CONVERSIONS BETWEEN U.S. CUSTOMARY UNITS AND SI UNITS (Continued)

U.S. Customary unit		Times conversion factor		Equals SI unit	
		Accurate	Practical		
Moment of inertia (area)					
inch to fourth power	in.^{4}	416,231	416,000	millimeter to fourth power	mm^4
inch to fourth power	in.^{4}	0.416231×10^{-6}	0.416×10^{-6}	meter to fourth power	m^4
Moment of inertia (mass)					
slug foot squared	slug-ft^2	1.35582	1.36	kilogram meter squared	kg·m^2
Power					
foot-pound per second	ft-lb/s	1.35582	1.36	watt (J/s or N·m/s)	W
foot-pound per minute	ft-lb/min	0.0225970	0.0226	watt	W
horsepower (550 ft-lb/s)	hp	745.701	746	watt	W
Pressure; stress					
pound per square foot	psf	47.8803	47.9	pascal (N/m^2)	Pa
pound per square inch	psi	6894.76	6890	pascal	Pa
kip per square foot	ksf	47.8803	47.9	kilopascal	kPa
kip per square inch	ksi	6.89476	6.89	megapascal	MPa
Section modulus					
inch to third power	in.^{3}	16,387.1	16,400	millimeter to third power	mm^3
inch to third power	in.^{3}	16.3871×10^{-6}	16.4×10^{-6}	meter to third power	m^3
Velocity (linear)					
foot per second	ft/s	0.3048*	0.305	meter per second	m/s
inch per second	in./s	0.0254*	0.0254	meter per second	m/s
mile per hour	mph	0.44704*	0.447	meter per second	m/s
mile per hour	mph	1.609344*	1.61	kilometer per hour	km/h
Volume					
cubic foot	ft^3	0.0283168	0.0283	cubic meter	m^3
cubic inch	in.^{3}	16.3871×10^{-6}	16.4×10^{-6}	cubic meter	m^3
cubic inch	in.^{3}	16.3871	16.4	cubic centimeter (cc)	cm^3
gallon (231 in.^{3})	gal.	3.78541	3.79	liter	L
gallon (231 in.^{3})	gal.	0.00378541	0.00379	cubic meter	m^3

*An asterisk denotes an *exact* conversion factor

Note: **To convert from SI units to USCS units, *divide* by the conversion factor**

Temperature Conversion Formulas

$$T(°C) = \frac{5}{9}[T(°F) - 32] = T(K) - 273.15$$

$$T(K) = \frac{5}{9}[T(°F) - 32] + 273.15 = T(°C) + 273.15$$

$$T(°F) = \frac{9}{5}T(°C) + 32 = \frac{9}{5}T(K) - 459.67$$

Materials Science and Engineering Properties

SI Edition

Charles M. Gilmore

George Washington University

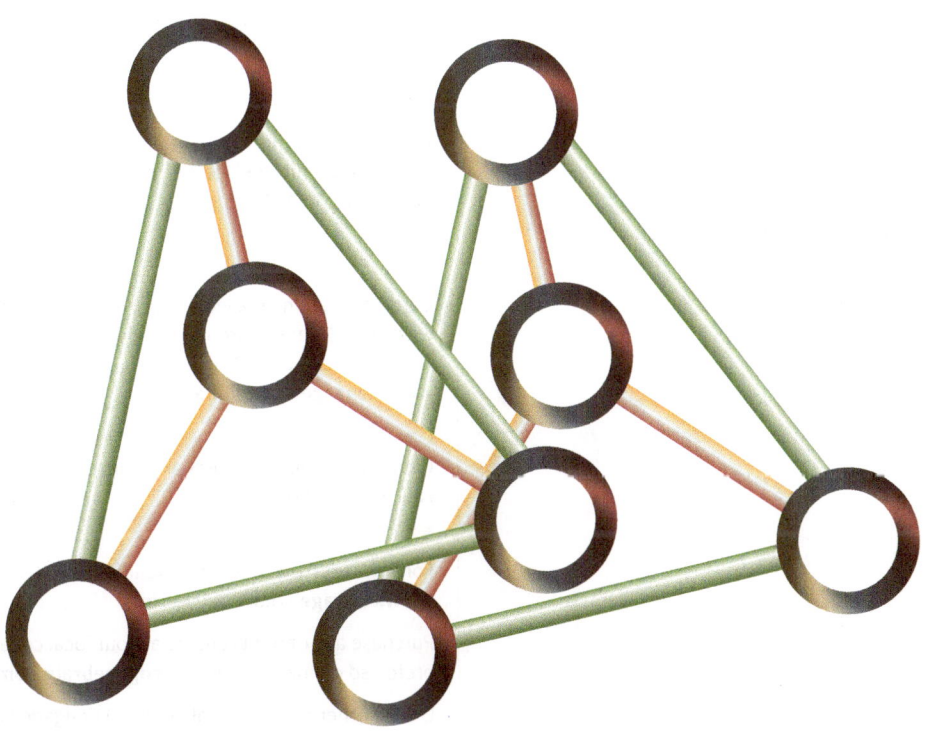

CENGAGE
Learning·

Australia · Brazil · Japan · Korea · Mexico · Singapore · Spain · United Kingdom · United States

Materials Science and Engineering Properties, SI Edition
Charles M. Gilmore

Publisher: Timothy Anderson

Senior Developmental Editor: Mona Zeftel

Senior Editorial Assistant: Tanya Altieri

Senior Content Project Manager: Jennifer Ziegler

Production Director: Sharon Smith

Media Assistant: Ashley Kaupert

Team Assistant: Samuel B. Roth

Rights Acquisition Director: Audrey Pettengill

Rights Acquisition Specialist, Text and Image: Amber Hosea

Text and Image Researcher: Kristiina Paul

Manufacturing Planner: Doug Wilke

Copyeditor: Patricia Daly

Proofreader: Erin Buttner

Indexer: Shelly Gerger-Knechtl

Compositor: MPS Limited

Senior Art Director: Michelle Kunkler

Internal Designer: MPS Limited

Cover Designer: Rose Alcorn

Cover Image: © ssuaphotos/Shutterstock

Library of Congress Control Number: 2013957160

ISBN-13: 978-1-111-98861-6

ISBN-10: 1-111-98861-7

Cengage Learning
200 First Stamford Place, Suite 400
Stamford, CT 06902
USA

Cengage Learning is a leading provider of customized learning solutions with office locations around the globe, including Singapore, the United Kingdom, Australia, Mexico, Brazil, and Japan. Locate your local office at: **www.cengage.com/global**

Cengage Learning products are represented in Canada by Nelson Education Ltd.

For your course and learning solutions, visit **www.cengage.com/engineering.**

Purchase any of our products at your local college store or at our preferred online store **www.cengagebrain.com.**

Unless otherwise noted, all items © Cengage Learning.

Printed at CLDPC, USA, 06-19

This book is dedicated to the important women in my life: To my wife Charlotte and to the memory of my mother Ruth.

About the Author

Charles M. Gilmore is Emeritus Professor of Engineering and Applied Science at the George Washington University. He obtained his B.S. and M.S. degrees in Engineering Mechanics at the Pennsylvania State University and his Ph.D. in Engineering Materials from the University of Maryland. He was employed by the Department of the Navy and the U.S. Naval Research Lab from 1963 to 1971. In 1971 he joined George Washington as an assistant professor. In addition to the position of professor, he was Chairman of the Department of Civil, Mechanical, and Environmental Engineering. In the School of Engineering and Applied Science he served as Associate Dean for Research and as Acting Dean. He developed the graduate program in materials science within the department and was responsible for the undergraduate courses in materials science and materials engineering. Dr. Gilmore was selected as an outstanding teacher by the ASCE student chapter. He is a member of the Sigma Tau and Tau Beta Pi honorary engineering fraternities, where he was an advisor for the George Washington chapter of Tau Beta Pi. He served as co-director of the George Washington University Institute for Materials Science along with Professor David Ramaker of the Department of Chemistry. Dr. Gilmore's research resulted in over 50 refereed publications on molecular dynamics simulation and experiments on the growth of thin films, fatigue and fracture of metals, and x-ray crystallography. Awards for his research include the George Kimball Burgess Award from the American Society for Metals Washington D.C. chapter and an Outstanding Paper award from the Materials Science and Technology Division of the Naval Research Lab. He served as an officer of the Washington D.C. chapter of the American Society for Metals and was the chapter chairman in 1984-1985.

CONTENTS

Chapter 3   The structure of real materials   101

Chapter 4   Temperature effects on atom arrangements and atom motion   135

Chapter 5   Phase transformations and phase diagrams   197

Chapter 6   Introduction to mechanical properties   261

Chapter 7   Making strong materials   321

Chapter 8   Engineering materials   363

Chapter 9   Time, temperature, and mechanical properties   419

Chapter 10   Oxidation, degradation, corrosion, electroprocessing, batteries, and fuel cells   457

Chapter 11　Fracture and fatigue　　　497

Chapter 12　Composite materials　　　551

Chapter 13   Materials processing   597

Chapter 14   Material selection   625

Chapters on the internet web site but not in the printed textbook   W-1

Chapter 16   Electrical properties of materials   W-3

Chapter 17   Magnetic materials   W-69

Chapter 18   Photonic materials   W-101

Purpose

The purpose of this textbook is to provide students and instructors with a materials science and engineering properties textbook with sufficient scientific basis that engineering properties of materials can be understood by students. For example, without an understanding of enthalpy, entropy, and Gibbs free energy students are not be able to understand why there are so many different metal microstructures, why water and alcohol mix, but water and gasoline do not mix, and why there are so many different types of phase diagrams. Internal energy, enthalpy, entropy, and Gibbs free energy are carefully developed so that a student without a course in thermodynamics can understand the discussion. However, it is recommended that students have completed an undergraduate course in university physics. The book discusses entropy as possible arrangements of atoms and molecules. In this way students can visualize entropy. The visualization of entropy as possible arrangements of atoms or molecules should help students understand entropy when it is used in thermodynamics courses to explain why heat engines are not 100% efficient. Energy is used as a unifying theme throughout the book to explain engineering properties such as melting temperature, thermal expansion, diffusivity, fracture, corrosion, and creep. The energy of electrons and holes, the Fermi energy, and energy level diagrams are used to explain the conductivity of materials and the operation of electronic and photonic devices such as diodes, lasers, solar cells, and light emitting diodes.

Organization

The textbook uses an integrated approach to treating the engineering properties of ceramics, metals, and polymers. The science of most engineering properties is the same for ceramics, metals, and polymers. For example, the equation for the entropy of mixing metal atoms is similar to the entropy equation for the mixing of polymer long chain molecules. Therefore, the mixing of metal atoms is covered in the same chapter as the mixing of polymers. The equations for the fracture of ceramics, metals, and polymers are all the same. Therefore, fracture of ceramics, metals and polymers is treated in the same chapter, and the differences in resistance to fracture are explained by the differences in chemical bonding. The change in the energy bands of both *pn*-junctions and a metal-polymer-metal junctions resulting from an applied voltage allows students to understand the operation of diodes and solar cells made from these different materials. Therefore, the electrical properties of metals, ceramics, and polymers are covered in the same chapter. If students understand the science behind the engineering properties and the differences in bonding between ceramics, metals, and polymers they can understand why these different materials have different engineering properties.

Textbook

The textbook focuses on materials science and applications to mechanical properties. Students and instructors interested in the mechanical properties of materials are also those most likely to be interested in a treatment of materials science that includes entropy and Gibbs free energy.

The introductory Chapter 1 presents a brief history of the development of materials and of the science necessary for the understanding of materials, the classes of materials, and an introduction to the experimental techniques available to analyze materials. Chapter 1 also presents several interesting case studies of materials applications, such as the use of shape memory alloys for coronary stents. Chapters 2 to 5 cover materials science subjects necessary for the understanding of the structure, microstructure, and engineering properties of materials. Chapters 6 to 14 cover mechanical properties of materials, how to make strong solids, mechanical properties of engineering materials, the effects of temperature and time on mechanical properties, electrochemical effects on materials including corrosion, electroprocessing, batteries, and fuel cells, fracture and fatigue, composite materials, material processing, and material selection for mechanical design. Chapter 15 is a more advanced treatment of experimental methods than is presented in the introduction.

Supplementary Web Content

Chapters 16, 17, and 18 on electrical, magnetic, and photonic properties of materials, respectively, are posted on the accompanying website. These chapters present more advanced coverage of these topics than is presently available in other materials science and engineering textbooks. There are also chapter appendices on the website that contain the derivations of equations and advanced subjects related to the textbook.

Chapter Organization

Each chapter begins with a photograph and description intended to arouse interest in the subject matter of the chapter. A list of goals tells the student and instructor what is to be accomplished. An introductory section describes why the subjects in this chapter are important and presents background and historical information. Scientific background necessary to understand the chapter is covered first followed by the discussion of engineering properties and applications. Example problems are included throughout the chapter. Figures and graphs are extensively included to provide students with a visual impression of the subjects. A summary covers the main points presented. There is a list of authors for supplemental reading, and a complete list of references in the back of the book.

Chapter Problems

Each chapter provides homework questions to test the student's grasp of the concepts in the chapter. Multiple choice questions are patterned after those in the Engineer in Training exam. An additional set of problems allows for analysis of concepts in the chapter. Design problems and questions are included where appropriate.

Audience and Prerequisites

The textbook is appropriate for a 3 credit course in materials science and engineering for sophomore or junior students in engineering or applied science with an emphasis on mechanical properties. The text with appendices of advanced subject material and chapters on electrical, magnetic, and photonic properties of materials is appropriate for more advanced study such as honors courses, higher credit courses, or an introductory graduate course for students that did not have an undergraduate course in materials science. It is assumed that students have completed university level chemistry and physics that includes chemical bonding, an introduction to quantum mechanics and quantum numbers, and an introduction to thermodynamics. These subjects are covered in sufficient detail in the text that a student can learn these subjects in this course, but it is recommended that students cover these subjects in prerequisite courses. A course in differential and integral calculus is essential.

Supplements for Students and Instructors

The website includes derivations of equations, additional advanced subject matter, and additional chapters on electrical, magnetic, and photonic properties of materials. The website also provides links to videos produced by the National Science Foundation and other organizations on subjects such as careers in materials science and engineering, the discovery of new materials, and the processing of materials to produce unique products.

Supplements for Instructors

In addition, the instructor's website provides a Solutions Manual with the answers to all questions and complete solutions to all problems, PowerPoint slides of all figures, and PowerPoint lecture slides.

MindTap Online Reader and Course

In addition to the print version, this textbook is also available online through MindTap, a personalized learning program. Students who purchase the MindTap version will have access to the book's MindTap Reader and will be able to complete homework and assessment material online, through their desktop, laptop, or iPad. If your class is using a Learning Management System (such as Blackboard, Moodle, or Angel) for tracking course content, assignments, and grading, you can seamlessly access the MindTap suite of content and assessments for this course.

In MindTap, instructors can:

- Personalize the Learning Path to match the course syllabus by rearranging content, hiding sections, or appending original material to the textbook content
- Connect a Learning Management System portal to the online course and Reader
- Customize online assessments and assignments
- Track student progress and comprehension with the Progress app
- Promote student engagement through interactivity and exercises

Additionally, students can listen to the text through ReadSpeaker, take notes and highlight content for easy reference, and check their understanding of the material.

Acknowledgments

Writing this textbook is the most significant accomplishment of my career as an engineer and teacher. The textbook would not have been possible without the contributions of many people. For as long as I can remember I planned to be an engineer because my grandfather Walter Brown and my father Charles E.M. Gilmore were both engineers, and my mother Ruth E. Brown Gilmore constantly encouraged my engineering studies. The textbook started as handout supplements to an assigned course textbook, because none of the textbooks available had the treatment I desired. Over many years the supplements evolved into chapters and finally into a draft textbook. I thank many years of engineering students at George Washington University for enduring the evolution of this textbook, for finding errors, criticizing confusing discussions and organization, and for suggesting alterations. During the development of the textbook many faculty at George Washington University reviewed chapters and made improvements including Professors David Ramaker, Douglas Jones, Martha Pardavi-Horvath, Mark Reeves, and Can Korman. I also thank the following for reviewing chapters or sections and making improvements: my former doctoral students Dr. M. Ashraf Imam and Dr. Wontae Chang, who are both now at the U.S. Naval Research Lab, Dr. Peter Matic of the U.S Naval Research Lab and Adjunct Professor at GWU, Dr. Catherine Cotell Adjunct Professor at GWU, and Ronald Reese, who is Emeritus Professor of Physics at Washington and Lee University. Ron Reese is the author of "University Physics" published by Cengage Learning. Whenever I have a question about physics I know I will get the correct answer from Ron's textbook. Ron and his wife Edith have been friends ever since we took a white water canoe course together nearly fifty years ago. Despite the best efforts of these friends and colleagues, any errors that remain are my responsibility. I thank Harold Adams, Fellow AIA, for assisting me with chapter and cover design and colors. I thank Mark Wagner, supervising engineering lab technician for the Department of Mechanical and Aerospace Engineering at George Washington, for helping me with experiments and photos for the text.

Independent outside reviewers contributed greatly to the evolution of the textbook. I thank the following reviewers for taking time out of their busy schedules to carefully review the book and to provide constructive comments.

Pranesh B. Aswath, *University of Texas at Arlington*
Amit Bandyopadhyay, *Washington State University*
Jeffrey Fergus, *Auburn University*

Gerhard Fuchs, *University of Florida*
Brian Grady, *University of Oklahoma*
Theodoulos Z. Kattamis, *University of Connecticut*
Leijun Li, *Utah State University*
Blair London, *Cal Poly State University, San Luis Obispo*
Lane W. Martin, *University of Illinois, Urbana-Champaign*
John R. Schlup, Kansas State University
Satya Shivkumar, *Worcester Polytechnic Institute*

Finally I thank Charlotte, my wife of 50 years, for enduring the many years of my sitting at my computer composing this book when we could have been on a cruise to an exotic sea. We will take that cruise once the book is published. The author invites comments and corrections to the textbook or solutions manual to be sent to cgilmore@gwu.edu.

PREFACE TO THE SI EDITION

This edition of *Materials Science and Engineering Properties* has been adapted to incorporate the International System of Units (*Le Système International d'Unités* or SI) throughout the book.

Le Système International d'Unités

The United States Customary System (USCS) of units uses FPS (foot−pound−second) units (also called English or Imperial units). SI units are primarily the units of the MKS (meter−kilogram−second) system. However, CGS (centimeter−gram−second) units are often accepted as SI units, especially in textbooks.

Using SI Units in this Book

In this book, we have used both MKS and CGS units. USCS units or FPS units used in the US Edition of the book have been converted to SI units throughout the text and problems. However, in case of data sourced from handbooks, government standards, and product manuals, it is not only extremely difficult to convert all values to SI, it also encroaches upon the intellectual property of the source. Some data in figures, tables, and references, therefore, remains in FPS units. For readers unfamiliar with the relationship between the FPS and the SI systems, a conversion table has been provided inside the front cover.

To solve problems that require the use of sourced data, the sourced values can be converted from FPS units to SI units just before they are to be used in a calculation. To obtain standardized quantities and manufacturers' data in SI units, the readers may contact the appropriate government agencies or authorities in their countries/regions.

Instructor Resources

The Instructors' Solution Manual in SI units is available through your Sales Representative or online through the book website at www.login.cengage.com. A digital version of the ISM and PowerPoint slides of figures, tables, and examples and equations from the SI text are available for instructors registering on the book website.

Feedback from users of this SI Edition will be greatly appreciated and will help us improve subsequent editions.

Cengage Learning

A photograph of liftoff of the space shuttle *Columbia* on January 16, 2003. The space shuttle provides many wonderful examples of recently developed materials allowing the development of a high-performance system.

However, approximately 82 seconds after liftoff, a suitcase-size piece of foam thermal insulation broke from the external fuel tank, striking the leading edge of *Columbia*'s left wing. The leading edge of the wing is covered with reinforced carbon-carbon (RCC) tiles that are part of the thermal protection system (TPS) that we will discuss in more detail in Section 1.3.1. The impact fractured the RCC tile, creating a hole as large as 25 centimeters in the wing. During reentry the leading edge of the wing reaches temperatures of 1650°C. Without the protection of the RCC tile, the interior metal structure of the wing was exposed to these temperatures. *Columbia* disintegrated upon reentry, with the loss of all seven on-board astronauts. This disaster shows the importance of the TPS for the space shuttle and the importance of avoiding fracture in engineering systems.

NASA

The goals of this chapter are to understand

- The importance of developing new materials that advance technology
- That the development of materials in the past was by trial and error, but the development of new materials is now a science
- The types of materials available to engineers, including ceramics, metals, polymers, and composites
- Some interesting applications of materials to engineering design, including
    - the reinforced carbon-carbon tiles used in space shuttles
    - high-temperature gas-turbine blade materials
    - composite materials for aircraft design
    - high-strength polymer filaments
    - applications of a smart alloy with a memory that can be trained
- The role of materials scientists, materials engineers, and engineers in the development of applications with new materials
- Some of the considerations necessary when selecting the best material available for a design

Introduction

1.1 MATERIAL CLASSES AND A BRIEF HISTORY OF THEIR EARLY DEVELOPMENT

The use of materials has been so important to human development that the advancements in technology are classified on the basis of the primary materials used: the stone, copper, bronze, and iron ages. The present age could be called the silicon age, because integrated circuits in electronic devices are primarily made from silicon. Initially humans utilized the materials that were naturally available, including stone, flint, wood, hides, shells, various fibers, clay, mud, and so on. The first human-made materials were ceramics, followed by metals, polymers, and composites.

1.1.1 Ceramics

Ceramic pottery figurines are the first known human-made materials. Figure 1.1 is an image of a ceramic figurine found in the Czech Republic that dates to approximately 27,000 before the common era (BCE). The word *ceramic* comes from the Greek word *keramos*, which means "potter's clay" or "pottery." Ceramic earthenware pottery is made from clay that is composed

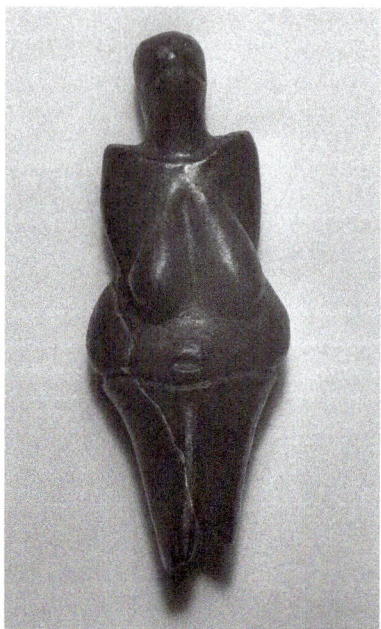

Figure 1.1   An image of a ceramic figurine that is one of the earliest known human-made materials, dating to approximately 27,000 BCE. (© *CTK / Alamy*)

of small particles of oxides of aluminum, silicon, and other elements such as potassium or sodium. When clay is mixed with water, it is easily formed into a shape. Once the shape is formed, it is heated to a high temperature at which the water is vaporized and the small solid particles bond with each other. In the case of the earliest pottery figurines, the heat was provided by an open wood fire. Prior to this time, humans had shaped materials available to them only by hammering, chipping, and performing other mechanical manipulations. Ceramic figurines were transformed from wet, pliable clay into a hard, solid shape with heat. The earliest ceramic pottery vessels date to approximately 16,000 BCE. The development of large, high-quality ceramic objects that could hold water required the development of furnaces that could reach temperatures approaching 1000°C.

Stoneware was developed in the fifteenth century BCE in China. Stoneware is also made from oxides of aluminum, silicon, and other elements, but the firing temperature is from 1100°C to 1300°C. Stoneware is dense and hard, and it is used to hold liquids and chemicals, such as acids. Porcelain was developed in China in approximately 200 of the common era (CE). Porcelain is primarily made from the alumino-silicate mineral kaolin, and it is fired at temperatures of 1200°C to 1400°C. Porcelain was not produced in Europe until 1708. Porcelain is used in fine dinnerware; in electrical high-voltage and high-frequency insulators; as a coating in sinks, dishwashers, and clothes washers; and in building materials such as floor tiles and wall panels.

The term *ceramic* now includes many more materials than pottery. **Ceramics** are inorganic nonmetallic materials with covalent or a combination of covalent and ionic bonding. The most frequently utilized ceramic for engineering applications, other than glass, is alumina (Al_2O_3). Alumina was first produced in the late 19th century. The present process for producing alumina from bauxite is the Bayer process, which was discovered in 1887. Alumina is used as a ceramic for high-temperature applications and as an electrical insulator. The white insulator on a gasoline engine spark plug, shown in Figure 1.2, is made from alumina.

Figure 1.2 An automobile gasoline engine spark plug. The white material is alumina ceramic. The threaded section is steel. (*Photo by C. M. Gilmore*)

Ceramics are now utilized for many technical applications. Many technical ceramics have carbon, nitrogen, or oxygen as a major constituent, along with an element from the left side of the periodic table. Technical ceramics include stoneware, porcelain, alumina (Al_2O_3), zirconia (ZrO_2), diamond, mica, silica (SiO_2), glass, silicon carbide (SiC), silicon nitride (Si_3N_4), and many others. One of the most important properties of ceramics is that they have very high melting points. For example, alumina (Al_2O_3) melts at 2045°C, and silicon carbide (SiC) melts at 2700°C. Because of their high melting temperatures, ceramics are not normally melted to fabricate a product. The figurine in Figure 1.1 is made of soft clay that was heated in a fire to produce a hard, solid object. In the fire the small particles of clay bond to each other without melting. The process of bonding small solid particles together at high temperature without melting them is called **sintering**. Sintering is used to produce parts of ceramics and metals that melt at very high temperatures. Although early humans used a furnace made of rocks, modern high-tech ceramic parts are sintered with computer-controlled heat cycles and ultra-pure environments.

Many of the modern applications of ceramics result from their high melting temperatures and their electrical insulating properties. The lining of high-temperature furnaces is made from foamed alumina. The thermal protection systems on the space shuttle, discussed in Section 1.3.1, include ceramic components. Ceramics such as Al_2O_3, SiC, and Si_3N_4 are relatively low in density because of the low atomic numbers of their constituents; however, ceramics such as tungsten carbide (WC) are higher in density. Ceramics in general have high stiffness, meaning that they resist deformation. Ceramics are weak in tension, which occurs when the two ends of a rod are pulled apart; but they are relatively strong in compression, which occurs when the two rod ends are pressed together. The strength of ceramics is dependent upon the production quality. In general, ceramics are electrical insulators. Ceramic superconductors are an exception to insulating ceramics. We discuss the conductivity of ceramics and superconductors in Chapter 16.

If ceramics are slowly cooled from the liquid state to the solid state, they are crystalline. In a **crystalline** material the atoms repeat in a regular arrangement. However, if a ceramic such as silica (SiO_2) is cooled rapidly from the liquid state, it forms amorphous solid silica glass. In an **amorphous** material the atoms are not in a regular arrangement. Glass is a metastable material that changes to a crystal over time, a process called devitrification. Glass is the only ceramic that is extensively recycled.

A negative mechanical property of ceramics is their brittleness. A **brittle** material is one that fractures with no permanent deformation. Glass, ceramic pottery, and ceramic plates are all brittle at room temperature. The brittle nature of ceramics limits their applications. For example, the designers of high-temperature gas turbines want to have the turbine temperature as high as possible for the maximum efficiency. Because of their high melting temperatures, ceramics are possible materials for a turbine blade. However, turbine blades are made with metal alloys primarily based upon nickel, which melts at 1453°C. The reason nickel alloys are used instead of ceramics is that ceramic turbine blades would break in a brittle manner like glass and cause catastrophic failure of the engine. Ceramics are used for nonrotating applications in gas turbines. Also, ceramics containing zirconia (ZrO_2) have been found to absorb a small but increased amount of energy before fracture. There are proposals to use mixtures of zirconia with other ceramics in high-temperature diesel and gasoline engines for applications such as cylinder liners.

1.1.2   Metals

Metals are inorganic materials with metallic bonding. Familiar elemental metals include iron, copper, silver, gold, aluminum, titanium, lead, and platinum. In metallic bonding, some of the outer electrons of the atom are free to move through the metal. Metallic bonding results in metals being good conductors of electricity.

The first use of metals by primitive humans involved the native metals copper and gold. Native metals are the elements found in a relatively pure state. Native iron is very rare because it easily oxidizes; however, some meteoric iron was available and utilized in tools, weapons, and jewelry as early as 35,000 BCE in Egypt.

It is thought that the first metal smelted from ore was lead, because of its low melting temperature of 327°C. A metal **ore** is the metal combined with other elements, such as oxygen or sulfur. **Smelting** of metals is the process of heating the metal ore to a high temperature in the presence of other chemicals, this process results in the production of the elemental metal. For example, when copper oxide (CuO) is heated in the presence of carbon monoxide (CO), the CO reacts with the oxygen in the ore (CuO), resulting in the production of copper and carbon dioxide (CO_2). Initially CO came from burning wood, and in later years charcoal provided the CO. It is thought that the first smelting occurred by accident when metal ores were used to decorate pottery. Lead oxide (PbO) provides black color and CuO is blue-green. If the temperature of the pottery kiln is sufficiently high and CO is present in the kiln, the decorative metal oxide is reduced to metal. Possibly the first metal workers were also potters.

The smelting of copper required furnaces capable of melting copper at 1085°C. Smelted copper contains impurity atoms such as arsenic, lead, silver, nickel, and iron; therefore it can be distinguished from native copper, which is purer. Some small items, such as pins, awls, and beads of smelted copper, have been dated as early as 6000 BCE. A cast copper axe dating to approximately 5500 BCE has been found in Serbia, and a cast copper mace dated to 5000 BCE has been found in Turkey. The earliest known furnace for smelting copper was found in the Sinai desert and dates to approximately 3500 BCE. Copper had relatively limited usage because it is a soft metal that cannot hold a sharp edge on an axe, knife, or spear point. Harder metals were needed.

Copper mixed with a metal other than zinc produces the alloy **bronze**. An **alloy** is a metal mixed with other elements. Copper mixed with zinc creates the alloy **brass**. Mixing copper with tin to produce bronze results in a much harder and stronger metal. Also, mixing copper with about 10 weight percent tin reduces the melting temperature of the bronze to approximately 950°C. By approximately 3500 BCE, bronze was extensively utilized in the Middle East. However, the source of tin is a mystery. No major tin deposits exist in the Middle and Near East, where large amounts of bronze were smelted. Sources of tin are in Malaya, China, Thailand, and England. It is thought that the use of bronze declined when the supply of tin was interrupted by war and invasion.

During the bronze age, iron was used; however, it was of poor quality; it did not have mechanical properties equal to bronze, and it corroded. The production of iron and steel also required higher-temperature processing than did copper and bronze. By 2000 BCE, smelted iron was produced in present-day Turkey, by 1800 BCE in northern India, and by 1500 BCE in sub-Saharan Africa. The beginning of the iron age corresponded with the peak of the bronze age. One of the oldest smelted iron artifacts is a knife blade dated to 2500 BCE, found in Turkey. In 2000 BCE iron was produced by layering iron ore with wood or charcoal in a furnace. The furnaces had clay blow tubes, and later bellows were made of animal hides to blow air into the furnace and increase the temperature. The CO produced by burning wood and charcoal reacted with the oxygen in the iron ore and reduced the ore to small particles of iron. However, the temperatures in the furnace did not reach 1538°C, the melting point of iron, and the iron did not melt. Rather, iron particles were mixed with the **slag**, which is all of the reaction products other than the iron, to produce a bloom, which is the iron plus slag mixture. The bloom was repeatedly heated in air and hammered to break the iron particles from the slag. Heating the iron in air removed carbon from the iron, resulting in relatively pure iron. The iron particles were then heated and hammered together to form a larger piece of wrought iron. When a metal is deformed into a shape, this metal is called **wrought**. Wrought iron is relatively pure iron that is soft and does not hold a sharp edge.

Very soon after wrought iron was being produced, it was discovered that if the iron is packed with charcoal and heated, the iron becomes stronger and can hold a sharp edge. Charcoal is primarily carbon (C), and the carbon is mixed with the iron to produce carbon steel. **Carbon steel** is iron plus approximately 0.2 to 1 weight percent C. Since the iron was a solid when it was carburized, how did the carbon get into the solid? In Section 4.7 we will discover that even in solids, the atoms are moving around in a process called **diffusion**. Albert Einstein was one of many people who studied the motion of atoms in both liquids and solids.

By 500 BCE, metal workers in China had achieved furnace temperatures of 1130°C. This is a sufficient temperature to melt iron that is mixed with 4.3% C and produce **cast iron**. In Section 5.14.1 we will discuss how the melting temperature of iron changes with the amount of carbon. With the development of cast iron, desired shapes could be made in sand molds. However, cast iron is brittle in comparison to steel. A **ductile** material is one that permanently deforms a significant amount before it fractures. Steel is more ductile than cast iron. It was also discovered that the amount of carbon in cast iron is reduced to the level desired in steel by heating the cast iron in air for a few days. The oxygen in air reacts with the carbon in the cast iron, producing carbon dioxide gas. The carbon atoms move to the surface of the iron by diffusion, and this decarburizes the cast iron to produce steel.

Nearly four thousand years after iron was first smelted in Turkey using the bloom process, iron was still made in Europe by this process. The furnaces in Europe were only capable of approximately 1000°C, and they could not melt iron. Several steps resulted in furnace temperatures capable of melting iron. Because all of the furnaces used charcoal made from wood, most of the trees in Europe had been cut down by 1600 CE. It was discovered that coke, which is made from coal by vaporizing the impurities with heat, is nearly pure carbon. The use of coke allowed furnaces in Europe to melt cast iron by the middle 1600s. Then it was discovered that if preheated air is forced into the furnace, the furnace temperature increased further. In the Bessemer blast furnace, developed in 1855, a blast of hot air reacts with the carbon mixed with the iron, heating the mixture to temperatures sufficiently high to melt relatively pure iron. This advancement allowed for the large-scale production of molten iron, and it contributed to the industrial revolution. The Bessemer furnace was replaced by the basic oxygen furnace, developed in 1952, that injects oxygen into the iron. This reduces the amount of nitrogen in the steel that came from the air used in the Bessemer blast furnace.

An important property of metals is that they can be extensively deformed into shape without breaking. The extensive deformation allows metals to be formed into products such as automobile bodies. The bodies of most automobiles are made of very-low-carbon steel, which is iron with less than 0.2 weight percent carbon. Low-carbon steel can be rolled into thin sheet and cold-stamped into shape. As we will discuss in Chapter 7 on strong solids, the rolling and stamping process can increase the strength of steel by up to ten times, resulting in a strong automobile body that maintains its shape under normal driving conditions. However, the steel permanently deforms on impact without breaking, such as in an accident. This is an important design safety consideration for automobiles. Automobile bodies are designed with crumple zones that crush in a controlled manner, as shown in Figure 1.3, to absorb the energy of a collision and to protect the occupants.

Fortunately, steel is also one of the cheapest metals available for engineering use. At the time this book was written, plain low-carbon steel is approximately $5 per kg, and high-strength alloy steel is $10 per kg. Aluminum alloys range from approximately $15 per kg to $30 per kg for a high-strength aluminum alloy. Titanium alloys are approximately $150 per kg.

Figure 1.3   A crash test of a sport utility vehicle showing how the steel in the front of the car deforms without breaking to absorb the energy of the impact and protects the passengers. (*Photo Courtesy of the National Crash Analysis Center of The George Washington University*)

In an effort to increase automobile mileage, aluminum is replacing iron and steel in some automobile applications. Aluminum is significantly lighter than iron, with densities of 2.70×10^3 kg/m^3 and 8.90×10^3 kg/m^3, respectively. Densities of the elements are available in Appendix B, and the density of metals is discussed in Section 2.9.2. Cast aluminum alloys have replaced cast iron in automobile engine blocks and cylinder heads. Some automobile bodies and parts are also being made out of stronger steels that are thinner and lighter.

The mechanical properties of metals result in their application to many designs. Metals can be made that are very strong both in tension and compression. However, pure metals such as iron and aluminum are inherently very weak. One of the achievements of materials science is to make these metals strong and to understand why the strength changes. We will cover the strengthening of metals in Chapter 7. Of the primary structural metals, steel has the highest density, stiffness, strength, and resistance to fracture; followed by titanium alloys; and then aluminum alloys. Metals are relatively stiff, but in general not as stiff as ceramics. Metals are much stronger than ceramics in tension, but they are of comparable strength in compression. Metals have a much higher resistance to fracture than ceramics. The melting temperature of metals varies widely. Mercury melts at $-38.9°C$, and tungsten melts at $3410°C$. The relatively high melting temperature of metals, such as nickel and cobalt, their resistance to fracture, and their oxidation resistance result in their use in high-temperature gas turbines, as discussed in Section 1.2.4.

Iron and steel are extensively recycled. Iron and steel products are magnetically separated from other materials. Aluminum in the form of beverage cans and other containers is extensively recycled, as are automotive aluminum parts. Titanium is recycled by industry, because most titanium usage is in industrial applications.

1.1.3 Polymers and Plastics

The word **polymer** means many ("poly") units ("mer"). Polymers are large molecules with many repeating units that are covalently bonded to each other. For example, in the polymer polytheylene, the mer unit is ethylene (C_2H_2). Organic polymers are based upon carbon, and inorganic polymers, such as silicones, are based upon other elements. Polymers can be liquids or solids at room temperature. In general, room-temperature liquid polymers have fewer mer units per molecule, such as 1000 mers, and solid polymers have a higher number of mer units, such as a million mers or more. Polymers can be natural or human-made. Latex is a natural polymer from the rubber tree that is used to make rubber, and wood is made from the two polymers lignin and cellulose. Cellulose is the most abundant polymer on earth because it is in plants. In the polymer deoxyribonucleic acid (DNA), which we humans are made from, the mers are nucleotides that are given the abbreviations A, G, C, and T. DNA is a polynucleotide that is made up of many repetitions of these four nucleotides, as shown in Figure 1.4.

Polyethylene (PE) is the most widely used synthetic polymer, and it was one of the first synthesized. PE is used in milk bottles, packaging, and many other applications. PE was first synthesized in 1898 by accident, and industrial production began in about 1940. PE is made from the mer ethylene (C_2H_4).

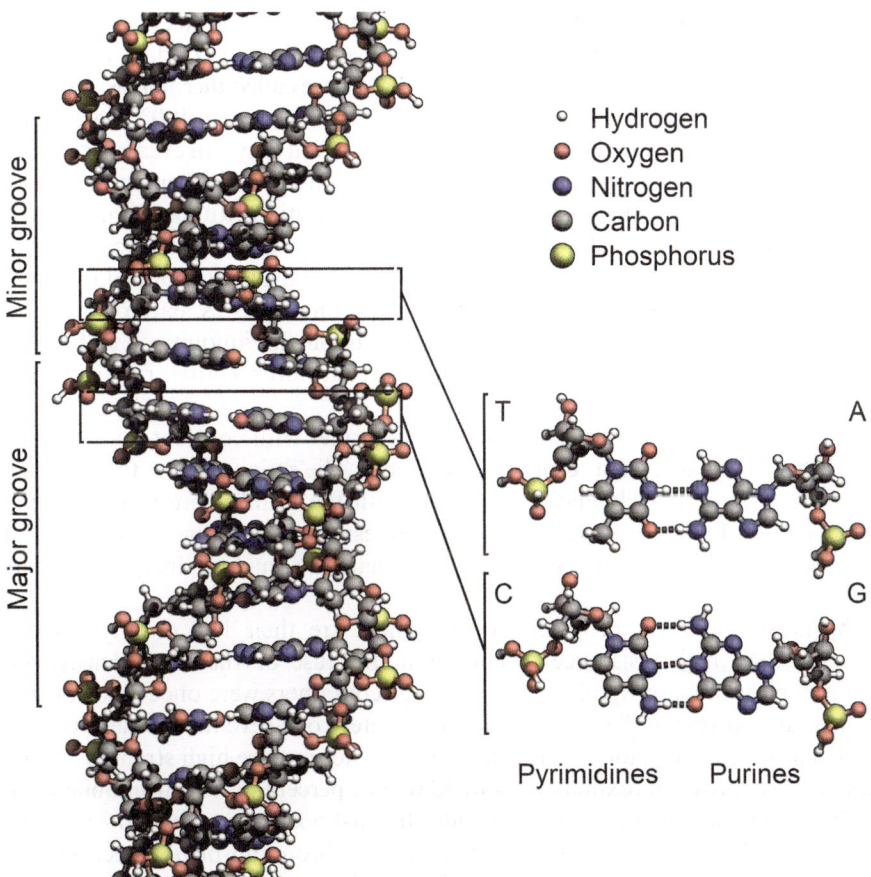

Figure 1.4 The structure of the polymer DNA consisting to the four molecules T, C, A, and G. (*Courtesy of Richard Wheeler*)

Ethylene is a gas at room temperature. In the polymerization of PE, ethylene gas is heated under pressure in a chamber along with a chemical that initiates the formation of long-chain molecules (LCMs) that can have hundreds to millions of repeating units of the ethylene mer. The mers are covalently bonded together in an LCM. An LCM of PE is shown in Figure 2.17.

Other polymers are produced by modifying the ethylene mer or by producing a completely different mer. For example, in polyvinylchloride (PVC) one of the hydrogen atoms in the ethylene mer is replaced by chlorine; and the mer is (C_2H_3Cl). Adding the chlorine to the polymer makes PVC much stronger than PE. PVC is used to make water and waste-water pipes, and wire insulation. Some other polymers that you may be familiar with include polyamide 66 (nylon™), polymethyl methacrylate (PMMA), polytetrafluoroethylene (PTFE, Teflon™), and polycarbonate (PC). Many of these polymers are known by their trade names, such as nylon and Teflon. PMMA is also known as Lucite™, Plexiglas™, Acrylite™, and Diakon™, depending upon the manufacturer. Silicones are polymers based upon a mer with silicon rather than carbon. Polymers are discussed in more detail in Section 2.8.

Plastics are polymers plus other materials that modify the properties of the polymer, such as pigments to change color, fillers such as silica powder to change mechanical properties, and plasticizers to decrease the viscosity. The term *plastic* applies to the total mixture, and the term *polymer* applies to the main structural molecules in the plastic, such as PVC. For example, if carbon is added to PVC to make black pipe, this product is a plastic pipe. However, many plastics are referred to by their polymer name. For example, plastic pipe made from PVC and carbon is referred to as black PVC.

In polymers, such as PE, PVC, and PMMA, different long-chain molecules are not bonded to each other with covalent bonds. They are held together in a solid with weak secondary bonds resulting from positive and negative charges on the molecules. These polymers are called **thermoplastic** polymers because they soften when heated, and they melt at relatively low temperatures, from 100°C to 300°C.

Thermoset polymers do not significantly soften when heated, and they can even harden. In a thermoset polymer there are covalent bonds between the molecules, such as in rubber and epoxy. When these polymers are heated above 300°C in air, they deteriorate by oxidation. Rubber is made from latex. It is the sap from the rubber tree, and it is a liquid polymer. Through the process of vulcanization, discovered by Charles Goodyear in 1839, the polymer molecules in latex are covalently bonded together with sulfur atoms. This results in rubber that is used in automobile and aircraft tires, battery casings, and rubber bands. The hardness and strength of the rubber depends upon the amount of sulfur added. In soft rubber, such as a rubber band, there is typically 3 weight percent sulfur. Harder rubbers have more sulfur and softer rubbers have less sulfur.

Epoxy is another material where there are polymer molecules with covalent bonds connecting long-chain molecules. Epoxy is made from a resin and a hardener. The resin contains the polymer long-chain molecules. The hardener contains chemicals that react with the resin to produce covalent bonds that connect different polymer LCMs. Epoxies are used as adhesives, automotive components, and molded parts. An important application of epoxy is in composite materials, such as fiberglass, that are discussed in Section 1.1.4.

Several of the most important properties of polymers are their low density, typically close to 1×10^3 kg/m³, and their high resistance to corrosion. Because of their low density, polymers are replacing metals and glass in many applications. Beverage containers were once made of glass or metal; now many of them are polymers. Plastics have approximately one-seventh the density of steel. Many automobile parts that were once made of sheet steel, but do not require high strength, are now made of plastics. Plastics now comprise approximately 14 to 18 weight percent of an automobile with an internal combustion engine. Some of the applications include the dashboard assembly, interior panels, the fan cowling, and air ducts. Automobile bumpers were made of chromium-plated steel, but now they are made of a steel beam covered with plastic. Automobile fenders are being made out of a plastic rather than steel; the change in material can decrease the fender weight by 50% and cost by 15% relative to a steel fender.

The stiffness, strength, and resistance to fracture of bulk polymers are much less than for structural metals. Polymers and ceramics have comparable resistance to fracture in tension and tensile strength; however, ceramics are stiffer than polymers and stronger in compression than polymers. The cost of polymers varies from $10 per kg for PE to $300 per kg for the high-performance bulk polymers.

Plastics are extensively recycled, and a numbering system, discussed in Section 14.4.3, has been developed to identify the type of polymer in the plastic for sorting. The numbering system allows the recycled plastic to have known polymer types.

1.1.4   Composite Materials

Composite materials are made of at least two separate materials that are combined into one material, but the separate materials maintain their identity within the combined material. For example, fiberglass is composed of glass fibers and epoxy. In fiberglass the glass fibers remain as glass fibers that are embedded in a matrix of epoxy. **Matrix** means that the material is continuous. Fiberglass is a human-made composite material, but wood is a natural composite material made of cellulose fibers and a matrix of lignin resin. Figure 1.5 shows steel reinforcing bars protruding from concrete. Concrete will be poured around these reinforcing bars to produce a steel-reinforced concrete column that will support a metro rail line. Some of the continuous concrete matrix is visible at the bottom of the photograph. In many composite materials, the main function of the matrix is to hold a strong material into the shape of the part. Matrix materials include metals, ceramics, plastics, and concrete. In fiberglass the glass fibers are a strong reinforcing phase. Other reinforcing phases include graphite fibers, boron fibers, high-strength polymer fibers, and steel wires and rods. With this variety of matrix and reinforcing materials, different materials can be combined to achieve the required properties from the composite material.

In composite materials with a brittle matrix, a second material is added to resist fracture. The concrete reinforced with steel rods in Figure 1.5 is an example of this. All buildings, bridges, and highways made

Figure 1.5   Steel reinforcing bars protruding from a continuous concrete matrix. (*Photo by C. M. Gilmore*)

from concrete use reinforcing steel rods to resist fracture. In an earthquake, a structure made only of concrete is likely to fracture. However, concrete reinforced with steel rods can be designed to resist severe earthquakes. Other high-performance ceramics are also reinforced with fibers to increase the resistance to fracture, such as the carbon matrix-graphite fiber composites used in the space shuttle, discussed in Section 1.3.1.

It is possible to make fibers that are much stronger than the same material in bulk form. For example, glass fibers are nearly a hundred times as strong as a bulk piece of glass. Graphite fibers are stronger than the strongest steel, even though bulk graphite has almost no strength. Fine fibers of polyethylene are more than 100 times as strong as bulk polyethylene and nearly as strong as the strongest steel. One way to use these strong fibers in a part is to make a composite material.

Composite materials are replacing metals in many applications that require high strength and light weight. For example, the use of graphite-fiber-reinforced epoxy composite materials for construction of the monocoque chassis of Formula 1 race cars started in 1981 by the McLaren racing team. *Monocoque* means single ("mono") shell ("coque"). On a race car, everything is attached to the monocoque chassis. Approximately 80% by volume of a modern Formula 1 race car is graphite-fiber-reinforced epoxy composite materials.

McLaren's experience with race car design and production has carried over to production of a sports car, shown in Figure 1.6, where graphite-fiber-reinforced epoxy composite material is used to replace a steel chassis. The carbon fiber chassis weighs only 75 kg. The chassis production time was reduced to four hours from the 3000 hours it took to produce a chassis in the 1990s. This has allowed McLaren to price this car at less than $250,000. A sports car with a carbon-fiber chassis can cost over a million dollars. A McLaren spokesman has predicted that in 30 years, general-production cars will have a carbon-fiber-composite chassis.

Aircraft utilize many different types of composite materials. For example, the body of the Boeing 787 discussed in Section 1.3.2 is made from graphite-fiber-epoxy composites rather than aluminum alloys.

The properties of composites depend upon the constituents in the composite. For example, graphite-fiber-reinforced epoxy has a density of approximately 1.67 g/cm^3, or approximately one-fifth the density of steel. It is as strong as high-strength steel. It is not as stiff as steel, but it is stiffer than aluminum and titanium. Its strengths in tension and compression are comparable, and its resistance to fracture is less

Figure 1.6   A McLaren sports car with a carbon-fiber chassis. *(Courtesy of Richard Betts)*

than that of steel but comparable to aluminum. At the time of this writing the cost of graphite-fiber-reinforced epoxy is approximately $250 per kg.

Graphite-reinforced epoxy is very difficult to recycle, because it is laborious to separate the graphite from the epoxy. Epoxy is a thermoset polymer and does not melt, and it is not easily dissolved in solvents. The composite can be ground into pellets or powder and used as a filler in a plastic, rubber, asphalt, or concrete. It is easier to recycle a composite material if the matrix and the reinforcement can be separated. If the matrix is a thermoplastic polymer, then the matrix can be melted to separate it from the reinforcement with a higher melting temperature. Composite materials are covered in more detail in Chapter 12.

1.2 THE DEVELOPMENT OF MATERIALS SCIENCE AND ENGINEERING

Prior to 1900, the development of new materials was primarily steered by the trial-and-error experiments conducted by craftspeople and a few scientists. By 1900 the foundation was present for an explosion in knowledge of science and in the development of new materials. To make the development of new materials a science required the development of the theory of the structure of matter and the development of experimental methods to observe materials in detail. **Materials science** is the application of the theory of the structure of matter to the understanding, development, and improvement of materials. **Materials engineers** specialize in the application of materials to designs. Materials scientists and engineers work in all of the engineering disciplines. In mechanical and aerospace engineering, materials scientists and engineers develop new applications, such as high-temperature metals for gas turbines, composite materials to replace steel in energy-efficient automobiles, and biomaterials for implantation in the body. In civil engineering, materials scientists and engineers develop higher-strength steels for skyscraper buildings that can withstand earthquakes, and higher-strength concrete and asphalt for roads. In electrical engineering, materials scientists have developed the solid-state materials that are used in consumer electronics. Mechanical, aerospace, civil, and electrical engineers apply the new materials to designs. Engineers are often responsible for both designing a new product as well as selecting the design materials.

1.2.1 The Development of the Theory of the Structure of Matter

Leucippus and Democritus, who were Greek philosophers living in the fifth century BCE, taught that the smallest unit of matter was an "atomos," even though they and their peers had no knowledge of what an atomos actually was. Even though the native elements copper and gold as well as meteoric iron were materials in use at that time, in about 330 BCE Aristotle proposed that everything is made of four "roots": earth, water, air, and fire. Plato later renamed "roots" as "elements." Many centuries closer to our own modern time, in 1661 Robert Boyle proposed that an **element** is a substance that cannot be broken down into a simpler substance by a chemical reaction. As time continued to pass and additional elements were discovered, a number of scientists noted periodic relationships between the elements, and Dimitri Mendeleev proposed a periodic table in 1869 that is similar to the present form of the periodic table, with eight groups. The highest atomic number in Mendeleev's periodic table was uranium. Now most periodic tables, such as the one at the end of this book, include elements to atomic number 118.

In 1896 J. J. Thompson showed that the atom was not the smallest unit of matter, but that atoms contained negatively charged particles, called electrons. By 1908 Ernest Rutherford proposed that an atom consisted of a small, positively charged nucleus surrounded by negatively charged electrons. In 1913 Neils Bohr proposed that the energy levels of the electrons in the hydrogen atom depended inversely upon the square of an integer quantum number (n). This model of hydrogen led to the quantum mechanical modeling of the electrons on atoms. Later it was found that there is an additional quantum number related to the angular momentum of the electron, which has a secondary effect on the electron energy. A more detailed discussion of the quantum numbers on atoms is presented in Sections 2.2.1 and 2.2.2. The model of an atom consisting of a small nucleus with protons and neutrons, surrounded by electrons with energies predicted by quantum mechanics, is appropriate for the understanding of materials.

1.2.2 The Development of Experimental Methods to Study Materials

New experimental techniques were developed that let material scientists observe materials from the macroscopic level to the atomic level. With these new techniques, scientists could test the new atomic theory of matter and apply it to the development of new materials. The following is a brief introduction to some of the most important experimental techniques used to characterize the structure and chemistry of materials. Chapter 15 provides a more complete presentation of all of these experimental techniques.

One of the first experimental tools used to study the structure of materials was the light optical microscope. A **light optical microscope** uses visible light as a probe with lenses to magnify the image. An early optical microscope dates to 1590. Galileo Galilei used a more advanced microscope by 1625. Early microscopes were used to study biological specimens, such as the structure of a fly's eye in 1644; and red blood cells, spermatozoa, and microorganisms in 1676. Improvements in illumination and optics resulted in extensive use of optical microscopes after 1900 in the characterization and development of metal alloys and in the study of biological materials. Figure 1.7 is a modern optical microscope image of an iron alloy with 20% nickel and 1.2% carbon. The lines in the figure are due to changes in the structure of the material during cooling that are explained in Section 5.14.2. Optical microscopes are limited to observing the structure of materials down to dimensions of approximately 10^{-6} m. They cannot observe materials at atomic dimensions, because the wavelength of light is thousands of times larger than the size of the atom. In a high-quality metallurgical microscope analyzing polished specimens, the maximum magnification is approximately 1000×. If specimens have irregular surfaces, lower magnifications are necessary to have the field of view in focus. Higher magnification microscopes that use probes other than light are discussed later in this section.

X-rays, discovered in 1895 by Wilhelm Roentgen, are electromagnetic radiation with wavelengths typically ranging from 0.1 nm to 1 nm. X-rays are like light waves, but with shorter wavelengths. The important properties of X-rays are that their wavelengths are comparable to the distance between atoms in materials, and that X-rays can penetrate through matter.

In 1912 Max von Laue discovered that X-rays are diffracted by crystals in a manner similar to the diffraction of light by a diffraction grating. The planes of atoms in a crystal act similar to a diffraction grating. A diffraction grating with a known line spacing diffracts light of a known wavelength to an angle that is calculated with the grating equation. In a crystal the planes of atoms with an unknown spacing diffract X-rays of a known wavelength to an angle that is measured in a device called a diffractometer. The angle of the diffraction intensity is used to determine the spacing between crystal planes with

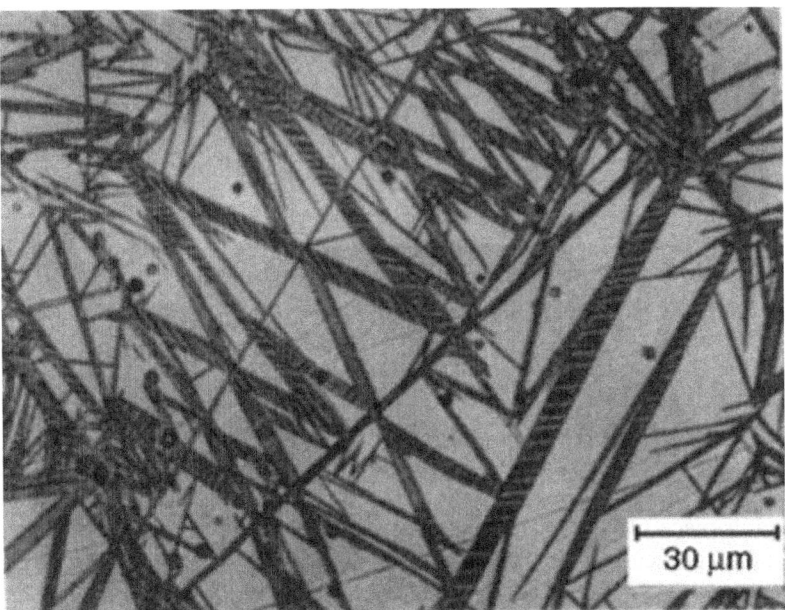

Figure 1.7   An optical-microscope image of the microstructure of an iron-20% nickel-1.2% carbon alloy cooled to 4 kelvin. The straight lines are due to structure changes in the iron alloy. (*ASM Handbooks Online, http://www.asmmaterials.info, ASM International, 2004*)

the X-ray diffraction equation. By measuring the spacing between different planes and the intensity of the diffraction peaks, the arrangement of atoms in the crystal is determined. The diffraction of the X-rays shows that in crystals, the atoms are in regular arrangements that repeat over large distances. For example, in some crystals the atoms are arranged in cubes, and the cubes in the crystal are all arranged in a regularly repeated pattern. One of the first crystal structures determined was salt, in 1914. The sodium and chlorine atoms in salt are in a cubic arrangement that repeats in three dimensions to produce individual salt crystals. X-rays were also used to determine the structure of DNA. Rosalind Franklin obtained a small crystal of DNA and determined the diffraction pattern. In 1953 James Watson and Francis Crick determined the structure of DNA, and for this they received a Nobel Prize. X-ray diffraction has allowed the determination of the atom arrangements in crystals of metals, ceramics, polymers, and biological materials.

The chemical analysis of materials is also possible with X-rays. Each atom type has characteristic energy levels for its electrons. The inner core electrons are strongly bound and have a low energy, and the outer electrons are higher in energy. An electron in an orbital on an atom can be excited out of its orbital by bombarding the atom with high-energy X-rays or electrons. Usually the empty electron orbital is filled by an electron from the same atom, one with higher energy than that of the removed electron. The energy change experienced by the electron changing to a lower energy is emitted from the atom as an X-ray. The emitted X-ray has an energy characteristic of the atom that emitted the X-ray. Atoms are identified by the energy of the characteristic emitted X-rays. Figure 18.10 shows the electron energy levels involved in X-ray absorption and emission. How is the energy of the emitted X-ray determined? Diffraction from a crystal of known planar spacing yields the wavelength of the X-ray and therefore the energy.

The light optical microscope discussed above had several limitations. For opaque materials it could only observe the surface, and it could only resolve features with dimensions of approximately a micron. The **transmission electron microscope** (TEM) uses high-energy electrons to create images through material. The first TEM was built in 1931. Some modern TEMs can resolve individual atoms in a solid. Section 15.4 covers the use in more detail of all the microscopes discussed in this section, and Figure 15.18 is a schematic of a TEM. In a TEM, electrons are created at the top of the microscope with a hot filament. The electrons are accelerated by a voltage of typically 100,000 volts (V), although some high-energy TEMs use 1,000,000 V or more. The electrons are focused into a small parallel beam by electromagnetic lenses. The electron beam is directed at a thin film or foil of the material under study. High-energy electrons can penetrate through matter. The higher the energy of the electron beam, the thicker the material that can be penetrated. A beam of electrons with an accelerating voltage of 100,000 V can penetrate through approximately 400 nm of nickel. As the electrons penetrate through matter, if they experience any changes in the material, such as defects, the electron path is altered by scattering or by changes in absorption. Figure 6.10 is a TEM image of molybdenum, and the dark lines are defects in the steel, called dislocations. The dislocations scatter the penetrating electrons out of the beam so they appear as dark. The electrons that penetrate through the material are observed on a fluorescent screen. A record of an image can be made on film or with electronic imaging. The TEM image is an image through the thin film of material. It is similar to looking through a piece of glass, where internal defects scatter or absorb light. However, in the case of a TEM, the magnification can be up to 2,000,000 times for a microscope with a high voltage and high resolution.

The TEM can also obtain an image of the surface contours of a bulk specimen. To produce this image, a bulk specimen is coated with a thin liquid polymer layer that assumes the contours of the surface. When the polymer is dried and stripped from the surface, it retains the surface contours, creating a surface replica. The polymer is then coated with a conducting material, such as carbon, and then it is viewed in the TEM.

Not only can the TEM produce images through the material, it can also be used for crystal-structure analysis and for chemical analysis. Electrons were first thought to be particles; however, crystals diffract electrons in the same way they diffract X-rays. Electrons behave both like particles and waves. The high-energy electron beam in the TEM is diffracted as it passes through a crystalline material, and the diffraction pattern of the material under study is used to determine the crystal structure and orientation of the material. Additionally, the high-energy electron beam is capable of exciting electrons out of the atoms under study with the subsequent emission of characteristic X-rays, as discussed above. The characteristic X-rays are used to identify the type of atoms in the volume of material under study. TEMs are used to study metals, ceramics, polymers, and biological materials.

A **scanning electron microscope** (SEM) scans an electron beam across the surface of a material to create an image of the surface. Figure 15.17 is a schematic of an SEM. The first commercial SEM was produced in 1965. Figure 1.11 is an SEM image, and the images of fracture surfaces in Figure 11.7 were created with a SEM. An SEM uses an electron accelerating voltage of 20 to 50 kilovolts (kV), and the SEM images only the surface of the specimen. In an SEM, the electron beam is created and focused in the same manner as in a TEM. However, in an SEM the electron beam is scanned across the surface of the material in a manner similar to the way that televisions with cathode-ray tubes scan the image across the television screen. The primary electron beam scanning across the surface of a material results in a number of emissions from the surface that are used to analyze the material. Most images from SEMs are from secondary electron emissions, which are electrons emitted from atoms when struck by the primary electron beam. The secondary electrons are collected at a detector as the primary beam scans the surface. If the primary beam is hitting a high spot on the specimen, this high spot emits more secondary electrons to the detector than does a cavity. High spots are bright, and a cavity looks dark. In this way a secondary electron image of the surface is created.

The primary electron beam exciting electrons from the atoms results in characteristic X-ray emissions from atoms that are struck in the specimen. The X-rays coming from the surface can be imaged with a detector in a manner similar to the secondary electron image. With the electronics available with solid-state detectors, the distribution of a particular type of atom on the surface is observed by accepting only the signal from X-rays of the characteristic energy of a particular atom type. In this way the distribution of a particular type of atom on the surface is determined.

Specimens for an SEM must conduct electrons; otherwise the surface of the specimen becomes electrically charged by the primary electron beam. Since metals conduct electricity, they are observed directly in an SEM. Materials that are not conductors, such as ceramics, plastics, and biological materials, are coated with a conductor, such as gold or carbon. You may have seen pictures of ants or other insects taken in an SEM. They are coated to make them conductive. One advantage of an SEM is that thick specimens can be studied with little or no specimen preparation; however, only the surface is analyzed. An SEM is also good for analyzing irregular surfaces, such as the fracture surfaces shown in Figure 11.7. The magnification in a typical SEM is 10,000; however, in a high-resolution SEM a magnification of 500,000 is possible.

The **scanning tunneling microscope** (STM) was invented in 1981. The image of atoms on the surface of nickel at the beginning of Chapter 2 was taken with an STM. Figure 15.19 is a schematic of an STM. In an STM, an atomically sharp probe is scanned across the surface to be investigated at a distance of a few tenths of a nanometer from the surface. A small constant voltage is established between the probe and the surface. A small electrical current, called the **tunneling current**, flows in the gap between the specimen and the probe. The tunneling current is a quantum mechanical effect where electrons have a probability of jumping across the gap between the probe and the sample even though classically no current should be present. An electronic feedback system keeps the tunneling current constant by positioning the probe at a constant distance from the surface. The image produced is from the vertical motion of the probe necessary to keep the tunneling current and distance from the specimen constant. As shown in the image of the nickel surface at the beginning of Chapter 2, the STM is capable of atomic resolution. In an STM, the specimen surface must be conductive. Only the surface is observed in an STM, and it is possible to observe the surface of a bulk specimen. An STM can be operated in air or in a vacuum. The atomic resolution in an STM, and the AFM discussed next, is made possible by very precise piezoelectric positioning devices. With a piezoelectric transducer (PZT), a controlled expansion or contraction of the material is produced by an applied voltage. We will discuss piezoelectric materials in Chapter 16.

In 1986 **the atomic force microscope** (AFM) was invented, and the AFM allows the surface of nonconductors to be viewed with atomic resolution. Figure 15.21 is a schematic of an AFM. The operation of an AFM in the contact mode is similar to the operation of record player, and of you sliding your finger over the bark of a tree to determine its roughness. On a vinyl record, changes in height in the grooves are sensed with a probe (the needle) and converted into sound as the vinyl record is rotated under the fixed needle. When you slide your finger over the bark of a tree, your finger moves up and down with the contour of the tree bark. From the up-and-down motion of your finger you sense the roughness of the bark surface.

In an AFM the sample is attached to a PZT, and the sample is scanned beneath an atomically sharp probe that is on the end of a thin cantilevered beam. In the contact mode of operation there is a constant applied force measured in nano newtons between the tip and the specimen. The up-and-down motion of the tip as the sample moves beneath the tip is measured with a laser light beam reflected off the end of the cantilever. The recorded image is of the up-and-down motion of the tip (z) as a function of the x and y positions of the probe on the surface. With electronic processing, different heights are converted into different colors, as shown in Figure 1.8. AFMs are used to study metals, ceramics, polymers, and biological materials. The specimen can be in air, in vacuum, and even in liquid solutions. Figure 1.8 is an image of individual carbon atoms in a nanocrystal of graphene.

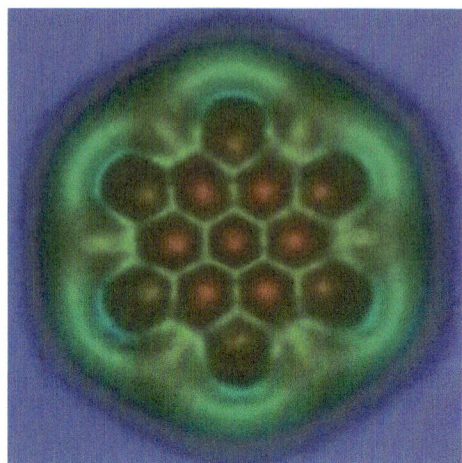

Figure 1.8   An AFM image of carbon atoms in a nanocrystal of graphene. Graphene is a single layer of graphite, as shown in Figure 2.16f. (*IBM Research - Zurich*)

1.2.3   The Development of Materials by Scientific Methods and Engineering Applications

Now materials scientists develop new materials using the atomic theory of matter and experimental tools, such as those discussed above. With the theory of matter and with modeling and simulation techniques, it is now possible to predict the behavior of materials before they are produced. Once the materials are produced, scientists can determine the physical properties and characterize the structure of new materials from the macroscopic level down to the atomic level. The development of high-temperature gas turbine blades provides an example of how the theory of matter and experimental methods were applied to improve the performance of jet aircraft engines.

1.2.4   An Example of Material Development: High-Temperature Gas-Turbine Blades

When aircraft propelled with gas turbines, shown in Figure 1.9, were first introduced during World War II, the operational lifetime of a turbine blade with a surface temperature of 650°C was about 10 to 25 hours. After that, chunks of metal would start flying out of the back of the engine. The turbine blade, shown in Figure 1.10, converts the energy of the burning fuel to power the compressor for the turbine. The expanding gas impinges upon the turbine blade, which turns the turbine. The stationary stator vane redirects the expanding gas onto the next stage of turbine blades. The first turbine blade materials were made from materials available at the time. Materials used in the turbine blades of modern jet engines allow the turbine blade to run at surface temperatures up to 1150°C, for operational lifetimes up to 50,000 hours for aircraft and up to 100,000 hours for land-based turbines. It is the turbine blade material that limits the turbine operating temperature. How was this improvement in material properties accomplished?

(a)

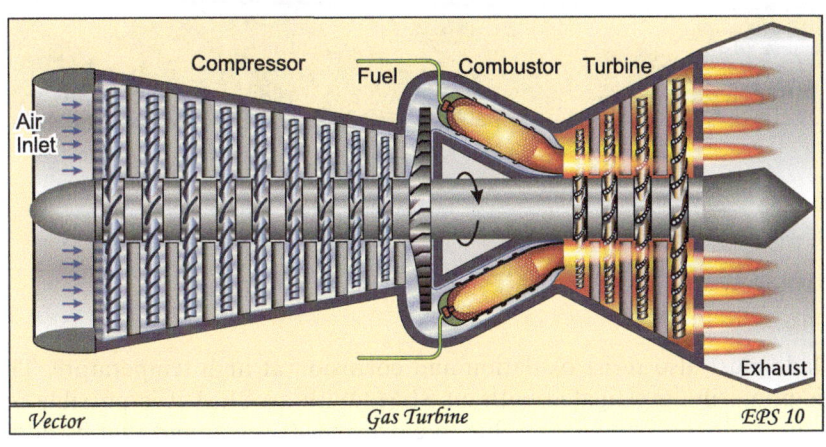

(b)

Figure 1.9 (a) A cutaway of a General Electric J85-GE-17A turbojet engine, showing the internal parts of the engine. This engine was used in the A-37 attack aircraft. (*Sanjay Acharya*) (b) A schematic of the sections of a jet engine. (*Fouad A. Saad / Shutterstock.com*)

Turbine blades are made of metals because some metals have sufficiently high melting temperatures for a gas turbine, and metals deform rather than break. Ceramics have high melting temperatures, but they are brittle and fracture without deforming. Nickel is the metal of choice for gas-turbine blades and vanes. It is known that metals with the atom arrangement of iron are more subject to oxidation than metals with the atom arrangement of nickel. Nickel has the same atom arrangement as gold, silver, and platinum. The arrangement of atoms in crystals is presented in Chapter 2.

Unfortunately, pure nickel is very low in strength. It was discovered that if a few percent aluminum is added to nickel, the strength increases considerably. Transmission electron microscopy (TEM) was used to determine that in nickel-aluminum mixtures, small particles of a compound Ni_3Al formed inside the nickel. The TEM was used to determine the shape, chemistry, and atom arrangement of the small Ni_3Al particles. It was found that very small Ni_3Al particles spaced closely together provided the highest strength. The diameter of the particles and their spacing is measured in nanometers. In Chapter 7 equations are developed to predict the strength of metal alloys. X-ray spectroscopy was used to determine the overall chemistry needed to obtain the desired strength.

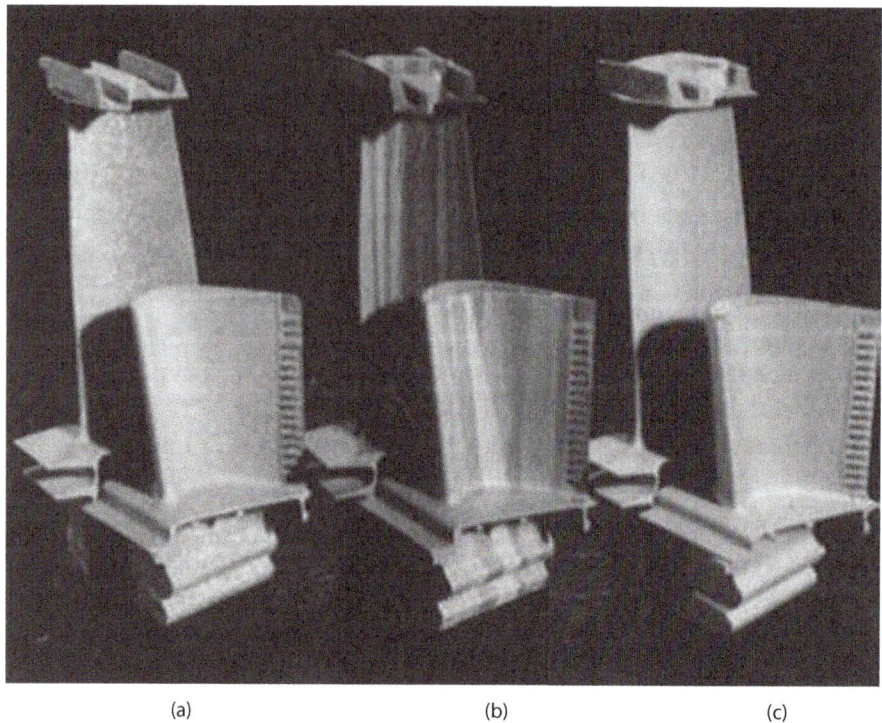

Figure 1.10   (a) A polycrystalline gas-turbine blade (front) and stator vane (behind) showing the grain structure. (b) A directionally solidified blade and vane. (c) A single crystal turbine blade and vane. (*Reprinted from Donachie, M.J. and Donachie, S.J, Superalloys: A Technical Guide, 2nd Edition, 2002, ASM International.*)

A turbine blade must also resist oxidation and corrosion at high temperature. The failure of a turbine blade is primarily investigated with an SEM. With an SEM it is possible to determine if the failure was due to loss of strength or due to corrosion. It was found that by adding chromium along with aluminum, the resistance to high-temperature corrosion was increased. By adding these elements a protective layer of chromium oxide or aluminum oxide formed on the metal. In Chapter 10 we discuss the high-temperature reaction of metals with oxygen and the corrosion of metals. The chemistry of the protective layers is determined by X-ray spectroscopy in an SEM. The metal in a modern turbine blade of a jet engine is called a superalloy. **Superalloys** are metal alloys based primarily upon nickel that are designed for high-temperature use, and some superalloys are based upon cobalt and iron. Aluminum, chromium, niobium, molybdenum, tungsten, and occasionally other elements are added to increase the high-temperature strength of superalloys. We discuss superalloys in Chapters 8 and 9.

The first turbine blades produced were polycrystalline metal, and polycrystalline superalloys are still used for most of the blades produced today. Figure 1.10a shows a polycrystalline turbine blade and a vane that were produced by casting liquid metal in a mold. A **polycrystalline** material is made up of many small crystals, called grains. Each grain has a regular arrangement of atoms. Grains of salt and grains of sand are also small crystals. The grains of a metal are joined together to form a polycrystalline metal. In a conventional casting, liquid metal is poured into a solid mold and the metal is cooled by contact with all of the surfaces of the mold. The grains in conventional cast turbine blade are randomly oriented, and they are visible in Figure 1.10a. Figure 1.10b shows a turbine blade and vane where the molten metal in the casting is cooled from one end, causing the grains to grow in a direction from the cool end to the hot end in a process called **directional solidification**. Directional solidification of turbine blades and vanes results in significantly improved high-temperature strength over conventional cast turbine blades and vanes with randomly oriented grains. Figure 1.10c shows a single crystal turbine blade and vane. In a single crystal of nickel, the atoms are arranged into small cubes that all line up in three dimensions throughout the entire crystal. Turbine blades and vanes made from single crystals have higher strength for extended times at high temperatures than polycrystalline and directionally solidified turbine blades and vanes. In Chapter 9 we discuss why single crystal turbine blades have a longer life at high temperature. Special crystal solidification procedures are necessary to produce single crystal turbine blades, and we will cover these in Chapter 13 on materials processing. Single crystal turbine blades are now used in high-performance jet gas turbines, such as those used for fighter aircraft.

This example of the development of superalloy turbine blades shows how the theory of matter and the experimental methods developed to study matter are used to develop a new material. The characterization of the structure of this material included

- The arrangement, positions, and types of atoms in the material
- The structure of the material from atomic size to macroscopic
- The nanostructure at dimensions of 10^{-9} m
- The microstructure at dimensions of 10^{-6} m
- The physical properties

Materials scientists determined how and why the strength changed when foreign elements, such as aluminum, niobium, and molybdenum, were added to nickel, and why a single crystal turbine blade had a longer life than a polycrystalline blade when operating at high temperature. This information is used by materials engineers to develop new materials for high-temperature gas-turbine blades that may be either single crystalline or polycrystalline. Engineers use superalloys to design new and improved gas-turbine blades, engines, and aircraft.

During the twentieth and twenty-first centuries, engineers have been able to produce an amazing array of products because of the new materials that have been developed by materials scientists. Strong, lightweight aluminum and titanium metal alloys allowed for the development of the body of jet aircraft. Polymers and composite materials allowed for the development of automobiles and jet aircraft that are lighter in weight and more fuel efficient. Corrosion-resistant metal alloys made possible the development of prosthetic devices to replace hip and knee joints.

1.2.5   The Discovery of New Nanomaterials: Buckyballs and Carbon Nanotubes

Nanomaterials are materials with whose size is measured in nanometers (10^{-9} m). A major materials science development in the twentieth century was the discovery in 1985 of spheres of 60 carbon atoms (C_{60}) naturally occurring in soot. C_{60} has been given the name **Buckyball** because the pattern of the atoms is similar to geodesic domes designed by R. Buckminster Fuller, and C_{60} looks like a soccer ball with a diameter measured in nanometers. The discovery of C_{60} was quickly followed by the discovery of **carbon nanotubes** (CNTs) in soot. A single-walled CNT is a sheet of graphite that is one atom layer thick wrapped into a tube that is a few nanometers in diameter and microns in length. There are also CNTs with multiple walls. Figure 2.16 shows the atom arrangements in C_{60} and in a CNT. Figure 1.11 is an SEM image of a bundle of aligned carbon nanotubes. C_{60} and CNTs are so small that they must be

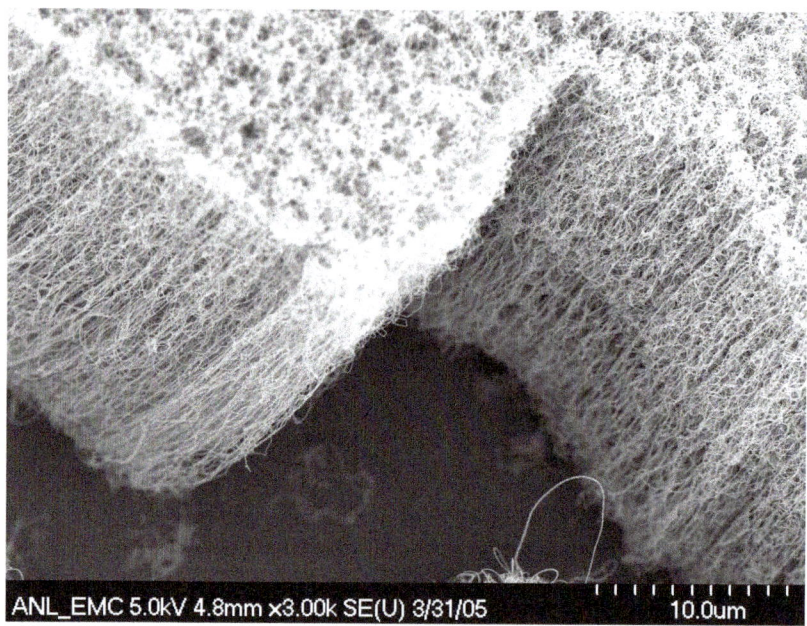

Figure 1.11   An SEM image of a bundle of aligned CNTs. *(Courtesy Argonne National Laboratory)*

viewed in microscopes with high magnification. Individual CNTs have been proposed as a replacement for silicon in transistors. CNTs are also one of the strongest materials ever found, and this has resulted in structural applications in composite materials for golf club shafts, hunting arrows, baseball bats, hockey sticks, surfboards, and many other applications.

The discovery C_{60} and of CNTs has stimulated great interest in nanotechnology that is applied to materials in electronics, medicine, structural materials, and biomaterials. **Nanotechnology** is the study, manipulation, and manufacture of materials with dimensions measured in nanometers.

Galileo Galilei was one of the first people to realize that a small specimen of a material is stronger than a large specimen of the same material. He was investigating why ships under construction collapsed under their own weight, but small-scale models were stable. The conventional reasoning was that a larger structure is stronger because it can support more force. Now we know that strength should be calculated as the force that can be supported by a structure divided by the cross-sectional area of the structure. A large structure is weaker based upon force per unit area because a large structure can have large defects. A small part cannot have a large defect, and nano-diameter filaments can have at the most nano-dimensioned defects.

1.2.6   Materials Science and Engineering Organizations

Prior to 1973, materials scientists belonged to various professional organizations, such as the American Society for Metals, the American Ceramic Society, the Society of Plastic Engineers, the American Chemical Society, the American Physical Society, the American Society of Mechanical Engineers, the American Institute of Aeronautics and Astronautics, the Institute of Electrical and Electronics Engineers, and the American Society of Civil Engineers. In 1973 the Materials Research Society (MRS) was established to provide a forum for the interchange of information across all of these disciplines. In addition, to keep the affiliation with their profession, most MRS members maintain membership in individual discipline professional societies. At the time this book was written the MRS has approximately 16,000 members in over 80 countries.

1.3 ENGINEERING APPLICATIONS OF MATERIALS: SOME INTERESTING CASE STUDIES

The development of new materials has made possible the creation of important products, such as high-performance automobiles, aircraft and aerospace vehicles, high-temperature gas turbines, high-performance sports equipment, and medical devices. In this book we develop the background necessary to understand the materials used in these products. This section presents a discussion of some interesting materials that in some cases are absolutely necessary for the application to function, and in other cases improve performance. The role that materials scientists, materials engineers, and mechanical and aerospace engineers play in developing these materials and products is briefly discussed.

1.3.1 Thermal Protection of Reentry Vehicles

When a reentry vehicle, such as the space shuttle, enters earth's atmosphere, the outer surfaces of the vehicle are heated due to the frictional interaction with the atmosphere. For example, the nose tip and the wing leading edges of the space shuttle reached temperatures of up to 1650°C. A thermal protection system (TPS) was required to prevent fires and equipment failure during reentry. Some of the materials utilized in the space shuttle TPS are shown in Figure 1.12. The surfaces that reach 1650°C to 1260°C are covered with reinforced carbon-carbon (RCC) tiles. The tiles have a carbon matrix that is reinforced with carbon (graphite) fiber. Carbon, in the form of graphite, is used in these areas because the melting temperature of graphite in the absence of oxygen is 3550°C. In graphite the atoms are arranged in a plane, and the planes of graphite slide over each other quite easily. This is why graphite is used as a solid lubricant.

How is a lubricant transformed into a super-strong fiber? Graphite is very strong parallel to the direction of the graphite planes. Materials scientists have learned to grow fibers of graphite where the planes are parallel to the fiber axis. Graphite fibers are stronger than the strongest high-strength steel wires, and they are one-quarter the density of steel. The graphite fibers are woven into a cloth for fabrication into RCC tiles.

The RCC tiles are made by first impregnating graphite fiber woven cloth with the polymer furan resin. Furan is a natural polymer found in oat hulls, and it is a byproduct of the manufacture of oatmeal. The graphite fiber cloth impregnated with furan resin is then carbonized by heating in an inert atmosphere to temperatures up to 1000°C. In carbonization elements in furan other than carbon are vaporized, resulting in amorphous chains of carbon. Additional furan is added to the composite to fill voids created by the vaporization of elements, and the composite is again carbonized. This process is repeated for up to four times to obtain a dense matrix of amorphous carbon. The RCC tiles are finally coated with silicon carbide to resist oxidation (burning) during reentry of space vehicles. RCC composites are utilized in other very high temperature applications, such as in rocket motors and in aircraft braking systems.

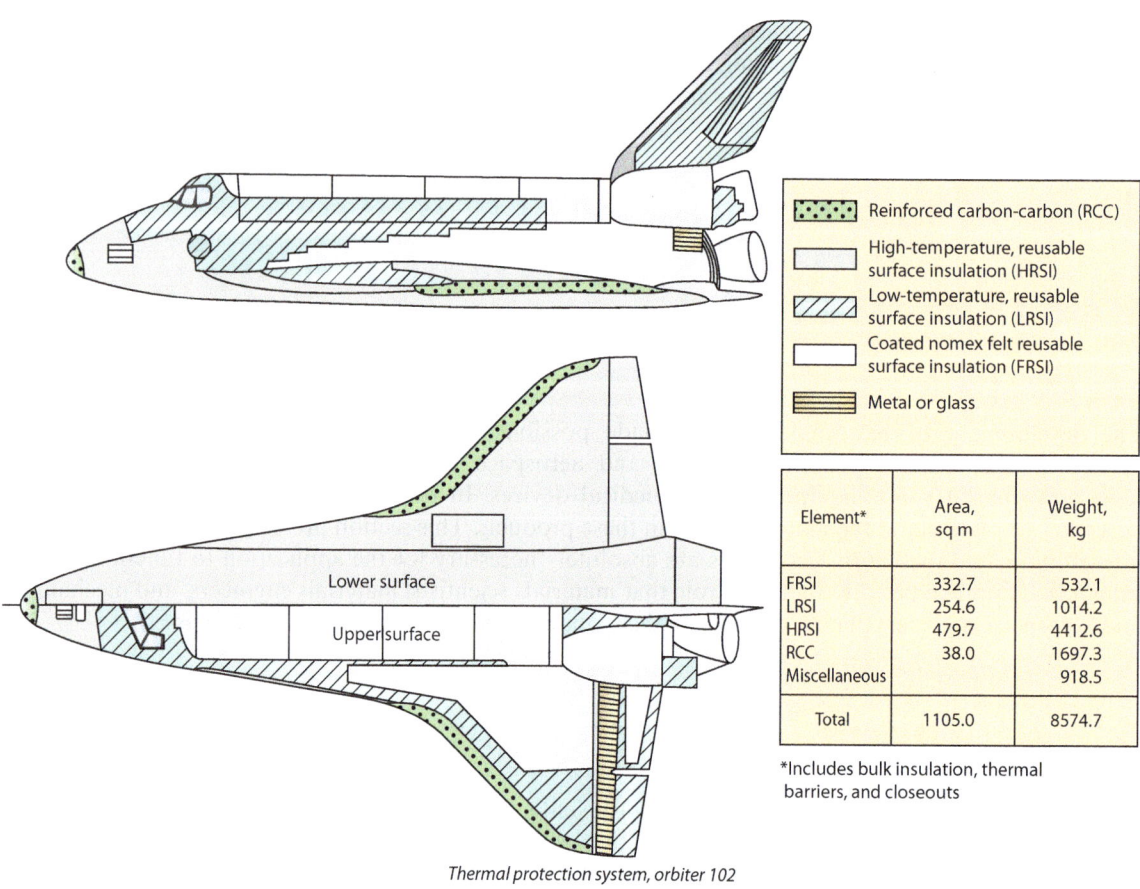

	Reinforced carbon-carbon (RCC)
	High-temperature, reusable surface insulation (HRSI)
	Low-temperature, reusable surface insulation (LRSI)
	Coated nomex felt reusable surface insulation (FRSI)
	Metal or glass

Element*	Area, sq m	Weight, kg
FRSI	332.7	532.1
LRSI	254.6	1014.2
HRSI	479.7	4412.6
RCC	38.0	1697.3
Miscellaneous		918.5
Total	1105.0	8574.7

*Includes bulk insulation, thermal barriers, and closeouts

Lower surface

Upper surface

Thermal protection system, orbiter 102

Figure 1.12   Materials used in the TPS in the space shuttle, and their locations. (*NASA*)

Even though the graphite fibers are strong, the RCC tiles are quite brittle and are subject to fracture, as occurred in the space shuttle *Columbia* disaster described in the caption of the introductory photograph of this chapter. If the carbon fibers in the RCC tiles are strong, why did the RCC tiles break on *Columbia* when hit by a piece of foam insulation? In Chapter 11 on the fracture of materials, we will find that high strength does not mean high resistance to fracture. In fact, resistance to fracture is general inversely related to strength. A piece of glass is stronger than a piece of lead. Lead is easily bent, but it does not fracture if we drop it on a concrete floor. If we drop glass on a concrete floor, it fractures.

Surface areas of the shuttle that reached temperatures from 1260°C to 650°C were covered with high-temperature reusable surface insulation tiles, as shown in Figure 1.12, made of silica (SiO_2) fibers. The outer surface of the tile was coated with a borosilicate glass. The tile was 95% space, because the density of the fibrous silica tile was only 0.14×10^3 kg/m³, compared with the 2.65×10^3 kg/m³ density for silica. The fibrous structure was similar to that of a sponge.

Surface areas that reached temperatures from 650°C and 370°C were also covered with a low-temperature reusable surface insulation of ceramic tiles made of SiO_2 fibers; however, the outer surface of the tile was coated with a silica-alumina (Al_2O_3) compound.

Surface areas that were at temperatures below 370°C were covered with a nomex™ felt reusable surface insulation. Nomex is a trade name for a flame-resistant polymer. Nomex can be made into a fiber, and the fibers can be made into a flexible felt. Firefighters wear fabric clothes made of nomex.

The development of new materials, such as graphite fibers and the RCC tiles, enabled the development of the TPS for the space shuttle. Materials scientists have developed high-strength graphite fibers, silica tiles, and nomex fiber. Materials scientists have characterized the structure and physical properties of these materials. Materials engineers have produced RCC tiles, and they have determined engineering properties of these products, such as the maximum operating temperature, the strength, and the resistance to fracture. Engineers designed the space shuttle TPS using these materials. The development of the TPS is just one example of the many unique materials in the space shuttle.

1.3.2 Composite Materials for the Boeing 787 Dreamliner

Composite materials are incorporated into many aerospace, civil, mechanical, and structural engineering applications, such as aircraft, automobiles, buildings, ships, and bridge reinforcement; and in large storage containers used to hold liquids, such as water or fuel. One of the most recent structural applications of composite materials is the Boeing Aircraft 787 *Dreamliner*, shown in Figure 1.13. Figure 1.14 shows the forward section of the airplane, which is made of a graphite-fiber-reinforced epoxy composite.

This is the first time that a large passenger plane is to use composite materials extensively. Eighty percent by volume of the construction materials are composites. Normally the fuselage (body) in a commercial passenger plane is made from an aluminum alloy. Graphite-fiber-reinforced epoxy composite has approximately three times the strength of the aluminum alloy metal it is replacing, and only 60% of the density. The graphite-fiber-reinforced epoxy is the gray-brown color shown in Figure 1.14. The use of graphite-fiber-reinforced epoxy in the fuselage will make the plane lighter, resulting in increased fuel efficiency, shorter takeoff and landing distances, and a longer flight range.

Materials scientists have developed high-strength graphite fibers and epoxy resins used in composite materials. Materials engineers have developed graphite-fiber-reinforced epoxy composite materials suitable for aircraft applications, and they have determined the engineering properties of these composite

Figure 1.13   Photograph of the Boeing 787 *Dreamliner*. (*© Robert Schlesinger/dpa/Corbis*)

Figure 1.14   The assembly of the forward section of the Boeing 787 *Dreamliner*, showing the composite material structure. (*markjhandel/Flickr*)

materials. Engineers use these composite materials in the design of parts for products, such as aircraft and automobiles.

1.3.3   Ultra-High-Strength Polymer Fibers

New lightweight high-strength materials have made possible the development of new sports, such as kitesurfing. Figure 1.15 shows a kitebsurfer coming off an ocean wave. In kitesurfing, the kiter controls a large kite that is typically 12 m² in area with four 25 m–long lines that are connected to a control bar on one end and the kite on the other. The kiter rides a small board, as shown in Figure 1.16. The high-strength lines are made from oriented ultra-high-molecular-weight polyethylene (OUHMWPE). The molecular weight is the average mass of the polymer molecules. Polyethylene (PE) is the same polymer that is in plastic milk bottles. However, materials scientists have learned to make polyethylene into a fiber that is as strong as high-strength steel and that has a density only one-eighth that of steel. In OUHMWPE, the PE long-chain molecules (LCMs) have 2 to 6 million mer units. Because the ultra-long-chain molecules are oriented parallel to the fiber axis, the fiber is incredibly strong. In a milk bottle, the molecules are only 2 to 5 thousand mer units long, and the LCMs are randomly oriented.

These high-strength lines allow kiters to be pulled across the water at high speed: At the time this book went to press, the kitesurf speed record was 94.38 km/h. Kiters also like to get some air time, as shown in Figure 1.16. Air time is when the kiter lifts off from the water using both the motion of the kite and the board, and then drifts down to the water by using the kite as a parachute. The record height for kiters getting air time has been set by kiters who have been unintentionally caught in winds that carried them up to heights of 50 meters or more. Fortunately their lines did not break, and they did come safely back to earth. Kitesurfing requires a kite and kite lines made of strong, lightweight polymers. The kite board,

Figure 1.15   A kitesurfer coming off an ocean wave. (*Eyalos / Shutterstock.com*)

shown in Figure 1.16, is primarily made of composite materials with strong reinforcing fibers, such as OUHMWPE, glass, and graphite; however, wood is often used in the core of the board.

OUHMWPE and other similar materials have found many other applications in addition to kiteboarding lines. Some bullet-proof vests are made of OUHMWPE. In bullet-proof vests, the fiber is woven into a fabric, and the fabric is layered to produce the thickness required to stop a bullet. OUHMWPE is used in fishing line, parachute and paraglider lines, bow strings, high-performance sails, cut-resistant gloves, and lines for tow boats. OUHMWPE is added to composite materials to provide both strength and resistance to fracture.

Materials scientists have developed high-strength OUHMWPE, and they have characterized the structure and physical properties of this material. Materials engineers incorporate these fibers into new composite materials to add strength and resistance to fracture. Engineers use OUHMWPE and new composite materials in the design of high-performance sporting goods and other products.

Figure 1.16   A kitesurfer getting some air time. (*Mai Techaphan / Shutterstock.com*)

1.3.4   A Smart Metal with a Trainable Memory: Biomedical Applications

The metal alloy 50 atomic percent titanium (Ti) and 50 atomic percent nickel (Ni) has found many applications because of its unique properties. You may have encountered TiNi orthodontic wires that pull your teeth into proper position, or eyeglass frames that if bent will return to shape by heating in hot water. TiNi is also used for variable-geometry chevrons for jet engines. The chevron is at the back of the jet engine, and it helps to control the direction of exhaust gases coming from the engine. Flexing the chevron during takeoff significantly reduces engine noise. Another application of TiNi is as a stent that is implanted inside the arteries of the heart to assist in blood flow. Fortunately, TiNi is compatible with the human body and is therefore suitable for implants.

TiNi is a smart metal that has a memory for a shape at specific temperatures, and it can be trained to assume a certain shape at specific temperatures. In a demonstration of the shape-memory effect, a straight wire of TiNi is bent into a new shape, such as a circle, at room temperature. Then the bent TiNi wire is placed in hot water, and it immediately straightens. Why does a wire of TiNi straighten in hot water? After finishing this book, you should be able to answer this question yourself, but the following paragraphs discuss the shape-memory effect without getting too involved in the science that is covered in Section 5.15.

The TiNi wire is normally formed straight at elevated temperatures, such as 500°C. At these temperatures the atoms in the TiNi are arranged in small cubes that regularly repeat over long distances. If the straight TiNi is cooled to room temperature, the square sides of the TiNi cubes are distorted into parallelograms by small shifts in position of the atoms. If the room-temperature TiNi wire is bent into a new shape, parallelogram distortions occur that accommodate the new bent shape of the wire. The sides of a cube can be distorted in many different directions to form parallelograms. When the TiNi is heated in hot water, the distorted atom arrangement goes back to the original cubic shape when the wire was first formed at high temperature, and the wire straightens out. This is the shape-memory effect.

It is not necessary that the TiNi always be in the shape of a straight wire at high temperature. By forming TiNi into some irregular shape at high temperature, the cubes of TiNi are formed in the irregular shape. If the TiNi is first cooled to room temperature and deformed into some new shape, and then the TiNi is heated to high temperature, the TiNi returns to the irregular shape. Also, by repeatedly forming TiNi into a certain shape at a low temperature, it can be trained so that the distortions occur to assume that shape at low temperature. The ability of TiNi to return to certain shapes at high and low temperatures is because at high temperature TiNi returns to the cubic arrangement of atoms, and at low temperature TiNi transforms into a distortion of the cubic structure. Most materials do not change their atom arrangement in this manner. Changes in atom arrangements in many different types of materials, including TiNi, are discussed in Chapter 5.

As mentioned previously, one application of TiNi is as a stent in coronary arteries. To produce a stent, a thin-walled TiNi tube of a diameter appropriate for insertion into arteries is formed at elevated temperature. The tube is then laser-cut at low temperature into the shape shown in Figure 1.17. After laser cutting, the stent is electropolished to reduce its interaction with blood. Electropolishing is discussed in Chapter 10. The stent is then heated to blood temperature, which will be the high temperature, and expanded to the shape it will assume in the artery. At this point, the expanded TiNi stent has the high-temperature cubic structure. The stent is then compressed at low temperature into a delivery tube that insulates it from warm blood and constrains it from expanding. At low temperature the TiNi has the distorted structure. The compressed stent in the delivery tube is placed at the end of a catheter, which is a medical tube inserted into the body, for

Figure 1.17   A photograph of a TiNi coronary stent. (*Alexey Kamenskiy / Shutterstock.com*)

insertion into the coronary artery. When the stent is in the proper location, it is pushed out of the catheter and into the coronary artery. The warm blood is sufficient to change the TiNi atoms into the cubic arrangement that expands the stent to the design size. The stent is sized so that it exerts a small pressure against the artery walls, opening the artery and allowing increased blood flow. By adjusting the ratio of titanium atoms relative to nickel atoms, or by substituting a small amount of cobalt for nickel, the temperature at which the TiNi assumes the cubic arrangement of atoms can be adjusted so that the blood temperature is sufficient to cause the high-temperature cubic arrangement of atoms to form.

Other applications of TiNi depend upon the observation that when TiNi transforms from its low-temperature distorted structure to its high-temperature cubic structure, its volume contracts by 3%. This contraction in orthodontic TiNi wires is what pulls teeth into position. The contraction is also utilized to make pipe couplings in submarines and other critical coupling applications. The coupling is cooled, and the two pipe ends to be joined are inserted into the coupling. The coupling is then warmed so that the 3% contraction occurs, clamping the pipe ends in a tight seal. Couplings of this sort are used where heating with a torch to fabricate or repair a piping system using a soldered joint is unsatisfactory, such as in a submarine, where oxygen is limited and a fire could be catastrophic.

The guide wires for a medical catheter can also be made of TiNi. The guide wire is a long wire attached to the catheter that guides the catheter from the insertion point in the thigh all the way up to the heart. Typically the TiNi guide wire is made of four different wires that can be individually heated by an electric current. When one of the four wires is heated, it can contract by up to 3%, but the other three wires retain their original lengths. If the single wire is heated just a small amount, the contraction of the wire is a fraction of 3%. This small contraction of one wire bends the guide wire in the direction of the shortened wire, shaping the guide wire into the desired curvature to guide it and the stent through the arteries from the thigh to the heart.

Materials scientists have analyzed the changes in the atom arrangements in TiNi to understand the shape-memory effect, and they have studied how changing the TiNi composition affects the temperature of the phase transformation. Materials engineers have developed techniques to manufacture TiNi alloys that have the correct temperatures for the change from the cubic arrangement to the distorted arrangement for applications, such as coronary stents. Engineers use the TiNi alloys in the design of products, such as coronary stents, pipe couplings, and jet-engine chevrons.

1.4   MATERIALS IN ENGINEERING DESIGN

A goal of this book is to teach engineers how to select the best materials available for a design. For example, what materials should we select for the chassis of a highly fuel-efficient, low-price automobile? To do this, engineers must know what materials are available to them. This book discusses all of the major classes of materials currently available, including metals, ceramics, plastics, and composite materials.

The engineer must know what physical properties are important for a design. An automobile chassis must be stiff and strong so that it does not change shape during use. It must absorb energy and protect its occupants in an accident. For fuel efficiency, a low density is important. Finally, the manufacturer must be able to sell the automobile at a profit. This book discusses the mechanical properties that relate to these considerations for each of type of material.

At the beginning of the automobile chassis design, all materials should be considered, but those with properties that make them unsuitable for the application should be eliminated. Are there any materials that are unsuitable for the chassis of an automobile? Ceramics are strong, but in a crash a ceramic chassis would shatter and provide little protection to the automobile occupants. For this reason, ceramics are eliminated at the beginning. It should also be possible to recycle the material. It would be desirable if the recycled material could be used for the same application or for another high-volume application. In Chapter 14, several methodologies are presented to compare all of the candidate materials for the important mechanical properties and for cost to determine the best material for the application. For an automobile chassis, the candidate materials are low-carbon steel; a high-strength, low-alloy steel; an aluminum alloy; a reinforced engineering polymer; and graphite-reinforced epoxy. Which material do you think best meets the design criteria? In Chapter 14 we will conduct material selections to answer this question.

1.5   THE CONTENTS OF THIS BOOK

It is not always sufficient to select materials based upon tables of strength and other physical properties. If engineers understand why a material has certain mechanical properties, they can more intelligently use that material in a design, and they are more likely to avoid selecting the wrong material. For example, the strength of a cold-formed steel automobile chassis results from the cold-forming process. If cold-formed steel is heated above a certain temperature in either fabrication or in service, the strength is decreased. The design engineer must know that this material cannot be heated by fabrication procedures, such as hot-working, and during service it cannot experience elevated temperatures. To provide this understanding, this book builds upon the studies of matter in chemistry and physics classes and applies this knowledge to engineering materials. This book uses a scientific approach to understanding engineering materials. Chapter 2 presents the chemical bonding and structure of materials, including metals, ceramics, and polymers. Important physical and mechanical properties, such as density, melting temperature, and hardness, are related to chemical bonding. This book uses an integrated approach to understanding physical and mechanical properties of materials by considering the similarities and differences of metals, ceramics, and polymers within the individual chapters. Chapter 3 discusses defects and alterations in materials. Many physical and mechanical properties, such as strength and fracture resistance, are directly related to the defects and alterations in materials. Chapter 4 discusses the effects that heat and temperature have on the structure of materials. In this chapter we will find that even in solids the atoms change their positions, and this can lead to the mixing of atoms that are initially separated. Chapter 5 discusses phase transformations and phase diagrams. Phase diagrams of two different elements or compounds show the phases present in a material as a function of temperature. Phase diagrams are important for understanding processes such as metals casting and strengthening. Chapter 6 introduces the mechanical properties of materials and focuses on their deformation and strength. Chapter 7 explains how to develop strong materials. Chapter 8 presents the different types of materials available to engineers and some of their mechanical properties. Chapter 9 presents the effect of temperature on the mechanical properties of materials. Chapter 10 presents the effect of oxidation and corrosion on the mechanical properties of materials. The theory of corrosion is electrochemical, and this theory also explains electroplating, electropolishing, electromachining, batteries, and fuel cells. Chapter 11 discusses the fracture and fatigue of materials. Chapter 12 covers composite materials. Chapter 13 discusses how products are produced from materials. Chapter 14 is devoted to procedures for selecting the best material for a design. The desired outcome for the engineering design is the selection of the most cost-effective material for the performance required. Chapter 15 presents experimental methods in more detail than does the present chapter. That concludes the printed textbook chapters of this publication; the remaining chapters are online. Chapter 16 is on the conductivity of materials. This chapter includes the conductivity of metals, ceramics, polymers, semiconductors, and superconductors. In addition, it covers dielectrics and junctions. Chapter 17 covers the magnetic properties of materials. Chapter 18 covers the photonic properties of materials. This includes the response of materials to electromagnetic radiation from infrared to X-rays.

Summary

- The use of materials has been so important to human development that the advancement in technology is classified on the basis of the primary materials used—for example, the stone, bronze, iron, and silicon ages.
- The major classes of materials are metals, ceramics, polymers, and composites.
- Metals are inorganic materials with metallic bonding.
- Ceramics are inorganic nonmetallic materials with covalent or a combination of covalent and ionic bonding.
- Ceramic pottery is the first known human-made material, dating to approximately 27,000 BCE.
- Smelted copper was in use by approximately 6000 BCE, and bronze was used extensively by 3500 BCE.
- Smelted iron was in use by approximately 2000 BCE, and the blast furnace to produce steel was invented in 1855.
- Polymers are large molecules with many repeating units. The units can be inorganic or organic.
- In some polymers, the large molecules are not covalently bonded to each other, resulting in thermoplastic polymers. In other polymers, the large molecules are covalently bonded to each other, resulting in thermoset polymers.
- Composite materials are made from at least two separate materials that maintain their identity in the composite.
- Materials science is the application of the theory of matter to the understanding, development, and improvement of materials.
- Materials engineers specialize in the application of materials to engineering designs.
- The development of materials by scientific methods is a result of the theory of matter and of experimental methods that allow detailed study of materials.
- Nanotechnology is a recent materials development that will affect human society significantly. Nanotechnology is the study, manipulation, and manufacture of materials with dimensions measured in nanometers.
- In the selection of a material for a design, the cost and the performance must be considered.

Supplemental Reading Subjects and Authors

Full references are listed at the end of the book.

General	*Askeland, Fulay, and Wright*
History of metals:	*Raymond*
Superalloys:	*Sims, Stoloff, and Hagel; Donachie and Donachie*
Shape-memory effect in TiNi:	*Wang*
Polymers:	*Osswald and Menges*

Homework

Concept Questions

1. If the atoms are in irregular arrangements, the material is _____.

2. In a _____ the atoms are in regular arrangements.

3. The structure at dimensions of 10^{-9} m is called _____ structure.

4. The materials that engineers have available for use in designs include metals, ceramics, plastics, and _____ materials.

5. Alumina (Al_2O_3) is a(n) _____ material.

6. Ceramics are often used in high-_____ applications.

7. The process of bonding small solid particles together at high temperature without melting them is called _____.

8. Iron is classified as a(n) _____.

9. _____ are mixtures of metal elements with other elements.

10. _____ is an alloy of copper and tin.

11. _____ is when a body is changed in shape.

12. A material that can permanently deform extensively before breaking is _____.

13. A _____ material is one that breaks without any significant permanent deformation and with little absorption of energy.

14. A negative mechanical property of ceramics is that they are _____.

15. The bodies of most automobiles are made of low-carbon _____.

16. The strength of low-carbon steel used in automobile bodies is achieved by _____ forming.

17. _____ are molecules with many repeating units.

18. The _____ is the basic building block of a polymer.

19. _____ are materials that are made out of polymers plus additives.

20. In the process of vulcanization the LCMs in latex are bonded together with _____ atoms to produce rubber.

21. The most important properties of polymers are their low _____ and their high corrosion resistance.

22. In high-strength OUHMWPE, the LCMs are oriented _____ to the fiber axis.

23. _____ materials are made of at least two separate materials that maintain their identity in the combined material.

24. The _____ in a composite material is continuous.

25. The function of the glass fibers in fiberglass is to increase the _____.

26. One of the most important mechanical properties of metals is their capacity for _____ deformation.

27. One of the most important physical properties of ceramics is their high _____ temperature.

28. One of the most important physical properties of polymers is their low _____.

29. A metal or ceramic composed of many small crystals is a _____.

30. The base metal in most jet-engine turbine blades is _____.

31. A major factor limiting the use of single crystal gas-turbine blades is _____.

Engineer in Training–Style Questions

1. The smelting of a metal
   (a) Raises the temperature above the melting point
   (b) Separates the metal from the ore
   (c) Deforms the metal into a shape
   (d) Cools the metal very rapidly from a high temperature

2. What is the first known human-made material?
   (a) Ceramic
   (b) Bronze
   (c) Flint arrowheads
   (d) Copper

3. The oldest-known smelted iron dates to approximately what year?
   (a) 27,000 BCE
   (b) 6000 BCE
   (c) 2500 BCE
   (d) 1000 BCE

4. Wrought iron is
   (a) Heated and then rapidly quenched into water
   (b) Packed in carbon to harden the iron
   (c) Heated in carbon monoxide
   (d) Deformed into shape

5. The maximum magnification of a typical light optical microscope is
   (a) 100,000 times
   (b) 10,000 times
   (c) 1000 times
   (d) 100 times

6. A microscope that can view the surface of insulators with atomic resolution is
    (a) A scanning electron microscope
    (b) A scanning tunneling microscope
    (c) A light optical microscope
    (d) An atomic force microscope

7. Which of the following is not a property of a ceramic such as alumina?
    (a) A high melting temperature
    (b) A high resistance to fracture
    (c) It is a good insulator.
    (d) It is inorganic.

8. Which of the following is not a property of a polymer?
    (a) Large molecules with repeating units
    (b) Can be either organic or inorganic
    (c) A high density
    (d) Good resistance to corrosion

9. The chassis of Formula 1 race cars is now made from
    (a) High-strength steel
    (b) High-strength aluminum
    (c) High-strength polymers
    (d) Graphite-fiber-reinforced epoxy

10. In the space shuttle, the areas exposed to the highest temperature are made of
    (a) Graphite-fiber-reinforced carbon-carbon composite
    (b) Silica tiles coated with silicon carbide
    (c) Graphite-fiber-reinforced epoxy composite
    (d) Alumina tiles

11. The highest-temperature aircraft gas-turbine blades are made from
    (a) Tungsten
    (b) Single crystals of nickel-based alloys
    (c) Polycrystals of nickel-based alloys
    (d) Alumina

12. Which of the following is not an appropriate use of OUHMWPE?
    (a) Milk bottles
    (b) Parachute lines
    (c) Bow string
    (d) Tug tow rope

13. A design project requires a TiNi wire to be shaped like the letter S at an operating temperature of 100°C. The selected TiNi atom arrangement starts to distort at a temperature of 90°C and the distortion is finished at a temperature of 60°C. The wire was drawn straight at a temperature of 500°C, and the wire supplied is straight. The wire is easily bent into the required S shape at room temperature. What will be the shape of the wire at the operating temperature of 100°C?
    (a) S shaped
    (b) C shaped
    (c) Straight
    (d) The shape is impossible to know.

14. A design requires a material that must have a high melting temperature, high stiffness, high compressive strength, low density, and low cost. Which of the following materials is most likely to best satisfy the design requirements?
    (a) Alumina ceramic
    (b) Carbon steel
    (c) Polyethylene
    (d) Graphite-reinforced-epoxy composite

15. A design requires a material to operate at normal atmospheric temperatures, be subjected to small applied forces, and have a low density and a low cost. Which of the following materials is most likely to best satisfy the design requirements?
    (a) Alumina ceramic
    (b) Carbon steel
    (c) Polyethylene
    (d) Graphite-reinforced-epoxy composite

16. A design requires a material that must operate up to a temperature of 300°C, have a high resistance to deformation, a high strength, a high resistance to fracture, and a low cost. Which of the following materials is most likely to best satisfy the design requirements?
    (a) Alumina ceramic
    (b) Low-carbon steel
    (c) Polyethylene
    (d) Graphite-reinforced epoxy

17. A design requires a material that must operate up to temperature of 200°C, have a high resistance to deformation, a high strength, a high resistance to fracture, and a low density. Highest performance is the most important design criteria. Which of the following materials is most likely to best satisfy the design requirements?
    (a) Alumina ceramic
    (b) Low-carbon steel
    (c) Polyethylene
    (d) Graphite-reinforced epoxy

Design-Related Question

Based on what you have learned in this chapter and what you have found in other sources, what are the current advantages and disadvantages of using low-carbon steel and graphite-reinforced epoxy as the chassis material for a low-cost automobile with a high fuel efficiency?

An STM image of the surface of nickel, showing the effect of individual atoms. The STM is one of several instruments recently developed that allow materials scientists to view materials at the atomic level—which helps in understanding how physical properties are related to the atoms in the material.

IBM

The goals of this chapter are to understand

- The electron structure of the elements of the periodic table
- The differences among vapors, liquids, and solids
- The packing of spherical atoms in solids
- Crystal structures and how to describe them
- The identification of planes and directions in crystals
- The calculations for determining the density of atoms in crystals, planes, and directions
- The types of primary and secondary chemical bonding
- The ways some physical properties are related to chemical bonding
- The reason why some materials are crystalline and others are amorphous
- The ways to model the energy of bonding and interatomic forces between atoms and ions

Chapter

2

Atoms, Chemical Bonding, Material Structure, and Physical Properties

2.1   INTRODUCTION

A goal of materials science is to understand the origin of the physical properties of materials, such as the hardness and the high melting temperature of diamond. In this chapter we will relate some physical properties to the types of chemical bonds in materials. We will start with the analysis of pure elements and then move on to materials with two types of atoms. Important classes of materials with two atom types include ceramics such as silicon nitride (Si_3N_4) that is used in high-temperature applications; polymers such as polyethylene that can be made into fibers as strong as high-strength steel; and intermetallic compounds such as nickel aluminide (Ni_3Al) that strengthen nickel for high-temperature gas turbines.

When atoms bond together to form a solid, some atoms form in a regular arrangement that repeats over long distances, which creates a crystal. For example, the diamond in an engagement ring is a single crystal of carbon atoms, and some high-performance jet aircraft use single-crystal turbine blades with nickel as the main atom type. We will study the arrangements of atoms in some important types of crystals and learn how to refer to planes and directions in crystals. In contrast, the atoms in some materials, such as glass, do not repeat in a regular arrangement over long distances. We will see why some materials form crystalline structures and others form amorphous structures. The answer lies in the type of chemical bonding and the complexity of the atom arrangements. Finally, we will develop models of bonding for inert-gas atoms and ionic bonding, allowing for the calculation of physical properties and dynamics of atomic motion.

37

2.2   ATOM ELECTRONIC STRUCTURE

We will begin this chapter with a review of basics from chemistry and physics that are necessary to understand chemical bonding. However, this treatment is not intended to replace the more complete treatment in introductory chemistry and physics courses. It is assumed that students have already studied these subjects in more detail. If you have not taken these courses, it is recommended that you consult textbooks on these subjects to obtain a more complete understanding than is presented here.

The chemical bonding in materials results from the electronic structure of the atoms. In this book, we will use an atomic model of a nucleus made of protons and neutrons that is surrounded by electrons in orbitals. Physicists have discovered that protons and neutrons in the nucleus are made of other particles, but details of that level are beyond the scope of this course.

2.2.1   Atom Electron Energies

The **binding energy** of the electron to the nucleus is the negative of the energy required to remove an electron from the atom to an infinite distance from the atom. With this definition the binding energy is a negative number. The binding energy is similar to the energy to remove an object from a well. The energy of the object at the bottom of the well is negative, and the energy of the object at the top of the well is zero. The deeper is the well, the more negative (lower) is the object energy at the bottom of the well.

The binding energy of electrons to a nucleus in an atom is described with quantum mechanics models. Each electron on an atom has a quantum state described by a set of quantum numbers that does not change as long as the atom is unperturbed. The bond energy of an electron is dependent upon the quantum numbers. The constant quantum numbers explain why an orbiting electron does not run out of energy with time, because the quantum numbers of the electron do not change on the orbiting electron unless the electron is perturbed in some way.

On an atom the **principal quantum number** (n) gives the shell of the electron, as shown in Figure 2.1. The magnitude of the electron binding energy is inversely related to the principal quantum number (n). Electrons with lower principal quantum numbers have more negative bond energies (deeper wells), and

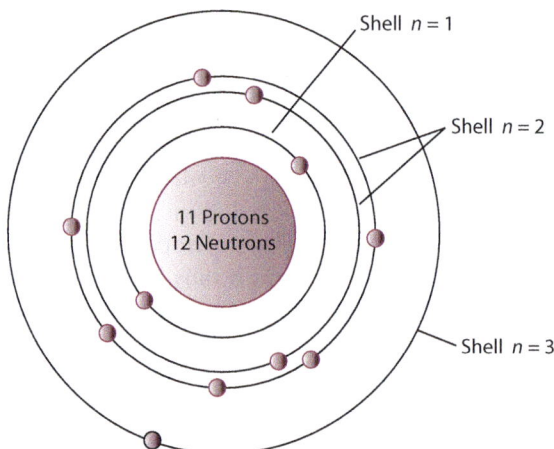

Figure 2.1   A quantum mechanical shell model of sodium (atomic number 11) showing the electron configuration $1s^2 2s^2 2p^6 3s^1$. (*Based on Askeland, D.R., and Phule, P.P., The Science and Engineering of Materials, 4th ed., Thomson-Brooks/Cole, Pacific Grove, Ca. (2003), p. 38.*)

electrons with higher principal quantum numbers have less negative bond energies. The possible principal quantum numbers are 1, 2, 3, 4, 5, ...

Hydrogen has only one electron, and its principal quantum number is $n = 1$. However, for atoms with two or more electrons, additional quantum numbers are necessary because of the **Pauli exclusion principle**: No two electrons with the same set of quantum numbers can occupy the same space at the same time, and therefore each electron on an individual atom must have a quantum state unique to that atom.

There are other quantum numbers in addition to the principal quantum number (n), such as the angular momentum quantum number (l). The **angular momentum quantum number** gives the subshell of the electron, and it determines the shape of the electron orbital. For example, the $l = 0$ electron orbital is spherically symmetric, and the $l = 1$ orbital is dumbbell shaped. The angular momentum quantum number l has a secondary effect on the electron energy. Electrons with increasing values of l have higher energies, and as before the lowest energy states fill first in the atoms of the periodic table. Possible values of l are $n - 1$, $n - 2$, . . ., 0. An s electron has $l = 0$, p electrons have $l = 1$, d electrons have $l = 2$, and f electrons have $l = 3$. Another quantum number is **the magnetic quantum number** (m_l); it has integer values of $-l$ to $+l$. The final quantum number is the spin s. Electrons spin about an axis, and the **spin quantum numbers** are $+\frac{1}{2}$ or $-\frac{1}{2}$, corresponding to spin up or spin down, respectively. In the absence of a magnetic field, the energy of an electron on an atom is independent of the magnetic and spin quantum numbers. In the absence of magnetic fields, the energy levels are defined only by the principal (n) and angular momentum (l) quantum numbers. Chapter 17, on magnetic materials, discusses the energy of the quantum states with magnetic fields. A quantum state is specified by all of the quantum numbers: the principal quantum number (n), the angular momentum quantum number (l), the magnetic quantum number (m_l), and the spin quantum number (s).

2.2.2 Atoms of the Periodic Table

The quantum numbers of the electrons on atoms are determined by using the Aufbau principle (that the electron orbitals with the lowest energy fill first), the Pauli exclusion principle, and the rules for the values of quantum numbers. For example, helium has two electrons. The lowest energy corresponds to electrons with the lowest principal quantum number of $n = 1$. According to the Pauli exclusion principle, this is possible if the two electrons have different quantum numbers. If the principal quantum number (n) is equal to 1, the only possible value of l is 0, and m_l must be 0. There can be two electrons with $n = 1$ if one has spin $-\frac{1}{2}$ and the other has spin $+\frac{1}{2}$. The electron configuration of helium is noted as $1s^2$, meaning that there are two $1s$ electrons. If the atom has three electrons, the third electron cannot have $n = 1$, because all quantum states available with $n = 1$ are filled.

Lithium has three electrons, two of which have $n = 1$ and one electron with $n = 2$. The lowest energy corresponds to the third electron having $n = 2$ and $l = 0$. The third electron is also an s electron, and the designation of the electron configuration is $1s^2 2s$, as shown in Appendix A.

For nonmagnetic applications, it is not important which values of the magnetic quantum number are filled first or which spin quantum number is filled first, for these quantum numbers have no effect on the electron energy. It is only important how many total magnetic quantum number states and spin quantum number states are available for a given value of the principal quantum number and angular momentum quantum number, and how many electrons there are to fill those states. For example, Figure 2.1 gives a shell model of the electrons around a sodium atom (atomic number 11) with 11 electrons. The 11 electrons are distributed in shells with principal quantum numbers $n = 1$, 2, and 3. There are two electrons in the $n = 1$ shell ($1s^2$), eight electrons in the $n = 2$ shell ($2s^2 2p^6$), and one electron in the $n = 3$ shell. In the $n = 1$ shell, according to rules for the quantum numbers, the only possible value for the l quantum number is $l = 0$, and m_l must also be 0. The possible values for the spin quantum numbers are $+\frac{1}{2}$ and $-\frac{1}{2}$. The $n = 1$ shell for sodium has the same quantum numbers as He ($1s^2$). In the $n = 2$ shell, according to rules for the quantum numbers, the possible values for the l quantum numbers are $l = 0$ and 1. There are two electrons ($2s^2$) with

$n = 2$ and a value for the l quantum number of 0 ($l = 0$), $m_l = 0$, and spin quantum number of $+\frac{1}{2}$ and $-\frac{1}{2}$. There are six electrons with $n = 2$ and a value for the l quantum number of one ($l = 1$) and magnetic quantum numbers (m_l) equal to -1, 0, and $+1$. Each of the quantum states with magnetic quantum number $m_l = -1$, 0, and $+1$ has two electrons with spin quantum numbers of $+\frac{1}{2}$ and $-\frac{1}{2}$, for a total of six $2p$ electrons ($2p^6$). With these electrons on the sodium atom, all of the quantum states in the $n = 2$ shell are filled. There are two electrons in the $n = 1$ shell, and eight electrons in the $n = 2$ shell for ten electrons. Sodium's eleventh electron must go into the $n = 3$ shell with $l = 0$ and $m_l = 0$, and the spin can be either $+\frac{1}{2}$ or $-\frac{1}{2}$, because there is no magnetic field. The electron configuration of sodium is therefore $1s^2 2s^2 2p^6 3s^1$.

Figure 2.1 shows that in a neutral sodium atom there are 11 protons, and in the most stable and abundant isotope there are 12 neutrons. The figure is not to scale, for the radius of a nucleus is typically 10^{-15} m and the radius of an atom is typically 10^{-10} m. The nucleus is a point in comparison to the atomic diameter. However, the mass of the atom is primarily contained in the nucleus. Each proton and neutron has a mass of 1.67×10^{-27} kg, and the electron mass is 9.11×10^{-31} kg. Almost all of the mass of the atom is in the nucleus; however, almost all of the volume of the atom is due to the size of the electron orbitals. Students interested in a more detailed discussion of the structure of the atom should consult the physics textbook by Reese.

Example Problem 2.1

Determine the electron configuration of the important engineering material carbon that is atomic number six.

Solution

The first three electrons go into the same electron quantum states as for lithium. The fourth electron can have the same quantum numbers as the third electron except that it must have opposite spin. It is a second $2s$ electron. The fifth electron has $n = 2$, but all of the $l = 0$ states are filled. The fifth electron must have $l = 1$. This is a $2p$ electron. The $2p$ quantum state can accommodate six electrons, and the sixth electron is another $2p$ electron. The electron configuration for carbon is $1s^2 2s^2 2p^2$.

2.2.3 Valence, Electronegativity, and Atom Stability

The periodic table gives each element's chemical symbol, group numbers, atomic number, molar mass, electronegativity, cohesive energy, melting temperature, and density of the atom at room temperature. The **atomic number** is the number of electrons or the number of protons in a neutral atom. The **molar mass** is the sum of the mass of the protons, neutrons, and electrons in a mole of average atoms with a single atomic number. Atoms can have different isotopes. Isotopes are atoms that have the atomic number of protons and electrons, but different numbers of neutrons. The molar mass is the mass in kilograms of a mole of average atoms. A mole of atoms contains **Avogadro's number** of atoms (6.02×10^{23}).

The **group number** at the top of the periodic table is the number of valence electrons for the atoms in that column. The **valence electrons** are the electrons in an incomplete electron shell that is outside of a spherically symmetric closed electron shell corresponding to an inert-gas atom. The inert-gas atoms are given the group number 0, because there are zero electrons outside of the spherically symmetric closed electron shell. The electron configuration of the inert-gas atoms provides a reference point for chemical bonding, because there is no primary chemical bonding between these atoms in the form of electron sharing or transfer. The electron configurations of the inert-gas atoms He, Ne, Ar, Kr, and Xe are given in Appendix A. Two s electrons in He complete the $n = 1$ shell of electrons, and this electron shell is spherically symmetric because s electrons have a spherically symmetric electron orbital. In the other inert-gas atoms the two s electrons (s^2) and six p electrons (p^6) form a spherically symmetric electron shell (s^2p^6).

Atoms, other than the transition elements, are placed in a group number from I to VII that is equal to the number of s plus p electrons in the valence electron shell. In the transition elements an electron subshell, such as the $3d$, $4d$, $4f$, $5d$, or $5f$, is incomplete. In general the group number of the transition elements is the number of electrons in the incomplete d or f subshell plus the number of s and p electrons in the outermost electron shell. For example, from Appendix A, the electron structure of chromium is $Ar + 3d^54s^1$. Chromium is in Group VI because of the five $3d$ electrons and the single $4s$ electron outside of the complete argon core, for a total of six valence electrons outside of the complete argon core. Iron is in Group VIII, because the electron structure is $Ar + 3d^64s^2$. Iron has six $3d$ electrons and two $4s$ electrons, for a sum of eight valence electrons outside of the complete argon core. Cobalt, nickel, rhodium, palladium, iridium, and platinum are all listed as being in Group VIII, even though they have more than eight valence electrons outside of a closed electron shell, because they have chemical properties that are similar to the elements iron, ruthenium, and osmium, which do have eight electrons outside of a closed electron shell. The rare earth and actinide series are also transition elements where the $4f$ and $5f$ subshells are incomplete, and the outer valence electron structure is similar in each series. The elements within each of these series of elements have similar chemical properties. These elements are located below the solid line in the periodic table, with the note [1].

The **electronegativity** of an atom is a relative measure of the atom's ability to attract electrons. The inert-gas atoms do not attract electrons, and their electronegativity is undefined. Atoms on the left side of the periodic chart have a low electronegativity; for example, lithium in Group IA has an electronegativity of 0.9, and fluorine in Group VII has the highest electronegativity in the periodic table of 4.0. In this row of the periodic chart, there is a nearly linear increase in electronegativity with atomic number.

The observation that atoms with eight s plus p electrons are inert indicates that this is a stable low-energy electron configuration. All atoms, other than the inert-gas atoms, have an incomplete shell of electrons outside of the inner inert-gas ion core. The incomplete shell of electrons is the valence electron shell. Atoms, other than the inert-gas atoms, chemically bond such that the atoms in the resulting material have an electron configuration as close as possible to the inert-gas atom electron configuration. When there is only one type of atom, an electron configuration similar to that of the inert-gas atoms results from sharing valence electrons. The way that the electrons are shared between atoms leads to the different types of chemical bonding. The types of chemical bonding are covered in introductory chemistry classes, and they are also reviewed in Sections 2.7 to 2.10, along with a discussion of how the type of bonding affects the arrangement of the atoms in materials and the physical properties of materials. The bonding of atoms to each other results in the formation of vapor, liquid, and solid materials.

2.3 VAPOR, LIQUID, AND SOLID MATERIALS

For this course we consider only three states of matter: vapor, liquid, and solid. We primarily discuss the properties of solids, but solids are usually produced from liquids or vapors, such as in casting a solid from a liquid or depositing a thin film of solid from a vapor. Therefore, it is important to understand these three forms of matter. In a **vapor** the moving units (either atoms or molecules) have weak interactions with each other and are not bonded to each other. The units can move unimpeded with kinetic energy over relatively large distances. The terms *gas* and *vapor* are essentially interchangeable.

When a vapor is lowered in temperature and condenses into a liquid, the units more strongly interact with each other. A **liquid** is a material that is easily changed in shape. In a liquid the units can move relative to each other. However, the velocity and distance of movement of the units in a liquid is significantly reduced relative to a vapor. In a liquid the bonds between the units are continuously being broken and re-formed with new neighbors. This allows the units to move relative to each other and for a liquid to flow. Because the units in a liquid move quite freely relative to each other, the units in a liquid are disordered. There is not a regular pattern to the positions of the units in a liquid, and we cannot predict the position of other units a large distance from an individual unit.

If the temperature of a liquid is lowered sufficiently, the liquid transforms to a solid. We are all familiar with water transforming to ice at temperatures below 0°C. A **solid** is a material that is not easily changed in shape, and a solid resists the application of nonhydrostatic forces. When nonhydrostatic forces are applied to a liquid, the units move relative to each other and the liquid easily changes shape permanently. A nonhydrostatic force applied to a material is not equal in all directions. A hydrostatic force is one that is equal in all directions. Pressure is a hydrostatic force. When you pull in one direction on a rubber band, the force you apply is nonhydrostatic. When nonhydrostatic forces are applied to a solid, the atoms do not easily move relative to each other, and the solid does not easily permanently change shape. A piece of steel is a solid. A rubber band is also solid because the rubber band snaps back into its original shape when the force is released, and the shape change during stretching is not permanent. A liquid does not return to its original shape when nonhydrostatic forces on it are reduced to zero.

2.3.1 Packing of Spherical Atoms in Solids

All elements when cooled sufficiently form solids, and this is true even for the inert-gas atoms. For example, argon forms a solid at 83.8 K. The inert-gas atoms have a spherical outer electron configuration, and we can think of inert-gas atoms as being like spherical billiard balls when considering how they pack together in a solid. If they are packed as closely as possible on a surface, they appear as in Figure 2.2a. This arrangement is **close-packing**, and each atom is surrounded by six other atoms in the same plane, except at the crystal edges. This is the most densely packed arrangement of equal-size spheres or atoms possible. A solid is three dimensional, and there are layers of close-packed atoms above those shown in Figure 2.2a. If the layer shown in Figure 2.2a is called the A layer, then atoms placed on top of the A layer naturally go to the low point among three atoms, as indicated by the B positions, as shown in Figure 2.2b. In Figure 2.2a there are also low points, labeled C on top of the A plane. If atoms are placed in the B positions on top of the A plane of atoms, there is not sufficient space to also place any atoms in C positions on the A plane. The next layer on top of the B layer can go on either the low points labeled C or on the A positions again, for both are low points on top of the B layer. The stacking of close-packed planes of spherical balls or atoms can go either ABA or ABC.

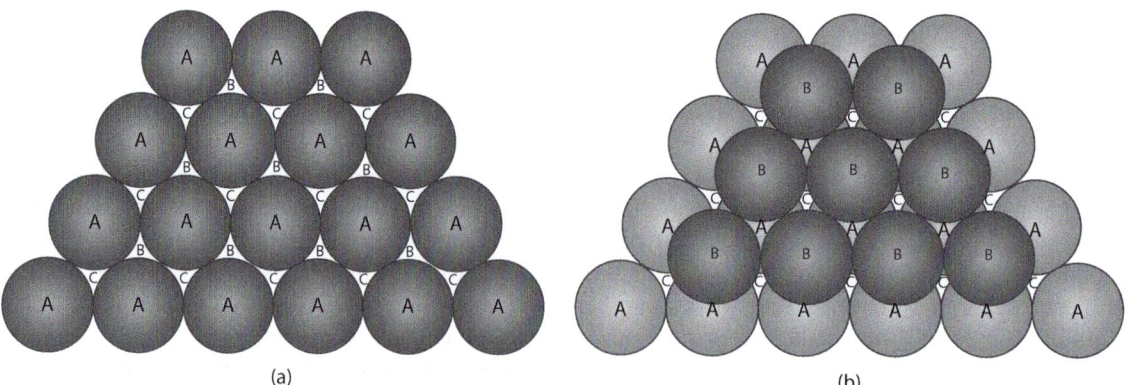

(a)   (b)

Figure 2.2   (a) Atoms in a close-packed plane labeled A, with the B and C positions indicated in low points between the A atoms. (b) Atoms placed in the B positions on top of the A layer. The next layer of atoms can go in either A positions or C positions.

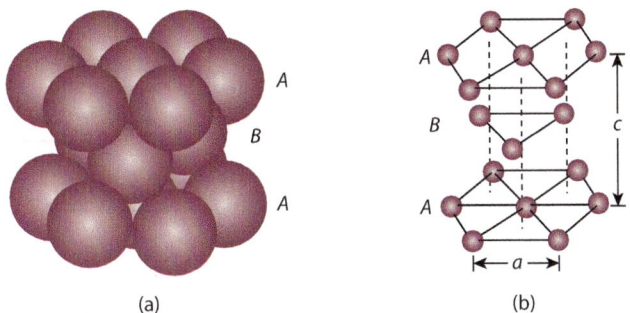

(a)   (b)

Figure 2.3 (a) Close-packed plane stacking $ABABA$ for the hexagonal close-packed (HCP) unit cell. (b) The hexagonal unit cell with reduced-size atoms, showing the atom positions and the lattice parameters a and c. ((a) Adapted from Cullity, B. D. Elements of X-Ray Diffraction, Addison-Wesley Pub. Co., Inc. Reading, MA (1956), p. 44. (b) Adapted from Askeland, D. R., Fulay, P. P., and Wright W. J., The Science and Engineering of Materials, 6th ed. Cengage Learning, Stamford, CT (2011), p. 71.)

In a perfect crystal, this stacking of atoms repeats over and over, so that the stacking is either $ABABABABABA$, which results in **a hexagonal close-packed** (HCP) crystal structure, as shown in Figure 2.3, or $ABCABCABCABC$, which results **in a face-centered cubic** (FCC) crystal structure, as shown in Figure 2.4. The close-packed plane in the FCC crystal structure is noted in Figure 2.4a. The atoms shown in Figures 2.3 and 2.4 are unit cells of atoms, and when they are repeated over and over in all directions, a crystal is produced. **A crystal** is an orderly array of atoms in space. Each of these crystal structures is close-packed, with twelve nearest neighbors to any individual atom, six nearest neighbors on the same plane, three nearest neighbors on the plane above, and three on the plane below. Twelve is the maximum number of nearest neighbors and the highest packing density of atoms. Materials that form in these crystal structures have the maximum density possible for atoms of that atomic mass. The inert-gas atoms when cooled to form a solid all have the FCC arrangement of atoms. Metals such as copper, silver, gold, nickel, and aluminum also have FCC structures. Metals that form in the HCP arrangement of atoms include titanium, cobalt, zinc, cadmium,

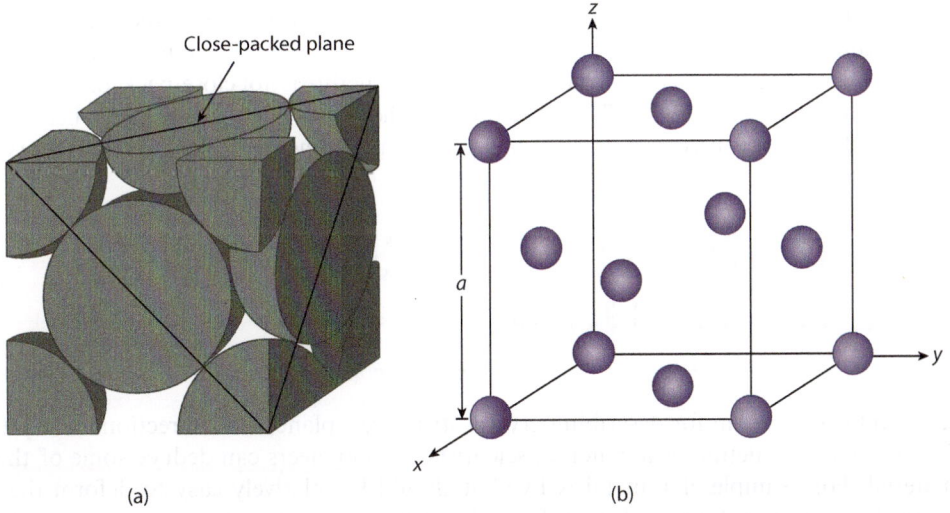

(a)   (b)

Figure 2.4 (a) The face-centered cubic (FCC) unit cell, with atoms touching along the close-packed directions. The close-packed planes $ABCABC$, cut diagonally through the FCC unit cell as shown. At the corners and face centers, only the amount of the atom in this unit cell is shown. (b) An FCC unit cell with reduced-size atoms, showing the atom positions and the lattice parameter a. (Adapted from Callister, W. D., Materials Science and Engineering, An Introduction 6th ed. John Wiley & Sons, New York (2003), p. 33.)

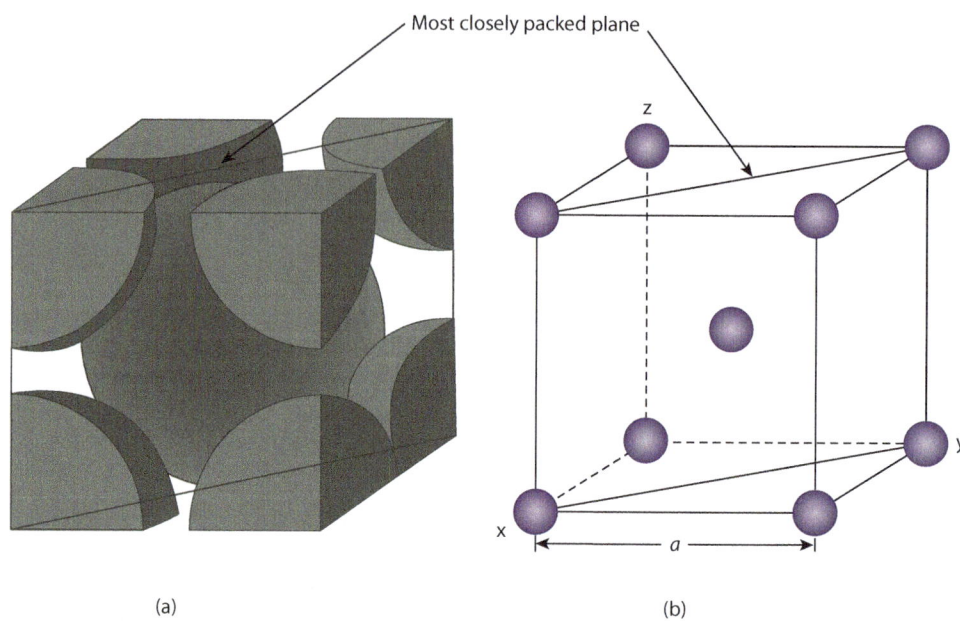

Most closely packed plane

(a)   (b)

Figure 2.5 (a) The body-centered cubic (BCC) unit cell, with atoms touching along the most closely packed direction through the body diagonal. At the corners only the fraction of the atom in this unit cell is shown. (b) The BCC unit cell with reduced-size atoms, showing the atom positions and the lattice parameter a. (*Adapted from Callister, W. D.,* Materials Science and Engineering: An Introduction, *6th ed. John Wiley & Sons, New York (2003), p. 35.*)

and zirconium. The reason that some elements form in the HCP structure and others in the FCC structure depends upon interaction forces between atoms that must be determined with quantum mechanical models.

The bonding in many materials is more complex than with the inert gases, which results in a variety of different crystal structures. For example, another common crystal structure for metals is the **body-centered cubic** (BCC) unit cell, shown in Figure 2.5. Metals such as sodium, iron, chromium, molybdenum, niobium, and tungsten form in the BCC crystal structure. The BCC structure is not close-packed; each atom has eight nearest neighbors rather than twelve. The structures of other materials are presented when these materials are discussed. But first it is necessary to develop a better understanding of how to describe crystals.

2.4 CRYSTAL STRUCTURES AND CRYSTALLOGRAPHY

Crystallography is a system for describing crystal structures, planes, and directions in crystals. Just by knowing the crystal structure of a material, scientists and engineers can deduce some of the properties of a material. For example, if a metal is FCC, it should be relatively easy to deform the material in processing through procedures such as rolling. The language and conventions of crystallography are necessary to understand because they allow scientists and engineers to communicate with each other about material structures with mutual understanding. Students desiring a more complete understanding of crystallography can consult books that specialize in that subject, some of which are listed at the end of this chapter.

2.4.1   Crystal Structures

The HCP, FCC, and BCC crystal structures, shown in Figures 2.3, 2.4, and 2.5, are examples of the unit cells of crystal structures. The **unit cell** is a small group of atoms or molecules that contains all of the necessary information about the crystal. The unit cell when repeated in space produces the entire crystal. There are three bits of information that are needed about the unit cell:

1. The type of the unit cell
2. The size of the unit cell
3. The positions of atoms in the unit cell

There are 14 different possible unit cell types for three-dimensional crystals, as shown in Figure 2.6; these are called the **Bravais lattice** types. The 14 Bravais lattices are variants of the seven crystal systems, which are listed in Table 2.1 along with their characteristics.

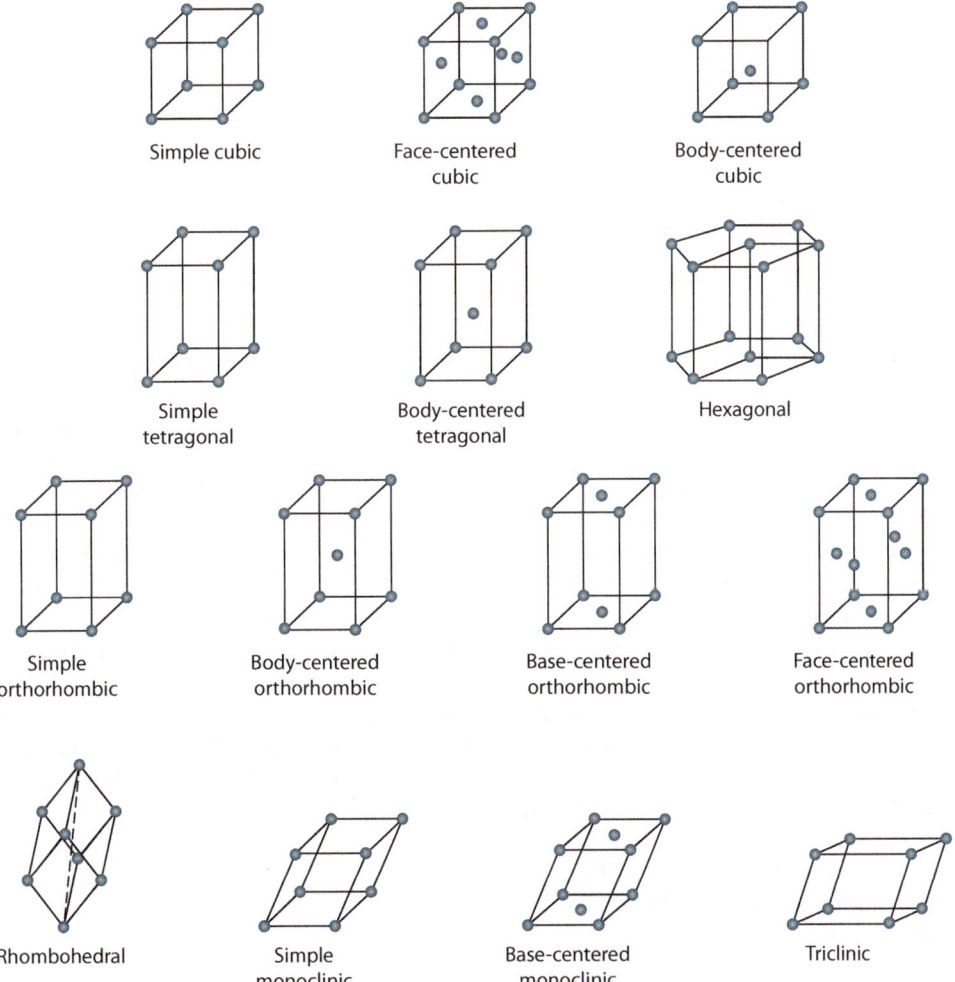

Figure 2.6   Unit cells of the 14 point lattices (Bravais lattices). (*Based on Askeland, D.R., Fulay, P.P., and Wright, W.J., The Science and Engineering of Materials, 6th ed. Cengage Learning, Stamford, CT. (2011), p. 62.*)

Table 2.1 **Characteristics of the Seven Crystal Systems**

Crystal System	Lattice Parameters	Lattice Angles
Cubic	$a = b = c$	Each angle equals 90°.
Tetragonal	$a = b \neq c$	Each angle equals 90°.
Orthorombic	$a \neq b \neq c$	Each angle equals 90°.
Hexagonal	$a = b \neq c$	Two angles each equal 90°, and the angle between a and b equals 120°.
Rhombohedral	$a = b = c$	All angles are equal to each other, and none of them equals 90°.
Monoclinic	$a \neq b \neq c$	Two angles each equal 90°, and one angle does not equal 90°.
Triclinic	$a \neq b \neq c$	No angles are equal to each other, and no angles equal 90°.

The *Bravais lattice* is made up of lattice points. The **lattice points** indicated in the unit cells are equivalent points. Lattice points are not atom locations until atoms are associated with a lattice point. Each lattice point can have one or more atoms associated with it. When the atom arrangement around one lattice point is specified, then each lattice point has the same atomic arrangement. The Bravais lattice type gives the shape of the unit cell, such as cubic, and it also indicates if there are additional lattice points in the unit cell. A primitive unit cell contains only one lattice point, such as the simple cubic unit cell. In a continuous crystal, the eight corners of a cubic unit cell are shared by eight surrounding unit cells, resulting in only one-eighth of a lattice point being contributed by each corner of a cube. The sharing at a corner is shown in Figures 2.4a and 2.5a, where only one-eighth of a sphere is shown in the unit cell at each corner. The BCC unit cell is cubic in shape, with a lattice point at the eight corners of the cube and a lattice point at the body center of the cube. The BCC unit cell, shown in Figures 2.5a and b, has two lattice points, one contributed by the eight corners and one contributed by the lattice point at the body-center position. If there is more than one lattice point in the unit cell, then the unit cell is nonprimitive. The BCC unit cell, for example, is nonprimitive.

The dimensions of the unit cell are given by the **lattice parameters** (a, b, and c). In unit cells without right angles at the corners, the **lattice angles** α, β, and γ are specified. The lattice parameters and crystal structure of some of the elements are presented in Appendix B, where the dimensions are in angstrom units (10^{-10} m). Lattice parameters in this book are normally presented in nanometers (1 nm = 10^{-9} m).

The coordinates of lattice points and of atoms are given in units or fractions of the lattice parameter, separated by commas. Using an x, y, and z coordinate system as shown in Figure 2.4b, the origin is given the coordinates 0, 0, 0. The lattice point a distance equal to the lattice parameter a along the x axis has the coordinates 1, 0, 0. The lattice parameter itself is not given. The lattice point at the body center has the coordinates $\frac{1}{2}, \frac{1}{2}, \frac{1}{2}$.

The atom positions only need be specified around the lattice point at the origin, because the arrangement of atoms around each lattice point is the same as it is around the origin. For example, as shown in Appendix B, nickel is FCC, with a lattice parameter (a) of 0.352 nm. The following specifies the location of all of the atoms in an FCC nickel unit cell:

1. Unit cell or Bravais lattice type: FCC
2. Lattice parameter: $a = 0.352$ nm
3. Atom positions: Ni at 0, 0, 0

Putting an atom at the lattice point 0, 0, 0 also puts an atom at all of the other lattice points in the FCC unit cell, since each lattice point has the same atomic arrangement. The resulting structure is shown in Figure 2.4.

Example Problem 2.2

Calculate the number of atoms in the FCC unit cell shown in Figure 2.4a.

Solution

In the FCC unit cell there are eight atoms at the eight corners of the cube; however, each of these atoms is shared with eight other cubes at each corner. Therefore, there is only one-eighth of an atom at each of the eight corners inside the unit cell, and the eight corners then contribute one total atom to the unit cell. The six faces of the unit cell also have an atom at the face center that is shared with another unit cell on the other side of the face. Each of the six faces contributes one-half of an atom, resulting in three atoms for the unit cell from the six faces. Therefore, the total number of atoms in the FCC metal unit cell is four.

$$8 \text{ corners} \times \tfrac{1}{8} \text{ atom per corner} = 1 \text{ atom at corners}$$

$$6 \text{ faces} \times \tfrac{1}{2} \text{ atom per face} = 3 \text{ atoms at face centers}$$

$$\text{Total number of atoms} = 1 \text{ atom at corners} + 3 \text{ atoms at face centers} = 4 \text{ total atoms}$$

Example Problem 2.3

For FCC copper, calculate (a) the number of atoms per unit volume using only the lattice parameter of 0.362 nm from Appendix B, and (b) the density of copper using the number of atoms per unit volume and the molar mass from the periodic table.

Solution

a) Since copper is FCC, the volume of a cubic unit cell is

$$a^3 = (0.362 \times 10^{-9} \text{m})^3 = 0.047 \times 10^{-27} \text{m}^3$$

In the FCC unit cell, there are four atoms. The number of atoms per unit volume (n_a) is

$$n_a = \frac{4 \text{ atoms}}{0.047 \times 10^{-27} \text{ m}^3} = 8.50 \times 10^{28} \frac{\text{atoms}}{\text{m}^3}$$

b) From the periodic table, the mass of one mole of copper atoms is 63.5 g or 0.0635 kg, and the units of density are kg/m^3. From dimensional analysis, the density of copper (ρ_{Cu}) can be obtained from the equation

$$\rho_{cu} = \frac{n_a M_{cu}}{N_A} = 8.50 \times 10^{28} \frac{\text{atoms}}{\text{m}^3} \left(\frac{\text{mole}}{6.02 \times 10^{23} \text{ atoms}} \right) 63.5 \times 10^{-3} \frac{\text{kg}}{\text{mole}}$$

$$= 89.7 \times 10^2 \frac{\text{kg}}{\text{m}^3} = 8.97 \times 10^3 \frac{\text{kg}}{\text{m}^3}$$

where M_{Cu} is the molar mass of copper and N_A is Avogadro's number.

From Appendix B, the density of copper is listed as 8.93×10^3 kg/m^3. Our calculated number is close. The difference is due to rounding-off errors.

Example Problem 2.3 shows that number of atoms per unit volume can be determined in two ways. One is by using the unit cell dimensions and the number of atoms in a unit cell, as in part (a). The other way

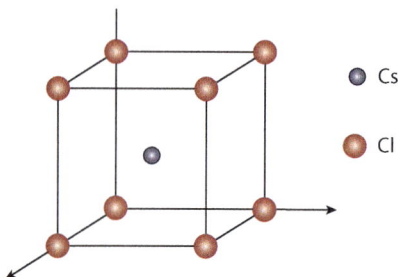

Figure 2.7   The crystal structure of CsCl. *(Based on Askeland, D.R., Fulay, P.P., and Wright, W.J., The Science and Engineering of Materials, 6th ed. Cengage Learning, Stamford, CT. (2011), p. 67.)*

is that if the density and the molar mass are known, then the equation in Example Problem 2.3 can be used to determine the number of atoms per unit volume (n_a).

Example Problem 2.4

The unit cell of cesium chloride (CsCl) has cesium ions at the eight corners of a cube and a chlorine ion at the body-centered position of the cube, as shown in Figure 2.7. What is the unit cell type, and what is the designation of the atom positions?

Solution

CsCl is simple cubic. If CsCl was BCC, the atoms at the corners and the body-centered position would have to be the same. In CsCl the corner atoms are different than the body-centered atom. The unit cell is simple cubic, and the atom positions are

$$\text{Cs at } 0, 0, 0.$$

$$\text{Cl at } \tfrac{1}{2}, \tfrac{1}{2}, \tfrac{1}{2}$$

This problem demonstrates the difference between lattice points and atom locations in a crystal.

2.4.2   Crystal Planes

In crystals it may be necessary to specify a certain plane. For example, the close-packed plane in an FCC crystal, indicated in Figure 2.4a, is the plane where permanent deformation of FCC metals occurs. A specific plane in a crystal is specified by the **Miller indices** (hkl), and Figure 2.8 shows several planes in a unit cell and the Miller indices. The Miller indices of a plane are found by using the following rules:

1. Choose an origin at a corner of a unit cell where the plane cuts through the x, y, and z axes within the distances of plus or minus the lattice parameters a, b, and c. If the plane cuts through the origin, move the origin to another lattice point of the unit cell so that the plane cuts through the axes at a distance equal to one of the lattice parameters from the origin. Remember that all lattice points are equivalent; therefore, the origin can be at any lattice point.
2. Determine the distance from the origin to the plane intercepts along the x, y, and z axes as fractions of the lattice parameters a, b, and c. Positive and negative intercepts are possible.
3. The Miller indices (hkl) are the inverse of the fractional intercepts determined previously in step 2.
4. For a specific plane, the Miller indices (hkl) are given in parentheses without any punctuation.

5. The Miller indices must be integers or 0. If there is a fractional index, multiply through by a factor that produces the smallest possible set of integers. If the origin is chosen properly, there should not be any fractions. Multiplication by a factor is equivalent to moving the origin.

6. Negative intercepts are indicated by a bar over the index, such as $(00\bar{1})$. This is called the "zero, zero, one bar" plane.

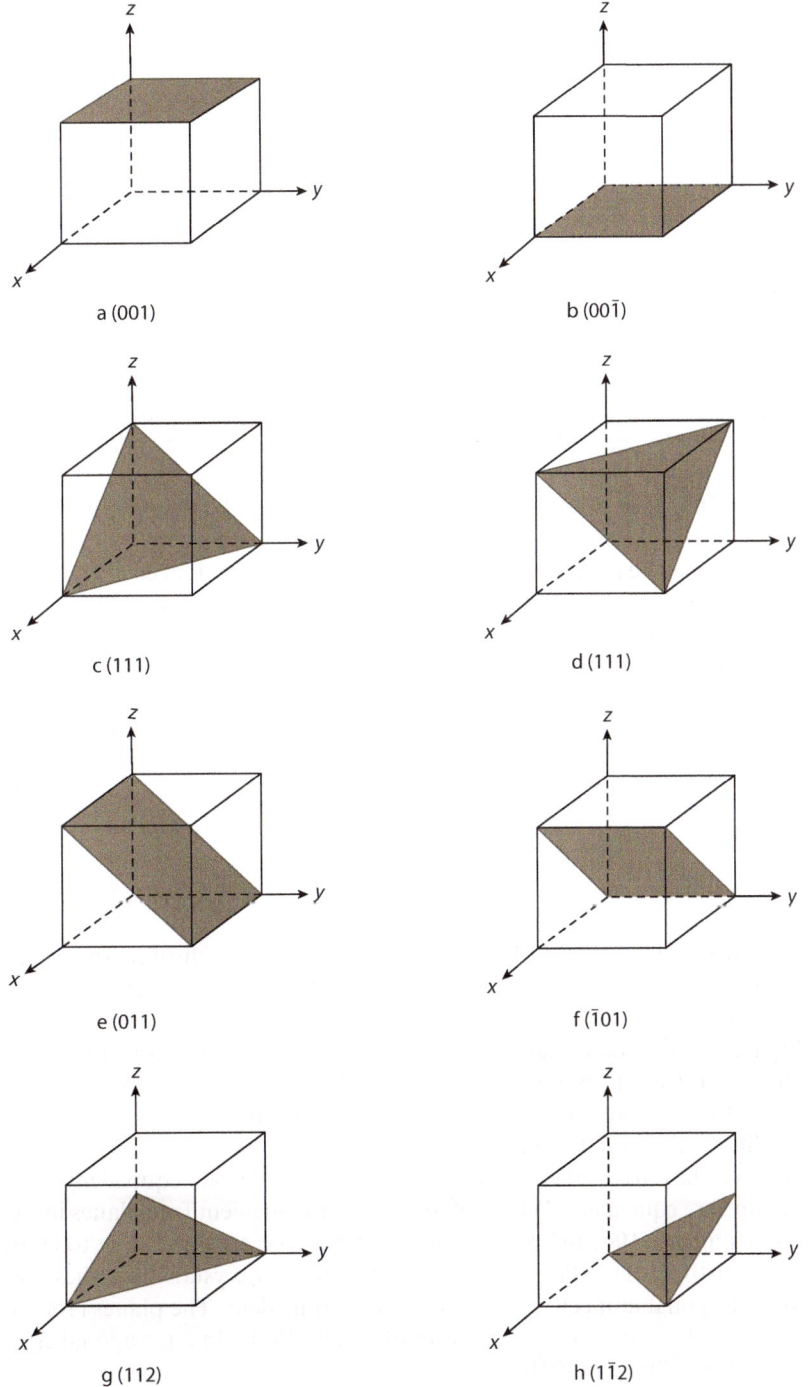

Figure 2.8   Several planes in a cubic unit cell.

The reason that the inverse of the intercepts is used for the Miller indices (hkl) is that it avoids the occurrence of infinite numbers. For example, the top face plane of the crystal shown in Figure 2.8a has fractional intercepts of ∞, ∞, and 1. Because the plane is parallel to the x and y axes, it has infinite intercepts along these axes, and it intercepts the z axis at $+1$. The Miller indices are then

$$h = 1/\infty = 0$$
$$k = 1/\infty = 0$$
$$l = 1/1 = 1$$

The plane is specified as (001). The bottom face of the unit cell in Figure 2.8b is the (00$\bar{1}$) plane. The bottom face of this cube passes through the origin. According to rule number 1 above, the origin must be moved. If the origin is moved a distance of one lattice parameter to the top of the cell, then the bottom face has a fractional intercept on the z axis of -1, and the Miller indices are (00$\bar{1}$). Note that we also could have moved the origin down to the coordinate 0, 0, -1. In that case the plane intercepts are ∞, ∞, and 1, and the Miller indices are (001). In all crystals the bottom and top planes of the unit cell are equivalent, because the top plane of one unit cell is the bottom plane of the unit cell on top of it.

The close-packed plane in the metallic FCC structure is the (111) plane, shown in Figure 2.8c. This plane has intercepts of 1, 1, and 1. The Miller indices are then (111). In Figure 2.4a the close-packed plane is indicated, and this same plane is shown in Figure 2.8d, and this plane is also a (111) plane. These Miller indices can be seen by placing the origin at the lattice point with coordinates 0, 0, 1 and extending the (111) plane into the unit cell that would sit on top of the one shown in Figure 2.8d. In the upper unit cell the plane in Figure 2.8d looks exactly like the plane shown in Figure 2.8c. Remember that each lattice point is equivalent, and the origin can be put at any lattice point. All lattice points are all equivalent to the 0, 0, 0 lattice point. Put the origin where it makes the geometric visualization of the plane intercepts easiest. The two (111) planes in Figures 2.8c and d are equivalent planes and parallel to each other, as is indicated by the same Miller indices.

In Figure 2.8e the intercepts are ∞, 1, and 1. The Miller indices are then

$$h = 1/\infty = 0$$
$$k = 1/1 = 1$$
$$l = 1/1 = 1$$

and the plane is specified as the (011). In Figure 2.8f the plane passes through the origin, and the origin must be moved. If the origin is moved to the coordinates 1, 0, 0, then the intercepts are -1, ∞, 1; and the Miller indices are then ($\bar{1}$01).

The plane in Figure 2.8g has fractional intercepts of 1, 1, and $\frac{1}{2}$. The inverses of the fractional intercepts are 1, 1, and 2, and the Miller indices are (112). In Figure 2.8h the plane shown passes through the origin. If the origin is moved to the coordinates 0, 1, 0 the plane intercepts the z axis at $\frac{1}{2}$, the x axis at 1, and the y axis at -1. The Miller indices are then (1$\bar{1}$2).

We stated earlier that the (001) and (00$\bar{1}$) planes in a cubic unit cell are equivalent. In a cubic crystal all of the faces of the cube are equivalent. When referring to the equivalent face planes in a cubic crystal, the symbol is {100}. The symbol {100} refers to all of the planes that are equivalent to (100), and in a cubic crystal, that is (100), (010), (001), ($\bar{1}$00), (0$\bar{1}$0), and (00$\bar{1}$). Not all crystals have these planes equivalent. For example, in the tetragonal unit cell, $a = b$, but c is not equivalent. The planes (100), (010), ($\bar{1}$00), and (0$\bar{1}$0) are equivalent, but these are not equivalent to (001) and (00$\bar{1}$). In a tetragonal crystal, {100} refers to the planes (100), (010), ($\bar{1}$00), and (0$\bar{1}$0).

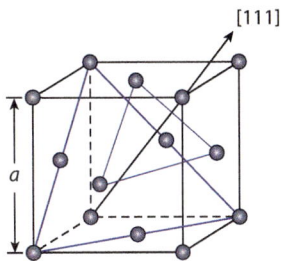

Figure 2.9 The (111) planes and the [111] direction in a cubic crystal.

In cubic crystals the **interplanar spacing** (d_{hkl}) between adjacent planes with the Miller indices (hkl) is given by Equation 2.1:

$$d_{hkl} = \frac{a}{(h^2 + k^2 + l^2)^{1/2}} \qquad \textbf{2.1}$$

where a is the cubic crystal lattice parameter. Equations for the interplanar spacing in noncubic crystals are presented in books specializing on crystallography, such as that by Cullity.

Example Problem 2.5

In a copper FCC crystal, the close-packed planes are the {111} planes. What is the spacing between the close-packed planes in copper with a lattice parameter of 0.362 nm?

Solution

The interplanar spacing is calculated from Equation 2.1.

$$d_{hkl} = \frac{a}{(h^2 + k^2 + l^2)^{1/2}} = \frac{0.362 \text{ nm}}{(1 + 1 + 1)^{1/2}} = \frac{0.362 \text{ nm}}{(3)^{1/2}} = \frac{0.362 \text{ nm}}{1.732} = 0.209 \text{ nm} = d_{111}$$

This result could have also been obtained by observing how the planes parallel to (111) cut the [111] direction in the FCC unit cell, as shown in Figure 2.9. In cubic crystals the [111] direction is perpendicular to the (111) planes. The distance between the (111) planes along the [111] direction is the interplanar spacing. Starting at the position 0, 0, 0 and going to the position 1, 1, 1 the [111] direction is sliced into three segments by (111) planes. There are also (111) close-packed planes that surround the atoms at 0, 0, 0 and at 1, 1, 1. The distance from 0, 0, 0 to 1, 1, 1 is $a3^{1/2}$, but the spacing between planes is one-third of this value, because the (111) planes cut this distance into three segments. The spacing between the (111) planes is $a/3^{1/2}$ in agreement with the result obtained from using Equation 2.1.

2.4.3 Planar Atom Density

There are applications that require knowledge of the planar atom density (*PAD*) as well as the linear atom density (*LAD*) in Section 2.4.5. For example, the permanent deformation of metal crystals occurs on the planes that have the highest planar atom density and in the direction with the maximum atom density. To determine the *PAD*, calculate the number of atoms whose centers lie in the plane of interest, with the plane's area given as *A*. Assume that the area (*A*) of interest is that enclosed by the atoms at 1, 3, 9, and 7 in Figure 2.10. If the atom center lies inside the area *A*, such as atom 5 in Figure 2.10, then

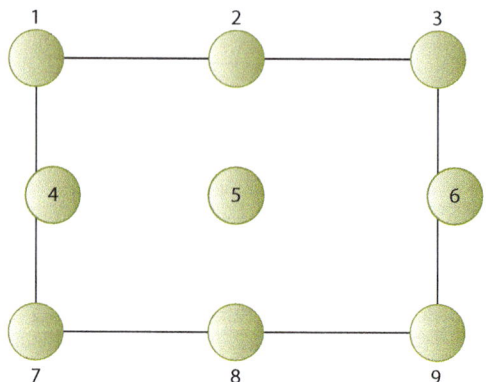

Figure 2.10 A hypothetical plane in a crystal, with atoms at positions 1 to 9.

the atom contributes one atom to the area A. If the atom is at a corner of an area A, such as atoms 1, 3, 9, and 7 in Figure 2.10, then the corner atom contributes only $\frac{1}{4}$ of an atom (90 degrees/360 degrees) to the area A. If there are atoms at the four corners of a rectangle, then the four atoms at the corners contribute a total of one atom. If the center of the atom lies along an edge of an area A, such as atoms 2 and 8 in Figure 2.10, then half of the atom is in the area, and the atom contributes $\frac{1}{2}$ of an atom to the area. If an atom's center does not lie in the plane of interest, it does contribute to the planar atom density. The atom 6 makes no contribution to the PAD for the area A, but the atom 4 contributes one atom to the area A. To determine the planar atom density, sum all of the atom contributions in the area A, and divide by the area A, as shown in Equation 2.2.

$$PAD = \frac{\text{Contribution of atoms with centers in area } A}{\text{Area of the plane } (A)} \qquad \textbf{2.2}$$

Example Problem 2.6

Compare the PAD of the (100) plane to the (110) plane in iron with a BCC structure; an atom at 0, 0, 0 and a lattice parameter of 0.286 nm.

Solution

For the (100) plane in the BCC structure of iron, there are atoms at the four corners of the (100) cube face; however, only $\frac{1}{4}$ of each of these atoms is in the area of the face. There is only one atom contributed by the four corners of the cube face (4 corners times $\frac{1}{4}$ atom per corner). The area of the cube face is a^2. The planar atom density is

$$PAD(100) = \frac{1 \text{ atom}}{a^2} = \frac{1 \text{ atom}}{(0.286 \times 10^{-9} \text{ m})^2} = \frac{1 \text{ atom}}{0.0818 \times 10^{-18} \text{ m}^2} = 12.2 \times 10^{18} \frac{\text{atoms}}{\text{m}^2}$$

The (110) plane in the BCC crystal of iron forms a rectangle that has sides that are the face diagonals of length $2^{1/2}a$ and height of a, where the lattice parameter is a. The area of the rectangle formed by the (110) plane is

$2^{1/2}a^2 = 0.116 \times 10^{-18}$ m². The four corners contribute $\frac{1}{4}$ atom each, for a total of 1 atom, and there is 1 atom at the body center. The total number of atoms in the rectangle is $1 + 1 = 2$. The atom density on the (110) plane is

$$PAD(110) = \frac{2 \text{ atoms}}{2^{1/2}a^2} = \frac{2 \text{ atoms}}{1.41(0.286 \times 10^{-9} \text{ m})^2} = \frac{2 \text{ atoms}}{1.41(0.0818 \times 10^{-18} \text{ m}^2)} = 17.2 \times 10^{18} \frac{\text{atoms}}{\text{m}^2}$$

The (110) plane has the higher planar atom density. If the (110) plane is compared with all planes in the BCC structure, it is found that the (110) plane has the highest atom density of any plane in the BCC structure, as shown in Figure 2.5. The permanent deformation of BCC crystals occurs along the planes equivalent to the (110) plane.

2.4.4 Crystal Directions

A **crystal direction** is specified by the vector components of the direction in fractions of the lattice parameters a, b, and c. Several directions in a cubic crystal are shown in Figure 2.11. The rules for determining a direction in a crystal are as follows:

1. Start a vector at the origin 0, 0, 0 of a unit cell with the desired direction. If the vector does not start at the origin, draw a line parallel to the desired direction that passes through the origin.
2. Locate where the vector leaves the boundary of the unit cell that has the origin located at 0, 0, 0. Express the coordinates of this intersection in fractions of the lattice parameters. Eliminate the fractions from the intersection coordinates by using a single multiplier to produce integers u, v, and w. If the vector exits the unit cell at the origin, move the origin one lattice parameter. The [0$\bar{1}$0] direction in Fig 2.11 is an example of this.
3. The direction is then specified as [uvw]. Negative components of a direction are indicated with a bar over the integer.

In Figure 2.11 the direction of the x axis is the [100] direction. The vector along the x axis crosses the boundary of the unit cell at a fractional distance of one from the origin, and this vector has zero components along the y and z axes. The values of u, v, and w are 1, 0, and 0, and the direction is specified as [100]. The [110] direction intercepts the unit cell with components 1, 1, and 0. This is the [110] direction. The [021] direction intercepts the unit cell with components of 0, 1, and $\frac{1}{2}$. If we eliminate fractions from

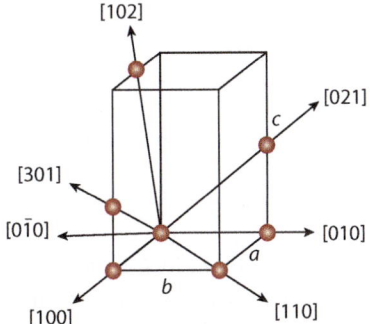

Figure 2.11 Several directions in an orthorhombic unit cell. The red dot indicates the point where the vector exits the unit cell relative to the origin at 0, 0, 0.

the components, the direction is specified as [021]. For the [301] direction, the fractional coordinates where the vector intersects the unit cell are 1, 0, and $\frac{1}{3}$. After clearing the fraction, the values of u, v, and w are 3, 0 and, 1, respectively. Multiplying the fractional components of a vector by a constant number does not change the direction of a vector; only the magnitude changes. The [102] direction passes through the point that has components of $\frac{1}{2}$, 0, and 1. If we eliminate fractions from these intercepts, the direction is [102]. The direction pointing in the negative y direction is [0$\overline{1}$0]. As with planes, this is called the "zero, one bar, zero" direction. The [0$\overline{1}$0] vector is shown exiting the unit cell at the origin. Move the origin for this vector to the position 0, 1, 0 relative to the position 0, 0, 0. Then relative to this origin it exits the unit cell at 0, −1, 0.

In a cubic crystal the directions along the sides of the cube are all equivalent directions. The directions [100], [010], [001], [$\overline{1}$00], [0$\overline{1}$0], and [00$\overline{1}$] in a cubic crystal are equivalent. The equivalent directions in a crystal are specified by $\langle uvw \rangle$. For example, the directions along all of the sides of a cube are specified by $\langle 100 \rangle$. In the orthorhombic crystal, shown in Figure 2.11, the [100] and [$\overline{1}$00] directions are equivalent. The [010] and [001] directions are different because in the orthorhombic unit cell $a \neq b \neq c$. In an orthorhombic unit, cell $\langle 100 \rangle$ refers only to the directions [100] and [$\overline{1}$00].

2.4.5   Linear Atom Density

To determine the linear atom density (LAD), calculate the number of atoms whose centers lie on a line. If an atom's center lies on a line and the line passes completely through the atom, then this atom contributes one atom to the LAD. Draw a line from atom 1 to atom 9 in Figure 2.10. The atom 5 contributes one atom to the LAD of this line. The atom 2 also makes a contribution of one atom to the line from atoms 1 to 3. If the line only goes to the atom's center, then the atom only makes a contribution of $\frac{1}{2}$ an atom to the linear density. Atoms 1 and 9 only contribute $\frac{1}{2}$ an atom each to the LAD of the line from atoms 1 to 9. If an atom's center does not lie on a line, it does not contribute to the linear atom density. Atom 6 makes no contribution to the LAD for the line from atoms 3 to 9, and atom 4 makes no contribution to the atom density to the line from atoms 1 to 7. To determine the LAD, sum the contribution of all of the atoms whose centers lie on a length of line (L), and divide by the line length, as shown in Equation 2.3.

$$LAD = \frac{\text{Contribution of atoms with centers on a length of line } L}{\text{Length of the line } (L)} \qquad \textbf{2.3}$$

Example Problem 2.7

Copper is FCC, with an atom at 0, 0, 0 and a lattice parameter of 0.362 nm. Assume that the copper atoms touch along the $\langle 110 \rangle$ type directions. (a) What is the LAD along the $\langle 110 \rangle$ type directions? (b) What is the radius of a copper atom?

Solution

a) The $\langle 110 \rangle$ type directions are the face diagonals of the FCC lattice shown in Figure 2.4a. Along a face diagonal from one corner of a face to the diagonally opposite corner, there are $\frac{1}{2}$ atom + 1 atom + $\frac{1}{2}$ atom = 2 atoms. The length of a face diagonal is 0.362 nm ($2^{1/2}$) = 0.512 nm.

The linear atom density is then

$$LAD = \frac{2 \text{ atoms}}{0.512 \text{ nm}} = 3.91 \frac{\text{atoms}}{\text{nm}}$$

If we checked the LAD in all directions of the FCC metal crystal, we would find that the $\langle 110 \rangle$ directions have the highest LAD.

b) To find the radius of a copper atom, we assume that the copper atoms touch along the direction that has the highest atom density, which is the $\langle 110 \rangle$ direction in the metal FCC crystal structure. This assumes that the copper atoms pack together like billiard balls. Along one face diagonal there are four atomic radii, and the length of this face diagonal is $2^{1/2}a$.

$$4R = 2^{1/2}a$$

$$R = \frac{a}{2^{3/2}} = \frac{0.362 \text{ nm}}{2.828} = 0.128 \text{ nm}$$

In Example Problem 2.7, the atomic radius is determined by assuming that the atoms touch along the most closely packed directions on the most closely packed planes in the crystal. This method of determining the atomic radius applies only when there is one type of atom, because in this case it is assumed that the atoms come together halfway between the two atoms. If there are two types of atoms, it is not appropriate to assume that the atoms come together halfway between them, because one atom may be larger than the other. Thus it is uncertain where two different atoms come together.

The designation of planes and directions as discussed above applies to all of the Bravais lattice types, except for the hexagonal type. All of the concepts that are covered in this book are explained without using hexagonal crystal Miller indices and directions. However, for those interested, the indices of planes and directions in hexagonal crystals are presented in Appendix 2A. Students looking for software and information on crystallography can find an alphabetical listing on http://www.iucr.org, which is maintained by the International Union of Crystallography.

2.5 CHEMICAL BONDING, ATOM ARRANGEMENTS, AND PHYSICAL PROPERTIES

Starting in this section we review the types of chemical bonding. Then we show how the chemical bonding determines the atomic arrangement in crystals and also influences the physical properties of materials. Physical properties that are related to chemical bonding in this chapter are density, melting temperature, and hardness. In later chapters we will relate chemical bonding to the thermal expansion coefficient, stiffness, and other properties of materials.

The **density** of a material is the mass of all of the atoms per unit volume of material. The number of atoms in a unit volume of the material is related to the size of the atoms and to the way that atoms are packed together. Some elements have different **allotropic** forms, meaning that they have different atom arrangements. For example, carbon is found in the form of graphite, diamond, and some newer forms that are discussed shortly. The density of graphite at room temperature is $2.25 \times 10^3 \text{ kg/m}^3$, and the density of diamond at room temperature is 3.52×10^3 kg/m^3. This difference in density is due to chemical bonding and the way that the atoms are packed together, as discussed below.

The **melting temperature** is the lowest temperature that causes a solid to transform to a liquid upon heating. The melting temperature of a material is directly related to the strength of the bonds that hold the atoms or molecules (units) together in a solid. When a material is heated, the atoms vibrate about their average atomic positions with increased amplitude. If a material is heated sufficiently, the units become more mobile because of the high amplitude of the thermal vibrations. When units move extensively such that neighboring units are continuously changing, the material melts to become a liquid. F. A. Lindemann proposed that materials melt when the amplitude of the lattice vibrations is approximately 10% of the nearest neighbor's interatomic spacing. If the material is heated to even higher temperatures, the amplitude of the atom vibrations becomes sufficiently large that the atoms vaporize. The stronger the chemical bonding, the higher the temperature necessary for the units to become mobile and form a liquid, and the higher the temperature necessary for the units to vaporize. The melting temperature is an important material property for mechanical applications at elevated temperatures. For example, the operating temperature of a high-temperature gas turbine is limited by the melting temperature of the metals that are suitable for use in the high-temperature section of turbines.

In the discussion of the relationship of melting temperature to chemical bonding throughout the book, the kelvin temperature scale is used, because with this scale zero chemical bonding energy results in a melting temperature approaching 0 degrees kelvin. With degrees celsius and fahrenheit temperature scales, negative melting temperatures result when there is still significant chemical bonding, and this is confusing. Also, in thermodynamic calculations it is necessary to use kelvin.

A material's hardness is an important mechanical property related to its chemical bonding. The **hardness** of a material is a measure of the resistance to permanent penetration by an indenter, as shown schematically in Figure 2.12. We run an informal hardness test when we push our finger into the sand at the beach, and we determine if the sand is soft or hard. In a formal hardness test, a hard indenter is driven into a material with a force, as shown in Figure 2.12b, that is applied through a mechanical system. Different hardness tests use different indenter materials, such as diamond or hard steel, and different indenter shapes, such as spherical balls, cones, and pyramids. For example, the Knoop hardness measurement uses a diamond pyramid indenter with various applied forces. The indenter is withdrawn from the specimen, and the depth of penetration (t) and the width (w) of the indentation, shown in Figure 2.12c, are measures of the hardness. In a hard material the penetration is small, and in a soft material the penetration is larger. The Knoop hardness is measured in units of force divided by area. Traditionally the units have been kilogram force per square millimeter, and these are the units used in this chapter. Other hardness measurements include Brinell, Mohs, Vickers, and Rockwell. All of the hardness measuring scales are discussed and compared in Chapter 6. The various hardness scales provide a relative number for the hardness, but the number is not a fundamental value such as the melting temperature of the material. Hardness tests are used extensively in manufacturing to provide a nondestructive test that gives a confirmation of the mechanical properties of the materials used in a part. Many parts can tolerate a small indentation without compromising the integrity of the part.

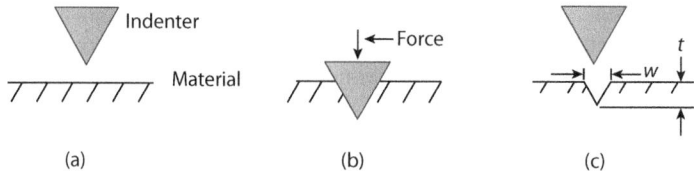

(a)   (b)   (c)

Figure 2.12   (a) A schematic of a hardness measurement system, showing both the indenter and the tested material. (b) The indenter penetrates into the material under an applied force. (c) The indenter is removed from the material; the depth of penetration (t) or the width of the indentation (w) is a measure of the hardness.

2.6 VAN DER WAALS BONDS

We start the discussion of bonding with van der Waals bonds. These bonds significantly influence the bonding characteristics of many polymers, plastics, adhesives, and paints. All materials have these weak secondary bonds, even though they are not the dominant bonds in many materials. In the inert-gas atoms they are indeed the dominant bonds, and these materials form solids at very low temperatures with the FCC crystal structure discussed in Section 2.3.1.

The **van der Waals bonds** result from electric dipoles formed on the atoms or molecules. If two charges of $+q$ and $-q$ are separated by a distance d, this forms an electric dipole. Positive and negative charges on different atoms or molecules attract each other, resulting in bonding. There are permanent-electric dipoles and fluctuating-electric dipoles on atoms and molecules. Permanent-electric dipoles exist on certain molecules, such as the water molecule, and all atoms have fluctuating electric dipoles.

2.6.1 Permanent-Electric Dipole Bonds

Permanent-electric dipole bonds are particularly important when hydrogen bonds are present. An example of a permanent-electric dipole bond exists in the water molecule. When hydrogen and oxygen covalently bond to form H_2O, as shown in Figure 2.13, the hydrogen atoms complete their $1s$ shells by sharing electrons across the molecule, and oxygen has eight electrons by sharing electrons with the two hydrogen atoms to complete its $2p$ shell. Because oxygen's electronegativity of 3.5 is larger than hydrogen's electronegativity of 2.1, all these covalently shared electrons are more attracted and physically closer to the oxygen than to the hydrogen. As shown in Figure 2.13, this results in a permanent-electric dipole on the water molecule. The ends of the hydrogen atoms that are furthest from the hydrogen-oxygen covalent bond are charged positive, and the oxygen is charged negative. This electronegativity-induced separation of the positive and negative charges on a water molecule forms a permanent-electric dipole. When two or more water molecules are within close proximity of each other, the positive ends of the hydrogen atoms on each water molecule are attracted to the negative area of the oxygen atoms on the neighboring water molecules, as shown in Figure 2.13. The attraction of the positive and negative charges bonds these molecules together to form liquid water at room temperature and solid ice at lower temperatures.

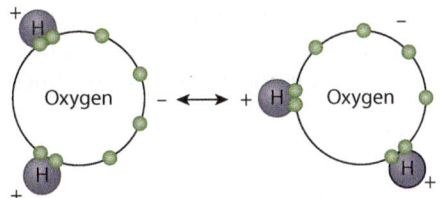

Figure 2.13 A water molecule (H_2O) carries a permanent-electric dipole; the hydrogen atoms are at the more positive end of the molecule, and the oxygen atom is the more negative end. The small green circles represent the electrons. Van der Waals bonds are generated between water molecules as the negatively charged area on one water molecule is attracted to a positively charged area on its neighbor, as indicated by the force arrow between the molecules. (*Based on Askeland, D.R., Fulay, P.P., and Wright, W.J., The Science and Engineering of Materials, 6th ed., Cengage Learning, Stamford, CT. (2011), pg. 39.*)

Many molecules that have permanent-electric dipoles involve hydrogen. As a result, permanent-electric dipole bonds are often called hydrogen bonds even if hydrogen is not present. However, the use of the term *hydrogen bond* for permanent-electric dipoles is confusing in a material that does not contain hydrogen. In this book we will only use the term *permanent-electric dipole bond*.

2.6.2 Fluctuating-Electric Dipole Bonds

In the inert-gas atoms the electron shells are spherically symmetric when averaged over time, and this is a low-energy configuration of the atoms. There is no tendency for these atoms to gain or lose electrons by chemically bonding with other atoms. This is why these atoms are gaseous at ambient temperatures. So why do the inert-gas atoms form liquids and solids if cooled sufficiently? For example, argon forms a solid at 83.8 K. This is a very low temperature for solidification, and this indicates that the bonding that forms the inert-gas atoms into solids is very weak. For example, from the periodic table the cohesive energy of argon is 0.08 electron-volts (eV). Compare this to diamond's cohesive energy of 7.43 eV. The **cohesive energy per atom** is the energy required per atom to move each atom in a solid at 0 kelvin to a point where all of the atoms are separated by an infinite distance. The cohesive energy of the elements is the bond energy per atom.

In the study of the cohesive energies of atoms and many other atomic level phenomena, the energy unit of **electron-volt** (eV) is often utilized, because the cohesive energies are often in the range from 0 to 10 eV. One eV is the kinetic energy that an electron acquires when it experiences a voltage drop of one volt, and $1\,eV = 1.6 \times 10^{-19}$ joules. However, the eV is not part of a unified system of units that is used for calculations of properties such as velocity; it is necessary to use the SI system (Système Internationale) of units for such calculations. The eV can be utilized in calculations where the dimensions cancel, such as when there is a ratio of energies. In general, the SI system of units is utilized for atomic calculations. If there is any doubt if the eV can be utilized in a problem, it is always safe to utilize the SI unit of energy (joule).

Even though the electron shells of the inert-gas atoms are spherically symmetric when averaged over time, they are not spherically symmetric at any particular instant. If two helium atoms are brought close to each other, the electrons in their outermost spherical shells will interact with each other, as shown in Figure 2.14. If there are electrons from atom 1 located between the two atoms, as shown in Figure 2.14, it is not likely that electrons with the same quantum numbers from atom 2 are also located between the two atoms, because of the Pauli exclusion principle and also from Coulomb's law that like charges repel each other. The absence of electrons in between the two atoms on atom 2 acts as a positive charge on the left side of atom 2, and this positive side is attracted to the negatively charged electrons on atom 1. This forms a temporary electric dipole between the two atoms. When inert-gas atoms are condensed into a liquid or a solid, the electron orbitals become coupled and form temporary electric dipoles throughout the liquid or solid. If we added a third atom to the right of atom 2, this atom would have fewer electrons on its left side, which faces atom 2, and more electrons on its right side, which is furthest from atom 2, because atom 2's negative side would repel them. However, the electrons on the atoms are in constant motion, and the orientation of the electric dipoles is continually changing. For this reason, the bond is called a fluctuating-electric dipole bond.

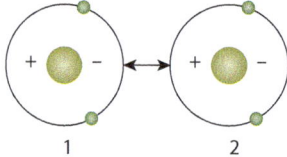

Figure 2.14 A schematic of possible locations of electrons around helium atoms in close proximity. The large outer circle indicates the size of the helium atoms. The two electrons on each helium atom are indicated by the green solid circles located on the outer large circle. The helium nucleus is the solid circle at the center of each large circle. The force of interaction between the two helium atoms is indicated by the double ended arrow.

The fluctuating-electric dipole bond is also called the London bond. This form of bonding is present in all materials, because electrical dipoles form due to the motion of electrons around all atoms. However, it is a weak form of bonding in comparison to covalent, metallic, or ionic bonding, and it does not significantly affect the properties of materials that have these stronger types of bonding.

The strength of the fluctuating-electric dipole bond increases with the number of electrons on the atom or molecule and with the length of the molecule. Observe in the periodic table that the melting temperature of the inert gas Xe (atomic number 54) is 161.4 K, and for Ne (atomic number 10) the melting temperature is 24.55 K. An element in Group VII (F, Cl, Br, and I) forms a diatomic molecule (a pair of identical elements), such as I_2. The bonding between molecules of these elements is the fluctuating-electric dipole bond, because there is no permanent-electric dipole between identical atoms. Iodine (atomic number 53) forms an I_2 molecule with 106 electrons and a melting temperature of 387 K. Also, the iodine molecule is elongated, not spherical. Later in this chapter we will continue the discussion about how the different molecules in polymers or plastics are bonded together with van der Waals bonds.

There are molecules that involve a hydrogen atom bonded to other atom types, such as carbon in methane, where the bonding that holds the molecules together in liquids and solids is not the hydrogen bond, but is instead the fluctuating-dipole bond. This is because hydrogen and carbon have similar electronegativities of 2.1 and 2.5, respectively, and the bonds are not strongly polar. Fluctuating-dipole bonds dominate the permanent-dipole (hydrogen) bonds in liquids and solids of methane. The hydrogen bond is not primarily responsible for the formation of liquids and solids of methane, even though the methane molecule has four hydrogen atoms.

2.6.3   Packing of Inert-Gas Atoms

The inert-gas atoms are solids at low temperatures. For example, argon is a solid at temperatures below 83.8 K. Since the inert-gas atom electron configuration is spherically symmetric when averaged over time, the inert-gas atoms pack together like they were spherical billiard balls. This packing results in the inert-gas atoms forming in the FCC crystal structure.

2.6.4   Physical Properties of Inert Gases

The fluctuating-electric dipole bond between inert-gas atoms is a relatively weak bond, as evidenced by the low melting and vaporization temperature of the inert gases. Neon (atomic number 10) melts at 24.55 K; however, radon (atomic number 86) melts at 202 K. Although the solids of the inert-gas atoms do not have engineering applications at present, the liquid and vapor forms do have engineering applications that result from their chemical stability and their melting and vaporization temperatures. Inert-gas atoms are used in applications where an environment that does not chemically react with a material is desired. For example, metals are heated to high temperatures in an inert-gas environment during various processing procedures to avoid oxidation, which is when the metal reacts with oxygen. Aluminum is very reactive with oxygen; therefore, aluminum is welded in an inert-gas environment, such as one filled only with argon gas. Also, inert-gas atoms are used in the vapor deposition process of sputtering to produce thin films of materials. In sputtering, ions of the inert-gas atoms are accelerated by an electrical potential toward a target material, and the target-material atoms are then blasted from the target toward a substrate, such as glass, where they land and form a thin film. The thin-metal films on sunglasses and on glass treated to reduce the transmission certain wavelengths of light are produced by sputtering.

Liquids of the inert gases are also used to cool materials. The vaporization temperature of helium is 4.1 K. If you want to run a test or process at a really low temperature, you can cool the material to temperatures approaching 4.1 K with liquid helium.

2.7   COVALENT BONDS

A **covalent bond** is the localized sharing of a pair of electrons between two atoms to complete the valence electron shell. Covalent bonding is present in many important materials, such as diamond, silicon, silicon carbide, polymers, and ceramics. Covalent bonding is present in materials that have high melting temperature and high hardness, such as diamond, and it is present in plastics that are relatively soft and have a low melting temperature. How can covalent bonding be present in high-hardness diamond as well as in soft plastics? This section and Section 2.8 answer that question.

If the valence shell of an atom is one-half filled or more, the atom acquires other electrons to complete the valence electron shell. If there is only one type of atom, the atoms acquire other electrons by sharing electrons with other atoms in a covalent bond, since each atom attracts the electrons equally. Carbon in the form of diamond provides an excellent example of covalent bonding; however, diamond is not the most stable form of carbon under ambient conditions. Graphite is the stable form of carbon, but buckyballs, nanotubes, and graphene can also form under ambient conditions. These forms of carbon are shown in Figure 2.16.

Carbon is in Group IV and has four valence electrons, as shown in Figure 2.15a. Therefore, its valence shell is one-half filled, and it requires four other electrons to complete its valence shell with eight electrons. By sharing electrons with four other carbon atoms, each carbon atom is surrounded by eight shared electrons, as shown in Figure 2.15b, and eight electrons is equal to the number of valence electrons in the inert-gas atoms. In Figure 2.15b only the central atom is shown with eight surrounding shared electrons; however, when all the carbon atoms are surrounded by four other carbon atoms, each carbon atom is encircled by eight shared electrons.

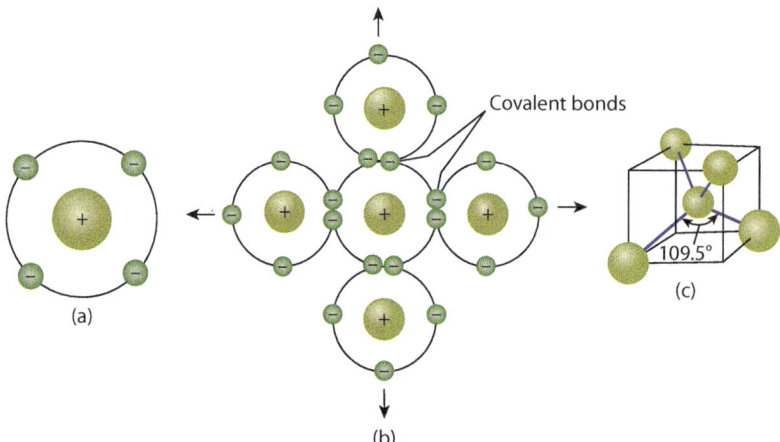

Figure 2.15   (a) A schematic of a carbon atom from Group IV, showing four valence electrons. (b) A planar schematic of carbon's covalently bonded electron sharing. The central atom now has eight valence electrons by sharing electrons with four of its nearest neighboring carbon atoms. Covalent bonding with other atoms in a crystal results in all the atoms sharing eight valence electrons. (c) In the diamond structure the four nearest neighboring carbon atoms are arranged in a tetrahedron. (*Reprinted from Askeland, D. R., Fulay, P. P., and Wright, W. J., The Science and Engineering of Materials, 6th ed., Cengage Learning, Stamford, CT (2011), p. 36.*)

2.7.1   Covalent Bonds, Atom Packing, and Crystal Structures

The type of atomic bonding determines how the atoms pack together to form a solid. For covalently bonded atoms of the same group number, the number of nearest neighbors (NNN) is given by 8 minus the group number (GN) as shown by Equation 2.4.

$$NNN = 8 - GN \qquad\qquad \textbf{2.4}$$

The NNN is the number of atoms closest to an atom. The number of nearest neighbors is also equal to the number of electrons needed in the valence shell, assuming that each atom pair forms a single covalent-pair bond. For Group VII the number of nearest neighbors is 1. As explained in Section 2.6.2, atoms such as chlorine combine with another identical atom to form diatomic molecules of chlorine gas, such as Cl_2. Equation 2.4 applies only when all of the valence electrons enter into covalent bonds and the nearest neighbors are equidistant. For carbon, which is in Group IV, NNN is 4 ($NNN = 8 - GN = 8 - 4 = 4$).

We use diamond as an example of covalent bonding because its covalent bonding and its allotropic forms result in interesting physical properties that have many engineering applications. The carbon atoms in diamond have the valence electron configuration $2s + 2p^3$. In diamond the s and the three p orbitals linearly combine to form four sp^3 **molecular orbitals** that point to the corners of a tetrahedron, as shown in Figure 2.16a. Each carbon atom must share electrons in covalent bonds with four other carbon atoms, so that each atom is effectively surrounded by eight valence electrons. The orientation of the atoms relative to each other is determined by the direction of the sp^3 molecular orbitals. The orientation of the atoms maximizes the overlap of the electrons forming the covalent bond between electrons in sp^3 molecular orbitals from different atoms. The carbon atoms in the diamond crystal structure, shown in Figure 2.16b, are arranged in the same tetrahedral arrangement shown in Figure 2.16a. The lattice parameter of diamond is 0.357 nm, with a carbon-carbon interatomic separation of 0.154 nm. The diamond cubic structure has an FCC unit cell; however, there are two atoms associated with each FCC lattice point. One atom is at the lattice point 0, 0, 0; this position also results in atoms at all lattice points of the FCC unit cell. In addition to each of these atoms, there is an associated atom located at a fractional distance of $\frac{1}{4}, \frac{1}{4}, \frac{1}{4}$ relative to 0, 0, 0 and each lattice point. The diamond cubic crystal structure is FCC, but it is not close-packed, because there are only four nearest neighbors to each atom.

Example Problem 2.8

Give all of the information necessary to describe the crystal structure of diamond.

Solution

The Bravais lattice type is FCC.
The atom positions are 0, 0, 0 and $\frac{1}{4}, \frac{1}{4}, \frac{1}{4}$.
The lattice parameter is 0.357 nm.

2.7.2   Physical Properties and Covalent Bonds

The strong covalent bond in diamond allows us to understand some of the physical properties of diamond presented in Table 2.2, along with physical properties of other materials with different types of bonding.

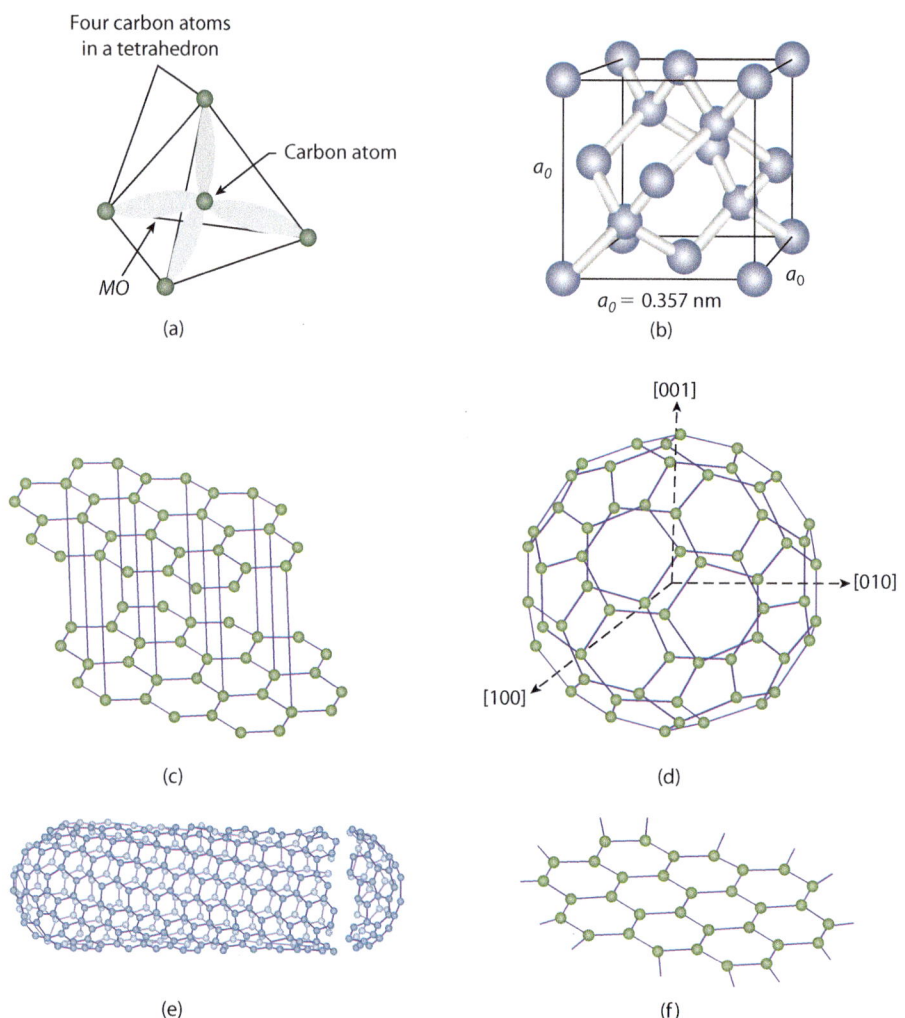

Figure 2.16 (a) The sp^3 molecular orbitals (MO—light gray area) of a carbon atom (dark solid circle of reduced size) at the center of a tetrahedron of carbon atoms in diamond. The allotropes of carbon are (b) the diamond cubic unit cell, (c) graphite, (d) the buckyball with the orientation shown in FCC crystals, (e) the carbon nanotube, and (f) graphene. (*(a) Adapted from Guy, A. G.,* Introduction to Materials Science, *McGraw-Hill, New York (1972) p. 11. (b) Adapted from http://math.ucr.edu/home/baez/diamond-conventional-unit-cell.gif (c) and (e) Reprinted from Askeland, D. R., Fulay, P. P., and Wright, W. J.,* The Science and Engineering of Materials, *6th ed., Cengage Learning, Stamford, CT (2011), pp. 92, 46, and 48. (d) Adapted from Haynes, W. M., ed.,* Handbook of Chemistry and Physics, *91st ed., CRC Press, Boca Raton, FL (2010–2011), pp. 12–78.*)

In diamond the valence electrons are bound between the carbon atoms in strong covalent bonds, resulting in a cohesive energy of 7.43 eV/atom. It is hard to deform a covalently bonded material that has strongly bonded electrons localized in specific orientations. Diamond is one of the hardest materials known, with a Knoop hardness of 7000 and a melting temperature of 3823 K. Compare these values with a melting temperature for argon of 83.8 K with a cohesive energy of 0.08 eV/atom resulting from Van der Waals bonding.

Table 2.2	Material; Bond Type, Including Covalent (C), van der Waals (VDW), Ionic (I), and Metallic (M); Density, Melting Temperature, and Knoop Hardness of Some Materials. The data are from a variety of sources. Hardness values can be affected by material preparation procedures, and values should be considered as typical.

Material (Bonding)	Density $(10^3\ kg/m^3)$	Melting Temperature (K)	Knoop Hardness $[kg_f/(mm)^2]$
Diamond (C)	3.52	3823	7000
Polyethylene (C+VDW)	0.95	403–410	5[1]
NaCl (I)	2.17	1070	16
SiC (C+I)	3.21	3003	2480
SiO_2 (I+C)	2.65	1983	820
Gold (M)	19.3	1336	69
High-strength steel (M)	7.85	1811	392

1 Brinell hardness

Table 2.2 shows that in general there is a correlation between a material's melting temperature and its hardness. Melting temperature indicates the strength of a material's chemical bonds: The higher the melting temperature, the stronger the bonding and the harder the material.

The hardness of diamond results in many industrial applications. Industrial diamonds are diamonds that are not gem quality or synthetic diamonds. Lathe cutting tools use diamond tips, and saws for cutting hard metals are imbedded with diamond chips. Small diamond particles are incorporated into a paste that is used to polish hard metals.

Despite diamond's high cohesive energy, diamond is not the ambient form of carbon. Graphite, buckyballs, nanotubes, and graphene are the ambient forms of carbon. Natural diamond was formed under conditions of high pressure and temperature in the early stages of the earth's formation. Synthetic diamonds are made under high pressure and temperature, such as 5×10^9 N/m² and 1500°C.

Recently it has been found that diamond-like thin films can be produced in the laboratory. Diamond-like coatings (DLCs) have a predominance of the sp^3 molecular orbitals of diamond, but they do not have the diamond crystal structure. They are amorphous; however, they are very hard. DLCs are used to coat cutting tools and other surfaces that need to resist wear. DLC films are normally produced by quenching a hot plasma containing carbon atoms onto a cold substrate (the part). The equipment required to produce DLCs is much simpler than that to produce synthetic diamond. There are commercial applications for these coatings on bearings and other surfaces to prevent wear.

2.7.3 The Structures and Properties of Allotropic Forms of Carbon: Diamond, Graphite, Buckeyballs, Nanotubes, and Graphene

The graphite structure, shown in Figure 2.16c, is the structure of carbon that normally forms at ambient temperature and pressure; buckyballs, nanotubes, and graphene can also form under ambient conditions. Under ambient conditions, three of the four valence electrons in carbon form in a planar sp^2 molecular orbital that has lobes in a plane that are separated by 120°. As a result of these molecular orbitals, the carbon atoms in graphite form rings, with each carbon atom surrounded by three carbon atoms at 120° from each other to maximize the overlap between electrons in covalent bonds, as shown in Figure 2.16c. The remaining p electron orbital is perpendicular to the plane of the sp^2 molecular orbital. In the plane of graphite, each carbon atom has three neighbors separated by 0.142 nm, and the planes of carbon atoms are separated by 0.34 nm. Two graphite planes are shown in Figure 2.16c. The structure is produced by these planes repeating in the sequence ABABAB . . . .

The bonding within the plane of the carbon atoms is a strong covalent bond, as evidenced by the interatomic separation of 0.142 nm, which is closer than the 0.154-nm interatomic separation in diamond. The large separation of 0.34 nm between the graphite planes results in a combination of very weak covalent and van der Waals type bonding between the p electron orbitals on different graphite planes. The bonding is so weak that graphite is used as a solid lubricant, because the planes of graphite easily slip over each other. The strong bonding within the plane of the sp^2 molecular orbitals results in strong graphite filaments where the graphite planes are aligned along the filament axis. These graphite filaments are utilized in high-strength composite materials, which are discussed in Chapter 13. These are the graphite filaments that are used in composite materials for aircraft such as the Boeing 787, in high-performance sporting equipment, and in high-performance automobiles and Formula 1 race cars.

Recently other forms of carbon have been found under ambient conditions, such as buckyballs, nanotubes, and graphene, as shown in Figures 2.16d, e, and f. Buckyballs, shown in Figure 2.16d, were first found in soot, and they are named after Richard Buckminster Fuller, an architect who designed geodesic domes of a similar geometric configuration. A **buckyball** has 60 carbon atoms (C_{60}) arranged in hexagons and pentagons. The carbon-carbon atom spacing of an atom pair between two hexagons is 0.140 nm, and the carbon atom pair spacing between a hexagon and a pentagon is 0.144 nm. The buckyballs can form into an FCC crystal with the center of each buckyball at an FCC position of a unit cell. Figure 2.16d shows the orientation of the buckyballs in the FCC lattice.

The **carbon nanotube** (CNT) shown in Figure 2.16e is a graphite sheet that is wrapped into a tube, with the ends terminated. CNTs are also found in soot, but commercial processes have been developed for producing CNTs. CNTs are typically a few nanometers in diameter, and they have lengths measured in microns. CNTs are one of the strongest filaments known. CNTs are already utilized in high-strength composite material applications such as in wind turbine blades, hunting arrows, ice hockey sticks, snow skis, racing bicycle frames, and surfboards. Much of the present interest in nanotechnology is a result of the discovery of CNTs.

In 2004 it was discovered that single layers of graphite can be mechanically isolated into two-dimensional crystals called **graphene**. It was expected that graphene crystals would be very strong; they could find applications in high-strength composite materials and in nanoelectronics.

2.7.4   Covalent Bonding with Two Types of Atoms

When two elements are combined that both have their valence shells half filled or more, then these two elements bond to each other with covalent bonds. One example of this is the compound silicon nitride (Si_3N_4). Silicon is in Group IV and nitrogen (N) is in Group V. This resulting material has very strong covalent bonding, and this in turn results in a solid that sublimes at 2173 K. Silicon nitride is utilized in high-temperature applications, such as for the bearings and the nonrotating parts of high-temperature gas turbines. Research is being conducted about using Si_3N_4 for rotating parts in land-based gas turbines, such as turbine vanes and blades. Presently rotors in turbochargers for gasoline and diesel engines are made of Si_3N_4. The turbocharger uses the hot exhaust gases to drive a turbine that compresses fresh air into internal-combustion engines. This allows a four-cylinder engine with a turbocharger to develop the power of a six-cylinder engine without a turbocharger. Si_3N_4 is also an excellent electrical insulator, because all of the valence electrons in Si_3N_4 are strongly bound to the atoms. Whenever there are two different atom types that are bonded to each other with different electronegativities, one atom attracts electrons more strongly than the other does, resulting in some ionic bonding. Other important engineering materials with two atom types that have both covalent and ionic bonding include alumina (Al_2O_3), zirconia (ZrO_2), quartz (SiO_2), and magnesia (MgO). There is further discussion of these materials and ionic bonding in Section 2.10.2. Another example of covalent bonding with at least two atom types is polymers.

2.8 POLYMERS

Polymers are large molecules with many ("poly") units ("mers"). Polymers are utilized in applications such as liquid containers, automobile bumpers, and high-performance fibers in body armor; and in composite materials for structural applications, such as the Boeing 787. **Plastics** are materials that are made from polymers that may also contain fillers, plasticizers, coloring agents, particles, and fibers. Plasticizers are compounds added to the polymer to make it flow more easily during processing. The word *plastic* also has another meaning when associated with the deformation of materials. Plastic deformation is permanent deformation.

2.8.1 Covalent Bonding in Polymer Molecules

One of the simplest polymers is polyethylene (PE). PE is made from the gas ethylene (C_2H_4), shown in Figure 2.17a. Hydrogen has a valence shell that is one-half filled with a single $1s$ electron, and in carbon the valence shell is one-half filled with four electrons. In ethylene each hydrogen atom acquires a second electron to complete the $1s$ shell by sharing one electron with a carbon atom in a covalent bond. Group IV carbon requires four additional electrons for each carbon to be surrounded by eight electrons, as in

Figure 2.17   (a) A schematic of the ethylene molecule (C_2H_4). (b) A schematic of the polymerization of ethylene molecules. The double bond in the ethylene molecule is broken, allowing each carbon atom on the ethylene molecule to bond to carbon atoms on other ethylene molecules. The ethylene molecule (C_2H_4) is the mer unit. (c) A planar schematic of polyethylene. Each bond is now a single covalent bond with two electrons. (d) A three-dimensional schematic of the arrangement of carbon and hydrogen atoms of reduced size on a polyethylene long-chain molecule. *(Based on Askeland, D.R., Fulay, P.P., and Wright, W.J., The Science and Engineering of Materials, 6th ed. Cengage Learning, Stamford, CT. (2011), (a) and (b) p. 606, (c) and (d) p. 604.)*

the inert-gas electron structure. The carbon atoms share eight electrons with two hydrogen atoms and another carbon atom in covalent bonds. The covalent bond between the carbon atoms in ethylene is a double covalent bond containing four electrons, as indicated by the double line between the carbon atoms in Figure 2.17a. A single line means two electrons in a covalent bond, and two lines indicates four electrons in a double covalent bond.

Polymerization is the process of converting monomers into polymers. In the polymerization process of PE the ethylene gas is heated to a high temperature and combined with an initiator, such as peroxide. The function of the initiator is to open up the double bond, as shown in Figure 2.17b, and start the polymerization reaction. The ethylene gas molecules react with each other to form long-chain molecules (LCMs) with single covalent bonds of two electrons between the two carbon atoms, as indicated in Figure 2.17b. Figure 2.17b shows the ethylene molecules with the double bond broken and replaced with a single covalent bond between the carbon atoms, and a single electron on the other side of the carbon atom, as indicated by the single dot at the end of a line on the carbon atoms. This line with a dot at the end does not indicate a covalent bond with two electrons; it indicates only one electron. Mer units combine with other mer units to form LCMs as shown schematically in Figure 2.17c. This chain formation happens when the single electron on one mer unit bonds with a single electron on another mer unit, and so on until the chain is formed. In the Figure 2.17c configuration, each carbon atom shares eight electrons, and each line between these atoms indicates a single covalent bond with two electrons. The PE molecule is not actually flat as shown in Figure 2.17c; rather, the atoms alternate from side to side, with the hydrogen atoms at an angle of 109.5° from each other, as shown schematically in Figure 2.17d.

The process of polymerization described for PE is addition polymerization, and it is the process for simple polymers. Addition polymerization is also used for polymers with the mer unit C_2H_3R, where R is another element, such as chlorine in polyvinylchloride, or a molecule, such as benzene in polystyrene. Addition polymerization is also used for polyvinylidenes that have the mer unit $C_2H_2R_1R_2$, where R_1 and R_2 are another element, such as $R_1=R_2=Cl$ in polyvinylidene-chloride, and it is used for polymers such as polymethyl-methacrylate (PMMA).

Condensation reactions produce more complex LCMs, such as the heterochain molecule of nylon that results from reacting hexamethylene diamene with adiptic acid. **Heterochain** molecules contain more than one mer unit. Condensation reactions can also produce three-dimensional polymers; for example, phenolformaldehyde (Bakelite™) is produced by the reaction of phenol and formaldehyde. Heterochain polymers and three-dimensional polymers are stronger polymers than polyethylene is, and these materials and processes are considered in more detail in the discussion of strong solids in Chapter 7.

The average number of mer units ($\bar{n}$) in the LCMs of a polymer is a measure of the degree of polymerization. The number-average molecular mass of the long-chain molecules ($\overline{M}_n$) in a polymer is equal to the total mass of the polymer (m) divided by the number of long-chain molecules (N), as shown in Equation 2.5a.

$$\overline{M}_n = \frac{m}{N} \qquad\qquad \textbf{2.5a}$$

The average number of mer units in long-chain molecules is the number-average molecular mass ($\overline{M}_n$) divided by the molecular mass of the mer unit (M_m), as shown in Equation 2.5b.

$$\bar{n} = \frac{\overline{M}_n}{M_m} \qquad\qquad \textbf{2.5b}$$

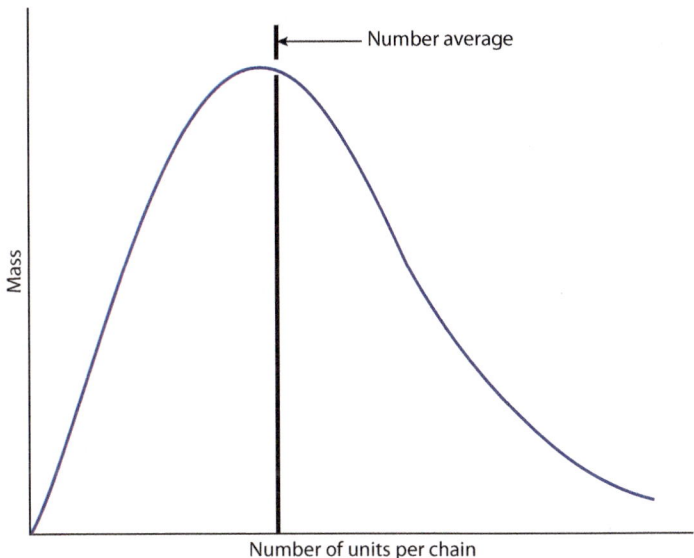

Figure 2.18   A hypothetical distribution of the mass of long-chain molecules in an interval of length as a function of the number of units per chain.

Mechanical properties of polymers, such as tensile strength, are related to the number-average molecular mass. As the molecular mass increases, the strength increases. For example, the polymer polystyrene (PS) is stiff and brittle at room temperature with an average of 1000 mer units in each LCM, and it is soft and sticky with only 10 mer units. In the discussion of polymers the term molecular weight is often utilized when what is meant is molecular mass.

A polymer has molecules with a distribution of molecular masses, as shown in Figure 2.18. To determine the distribution of molecular masses, the molecular masses are divided into intervals of length. With PS the intervals could be divided into intervals of 10 mers. The number of long-chain molecules with mer units from 0 to 10 is the first interval, and the mass plotted in Figure 2.18 is the total mass of all molecules in this interval. The second interval is long-chain molecules with 10 to 20 mer units, and the mass plotted in Figure 2.18 is the total mass of all molecules of this length. In polymers, such as the polyethylene in milk bottles that has ethylene [C_2H_4] mers, the number of mer units on the molecules range from 200,000 to 500,000. Ultra-high-molecular-weight polyethylene (UHMWPE) has molecules with 2 million to 6 million mer units.

The name, abbreviation, common name if applicable, mer structure, and applications of some common polymers are shown in Figure 2.19, and some more complex polymers are shown in Figure 2.20. In some of these long-chain molecules, an element is substituted for hydrogen. For example, in polyvinyl chloride (PVC), one of the four hydrogen atoms is replaced by a chlorine atom. In vinyl polymers one of the hydrogen atoms in each polyethylene mer unit is replaced by another atom. If two of the hydrogen atoms in each mer unit of a vinyl polymer are replaced by other atoms, the compound becomes a vinyldiene. Polyvinyldiene chloride has two chlorine atoms and two hydrogen atoms on each mer unit. In polytetraflouroethylene (Teflon™) all four (tetra) of the hydrogen atoms are replaced by fluorine atoms. In other cases a molecule will replace a hydrogen atom. For example, in polypropylene one hydrogen atom is replaced by a CH_3 molecule. In Figure 2.19, the hexagon with an enclosed circle on the polystyrene mer is benzene.

Polymer	Repeat Unit	Application	Polymer	Repeat Unit	Application
Polyethylene (PE)	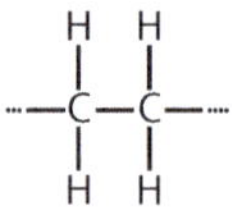	Packing films, wire insulation, squeeze bottles, tubing, household items	Polyacrylonitrile (PAN)	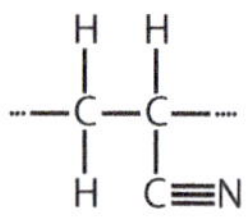	Textile fibers, precuesor for carbon fibers, food containers
Polyvinyl chloride (PVC)	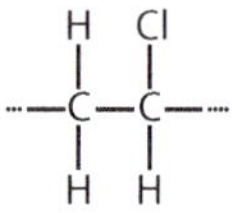	Pipe, valves, fittings, floor tile, wire insulation, vinyl automobile roofs	Polymethyl methacrylate (PMMA) (acrylic-Plexiglas)	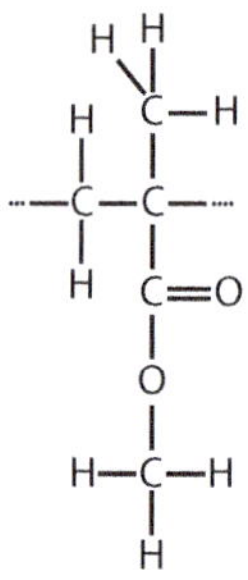	Windows windshields, coatings lenses, lighted signs
Polypropylene (PP)	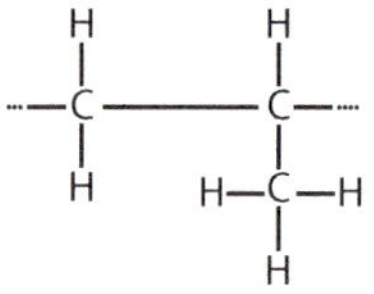	Tanks, carpet fibers, rope, packaging			
Polystyrene (PS)	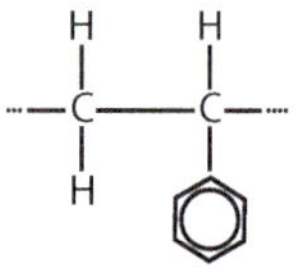	Packaging and insulation foams, lighting panels, appliance components, egg cartons	Polychlorotri-fluorcethylene		Valve components, gaskets, tubing, elecrical insulation
Polyacetylene	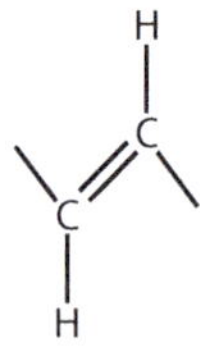	Conductor Semiconductor	Polytetrafluoro-ethylene (Teflon) (PTFE)		Seals, valves, nonstick coatings

Figure 2.19 A listing of the name, abbreviation, common name if applicable, mer structure, and applications of some common thermoplastic polymers. (*Based on Askeland, D.R., Fulay, P.P., and Wright, W.J., The Science and Engineering of Materials, 6th ed. Cengage Learning, Stamford, CT. (2011), p. 613.*)

2.8.2   Molecular Arrangements in Thermoplastic Polymers

A **thermoplastic** polymer is one that softens when it is heated. Thermoplastic polymers have only van der Waals bonds between different LCMs. The arrangement of the long-chain molecules in a liquid thermoplastic polymer is amorphous, and it looks like a bowl of spaghetti noodles, as shown in Figure 2.21a. An amorphous structure is one that does not have long-range order. Crystals have long-range order, because the crystal structure repeats over large distances. However, if the material is amorphous, it is impossible to predict the location of atoms over large distances. If an amorphous liquid polymer is cooled rapidly to room temperature to form a solid, the solid structure is still amorphous. The reason that the solid structure is still amorphous after a rapid cool is that the long-chain molecules do not have

Polymer	Repeat Unit	Applications
Polyoxymethylene (acetal)(POM)	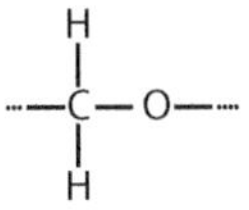	Plumbing fixtures, pens, bearings, gears, fan blades
Polyamide (nylon) (PA)	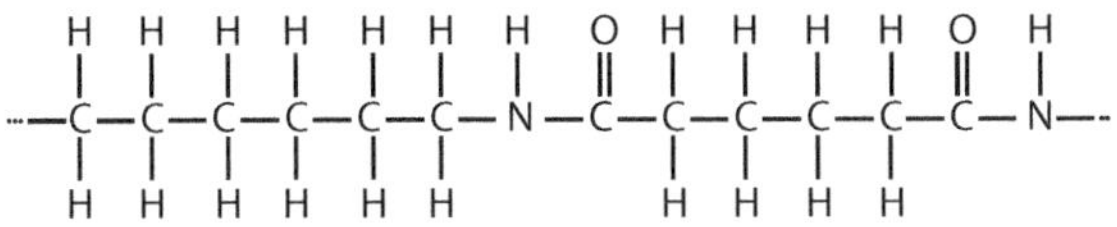	Bearings, gears, fibers, rope, automotive components, electrical components
Polyester (PET)	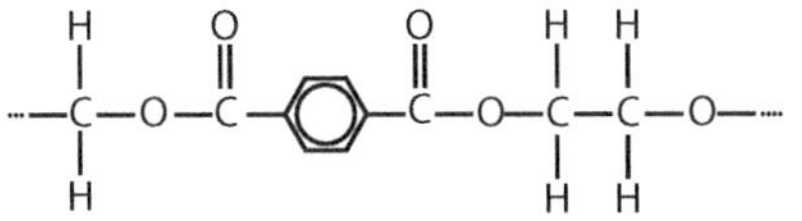	Fibers, photographic film, recording tape, boil-in-bag containers, beverage containers
Polycarbonate (PC)	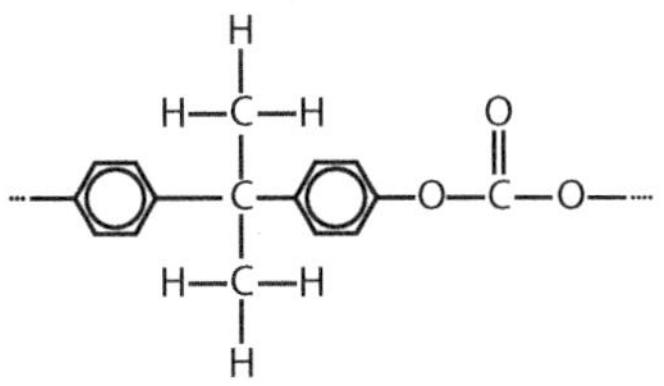	Electrical and appliance housings, automotive components, football helmets, returnable bottles
Polyimide (PI)	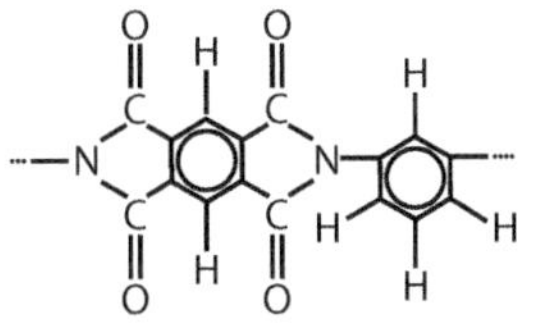	Adhesives, circuit boards, fibers for space shuttle
Polyetheretherketone (PEEK)	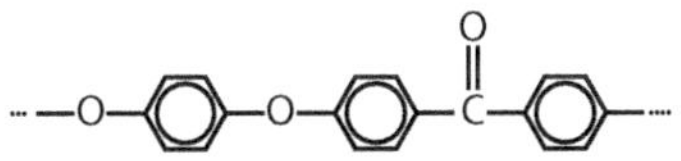	High-temperature electrical insulation and coatings
Polyphenylene sulfide (PPS)	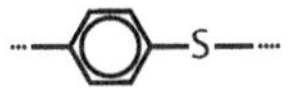	Coatings, fluid-handling components, electronic components, hair dryer components
Polyether sulfone (PES)	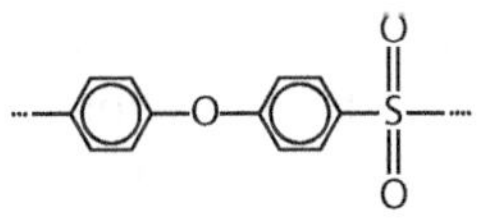	Electrical components, coffeemakers, hair dryers, microwave oven components
Polyamide-imide (PAI)	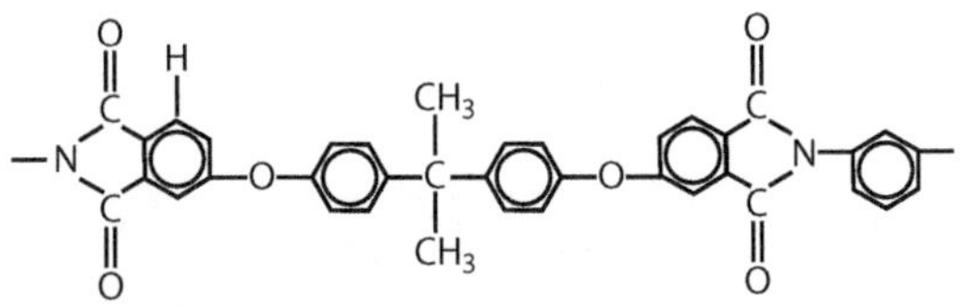	Electronic components, aerospace and automotive applications

Figure 2.20   A listing of the name, abbreviation, common name if applicable, mer structure, and applications of some more complex thermoplastic polymers. (*Based on Askeland, D.R., Fulay, P.P., and Wright, W.J., The Science and Engineering of Materials, 6th ed. Cengage Learning, Stamford, CT. (2011), p. 614.*)

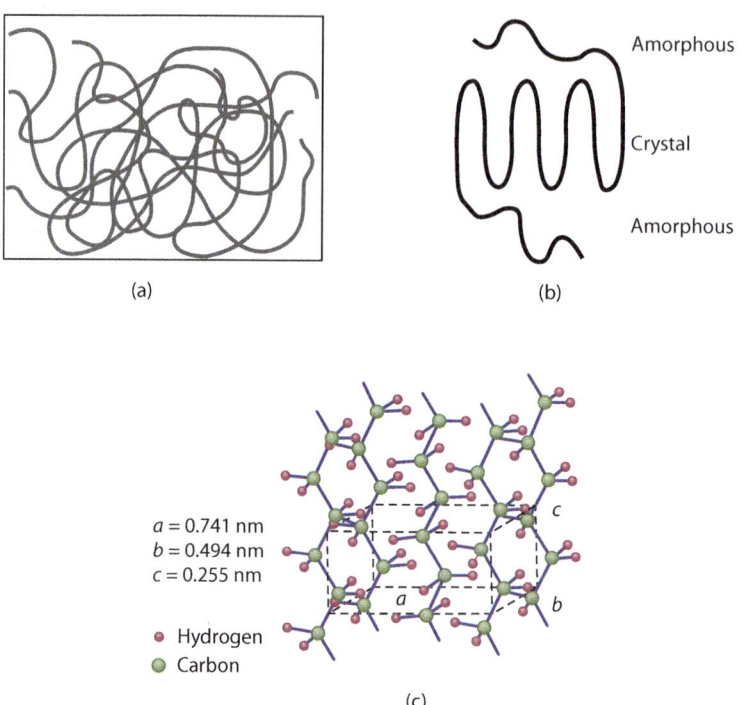

a = 0.741 nm
b = 0.494 nm
c = 0.255 nm

● Hydrogen
● Carbon

(c)

Figure 2.21   (a) A schematic representation of an amorphous polymer made from long-chain molecules. (b) A long-chain molecule showing a crystalline region formed by looping of the molecule, as well as amorphous regions. (c) The unit cell of polyethylene. *((a) and (c) based on Askeland, D.R., Fulay, P.P., and Wright, W.J., The Science and Engineering of Materials, 6th ed. Cengage Learning, Stamford, CT. (2011), p. 95.)*

enough time to untangle themselves and form an ordered crystal structure before the cooling is completed. If the polymer is a simple one, such as polyethylene, and it is cooled slowly, some crystalline areas will form as the polyethylene molecule loops back on itself, as shown in Figure 2.21b, or by alignment of adjacent molecules. A crystalline region of polyethylene has the unit cell shown in Figure 2.21c.

When a large molecule replaces a hydrogen atom, this inhibits crystallization. The large molecules make it difficult for molecules to line up. The polymer PMMA has large methyl and methacrylate molecules that replace the hydrogen atoms. PMMA is normally completely amorphous. Completely amorphous polymers are clear unless a coloring agent is added. PMMA is a clear polymer that is used to replace glass in many applications. Polymers that have crystalline areas are translucent because the crystals scatter light. Semicrystalline PE in milk bottles is not clear.

2.8.3   Properties of Thermoplastic Polymers

Since the covalent bonds in a polymer chain are strong, why is PE very soft in comparison to a material like diamond that also has covalent bonds and is hard, as shown in Table 2.2? Note that in Table 2.2 the PE is measured with the Brinell hardness. There is not a direct conversion from one hardness scale to the other, but a Knoop hardness of 100 corresponds to approximately 95 on the Brinell scale. The softness of PE is because the bond between different long-chain molecules in solid PE is the weak van der Waals bond. Figure 2.22 shows the bonds between LCMs of PVC.

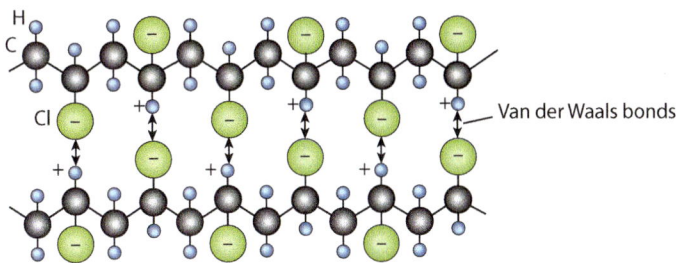

Figure 2.22 A schematic of the van der Waals bonds between polymer chains in PVC. *(Based on Askeland, D.R., Fulay, P.P., and Wright, W.J., The Science and Engineering of Materials, 6th ed. Cengage Learning, Stamford, CT. (2011), p. 40.)*

In thermoplastic polymers the thermal vibrations of the molecules at elevated temperatures are sufficient to break the van der Waals bonds along the molecules, allowing the long-chain molecules to slide past each other when a force is applied. Once the van der Waals bond is broken, the long-chain molecules are permanently deformed relative to each other. All of the polymers in Tables 2.19 and 2.20 are thermoplastic polymers. Note that the word *plastic* in the term *thermoplastic* refers to the polymer's "plasticity," or "deformability"; the word *plastic* in this case does not mean a polymer with fillers, pigments, plasticizers, and so on.

2.8.4   Thermosetting Polymers

Thermoset polymers do not significantly soften when heated, and they can harden. Polymers with three-dimensional covalent bonds are thermosetting polymers. One way to form three-dimensional covalent bonding with polymers is to cross-link LCMs. Different LCMs are **cross-linked** by adding atoms or molecules that covalently bond the different LCMs to each other, as shown in Figure 2.23. An example of a cross-linked polymer is rubber. The molecular structure of latex (*cis* isoprene) is shown in Figure 2.24a. Latex is a milky sap that is extracted from the rubber tree. The process of vulcanization adds the sulfur cross-link atoms between the latex long-chain molecules of latex, as shown in Figure 2.24b, creating solid rubber. The sulfur cross-links are located at sites where previously there was a carbon double bond. The hydrogen atoms bond with the second electron of the broken carbon-carbon double covalent bond.

Solid rubber typically has 0.5 to 5 weight percent sulfur. Low amounts of sulfur produce rubber that is very flexible, such as rubber bands and gloves. High amounts of sulfur produce a material that is rigid. Automobile tires typically contain 3 weight percent sulfur. In the undeformed state the long-chain

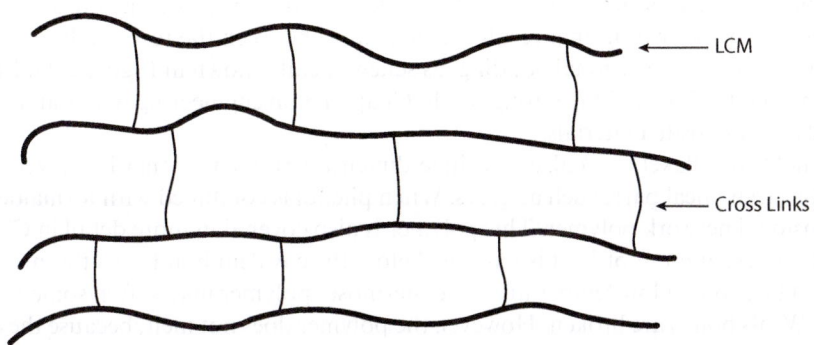

Figure 2.23 A three-dimensional molecular structure formed by cross-linking long-chain molecules (LCMs).

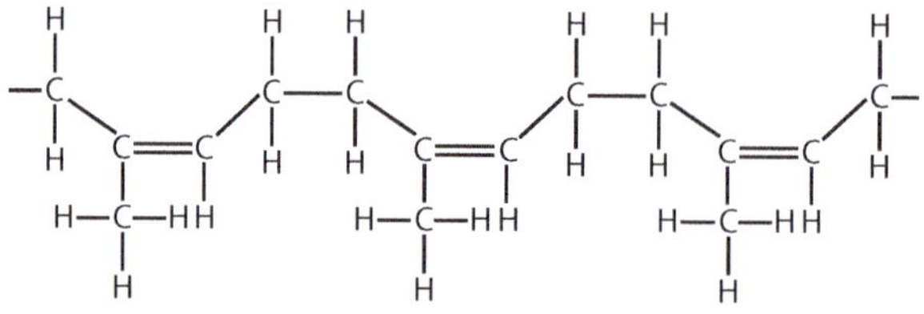

(a)

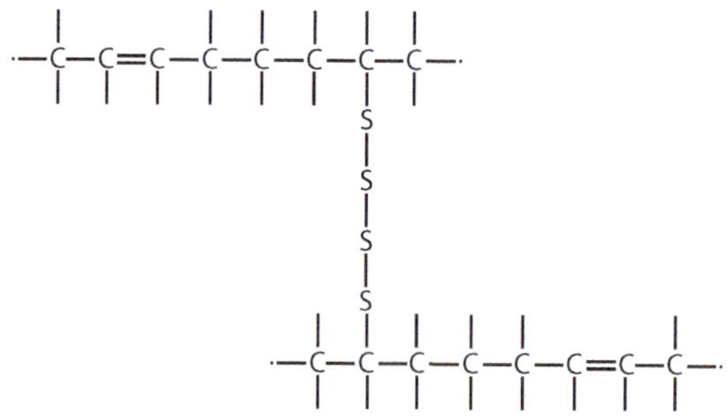

(b)

Figure 2.24   (a) A planar schematic of the long-chain molecular structure of liquid latex before the vulcanization process. The long-chain molecule is bent due to the presence of the CH_3 side groups. (b) A schematic of the latex long-chain molecules cross-linked with sulfur (S) by vulcanization to form rubber. (*Based on Askeland, D.R., Fulay, P.P., and Wright, W.J., The Science and Engineering of Materials, 6th ed. Cengage Learning, Stamford, CT. (2011), (a) p. 630 and (b) p. 632.*)

molecules in rubber are coiled and bent, because the bulky side-group molecules bend the long-chain molecules, as shown in Figure 2.24a. When the rubber is stretched, these coils and bends are straightened. This allows the rubber to stretch as much as ten times its original length without breaking. Then if the stretched rubber is released, the sulfur cross-links realign the rubber molecules, the bends and coils return, and the rubber returns to its previous shape.

Epoxy is another cross-linked polymer. Epoxy is utilized as the matrix in many composite materials, such as fiberglass, because of its ease of fabrication. Liquid epoxy resin composed of LCMs that are not cross-linked is mixed with a liquid hardener. The hardener produces a chemical reaction with the long-chain molecules of the epoxy resin that results in cross-links between the resin molecules. This produces a hard solid that has three-dimensional bonding, as schematically shown in Figure 2.23. Epoxy is covered in more detail in Chapter 7 on making strong solids, Chapter 8 on engineering materials and applications, and Chapter 12 on composite materials.

Phenolformaldehyde (Bakelite™) also is a three-dimensional polymer that is utilized as an electrical insulator and for mechanical parts such as gears. When phenol is combined with formaldehyde the result is a three-dimensional network polymer. This polymer is also covered in more detail in Chapters 7 and 8.

More than a modest amount of heat is required before the covalent bonds in a thermoset polymer can be broken. When heated to a low temperature, the thermoset polymer may soften some because the ever-present van der Waals bonds are broken. However, the polymer does not melt, because the covalent bonds remain. When thermosetting polymers are heated in air, they react with oxygen, harden, and deteriorate.

2.8.5 Physical Properties of Polymers and Applications

Some physical properties of polymers result from the strong covalent bonding of the atoms in the long-chain molecular structure, and other properties result from the weak van der Waals bonds that hold the long-chain molecules together. Even though the carbon-carbon and hydrogen-carbon covalent bonds in polyethylene are very strong, bulk polyethylene is easy to deform at room temperature, and polyethylene typically melts at a low temperature of 403 K (130°C), as shown in Table 2.2. In polymers the melting temperature is defined as the maximum temperature where crystalline structures can exist.

The density of polymers is very low because the long-chain molecules are primarily made of carbon and hydrogen. Also, the bonding between these chains is weak, resulting in relatively large distances between chains. The density of polymers is typically 1g/cm^3, in comparison to aluminum, with a density of 2.69 g/cm^3; and iron, with a density of 7.87 g/cm^3. This explains why polymers are replacing metals in applications where low weight is important but high strength is not. As other atoms or molecules are substituted for hydrogen, the density increases; for example, PVC has one hydrogen out of four replaced by chlorine, and the density increases from 0.95 g/cm^3 for PE to typically 1.46 g/cm^3 for PVC.

It is possible to produce polymer fibers that are as strong as the highest strength steel, as discussed in Section 1.3.3. Strong polymer fibers are produced by orienting the long-chain molecules along the fiber axis, so that the strong covalent carbon-carbon bond carries the load, rather than the weak van der Waals bonds between long-chain molecules. Oriented ultra-high molecular-weight polyethylene (OUHMWPE) is one of the commercial fibers. OUHMWPE has been engineered into a high-performance material and is used to make materials such as high-performance sporting equipment, bullet-proof vests, tug tow lines, bow strings, and high-performance sails. OUHMWPE is an example of an engineered material where the material is designed to maximize the strength of the molecular structure and also minimize the effects of the weakness in the bonding. By understanding the nature of the chemical bonding in these materials, it is possible to produce polymers of amazing strength. Spectra™ is one of the trade names for OUHMWPE. The production and properties of OUHMWPE are covered in more detail in Section 7.6.1.4.

2.9 METALLIC BONDING

Metals are inorganic materials with metallic bonding. In the **metallic bond** the valence electrons are collectively shared by many positive-ion cores, and the valence electrons are free to move throughout the metal. Metals are found in many products, from the steel in automobiles and bridges to the nickel in high-temperature gas-turbine blades. The nature of the metallic bond gives metals physical properties that are different from those of covalently bonded materials and polymers. First we consider pure elements, such as iron. Then at the end of this section we consider **alloys**, which are a mixture of metallic elements with other elements. For example, the steel in an automobile or a bridge is a mixture of iron and a small amount (such as 0.2 weight percent) of carbon.

Atoms of pure elements whose valence electron shells are less than half filled (Groups I to III), when brought together, give up their valence electrons to a nonlocalized electron gas to achieve an inert-gas electron configuration. Aluminum (in Group III of the periodic table) has the valence-electron configuration of three valence electrons ($3s^2 3p$) outside of the neon ion core, as shown in the top part of

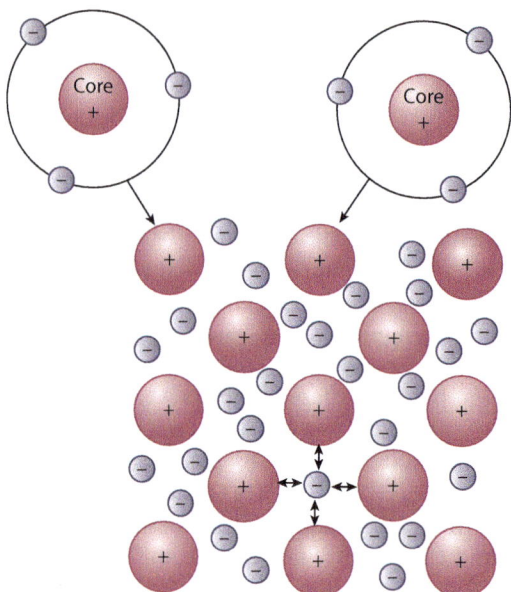

Figure 2.25   The upper part of the figure shows atoms with a valence of 3 (such as aluminum) coming together, indicated by the arrows, to form a solid with metallic bonding. The schematic of metallic bonding has positive-ion cores in regular locations, and the free valence electrons (negative charges) are randomly located. The force arrows in the lower part of the figure indicate the attractive bonding between the positive-ion cores and the negative electrons. (*Based on Askeland, D.R., Fulay, P.P., and Wright, W.J., The Science and Engineering of Materials, 6th ed. Cengage Learning, Stamford, CT. (2011), p. 35.*)

Figure 2.25. Aluminum has a spherically symmetric ion core of the inert-gas neon if the valence electrons are removed from the atom. If aluminum atoms are brought together to form a solid, the valence electrons become separated from the aluminum positive-ion cores and are free to move about in the metal, as shown in the lower part of Figure 2.25. However, the electrons are confined to the solid metal and neutralize the positive charge of the aluminum ions. In the lower part of Figure 2.25, the positive-ion cores (large circles) are regularly spaced in the crystal, and the negatively charged free electrons (small circles) are shown randomly positioned in the metal to indicate that they are not localized to a particular ion core. The free electrons are often referred to as a sea of electrons or as an electron gas to indicate their freedom of motion. In reality they are not a liquid or a gas. Because the electrons are free to move anywhere in the metal, the electrical and thermal conductivity of metals is high. The electrostatic attraction of the positive-ion cores for the free electrons bonds the atoms together into a solid, as indicated by the force arrows in the bottom part of Figure 2.25.

The transition-series elements, such as iron, where the $3d$ subshell is being filled, are also metals because the outer $4s$ electrons become free electrons when a solid is formed. The electrons in the incomplete $3d$ subshell are localized onto the atoms and covalently bond with other metal atoms. Metals also occur in Groups IV, V, and VI when the atomic number is large, such as Group IV tin (atomic number 50) and lead (atomic number 82), Group V antimony (atomic number 51), and Group VI polonium (atomic number 84). In these elements the atomic radius is large, and there are many inner electrons shielding the valence electrons from the nucleus. As a result, the valence electrons are loosely bound, and in a solid they become free electrons surrounding ion cores in a metallic bond, rather than forming the covalent bonds usually expected of Group IV, V, and VI elements.

2.9.1　Metal Atom Packing

The spherical positive-ion cores in many metals pack together as though they were billiard balls, as shown in Figures 2.3 and 2.4. Metals that solidify in a hexagonal close-packed crystal structure, as shown in Figure 2.3, include titanium, zirconium, cobalt, magnesium, beryllium, and zinc. Metals that form in the FCC crystal structure, as shown in Figure 2.4, include copper, silver, gold, nickel, platinum, and palladium. Metals such as sodium, iron, chromium, molybdenum, niobium, and tungsten form in the BCC crystal structure, shown in Figure 2.5. Most metals form in one of these three structures. However, there are a few metals that have other crystal structures; for example, polonium is simple cubic and tin is tetragonal. To predict the structure of a metal requires a quantum mechanical analysis of the bonds between the atoms. However, there are some trends. The group IB metals Cu, Ag, and Au form in the FCC structure, as do the elements with ten electrons in the outermost d plus s subshells (Ni, Pd, and Pt). The BCC structure is favored in the transition elements where there is significant covalent bonding present between electrons on different atoms with an incomplete inner electron subshell, such as the $3d$ subshell in iron, the $4d$ subshell in molybdenum, and the $5d$ subshell in tungsten.

2.9.2　Metal Physical Properties

Metals are good conductors of electricity, because the free electrons are not bonded to a particular atom and they can therefore flow through the metal, as discussed in Section 16.2. The free electrons in metals also result in a high thermal conductivity, as discussed in Section 4.4. Most pure metals are ductile, meaning that they can be permanently deformed without breaking. An example of ductility is when you permanently bend a paper clip. The permanent deformation of metals is discussed in Chapter 6. We will find that the ductility of metals is due to the close-packing of metal atoms and because the metallic bond is not localized between two atoms. Metals have a high density relative to their atomic mass because the packing of metal atoms efficiently fills the space in the metal crystal. Compare the density of silicon (atomic number 14) to aluminum (atomic number 13). Silicon is in Group IV and has the same crystal structure as diamond. Aluminum is in Group III of the periodic table and is a FCC metal with close-packing of atoms. Silicon is less dense (2.33×10^3 kg/m³) than aluminum (2.7×10^3 kg/m³), even though silicon has a higher atomic mass. The reason is because each silicon atom has only four nearest neighbors, whereas aluminum has twelve nearest neighbors. Therefore, the silicon crystal structure is more open and less dense than the close-packed aluminum structure.

In general, the melting temperatures for materials with metallic bonding are lower than for materials with covalent bonding, and melting temperatures are in direct relationship to the cohesive energy. For example, covalently bonded silicon has a melting temperature of 1683 K, whereas metallic aluminum has a melting temperature of 933 K. These melting temperatures are directly related to the cohesive energies in the periodic table of 4.64 and 3.34 eV/atom, respectively, for silicon and aluminum.

There is a significant range of cohesive energies for metals, as shown in the periodic table. For example, the metals aluminum, nickel, and tungsten have cohesive energies of 3.34, 4.435, and 8.66 eV per atom. These cohesive energies are in direct proportion to their respective melting temperatures of 933, 1726, and 3683 K. High melting temperature metals are usually transition series metals that have some covalent bonding between the unfilled inner electron subshells, such as titanium (1941 K), iron (1811 K), and tungsten (3683 K). If the bonding is primarily metallic, then the melting temperature is lower, as for aluminum (933 K), copper (1358 K), and silver (1235 K).

Different metals also have a wide range of hardness at room temperature. From Table 2.2, gold has a Knoop hardness of 69 and high-strength steel has a hardness of 392. Materials with primarily metallic bonding, such as gold, have a lower hardness than materials with some covalent bonding in addition to metallic bonding, such as steel. In metals the hardness is sensitive to mechanical and thermal treatments. This sensitivity allows for very interesting changes in their mechanical properties. For example, if soft pure gold is hammered, it becomes harder. Then if this same hammered gold is put in a furnace at 400°C for an hour, it softens. The reasons for this behavior are discussed in Chapter 7. There is not a single number for the hardness of a metal. The hardness depends upon the deformation and temperature history of the metal. Metal alloys that are deformed and given thermal treatments can have much higher hardness; for example, high-strength steel can have a Knoop hardness as high as 1000.

Diamond has a Knoop hardness of approximately 7000, whereas gold's Knoop hardness is 69. This large difference is due to diamond's covalent bonds, which resist permanent deformation. As we shall see in Chapter 6, pure metals, such as gold, are easy to deform permanently.

2.9.3   Metal Alloys

Most applications of metals are alloys; the metal is not pure. If we mix two similar metallic elements, we normally obtain a resulting metal alloy that is similar to the two original metals. For example, consider a mixture of copper and nickel, which are similar to each other. If copper is mixed with nickel, the resulting solid has properties that are an "average" of the properties of copper and nickel.

Compounds are often created from two metals with significantly different properties, and this can produce some very interesting engineering properties. For example, if nickel is mixed with 25 atomic percent aluminum, the intermetallic compound Ni_3Al is formed. Ni_3Al is called an **intermetallic compound** because it is composed of at least two metals and there is a fixed number of atoms of each type; three nickel atoms and one aluminum atom. The relative numbers of the types of atoms in a compound is the **stoichiometry**. The intermetallic compound is also a metal. When two metals form a compound, this is an indication of strong bonding. Many intermetallic compounds are very hard and have high melting temperatures. The intermetallic compound Ni_3Al has a relatively high melting temperature of 1669 K and is used to provide the strength in the nickel-based alloys used in high-temperature gas turbines.

If a metal is mixed with a nonmetallic element, a compound often results. For example, if 25 atomic percent carbon is mixed with iron, the compound iron carbide (Fe_3C) results. This compound does not have the properties of a metal. It is not a good conductor of electricity. It is brittle, not ductile, and it melts at a high temperature of approximately 2523 K. If iron is mixed with a small amount of carbon, such as 0.2 weight percent, the result is a mixture of iron and a small amount of Fe_3C. This alloy of carbon steel is ductile, and it is used in automobiles and bridges. Other metal-alloy applications include aluminum alloys used in aircraft fuselage and frames, nickel alloys used in high-temperature gas turbines, titanium alloys used in supersonic aircraft fuselages, and copper alloys used in heat exchangers. Phase diagrams in Chapter 5 show us the alloys that result from mixing metals with other elements.

2.10   IONIC BONDING

An ionic bond is the bond that occurs between the positive and negative ions of different elements with significantly different electronegativities. A classic example of an ionic bonding is sodium chloride (NaCl), where sodium (Na, Group IA, electronegativity 0.9) bonds ionically with chlorine (Cl, Group VII, electronegativity 3.0). Ionic bonding occurs in NaCl by electron transfer of the $3s$ electron from the sodium atom to the chlorine atom, as shown in Figure 2.26a, to create the cation Na^+ and the anion

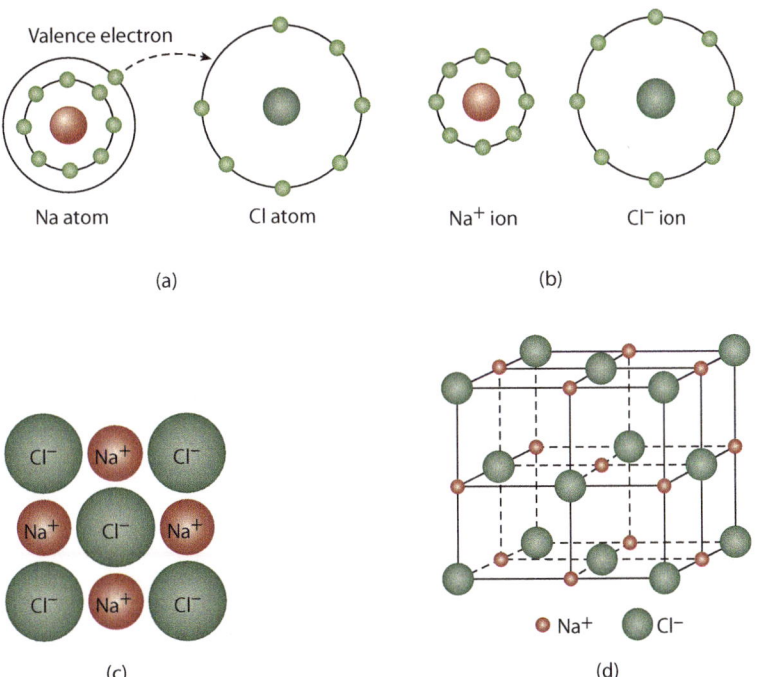

Valence electron

Na atom   Cl atom   Na⁺ ion   Cl⁻ ion

(a)   (b)

(c)   (d)

● Na⁺   ● Cl⁻

Figure 2.26 (a) A schematic of electron transfer from a sodium (Na) atom to a chlorine (Cl) atom. (b) Sodium (+) and chlorine (−) ions, each with eight valence electrons. (c) A two-dimensional schematic of the packing of positive and negative ions. (d) A unit cell of the sodium chloride (NaCl) crystal structure. ((a), (b), and (c) adapted from Askeland, D. R., Fulay, P. P., and Wright, W. J., The Science and Engineering of Materials, 6th ed., Cengage Learing, Stamford, CT (2011) p. 38. (d) Adapted from Callister, W. D., Materials Science and Engineering: An Introduction, 6th ed. John Wiley & Sons, New York (2003), p. 388.)

Cl^-. Na^+ has the inert-gas electron structure of neon, and Cl^- has the electron structure of argon, as shown in Figure 2.26b. The solid NaCl is held together by the electrostatic attraction of the Na^+ and Cl^- ions for each other, as shown in Figure 2.26c. Other materials of significant scientific and engineering importance that have ionic bonding are MgO, Al_2O_3, and ZrO_2. These materials have a mixture of ionic and covalent bonding, as shown in Table 2.3. A mixture of ionic and covalent bonding means that there is some transfer of electrons from one atom to another, and there is some localized electron sharing. The greater the difference in electronegativity, the greater the tendency to form ions and an ionic bond, such as in NaCl. If the electronegativity of two different atoms is similar, there is more of a tendency to share electrons in a covalent bond. When there is an electronegativity difference between two different atoms, the electrons are more attracted to the element with the higher electronegativity. An equation proposed by Linus Pauling is used to calculate the percent ionic bonding from the electronegativity difference (Δ), as shown in Equation 2.6.

$$\% \text{ ionic} = (1 - e^{-0.25\Delta^2})100 \qquad \textbf{2.6}$$

Similar electronegativity results in more covalent bonding, such as 2.5 for carbon and 1.8 for silicon in SiC; and a electronegativity difference results in more ionic bonding, such as 1.2 for Mg and 3.5 for O in MgO.

Table 2.3   **Percent Ionic and Covalent Bonding in Some Ceramic Compounds, the Electronegativity Difference (Δ), and the Melting Temperature (T_m)**

Ceramic	Δ	% Ionic	% Covalent	T_m K
ZrO_2	2.1	67	33	2953
MgO	2.3	73	27	3062
Al_2O_3	2.0	63	37	2300
SiO_2	1.7	51	49	1983
Si_3N_4	1.3	30	70	2173[1]
SiC	0.7	11	89	3003

1   Sublimes

Example Problem 2.9

Calculate the percent ionic and covalent bonding for MgO.

Solution

From the periodic table, the electronegativities of Mg and O are 1.2 and 3.5, respectively.

$$\Delta = 3.5 - 1.2 = 2.3.$$

Inserting Δ into Equation 2.6 gives

$$\% \text{ ionic} = (1 - e^{-0.25\Delta^2})100 = (1 - e^{-0.25(2.3)^2})100 = (1 - e^{-1.32})100 = (1 - 0.27)100 = 73\%$$

2.10.1   Atom Arrangements with Ionic Bonding

Generally, within an ionic bond, the anion's radius is larger than the cation's radius, since the anion has "grown" from gaining the electron that the "shrunken" cation has lost, relative to the neutral atoms. Each cation attracts as many anions around it as possible; however, the anions cannot overlap each other, and there should not be extra space between the anions. The larger the cation, the more anions that can surround the cation without overlapping. The number of anions that surround a cation is called the **coordination number**, and it is controlled by the radius of the cation (r_c) relative to the anion (r_a), as shown in Figure 2.27. For example, from Appendix C the ionic radius ratio (r_c/r_a) in NaCl is 0.54 (0.181 nm/0.181 nm). Figure 2.27 indicates that the Na^+ cation should have six neighboring Cl^- anions. Six spheres of radius 0.181 nm can be packed around one sphere of radius 0.097 nm. The crystal structure of NaCl with six nearest neighbors is shown in Figure 2.26d. In the NaCl unit cell there are four Cl^- ions surrounding the Na^+ at the body center position, and there is one Cl^- ion above and one Cl^- ion below for a total of six Cl^- ions surrounding the Na^+ at the body center position.

To describe NaCl, it is FCC with a possible set of atomic positions of chlorine at 0, 0, 0 and sodium at 0, $\frac{1}{2}$, 0. For every chlorine at the FCC positions, there is a sodium located 0, $\frac{1}{2}$, 0 away from the FCC

Coordination Number	Location of Smaller Atom or Ion	Radius Ratio	Coordination Geometry
2	Linear	0–0.155	
3	Center of triangle	0.155–0.225	
4	Center of tetrahedron	0.225–0.414	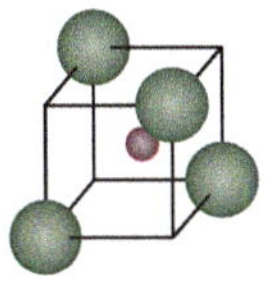
6	Center of octahedron	0.414–0.732	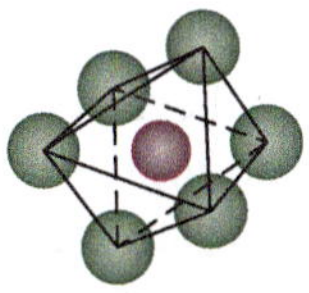
8	Center of cube	0.732–1.000	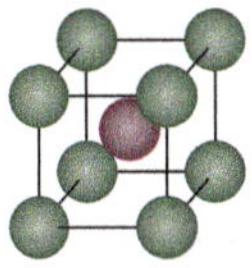

Figure 2.27 Coordination numbers and geometries for various atom or ion radius ratios. (*Based on Askeland, D.R., Fulay, P.P., and Wright, W.J., The Science and Engineering of Materials, 6th ed. Cengage Learning, Stamford, CT. (2011), p. 85.*)

atom. The sodium atom could also have been chosen at $\frac{1}{2}, 0, 0$ or at $0, 0, \frac{1}{2}$ and the same structure would have resulted, but all of the atom positions in a unit cell are established by selecting only one of these positions. And for proper stoichiometry, there is only one Na for one Cl. Other ionic materials that form in the NaCl structure include MgO, CaO, and FeO.

Example Problem 2.10

MgO has the NaCl crystal structure with a lattice parameter of 0.4212 nm. Calculate the expected density of MgO.

Solution

This problem is similar to Example Problem 2.3. However, now there are two atom types per lattice point (LP). This problem can be worked in the same way as Example Problem 2.3, except that we deal with LPs and the atoms associated with each LP. Using the LPs will do the accounting for us. Trying to deal with all the individual atoms can get confusing. In the FCC lattice there are four LPs, and each LP has one oxygen (O) and one Mg atom. The molar mass of oxygen is 16.00 g/mole, and for Mg it is 24.31 g/mole, and each mole of LPs contributes 40.31 g/mole (M_L).

The volume of the MgO unit cell is

$$a^3 = (0.4212 \times 10^{-9}\,\text{m})^3 = 0.0747 \times 10^{-27}\,\text{m}^3$$

The number of LPs per unit volume (n_L) is

$$n_L = \frac{4\,\text{LPs}}{0.0747 \times 10^{-27}\,\text{m}^3} = 5.35 \times 10^{28}\,\frac{\text{LPs}}{\text{m}^3}$$

The density of MgO (ρ_{MgO}) is then calculated from

$$\rho_{Mgo} = \frac{n_L M_L}{N_A} = 5.35 \times 10^{28}\,\frac{\text{LPs}}{\text{m}^3}\left(\frac{\text{mole of LPs}}{6.02 \times 10^{23}\,\text{LPs}}\right) 40.31 \times 10^{-3}\,\frac{\text{kg}}{\text{mole of LPs}} = 3.58 \times 10^3\,\frac{\text{kg}}{\text{m}^3}$$

This is the same equation as in Example Problem 2.3, except that here it is the number of lattice points per unit volume (n_L) that is used rather than the number of atoms per unit volume. In Example Problem 2.3 there is one atom per lattice point, and the number of atoms per unit volume is equal to the number of lattice points per unit volume in that problem. The answer is equal to the handbook value of the density of MgO.

Another ionic material is CsCl. The structure of CsCl is shown in Figure 2.7. From Appendix C the ionic radius ratio (r_c/r_a) is equal to 0.923 (0.167 nm/0.181 nm). From Figure 2.27 the coordination number should be 8, and in Figure 2.7 there are eight Cs atoms around the Cl atom. In CsCl the atom positions are cesium at 0, 0, 0 and chlorine at $\frac{1}{2}, \frac{1}{2}, \frac{1}{2}$.

If an anion and a cation are pushed too close together, the plus-charged and minus-charged ions repel each other rather than attracting each other, as would two separated positive-point and negative-point charges. The ions do not behave as point charges when pushed together, for if they did the ionic crystal would collapse, and the ionic charges would be neutralized. What keeps the ionic crystal from collapsing? Because of the Pauli exclusion principle, the system energy increases rapidly if the $2p$ electrons from the sodium and chlorine ions start to overlap, and this keeps the NaCl crystal from collapsing.

2.10.2 Physical Properties and Ionic Bonding

Ionically bonded materials are not good conductors of electricity, because there are no free electrons. Because the ionic bond is a relatively strong bond, ionically bonded materials have relatively high melting temperatures. The melting temperatures (T_m) of ionic materials are higher than polymers and similar to those of metals. As shown in Table 2.2, NaCl melts at 1070 K, PE melts at 403 K, and gold melts at 1336 K. Materials such as SiC (T_m = 3003 K) and SiO$_2$ (T_m = 1983 K) are in part bonded covalently, as shown in Table 2.3, which explains why they have higher melting temperatures than do materials with primarily ionic bonding, such as NaCl. The Knoop hardness of 16 for NaCl is very low. The hardness of NaCl is lower than even that of gold (69). In a hardness test of NaCl, the material under the indenter fractures rather than permanently deforms as it does in a metal. Ionic materials fracture with very low energy input. Also, when more covalent bonding is present in materials, such as 89% in SiC in comparison to 49% for SiO_2, the hardness of the SiC is much higher at 2480 in comparison to 820 for SiO_2. The hardness and melting temperature of materials with ionic plus covalent bonding in general increases with an increasing amount of covalent bonding.

2.10.3 Ceramics

Ceramics are normally oxide, carbide, or nitride compounds that include a metal or silicon. Ceramics have a combination of covalent and ionic bonding. Examples of common ceramics are Al_2O_3, SiC, Si_3N_4, and TiO_2. Ceramic parts are made by pressing powders of the ceramics together at high temperatures. These powder-processing techniques are discussed in Chapter 13 on materials processing. Ceramic materials have relatively high melting temperatures and hardness, as shown in Table 2.2. Ceramics are utilized in very-high-temperature applications such as furnaces, rocket motors, and for some nonrotating components of high-temperature gas turbines. It has been proposed to use ceramic materials, such as ZrO_2, in high-temperature gasoline and diesel engines.

2.10.4 Crystalline and Amorphous Materials

Some materials can be either crystalline or amorphous depending upon the cooling rate of a liquid to a solid. Silica (SiO_2) is a material that can be crystalline quartz, shown in Figure 2.28, or amorphous silica glass, shown in Figure 2.29. Solid SiO_2 can take on either of these forms, for two reasons. It has a complex atomic arrangement, and the bonding between silicon and oxygen is strong. In quartz each silicon atom is surrounded by four oxygen atoms, in the shape of a tetrahedron, as shown in Figure 2.28a. Each oxygen atom is shared by two silicon atoms, as shown in the quartz crystal structure in Figure 2.28b. Thus the chemical composition is two oxygen atoms for each silicon atom. In crystalline quartz the atom arrangement, shown in Figure 2.28b, is repeated over long distances; this is long-range order. When quartz is heated to form a liquid, some of the silicon-oxygen bonds are momentarily broken, and the silicon atoms move relative to the oxygen atoms. Then the silicon atoms re-form bonds with other oxygen atoms so that the movement is permanent. In the liquid state, each silicon atom is still surrounded by approximately four oxygen atoms, but all the paired silicon and oxygen atoms are continually changing partners. Also, the silicon-oxygen tetrahedra, shown in Figure 2.28a, can rotate when the silicon-oxygen

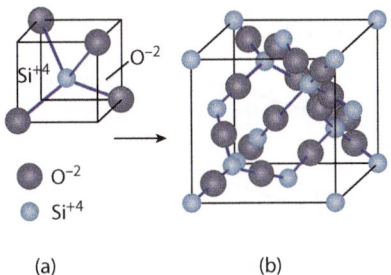

(a)   (b)

Figure 2.28 (a) The tetrahedral arrangement of four oxygen (O) ions around a single silicon (Si) ion. (b) The arrangement of Si and O ions in a unit cell of crystalline SiO_2 (cristobalite quartz). (*Based on Askeland, D.R., Fulay, P.P., and Wright, W.J., The Science and Engineering of Materials, 6th ed. Cengage Learning, Stamford, CT. (2011), p. 95.*)

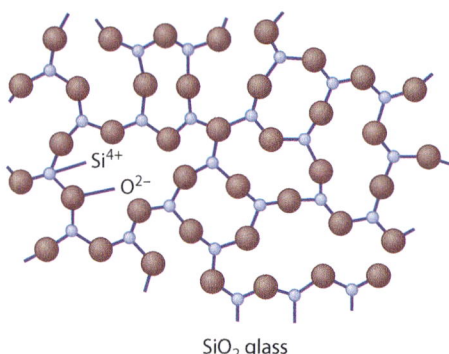

SiO₂ glass

Figure 2.29   A two-dimensional schematic of the disordered ion arrangements in SiO₂ glass. The silicon ions are surrounded by a tetrahedron of four oxygen ions. (*Based on Askeland, D.R., Fulay, P.P., and Wright, W.J., The Science and Engineering of Materials, 6th ed. Cengage Learning, Stamford, CT. (2011), p. 584.*)

bonds are broken, so that the long-range order, shown in Figure 2.28b, does not exist in the liquid. A schematic of the ion arrangements in liquid silica is shown in Figure 2.29. A structure without long-range order, such as that in Figure 2.29, is amorphous. However, each silicon atom is still surrounded by four oxygen atoms in the liquid; thus there is **short-range order**. Short-range order is order over interatomic distances.

  If the liquid (amorphous) SiO₂ is rapidly cooled to room temperature, the amorphous structure is frozen in place, and the resulting solid glass retains this amorphous structure. You may have heard that glass at room temperature is a liquid. You can forget that, for it is wrong. A liquid is easily changed in shape, as discussed in Section 2.3. In solid glass, the silicon-oxygen neighbors are constant in time. If the cooling of liquid silica is very slow, so that each tetrahedron of silicon and oxygen atoms has time to move into the lowest-energy position, then quartz forms. Quartz, which is extremely abundant on the earth's surface, formed crystals naturally as our planet slowly cooled. Glass also forms naturally when molten SiO₂ erupts from a volcano and then cools rapidly. The transformation from a liquid to

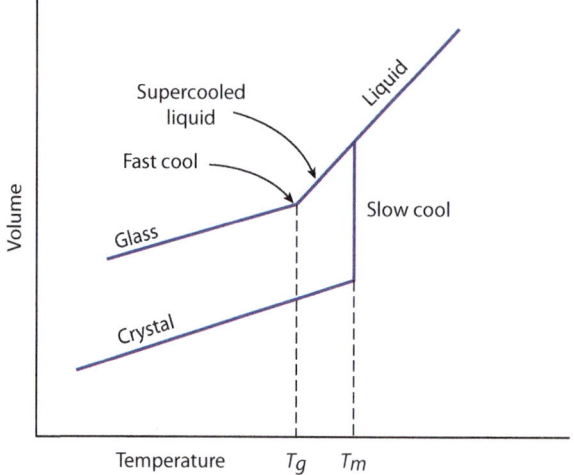

Figure 2.30   The volume of a glass-forming material as a function of temperature for rapid and slow cooling rates.

a solid can be followed by observing the volume of the material, as shown in Figure 2.30. If a material is cooled slowly from above the liquid-crystalline solid melting temperature (T_m), a plot of the volume (V) versus the temperature (T) usually has a stepdown when the crystalline solid forms, as shown in Figure 2.30. The water-ice transformation is an exception: When liquid water transforms to ice upon cooling, it expands, which is why ice floats upon water. The change in volume is due to the difference in density of the liquid and crystalline solid phases. If a material that forms glass is cooled rapidly to temperatures below the melting temperature (T_m), the transformation to a crystal does not occur, and the liquid is super-cooled. As the temperature is reduced below T_m, the resistance to deformation of the super-cooled liquid rapidly increases, and below the glass transition temperature (T_g), the material is a solid.

At temperatures below the glass transition temperature, amorphous materials, such as glass, are solids without long-range order, but with short-range order. In glass, there is short-range order because silicon atoms are surrounded by four oxygen atoms in the tetrahedral arrangement shown in Figure 2.28a. Whether a material is a solid or a liquid is not based upon the atomic arrangement; it is based upon the resistance to deformation by nonhydrostatic forces.

There is no abrupt change in the volume of an amorphous material at the glass transition temperature, because the structure of the liquid and amorphous solid is the same. This is a second-order transformation. The liquid to crystalline transformation is first-order, because there is a discontinuity in the volume-versus-temperature plot at T_m. In a second-order transformation, there is a discontinuity in the derivative (dV/dT) of the volume-versus-temperature plot. When the temperature of solid glass is increased from a low temperature, the glass starts to soften at the glass transition temperature (T_g); in contrast, a crystalline material remains solid up to the melting temperature (T_m).

Polymers also form amorphous solids when rapidly cooled from the liquid, as discussed in Section 2.8.2, because of the difficulty of aligning LCMs. Pure elemental metals cannot be cooled sufficiently fast from the liquid to form amorphous solids. However, metal alloys, such as $Nd_2Fe_{14}B$, have been produced that can be cooled sufficiently fast that they are amorphous.

2.11 INTERATOMIC POTENTIALS

2.11.1 Introduction

Interatomic potentials permit the calculation of physical properties of a crystal, such as the equilibrium lattice parameter, cohesive energy, stiffness, and thermal expansion coefficient. Also, interatomic potentials allow for simulations of dynamic processes in materials that cannot be observed experimentally because they happen too quickly, such as the atom displacements resulting from the impact of a high-energy atom onto a crystal. The **interatomic potential** is the potential energy due to the bonding between an atom at the origin ($r = 0$) and a second atom at a variable interatomic separation (r), as shown by the blue line in Figure 2.31. The plot of the interatomic potential as a function of interatomic separation also provides a visual representation of the bonding in addition to providing an analytical model. As shown in Figure 2.31, at an infinite separation there is no bonding, and the potential energy is 0. As the atoms come closer together, they bond and the potential energy decreases, because it would require an input of energy to separate the atoms. The potential energy reaches a minimum at the equilibrium interatomic separation (r_0). If the atoms are pushed together more than the equilibrium interatomic separation, the potential energy increases rapidly because the ion cores of the atoms overlap.

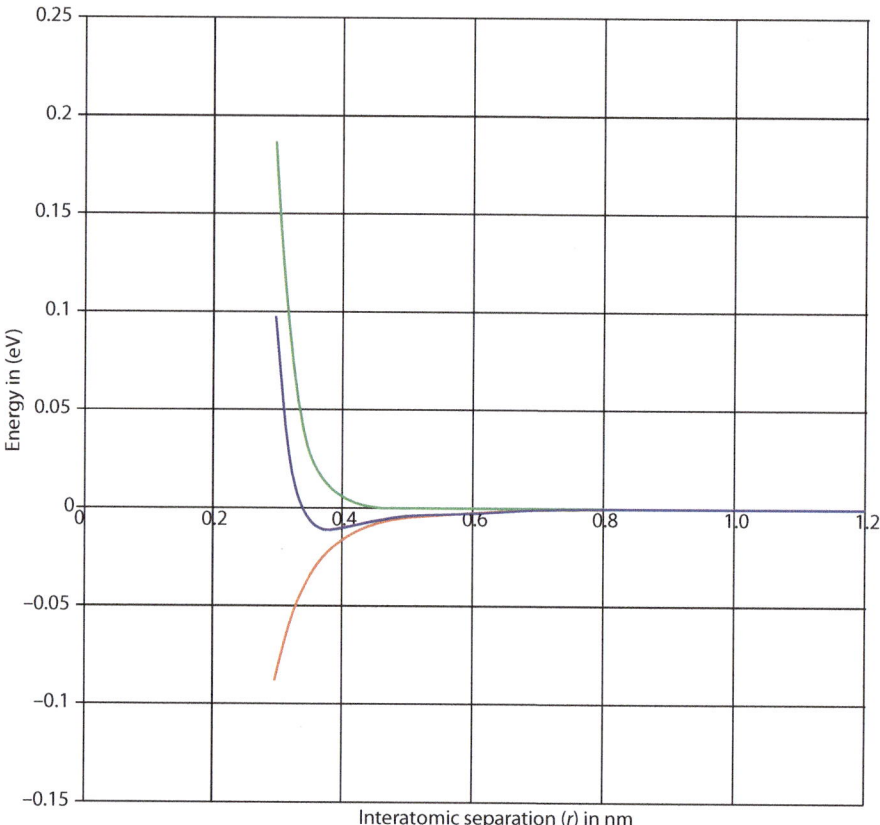

Figure 2.31   The interatomic potential energy in energy electron-volts (eV) between two argon atoms (blue), the attractive energy (red), and the repulsive energy (green) as a function of interatomic separation in nanometers (nm).

2.11.2   Inert-Gas Atom Potentials

If an inert gas is cooled to a very low temperature, it forms a liquid, and at an even lower temperature, its atoms will solidify into the FCC crystal structure. For example, argon solidifies at 83.8 K. To avoid the "solid gas" oxymoron, these elements are called Group 0 solids in this section of the book. The weak bonding that holds the Group 0 atoms together is the fluctuating-electric dipole bond that we discussed in Section 2.6. The attractive potential of the fluctuating-electric dipole bonds is modeled with an empirical potential $V_A(r)$ between pairs of atoms of the form given by Equation 2.7,

$$V_A(r) = \frac{-B}{r^n} \qquad\qquad \textbf{2.7}$$

where B and n are constants evaluated from physical properties or by theoretical analysis. The attractive potential is negative; a negative energy indicates bonding. For the inert-gas atoms, n is approximately equal to 6. The attractive potential is the negative red line in Figure 2.31.

   If the atoms are pushed together so that the atom cores start to overlap, there is a very strong repulsive interaction. This repulsion is much stronger than normal coulombic repulsion of like charges that have a

$1/r$ dependence. The repulsive energy of interaction has a dependence upon interatomic separation (r) of approximately $1/r^{12}$. This repulsion is due to the Pauli exclusion principle. If two argon atoms are brought together so that the $3p$ electrons on the different argon atoms occupy some of the same space, there is a repulsive energy (V_R), as shown by the positive green line in Figure 2.31. An empirical equation for the positive repulsive interaction of a pair of atoms (V_R) is given by Equation 2.8,

$$V_R(r) = \frac{A}{r^m}$$

2.8

where A and m are constants evaluated from theoretical analysis or from experimental data. Repulsive energies are positive. The value of m is typically from 8 to 12; thus V_R is positive and very large for a small interatomic separation, and V_R is very small for a large interatomic separation (r), as shown in Figure 2.31. The sum of $V_A(r)$ and $V_R(r)$ is $V_P(r)$, and $V_P(r)$ is the total potential energy for a pair of inert-gas atoms, as shown in Equation 2.9 and by the blue line in Figure 2.31.

$$V_P(r) = V_R(r) + V_A(r) = \frac{A}{r^m} - \frac{B}{r^n}$$

2.9

If $m = 12$ and $n = 6$ in Equation 2.9, this is called a Lennard-Jones pair potential, which is discussed in more detail in Appendix 2B.

An applied force [($F_P(r)$] is required to move an atom by an amount dr from the minimum energy interatomic separation at r_0 relative to an atom fixed at the position $r = 0$. The amount of energy $dV_P(r)$ required to change the interatomic separation by dr is then given by Equation 2.10.

$$dV_P(r) = F_P(r)\, dr$$

2.10

From Equation 2.10, if the potential is given by Equation 2.9, the applied force required to move the atom to the position r is given by Equation 2.11.

$$F_P(r) = \frac{dV_P(r)}{dr} = \frac{A(-m)}{r^{m+1}} - \frac{B(-n)}{r^{n+1}}$$

2.11

The applied force required to have two argon atoms at an interatomic separation of r is shown in Figure 2.32. A positive applied force means that the force applied to the atom at position (r) is in the positive direction, and the atoms are pulled apart. A negative applied force means that the force applied to the atom at position (r) is in the negative direction, and the atoms are pushed together.

Note in Figure 2.32 that the force is 0 at the interatomic separation corresponding to the minimum of the potential-energy plot. This corresponds to the equilibrium interatomic separation (r_0) of 0.381 nm for argon. The **equilibrium interatomic separation** is the separation that is present after a long time and that minimizes the interatomic potential. Put a small ball on a surface with the shape shown in Figure 2.31, and the ball will eventually come to rest at the minimum energy position at r_0. The equilibrium interatomic separation is determined from Equation 2.12.

$$F_P(r_0) = \frac{dV_P(r_0)}{dr} = 0$$

2.12

The general form of Equation 2.12 is a very important equation in materials science. The equilibrium value of a variable, such as interatomic separation, is the value that minimizes the energy.

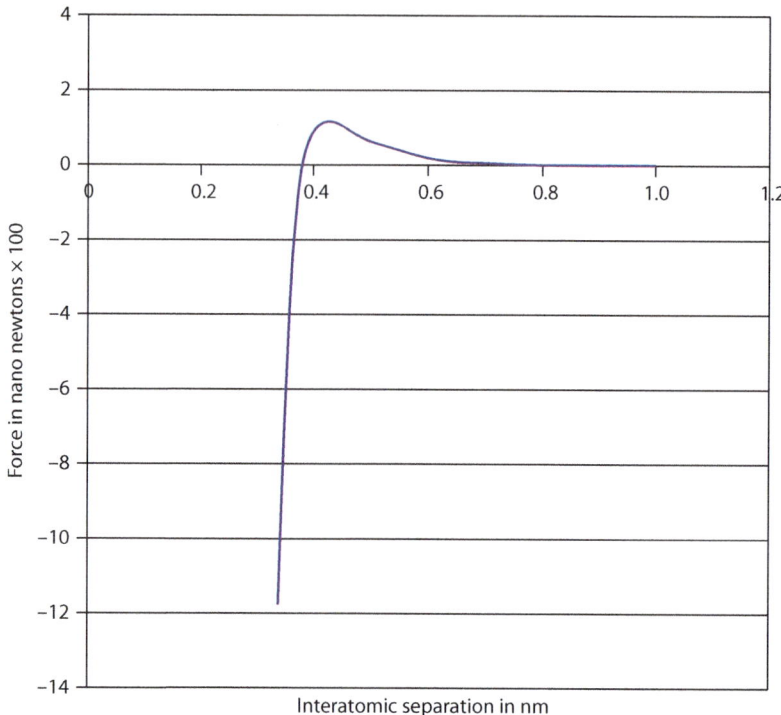

Figure 2.32   The applied force, in nano newtons times 100, required to move a pair of argon atoms to the indicated interatomic separation, in nanometers.

Several other observations are made from Figure 2.32. First a large negative applied force is required to push the second atom to a small interatomic separation. This large force is a result of the Pauli exclusion principle, because electrons with the same quantum numbers on different atoms are overlapping at small interatomic separations. Secondly, when the atom at the equilibrium interatomic separation is pulled away from the atom at the origin, the force increases, reaches a maximum, and then decreases. In a solid, the maximum applied force per unit of cross-sectional area of material required to pull the atoms apart is the theoretical strength of the solid. The applied force is equal and opposite to the interatomic force.

Pair potentials are often expressed in energy units of electron-volts and nanometers, and this is satisfactory for calculations of energy alone. However, to obtain a correct answer with Equations 2.10, 2.11, and 2.12, where a derivative is present, all terms must be in a unified system of units, such as the SI system.

The cohesive energy of a crystal is the bond energy per atom in a crystal. In the FCC crystal structure of the Group 0 atoms, there are 12 nearest neighbors to each atom, which is called the coordination number (C_N). The number of pair bonds per atom is equal to the coordination number divided by 2, which for the FCC crystal structure would equal six pair bonds per atom. It is necessary to divide the number of neighbors by 2 to get the number of pair bonds, because if there are N atoms, there are $N/2$ pairs of atoms. The pair potentials for the Group 0 atoms are short range, and the pair potential is primarily determined by the interaction with the nearest-neighbor atoms. This is observed in Figure 2.31, where the bond energy has decreased to a relatively small value by the next nearest interatomic separation of 0.54 nm (for argon). In many computations with the Group 0 atoms, only the nearest-neighbor interactions are determined. If only nearest-neighbor interactions are considered for a Group 0 crystal, the cohesive energy per atom $[V_{coh}(r)]$ as a function of interatomic separation (r) is then given by Equation 2.13.

$$V_{coh}(r)\frac{energy}{atom} = \frac{C_N \text{ pair bonds}}{2 \text{ atoms}} V_p(r)\frac{energy}{\text{pair bond}} \qquad \textbf{2.13}$$

Example Problem 2.11

An empirical interatomic pair potential for argon atoms, in units of eV and nanometers, is:

$$V_p(r) = \frac{1.0 \times 10^{-7}}{r^{12}} - \frac{0.64 \times 10^{-4}}{r^6}$$

The lattice parameter of FCC argon is equal to 0.539 nm.
(a) Determine the equilibrium pair bonding energy.
(b) Determine the cohesive energy at 0 kelvin using only the nearest-neighbor interactions, and compare this with the value in the periodic table.

Solution

a) From the lattice parameter (a) of FCC argon, the equilibrium interatomic separation (r) is equal to

$$r = \frac{a}{2^{1/2}} = \frac{0.539 \text{ nm}}{1.414} = 0.381 \text{ nm}$$

Substituting the interatomic separation of 0.381 nm into the interatomic potential for argon results in

$$V_p(r) = \frac{1.00 \times 10^{-7}}{(0.381)^{12}} - \frac{0.64 \times 10^{-4}}{(0.381)^6} = \frac{1.00 \times 10^{-7}}{9.4 \times 10^{-6}} - \frac{0.64 \times 10^{-4}}{3.1 \times 10^{-3}} = 0.0106 \text{ eV} - 0.0206 \text{ eV} = -0.01 \text{ eV}$$

Experimentally the pair bond energy has been determined to be -0.0104 eV, for excellent agreement.
b) Considering only nearest-neighbor interactions, the cohesive energy per atom is

$$V_{coh}(r) = \frac{C_N \text{ pair bonds}}{2 \text{ atoms}} V_P(r)\frac{energy}{\text{pair bond}} = \frac{12 \text{ pair bonds}}{2 \text{ atoms}}\left(-0.01 \frac{\text{eV}}{\text{pair bond}}\right) = -0.06 \frac{\text{eV}}{\text{atom}}$$

The experimental value in the periodic table is -0.08 eV. The difference is due to considering only nearest-neighbor interactions.

2.11.3   Interionic Potentials

For the computation of ionic-bond energy, a pair of ions, such as Na^+ and Cl^-, is initially assumed to exist with an infinite separation before being brought together. Starting with ions at infinite separation avoids the complicated problem of when and how the $3s$ electron on a sodium atom transfers to the chlorine atom to create the two ions. For ionic bonding, an interionic potential is utilized, and the energy to separate the ions from their equilibrium positions to infinity is the separation energy. The coulombic attraction $[V_c(r)]$ between a cation and anion as a function of interionic separation (r) is given by the standard equation for the electrostatic interaction of a positive and negative point charge given in Equation 2.14,

$$V_c(r) = -\frac{(Ze)^2}{4\pi\varepsilon_0 r} \qquad\qquad \textbf{2.14}$$

where Ze is the electronic charge on each ion and ε_0 is the permittivity of a vacuum (8.85×10^{-12} farads/m). Each ion in an ionic crystal has the electron configuration of an inert-gas atom; for example, Na^+ has the electron configuration of neon, and Cl^- has the electron configuration of argon. Thus the repulsive interaction can be represented by a power-law relationship similar to Equation 2.8 for the inert-gas atoms. A possible interionic potential $[V_{pion}(r)]$ is the attractive coulombic potential plus a repulsive power-law term, as presented in Equation 2.15.

$$V_{pion}(r) = -\frac{(Ze)^2}{4\pi\varepsilon_0 r} + \frac{B}{r^m} \qquad\qquad \textbf{2.15}$$

Since the attractive coulombic potential is a function of $1/r$, it extends over more ions in the crystal than the repulsive potential that is a function if $1/r^m$, where m is typically a number from 8 to 12. For this reason, to calculate the **crystal separation energy**, which is the energy needed to separate all of the crystal ions to infinity, it is necessary to sum the coulombic potential over all of the ions in the crystal. This calculation is discussed in Appendix 2C. To calculate the cohesive energy of an ionic crystal, the energy to create the positive and negative ions from the neutral atoms must be added to the crystal separation energy.

2.11.4 Interatomic Potentials and Physical Properties

Some physical properties of materials can be developed from interatomic potentials. The equilibrium interatomic separation (r_0) of a pair of atoms is found by using Equation 2.12. The cohesive energy of a crystal (V_{coh}) is the energy per atom required to increase the atom separations from the equilibrium value to infinity. The cohesive energy is found by substituting the equilibrium interatomic separation (r_0) into an equation for the crystal cohesive energy, such as Equation 2.13. Thermal expansion of a material is the increase in dimensions resulting from an increase in temperature. Thermal expansion is related to the asymmetry of the interatomic potential as a function of interatomic separation (r), as we will discuss in Chapter 4. The stiffness of a material is calculated from the second derivative of the interatomic potential with respect to interatomic separation, or the curvature of the potential energy, at the equilibrium interatomic separation. And finally the theoretical strength of the material is related to the maximum in the applied force as a function of interatomic separation, as shown in Figure 2.32.

2.11.5 Models for Metallic and Covalent Bonding

Van der Waals bonding in inert-gas atoms and ionic bonding can be modeled with pair potentials. However, pair potentials do not properly model the interactions between the positive-ion cores in metals and the free-electron gas. This issue arises because many electrons simultaneously interact with a single ion core, and a single electron interacts simultaneously with more than one ion core. It is not possible to properly model these many-bodied interactions with a pair potential. Computationally efficient techniques have been developed to calculate the local density of electrons in a metal. In the local density approximation, the local energy of the crystal is assumed to be a function of the local electron density. The local electron density approximation has permitted the calculation of the bond energy and properties of metals.

Pair potentials also cannot account for the observation that some materials, such as diamond, have an angle of 109.5 degrees among three carbon atoms. Pair potentials are a function of interatomic separation (r), but not of angle. Covalently bonded materials, such as diamond, have been modeled by including three body or angular terms into the interatomic potentials and by including terms that depend upon the number of nearest neighbors.

Summary

- A goal of materials science is to understand the origin of the physical properties of materials, such as the density, hardness, and the high melting temperature of diamond.
- The binding energy of electrons to the nucleus on atoms is described with quantum mechanical models. Each electron on an atom has a quantum state that is a set of quantum numbers that does not change as long as the atom is unperturbed. The magnitude of binding energy of an electron to the nucleus is inversely related to its principal quantum number.
- The Pauli exclusion principle states that no two electrons with the same set of quantum numbers can occupy the same space at the same time.
- The electron configuration of an atom is determined using the Aufbau principle that the electron orbitals with the lowest energy fill first, the Pauli exclusion principle, and the rules for the values of quantum numbers.
- The electronegativity of an atom is a relative measure of the atom's ability to acquire electrons.
- A material can be in a vapor, liquid, or solid phase. In a vapor, the atoms or molecules are not bonded to each other, and they move with kinetic energy relative to each other. In a liquid, the atoms are bonded to each other, but the atoms are in constant motion. When nonhydrostatic forces are applied to a liquid, the atoms easily move relative to each other, and the liquid easily changes shape permanently. When nonhydrostatic forces are applied to a solid, the atoms do not easily move relative to each other, and the solid does not easily change shape.
- All elements when cooled sufficiently will form solids, even the inert-gas atoms.
- Most atoms of elemental solids with a spherical outer electron configuration are in a close-packed arrangement like spherical billiard balls. The close-packed arrangement is the most densely packed arrangement of equal-size spheres or atoms possible. The HCP and FCC crystal structures result from the close-packing of atoms.
- Crystallography is a system for describing crystal structures, planes, and directions in crystals.
- The unit cell is a small group of atoms or molecules that contains all of the necessary information about the crystal, and the unit cell when repeated in space produces the entire crystal.
- The lattice points in unit cells are equivalent points.
- The dimensions of the unit cell are given by the lattice parameters a, b, and c. In unit cells without right angles at the corners, the angles are α, β, and γ.

- The positions of all of the atoms in a perfect crystal are specified by giving the unit cell or Bravais lattice type, the lattice parameters and lattice angles, and the atom positions around one lattice point, and by repeating the unit cell in space.

- A specific plane in a crystal is specified by the Miller indices (hkl); and $\{hkl\}$ refers to all of the planes that are equivalent to (hkl).

- The planar atom density (PAD) is the number of atoms whose centers lie in a plane of area A, divided by the area A. The linear atom density (LAD) is the number of atoms whose centers lie on a line with length L, divided by the length L.

- A direction in a crystal [uvw] is specified by the vector components of the direction in fractions of the lattice parameters a, b, and c, and <uvw> refers to all of the directions that are equivalent to [uvw].

- The valence electron shell is the incomplete shell of electrons on an atom.

- Chemical bonding determines how the atoms are arranged in crystals, and it influences the physical properties of materials. Physical properties that are related to chemical bonding include density, melting temperature, and hardness.

- Van der Waals bonds are weak secondary bonds between atoms or molecules that result from electric dipoles formed on the atoms or molecules. There are permanent-electric dipole bonds created when chemical bonds between atoms cause a permanent asymmetric distribution of electrons; and fluctuating-electric dipole bonds, created by the temporary asymmetric distribution of electrons around an individual atom.

- The inert-gas atoms only have fluctuating-electric dipole bonds. The inert-gas atoms have a spherical electron distribution when averaged over time, and in solids the inert-gas atoms pack like billiard balls into the FCC crystal structure. The weak secondary bonding results in very low melting temperatures and insignificant hardness.

- The inert-gas atoms are important for providing an inert atmosphere in many industrial processes, including the welding of reactive metals, sputter deposition of metals, and sintering of metals. Liquids of the inert-gas atoms are used for cooling materials.

- A covalent bond is the localized sharing of a pair of electrons between two atoms, to complete the valence electron shells of both atoms.

- For covalently bonded materials, the number of nearest neighbors is also equal to the number of electrons needed to complete the valence shell, assuming that each atom pair forms a single covalent pair bond.

- Covalent bonding is present in many important materials, such as diamond, silicon, silicon carbide, and ceramics. These materials have a high melting temperature and high hardness because of the strength of the covalent bond, and a relatively low density because of the low number of nearest neighbors.

- The high hardness of covalently bonded materials results in applications in cutting tools, grinding, and abrasives. The relatively low density and high melting temperature results in applications in nonrotating parts of gas turbines and rockets, and in vehicle turbochargers.

- The most prevalent ambient form of carbon is graphite, but carbon is also found as diamond, buckyballs, nanotubes, and graphene. Nanotubes and graphene have many potential nanotechnology engineering applications because of their high strength and electrical properties.

- Polymers are large molecules with many (poly) units (mers).

- LCMs in a liquid polymer have an amorphous arrangement.

- If an amorphous liquid polymer is cooled rapidly to room temperature to form a solid, the solid structure is still amorphous. If the polymer is a simple one, such as polyethylene, and it is cooled slowly, some crystalline areas form.

- Thermoplastic polymers have only van der Waals bonds between different molecules, and they soften when heated. Thermoplastic polymers melt at relatively low temperatures, and they have low hardness.

- Polymers with three-dimensional covalent bonding are thermosetting polymers; they do not soften significantly when heated, and they can harden. Thermoset polymers do not melt when heated; they deteriorate in the presence of oxygen.

- Plastics are materials that are made from polymers that may also contain fillers, plasticizers, coloring agents, particles, and short fibers.

- An important physical property of plastics is their low density. This has resulted in applications of plastics replacing steel in many products to reduce weight if the strength of steel is not required.

- Extremely strong polymer fibers are produced when the LCMs are oriented along the fiber axis.

- Metals are inorganic materials with metallic bonding. In the metallic bond, valence electrons are collectively shared by many positive-ion cores, and valence electrons are free to move throughout the metal. The free electrons in metals result in high electrical and thermal conductivity.

- Many metals form in the close-packed FCC or the HCP crystal structures. Transition-series metals that have incomplete ion cores often form in the BCC crystal structure.

- Metals have high densities relative to those of covalently bonded materials, when normalized for molar mass. Metals with only metallic bonding (such as Al, Cu, Ag, and Au) have a lower melting temperature and hardness than do covalently bonded materials, such as diamond, silicon carbide (SiC), and boron nitride (BN). Most metals have higher melting temperatures and hardness than polymers. Transition-series metals have some degree of covalent bonding between atoms, and the covalent bonds result in higher melting temperatures and hardness than for metals with only metallic bonding. The hardness of metals is also dependent upon the deformation and temperature history of the metal.

- An alloy is a metal element mixed with another element.

- An intermetallic compound is a compound between two metals, and an intermetallic compound is also a metal.

- Metal-alloy applications include the steel in an automobile chassis and in bridges, aluminum alloys used in aircraft fuselages and frames, nickel alloys used in high-temperature gas turbines, titanium alloys used in supersonic aircraft fuselage and gas-turbine compressors, and copper alloys used in heat exchangers.

- An ionic bond forms between positive and negative ions, which are attracted to each other because of their opposite charges.

- The packing in ionic materials depends upon the relative size of the ions. Each cation attracts as many anions around it as possible; however, the anions cannot overlap each other, and there cannot be extra space between the anions.

- Many ceramic materials have a mixture of ionic and covalent bonding. If a mix of bonds is present, some electrons are transferred from one atom to another (the ionic bonds), and some electrons are shared locally (the covalent bonds). Ceramics are normally oxide, carbide, or nitride compounds with a metal or with silicon.

- Ceramics have relatively high melting temperatures and hardness; however, they are brittle.
- Ceramics are utilized in high-temperature applications such as furnaces and rocket motors, for some nonrotating components of high-temperature gas turbines, and in vehicle turbochargers. It is also proposed to use ceramic materials, such as ZrO_2, in high-temperature gasoline and diesel engines for applications such as cylinder liners.
- Interatomic potentials permit the calculation of physical properties of a crystal, such as the crystal structure, lattice parameter, cohesive energy, stiffness, and thermal expansion coefficient. Also, they allow for simulations of dynamic processes in materials that cannot be observed experimentally because they happen too fast. The interatomic potential is the potential energy due to the bonding between an atom at the origin ($r = 0$) and a second atom with an interatomic separation (r).
- Interatomic force is the negative of the derivative of the interatomic potential with respect to interatomic separation.
- Interatomic pair potentials have been developed to model the fluctuating-electric dipole bond in inert-gas atoms, and interionic pair potentials have been developed to model ionic bonds. Modeling of covalent bonds requires three body potentials with angular energy dependence, and modeling of metals requires evaluating the interaction of positive-ion cores with many electrons.

Supplemental Reading Subjects and Authors

Full references are listed at the end of the book.

General:	Askeland, Fulay, and Wright
Structure of the atom:	Reese
Crystallography:	Cullity; Barrett and Massalski
Chemical bonding:	Harrison
Polymers:	Oswald and Menges; van Krevelen
Ceramics:	Chaing, Birnie, and Kingery
Metals:	Barrett and Massalski
Interatomic potentials:	Harrison

Homework

Concept Questions

1. The _____ _____ principle says that no two electrons that occupy the same space can have the same quantum numbers.

2. The quantum numbers of electrons on an atom include the principal quantum number, the angular-momentum quantum number, the magnetic quantum number, and the _____ quantum number of $\pm\frac{1}{2}$.

3. The _____ electrons of an atom are the electrons in an incomplete electron shell that are outside of a spherically symmetric closed electron shell corresponding to an inert-gas atom.

4. The _____ number in the periodic table is equal to an element's number of valence electrons.

5. The inert-gas atoms have the group number of _____.

6. Atoms chemically bond to achieve an electron configuration as similar as possible to that of the _____ _____ atoms.

7. If a ceramic has strong chemical bonding, the ceramic has a(n) _____ melting temperature.

8. Hardness is a measure of a material's resistance to mechanical _____.

9. If a metal has a low cohesive energy, the metal is expected to have a _____ hardness.

10. The bonding between different H_2O molecules in water is a _____ electric dipole bond.

11. The bond between different inert-gas atoms in liquids and solids is the _____ electric dipole bond.

12. An element has metallic bonding when the valence electron shell is _____ than half filled.

13. In metallic bonding, the valence electrons are _____ electrons.

14. Ideal metals form close-packed crystal structures because the positive _____ cores pack like spherical billiard balls.

15. The density of metals is _____ relative to their molar mass in comparison to covalently bonded materials with the same interatomic spacing.

16. The _____ cell is a small group of atoms that contains all of the necessary information about the crystal that when repeated in space produces the crystal.

17. Lattice points in a Bravais lattice are _____ points in space.

18. A primitive unit cell contains _____ lattice point(s).

19. The eight corners of a cubic lattice unit cell contribute _____ lattice point(s) to the unit cell.

20. The six face-centered lattice points of a FCC lattice unit cell contribute _____ lattice point(s) to a unit cell.

21. A BCC unit cell contains _____ lattice point(s).

22. The Miller indices of a plane are the _____ of the intercepts of the plane along the unit cell axes.

23. The {100} planes of a cubic unit cell form the _____ of the cube.

24. The close-packed planes in a FCC metal are the _____ family of planes.

25. The planar atom density of the (100) plane of a FCC metal is _____ atoms divided by the square of the lattice parameter.

26. In a BCC metal, the _____ family of directions is the most closely packed.

27. The linear atom density of the [110] direction for a FCC metal is _____ atoms divided by the length of the face diagonal.

28. If there is only one atom type and the valence shell is one-half filled or more, then the atoms share electrons in a _____ bond.

29. An atom in face centered cubic aluminum has _____ nearest neighbors.

30. A carbon nanotube is a single sheet of _____ rolled into a tube.

31. The mer of polyethylene is made from two atoms of _____ and two atoms of hydrogen.

32. In a polyethylene long-chain molecule, the carbon-carbon and hydrogen-carbon bonds are all _____ bonds.

33. If liquid PMMA is rapidly cooled to room temperature, the structure is _____.

34. In vulcanized rubber, sulfur atoms _____ the latex long-chain molecules.

35. Polyethylene is mechanically soft because different long-chain molecules in polyethylene are held together with weak _____ ____ _____ bonds.

36. OUHMWPE fibers are high in strength because the LCMs are oriented _____ to the fiber axis.

37. In ionic bonding, electrons are transferred from an atom with its valence shell less than half filled to an atom with the valence shell _____ than half filled.

38. The atom arrangement in liquid copper is _____, with no long-range order.

39. It is possible to cool liquid copper sufficiently fast to form an amorphous structure. True or false?

40. When polyethylene is melted, the only bonds that are broken are the weak ____ ____ _____ bonds between the long-chain molecules.

41. In silica glass (SiO_2), there is no long-range order, but there is _____-range order.

42. Glass at room temperature is not a liquid, it is a solid because it can resist a change in _____.

43. During a rapid cool, liquid SiO_2 solidifies into _____ because of its complex structure.

44. During heating, an amorphous material starts to soften at the _____ transition temperature.

45. The atom pair bond energy as a function of interatomic _____ is the interatomic pair potential.

46. The equilibrium interatomic separation between two atoms is determined by setting the derivative of the interatomic pair potential with respect to separation equal to _____.

47. In an ionic material, the _____ potential is the attractive energy between ions.

48. The bond energy of a pair of atoms is equal to the depth of the interatomic potential at the _____ interatomic separation.

49. Covalently bonded materials cannot be modeled with pair potentials, because the pair potential energy is only a function of interatomic _____.

50. The element rubidium (Rb, atomic number 37) is in Group IA of the periodic table; thus the chemical bonding should be _____.

51. The element rubidium (Rb, atomic number 37) is in Group IA of the periodic table, and the element chlorine (Cl, atomic number 17) is in Group VIIB of the periodic table. RbCl should have _____ bonding.

52. The molecular mass of a long-chain molecule is equal to the molecular mass of a _____ unit times the number of _____ units.

Engineer in Training–Style Questions

1. Which of the following types of chemical bond is not a primary bond type?
   - (a) Covalent
   - (b) Fluctuating-electric dipole
   - (c) Metallic
   - (d) Ionic

2. If the principal quantum number (n) is equal to 3, which of the following angular momentum quantum numbers (l) is not allowed?
   - (a) 0
   - (b) 1
   - (c) 2
   - (d) 3

3. The energy of an electron on an atom that is not in a magnetic field is not dependent upon which of the following?
   - (a) Charge on the nucleus
   - (b) Spin quantum number
   - (c) Angular momentum quantum number
   - (d) Principle quantum number

4. The radius of an atom is typically:
   - (a) 10^{-6} m
   - (b) 10^{-10} m
   - (c) 10^{-15} m
   - (d) 10^{-18} m

5. The radius of a nucleus is typically:
   - (a) 10^{-6} m
   - (b) 10^{-10} m
   - (c) 10^{-15} m
   - (d) 10^{-18} m

6. The electronegativity of the inert-gas atoms is equal to:
   - (a) undefined
   - (b) 0
   - (c) 1
   - (d) 2

7. In a tetragonal unit cell, $a = b$ but c is not equivalent. Which plane is therefore not equivalent?
   - (a) (100)
   - (b) (010)
   - (c) (0$\bar{1}$0)
   - (d) (001)

8. Which of the following is not associated with a solid that has covalent bonding?
   - (a) Localized valence electrons
   - (b) High cohesive energy
   - (c) Close-packing of atoms
   - (d) Low density relative to molar mass

9. Which of the following physical properties is not associated with a solid that has covalent bonding between all atoms?
    (a)   Low melting temperature
    (b)   High hardness
    (c)   Electrical insulator
    (d)   Brittle fracture

10. Which of the following is not associated with van der Waals bonds?
    (a)   Fluctuating-electric dipoles
    (b)   Electron transfer
    (c)   Permanent-electric dipoles
    (d)   Relatively low melting temperature

11. Which of the following is not an allotropic form of carbon?
    (a)   Ethylene
    (b)   Diamond
    (c)   Buckyball
    (d)   Graphene

12. Which of the following is not a thermoplastic polymer?
    (a)   Polyethylene
    (b)   Epoxy
    (c)   Polyvinylchloride
    (d)   Polypropylene

13. Which of the following does not increase the strength of a polymer?
    (a)   Orienting molecules
    (b)   High molecular mass
    (c)   Cross-links
    (d)   Plasticizers

Design-Related Questions

1. If you have to select a material and the primary requirement is a high melting temperature, what class of material would you investigate first for suitability?

2. If low density is the primary design requirement, what class of material discussed in this chapter would you first investigate for suitability?

3. You have to select a material as a coating on an aluminum part that improves the wear and abrasion resistance of the part. What class of material would you investigate first for suitability?

4. You are asked to select a material for a barge tow line that must be as strong as steel cable, but can float on water and is not corroded by salt water. What material discussed in this chapter might be suitable?

5. Aluminum (Group III) and silicon (Group IV) are adjacent to each other in the periodic table. Relative to aluminum, silicon is less dense, has a higher melting temperature, is harder, and is very prevalent in the sand and rocks of the earth's crust. And yet aluminum has many more mechanical applications, such as in the structure and skin of aircraft, the cylinder heads in automobile engines, in small boats, and in marine engines. We will cover this later in the book, but from what you know about metals, such as aluminum, what property results in the use of aluminum in these applications rather than silicon?

Problems

Problem 2.1: How many atoms are there in a FCC unit cell that has an atom at 0, 0, 0?

Problem 2.2: The compound Ni_3Al is used to strengthen nickel-based alloys used in high-temperature gas-turbine materials. The crystal structure of Ni_3Al is a cube with Al atoms at the eight cube corners and Ni at all of the cube face centers. (a) What is the Bravais lattice type for Ni_3Al, and (b) what are the atom positions?

Problem 2.3: Calculate the number of atoms per unit volume in FCC silver (Ag), assuming that the lattice parameter for silver is 0.407 nm.

Problem 2.4: Calculate the number of atoms per unit volume in BCC solid sodium (Na), assuming that the lattice parameter for sodium is 0.428 nm.

Problem 2.5: The density of silver at room temperature is 10.49 g/cm³. You need to know the density of solid silver just below the melting temperature. At 960°C, the lattice parameter was measured to be 0.4176 nm. Compare the theoretical density of silver at 960°C to that at room temperature.

Problem 2.6: The density of BCC iron at room temperature is listed as 7.87 g/cm³ in Appendix B. At temperatures above 912°C, iron has a FCC (γ) structure, but the density is not listed. (a) Calculate the density of FCC iron based upon the listed lattice parameter of 0.3589 nm. (b) Explain any difference observed in the density of the FCC phase relative to the BCC phase.

Problem 2.7: Nanoparticles are finding many applications, including medicine, magnetic permanent memory, and high-strength materials. Assume that a high-strength nickel alloy is to be made out of nanoparticles, and that the size of the nanoparticles is a cube 10 nm on each side. Calculate the number of atoms in these particles in two ways: (a) For FCC nickel, calculate the number of atoms using only the lattice parameter of 0.352 nm, and (b) using the density of nickel and the atomic mass from Appendix B.

Problem 2.8: Silicon is FCC with an atom at 0, 0, 0 and an atom at $\frac{1}{4}, \frac{1}{4}, \frac{1}{4}$ and a lattice parameter of 0.543 nm. (a) How many atoms are there in this FCC unit cell? (b) Calculate the number of atoms per unit volume based upon the unit cell. (c) Calculate the number of atoms per unit volume based upon the density of 2.33×10^3 kg/m³.

Problem 2.9: Diamond is FCC with a lattice parameter of 0.357 nm and atoms located at 0, 0, 0 and at $\frac{1}{4}, \frac{1}{4}, \frac{1}{4}$, and these two atoms are nearest neighbors. Calculate the atomic radius of a carbon atom in diamond, assuming that the radii touch between nearest neighbors.

Problem 2.10: In an orthorhombic unit cell with $a < b < c$, draw the following planes: (001), (101), $(1\bar{1}0)$, (132), $(11\bar{2})$, and $(1\bar{1}0)$. Label each plane, and show the intercepts of the plane with the x, y, and z axes in the unit cell.

Problem 2.11: In the BCC metal iron, the (110)-type planes are the most closely packed planes.

    (a) Calculate the interplanar spacing between the (110)-type planes.

    (b) Draw a BCC unit cell and show the (110)-type planes and a [110] direction that is perpendicular to the (110) planes.

    (c) Calculate the segment lengths where the (110) planes cut the [110] direction.

Problem 2.12: Compare the planar atom density of the {100}-type planes with the {111}-type planes in the FCC structure of copper that has a lattice parameter of 0.361 nm.

Problem 2.13: Draw the following directions in a cubic unit cell: [001], [00$\bar{1}$], [110], [111], [11$\bar{1}$], [112], and [123]. Label each direction, and show the coordinates of where each direction intersects the boundary of the unit cell.

Problem 2.14: (a) Compare the linear atom density of the [100] and [111] directions in the BCC metal iron, with a lattice parameter of 0.286 nm. (b) Which is the most closely packed direction in the BCC structure? (c) What is the radius of an iron atom if it is assumed that the atoms touch along the most closely packed direction?

Problem 2.15: Pure iron at room temperature has the BCC structure; however, iron can also be found in the FCC structure at higher temperatures. Predict the lattice parameter of FCC iron if it did form at room temperature, assuming that atoms touch only along the most closely packed directions in both the FCC and BCC structures. The lattice parameter of BCC iron at room temperature is 0.286 nm.

Problem 2.16: If sufficient force is applied to a crystal, it can be permanently deformed. Permanent deformation in metal crystals occurs due to atomic displacements on the planes that are most closely packed, and it happens in the most closely packed directions. Assume that a force is applied to a crystal of FCC copper that permanently stretches the crystal in the [001] direction. The (111) plane is one of the close-packed planes in FCC copper on which permanent deformation could occur.

    (a)   What is the angle between the [001] direction and the normal to the (111) plane?

    (b)   Atom displacements along which close-packed directions within the (111) plane could contribute to the permanent deformation of the crystal in the [001] direction?

Problem 2.17: Permanent deformation in BCC crystals occurs on the most closely packed planes, and it happens in the most closely packed directions that are {110} planes and <111> directions in the BCC crystal. Assume that a force is applied to a BCC crystal that permanently deforms it in the [001] direction. For permanent deformation that is on a (011) plane:

    (a)   What is the angle between the normal to the (011) plane and the [001] direction?

    (b)   Atomic displacements along what most closely packed directions could contribute to permanent deformation of the crystal in the [001] direction?

Problem 2.18: Briefly explain how a fiber made of oriented ultra-high-molecular-weight polyethylene can be much stronger than structural steel in tension, but perpendicular to the fiber axis, its hardness is much less than that of steel.

Problem 2.19: How many carbon and hydrogen atoms are there in the unit cell of polyethylene shown in Figure 2.21c, assuming that the density of crystalline polyethylene is 0.996 g/cm³?

Problem 2.20: MgO is a high-temperature ceramic material that has mixed ionic and covalent bonding. Should MgO have the NaCl structure or the CsCl structure based upon ionic radii?

Problem 2.21: Use a spreadsheet or write a short computer program to confirm the percent ionic and covalent character of the bonding in Table 2.3.

Problem 2.22: The lattice parameter of CsCl is 0.4123 nm. Calculate the density of CsCl.

Problem 2.23: You are trying to identify a mineral that has been brought to your laboratory. The density is measured to be 1.984×10^3 kg/m³. From the density you think it might be KCl. From data available, you know the KCl has the same crystal structure as NaCl, but you cannot find the lattice parameter of KCl. Determine the lattice parameter of KCl from the density that you can then confirm with X-ray diffraction.

Problem 2.24: An empirical interatomic pair potential for xenon atoms, in units of eV, and nm, has been determined to be

$$V_p(r) = \frac{12.6 \times 10^{-7}}{r^{12}} - \frac{31.8 \times 10^{-4}}{r^6}$$

The lattice parameter of FCC xenon is equal to 0.630 nm.

(a) Determine the equilibrium pair bonding energy.

(b) Determine the cohesive energy of a xenon crystal at 0 kelvin using only the nearest-neighbor interactions, and compare your result with the value in the periodic table. Comment on any observed difference.

Problem 2.25: Assume that an interionic pair potential between K^+ and Cl^- ions can be approximated by Equation 2.15. Assume that the repulsive ion core interactions can be modeled with a power of $m = 12$. Experimentally it has been determined that the energy to separate one pair of K^+ and Cl^- ions is 5.0 eV, and that the equilibrium interionic separation is equal to 0.266 nm. Use this experimental data to determine the value of B for an ion pair. Use the SI system of units for this problem.

Problem 2.26: Test the K^+ and a Cl^- interionic potential that you developed for Problem 2.25 by determining the equilibrium interionic separation and comparing this value to the experimental value of 0.266 nm.

A photograph of a naturally occurring ruby crystal. A ruby is an aluminum oxide (Al_2O_3) crystal with chromium impurities. Without the chromium impurities, the aluminum oxide crystal is a clear white sapphire that has much less value. Blue sapphires are aluminum oxide with impurity atoms of iron and titanium. In many materials the properties are due to impurities, defects, or modifications that are present in the material. When carbon atoms are added to iron, it increases the material's strength and results in steel. The strength of nickel-based high-temperature gas-turbine blades is increased by the addition of aluminum. The semiconductors in solid-state electronics operate because atoms such as arsenic and gallium are added to silicon. The high-temperature superconducting material $YBa_2Cu_3O_{7-\delta}$ depends upon the missing oxygen atoms (δ).

© Rhea Eason/Alamy

The goals of this chapter are to understand

- The difference in the structure of perfect materials and real materials
- Point defects in metals and ceramics
- Atomic and molecular modifications to the individual mers of polymers
- Edge and screw dislocations in metals and ceramics
- Copolymers and branching of polymer chains
- Surfaces, grains, grain boundaries, and phase boundaries
- Lamella and spherulites in polymers
- The structure of composite materials

Chapter 3

The Structure of Real Materials

3.1 INTRODUCTION

In Chapter 2, we assumed crystals to be perfect and long-chain molecules to be uniform and continuous. However, all real crystals have defects, and long-chain molecules can have various modifications. In many materials the deviations from perfection control their physical properties. For example, the steel for an automobile body is not pure iron. Impurity atoms of carbon, nitrogen, and manganese are added to iron that increase the material's strength and result in steel, and additions of nickel and chromium to iron provide corrosion resistance in stainless steel. To control a material's physical properties, engineers and scientists must know how alterations to perfect crystals and modifications to uniform LCMs affect the physical properties.

The modifications to perfection discussed above are atomic in size. For example, when a small amount of nickel and chromium is added to iron to produce stainless steel, the nickel and chromium atoms replace iron atoms in the iron crystals. This modification is of atomic dimension, and it is classified as zero-dimensional (0-D). In materials there are also one-dimensional (1-D), two-dimensional (2-D), and three-dimensional (3-D) modifications. In the literature for metals and ceramics, modifications to a perfect crystal are generally called defects, even if the modification is intentional and it produces an improved material. In polymer science, modifications to long-chain molecules are not called defects.

3.2 ZERO-DIMENSIONAL DEFECTS AND MODIFICATIONS TO MATERIALS

3.2.1 Point Defects in Crystals

Substitutional Atoms When a nickel atom replaces an iron atom in an iron crystal, it is called a **substitutional** atom, as schematically shown in Figures 3.1c and d, and the resulting mixture of atoms is called a substitutional **solid solution**. Atoms in a solid can mix together to form a solid solution, just as different liquids such as alcohol and water mix together to form a liquid solution. If the foreign atom is similar in size and electronegativity to the atoms in the host crystal, the foreign atom substitutes into the host crystal. If the foreign atom is more than 15% different in radius than the host atoms, it normally does not substitute into the host crystal in large concentrations, because of the lattice deformation that is indicated in Figures 3.1c and d.

Vacancies A **vacancy**, shown in Figure 3.1a, is an atom missing from the crystal. All crystals at temperatures above absolute zero kelvin contain vacancies, and in Chapter 4 we will calculate the equilibrium number of vacancies. The presence of vacancies in crystals has several important consequences. They allow for the movement and mixing of atoms in crystalline solids, and vacancies also contribute to high-temperature permanent deformation of crystals. Vacancies in photonic crystals are used to guide light through the crystal; vacancies can be used to guide light through a

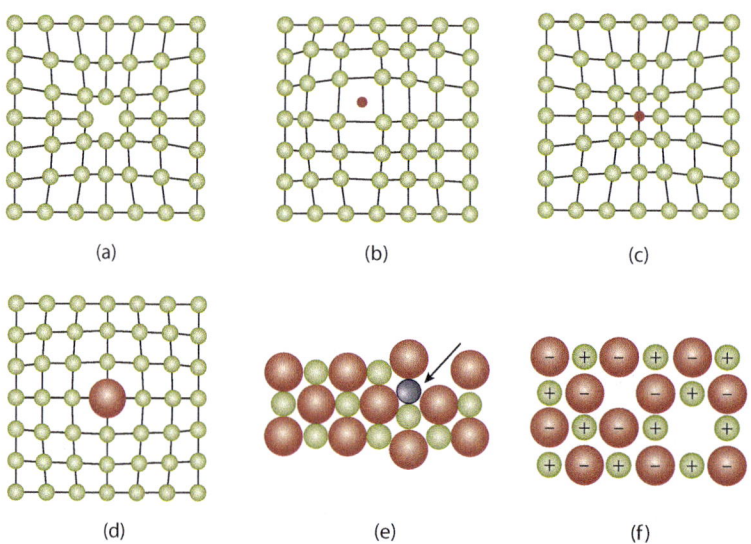

(a)   (b)   (c)

(d)   (e)   (f)

Figure 3.1 A schematic of point defects in crystals: (a) a vacancy, (b) an interstitial atom, (c) a small substitutional atom, (d) a large substitutional atom, (e) a Frenkel defect, and (f) a Schottky defect. (*Based on Askeland, D.R., Fulay, P.P., and Wright, W.J., The Science and Engineering of Materials, 6th ed. Cengage Learning, Stamford, CT. (2011), p. 114.*)

Substitutional Snoopy

90-degree turn in a photonic crystal. Photonic crystals are used in electro-optical systems to control the path of light.

Interstitial Atoms

If a small foreign atom is mixed with a much larger host atom, the smaller atom becomes a foreign **interstitial**, as shown in Figure 3.1b. Interstitial atoms go into the spaces between the host crystal atom positions. The easiest interstitial site to visualize is for the simple-cubic lattice, as shown in Figure 3.2, where the interstitial site is at the cube center in the position $\frac{1}{2}, \frac{1}{2}, \frac{1}{2}$. The metal polonium is the only element to form in the simple-cubic structure. In the BCC metal structures there are two types of interstitial sites possible, as shown in Figure 3.2. The interstitial sites at positions such as $1, \frac{1}{2}, \frac{1}{4}$ have four nearest neighbors and are tetrahedral sites, because they are at the center of a tetrahedron. Also in the BCC structure there are interstitial sites at positions such as $\frac{1}{2}, 1, \frac{1}{2}$. These are octahedral sites, because the interstitial atom is at the center of an eight-sided octahedron that is created by the six surrounding atoms. Interstitial atoms preferentially fill the interstitial sites that best satisfy the atom-packing criteria developed for ionic materials, shown in Figure 2.27, with the radius ratio of the interstitial to the host (r_i/r_h) replacing the radius of the cation to the anion. There are octahedral interstitial sites in an FCC metal at the midpoint along each side of the unit cell; in positions of the type $\frac{1}{2}, 0, 0$; and in the position $\frac{1}{2}, \frac{1}{2}, \frac{1}{2}$ at the center of the unit cell. In the FCC lattice there are tetrahedral interstitial sites at positions such as $\frac{1}{4}, \frac{3}{4}, \frac{1}{4}$.

Foreign interstitial atoms mixed into a host crystal create an **interstitial solid solution**. For example, small amounts of interstitial carbon are present in iron to produce low-carbon steel (Fe + C). Interstitials are more easily incorporated into BCC transition metals, such as Fe, Cr, V, Zr, Mo, and W, than in close-packed noble FCC metals, such as Cu, Ag, and Au. One proposal for

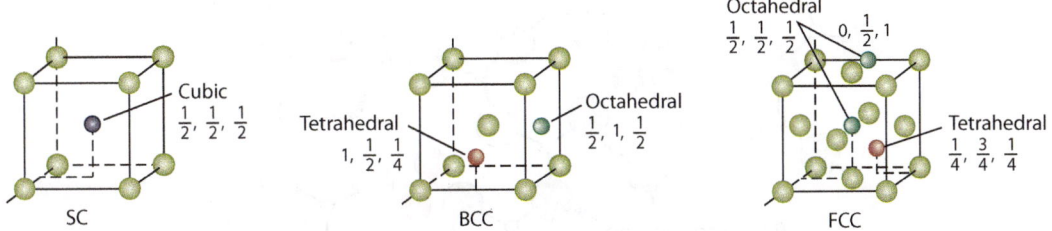

Figure 3.2 Interstitial sites in simple-cubic (SC), body-centered cubic (BCC), and face-centered cubic (FCC) metals. *(Based on Askeland, D.R., Fulay, P.P., and Wright, W.J., The Science and Engineering of Materials, 6th ed. Cengage Learning, Stamford, CT. (2011), p. 84.)*

safely storing hydrogen for use in fuel cells for automobiles is as interstitial ions in metals such as palladium.

It is also possible to have interstitials of the crystal material (for example, interstitials of iron in an iron crystal). Self-interstitials, also called interstitialcies, are found in materials that have been bombarded with high-energy particles, such as electrons, neutrons, or atoms; and in metals that have been extensively deformed permanently, such as by rolling or extrusion. Self-interstitials affect the physical properties of materials. For example, high-energy particles bombard nuclear reactor materials during reactor operation, producing self-interstitials and making the reactor metals more brittle. And during the rolling of a metal, self-interstitials contribute to the hardening of the rolled metal.

Materials with ionic bonding have all of the point defects discussed above. For example, chromium atoms substitute for aluminum atoms in the Al_2O_3 lattice of a ruby, which is an example of a cation foreign substitutional atom. There can be either cation or anion vacancies. Ionic materials have additional defects that preserve electrical neutrality. Figure 3.1e is a Frenkel defect (cation vacancy and interstitial). The dark atom in Figure 3.1e is the cation interstitial, and it is the same atom type as the other cations. Figure 3.1f shows a Schottky defect (cation and anion vacancy).

3.2.2   Point Defects in Glass

The atom arrangement in pure amorphous silica (SiO_2) is shown in Figure 2.29. When Na_2O (soda) is added to silica glass, the Na^{+1} ions occupy interstitial spaces in the glass structure close to the oxygen ion, to provide electrical neutrality, as shown in Figure 3.3. In pure silica each oxygen bonds with two silicon atoms, and the connecting oxygen is called a bridging oxygen. The oxygen ion associated with the Na_2O forms a bond with a single silicon atom, and the two sodium ions are not bonded to the silica structure, as shown in the box in Figure 3.3. This terminates the SiO_2 structure at the nonbridging oxygen atom,

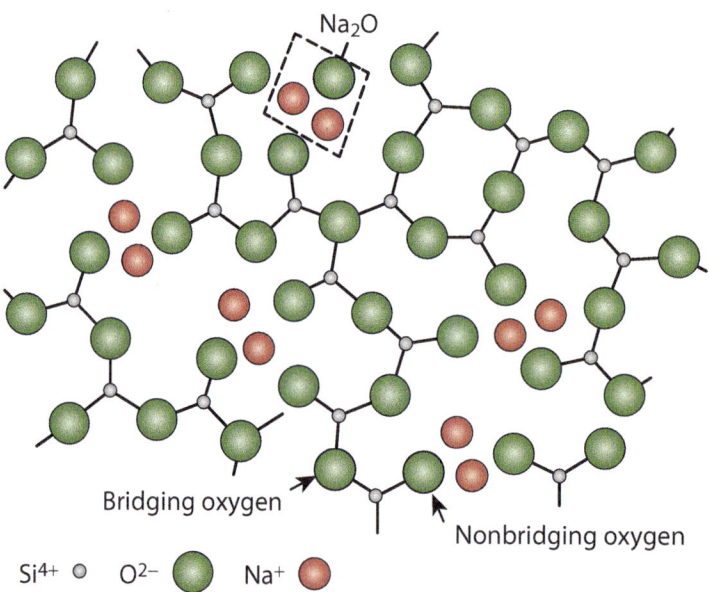

Figure 3.3   A schematic of the effect of adding Na_2O (soda) to silica (SiO_2) glass.

which weakens the SiO_2 structure. The addition of Na_2O to SiO_2 can reduce the viscosity of SiO_2 by as many as ten orders of magnitude.

3.2.3 Zero-Dimensional Modifications to Polymer Chains

The mer structure of polyethylene is shown in Figure 2.17b. When a chlorine atom regularly substitutes for a hydrogen atom in the mer structure, polyvinyl chloride (PVC) is produced, as shown in Figure 2.19. Other atoms or groups of atoms substituting for the hydrogen in the mers, shown in Figure 2.19 and 2.20, result in other polymers. It is possible to have several possible arrangements of mers to produce the polymer, as shown in Figure 3.4. Using Cl as an example of an atom or a molecule that replaces a hydrogen atom on the mer, if the Cl always repeats in the same position in the mer, this is an **isotactic** arrangement. If the Cl alternates from one side of the chain to the other, this is a **syndiotactic** arrangement, and if the Cl appears at random locations, this is an **atactic** arrangement. The arrangement of the atoms on the polymer chains can be affected by growing the long-chain molecules (LCMs) on the surface of a metal, such as titanium or platinum. The metal surface preferentially attracts one side of the mers that are attaching to the growing polymer chain. In this way the LCM has mers with an atom, such as chlorine, that prefer one side of the polymer chain. A polymer with an atactic arrangement is more difficult to crystallize than polymer chains with regular atom arrangements, because the chains are difficult to align when the substituting atom or group of atoms is in irregular positions along the chain. A more crystalline polymer has a higher strength than does the same polymer that is less crystalline.

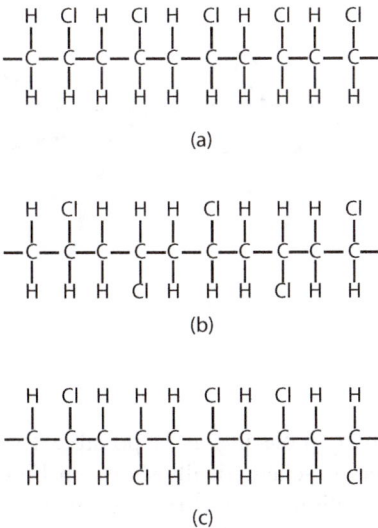

Figure 3.4 (a) An isotactic arrangement of the chlorine atoms on a PVC long-chain molecule. (b) A syndiotactic arrangement. (c) An atactic arrangement. (*Based on Askeland, D.R., Fulay, P.P., and Wright, W.J., The Science and Engineering of Materials, 6th ed. Cengage Learning, Stamford, CT. (2011), p. 617.*)

3.3   ONE-DIMENSIONAL OR LINEAR DEFECTS AND MODIFICATIONS TO MATERIALS

3.3.1   Linear Modifications to Polymers

Side Branches   During the polymerization process, discussed in Section 2.8.1, side branches can grow on the long-chain molecule, as shown in Figure 3.5. This happens when a carbon-hydrogen bond along a chain is broken, and a mer attaches where there had previously been a hydrogen atom. Other mers then attach to this one, and a side branch grows as did the original LCM. This happens particularly at higher polymerization temperatures. Polyethylene has a side branch attached to every 100th to 200th carbon atom along the LCM. Low-density PE (LDPE) has a lower average molecular mass and longer side branches than does high density PE (HDPE), as shown in Figure 3.5. Long side branches result in polymers that are more difficult to crystallize, because the side branches make it difficult for atoms on different chains to line up with each other. LDPE is typically 40 to 50% crystalline, and HDPE can be up to 80% crystalline. Side branches on adjacent long-chain molecules keep different long-chain molecules from coming close to each other. This weakens the van der Waals bonding between the different long-chain molecules, and it lowers the melting temperature, density, and hardness (as shown in Table 3.1). In Chapter 7, we will learn more about how the presence of side chains on LDPE make it less strong than HDPE.

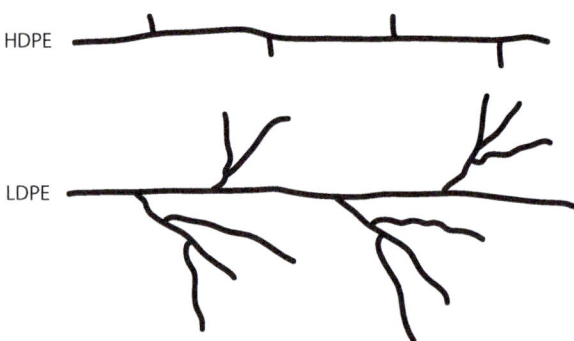

Figure 3.5   A schematic of the side branches on high-density polyethylene (HDPE) and low-density polyethylene (LDPE).

Copolymers   In a long-chain molecule, it is possible to have different polymer structures along the chain to form a **copolymer**, as schematically shown in Figure 3.6. Copolymers are formed by introducing different mers into the reaction vessel. If there are two mers, the molecule is a bipolymer, and if there are three mers, it is a terpolymer. Depending upon how the different mers react with each other, as well as the temperature, pressure, presence of catalysts, and so on, different structures are possible. The mers can regularly alternate as individual or multiple mers, as in the regular copolymer in Figure 3.6a; they can be randomly distributed along the LCM as in 3.6b, in longer blocks as in Figure 3.6c, or as

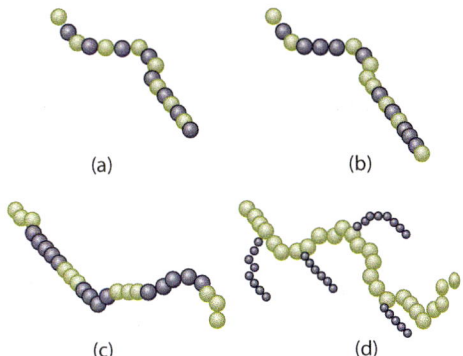

Figure 3.6 A schematic of some possible polymer solutions: (a) A regular copolymer, (b) random copolymer, (c) block copolymer, and (d) graft copolymer. The spheres represent mers, and different colors represent different mers. (*Based on Askeland, D.R., Fulay, P.P., and Wright, W.J., The Science and Engineering of Materials, 6th ed. Cengage Learning, Stamford, CT. (2011), p. 618.*)

grafts as in Figure 3.6d. In general, the properties of regular and random copolymers are weighted averages of the properties of the individual polymers. Block and graft copolymers tend to have properties of the primary polymer; however, certain properties are affected by the addition of blocks or grafts of another polymer.

Heat shrink wrap is a bipolymer of ethylene and approximately 10 weight percent acetate. ABS plastic is a terpolymer of polyacrylonitrile, polybutadiene, and polystyrene that is used in many applications such as helmets, telephone housings, luggage, small appliance housings, and so on. The properties of ABS can be varied depending upon the amounts of the individual polymers. Some of the physical properties of ABS are presented in Table 3.1. In comparison to PE, ABS can be up to twice as hard as HDPE; otherwise the properties are similar.

Cross-Linking of Polymers
Cross-links are atoms or molecules that form covalent bonds between long-chain molecules, as discussed in Section 2.8.4. The cross-link chains, shown in Figure 2.23, significantly strengthen a polymer by reducing the sliding of the long-chain molecules. In rubber the latex (*cis* isoprene) is cross-linked with sulfur, as discussed in Section 2.8.4.

Table 3.1   Density, Melting Temperature, and Brinell Hardness of Some Polymers

Polymer Name	Density[1] $10^3 \, kg/m^3$	Melting Temp.[2] °C	Brinell Hardness[1] $kg_f/(mm)^2$
LDPE	0.914–0.928	98–115	1.3–2.0
HDPE	0.94–0.96	130–137	4.1–6.6
ABS	1.04–1.06	110–125	8.0–12.0

1   Osswald, T.A., and Menges G., Materials Science of Polymers for Engineers, Hanser Publishers, Munich (2003), Table I.

2   Askeland, D.R., Fulay, P.P., and Wright, W.J., The Science and Engineering of Materials, 6th ed. Cengage Learning, Stamford, CT. (2011), p. 621.

3.3.2   Line Defects in Crystalline Metals

The line defect shown in Figure 3.7 is an **edge dislocation**, and the line defect shown in Figure 3.8 is a **screw dislocation**. Dislocations form in crystals during the solidification process as a result of temperature and interatomic distance differences between the liquid and solid. Dislocations always increase the crystal energy relative to that of a perfect crystal, and therefore theoretically they can be eliminated from a crystal. However, metals that are dislocation free have not yet been produced, because dislocations form and move easily in a metal at its melting temperature, which is when crystal growth occurs. Dislocation-free silicon has been produced because dislocations in silicon have a high energy.

In Chapters 6 and 7 we will find that dislocations are responsible for many mechanical properties of metals. The motion of dislocations results in room-temperature permanent (plastic) deformation of metals. Also, dislocations can strengthen a metal, which may sound counterintuitive; we will learn more about this in Chapter 7. For steel a high dislocation density is often desirable, because it results in a stronger steel. Conventional steel has a much higher dislocation density than a silicon crystal does, because the solidification process in steel production is not as well controlled as in semiconductor production, and it is easier for dislocations to form in a metal like steel than in a covalently bonded material like silicon. If steel is permanently deformed by processes such as cold rolling, extruding, drawing, forging, or swaging, this increases the dislocation density, thereby increasing the strength of the steel by up to a factor of 10. These processes for metals are discussed in Chapter 13.

However, the strongest metal would result if there were no dislocations, which at present it is not possible. It is possible to produce **whiskers** of metals and ceramics with very small diameter (1 μm) that are dislocation free except for one screw dislocation up the center. Whiskers of iron have strengths 500 times that of a typical steel. The whisker with its single screw dislocation looks like the material in Figure 3.8.

The edge dislocation in Figure 3.7 looks like it has an extra plane of atoms in the top half of the crystal; so why is the edge dislocation considered to be a line defect? In the edge dislocation the atoms are not properly bonded along a line that is perpendicular to the page at the symbol $\perp$. All of the atoms in the remainder of the simple-cubic crystal, shown in Figure 3.7, form bonds with six nearest-neighbor atoms: four in the plane of the page, one in front, and one behind the plane of the page. However,

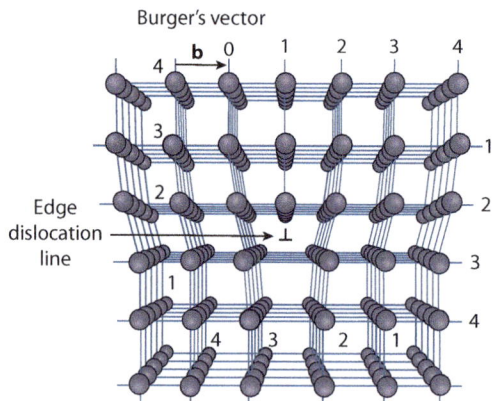

Figure 3.7   An edge dislocation with a Burger's vector **b** in a simple-cubic material, with a Burger's circuit indicated by the numbers. (*Adapted from Callister, W. D., Materials Science and Engineering: An Introduction, 6th ed., John Wiley & Sons, New York (2003), p. 74.*)

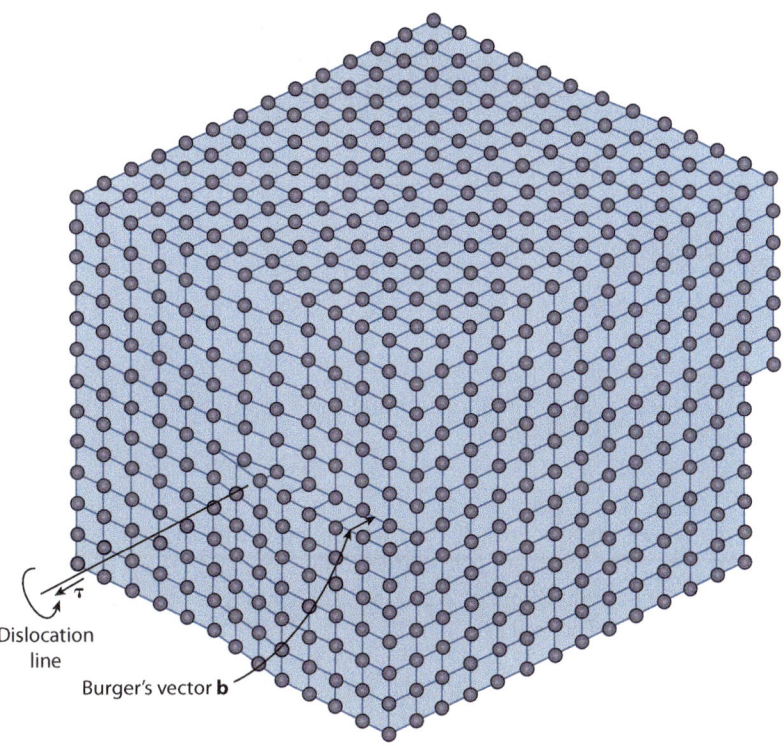

Dislocation
line

Burger's vector **b**

Figure 3.8   A screw dislocation with Burger's vector **b** and tangent vector **τ**. (*Adapted from Callister, W. D., Materials Science and Engineering: An Introduction, 6th ed., John Wiley & Sons, New York (2003) p. 75.*)

the atoms along the line that is perpendicular to the page at the symbol ⊥ have only five nearest-neighbor atoms: three in the plane of the page, one in front of the page, and one behind the plane of the page. It is only along the line marked by ⊥ where the bonding is incorrect. That is why the dislocation is a line or 1-D defect. The measure of the dislocations in a crystal is the dislocation density or the dislocation line length per unit volume (m/m^3).

A dislocation is characterized by two vectors: a unit vector (**τ**) tangent to the dislocation line, and the **Burger's vector** (**b**) that gives the magnitude and direction of the displacement in the crystal caused by the dislocation. The tangent vector (**τ**) for the edge dislocation in Figure 3.7 is normal to the page, pointing into the page and located at position ⊥. The direction of the tangent vector of the dislocation line is arbitrary, but the tangent vector cannot reverse along the dislocation line. The Burger's vector (**b**) is determined by performing a Burger's circuit in the crystal. A Burger's circuit consists of measuring equal numbers of lattice spacings in each direction around a lattice point in a crystal. The Burger's vector closes the Burger's circuit. A Burger's circuit performed in a dislocation-free crystal begins and ends at the same location; the net result of a Burger's circuit would be that **b** = 0. If the Burger's circuit surrounds a dislocation, it does not begin and end at the same lattice point, and the vector that connects the ending lattice point to the beginning lattice point of the circuit is the Burger's vector (**b**). To determine the direction of rotation for the Burger's circuit, use the right-hand rule: Point the thumb of your right hand in the direction of the tangent vector of the dislocation, and your other fingers will point in the correct rotational direction for the Burger's circuit. For example, in Figure 3.7, start at the lattice point at the pointed end of the Burger's vector and go to the right (clockwise) four lattice spaces, then go down four lattice spaces, then go to the left four lattice spaces, and finally go up four lattice spaces. The result is that the end of the Burger's circuit is at the beginning of the vector shown in Figure 3.7. The Burger's vector (**b**) is the magnitude and direction of the vector needed to close the circuit. The Burger's vector connects two lattice points in the crystal. For an edge dislocation the tangent vector (**τ**) and the Burger's vector (**b**) are perpendicular to each other.

Example Problem 3.1

Assume that there is a Burger's vector for a dislocation in a BCC iron crystal that goes from the origin at $0, 0, 0$ to the nearest lattice point in the unit cell, with all positive coordinates. For BCC iron, the lattice parameter is 0.286 nm.
(a) What is the Burger's vector of the dislocation, in vector notation?
(b) What is the magnitude of this Burger's vector?

Solution

a) In a BCC crystal, the nearest lattice point to the $0, 0, 0$ lattice point with all positive coordinates is the body-centered position at $\frac{1}{2}, \frac{1}{2}, \frac{1}{2}$. The position $\frac{1}{2}, \frac{1}{2}, \frac{1}{2}$ relative to the origin is in the [111] direction. The Burger's vector is the sum of $a/2$ in the x direction, of $a/2$ in the y direction, and of $a/2$ in the z direction. In vector notation the Burger's vector is expressed as

$$\mathbf{b} = \frac{a}{2}[111]$$

b) The magnitude of this vector is

$$b = \frac{0.286 \text{ nm}}{2}(1^2 + 1^2 + 1^2)^{1/2} = 0.143 \text{ nm } (1.732) = 0.248 \text{ nm}$$

In the screw dislocation shown in Figure 3.8, the tangent vector (**τ**) is shown coming out the crystal. The tangent vector could have also been chosen to point into the crystal. The reason for this selection of the direction of the tangent vector is to be consistent with Figure 3.9. Using the right-hand rule and taking a Burger's circuit results in Burger's vector (**b**), which is parallel to the dislocation line. The screw-dislocation name comes from the similarity to a screw thread on a bolt. When a Burger's circuit is taken around the screw-dislocation line, the resulting displacement is an increment along the dislocation line by an amount **b** parallel to the dislocation line. This is the same result as when a nut on a bolt is rotated 360°; it advances along the bolt by the spacing between the screw threads along the bolt axis.

The edge and the screw dislocations are two extremes. Most dislocations in a crystal are between an edge and screw orientation. Figure 3.9 shows how a dislocation line with Burger's vector (**b**) changes from an edge dislocation to a mixed dislocation to a screw dislocation, depending upon the orientation of the Burger's vector relative to the dislocation line. Note that the Burger's vector of the edge and screw components is the same. The Burger's vector of a dislocation line is constant irrespective of the orientation of the dislocation line. The orientation of the dislocation line can change relative to the Burger's vector. If the tangent vector is pointing in to the page for the edge part of the dislocation line, then to be consistent along the dislocation line, the tangent vector must point out of the crystal where the dislocation line emerges as a screw dislocation.

The dislocations shown in Figures 3.8 and 3.9 are in simple-cubic crystals; however, most cubic metals are either FCC or BCC. To see an edge dislocation in a FCC crystal, see Appendix 3A.

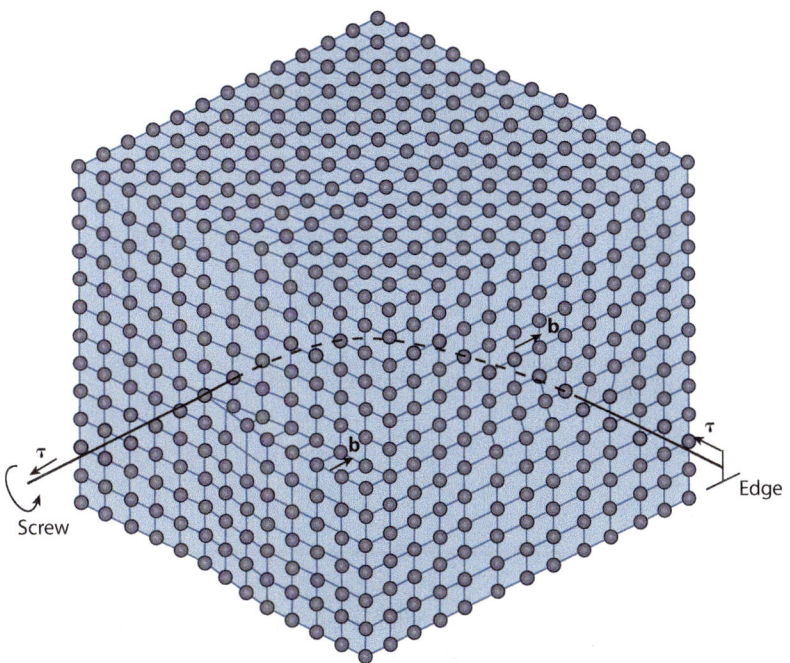

Figure 3.9 A mixed dislocation showing the change from a screw dislocation, where the dislocation line is parallel to the Burger's vector, to an edge dislocation, where the Burger's vector is perpendicular to the dislocation line. (*Based on W. T. Read, Dislocations in Crystals. McGraw-Hill, 1953.*)

3.3.3 Line Defects in Materials with Ionic and Covalent Bonding

Dislocations are also present in most materials with covalent and ionic bonding. Silicon crystals have been produced that are free of dislocations; however, dislocation-free silicon crystals are difficult to produce, and most silicon single crystals have a few dislocations. It requires a high energy to form dislocations in materials with covalent bonding because the covalent bond is directional and resists bending. In the vicinity of a dislocation the bonds must be bent. In semiconductors it is desired to have the minimum dislocation density, because dislocations trap electrons, and this decreases the conductivity of the semiconductor.

One important characteristic of dislocations in materials with ionic bonding is that they have relatively large Burger's vectors. This is because a Burger's vector connects lattice sites and not atoms. MgO has the same FCC crystal structure as NaCl does, shown in Figure 2.26d. If the O^{2-} ions are placed at the lattice sites of the FCC structure, the Mg^{2+} ions are not at lattice sites. The Burger's vector (**b**) must go from an O^{2-} ion at a lattice site to another O^{2-} ion also at a lattice site, and not from the O^{2-} ion at a lattice site to the closer Mg^{2+} ion that is not at a lattice site. In Figure 3.10 the Burger's circuit starts at the lattice point x and finishes at the lattice point y.

In Figure 3.10 the front face shown is the $(1\bar{1}0)$ plane of MgO. Compare Figure 3.10 with Figure 2.26d, and observe that the [110] direction lies in the $(1\bar{1}0)$ plane. Also recall that in the FCC structure the closest lattice point to 0, 0, 0 is at a position such as $\frac{1}{2}, \frac{1}{2}, 0$ that is in the [110] direction.

The magnitude of the Burger's vector is important because the force necessary to move a unit length of dislocation is proportional to the magnitude of the Burger's vector. In Chapter 6, we will show that

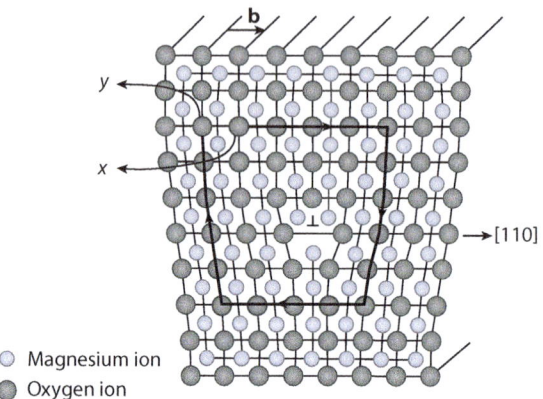

Figure 3.10   An edge dislocation in ionic MgO showing a Burger's circuit starting at *x* and ending at *y,* with the resulting Burger's vector (**b**). (*Based on W. D. Kingery, H. K. Bowen, and D. R. Uhlmann, Introduction to Ceramics, John Wiley, 1976.*)

the movement of dislocations produces the permanent deformation of crystals. The large Burger's vector in materials with ionic bonding makes it difficult to move dislocations, and this contributes to their brittleness.

3.4   TWO-DIMENSIONAL DEFECTS IN MATERIALS

3.4.1   Surfaces and Phase Boundaries

The crystal surface shown in Figure 3.11 is a planar defect, because atoms at the surface are not bonded in the same way as the atoms in the interior of the crystal. In the section of a close-packed plane in Figure 3.11, each atom in the interior has six nearest neighbors in the plane; however, an atom at the surface has only four nearest neighbors in the plane. The energy of any atom at the surface is increased relative to that in the bulk, because surface atoms do not have the proper number of neighbors sharing bonding electrons. Proper chemical bonding in the bulk results in the lowest energy. The increase in energy of the atoms at the surface is expressed as an energy per unit area or surface energy (γ) in joules/m^2. Table 3.2 presents the surface energy of some materials, the temperature of the surface energy measurement when available, and surface crystal planes if appropriate. When no crystal plane is given, assume that the specimen is a polycrystal. Polycrystals are explained in Section 3.4.2. When a polycrystal grows, the surfaces with low energies grow in preference to surfaces with high energies. In general, surface energy is proportional to bond energy, because the surface energy results from bonds that are broken at the surface. Materials that have a high melting temperature, such as tungsten and MgO, have a high surface energy, and materials with weak bonding, such as polyethylene, have a low surface energy. For polymers, such as polyethylene, it is primarily the weak van der Waals bonds between LCMs that are broken at the surface; thus polymers have a low surface energy.

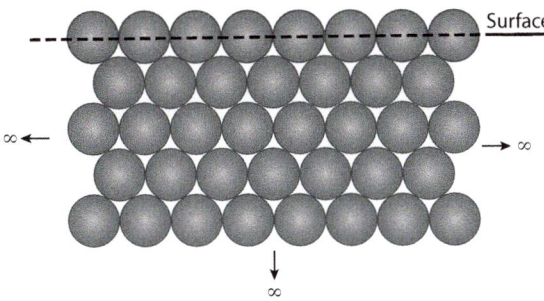

Figure 3.11 A schematic of a section through a crystal, showing a surface. The crystal is assumed to extend a large distance in all directions except above the surface. The atoms at the surface have no atoms to bond to above the surface.

Table 3.2 The Surface Energy of Some Selected Materials at the Indicated Temperatures

Material	Surface Energy (J/m²)
Metals	
Sn	0.680[1]
Al	1.08[1]
Ag	1.12[1]
Au	1.39[1]
Cu	1.72[1]
Cu	1.43[2]
δ-Fe	2.08[1]
γ-Fe	2.1[2]
Pt	2.28[1]
W	2.65[1]
Covalent bonding	
Ge(111)	1.06[2]
Si(111)	1.23[2]
Ionic bonding	
LiF (100)	0.34[2]
NaCl (100)	0.30[2]
Ceramics[2]	
Al_2O_3	0.905
TiC	1.19
MgO (100)	1.5
0.8 SiO_2—0.2 NaO	0.38

(Continued)

Table 3.2 *Continued*

Material	Surface Energy (J/m²)
Solid polymers[3]	
PE	0.031–0.036
PP	0.029–0.030
PVC	0.039–0.042
PTFE	0.019
PS	0.033–0.043
PMMA	0.039–0.040
Nylon 6, 6	0.039
Liquids[4]	
Water	0.0720
Mercury	0.485
Methanol	0.022
Ethanol	0.022
Ethylene glycol	0.048
Acetone	0.023
Benzene	0.028

1 Jones, H. Metal Science Journal, 5:15 (1971) and Porter, D.A., and Easterling, K.E., Phase Transfromations in Metals and Alloys, Chapman and Hall (1993), p. 113.

2 Chaing, Y-M., Birnie D.B., and Kingery, W.D., Physical Ceramics-Principles for Ceramic Science and Engineering, John Wiley & Sons, N.Y. (1997), p. 359.

3 van Krevelen, D. W., Properties of Polymers, Correlations with Chemical Structure, Elsevier, Amsterdam (1972), p. 100.

4 CRC Handbook of Chemistry and Physics, Lide D.R. ed. 90th ed. Boca Raton, FL. (2009), pp. 6-161.

Example Problem 3.2

What is the internal energy difference between one piece of pure Al_2O_3 of dimension 0.02 m × 0.01 m × 0.01 m and two pieces of Al_2O_3 of dimensions 0.01 m × 0.01 m × 0.01 m?

Solution

The only difference between the single piece of Al_2O_3 and the two pieces is that the two pieces have two additional surfaces of area 0.01 m × 0.01 m. The surface energy of Al_2O_3 is 0.905 J/m², from Table 3.2. The difference in internal energy (ΔE) between the single and two pieces is then

$$\Delta E = \gamma A = 0.905 \, \frac{J}{m^2} (2) (0.01 \text{ m})^2 = 1.81 \times 10^{-4} \text{J}$$

The two pieces of Al_2O_3 have this much more internal energy than a single piece does. A similar result happens when a piece of a brittle material breaks, such as glass. New surface area is created, and this increase in the internal energy must be provided when breaking the part.

Although surface energy is justified for a crystalline solid in the discussion above, all liquids and solids whether crystalline or amorphous also have surfaces; and the atoms at the surface have fewer neighbors for bonding than the interior atoms do, no matter what the structure of the liquid or solid. The effect of surface energy is visible in a drop of water on the waxed surface of a car. The water forms a shape approximating a hemisphere, because this shape minimizes the amount of surface area of the water, and it minimizes the total surface energy of the water (γ) times the surface area of the water droplet. Water has a higher surface energy (0.072 J/m^2) than wax does (≈ 0.030 J/m^2). The water does not spread over the wax, because the water with a higher surface energy would result in a large area of higher surface energy. When the wax disappears with time, the paint contains materials that have a higher surface energy than water does, such as metal oxides (TiO_2 is used to make white paint). Then the water spreads over the old paint surface, because now the water layer spread over the old paint results in a lower-energy surface exposed to the air. For a material that has mobility to uniformly spread over another material, it must have a lower surface energy than the base material. Uniform spreading of one material over another is called wetting. One material wetting another is important in many manufacturing processes, such as coatings and plating, and in the production of composite materials where a matrix must bond to a strong fiber.

Manufacturing problems can arise when a material with a high surface energy is coated onto a material with a low surface energy. For example, polymer sunglass lenses are coated with metals such as iridium that absorb the sun's light. However, as can be seen from Table 3.2, metals have a much higher surface energy than do polymers. A metal deposited from a vapor would form three-dimensional (3-D) islands (droplets) on the polymer, rather than a smooth thin coating, to minimize the surface area of the metal, and to maximize the surface area of the low surface energy polymer. To obtain a smooth thin film of metal, the metal atoms are deposited in vacuum with a moderately high kinetic energy, approximately 50 eV/atom, so that the metal atoms move across the surface of the polymer with some kinetic energy. Then the metal atoms become frozen on the surface of the polymer in a thin uniform layer before the 3-D islands can form.

The term **surface tension** is used in the discussion of surfaces, particularly in the discussion of liquids. A good example of surface tension is observed in a blown-up balloon or a soap bubble. We can hypothetically cut the inflated balloon in half. To keep the two halves of the balloon from blowing apart when we cut it, we imagine applying a force per unit length (newtons/meter) along the cut surface, as shown in Figure 3.12. The balloon separates the high-pressure inside air from the outside air at a pressure of 1 atmosphere. The magnitude of the distributed force per unit length along the cut required to keep the two halves of the balloon together is the surface tension. The units of newtons/meter for surface tension and joules/m^2 for surface energy are the same. A joule is a newton-meter, and by canceling the meters in the numerator and denominator of joules/m^2, the result is newtons/meter.

For a liquid-vapor interface the liquid surface energy is numerically equal to the surface tension of the liquid, and this tension is what keeps the water droplet on the waxed surface of a car in the droplet shape. In solids the surface tension and the surface energy are not necessarily equal, because in solids there can be displacements of the atoms from their equilibrium atomic positions at the surface that contribute to

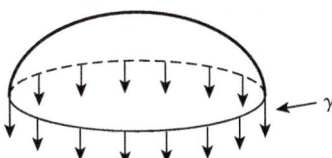

Figure 3.12 A schematic of a section of an inflated balloon that separates high-pressure air from atmospheric-pressure air. A cut through the center of the balloon is replaced by a surface tension or a distributed force per unit length (γ).

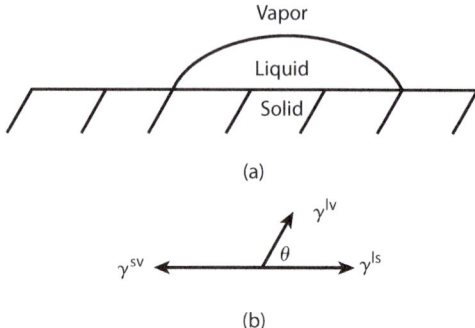

(b)

Figure 3.13   (a) A schematic of a cross section through a liquid droplet on a surface. (b) The surface tensions at the left side of the liquid droplet.

the surface energy that do not contribute to surface tension. However, usually surface energy and surface tension are considered equal in both liquids and solids for most applications, and the terms are used interchangeably in many discussions.

When a liquid forms a droplet on a surface, such as in Figure 3.13, it is surface tension that determines the shape of the droplet on the surface. The concept of surface tension allows the analysis of the shape of liquids on a surface, because surface tension is a force per unit length, and it has magnitude and the direction of the force as shown in Figure 3.13. Surface energy does not have a direction. When a liquid droplet forms an equilibrium shape on a surface, as shown in Figure 3.13a, the surface tensions are taken as vector forces for analysis of the shape. In Figure 3.13b the vector forces are shown; γ^{sv} is the surface tension between the solid and the vapor phase. The surface tension between the liquid and the vapor (γ^{lv}) is at an angle of θ relative to the solid surface, and this is the contact angle between the solid surface and the liquid surface. The interfacial tension between the liquid and solid (γ^{ls}) is parallel to the solid surface. Because the liquid and solid are two different phases, the interface between them is a phase boundary. A phase boundary has an interface energy just as a surface has an energy. Actually, surfaces are a special case of phase boundaries because they separate the liquid or solid phase from the vapor phase. In this book phases are indicated by superscripts.

If the liquid droplet is at equilibrium and not changing with time, then the surface-tension sum in Figure 3.13b is 0. Summing the forces in the x direction in Equation 3.1 results in (Young's) Equation 3.2 and Equation 3.3.

$$\sum F_x = \gamma^{sv} - \gamma^{ls} - \gamma^{lv} \cos \theta = 0 \tag{3.1}$$

$$\gamma^{sv} = \gamma^{ls} + \gamma^{lv} \cos \theta \tag{3.2}$$

$$\cos \theta = \frac{\gamma^{sv} - \gamma^{ls}}{\gamma^{lv}} \tag{3.3}$$

Example Problem 3.3

The contact angle for a water droplet on polystyrene (PS) is 91°. Assume that the surface tension for PS is 0.033 N/m. What is the value of the interfacial tension between water and PS at 25°C?

Solution

From Table 3.2 the surface tension of water is 0.072 N/m. Inserting the known values into Equation 3.2 results in

$$\gamma^{sv} = \gamma^{ls} + \gamma^{lv} \cos \theta = 0.033 \text{ N/m} = (\gamma^{ls} + 0.072 \cos 91°) \text{ N/m} = (\gamma^{ls} + 0.072 \, (-0.017)) \text{ N/m}$$

$$\gamma^{ls} = 0.033 \text{ N/m} + 0.001 \text{ N/m} = 0.034 \text{ N/m}$$

The interfacial tension between PS and water is 0.034 N/m. This high value of interfacial tension indicates that there is a strong interaction between water and polystyrene.

Interfaces between any two phases have an interfacial tension. The interface between two solids, such as epoxy and glass fibers, has an interfacial tension. The interface between a solid and a liquid has tension, and the interface between and solid and a vapor has an interfacial tension. Normally solid-vapor and liquid-vapor phase interfaces are referred to as surfaces rather than as interfaces, even though they are the interface between two phases. Table 3.3 presents the interfacial tensions for some polymer combinations. These values are determined at temperatures where the first listed polymer is a liquid and extrapolated back to 20°C unless otherwise noted.

Example Problem 3.4

Determine the contact angle between liquid polyethylene (PE) and a solid slab of polystyrene. Assume that the surface tension of the PE is 0.031 N/m and for the PS it is 0.036 N/m.

Solution

Inserting known values into Equation 3.2,

$$\gamma^{sv} = \gamma^{ls} + \gamma^{lv} \cos \theta$$

$$0.036 \text{ N/m} = 0.0044 \text{ N/m} + 0.031 \text{ N/m} \cos \theta$$

$$\cos \theta = \frac{(0.036 - 0.0044) \text{ N/m}}{0.031 \text{ N/m}} = \frac{0.0316}{0.031} \approx 1$$

$$\theta = 0$$

Table 3.3   Interfacial Tension in Newtons/m for Selected Polymer Combinations at 20°C (68°F) or at the Indicated Temperature

Polymer Pair	γ
PMMA/PS	0.0032
PE/PS	0.0044[1]
PE(linear)/PS	0.0083
PE(linear)/PMMA	0.0119

Adapted from Owen, M. J., "Surface and Interfacial Properties," in *Physical Properties of Polymers Handbook*, Mark, J. E., ed., AIP Press, Woodbury, New York (1996), p. 669.

1   Measured at 200°C.

If $\theta = 0$, the liquid wets the surface and covers it with a material of lower surface tension or surface energy. For the calculations of cos θ from Equation 3.3, it is possible for the ratio on the right-hand side of the equation to be greater than 1; in that case the liquid wets the surface because the solid vapor surface tension is much larger than the liquid vapor surface tension. If the right-hand side of Equation 3.3 is greater than 1, the equilibrium assumed in Equation 3.1 is not present. It is also possible for the contact angle to be greater than 90°, in which case the liquid-solid interface tension is greater than the solid vapor surface tension, and the liquid does not wet the surface.

Surface tension is a function of temperature, which is why the temperature is given in Table 3.2. For example, the surface tension of water is 0.7423 N/m at 10°C and 0.5891 N/m at 100°C. For crystalline materials the surface energy also depends upon the crystal plane, as indicated in Table 3.2. For polymers, the surface tension depends upon the molecular weight, the degree of side branching, and the crystallinity.

3.4.2   Grain Boundaries in Crystalline Materials

In Chapter 2 crystal structures were introduced. Only in a few applications are materials made of single crystals, such as the silicon in an integrated circuit, some high-performance gas-turbine blades, the crystal in a laser, and the diamond in an engagement ring. However, the aluminum in an automobile cylinder head and the titanium in the fan blades of a jet engine are polycrystalline. Polycrystalline materials are made from many (poly) small grains (crystals) that are joined together along grain boundaries. A **grain boundary** separates two crystals of the same phase that are in different orientations, as shown in Figure 3.14a. Figure 3.14b shows a polished surface of stainless steel. The dark lines on the surface are where the grain boundaries intersect the surface. The different grains in this specimen are different shades of gray, because crystals of different orientations reflect light with different intensity.

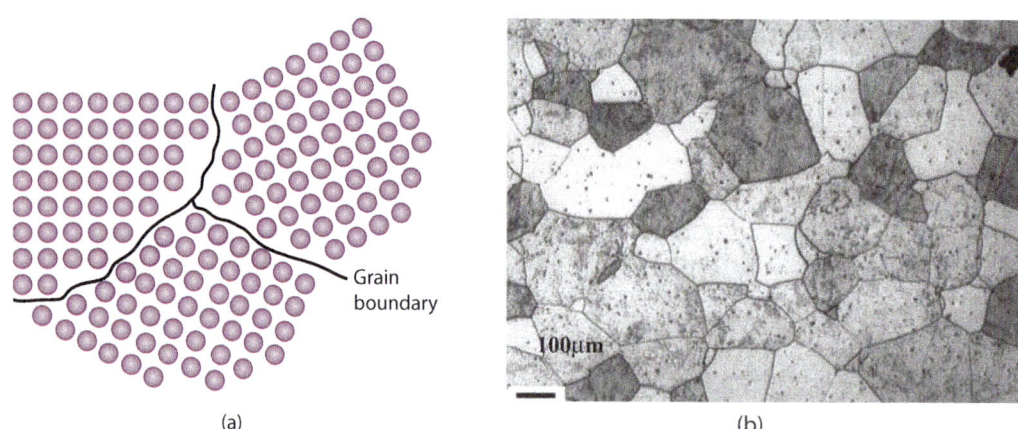

(a)   (b)

Figure 3.14   (a) A schematic of grains of different orientations coming together at grain boundaries. (b) A photomicrograph of grains and grain boundaries in stainless steel. *((a) Based on Askeland, D.R., Fulay, P.P., and Wright, W.J., The Science and Engineering of Materials, 6th ed. Cengage Learning, Stamford, CT. (2011), p. 136. (b) Anthony J. DeArdo)*

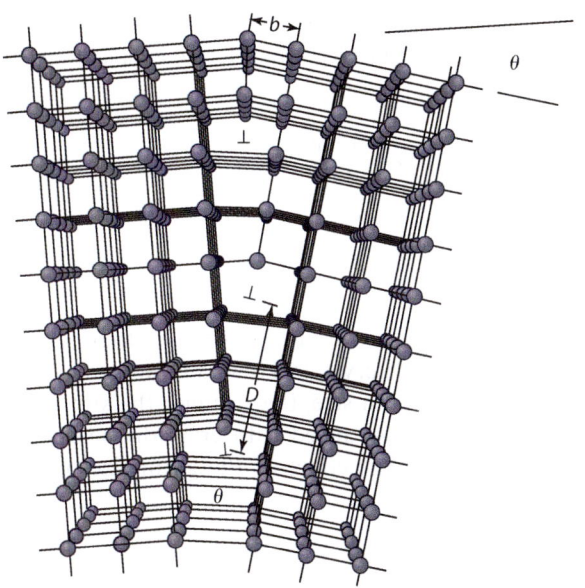

Figure 3.15 A schematic of edge dislocations in a simple-cubic crystal of Burger's vector **b** and spacing *D* forming a low-angle-tilt grain boundary of angle *θ*. (*Adapted from Guy, A. G., Introduction to Materials Science, McGraw-Hill, New York (1972), p. 184.*)

In Figure 3.15 a low-angle-tilt grain boundary is schematically shown in a crystal. In a low-angle-tilt grain boundary, the two grains are tilted relative to each other by a small angle (θ), and the grain boundary is a stack of edge dislocations of Burger's vector **b** and a spacing of *D* with the relationship given in Equation 3.4, where the magnitude of the Burger's vector is represented by *b*.

$$\tan \theta = \frac{b}{D} \qquad\qquad \textbf{3.4}$$

A stack of edge dislocations creates a tilt between the grains, and a series of screw dislocations produce a twist between the grains. Grain boundaries can be pure tilt, pure twist, or anything in between.

Experimentally it has been found that high-angle grain boundary energies are approximately 0.25 to 0.40 times the surface energy. The ratio depends upon the orientation of the grain boundary. Low-angle grain boundaries have a lower energy than high-angle grain boundaries do.

Example Problem 3.5

A simple-cubic crystal with a lattice parameter of 0.3 nm has a low-angle tilt boundary of 10 degrees. Calculate the spacing (*D*) between edge dislocations in the tilt boundary.

Solution

The relationship among the tilt angle ($\theta = 10°$), the magnitude of the Burger's vector (b), and the dislocation spacing (D) is given by

$$\tan \theta = \tan 10° = 0.18 = \frac{b}{D} = \frac{0.3 \text{ nm}}{D}$$

In a simple-cubic crystal, the magnitude of the lattice parameter (a) is equal to the magnitude of the Burger's vector (b). Solving for D,

$$D = \frac{0.3 \text{ nm}}{0.18} = 1.7 \text{ nm}$$

On average there is a edge dislocation every 5.7 unit cells in this example.

Grain boundaries are not desirable in semiconductors, because they scatter and trap electrons. This is why semiconductors are made from single crystals. However, for steel structural applications a small grain size is desired, and the smaller the grain size, the stronger the steel. A small grain size results in more grain boundaries. Materials are now produced with grain diameters of less than 100 nm. These nanostructured materials are very strong. The effect of the grain size on the strength of materials is covered in Chapter 7, on strong solids.

3.4.3   Spherulites in Polymers

When uniformly cooled from the liquid phase, polymers that form crystalline and amorphous regions will form **spherulites** like those shown in the photomicrograph in Figure 3.16. Figure 2.21b illustrates how LCMs loop back and forth to form crystalline regions. The crystalline regions form lamellae that are 20 to 60 nm in size, depending upon the polymer and the cooling rate. Different lamellae surrounded by amorphous material come together to form a spherulite that is typically 50 to 500 μm

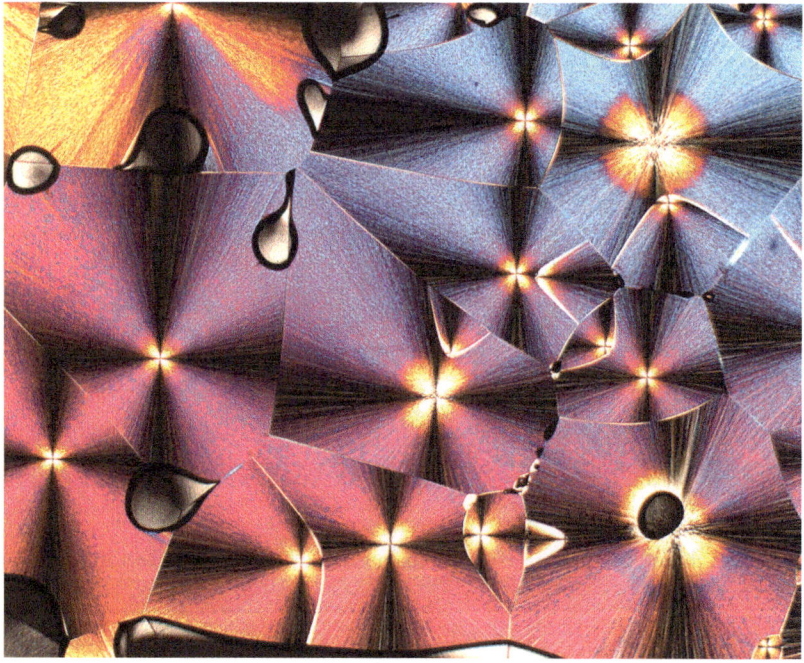

Figure 3.16   Spherulites in polystyrene. *(US NSF)*

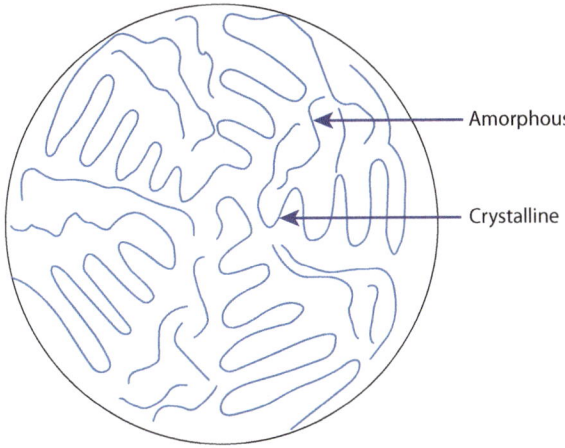

Figure 3.17   A schematic of a spherulite showing lamellae that form the crystalline regions and amorphous regions.

in diameter, as shown schematically in Figure 3.17. The spherulite is a combination of crystalline and amorphous regions that grows radially—it is semicrystalline—whereas in a metal the grain is a small single crystal.

3.4.4   Stacking Faults and Twin Boundaries

In Figure 2.4 the layers of metal atoms in a FCC crystal are shown to be ABCABCABC and so on. However, faults in the layers of a crystal occur during the formation and deformation of the crystal. If the layers of a FCC crystal are ABCABCA/CABCABC, there is a **stacking-fault** where the slash is located and the B layer of atoms is missing. There is a stacking-fault energy associated with this defect, because the stacking of the layers of the atoms is incorrect. Stacking-fault energies for FCC metals range from approximately 250 millijoules per square meter (mJ/m^2) for aluminum to less than 10 mJ/m^2 for brass and FCC stainless steel. The low stacking-fault energy in cartridge brass allows for the extensive deformation of brass into the shape of a bullet cartridge. The effects of stacking-fault energy on plastic deformation are discussed in more detail in Section 7.2.1. Stacking-faults are more likely to occur in metals with a low stacking-fault energy, such as brass, than in metals with a high stacking-fault energy, such as aluminum.

A coherent **twin boundary** in a FCC crystal is also an error in the stacking of layers of atoms where the sequence in the layers is ABCAB/C/BACBA. To the right of the /C/ plane between the slashes is a mirror image or twin of the stacking of layers to its left. The term *coherent* means that there is continuity in the arrangement of atoms within the planes at the interface. In contrast, the grain boundaries shown in Figure 3.14a are not coherent, because the arrangement of atoms within the grain boundary is not continuous. Twins form in crystals as a result of growth and of crystal deformation. Growth twin boundaries are shown in the photomicrograph of brass in Figure 3.18; they are the straight boundaries within a grain that separate regions of different shades of gray. Twins form in brass because a twin is a low-energy defect. The coherent twin-boundary energies are approximately twice the stacking fault energy, because a twin boundary is similar to two stacking faults adjacent to each other. For example, the coherent twin-boundary energy in FCC stainless steel is approximately 20 mJ/m^2, and the stacking-fault energy of the same stainless steel is 10 mJ/m^2. Stacking faults and twins occur in materials other than FCC metals; however, the layer sequence is most easily demonstrated for FCC metals. The energy of high-angle grain boundaries is approximately 300 to 400 times the twin-boundary energy in FCC metals.

Figure 3.18 An optical photomicrograph of polycrystalline cartridge brass showing grains and twins within the grains. The twin boundaries are straight. The inset schematic of a slab of material in the lower left of the figure shows the rolling direction. (*ASM Handbooks Online, http://www.asmmaterials.info, ASM International, 2004.*)

3.5   THREE-DIMENSIONAL OR BULK DEFECTS AND MODIFICATIONS TO MATERIALS

3.5.1   Three-Dimensional or Bulk Defects and Modifications to Metals

An example of bulk modification to a metal is the formation of a second phase inside the metal. The formation of the precipitate Ni_3Al (γ'), shown in Figure 3.19, in a nickel-aluminum alloy is a three-dimensional (3-D) modification that strengthens the alloy for use as a high-temperature gas-turbine material. However, the formation of iron sulfide particles (inclusions) in steel is a 3-D defect, and they decrease the steel's resistance to fracture. All bulk defects or modifications have a phase boundary separating the material from the bulk defect, because the bulk defect is a different phase.

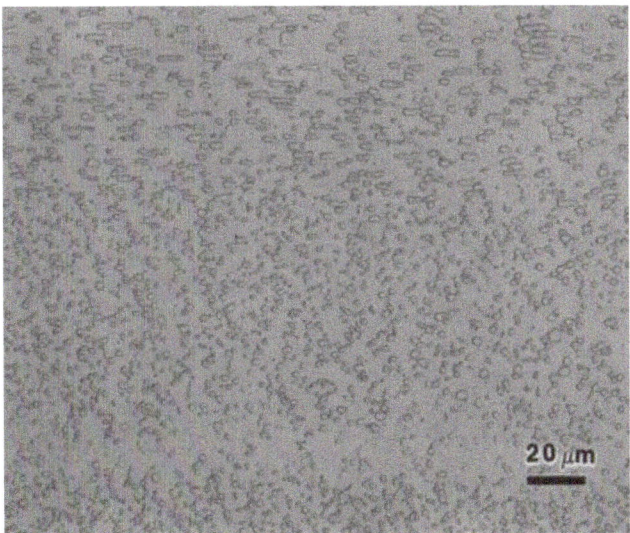

Figure 3.19 A photomicrograph of a nickel-based alloy U-720 as forged, containing 2.5 weight percent aluminum, showing precipitate particles of Ni$_3$Al (darker particles). (*ASM Handbooks Online, http://www.asmmaterials. info, ASM International, 2004.*)

3.5.2 Three-Dimensional or Bulk Defects in Ceramics

Voids are the main 3-D defect of concern in ceramics. Most of the processing procedures for producing ceramic parts utilize small particles of ceramic material that are bonded together by chemical reactions, as in the formation of concrete, or by heating to a high temperature, as in the firing of clay pottery and the sintering of ceramics. These procedures bond the small particles together, but the particles do not melt and they retain their original shapes. As a result, the particles in the final product do not fit perfectly together, and there are usually voids in ceramic materials, as shown in Figure 3.20. Voids significantly reduce the fracture strength of ceramics.

3.5.3 Three-Dimensional Modifications to Polymers

There are many 3-D additives to polymers. In Chapter 4 we will find that LCMs do not mix well with other molecules unless they are solvents. Most additives to polymers are second phases that are 3-D. Some of the additives to polymers include flame retardants, thermal stabilizers, plasticizers, and fillers. For example, carbon particles are added to rubber to make automobile tires. **Plasticizers** are added to polymers to make them flow more easily into shape during processing. **Fillers** are added to polymers to increase volume (extenders) and to increase strength or to modify other properties. Fillers include silica flour (fine powder) and wood flour or fibers. Particles or fibers added to polymers produce a composite material. Composite materials are discussed next, and in more detail in Chapter 12.

2 μm

Figure 3.20 A photomicrograph of sintered ceramic barium magnesium tantalite, showing the grain structure with voids between the grains. (*Heather Shirey*)

3.6   COMPOSITE MATERIALS

Composite materials are mixtures of at least two different material phases, and the two phases are separate. An example of a composite material is concrete reinforced with steel rods, shown in Figure 1.5. Composite materials are produced because the resultant material has unique properties that are not possible with any of the original single materials. Composite materials are normally made of a matrix phase that holds the material together, and a second phase that adds unique physical properties to the composite, such as strength, conductivity, hardness, or fracture resistance. The second phase can be in the form of small particles, fibers, sheets, or a three-dimensional bulk phase. A familiar example of a fiber second phase is fiberglass. Fiberglass has a matrix of epoxy resin that surrounds the second phase of glass fibers. The glass fibers provide the strength to the composite. In advanced composites, the glass fibers are replaced with fibers such as graphite, boron, or high-strength polymer fibers, such as oriented UHMWPE. These fibers are used in composite materials because of their extremely high strength that is many times stronger than that of their corresponding bulk materials.

Metals such as aluminum, cobalt, and nickel can also be the matrix material with both particle and fiber second phases. Figure 3.21 shows a composite material with an aluminum-alloy matrix that is reinforced with boron fibers coated with silicon carbide (SiC). The SiC suppresses the reaction between the boron and the aluminum matrix. Composites such as this are used in high-performance aircraft, for which aluminum alloys alone do not have sufficient strength and stiffness. A dentist's drill is made of a cobalt metal matrix and tungsten carbide particles, as shown in Figure 3.22. The cobalt matrix material is the light continuous phase, and the tungsten carbide particles are dark particles with sharp edges that cut the teeth. The cobalt metal matrix holds the tungsten carbide second-phase particles in place. Many carbide cutting tools are made of similar materials.

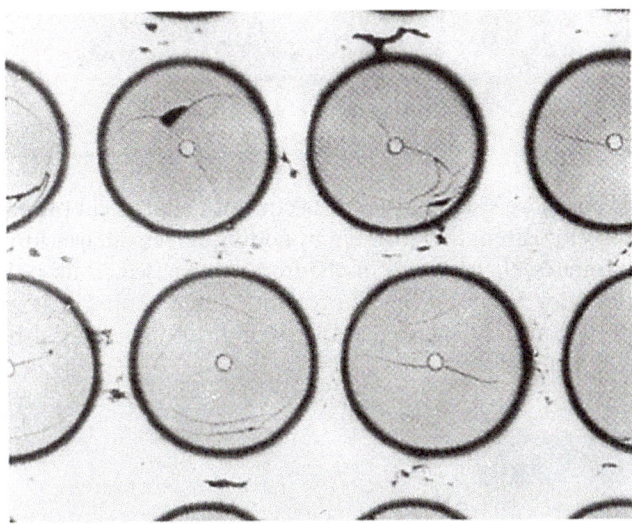

Figure 3.21   An optical photomicrograph of a cross section of a composite material: an aluminum matrix reinforced with boron fibers coated with silicon carbide (SiC). The cross section of the fibers is circular. Each boron fiber is grown on a tungsten wire that appears as the small circle at the center of the fiber's cross section. The dark circle around the boron fiber is a coating of SiC. (1000× magnification.) (*ASM Handbook Volume 9, Metallography and Microstructures, 2004, ASM International, in ASM Handbooks Online, http://www.asmmaterials.info, ASM International, 2004.*)

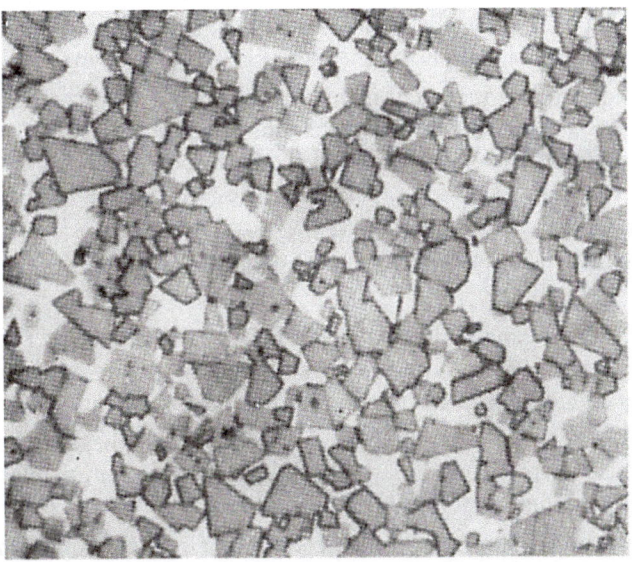

Figure 3.22   An optical photomicrograph of a composite material with 80 weight percent tungsten carbide particles (dark phase) in a 20 weight percent cobalt matrix. (1300× magnification.) (*ASM Handbook Volume 7, Powder Metal Technologies and Applications, 1998, ASM International in ASM Handbooks Online, http://www.asmmaterials.info, ASM International, 2004.*)

3.7   DEFECTS, MODIFICATIONS, AND PHYSICAL PROPERTIES

For the remainder of this book, we study how defects control the physical properties of materials. Pure, well-annealed iron has a very low strength. However, by adding defects such as foreign atoms, dislocations, grain boundaries, or precipitates, the strength of the iron is increased. Bulk polyethylene is a naturally weak material, but by orienting the long-chain molecules along a fiber axis, extremely high-strength filaments are produced. The science of increasing the strength of materials is the science of controlling the structure and defects in materials.

Summary

- Point defects present in crystalline solids include vacant atom sites, foreign substitutional atoms, and foreign and self-interstitial atoms. In addition, ionic solids have Frenkel defects (cation vacancy and interstitial) and Schottky defects (cation and anion vacancies) that are neutral in charge.

- Foreign substitutional atoms in a host crystal create a substitutional solid solution, and foreign interstitial atoms in a host crystal create an interstitial solid solution.

- The addition of Na_2O to SiO_2 reduces the viscosity of SiO_2 glass by as many as ten orders of magnitude.

- Polyvinylchloride has a mer that is ethylene with one of the hydrogen atoms replaced by chlorine (Cl). If the Cl always repeats in the same position in the mer, this is an isotactic arrangement. If the Cl alternates from one side of the chain to the other, this is a syndiotactic arrangement, and if the Cl appears at random locations on the mer, this is an atactic arrangement.

- A polymer with an atactic arrangement is more difficult to crystallize than are polymer chains with regular atom arrangements, because the chains are difficult to align when the substituting atom or group of atoms is in irregular positions along the chain. A more crystalline polymer has a higher strength than does a similar polymer that is less crystalline.

- Side branches in polymers result in polymers that are more difficult to crystallize; have weaker van der Waals bonding between LCMs; and have lower melting temperatures, densities, and hardness.

- A copolymer has different mers along a single long-chain molecule. The different mers can alternate in a regular manner or at random, be in blocks along the polymer chain, or be in side branches (grafts). Polymer properties are altered by producing a copolymer.

- Cross-links are atoms or molecules that form covalent bonds between long-chain molecules and produce a thermoset polymer. Cross-links decrease the amount of plastic deformation and increase the stiffness, strength, and hardness of polymers.

- Edge, screw, and mixed dislocations are line defects present in all metals and in most materials with covalent and ionic bonding. Dislocations form in crystalline materials during the solidification process as a result of forces in the solid caused by the temperature difference between the liquid and solid and by forces that produce permanent deformation.

- A dislocation is characterized by two vectors: a unit vector (τ) tangent to the dislocation line, and the Burger's vector (**b**), which gives the magnitude and direction of the displacement in the crystal caused by the dislocation. For an edge dislocation the Burger's vector is perpendicular to the tangent vector, and for a screw dislocation the Burger's vector is parallel to the tangent vector.

- A crystal's surface is a planar defect, because atoms at the surface are not bonded in the same way as the atoms in the interior of the crystal. The energy of any atom at the surface is increased relative to that in the bulk, because surface atoms do not have the proper number of neighbors sharing bonding electrons. This results in surfaces having a positive surface energy, measured as energy per unit area.

- Surface tension is observed in an inflated balloon or a soap bubble. Surface tension is the force per unit length necessary to keep the two halves of the balloon or soap bubble from blowing apart if it is cut in half. In liquids surface tension and surface energy are numerically equal, and in solids they are normally assumed to be equal even though they can be different.

- Polycrystalline materials are made from many (poly) small grains (crystals) that are joined together along grain boundaries. Grain boundaries are interfaces between different grains of the same material.

- Polymers with uniform LCMs, such as polyethylene, when uniformly cooled from the liquid phase will form spherulites. Spherulites contain crystalline regions surrounded by amorphous material. Spherulites are typically 50 to 500 μm in diameter.

- Stacking faults are errors in the stacking of planes of a crystal that occur during the formation and deformation of a crystal. There is an energy associated with a stacking fault, because the bonding of the layers of atoms is incorrect. The low stacking-fault energy in cartridge brass allows for the extensive deformation of brass into the shape of a bullet cartridge.

- A twin boundary in a crystal is a plane that has mirror images of the layers of atoms on either side of the twin boundary.

- Bulk modifications to metals include the formation of second-phase particles or precipitates, such as $Ni_3Al\,(\gamma')$, in nickel-based superalloys. The formation of iron sulfide particles (inclusions) in steel is a 3-D defect, and they decrease the steel's resistance to fracture. All bulk defects or modifications have a phase boundary separating the material from the bulk defect, because the bulk defect is a different phase.

- The main 3-D defect of concern in ceramics is voids. Most of the processing procedures for producing ceramic parts utilize small particles of ceramic material that are bonded together by chemical reactions, as in concrete production, or by heating to a high temperature, as in the firing of clay pottery and sintering of ceramics. These procedures bond the small particles together, but the particles do not melt, and voids occur between the particles. Voids significantly decrease the strength and resistance to fracture of ceramics.

- Many additives to polymers are second phases that are three dimensional. Fillers are added to polymers to increase volume (extenders) and to increase strength or other properties. Fillers include silica flour (fine powder) and wood flour or fibers.

- Composite materials are mixtures of at least two different material phases that are separate. Composite materials are usually produced because the resultant material has unique properties that are not possible with a single material. Composite materials are normally made of a matrix phase that holds the composite material together, and a second phase adds unique physical properties to the composite, such as strength, conductivity, hardness, or fracture resistance. The second phase can be in the form of particles, fibers, sheets, or a three-dimensional bulk phase.

- The science of increasing the strength of materials is the science of controlling the structure and defects in materials.

Supplemental Reading Subjects and Authors

Full references are listed at the end of the book.

General:	*Askeland, Fulay, and Wright*
Crystal defects:	*Allen and Thomas*
Dislocations:	*Weertman and Weertman*
Surfaces:	*Hudson*
Polymers:	*Oswald and Menges; Mark; van Krevelen*
Ceramics:	*Chaing, Birnie, and Kingery*
Metals:	*Porter and Easterling; Barrett and Massalski*
Metal microstructures:	*ASM Handbook*

Homework

Concept Questions

1. A ruby's red color results from the substitution of _____ atoms for aluminum atoms in a crystal of Al_2O_3.

2. In stainless steel, nickel and chromium atoms form a _____ solid solution in iron.

3. If a foreign atom is more than _____ percent larger than its host, it normally does not form a continuous substitutional solid solution.

4. In a simple-cubic lattice, the interstitial atoms are positioned at the coordinates _____.

5. In the BCC lattice, the interstitial site with coordinates $\frac{1}{2}$, 1, $\frac{1}{2}$ is called a(n) _____ site.

6. Hydrogen atoms in palladium form a(n) _____ solid solution.

7. Self-interstitials are produced when metals are _____ deformed.

8. A Frenkel defect in an ionic material is a cation _____ and vacancy.

9. A Schottky defect in an ionic material is a cation and anion _____.

10. The addition of Na_2O to SiO_2 to form soda glass reduces the _____ of SiO_2.

11. In PVC, if the chlorine atoms are all on the same side of the LCM, this is a(n) _____ arrangement.

12. PVC with a(n) _____ arrangement of Cl substituting for hydrogen on the LCM is difficult to crystallize.

13. A polymer with a high percent of crystallinity has a _____ hardness than does the same polymer that is less crystalline.

14. Polyethylene LCMs with long side branches results in _____ -density polyethylene.

15. An increase in the length of side branches on polyethylene LCMs results in a(n) _____ in the hardness of polyethylene.

16. The physical properties of a regular copolymer are normally a weighted _____ of the physical properties of the component polymers.

17. Polymers, such as epoxy, with cross-links are _____ polymers that do not soften when heated.

18. A one-dimensional or line defect in a crystalline material is a(n) _____.

19. If a Burger's circuit is performed around a dislocation, the Burger's _____ closes the circuit.

20. For an edge dislocation, the Burger's vector is _____ to the dislocation line.

21. Dislocations always _____ the energy of a crystal.

22. The Burger's vector in ionic materials is larger than in elemental metals, because the Burger's vector must connect two_____ points.

23. Polymers have a lower surface energy than metals and ceramics do, because at the polymer surface the _____ _____ _____ bonds between LCMs are broken.

24. The interface between epoxy and glass fibers in fiberglass is a(n) _____ boundary.

25. The interface between two crystals in a polycrystalline single phase solid is a(n) _____ boundary.

26. In uniformly cooled polymers crystalline lamellae come together with amorphous regions to form _____.

27. If the layering of the close-packed planes in a FCC metal is ABCABCACABC, this is a _____ _____.

28. If the layering of the close-packed planes in a FCC metal is ABCABCBACBA, this is a _____ _____.

29. One of the most serious 3-D defects that affects the strength of ceramics is the presence of _____.

30. Fine silica flour (fine powder) added to a polymer is a(n) _____.

31. A composite material is made from at least _____ phases that are separate.

32. In a composite material the _____ phase is continuous and is responsible for holding the composite material together.

33. The science of increasing the strength of materials is the science of controlling the structure and _____ in materials.

Engineer in Training–Style Questions

1. Which of the following is not a point defect?
   - (a) Dislocation
   - (b) Interstitial
   - (c) Vacancy
   - (d) Frenkel

2. Which of the following is not an arrangement of atoms or molecules on the mers of a LCM?
   - (a) Isotactic
   - (b) Syndiotactic
   - (c) Atactic
   - (d) Polytactic

3. Which of the following is not caused by side branches on a polymer?
    (a) Decreased density
    (b) Decreased transparency
    (c) Decreased crystallinity
    (d) Decreased hardness

4. Which of the following is not a form of copolymer?
    (a) Uniform
    (b) Random
    (c) Block
    (d) Graft

5. Which of the following is not a result of cross-linking polymer chains?
    (a) Increased hardness
    (b) Increased melting temperature
    (c) Decreased plastic deformation
    (d) Increased brittleness

6. For a metal with the indicated structure, which of these is not the shortest Burger's vector in the indicated metal crystal structure?
    (a) Simple cubic $\mathbf{b} = a\,[100]$
    (b) Simple tetragonal $\mathbf{b} = a\,[100]$
    (c) Body-centered cubic $\mathbf{b} = \dfrac{a}{2}\,[111]$
    (d) Face-centered cubic $\mathbf{b} = \dfrac{a}{2}\,[100]$

7. Which of the following is not a planar defect?
    (a) Edge dislocation
    (b) Stacking fault
    (c) Twin
    (d) Grain boundary

Design-Related Questions

1. You are asked to select a polymer for a design where the temperature could vary between 0°C and 100°C, and it is desired that the polymer's mechanical properties should not significantly change over this temperature range. What type of polymers would you initially investigate for this design, and why would you make this selection?

2. For a high-temperature gas-turbine application, you have chosen to use alumina ceramic material for a nonrotating part that is subject to modest forces. You find that you can save a significant amount of money by buying alumina that is of 1% lower density than more expensive alumina, and in aircraft design saving weight is always desired. What aspect of the defect structure of the cheaper alumina should you investigate to determine its possible effects on mechanical properties, and why may this be important?

3. You are asked to produce a part that will be shaped by forging from a thick block of metal. Forging is a form of controlled hammering the part into shape that is covered in Section 13.2.1. The final part will have some thin sections that will experience extensive deformation and some thick sections that will experience minimal deformation. It is desired that the mechanical properties of the as forged metal be relatively uniform throughout the part. It is planned to use the part in the as-forged condition without subsequent treatment. What defect property of metals may be important in the metal selection, and why is this property important for this design?

Problems

Problem 3.1   Draw the octahedral shape with the atoms shown that surrounds the interstitial site at $\frac{1}{2}, \frac{1}{2}, \frac{1}{2}$ in the FCC metal structure.

Problem 3.2   Draw the octahedral shape with the atoms shown that surrounds the interstitial site at $\frac{1}{2}, 1, \frac{1}{2}$ in the BCC metal structure. You will have to put two unit cells side by side to show this.

Problem 3.3   (a) For BCC iron, calculate the diameter of the minimum space available in an octahedral site at the center of the (010) plane, and compare this to the diameter of a carbon atom. Assume that the iron atoms in the BCC structure are hard spheres that touch along the [111] direction.

   (b) Compare the space available with the size of a carbon atom from Appendix C, and comment about the potential solubility of carbon in BCC iron in octahedral interstitial sites.

Problem 3.4   Iron has an FCC structure at temperatures just above 912°C, and the lattice parameter is 0.3589 nm. Assume that BCC iron's atomic radius is equal to 0.1241 nm as given in Appendix C, and for carbon it is 0.0077 nm. The atomic radius of iron should not increase with temperature and with changes in crystal structure; only the size of the lattice changes.

   (a) Calculate the radius ratio of carbon to iron to see if it is appropriate for octahedral coordination.

   (b) Calculate the diameter of the space available for the octahedral site at the center of the FCC unit cell, and compare this to the diameter of a carbon atom.

   (c) Comment on the diameter available relative to the diameter of the carbon atom and upon the possibility of solubility of carbon in FCC iron at octahedral interstitial sites. Also comment on the expected relative solubility of carbon in octahedral sites in FCC and BCC iron (see Problem 3.3).

Problem 3.5   There is a dislocation with a Burger's vector between the origin of an FCC copper (Cu) crystal at 0, 0, 0 and the nearest lattice point in the $(00\bar{1})$ plane. The Cu lattice parameter is 0.362 nm.

   (a) What is the Burger's vector of this dislocation, in vector notation?

   (b) What is the magnitude of the Burger's vector?

Problem 3.6   Tungsten is a BCC metal with a lattice parameter of 0.3165 nm. What is the direction and magnitude of the Burger's vector in tungsten, corresponding to the plane and direction of maximum packing density?

Problem 3.7   Determine the magnitude of the Burger's vector in MgO, shown in Figure 3.10, that has the NaCl crystal structure with a lattice parameter of 0.4212 nm, and compare the magnitude of this Burger's vector to that of copper. Note that the Burger's vector is in the [110] direction in MgO.

Problem 3.8   The crystal structure of CsCl is shown in Figure 2.7, and the lattice parameter of CsCl is 0.4123 nm. Determine the magnitude and the direction of the Burger's vector in CsCl.

Problem 3.9   Nanostructured materials can have unique properties. For materials with dimensions measured in nanometers, the amount of surface energy is comparable to the bonding energy in the material. (a) Compare the magnitude of the total surface energy to the total bond energy for a sphere of Ni atoms with one atom at the center surrounded by 12 nearest neighbors. The surface energy of Ni is 1.90 J/m^2, the cohesive energy of Ni is -4.435 eV/atom, and the lattice parameter of FCC Ni is 0.352 nm. Assume that the total radius of this sphere is equal to 1 atomic diameter. (b) Is this 13-atom cluster of atoms stable?

Problem 3.10   The contact angle for a water droplet on polymethyl-methacrylate (PMMA) is 80°. Assume that the surface tension for PMMA is 0.039 N/m. What is the value of the interfacial tension between water and PMMA at 25°C?

Problem 3.11   The contact angle for a water droplet on wax is 112°. Assume that the surface tension for wax 0.023 N/m. What is the value of the interfacial tension between water and wax at 25°C?

Problem 3.12   Determine the contact angle between liquid polyethylene (PE) and a solid slab of polystyrene (PS) at 200°C. Assume that the surface tension of the PE is 0.036 N/m, and for the PS it is 0.035 N/m. The interfacial tension between PE and PS is presented in Table 3.3.

Problem 3.13   A simple-cubic crystal with a lattice parameter of 0.2 nm has a low-angle-tilt boundary of 5°. Calculate the spacing (D) between edge dislocations in the tilt boundary.

Problem 3.14   A cube of copper 0.10 meters on each side has an average grain size of 10×10^{-6} m. The average grain boundary energy of copper is 625×10^{-3} J/m^2. Assume that all of the grains are cubes 10×10^{-6} m across.

   (a)   How many grains are there in the 0.1-m cube of metal?

   (b)   What is the total grain boundary area? You can assume that for the grains at the surface, the surface energy is equal to the grain boundary energy.

   (c)   What is the total amount of energy present in the 0.1-m cube of copper due to the grain boundaries?

   (d)   The density of copper is 8.96×10^3 kg/m^3. To what height would the copper cube have to be raised to increase the energy by an amount equal to the energy in the grain boundaries? Assume that the acceleration of gravity, $g = 9.81 m/s^2$, is constant.

A C-130 Hercules aircraft spreads an oil-dispersing chemical into the Gulf of Mexico in an effort to dissolve the leaked oil at the Deep Water Horizon drilling unit. Why is it that oil and water do not mix, but alcohol and water do mix? Why do nickel and copper mix to form a solid alloy, but nickel and silver do not mix? If solid pieces of nickel and copper are joined together and heated to a temperature of only half the melting temperature of copper, the atoms mix. How do atoms in solids mix? This chapter addresses these questions and others that are related to the effect that temperature and heat have on materials and the arrangements of the atoms in materials.

The goals of this chapter are to understand

- How heat is absorbed in materials
- Why most materials expand when heated
- How the coefficient of thermal expansion is related to chemical bonding
- How heat is conducted through materials
- How thermal conductivity is related to chemical bonding and defects in materials
- How and why atoms in liquids and solids mix in some cases but not in others
- How entropy relates to the configuration of atoms in materials
- What energy is minimized at equilibrium at constant temperature and pressure
- The entropy and Gibbs free energy of atom mixtures
- The important parameters in the rates of change in materials, and how to determine them
- How to determine rates of atom diffusion and the chemical compositions resulting from diffusion
- The entropy and Gibbs free energy of polymer blends
- The parameters that control the solubility of solvents in polymers
- The processes involved in the permeation of gases and liquids into polymers
- How catalytic processes increase the rates of reactions

Chapter 4

Temperature Effects on Atom Arrangements and Atom Motion

4.1   INTRODUCTION

In this chapter we will study the effects of temperature on the structure of materials. Throughout this book we will learn how the properties of materials change as the temperature changes. As the temperature of a solid increases, its atoms move more rapidly, and as a result the solid melts, weakens, expands, creeps, and/or deforms. In this chapter we will learn how increased temperatures affect the atoms of a material to create these changes, and learn how to calculate the amount of heat required to change the temperature. When the energy required to heat or cool a building is calculated, the heat-absorbing capacity of the building and all the materials in it must be known. We will see why materials expand as they are heated, and why this expansion is an important design factor. For example, an engine designer must consider the expansion of the materials that occurs when the engine runs at a high temperature. We will explore how heat is transferred through materials, and we will examine how even in a solid material the atoms can still move around. The movement of atoms in solids is called diffusion, and the metal-joining technique of diffusion bonding has been developed to bond titanium parts for aircraft structures.

Section 1 Thermal Properties of Materials

4.2   HEAT AND TEMPERATURE

When a pot of cold water is placed on a hot stove, heat (Q) is transferred to the water, and the temperature (T) of the water increases. Heat (Q) is the amount of energy that is transferred from the hot system to the cold system because they are at different temperatures. The units of heat are the same as those used for energy. A **thermodynamic system** is the collection of atoms or molecules that we are studying. In this example the hot stove and the pot of cold water is the thermodynamic system.

We all know what temperature is; we see it displayed on the weather forecast. The kelvin (K) temperature scale is used for most thermodynamic calculations. With the kelvin temperature scale, the freezing temperature of water at 1 atmosphere of pressure is 273.15 K. Zero kelvin is the lowest temperature of the scale, but it cannot actually be reached in any material. The reason is that to lower the temperature of a system approaching 0 K, it is necessary to remove heat and transfer it to a system that is at an even lower temperature, and no system can be at a temperature of less than 0 K. In the celsius temperature scale, the freezing temperature of water at 1 atmosphere of pressure is 0°C, and absolute zero is −273.15°C. A temperature increment of 1 degree on the celsius scale is equal to 1 degree on the kelvin temperature scale. In the fahrenheit temperature scale, the freezing temperature of water at 1 atmosphere of pressure is 32°F, and absolute zero is −459.67°F. A temperature increment of 1 degree on the celsius scale is equal to 1.8 degree on the fahrenheit scale.

The temperature in kelvin is directly related to the kinetic energy of the atoms and molecules in the material, and measuring the temperature is much easier than measuring the kinetic energy in a material. In a solid, when the temperature is increased, the amplitude of the vibrations of the atoms about their atomic position increases. All atoms in solids vibrate even at the lowest attainable temperatures, where the amplitude of the atomic vibrations is at a minimum. The atomic vibrations in solids and liquids form waves called **phonons**. Figure 4.1 shows schematically the vibration of a line of atoms in a solid. The small solid dots along the straight line are the positions of the atoms if they are at rest. The large circles are the positions of the atoms as a result of the atomic vibrations. The line connecting the two circles indicates the atom displacement resulting from the atomic vibrations. The vibrations of the atoms are not at random; rather they are coupled into phonons that are like sine waves of displacement that move through the solid. The thermal vibrations shown in Figure 4.1 form a transverse wave because the displacements are transverse (perpendicular) to the direction of propagation of the wave. There are also longitudinal waves, where the displacements are in the same direction as the direction of propagation of the wave. In longitudinal waves the atoms are either squeezed together or pulled apart as the wave propagates.

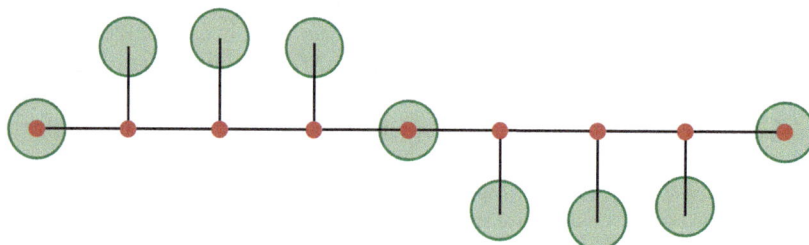

Figure 4.1   A schematic of the thermal vibration of a line of atoms in a crystal. If the atoms were at rest, they would sit along the straight line at the points indicated by the dots. Thermal vibrations displace the atoms in waves or phonons to the positions shown by the circles.

4.2.1   Heat Capacity and Specific Heat

For a mole of material, the increment of heat (dQ) added to the material divided by the change in temperature (dT) is the **heat capacity** (C), or the heat capacity is the amount of heat required to increase the temperature of 1 mole of material by 1 degree. It is possible to transfer the heat under conditions of constant pressure (C_P) as shown in Equation 4.1, or under conditions of constant volume (C_V) as shown in Equation 4.2.

$$C_P = \left(\frac{dQ}{dT}\right)_{P=\text{const}}$$

4.1

$$C_V = \left(\frac{dQ}{dT}\right)_{V=\text{const}}$$

4.2

The SI units of heat capacity are joules per kelvin-mole, and the British thermal units (Btu's) are calories per kelvin-mole. The heat capacity in Equations 4.1 and 4.2 is also known as the molar specific heat.

From the kinetic theory of gases, the heat capacity at constant volume for an ideal gas, such as the inert gas atoms, is equal to $1.5R$, as shown in Appendix 4A, where R is the universal gas constant (8.31 J/mole · K). The value of $1.5R$ for the heat capacity of ideal gases is valid for temperatures where bonding and quantum mechanical effects are not present.

For solids, quantum mechanical effects are not present above a temperature called the **Debye temperature** (θ_D). For solids at temperatures above θ_D the heat capacity at constant volume is equal to $3R$ or 24.9 J/mole · K, as shown in Figure 4.2 and as derived in Appendix 4B. For a pendulum the average kinetic energy is equal to the average potential energy, and these energies are also equal for a solid. An ideal gas has kinetic energy but no potential energy. In a solid both the kinetic energy and potential energy must be provided to heat the solid, and as a result the classical heat capacity of a solid is double that of an ideal gas. The decrease in heat capacity as temperature is decreased below θ_D in Figure 4.2 is due to quantum mechanical effects.

Specific heat (c) is the heat capacity per unit mass (kg), and for elements it is equal to the heat capacity in Equations 4.1 and 4.2 divided by the molar mass (M).

$$c_P = \frac{C_P}{M}$$

4.3

$$c_V = \frac{C_V}{M}$$

4.4

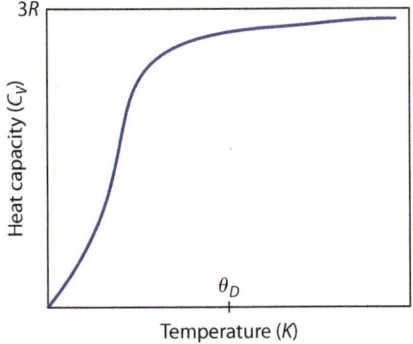

Figure 4.2   The heat capacity at constant volume for a solid with no phase transformations, as a function of temperature. Above the Debye temperature (θ_D) the material behaves classically and has a heat capacity of approximately $3R$.

Table 4.1 presents values of the specific heat for some materials. The SI units of specific heat are joules per kilogram-kelvin (J/kg · K), and presented in Table 4.1 are calories per gram-kelvin; the latter units are a mixture of British thermal and metric units. The most generally available data is for specific heat at constant pressure. It is easier to weigh specimens than to determine the number of moles of atoms. Also, it is much easier to conduct experiments at 1 atmosphere of pressure than to maintain constant volume. For solids the specific heats at constant volume and constant pressure are nearly equal to each other, because the work change (PdV) of a solid associated with a 1-degree temperature change at 1 atmosphere of pressure is very small.

Table 4.1	The Specific Heats of Selected Materials, at 1 Atmosphere of Pressure and at 300 K	
Material	c_p (J/kg · K)	c_p (calories/kg · K)
FCC Metals		
Ag	235	0.056
Al	900	0.215
Au	128	0.031
Cu	385	0.092
Ni	444	0.106
Pb	159	0.038
BCC Metals		
Fe	444	0.106
W	134	0.032
HCP Metals		
Mg	1017	0.243
Ti	523	0.125
Zn	389	0.093
Alloys		
1025 steel	486	0.116
316 stainless	502	0.120
Semiconductors		
B	1025	0.245
Si	703	0.168
Ceramics		
Al_2O_3	837	0.200
Diamond	519	0.124
SiC	1046	0.250
Si_3N_4	711	0.170
SiO_2	1108	0.265
Polymers		
HDPE	1841	0.440
LDPE	2301	0.550

Table 4.1 *Continued*		
Material	c_P (J/kg $\cdot$ K)	c_P(calories/kg $\cdot$ K)
PTFE	1000	0.239
Nylon 6, 6	1674	0.400
PS	1172	0.280
Liquids		
Water	4184	1.0

Based on Askeland, D.R., Fulay, P.P., and Wright, W.J., The Science and Engineering of Materials, 6th ed. Cengage Learning, Stamford, CT. (2011), p. 833.

Example Problem 4.1

From the specific heats given in Table 4.1 and the molar masses provided in Appendix B, calculate the heat capacity at constant pressure for aluminum and tungsten, and compare these two results.

Solution

Tungsten has a specific heat of 134 J/kg $\cdot$ K, and aluminum has a specific heat of 900 J/kg $\cdot$ K. The molar mass for each is obtained from Appendix B. From Equation 4.3,

$$C_P = c_P M$$

For tungsten,

$$C_P = c_P M_W = 134 \frac{J}{kg \cdot K} \left(183.86 \times 10^{-3} \frac{kg}{mole}\right) = 24.6 \frac{J}{K \cdot mole}$$

For aluminum,

$$C_P = c_P M_{Al} = 900 \frac{J}{kg \cdot K} \left(26.98 \times 10^{-3} \frac{kg}{mole}\right) = 24.3 \frac{J}{K \cdot mole}$$

This calculation shows that on a molar basis, the heat capacity of the two materials is very similar, even though the specific heats are very different. As can be seen from these results, the heat capacity at constant pressure is nearly equal to the heat capacity at constant volume ($3R = 24.9$ J/mole $\cdot$ K) at atmospheric pressure.

Example Problem 4.2

The specific heat of water at constant pressure in the temperature range from 300 K to 373 K is 4184 J/kg $\cdot$ K. Compare the magnitude of the heat necessary to raise the temperature of 1 kg of water from 300 K to 373 K (the normal boiling temperature), to the latent heat of fusion (3.3×10^5 J/kg) and the latent heat of vaporization (22.75×10^5 J/kg) of water.

Solution

Since the heating and vaporization are both conducted at 1 atmosphere of pressure, the heat capacity at constant pressure in Equation 4.3 is utilized to calculate the heat required to raise the temperature of the water from 300 to 373 K.

$$c_P = \left(\frac{dQ}{dT}\right)_P = 4184 \frac{J}{kg \cdot K} = \frac{dQ}{73 \text{ K}}$$

Solving for dQ,

$$dQ = 4184 \frac{J}{kg \cdot K} (73 \text{ K}) = 3.05 \times 10^5 \frac{J}{kg}$$

The heat necessary to raise the temperature from 300 K to the boiling temperature of 373 K is approximately equal to the heat necessary to transform ice to water (3.3×10^5 J/kg); however, it is only one-seventh of the heat necessary to transform water to water vapor at 373 K (22.57×10^5 J/kg). The heat required to boil water is much greater than the heat required to melt ice at 273 K or to bring water from 300 K to the boiling temperature.

4.2.2 Heat Capacity and Specific Heat of Polymers

Observe in Table 4.1 that the specific heat of polymers is higher than for any other materials. However, if we multiply the specific heat of high-density polyethylene times the molar mass of a carbon atom and two attached hydrogen atoms (14×10^{-3} kg/mole) to obtain the heat capacity or molar specific heat, the result is 25.8 J/K · mole. This result is just a little higher than the results for aluminum and tungsten in Example Problem 4.1. The heat capacity of the polyethylene is expected to be higher than for a solid made of atoms of the same atomic mass, because the vibration of the hydrogen and carbon atoms relative to each other absorbs heat in addition to the vibration of the carbon atom with attached hydrogen atoms. The primary reason for the high specific heat of polymers is their low monomer mass. Low-density polyethylene (LDPE) has a higher heat capacity than does high-density polyethylene (HDPE) because of the vibration of side branches. The specific heats of polymers with a higher monomer mass, such as PTFE, are lower than for polyethylene; however, the heat capacity or molar specific heat of these polymers is similar to that of polyethylene.

For amorphous polymers, such as the polyvinyl chloride (PVC), polystyrene (PS), and polycarbonate (PC) in Figure 4.3a, the specific heat gradually increases with temperature in this temperature range, because more modes of vibration in the LCMs are excited as the temperature increases. For semicrystalline polymers, such as ultra-high-molecular-weight polyethylene (UHMWPE) in Figure 4.3b, there is a spike in the specific heat when the crystallites start to melt. All materials have an increase in the specific heat at the temperature of phase transformations, because of the latent heat of transformation. UHMWPE has more of a spike in the specific heat than LDPE does, because UHMWPE is more crystalline. The crystallinity of HDPE is between that of UHMWPE and LDPE, and so is its spike in specific heat. In amorphous polymers, there is no discontinuity in the specific heat when the temperature passes through the glass transition temperature. In thermoset polymers, if there are chemical reactions that occur during heating, they affect the specific heat because of heats of reaction, as shown in Figure 4.3c. Once the thermoset polymer is cured and the chemical reactions are complete, the specific heat has a gradual increase with temperature, because there are no further chemical reactions or phase transformations.

For polymers with fillers, such as carbon black or glass particles, the specific heat follows the rule of mixtures based upon weight fraction. Weight fraction is the weight of each component, such as the weight of just the glass particles, divided by the total weight. In the rule of mixtures, the total property value is the sum of the fractional presence of each constituent times its property value. We will cover the rule of mixtures in more detail in Chapter 12, on composite materials. For copolymers, the specific heat follows the rule of mixtures based upon mole fraction.

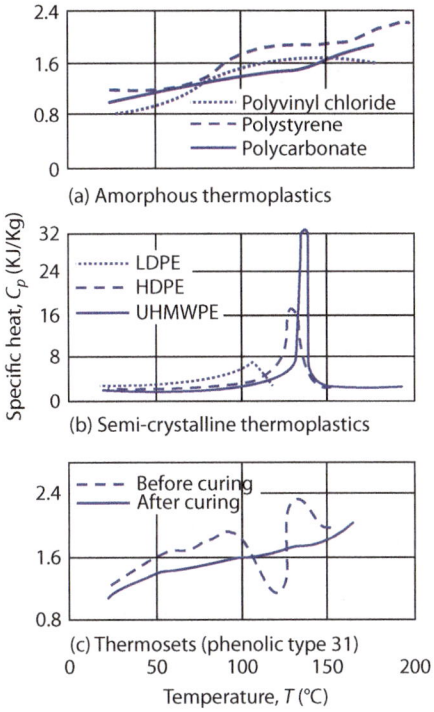

Figure 4.3 (a) Specific heat as a function of temperature for the amorphous polymers polyvinyl chloride (PVC), polystyrene (PS), and polycarbonate (PC). (b) Specific heat as a function of temperature for the semicrystalline polymers low-density polyethylene (LDPE), high-density polyethylene (HDPE), and ultra-high-molecular-weight polyethylene (UHMWPE). (c) Specific heat as a function of temperature for a thermoset polymer. (*Based on Osswald, T.A., and Menges G., Materials Science of Polymers for Engineers, Hanser Publishers, Munich (2003), p. 108.*)

4.3 THERMAL EXPANSION

When we cannot remove a metal lid screwed onto a glass jar, we pour hot water on the metal lid, and it will loosen. Most metals expand more than glass does for the same temperature change; thus the metal lid expands more than the glass jar does, breaking the seal between the lid and the glass. Most materials expand when they are heated. An exception is if there is a phase transformation in the material. For example, water is more dense than ice, and the shape-memory alloy TiNi contracts when heated through the phase transformation that results in the shape-memory effect. Also, there are a few materials such as graphite filaments and oriented polymer fibers that contract in the axial direction when heated. These fibers and filament materials are covered later in this section.

We must consider thermal expansion in designs, such as for engines, that have large temperature changes; and for large structures, such as bridges, that can have significant dimensional changes because of their large dimensions. Also, in composite materials, such as graphite-reinforced epoxy, we must consider the difference in the thermal expansion of the two components of the composite, if the composite is to undergo temperature changes during fabrication or use. We will consider the effects of temperature change on composite materials in Chapter 12.

A solid rod has an original length L_0 at temperature T_0. When the temperature of the rod is raised (or cooled) by an amount dT to a final temperature (T_f), the length of the rod increases (or decreases) in length by dL to L_f, where dL is given by Equation 4.5:

$$dL = L_f - L_0 = L_0 \alpha_L dT = L_0 \alpha_L (T_f - T_0) \qquad \textbf{4.5}$$

where α_L is the **linear coefficient of thermal expansion**. Values of the linear coefficient of thermal expansion are given in Table 4.2. Equation 4.6 expresses the linear coefficient of thermal expansion (α_L) from Equation 4.5 as

$$\alpha_L = \frac{1}{L_0} \left(\frac{dL}{dT} \right) \qquad \textbf{4.6}$$

The units of the linear coefficient of thermal expansion are (degree)$^{-1}$. The coefficient of thermal expansion can be listed as degrees celsius or kelvin; the value is the same because the temperature appears as dT. A one-degree difference in celsius is equal to a one-degree difference in kelvin. Not only does the linear coefficient of thermal expansion apply to the length of the specimen; it applies to all of the other dimensions. In a cylindrical rod of isotropic material, the linear coefficient of thermal expansion applies to the diameter, and for a hollow cylinder, the same linear coefficient of thermal expansion applies to the inside and outside diameter.

In isotropic materials, the linear expansion coefficient is the same in all directions. An isotropic material is one that has the same properties in all directions. The properties of polycrystalline metals with randomly oriented grains are usually isotropic, and polymers with randomly oriented long-chain

Table 4.2 Linear Coefficient of Thermal Expansion (α) for Some Selected Materials

Material	Linear Coefficient of Thermal Expansion (α) ($\times 10^{-6}/°C$)	Material	Linear Coefficient of Thermal Expansion (α) ($\times 10^{-6}/°C$)
Al	25.0	Nylon 6,6	80.0
Cu	16.6	Nylon 6,6—33% glass fiber	20.0
Fe	12.0	Polyethylene (HDPE)	200.0
Ni	13.0	Polyethylene—30% glass fiber	48.0
Pb	29.0	Polystyrene	70.0
Si	3.0	Al_2O_3	6.7
W	4.5	Fused silica	0.55
1020 Steel	12.0	Partially stabilized ZrO_2	10.6
3003 Aluminum alloy	23.2	SiC	4.3
Gray iron	12.0	Si_3N_4	3.3
Invar (Fe–36% Ni)	1.54	Soda-lime glass	9.0
Stainless steel	17.3		
Yellow brass	18.9		
Epoxy	55.0		

Data primarily from Askeland, D.R., Fulay, P.P., and Wright, W.J., The Science and Engineering of Materials, 6th ed. Cengage Learning, Stamford, CT. (2011), p. 835. The data for HDPE is from Osswald, T.A., and Menges G., Materials Science of Polymers for Engineers, Hanser Publishers, Munich (2003), Table I.

molecules are usually isotropic. However, when a polycrystalline metal or a polymer is processed into a nonisotropic shape, such as a thin sheet, the properties of the sheet in the plane of the sheet can be much different than the properties normal to the sheet.

If the volume (V) of material is measured rather than a linear dimension, then the volume coefficient of expansion (α_V) is as shown in Equation 4.7:

$$\alpha_V = \frac{1}{V_0}\left(\frac{dV}{dT}\right)$$

4.7

The relationship between α_L and α_V for an isotropic material is that $\alpha_V = 3\alpha_L$, but for nonisotropic materials, this relationship cannot be used, because there are different coefficients of expansion in different directions.

The origin of thermal expansion is explained with the interatomic potential shown in Figure 4.4. The equations for the interatomic potential in Chapter 2 do not contain temperature; they calculate the potential energy at zero kelvin. Kinetic energy must be added to the potential energy if the material is at a temperature above zero kelvin. Kinetic energy results in the vibration of the atoms about their equilibrium atomic positions. As the temperature increases, the kinetic energy of the atoms in the material increases, and the amplitude of the lattice vibrations increase. At high temperatures, the average kinetic energy of each atom in the crystal is approximately $1.5kT$. The average kinetic energy (KE_1) of a pair of atoms at temperature T_1 is then equal to $3kT_1$. If the atoms vibrate toward each other, they can do so until all of the kinetic energy is converted to potential energy. This occurs at the interatomic distance $r_{<1}$, as shown in Figure 4.4. If the atoms move apart, they can do so until all of the kinetic energy is converted to potential energy at $r_{>1}$. At temperature T_1 the atoms vibrate between the equal energy positions given by the intersection of KE_1 with the potential energy curve at $r_{<1}$ and $r_{>1}$. The average interatomic separation at temperature T_1 is r_1, and r_1 is given by Equation 4.8:

$$r_1 = \frac{1}{2}(r_{<1} + r_{>1})$$

4.8

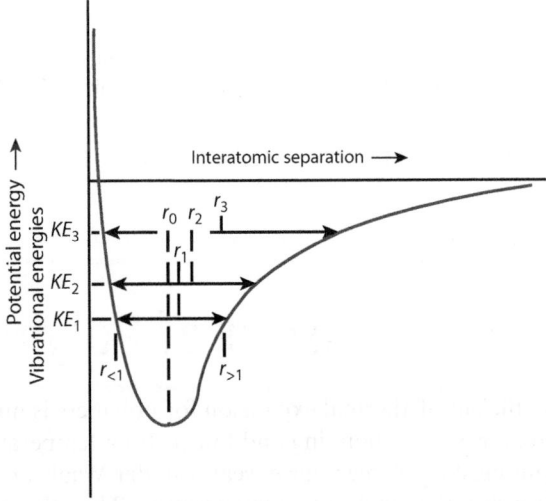

Figure 4.4 Interatomic potential energy as a function of interatomic separation. As the material is heated, the kinetic energy increases to KE_1, KE_2, and KE_3, and the average interatomic separation also increases to r_1, r_2, and r_3.

If the temperature is further increased to T_2, the increase in thermal energy results in an average value of $r_2 > r_1 > r_0$ as shown in Figure 4.4. As the interatomic separation increases between a pair of atoms, the dimensions of the solid increase, and the material expands. At the atomic level, the linear coefficient of thermal expansion is taken from Equation 4.6, as shown in Equation 4.9:

$$\alpha_L = \frac{1}{r_i}\left(\frac{dr_i}{dT}\right)$$

    4.9

where r_i is the interatomic separation at the temperature T_i.

As shown by Figure 4.4, the thermal expansion coefficient is directly dependent upon the interatomic bonding, and this explains many of the characteristics of the thermal expansion coefficient. Stronger bonding results in smaller-amplitude lattice vibrations, which in turn results in less thermal expansion. Stronger bonding also results in a higher melting temperature, as discussed in Chapter 2. For example, tungsten melts at a temperature of 3683 K, and it has a thermal expansion coefficient of only $4.5 \times 10^{-6}\,°C^{-1}$, whereas aluminum melts at a temperature of 933 K, and it has a thermal expansion coefficient of $25 \times 10^{-6}\,°C^{-1}$. The thermal expansion coefficient is not constant with temperature, because as the temperature increases, the vibrating atoms or molecules experience a different portion of the interatomic potential. In general, the thermal expansion coefficient increases with increasing temperature, because as the material is heated the atoms are more separated and the bonding is weaker. If there is a change in crystal structure, and therefore a change in bonding, there is an abrupt change in the coefficient of thermal expansion. For example, when pure iron transforms upon heating from the BCC phase to the FCC phase at 912°C, the coefficient to thermal expansion changes from $17 \times 10^{-6}\,°C^{-1}$ to $23 \times 10^{-6}\,°C^{-1}$.

Example Problem 4.3

In the high-temperature section of a high-performance gas turbine, the temperature can reach 1100°C. The turbine blades are usually nickel-based alloys. In the design of the turbine, allowances must be made for the expansion of the turbine blades. Assume that the turbine blades are 7 cm long, and the turbine can be as cold as 0°C. Calculate the expansion of the turbine blades if they were to reach 1100°C, using the thermal expansion coefficient of pure nickel.

Solution

The total length change (dL) for this temperature range is given by Equation 4.5, the total temperature change (dT) is 1100°C, and the coefficient of thermal expansion for nickel is given in Table 4.2 as $13 \times 10^{-6}\,°C^{-1}$.

$$dL = L_0 \alpha_L dT = 7\ \text{cm}\left(13 \times 10^{-6}\,\frac{1}{°C}\right)1100°C = 100 \times 10^{-3}\ \text{cm} = 0.1\ \text{cm}$$

4.3.1   Thermal Expansion of Polymers

Note in Table 4.2 that the coefficient of thermal expansion for polymers is much greater than for metals or for ceramics. As discussed above, weak bonding and low melting temperatures result in high thermal expansion coefficients. Thermoplastic polymers have weak van der Waals bonding holding the different long-chain molecules together, and they melt a low temperature. Thus thermoplastic polymers such as polyethylene have a high coefficient of thermal expansion. In polymers, if the long-chain molecules are oriented, such as in fibers, the coefficient of thermal expansion in the fiber direction is lower than in the radial direction, because the strong covalent bonds are oriented in the fiber direction. Thermosetting polymers, such as epoxy, have a lower coefficient of thermal expansion than thermoplastic polymers, because stronger covalent bonds cross-link the long-chain molecules.

If fillers or particles, such as glass, are added to the polymer, the coefficient of thermal expansion follows the rule of mixtures based upon volume fraction. For polymers that are reinforced with uniaxial continuous strong fibers (composite materials), the coefficient of thermal expansion in the direction of the fibers is controlled by the coefficient of thermal expansion of the strong fibers. The coefficient of thermal expansion perpendicular to the fiber direction follows the rule of mixtures based upon volume faction. We will cover the coefficient of thermal expansion of composite materials in more detail in Chapter 12, on composite materials.

4.3.2 Negative Coefficients of Thermal Expansion

In general, the separation of atoms increases in materials when they are heated. Why do some materials, such as graphite fibers and Kevlar™ fibers, have negative axial coefficients of thermal expansion? Kevlar is an oriented polymer fiber, as is oriented UHMWPE. High-modulus graphite fibers have an axial thermal expansion coefficient of -0.7×10^{-6} K^{-1} and a radial expansion coefficient of 10×10^{-6} K^{-1}. Kevlar polymer fibers have an axial thermal expansion coefficient of -6×10^{-6} K^{-1} and a radial expansion coefficient of 54×10^{-6} K^{-1}. The negative coefficient of expansion of these materials is related to their structures and bonding. Graphite has two-dimensional sheets of carbon atoms, as shown in Figure 2.16c. When graphite is heated, all of the interatomic separations shown in Figure 2.16c increase in agreement with the discussion above that relates the interatomic separation to the interatomic potential. In graphite, the separation between the graphite sheets increases more than the separation between the carbon atoms in the sheets, because the bonding between the graphite sheets is much weaker than the bonding between the atoms within the sheets. Thermal vibrations in materials not only make the atoms vibrate about the equilibrium interatomic separation, but the vibrations form waves, or phonons, that propagate through the material, as shown in Figure 4.1. These phonons result in the graphite sheets becoming wavy as the graphite temperature increases. As the graphite sheet becomes wavy, the straight-line distance from the beginning to the end of the sheet is lessened, even though the length of the sheet is greater if it is flattened out. The higher the temperature, the wavier the sheet becomes, and the shorter it becomes even though all of the interatomic separations are increasing. The negative coefficient of axial expansion in Kevlar filaments has a similar origin, because in Kevlar there are long-chain molecules that are oriented along the axis of the filaments, and the bonding between these long-chain molecules is weak-dipole bonding. The long-chain molecules oriented along the filament axis increase in length, but they become wavy as the temperature increases. The filament decreases in length, even though the long-chain molecules increase in length when straightened out.

4.4 THERMAL CONDUCTIVITY

If you take a bar of metal that is at room temperature and you put one end in a hot furnace, the end of the bar that is not in the furnace increases in temperature. This is due to thermal conductivity. Heat (Q) is conducted from the hot end of the bar to the cold end. We know from experience that the hotter the furnace relative to room temperature (dT) and the shorter the bar (dx), the hotter the bar when we

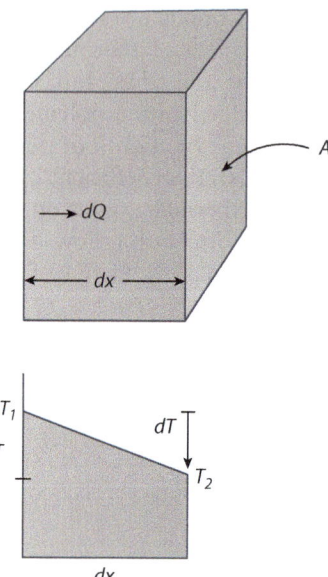

Figure 4.5 The top figure is a block of material of thickness dx that is at temperature T_1 on the left side, as indicated in the lower plot, and at a lower temperature of T_2 on the right side. This results in the transfer of an amount of heat dQ in the time dt across the area A.

touch it. Thus more heat must have been transferred to the end that we touch. These observations are schematically shown in Figure 4.5 and are stated in Equation 4.10 as

$$q = \frac{1}{A}\frac{dQ}{dt} = -k_T\frac{dT}{dx}$$

4.10

where q is the heat flux or the increment of heat (dQ) that crosses a unit area (A) per unit time (dt), k_T is the **thermal conductivity** measured in units of watts/meter $\cdot$ kelvin ($Wm^{-1} \cdot K^{-1}$), and dT/dx is the temperature gradient. A watt is a joule per second. The greater the temperature gradient, the higher is the heat flux. The negative sign is necessary because dT is the low temperature minus the high temperature, and this is always negative. The heat flows from the high temperature to the low temperature, and the heat flux (q) is always positive. Values of the thermal conductivity are given in Table 4.3 for some selected materials. Equations similar to 4.10 are also found in the movement of atoms in solids (Fick's first law) and in the conduction of electrons in an electric field (Ohm's law).

4.4.1   Thermal Conductivity in Ceramics

As shown in Table 4.3, diamond has the highest thermal conductivity (2000 $Wm^{-1} \cdot K^{-1}$) of any bulk material at room temperature, and ceramics, such as silicon carbide and silicon nitride, have a relatively high thermal conductivity. Diamond and graphite have high thermal conductivity because of their strong bonding. In all materials, heat is conducted by atoms that are vibrating at a high amplitude in the hot location, exciting neighboring atoms at a lower temperature and a smaller vibration amplitude. This produces higher-amplitude vibrations of the lower-temperature atoms, increasing their temperature, and the heat is conducted. This mechanism results in materials with strong bonding having high thermal conductivity, because the strong coupling between atoms transfers the lattice vibrations from one atom to another. Graphene is reported to have a thermal conductivity of approximately 5000 $W \cdot m^{-1} \cdot K^{-1}$,

Table 4.3 The Thermal Conductivity of Some Selected Materials at Room Temperature

Material	Thermal Conductivity (k_T) (W · m^{-1} · K^{-1})	Material	Thermal Conductivity (k_T) (W · m^{-1} · K^{-1})
Pure Metals		**Ceramics**	
Ag	430	Al_2O_3	16–40
Al	238	Carbon (dimond)	2000
Cu	400	Carobon (graphite)	335
Fe	79	Fireclay	0.26
Mg	100	Silicon carbide	up to 270
Ni	90	AIN	up to 270
Pb	35	Si_3N_4	up to 150
Ti	22	Soda-lime glass	0.96−1.7
W	171	Vitreous silica	1.4
Zn	117	Vycor™ glass	12.5
Zr	23		
Alloys		**Polymers**	
1020 steel	51.9	6,6-nylon	0.25
3003 aluminum alloy	280	Polyethylene	0.33
316 stainless steel	15.9	Polyimide	0.21
Cementite	50	Polystyrene	0.13
Cu-30% Ni	50	Polystyrene foam	0.029
Ferrite	75	Teflon	0.25
Gray iron	79.5	PMMA	0.18
Brass (70% Cu-30% Zn)	120		
		Semiconductors[1]	
		Si	124
		Ge	64

Based on data from Askeland, D.R., Fulay, P.P., and Wright, W.J., The Science and Engineering of Materials, 6th ed. Cengage Learning. Stamford, CT. (2011), p. 840.

1   CRC Handbook of Chemistry and Physics, ed.D.R. Lide, 90th ed. 2009–2010 pg. 12–80.

which is much higher than the value for diamond. Graphene is a sheet of graphite one atomic layer thick. Strong bonding results in effective conduction of the lattice vibrations, or phonons, through the material.

4.4.2   Thermal Conductivity in Metals

Note that the thermal conductivity of metals in Table 4.3 is quite high relative to that of other materials, with silver having the highest conductivity of the metals. However, silver does not have strong bonding, as indicated by its relatively low melting temperature of 962°C. Metals have free electrons that transfer energy when they are excited. The kinetic energy of free electrons is increased by interactions with the thermal vibrations of atoms. The high-kinetic-energy free electrons transfer heat by moving through the lattice and

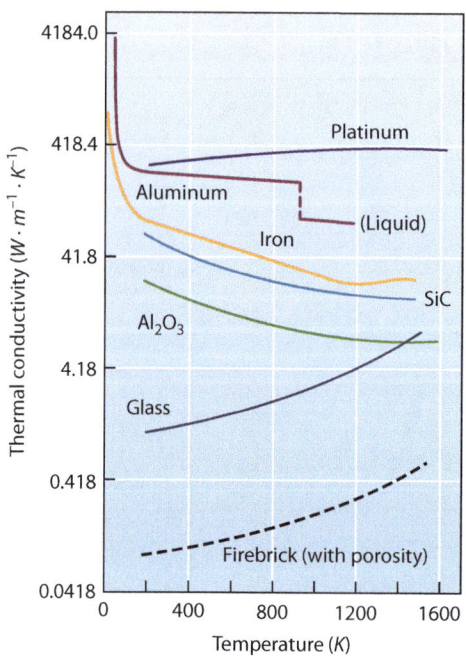

Figure 4.6   The thermal conductivity of some metals and ceramics as a function of temperature. (*Based on Askeland, D.R., Fulay, P.P., and Wright, W.J., The Science and Engineering of Materials, 6th ed. Cengage Learning, Stamford, CT. (2011), p. 841. 1. CRC Handbook of Chemistry and Physics, ed.D.R. Lide, 90th ed. 2009–2010 pg. 12–80.*)

then colliding with crystal atoms. In the collision the electrons transfer their energy to the atoms. Metals with a high thermal conductivity, such as silver and copper, also have a high electrical conductivity. At temperatures below 100 K, the thermal conductivity of metals is very high and increases rapidly with decreasing temperature, as shown in Figure 4.6. At low temperatures, the conductivity of electrons is very high, because electrons can move through the crystal without being scattered. However, as the temperature increases, the electrons are increasingly scattered by atom vibrations, and the free electrons travel over shorter distances. Once an electron is scattered, it no longer transfers energy in its original direction of propagation. Free electron scattering limits the free-electron heat-transfer mechanism at high temperatures. Tungsten is an example of a metal with strong bonding, as indicated by a melting temperature of 3683 K, and a relatively high thermal conductivity at room temperature of 171 W · m^{-1} · K^{-1}.

At higher temperatures, such as room temperature, there is no general trend for the thermal conductivity of metals as a function of temperature, because there are several different mechanisms acting to both increase and decrease the thermal conductivity as temperature increases. For example, the thermal conductivity of platinum increases with temperature, and that of iron decreases with temperature, as shown in Figure 4.6.

4.4.3   Thermal Conductivity in Semiconductors

In semiconductors, such and silicon and germanium, the thermal conductivity at low temperatures is primarily due to the phonons conducting heat through the crystal lattice; therefore, at low temperature semiconductors behave like ceramics. However, in semiconductors at higher temperatures, covalently bonded electrons are excited to become free conduction electrons, and then the thermal conductivity is also due to conduction electrons in addition to phonons. In ceramics that are not semiconductors, the valence electrons are localized in very strong covalent and ionic bonds, and these electrons are not excited to become free electrons, even at high temperatures. Covalently bonded electrons and electrons on ions cannot have their kinetic energy increased by collisions with vibrating atoms, because the electron energies in these types of bonding are determined by quantum numbers. The quantum numbers of electrons in covalent bonds and on ions do not change unless the electrons are excited by a large amount of energy, and thermal vibrations only provide a small amount of energy.

4.4.4   Thermal Conductivity in Polymers

In Table 4.3, the thermal conductivity of polymers is less than that of metals or ceramics. Unlike metals, polymers have no free conduction electrons. The valence electrons in polymers are localized in covalent bonds, and they are so strongly bound that they cannot have their kinetic energy increased by collisions with vibrating atoms. In polymers the thermal conductivity is particularly low, because the weak van der Waals bonding between long-chain molecules makes it difficult to transfer thermal energy through atomic or molecular vibrations from one long-chain molecule to another.

The thermal conductivity in polymers depends upon the polymer configuration. The thermal conductivity along long-chain molecules, such as along oriented fibers, is higher than between long-chain molecules. The thermal conductivity is higher in polymers with high crystallinity and high molecular mass. For example, HDPE has thermal conductivities from 0.33 to 0.51 W $\cdot$ m^{-1} $\cdot$ K^{-1}, and LDPE has thermal conductivities from 0.32 to 0.40 W $\cdot$ m^{-1} $\cdot$ K^{-1}. These are some of the highest thermal conductivities of polymers in Table 4.3. HDPE is of a high molecular mass and crystallinity. Thermal conductivity is lower in polymers that are amorphous, such as PMMA with 0.18 W $\cdot$ m^{-1} $\cdot$ K^{-1}, and polymers that have extensive side branching.

4.4.5   Thermal Conductivity and Defects

Defects affect the thermal conductivity, because defects scatter the phonons that provide the thermal conductivity. Single crystals have the highest thermal conductivity, because the phonons propagate along the crystal planes over large distances. Porous materials have lower thermal conductivity than 100% dense materials, because heat is not effectively transferred across the pores. Materials with pores have less than the theoretical (100%) density, because the air in the pores is less dense than the material. Firebrick is a highly porous ceramic insulator used to line furnaces, and its thermal conductivity is very low, as shown in Figure 4.6. Polystyrene has a thermal conductivity of 0.13 W $\cdot$ m^{-1} $\cdot$ K^{-1}, whereas polystyrene foam has a thermal conductivity of only 0.029 W $\cdot$ m^{-1} $\cdot$ K^{-1}. The low thermal conductivity makes foamed polymers excellent thermal insulators.

Glasses have a low thermal conductivity relative to that of crystalline material, because in glasses the phonons cannot propagate large distances before they are scattered by the disordered atoms. For example, single crystal quartz (SiO_2) has a thermal conductivity of 10.7 W $\cdot$ m^{-1} $\cdot$ K^{-1} along the [100] direction at 323 K, whereas vitreous (amorphous) fused silica (SiO_2) has a thermal conductivity at room temperature, according to Table 4.3, of 1.4 W $\cdot$ m^{-1} $\cdot$ K^{-1}.

Defects in a metal decrease the thermal conductivity, because defects scatter the phonons and the free electrons that transfer energy through the metal. Note in Table 4.3 that the thermal conductivity of pure iron is 79 W $\cdot$ m^{-1} $\cdot$ K^{-1}, whereas that of iron plus 0.20 weight percent carbon (1020 steel) is 51.9 W $\cdot$ m^{-1} $\cdot$ K^{-1}, and 316 stainless steel, which is iron with substitutional atoms of nickel and chromium, has a thermal conductivity of 15.9 W · m⁻¹ · K⁻¹. The low thermal conductivity of stainless steel make it the metal of choice for producing devices to transfer fluids, such as liquid hydrogen or liquid helium, that have a low heat of vaporization and a low boiling temperature.

Example Problem 4.4

A 316 stainless-steel pan that is 1 mm thick is placed on a 500°C stove and contains boiling water at 100°C. (a) Calculate the heat flux into the water. (b) If the pan is made of copper rather than 316 stainless steel that is 1 mm thick, calculate the temperature of the stove that would result in the same heat flux as in part (a).

Solution

a) Use Equation 4.10 to determine the heat flux. It is given that $dT = -400\,\text{K}, dx = 0.001\,\text{m}$, and $k_T = 15.9\,\text{W/m} \cdot \text{K}$. Inserting these values into Equation 4.10 yields

$$q = -k_T \frac{dT}{dx} = -15.9 \frac{\text{W}}{\text{m} \cdot \text{K}} \left(\frac{373\,\text{K} - 773\,\text{K}}{1 \times 10^{-3}\,\text{m}} \right) = 6.36 \times 10^6 \frac{\text{W}}{\text{m}^2}$$

The units of k_T are $\text{W} \cdot \text{m}^{-1} \cdot \text{K}^{-1}$ indicating that degrees kelvin should be used. However, since temperature enters the equation as dT, the increment in degrees kelvin is the same as the increment in degrees celsius. If the actual temperature enters an equation, then the temperature in degrees kelvin must be utilized.

b) We want to achieve the same heat flux (q) as in part (a), and the unknown is now the stove temperature (T_H) and dT. The water in the pan is still at 100°C, and the thermal conductivity of copper is 400 $\text{W} \cdot \text{m}^{-1} \cdot \text{K}^{-1}$.

$$q = 6.36 \times 10^6 \frac{\text{W}}{\text{m}^2} = -k_T \frac{dT}{dx} = -400 \frac{\text{W}}{\text{m} \cdot \text{K}} \left(\frac{373\,\text{K} - T_H}{1 \times 10^{-3}\,\text{m}} \right)$$

$$373\,\text{K} - T_H = -\left(\frac{1 \times 10^{-3}\,\text{m}}{400\,\text{W/m} \cdot \text{K}} \right) 6.36 \times 10^6 \frac{\text{W}}{\text{m}^2} = -15.9\,\text{K}$$

$$T_H = 373\,\text{K} + 15.9\,\text{K} = 389\,\text{K} = 116\,°\text{C}$$

Because of the difference in thermal conductivity, the same heat flux can be delivered with a copper pan with a temperature increment of 16 K, as is delivered by a stainless-steel pan with a temperature difference of 400 K. Stainless steel is most effectively used in applications that require a minimum amount of heat transfer, such as in transport tubes for hot or cold liquids. Copper is most effectively used in applications that require a large amount of heat transfer, such as in cooking utensils and heat exchangers. Stainless steel is used in cookware because it is stronger than copper and can be made into thinner and lighter products requiring less material, and it is corrosion resistant.

Section 2 An Introduction to Thermodynamics

4.5   THE MIXING OF ATOMS

The previous section was about the conductivity of heat; this section and following sections could be titled the conductivity of atoms and molecules. Why is it that some atoms or molecules mix and others do not? For example, alcohol and water mix easily, but oil and water do not. We will find the same is true

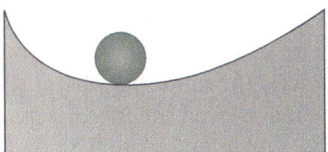

Figure 4.7 A ball on a curved surface always moves toward the lowest energy position.

of solids. At 1000°C nickel and copper are solids whose atoms can mix completely; however, at 1000°C only a few percent of nickel and silver can mix into each other.

When a thermodynamic system changes, it should always be to decrease its energy toward the minimum possible, assuming that we use the correct energy. For example, a ball on a curved surface always goes to the minimum internal energy position, as shown in Figure 4.7. The internal energy is the sum of the kinetic energy of the ball and the potential energy of the ball due to gravity. The ball should not roll uphill on its own. The same principle should apply if atoms mix; the appropriate energy should decrease. In this section we determine the appropriate energy that we should use to understand why atoms in a solid or liquid either mix or do not mix. We will find that it is not sufficient to know only the kinetic and potential energy of the atoms. Another contribution to the energy, called entropy, is necessary to understand why some atoms or molecules mix and others do not mix.

4.5.1 Entropy

To observe the effect of entropy, we will first study the mixing of two ideal gases. Ideal gases are chosen because we can do the calculations for them, and they demonstrate why the internal energy is not sufficient to explain the mixing of atoms. We start with two sealed containers with inert gases, such as helium (He) in the left container and argon (Ar) in the right container, as shown in Figure 4.8a. If the atoms in the container are at a constant temperature (T), then the internal energy (E) is equal to the kinetic energy of the atoms in the container. We assume that the atoms in a gas have no potential energy because of the large distance of separation of the atoms. The kinetic energy of 1 mole of gas atoms in each container is

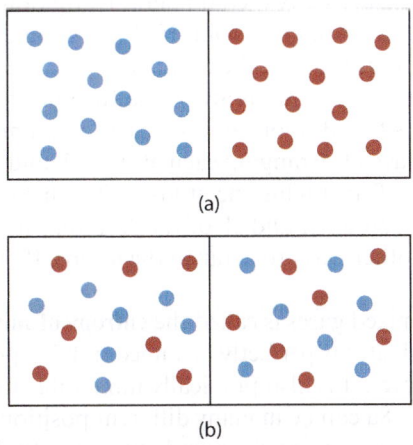

(a)

(b)

Figure 4.8 (a) Inert gases, such as He and Ar, in separate insulated containers, each at the same temperature. (b) Two separate, insulated containers with a mixture of He and Ar atoms, at the same temperature.

$1.5RT$. In these same two containers with the same number of He and Ar atoms, the atoms are mixed so that each container now is a 50% mixture of He and Ar, as shown in Figure 4.8b. The internal energy of the atoms in Figures 4.8a and 4.8b is exactly the same, but obviously there is an energy difference between these two arrangements of atoms. If you are given the job of separating the mixture of atoms shown in Figure 4.8b to obtain the configuration shown in Figure 4.8a, so that He is in one container and Ar is in the other container, a significant amount of energy must be expended to do this. The container with the mixed gases must be at lower energy, and it is when we use the correct energy that includes **entropy**.

We also know that if we drill a hole in the wall between the two containers in Figure 4.8a, the He and Ar atoms will mix, and given sufficient time there would be a mixture of atoms in the two containers similar to that in Figure 4.8b. But if we drill a hole in the partition between the two separate containers in Figure 4.8b where the He and Ar atoms are mixed, it is not very probable that the atoms will separate into pure He in one container and pure Ar in the other container as in Figure 4.8a. Atoms will move through the hole between the two containers, but the most likely state is for the atoms to remain mixed, as they are in Figure 4.8b. Referring to Figure 4.7, the system with the mixed atoms must be at a lower energy, even though the internal energy with the atoms mixed or separated is the same. What is it that makes the mixed arrangement in Figure 4.8b lower in energy than Figure 4.8a?

The atoms in Figure 4.8a are ordered in the sense that He is only found in one container, and those in Figure 4.8b are disordered because He can be found in either container. Boltzmann was the first to propose that **entropy** is related to disorder in materials. The concept of entropy was developed in association with the **second law of thermodynamics,** which states it is impossible for a heat engine to be 100% efficient. The **first law of thermodynamics** is that energy is conserved; it cannot be created or destroyed. The **third law of thermodynamics** is that a perfect crystal has 0 entropy at 0 kelvin, and no material can be at 0 kelvin.

Planck proposed Equation 4.11 to relate entropy (S) to the arrangement of atoms or molecules:

$$S = k\ln\omega \qquad\qquad\qquad \textbf{4.11}$$

where ω is the number of different distinguishable ways to arrange the atoms or molecules in the material and k is Boltzmann's constant (1.38×10^{-23} J/atom $\cdot$ K). In the example of the gases in Figure 4.8a, the number of ways of arranging the atoms is 1; all of the He atoms must go in the container on the left, and all of the Ar atoms must go in the container on the right. Inserting 1 into Equation 4.11 results in an entropy of 0, because the natural log of 1 is 0.

To calculate the entropy when the atoms are mixed, assume that we have a machine that can place individual He or Ar atoms into one of the two containers. To randomize the arrangement process, we decide the container for the atoms by flipping a coin. If the coin lands heads up, the atom goes in the container on the left, and if the coin is tails up, the atom goes into the container on the right. The most likely outcome from this process is that the atoms are mixed together, as shown in Figure 4.8b. When we place the atoms into the containers based upon coin tosses, each atom has an equal chance of being in either container. The number of ways of arranging each atom is 2, and since there are N total atoms, the number of different possible ways of arranging the atoms is 2^N. One of the possible arrangements is to have all of the He atoms in the left container and all of the Ar atoms in the right container. However, this is not very likely. There are many other ways to arrange the atoms. If there is a mole of each atom type, this gives a large entropy indeed.

The entropy calculated for the mixed gases is called the **entropy of mixing**. The entropy of pure helium is 0 only if the helium is at 0 K with atoms perfectly arranged in FCC positions. There is only one way to arrange these helium atoms. However, it is also physically impossible for a material to be cooled to 0 K. In helium gas the atoms in Figure 4.8a can be in many different positions or arrangements, and the many possible arrangements result in a large entropy for pure helium gas. In this chapter we will only calculate the entropy contribution due to mixing of atoms.

Energy and entropy are both **state variables**. A state variable is one that depends only upon the condition of the thermodynamic system at the time it is observed. For example, the internal energy of the thermodynamic system in Figure 4.8a is $1.5RT$, and the entropy of mixing is zero. The internal energy and entropy of mixing do not depend upon how this state was achieved.

The study of entropy brings meaning to the quote attributed to Arnold Sommerfeld, one of the great scientists of the twentieth century: "Thermodynamics is a funny subject. The first time you go through it, you don't understand it at all. The second time you go through it, you think you understand it except for one or two points. The third time you go through it, you know you don't understand it, but by that time you are so used to the subject, it doesn't bother you any more." The classical interpretation of entropy as the reason that 100% efficiency cannot be attained in a heat engine is difficult to visualize. However, the statistical interpretation of entropy relating to the arrangements of atoms is something that can be visualized. We can determine the separation or the mixing of atoms with experiments, and this is something that can be visualized. Hopefully the statistical study of entropy will help in the understanding of entropy and how it applies to many thermodynamic systems.

Example Problem 4.5

What is the entropy of mixing of a iron crystal with 10^{23} atom sites that has a single chromium substitutional atom that can be located at any one of the atom sites?

Solution

The chromium atom can occupy any one of the 10^{23} atom sites, and the chromium atom at each different site in the crystal provides a different distinguishable arrangement of atoms. Scanning tunneling microscopes can image individual atoms on a surface, and transmission electron microscopes can image individual atoms in a thin film of material; therefore, we can experimentally distinguish arrangements of atoms. There are 10^{23} distinguishable ways to arrange the chromium atom. We assume that the iron atoms are indistinguishable from one another, and therefore if two iron atoms are interchanged, this is the same arrangement, because no experiment can tell the difference between two iron atoms.

From Equation 4.11,

$$S = k \ln \omega = 1.38 \times 10^{-23} \frac{J}{K \cdot atom} (\ln 10^{23}) = 73 \times 10^{-23} \frac{J}{K \cdot atom}$$

This is the entropy of mixing for only one substitutional atom in 10^{23} atoms. If the number of substitutional atoms becomes a significant fraction of the total number of atoms, then the magnitude of the entropy due to mixing is significantly increased.

In a perfect crystal made of one atom type, there is only one way to arrange the atoms ($\omega = 1$). Each atom is placed in a crystal atom site, and it is assumed one atom cannot be distinguished from another. The entropy of mixing of this perfect crystal is 0, according to Equation 4.11. When atoms mix (for example, when copper and nickel atoms mix in solids) the entropy is increased because there are many ways to arrange the different atoms in the alloy. If we generalize the problem to the mixing of B-type atoms into an A-type crystal with N total atoms, where the A-type and B-type atoms mix ideally, the entropy change (ΔS_m) relative to two pure crystals is given in Equation 4.12:

$$\Delta S_m = -kN(C_A \ln C_A + C_B \ln C_B) \qquad \textbf{4.12}$$

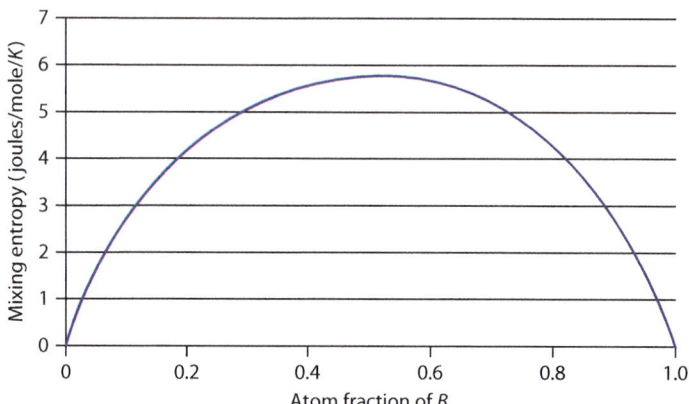

Figure 4.9   The entropy of mixing (ΔS_m) for the atom fraction C_B of B-type atoms into an A-type crystal with a total of 1 mole of atoms.

where C_B is the atom fraction of B-type atoms (N_B/N) and C_A is the atom fraction of host A-type atoms (N_A/N). Ideal mixing occurs when there is no reaction between the A-type and B-type atoms, and they mix as though they are an ideal gas. Equation 4.12 is derived from Equation 4.11 in Appendix 4C.

In Equation 4.12, k can be replaced with the universal gas constant (R); however, N then becomes the total number of moles of atoms. The universal gas constant is Boltzmann's constant times Avogadro's number. If working a problem on a per atom basis, use k; and if working a problem on a per mole basis, use R.

Figure 4.9 is a plot of the entropy change (ΔS_m) due to mixing atom fractions of B-type atoms (C_B) into a crystal made of A-type atoms with a total of one mole of atoms. Note that the entropy of mixing is 0 when C_B is equal to 0 or 1, and it is a maximum when C_B is equal to 0.5. Now we have to see how entropy is related to energy and how the mixing entropy can decrease energy.

Example Problem 4.6

Calculate the entropy of mixing for an alloy that has a total of 1 mole of atoms and is 40 atomic percent copper and 60 atomic percent nickel and the atoms are mixed randomly.

Solution

Writing Equation 4.12 on a molar basis with the universal gas constant and N equal to the number of moles results in

$$\Delta S_m = -RN(C_{Cu} \ln C_{Cu} + C_{Ni} \ln C_{Ni})$$

Inserting all of the known values results in

$$\Delta S_m = -8.31 \frac{J}{mole \cdot K} (1 \text{ mole})[0.4 \ln 0.4 + 0.6 \ln 0.6] = -8.31 \frac{J}{K}[0.4(-0.92) + 0.6(-0.51)]$$

$$\Delta S_m = -8.31 \frac{J}{K}[-0.37 - 0.31] = -8.31 \frac{J}{K}[-0.68] = 5.62 \frac{J}{K}$$

The entropy of mixing is always a positive value.

4.5.2   The Gibbs Free Energy

In his study of the possible energy output of heat engines, J. W. Gibbs related entropy (S) to the **Gibbs free energy** (G), internal energy (E), pressure (P), volume (V), absolute temperature (T), and enthalpy (H) through Equation 4.13.

$$G = E + PV - TS = H - TS \qquad\qquad \textbf{4.13}$$

The Gibbs free energy is the energy that is available or free to do work under conditions of constant pressure and temperature; this is proven in Appendix 5A after the introduction of an additional equation for entropy calculations. When a system at constant pressure and temperature is at **equilibrium**, the Gibbs free energy is a minimum, similar to Figure 4.7, and other variables, such as composition, do not change with time. The internal energy (E) for a material is the sum of the kinetic energy (KE) and potential energy (PE) for all of the atoms. The kinetic energy is the energy of motion of all of the atoms, including the translation of atoms in a gas or liquid and the vibration of atoms in liquids and solids. The potential energy includes the energy of bonding of all of the atoms, the energy of all defects, and the energy associated with the interaction of the material with external fields such as gravitational, electrical, and magnetic. We normally assume that the mechanical energy is zero. The mechanical energy of a material is the energy due to the macroscopic motion of the material, such as translation and rotation. We assume that center of gravity of the material is at rest, there is no rotation, and we assume that the gravitational potential energy of the body is zero. The **enthalpy** (H) is the sum of the internal energy (E) and the product of the pressure times the volume (PV). All of the variables in Equation 4.13 are state variables, these variables define the state of the system. Heat Q and work W are not state variables. The state of a system is changed when heat is transferred to a system or if work is done on it, but the amount of heat or work does not define the state of the system.

If a material at constant temperature and pressure is changed from a state with an initial Gibbs free energy of G_0 to a state with G, such as when atoms mix, then the change in the Gibbs free energy (ΔG) is given by Equation 4.14:

$$\Delta G = G - G_0 = \Delta E + P\Delta V - T\Delta S = \Delta H - T\Delta S \qquad\qquad \textbf{4.14}$$

If ΔG is equal to 0, the change occurs under equilibrium conditions and is reversible. If ΔG is negative, the change is spontaneous and not reversible. If a change is spontaneous, it should happen; however, the change may take time, and it is possible for the time to be very large. Diamond provides an example of a change that should occur, but fortunately takes a long time. Natural diamonds were formed under high pressure and temperature deep within the earth, where diamond was the form of carbon with the lowest Gibbs free energy. On earth's surface at one atmosphere of pressure and room temperature, graphite has a lower Gibbs free energy than diamond does. Fortunately, diamond can persist at room temperature and 1 atmosphere of pressure for a long time. Changes that produce a positive ΔG can occur, but this produces a metastable system that eventually changes to a system with a lower Gibbs free energy, although systems can persist in metastable states for a long time.

The entropy change resulting from the mixing of atoms at constant temperature and pressure makes a negative contribution to the change in the Gibbs free energy. Mixing increases the entropy, and the entropy enters the change in the Gibbs free energy as $-T\Delta S$. The greater the entropy, the lower the Gibbs free energy, and the lowest value of the Gibbs free energy is the equilibrium value. The term in the Gibbs free energy change that drives mixing at constant temperature and pressure is $-T\Delta S$.

Why do some liquids and solids not mix? If the change in enthalpy (ΔH) due to mixing is large and positive, then this term can dominate the $-T\Delta S$ term. If the total change in the Gibbs free energy (ΔG) due to mixing is positive, then the atoms or molecules do not mix. Copper and nickel mix readily at 1000°C because the atoms are chemically similar, with similar lattice parameters of 0.352 nm and 0.362 nm, respectively; thus the change in mixing enthalpy (ΔH) is smaller than the term $-T\Delta S$. However, copper and silver have much different lattice parameters of 0.362 nm and 0.409 nm, respectively. When a large silver atom is mixed into the smaller nickel lattice, it results in a positive value of ΔH that is larger in magnitude than the term $-T\Delta S$, and they do not mix. The change in the Gibbs free energy per mole (ΔG) of an A-type crystal mixed with an atom fraction (C_B) of type-B atoms is calculated with Equation 4.15:

$$\Delta G = G - G_0 = C_B N \Delta H_{BA} - T\Delta S_m \qquad \textbf{4.15}$$

where G is the Gibbs free energy per mole of the A-type crystal mixed with the chemical composition (C_B) of type-B impurity atoms. G_0 is the Gibbs free energy of a pure A-type crystal at the temperature of the calculation, and G_0 can be taken as 0. N is the total number of moles of atoms, and $C_B N$ is the number of moles of B-type atoms. ΔH_{BA} is the enthalpy of mixing 1 mole of B-type atoms into an A-type crystal with an infinite number atoms, and ΔS_m is the entropy of mixing 1 mole of B-type atoms into an A-type crystal resulting in a chemical composition of C_B.

Example Problem 4.7

Calculate the entropy change of mixing 20 atom percent copper atoms into a silver crystal for a total of 1 mole of atoms. The silver-copper alloys form several technical alloys such as sterling silver (approximately 15 atom percent copper) and silver solder.

Solution

To determine the entropy of mixing the copper into the silver to obtain a total of 1 mole of atoms, the chemical composition values of 20 atom percent copper and 80 atom percent silver are inserted into Equation 4.12. Since the problem is on a per-mole basis, R replaces k in Equation 4.12:

$$\Delta S_m = -RN(C_{Ag} \ln C_{Ag} + C_{Cu} \ln C_{Cu})$$

$$\Delta S_m = -8.31\frac{J}{\text{mole} \cdot K}(1 \text{ mole})[0.8 \ln 0.8 + 0.2 \ln 0.2] = -8.31\frac{J}{K}[0.8(-0.223) + 0.2(-1.61)]$$

$$\Delta S_m = -8.31\frac{J}{K}[-0.179 - 0.322] = -8.31\frac{J}{K}[-0.500] = 4.16\frac{J}{K}$$

Example Problem 4.8

(a) If the mixing of 20 atom percent copper into a silver crystal occurs at a temperature of 727°C, determine the change in the Gibbs free energy on a per-mole basis if the enthalpy of solution for copper into silver is 0.25 eV/atom, according to data in Hultgren et al. (b) Based upon the result, should it be possible to mix 20 atom percent copper into silver at 1000 K? Explain your answer.

Solution

a) From Example Problem 4.7, the entropy change per mole is 4.16 J/K-mole. The enthalpy of solution of copper substituted into silver is 0.25 eV per copper atom.

Converting 0.25 eV per copper atom to joules per mole of copper atoms results in

$$\frac{0.25 \text{ eV}}{\text{Cu atom}}\left(\frac{1.602 \times 10^{-19} \text{ J}}{1 \text{ eV}}\right)\frac{6.023 \times 10^{23} \text{ Cu atoms}}{\text{mole of Cu}} = \frac{2.41 \times 10^4 \text{ J}}{\text{mole of Cu}}$$

However, in a mole of atoms only 20 atom percent are copper. For 20 atom percent of a mole of copper atoms, the enthalpy change is

$$C_{Cu}N\Delta H_{CuAg} = 0.20 \frac{\text{mole of Cu}}{\text{mole}} (1 \text{ mole}) \frac{2.41 \times 10^4 \text{ J}}{\text{mole of Cu}} = 0.482 \times 10^4 \text{ J}$$

The change in the Gibbs free energy due to substituting 20 atom percent copper atoms into silver is then calculated with Equation 4.15. The temperature must be in kelvin.

$$727°C = 1000 \text{ K}$$

$$\Delta G = G - G_0 = C_{Cu}N\Delta H_{CuAg} - T\Delta S_m = 0.482 \times 10^4 \text{ J} - (1000 \text{ K})4.16 \frac{\text{J}}{\text{K}}$$

$$\Delta G = 4820 \text{ J} - 4160 \text{ J} = 660 \text{ J}$$

b) The Gibbs free energy required to mix 20 atom percent copper into silver is positive, and this amount of mixing should not happen. In Chapter 5 we will see in more detail that 20 atom percent copper does not mix into silver at 1000 K (727°C). A calculation for 10 atomic percent copper substituted into silver at 1000 K results in a negative Gibbs free energy change, as shown in Homework Problem 4.14. Ten atomic percent copper can be substituted into silver at 1000 K (727°C). In Chapter 5 we will see that these calculations are consistent with experimental results presented in phase diagrams.

Why is it that for mixing atoms at constant temperature and pressure, we have to use the Gibbs free energy to determine the energy changes? However, with a ball on a surface we can use the internal energy even though the ball is also at constant temperature and pressure? The reason is that with the ball on a surface, there is no change in entropy, because there is no change in the arrangement of the atoms in the rolling ball. Therefore, the term $-T\Delta S$ is 0 in Equation 4.14. Also, there is no change in volume; the term $P\Delta V$ is 0. From Equation 4.14, for the ball on a surface, the change in the Gibbs free energy is equal to the change in internal energy. However, when there is an entropy change at constant temperature and pressure, it is necessary to use the Gibbs free energy to determine the arrangement of the system that gives the minimum energy.

4.5.3 Equilibrium Point Defect Concentrations in Metals and Ceramics

Is there a limit to the mixing of atoms, or do atoms just mix or not mix with no in-between degree of mixing? Above we determined that 20 atom percent copper should not mix into a silver crystal at 1000 K, but in Homework Problem 4.14, we calculate that 10 atom percent copper should mix into a silver crystal at 1000 K. This indicates that there is limited mixing of copper into silver. In this example some mixing appears to decrease the Gibbs free energy, but too much mixing appears to increase the Gibbs free energy. Is there an amount of mixing that produces a minimum in the Gibbs free energy?

We will consider substituting B-type atoms into an A-type crystal, and for now we will assume that the mixing entropy is the only entropy change that results from this substitution. In Figure 4.10 the term $-T\Delta S_m$ is plotted as a function of the number of B-type atoms (N_B). ΔH_{BA} is the positive enthalpy of

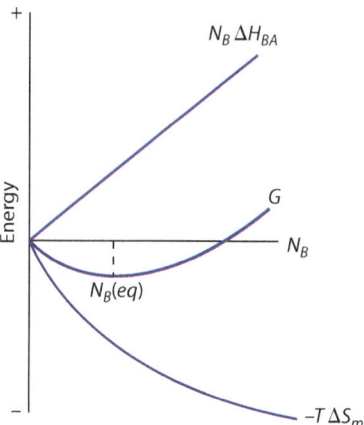

Figure 4.10   A hypothetical plot of the negative of temperature times mixing entropy ($-T\Delta S_m$), the number of B-type atoms times the enthalpy change of substituting a B-type atom into the A-type crystal ($N_B \Delta H_{BA}$), and the Gibbs free energy (G) as a function of the number of B-type atoms (N_B).

mixing a single B-type atom into an A-type crystal. The enthalpy of mixing is the enthalpy required to remove an A-type atom from the A-type material and move it to infinity, creating a vacancy in the A-type material, and to take a B-type atom that is at infinity and to place it in the vacancy. The enthalpy of mixing can be positive or negative. A negative enthalpy of mixing leads to compound formation, which we will discuss in Chapter 5. Substituting N_B type B atoms into an A-type crystal requires an input enthalpy of $N_B \Delta H_{BA}$; this term is also plotted in Figure 4.10. The number of substitutional B-type atoms in an A-type crystal that minimizes the Gibbs free energy (G) at constant temperature and pressure is found by taking the derivative of G with respect to N_B, setting the derivative equal to 0, and solving for the resulting value of N_B. The result is shown in Equation 4.16:

$$N_B(\text{eq}) = N \exp -\left(\frac{\Delta H_{BA}}{kT}\right)$$ **4.16**

where N_B (eq) is the equilibrium number of B-type atoms that minimizes the Gibbs free energy if there are N total atoms, as shown in Figure 4.10. Equation 4.16 is derived in Appendix 4D.

If the substitution of a B-type atom into the A-type crystal creates disorder in the form of atomic displacements in addition to the substitution, as shown in Figure 3.1, then this additional disorder results in a configurational entropy change (ΔS_{BA}). Figure 3.1 shows the displacement, or disorder, around several types of point defects, including substitutional atoms, interstitial atoms, and vacancies. For N_B substitutional B-type atoms a term $-TN_B\Delta S_{BA}$ is inserted into Equation 4.15 to account for the displacement disorder, as shown in Equation 4.17:

$$\Delta G_{BA} = \Delta H_{BA} - T\Delta S_{BA}$$ **4.17**

where ΔG_{BA} is the Gibbs free energy change resulting from mixing a single B-type atom into an A-type crystal. Multiply ΔG_{BA} by the number of B-type atoms (N_B) and add the term $-T\Delta S_m$; then G in Equation 4.15 is replaced by G in Equation 4.18.

$$G = G_0 + N_B\Delta G_{BA} - T\Delta S_m$$ **4.18**

The total Gibbs free energy change given by Equation 4.18 must be negative for the B-type atoms to substitute into the A-type crystal. Taking into account configurational entropy due to displacement of atoms around defects results in the equilibrium number of B-type atoms that minimizes the Gibbs free energy [N_B(eq)], as shown in Equation 4.19:

$$\frac{N_B(\text{eq})}{N} = \exp - \left(\frac{\Delta G_{BA}}{kT}\right) \qquad \textbf{4.19}$$

Normally you will see the form of Equation 4.16, because it is very difficult to determine the configurational entropy change ΔS_{BA}; thus it is usually neglected. For consistency with concepts to follow, you should be aware of Equation 4.19.

Equations 4.16 and 4.19 show that the equilibrium number of foreign atoms in the crystal increases exponentially with temperature. This is why in steel intended for submarine use that must resist fracture down to low temperatures, the high-temperature processing is conducted in a vacuum. Otherwise the steel picks up foreign atoms from the environment. These equations also give the amount of sugar that you can dissolve in your coffee while you stay up late to study for your materials science exam. The hotter the coffee, the larger is the amount of dissolved sugar. If the number of foreign atoms in solution is less than the number given by Equation 4.16, then all of the foreign atoms are in solution. If there are more foreign atoms present than given by Equation 4.16, the extra atoms in solution would increase the Gibbs free energy, as shown in Figure 4.10; thus they are rejected from the solution. This is why you find the sugar in the bottom of your coffee cup after you put three spoons of sugar into your coffee and let it cool.

Foreign atoms must be added to a material, but vacancies form spontaneously if they lower the Gibbs free energy of the crystal. Thus vacancies are always present in crystals at ambient temperatures. We discuss vacancies now because vacancies are critical to the mixing of atoms of similar size in solids. The presence of an atom fraction of vacancies of N_v/N adds disorder to a crystal, just as adding foreign atoms to a crystal adds disorder. The entropy of a crystal with an atom fraction of N_v/N vacancies is calculated with Equation 4.12 by replacing C_B with C_v, which is equal to N_v/N. The equilibrium atom fraction of vacant sites (N_v/N) is given by Equation 4.19 if the Gibbs free energy to create a vacancy is given by ΔG_v. The fraction of occupied interstitial sites that minimize the Gibbs free energy is also expressed by equations similar to Equations 4.16 and 4.19; however, geometrical factors must be introduced, because the number of interstitial sites is not equal to the number of atom sites N.

If Equation 4.19 is written for vacancies, then the atom fraction of vacant sites is also the probability (P_v) that a particular site is vacant. Equation 4.19 is similar in appearance to the **Maxwell-Boltzmann probability distribution** (MBPD) that applies to classical particles, such as atoms and molecules. The MBPD gives the probability [$P(G + \Delta G)$] that the increment in the Gibbs free energy (ΔG) is available at a site in excess of the average Gibbs free energy (G) at constant temperature and pressure, as shown in Equation 4.20.

$$P(G + \Delta G) = \exp - \left(\frac{\Delta G}{kT}\right) \qquad \textbf{4.20}$$

The increment in Gibbs free energy is provided by thermal vibrations. There is an average Gibbs free energy per atom in a material; however, every atom does not vibrate with exactly the same energy. Some atoms have more thermal energy than others, and the extra thermal energy of some atoms above the average provides the local energy for the increment in Gibbs free energy (ΔG). Comparing Equation 4.20 to Equation 4.19, which is written for vacancies, the conclusion is that the probability of the presence of a vacancy is equal to the probability that the Gibbs free energy to form the vacancy is available in excess of the average Gibbs free energy. We will use the MBPD in Equation 4.20 when we consider the kinetics of changes in materials, such as the mixing of atoms in solids.

In Equations 4.16, 4.19, and 4.20, Boltzmann's constant (k) can be in units of Joules per atom · K or in units of eV per atom · K. The units of kT must be consistent with the units of ΔH or ΔG; the units for the terms in the exponent in these equations must cancel so that the entire exponential is dimensionless. Equation 4.16 can also be written as Equation 4.21:

$$N_B(\text{eq}) = N \exp - \left(\frac{\Delta H_{BA}}{RT} \right)$$ **4.21**

where R is the universal gas constant in joules per mole · K, and ΔH_{BA} is in units of joules per mole of B-type atoms. This discussion applies to any equation with an exponential term similar to that in Equation 4.16.

Example Problem 4.9

Calculate the atom fraction of vacancies in solid gold at the melting temperature of 1336 K, knowing that the vacancy-formation enthalpy in gold is 0.94 eV per vacancy.

Solution

Since ΔH_v is 0.94 eV per vacancy, it is possible to calculate the equilibrium atomic fraction of vacancies ($N_v(eq)/N$) at 1336 K by using Equation 4.16 and writing the equation for vacancies rather than for B-type atoms. Boltzmann's constant must be in units of eV/atom · K for the exponential term to be dimensionless.

$$\frac{N_v}{N} = \exp - \left(\frac{\Delta H_v}{kT} \right) = \exp - \left(\frac{0.94 \text{ eV/vacancy}}{8.62 \times 10^{-5} \dfrac{\text{eV}}{\text{atom} \cdot \text{K}} \, 1336 \text{ K}} \right) = \exp(-8.16) = 2.9 \times 10^{-4} \frac{\text{vacancies}}{\text{atom}}$$

Another term that is used for the fraction of defects is parts per million. If, for example, there are one million atom sites, then there are 290 vacancies, or there are 290 parts per million vacancies.

    In the production of thin films for conductors in satellites, gold is often used as the conductor. If gold is vapor deposited onto a silicon wafer in an integrated circuit, it is rapidly cooled, or quenched, from a high temperature. The solid gold would contain more than 290 parts per million vacancies, because the initial temperature of the gold vapor is greater than the melting temperature, and then the gold atoms are rapidly cooled (quenched) when they stick to gold thin film. For electronic systems, this is a high vacancy concentration, and the vacancies increase the resistivity of the gold. How would you reduce the vacancy concentration? Equation 4.16 gives the solution. Hold the solid gold at a lower temperature, such as 600 K, and the equilibrium atomic fraction of vacancies adjusts to this temperature. There are a number of ways that vacancies can disappear from a crystal. Vacancies can migrate to a surface or to a grain boundary where they disappear, and they can be adsorbed along edge dislocation lines.

Section 3 Thermal Activation of Changes in Materials

4.6   RATES OF CHANGE IN MATERIALS

Above we showed that a system at constant temperature and pressure always changes toward the equilibrium state that minimizes the Gibbs free energy. Now we are going to address how fast the system changes. In many cases the change to the minimum Gibbs free energy state does not occur instantly;

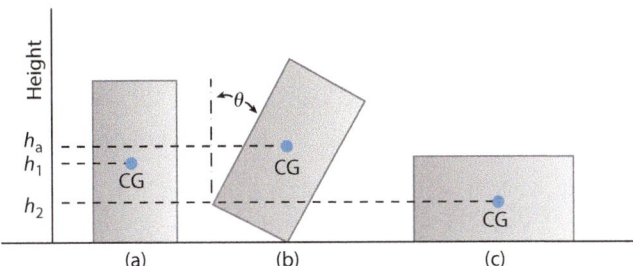

Figure 4.11 A block in various positions on a table top: (a) The block in a metastable state. (b) The block in an activated state, tilted by an angle θ, raising the center of gravity (CG) to a height h_a. (c) The block in a stable state, lowering the center of gravity to h_2.

rather, the system moves toward the equilibrium state at a finite rate. For example, iron reacts with oxygen at room temperature to form iron oxide with a decrease in the Gibbs free energy; fortunately this reaction does not occur instantly throughout the iron. The iron ions must move through the oxide layer that forms; then the oxygen reacts with iron at the oxide-environment interface. Thus our cars do not instantly turn to rust. In this section we develop equations for the rate of change in materials. The rate of change of many physical systems follows the same dependence upon temperature, including the viscosity of fluids, the high-temperature deformation of gas-turbine blades, the rate of strengthening an alloy by precipitation hardening, the transformation of amorphous glass to a crystal, and the mixing of carbon into steel to case-harden the steel.

From an energy viewpoint, iron in the presence of oxygen is like the block shown in Figure 4.11a: The block is in a metastable state. The block can remain in a metastable energy state for a long time, but it is possible for the block to change to a lower-energy state, with a lower center of gravity (CG) if the block falls over to the state shown in Figure 4.11c. The block in Figure 4.11a does not instantly fall over to the lower-energy state in Figure 4.11c; the block could stay in the state shown in Figure 4.11a for an infinite amount of time, if the system is not activated by vibrating the table top to tilt the block by the angle θ, as shown in Figure 4.11b. In Figure 4.12a the potential energy of the block is plotted as a function of the angle θ. The angle θ is the **reaction coordinate**; it is a measure of the progress of the change from state 1, the metastable state in Figure 4.11a, to the stable state 2 in Figure 4.11c. For the block to go from state 1 to state 2, the block must pass through the higher-energy activated state (a) shown in Figure 4.11b. To tip the block, it is necessary to raise the CG of the block by an increment $\Delta h = h_a - h_1$.

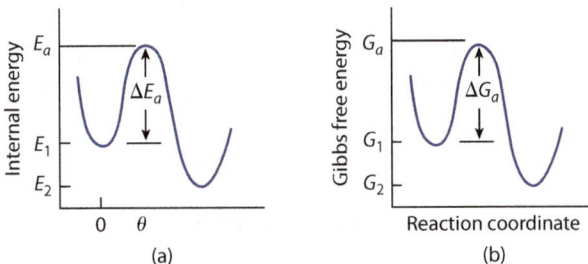

Figure 4.12 (a) The internal energy change for the block in Figure 4.11, moving from the metastable state with energy E_1 through the activated state of energy E_a to the stable state of energy E_2. (b) The Gibbs free energy change for a material at constant temperature and pressure, moving from a metastable state with the Gibbs free energy (G_1) through the activated state of energy G_a to the stable state of the Gibbs free energy G_2.

adatom

1   a   2

surface

Figure 4.13   The diffusion of an adatom on a surface. The adatom is initially in position 1. To move to position 2, the adatom must pass through the activated state (*a*).

This increases the internal energy of the block by $\Delta E_a = mg\,\Delta h$. This position is called the tipping point in many applications, including economics, but here it is called the activated state (*a*). The activation internal energy (ΔE_a) required to tip the block is provided by vibrating the table top; in a material the vibrations that cause this change are the thermal vibrations of the atoms at the temperature (*T*). The internal energy (*E*) is used to define the equilibrium configuration of the block in Figure 4.11, because there is no volume or entropy change for tipping the block. For a material at constant temperature and pressure, the equilibrium state corresponds to the minimum Gibbs free energy, as shown in Figure 4.12b and as discussed above.

A Gibbs free energy plot as a function or reaction coordinate for a process that involves a change in entropy is shown in Figure 4.12b. An example of atom motion that requires an activation Gibbs free energy (ΔG_a) is surface diffusion, shown in Figure 4.13, where an atom is added (an adatom) on top of a surface in position 1. The Gibbs free energy required to break the atomic bonds in position 1 and move the atom to position *a* is the activation Gibbs free energy (ΔG_a). Once the atom moves to the activated state *a*, then it can continue to move to state 2. If the activation Gibbs free energy (ΔG_a), shown in Figure 4.12b, is provided by thermal vibrations of atoms at the surface, then the probability of this energy being present in excess of the energy G_1 during a thermal vibration is given by $P_a(G_1 + \Delta G_a)$ in the Maxwell-Boltzmann probability distribution in Equation 4.20, as shown in Equation 4.22:

$$P_a(G_1 + \Delta G_a) = \exp -\left(\frac{\Delta G_a}{kT}\right) \qquad\qquad \textbf{4.22}$$

If the adatom is vibrating at a frequency of ν, then ν is the frequency of the atom attempting to reach the activated energy state ($G_1 + \Delta G_a$). The rate at which the system reaches the activated state (R_a) is then the product of the vibration frequency and the probability that the activation Gibbs free energy (ΔG_a) is present, as shown in Equation 4.23.

$$R_a = \nu P_a(G_1 + \Delta G_a) = \nu \exp -\left(\frac{\Delta G_a}{kT}\right) \qquad\qquad \textbf{4.23}$$

Equation 4.24 shows that the activation Gibbs free energy is equal to the enthalpy change to reach the activated state (ΔH_a) minus the temperature times the entropy change from the state 1 to the activated state (ΔS_a).

$$\Delta G_a = \Delta H_a - T\Delta S_a \qquad\qquad \textbf{4.24}$$

The rate of the change (R_c) is then given by Equation 4.25:

$$R_c = \frac{R_a}{2} = \frac{\nu}{2}\exp -\left(\frac{\Delta H_a - T\Delta S_a}{kT}\right) \qquad\qquad \textbf{4.25}$$

The factor of $\frac{1}{2}$ in Equation 4.25 results from an atom in the activated state (*a*) having the probability of moving to the new position being equal to the probability that the atom moves back to the original position. Rewriting Equation 4.25 and separating out the temperature dependent terms results in Equations 4.26 and 4.27.

$$R_c = \frac{\nu}{2} \exp\left(\frac{\Delta S_a}{k}\right) \exp - \left(\frac{\Delta H_a}{kT}\right)$$

4.26

$$R_c = C \exp - \left(\frac{\Delta H_a}{kT}\right)$$

4.27

The vibrational frequency and entropy change term are combined into the constant C. Equation 4.27 is the **Arrhenius rate equation**. Note that the rate of change from state 1 to state 2 does not depend upon the enthalpy of state 2, as shown in Equation 4.27. The rate of going from state 1 to state 2 depends only upon the activation enthalpy (ΔH_a). An example of this is given by the block in state 1, which is on a table top. If the block is placed at the edge of the table, so that the block falls off the table to the floor if it reaches the activated state (a), the rate of falling to the floor is the same as the rate of falling onto the table. If the table is placed at the edge of the Grand Canyon, so that the block would fall to the bottom of the Grand Canyon, it would fall into the canyon at the same rate as onto the table.

The enthalpy of activation (ΔH_a) is determined experimentally by measuring the rate of change (R_c) in a material as a function of temperature. Taking the natural log of each side of Equation 4.27 results in Equation 4.28:

$$\ln R_c = \ln C - \frac{\Delta H_a}{kT}$$

4.28

This is the equation of a straight line with a slope of $\Delta H_a / k$ when $\ln R_c$ is plotted as a function of the inverse of the absolute temperature $1/T$. It is the activation enthalpy (ΔH_a) that is experimentally determined and not the activation Gibbs free energy. Therefore, activation enthalpy is presented in thermodynamic data tables rather than the Gibbs free energy of activation. The knowledge of the activation enthalpy also explains why Equation 4.16 appears in most textbooks rather than Equation 4.19. The activation Gibbs free energy is not known because the entropy times temperature contribution is not experimentally determined, and it is very difficult to calculate the total entropy.

Example Problem 4.10

The time required for honey to drain through a small exit hole in a graduated container was measured as a function of temperature. Three temperatures were selected. The honey was in a warm room at 295 K, in a cold room at 285 K, and in a refrigerator at 278 K. At 295 K, 3.5 milliliters (mL) of honey drained from the container in 17 minutes; at 285 K, 6.0 mL required 626 minutes to drain; and at 278 K, 5.4 mL required 2177 minutes to drain. Determine the activation enthalpy for the flow of honey.

Solution

Taking these temperatures of 295, 285, and 278 K as T_1, T_2, and T_3, respectively, the rates are calculated as

$$R_1 = 3.5 \text{ mL}/17 \text{ minutes} = 0.2 \text{ mL/minute}$$

$$R_2 = 6.0 \text{ mL}/626 \text{ minutes} = 0.0096 \text{ mL/minute}$$

$$R_3 = 5.4 \text{ mL}/2177 \text{ minutes} = 0.0025 \text{ mL/minute}$$

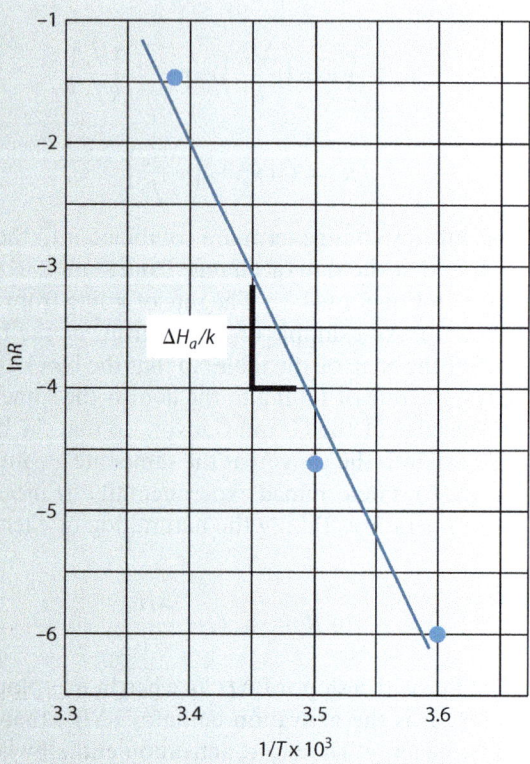

Figure 4.14   A plot of the natural log of the rate of honey flow as a function of the inverse of the temperature multiplied by 10^3.

In Figure 4.14, the natural log of the rate of honey flow is plotted as a function of the inverse of the absolute temperature, and a straight line is drawn through the three data points. The activation enthalpy (ΔH_a) is solved for analytically by taking two points that are in the linear portion of the plot of $\ln R_c$ versus $1/T$. In this example R_1 and R_3 are utilized, and there are two unknowns (ΔH_a and C). By determining the ratio R_1/R_3, as shown in Equation 4.29, the constant C cancels assuming that the constant C does not change significantly over these small temperature ranges.

$$\frac{R_1}{R_3} = \frac{\exp-\left(\dfrac{\Delta H_a}{kT_1}\right)}{\exp-\left(\dfrac{\Delta H_a}{kT_3}\right)} = \exp-\left(\frac{\Delta H_a}{k}\right)\left(\frac{1}{T_1} - \frac{1}{T_3}\right) \tag{4.29}$$

Taking the natural log of each side of Equation 4.29 results in Equation 4.30:

$$\ln\frac{R_1}{R_3} = -\frac{\Delta H_a}{k}\left(\frac{1}{T_1} - \frac{1}{T_3}\right) \tag{4.30}$$

Solving for ΔH_a results in Equation 4.31:

$$\Delta H_a = -\frac{k \ln \dfrac{R_1}{R_3}}{\dfrac{1}{T_1} - \dfrac{1}{T_3}}$$

4.31

Substituting the data from tests number 1 and 3 into Equation 4.31 and inserting Boltzmann's constant (k) of 8.62×10^{-5} eV/K-molecule lets us solve for ΔH_a:

$$\Delta H_a = -\frac{8.62 \times 10^{-5} \dfrac{\text{eV}}{\text{molecule} \cdot \text{K}} \left(\ln \dfrac{0.2 \text{ mL/min}}{0.0025 \text{ mL/min}} \right)}{\dfrac{1}{295 \text{ K}} - \dfrac{1}{278 \text{ K}}}$$

$$\Delta H_a = -\frac{8.62 \times 10^{-5} \dfrac{\text{eV}}{\text{molecule} \cdot \text{K}} (\ln 80.65)}{-2.075 \times 10^{-4} \text{K}^{-1}} = 1.82 \frac{\text{eV}}{\text{molecule}} = 1.76 \times 10^5 \frac{\text{J}}{\text{mole}}$$

The activation enthalpy (ΔH_a) is the energy required for the molecules in honey to slide past each other as the honey flows out of the restricted orifice. The constant (C) is solved for by selecting one of the temperatures, such as 295 K, and substituting into Equation 4.28 the rate $R_1 = 0.2$ mL/minute, and for ΔH_a the value of 1.82 eV/molecule.

From a very simple test, such as measuring the flow rate of honey out of an orifice, fundamental information about the motion of molecules in honey is obtained.

4.7 DIFFUSION IN METALS AND CERAMICS

It is easy to visualize the mixing of the atoms in two gases that are first separated and then released so that they can mix. Also, we are familiar with the mixing of two liquids, such as alcohol and water. Because atoms and molecules are mobile in liquids, it is easy to visualize how liquid atoms or molecules mix. We can observe ink mixing into water. However, it may come as a surprise that the atoms in two different solids with clean polished surfaces that are pressed together can also mix. For example, if a block of copper and a block of nickel are polished and pressed together at an elevated temperature to form what is called a **diffusion-couple**, as shown in Figure 4.15a, the atoms mix together by the process of diffusion. **Diffusion** is the motion of individual atoms or molecules in steps comparable to interatomic distances. There are interesting applications of diffusion, such as the diffusion bonding of titanium alloys in aircraft. In diffusion bonding the atoms of two different parts mix and are joined together without melting either part. The unexpected diffusion bonding of metal and ceramic parts in contact in the ultra-clean and low-temperature environment of outer space was one of the surprise discoveries of early space exploration. It is possible that the clean environment of outer space could be used as a location for intentionally diffusion bonding materials together. In addition, we find that diffusion is important in many processes in materials, including oxidation, case hardening, phase transformations, hydrogen gas purification, and fuel-cell operation.

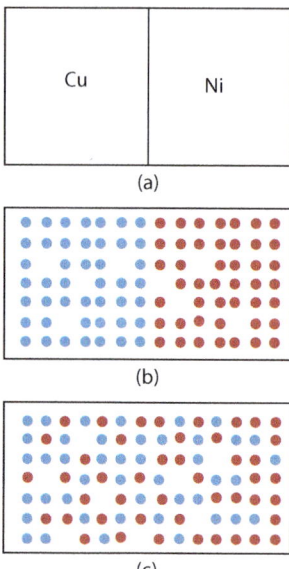

Figure 4.15   A schematic of a diffusion-couple of copper and nickel, at various stages of mixing by diffusion: (a) Polished copper and nickel blocks pressed together to form a diffusion-couple before diffusion has begun. (b) The copper and nickel atoms and some vacancies, before diffusion has begun. (c) The copper and nickel atoms and some vacancies, after a significant amount of diffusion.

4.7.1   Vacancy Diffusion

How do similar sized atoms, such as Cu and Ni, move inside a solid? We normally think of the atoms or molecules of a solid as fixed in place. The atoms of two crystals, such as Cu and Ni, are schematically shown in Figure 4.15b before mixing, and the mixed atoms are shown in Figure 4.15c after a significant amount of diffusion. The problem of figuring out how nickel and copper atoms mix in a solid had challenged many of the outstanding scientists, including Albert Einstein, in the early part of the twentieth century. There were proposals that the atoms mixed by rotating in little circles. Figure 4.15b shows that there are vacancies in every material. The equilibrium concentration of vacancies is given by Equation 4.16 rewritten for vacancies rather than B-type atoms. The presence of only one vacancy in the crystal allows the nickel and copper to mix. You have experienced this if you have played the game where the sliding numbered tiles are rearranged through one missing (vacant) tile, as shown in Figure 4.16. In Figure 4.16a the red tiles are separated from the white tiles. The numbers are meaningless in this example of atom mixing, for we cannot place numbers on atoms. For our purposes, red represents one type of atom and white represents another. In Figure 4.16b, the red and white tiles have been completely mixed because of the presence of one vacant site. Without a vacant site, the atoms (tiles) cannot be moved. The same process is observed in a crystal. If a vacancy is present in a crystal, as shown in Figure 4.17a, an adjacent atom can jump into the vacant site, and the vacancy moves to the former site of the jumping atom. The diffusing atom is most likely to jump when the atoms between the diffusing atom and the vacancy vibrate away from each other, creating an easy pathway for movement of the diffusing atom or molecule into the vacancy. Figure 4.17b shows vacancy diffusion in an ionic material, where positive ions diffuse to positive-ion sites and negative ions diffuse to negative-ion sites.

Figure 4.16 (a) The red and white tiles are color separated, with one vacant site. (b) The red and white tiles are mixed, with a vacant site still present but in another location. (*Photo by C. M. Gilmore.*)

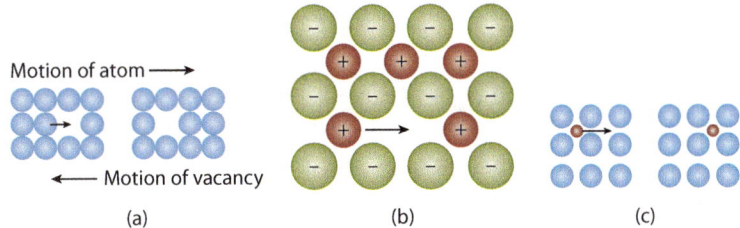

Figure 4.17 (a) The vacancy diffusion mechanism in a crystalline metal or covalently bonded solid. (b) The vacancy diffusion mechanism in an ionic material. (c) Interstitial diffusion. (*Based on Askeland, D.R., Fulay, P.P., and Wright, W.J., The Science and Engineering of Materials, 6th ed. Cengage Learning, Stamford, CT. (2011), (a) and (c) p. 162, (b) p. 170. 1. CRC Handbook of Chemistry and Physics, ed.D.R. Lide, 90th ed. 2009–2010 pg. 12–80.*)

Based upon the vacancy model of diffusion, the greater the amplitude of the lattice vibrations, the larger the diffusion rate. A high temperature results in higher-amplitude lattice vibrations, and therefore the diffusion rate should increase with temperature, as has been experimentally confirmed. Also, as temperature increases the concentration of vacancies increases, as shown by Equation 4.16. The increase in the number of vacancies increases the number of sites where vacancy diffusion can occur.

4.7.2 Interstitial Atom Diffusion

For the diffusion of interstitial atoms in interstitial solid solutions, such as carbon in iron, vacancies are not required. The interstitial sites are not occupied by atoms. Therefore, interstitial atoms move from interstitial site to site, as shown in Figure 4.17c. This usually results in a more rapid diffusion for interstitial solid solutions than for vacancy diffusion in substitutional solid solutions at the same temperature.

4.7.3   Vacancy Diffusion Rates in Metals and Ceramics

The rate for vacancy diffusion requires the presence of vacancies. In a crystal at a constant temperature (T) and pressure (P), the equilibrium vacancy concentration is given by inserting the enthalpy for vacancy formation into Equation 4.16, yielding Equation 4.32:

$$\frac{N_v(\text{eq})}{N} = \exp - \left(\frac{\Delta H_v}{kT}\right) \qquad \textbf{4.32}$$

The term $N_v(\text{eq})/N$ is the probability (P_v) that a vacancy exists at any individual site. The probability that an atom will move into an adjacent vacancy (P_m) is given by Equation 4.33:

$$P_m = \exp - \left(\frac{\Delta G_m}{kT}\right) = \exp - \left(\frac{\Delta H_m - T\Delta S_m}{kT}\right) = A\exp - \left(\frac{\Delta H_m}{kT}\right) \qquad \textbf{4.33}$$

where ΔG_m is the activation Gibbs free energy for atom migration, and A contains the entropy term. The total probability of vacancy diffusion (P_D) is then given by the product of the formation (P_v) and migration (P_m) probabilities, as shown in Equations 4.34 and 4.35:

$$P_D = P_v P_m = \exp - \left(\frac{\Delta H_v}{kT}\right) A\exp - \left(\frac{\Delta H_m}{kT}\right) = A\exp - \left(\frac{\Delta H_v + \Delta H_m}{kT}\right) \qquad \textbf{4.34}$$

$$P_D = P_v P_m = A\exp - \left(\frac{\Delta H_D}{kT}\right) \qquad \textbf{4.35}$$

where $\Delta H_D = \Delta H_v + \Delta H_m$ is the activation enthalpy for diffusion. The rate of diffusion (R_D) from Equation 4.27 is then given by Equation 4.36:

$$R_D = C\exp - \left(\frac{\Delta H_D}{kT}\right) \qquad \textbf{4.36}$$

where the constant (C) contains the vibrational frequency and entropy terms. If we had used the Gibbs free energy of vacancy formation in Equation 4.32, the entropy would be contained in C in Equation 4.36, and the resulting equation would have been the same. The activation enthalpy for vacancy diffusion (ΔH_D) of some atoms in a host lattice is presented in Table 4.4. The pre-exponential constant (D_{0BA}) is explained in Section 4.7.6.

Self-diffusion in metals is primarily by vacancy diffusion. Self-diffusion is not particularly significant for engineering applications, because it refers to the diffusion of atoms in a pure material, and there is no composition change. Even in a pure metal the atoms move in random jumps. Self-diffusion is measured by following the movement of radioactive isotopes of elements in a nonradioactive host of the same element. Self-diffusion is studied because it is diffusion without any chemical driving force, and it relates to a homogenous material without chemical interactions. Therefore, self-diffusion reveals information purely about diffusion without the complication of differences in atoms or molecules. In Table 4.4, observe that the activation enthalpy for self-diffusion is directly related to the melting temperature for metals with the same crystal structure. For example, of the FCC metals, lead melts at 600 K and copper melts at 1356 K, and the activation enthalpy for self-diffusion in copper is nearly double that of lead. As we noted in Chapter 2, the melting temperature is related to the strength of the bonding; thus the activation enthalpy for self-diffusion is also related to the strength of the bonding.

Table 4.4 The Activation Enthalpy (ΔH_D) and Pre-Exponential Constant (D_{0BA}) for the Diffusion of Atoms in Some Elements and Compounds. D_{0BA} is defined in Equation 4.39.

Diffusion Couple	ΔH_D kcal/mole	ΔH_D kJ/mole	ΔH_D eV/atom	D_{0BA} m²/s
Interstitial Diffusion				
C in FCC Fe[1]	35.3	148	1.53	2.3×10^{-5}
C in BCC Fe	20.9	87.5	0.907	1.1×10^{-6}
N in FCC Fe	34.6	145	1.50	3.4×10^{-7}
N in BCC Fe	18.3	76.6	0.794	4.7×10^{-7}
H in FCC Fe	10.3	43.1	0.447	6.3×10^{-7}
H in BCC Fe	3.6	15.1	0.156	1.2×10^{-7}
Self-Diffusion				
Pb in Pb	25.9	108	1.12	1.27×10^{-4}
Al in Al	32.2	135	1.40	1.0×10^{-5}
Cu in Cu	49.3	206	2.14	3.6×10^{-5}
Fe in FCC Fe	66.7	279	2.90	6.5×10^{-5}
Fe in BCC Fe	58.9	247	2.56	4.1×10^{-4}
Zn in Zn	21.8	91.3	0.946	1.0×10^{-5}
Mg in Mg	32.2	135	1.40	1.0×10^{-4}
W in W	143	600	6.22	4.1×10^{-4}
Si in Si	110	460	4.77	1.8×10^{-1}
C in C (graphite)	163	657	7.08	5.0×10^{-4}
Heterogeneous Diffusion				
Ni in Cu	57.9	242	2.51	2.3×10^{-4}
Cu in Ni	61.5	257	2.67	6.5×10^{-5}
Zn in Cu	43.9	184	1.91	7.8×10^{-5}
Ni in FCC Fe	64	268	2.78	4.1×10^{-4}
Au in Ag	45.5	190	1.98	2.6×10^{-5}
Ag in Au	40.2	168	1.74	7.2×10^{-6}
Al in Cu	39.5	165	1.71	4.5×10^{-6}
Al in Al_2O_3	114	477	4.95	2.8×10^{-3}
O in Al_2O_3	152	636	6.60	1.9×10^{-1}
Mg in MgO	79	331	3.43	2.5×10^{-5}
O in MgO	82.1	344	3.46	4.3×10^{-9}

Based on data from Askeland, D.R., Fulay, P.P., and Wright, W.J., The Science and Engineering of Materials, 6th ed. Cengage Learning, Stamford, CT. (2011), p. 164 and from (1) Smithells Metals Reference Book, eds. Gale, W.F., and Totemeier, T.C., Elsevier Butterworth-Heinemann, Oxford (2004), p. 13–21.

In metal oxide ceramics, such as MgO and Al_2O_3, diffusion of both the metal and the oxygen ions occurs primarily by vacancy diffusion, as indicated by the magnitude of the activation enthalpy. Note also that the activation enthalpy of oxygen diffusion in each ceramic is greater than the activation enthalpy for the metal, even though the atomic numbers of Mg and Al are higher than that of oxygen. This results from the oxygen ion being larger than the metal ion because of electron transfer to the oxygen ion.

4.7.4   Interstitial Diffusion Rates in Metals and Ceramics

For interstitial diffusion, the enthalpy of diffusion is equal to the enthalpy for migration ($\Delta H_D = \Delta H_m$) because it is not necessary to create vacant interstitial sites. Therefore, the activation enthalpy for interstitial diffusion is normally less than that for vacancy diffusion where vacancies must first be created before migration can occur. As shown in Table 4.4, the activation enthalpy for interstitial diffusion of carbon in iron is less one-half the value for self-diffusion of iron in iron. In many metals hydrogen forms a positive ion that diffuses by the interstitial mechanism.

4.7.5   Diffusion Rates at Defects in Metals and Ceramics

It is easier for atoms to move on a surface than through the bulk of the material, because a vacancy need not be present on the surface for diffusion to occur, and the surface atoms are less strongly bonded than are bulk atoms. Thus the velocity of diffusion of an atom on a surface is greater than through the bulk. Surface diffusion is an important consideration in the growth of thin-film materials by vapor deposition. Relative to the bulk, diffusion occurs more rapidly along surfaces, grain boundaries, phase boundaries, and edge dislocations. Table 4.5 shows that the activation enthalpy for self-diffusion in bulk silver is twice the value for grain-boundary diffusion and five times the value for surface diffusion. Atom

Table 4.5   The Activation Enthalpy (ΔH_D) and Pre-Exponential Constant (D_0) for Self-Diffusion in Silver in the Bulk, along Grain Boundaries, and at Surfaces. D_0 is Defined in Equation 4.39.

Diffusion Type	ΔH_D kcal/mole	ΔH_D kJ/mole	ΔH_D eV/atom	D_0 m²/s
Bulk	45.7	191	1.98	1.00×10^{-4}
Grain Boundary	22.8	95.2	0.99	0.74×10^{-4}
Surface	8.9	37.3	0.39	0.47×10^{-4}

Based on data from Askeland, D.R., Fulay, P.P., and Wright, W.J., The Science and Engineering of Materials, 6th ed. Cengage Learning, Stamford, CT. (2011), p. 174.

diffusion is faster along edge dislocations than through a dislocation-free crystal. The atoms diffuse faster in the region of the edge dislocation where the spacing between atoms is greater. Diffusion along edge dislocations is referred to as pipe diffusion. Similarly, it is easier for atoms to move through the spaces in a tilt-grain boundary than through the bulk of the material. Grain boundary diffusion has important implications in the design of high-temperature materials such as gas-turbine blades. The blades are sometimes made from single crystals to avoid grain-boundary diffusion. We will cover this subject in more detail in Chapter 9, on the temperature and time dependence of material response to applied stresses.

4.7.6 Analysis of Steady-State Diffusion

In **steady-state diffusion**, the concentration of atoms does not change with time. An example of steady-state diffusion is the purification of hydrogen, as shown in Figure 4.18. High-purity hydrogen is required for many applications, such as in fuel cells. In a hydrogen-purification system, low-purity hydrogen at a constant and high pressure of approximately 14 atmospheres is on the inside of a palladium tube at a temperature of approximately 773 K. The partial pressure of the hydrogen on the high-pressure side of the palladium tube (P_{Hhi}) must be higher than the hydrogen partial pressure on the low-pressure side (P_{Hlo}), because then the concentration of hydrogen is higher on the high-pressure side of the palladium (C_1 hydrogen atoms/m³) than on the low-pressure side (C_2 hydrogen atoms/m³), as shown in the plot in the lower part of Figure 4.18. The hydrogen molecules (H_2) are **adsorbed** on the palladium surface on

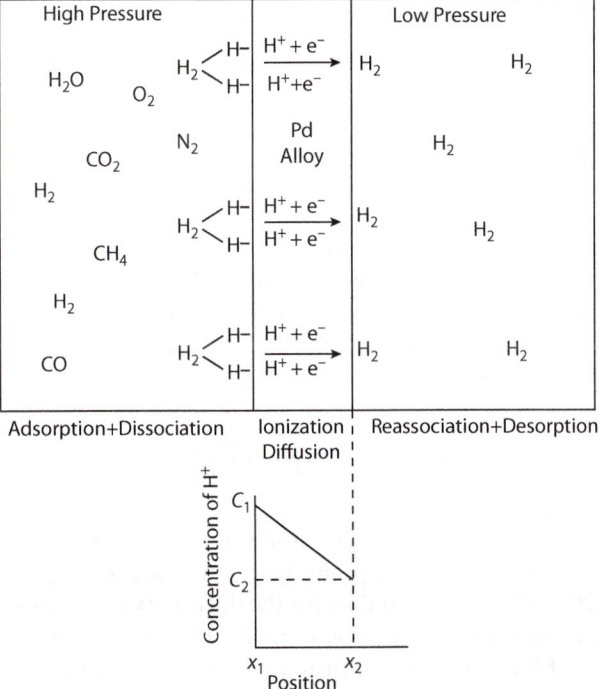

Figure 4.18 The top figure is a schematic of a hydrogen purification cell. The lower part of the figure shows the concentration of hydrogen ions in the palladium metal as a function of position through the metal.

the high-pressure side, where H_2 then disassociates into two separate H atoms. An atom or molecule is adsorbed when it attaches to a surface. Each hydrogen atom is ionized and the electron becomes one of the palladium metal conduction electrons, and the much smaller H^+ ion is only a nucleus. The small H^+ ion nucleus diffuses rapidly through interstitial sites in the palladium, but the larger-size ions of carbon, oxygen, and nitrogen in the air cannot as easily diffuse through the palladium. The palladium tube acts as a filter that allows only the rapid diffusion of hydrogen ions. Hydrogen with one impurity atom in 10^8 hydrogen atoms is accumulated on the outside of the tube in a chamber at a lower hydrogen partial pressure.

Equation 4.37 is the most general equation predicting the steady-state diffusion rate of atoms

$$J_B = -C_B B_B \frac{\partial G_B}{\partial x} \qquad \qquad \textbf{4.37}$$

where J_B is the number of B-type atoms, hydrogen in the above example, that cross a square meter of the A-type material per second; C_B is the concentration of the B-type atoms at position x in atoms per cubic meter; and B_B is the mobility of the B-type atoms in units of velocity per unit of force. J_B is also called the B-type atom flux. The term $\partial G_B/\partial x$ is the gradient with respect to position (x) of the Gibbs free energy of the B-type atoms inside the A-type material, and it is the chemical force on the B-type atoms. The atoms always move in a direction to minimize the Gibbs free energy of the system (G_B); thus the gradient is always negative. Equation 4.37 for the diffusion of atoms is similar to Equation 4.10 for the heat flux, and it is similar to Ohm's law for electrical current density.

Unfortunately, there is no way to measure the Gibbs free energy or its gradient in a material. We can measure the chemical concentration (C_B) of the B-type atoms as a function of position, as discussed in Chapter 15, on experimental methods. In many cases, such as for the Ni and Cu shown in Figure 4.15, atoms move from a high concentration to a low concentration. This is true if the diffusing atoms mix into the host atoms. For atoms that mix ideally, the gradient of the concentration is directly related to the gradient of the Gibbs free energy. For this case the steady-state diffusion equation is written as Equation 4.38:

$$J_B = -D_{BA} \frac{dC_B}{dx} \qquad \qquad \textbf{4.38}$$

where J_B is the rate of diffusion of B-type type atoms in the A-type material, D_{BA} is the diffusivity or the diffusion coefficient in units of square meters per second of the B-type material in the A-type host, and dC_B/dx is the concentration gradient with respect to position (x) of the B-type atoms in the A-type host. This form of the steady-state diffusion equation is Fick's first law. The temperature dependence of the diffusivity is given by Equation 4.39:

$$D_{BA} = D_{0BA} \exp - \left(\frac{\Delta H_{DBA}}{kT} \right) \qquad \qquad \textbf{4.39}$$

where D_{0BA} is a constant for the diffusion of B-type atoms into A-type atoms, and ΔH_{DBA} is the activation enthalpy for diffusion of B-type atoms in A-type atoms. Some values of D_{0BA} and ΔH_{DBA} are presented in Table 4.4. Figures 4.19, 4.20, and 4.21 present data for the diffusivity of various materials as a function of $1/T$. In Figure 4.19 the diffusion coefficient of iron (Fe) into BCC iron is plotted; this is the self-diffusion that we discussed above. In Figure 4.20 the diffusion coefficient for gold in silicon is presented in the plots of Au_s^1 and Au_s^2. Gold is an interstitial solid solution in silicon. For Au_s^1 the concentration of gold is supersaturated, and for Au_s^2 the dislocation density is increased in the silicon and the concentration of gold is not saturated.

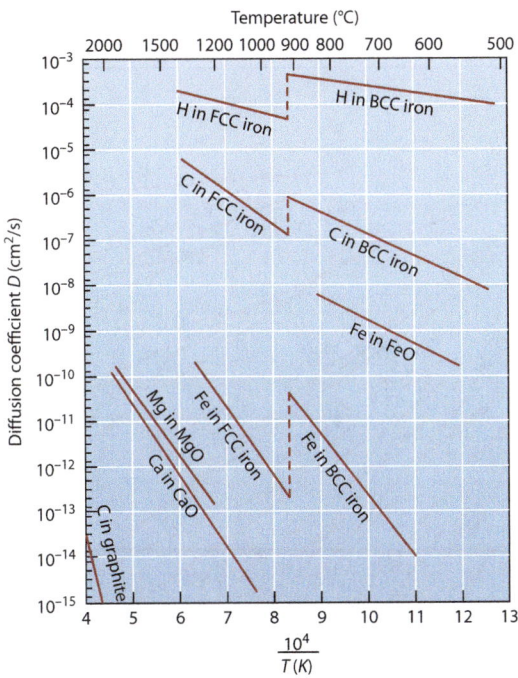

Figure 4.19 The diffusion coefficient for atoms in some metals and ceramics as a function of $10^4/T$. *(Based on Askeland, D.R., Fulay, P.P., and Wright, W.J., The Science and Engineering of Materials, 6th ed. Cengage Learning, Stamford, CT. (2011), p. 169.)*

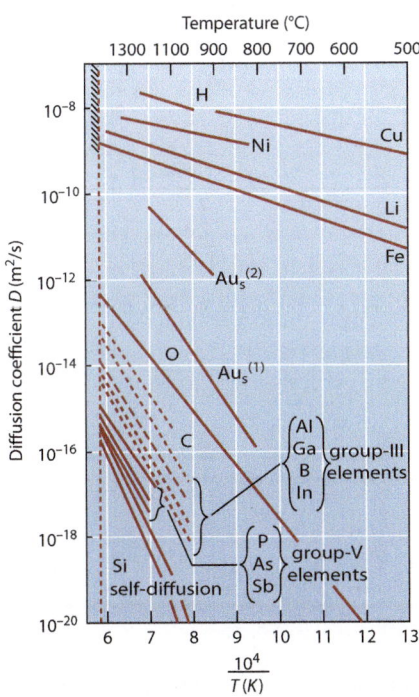

Figure 4.20 The diffusion coefficient for atoms in silicon as a function of $10^4/T$. (Based on "Diffusion and Diffusion Induced Defects in Silicon," by U. Gösele. In R. Bloor, M. Flemings and S. Mahajan (eds.), Encyclopedia of Advanced Materials, Vol. 1, 1994, p. 631.)

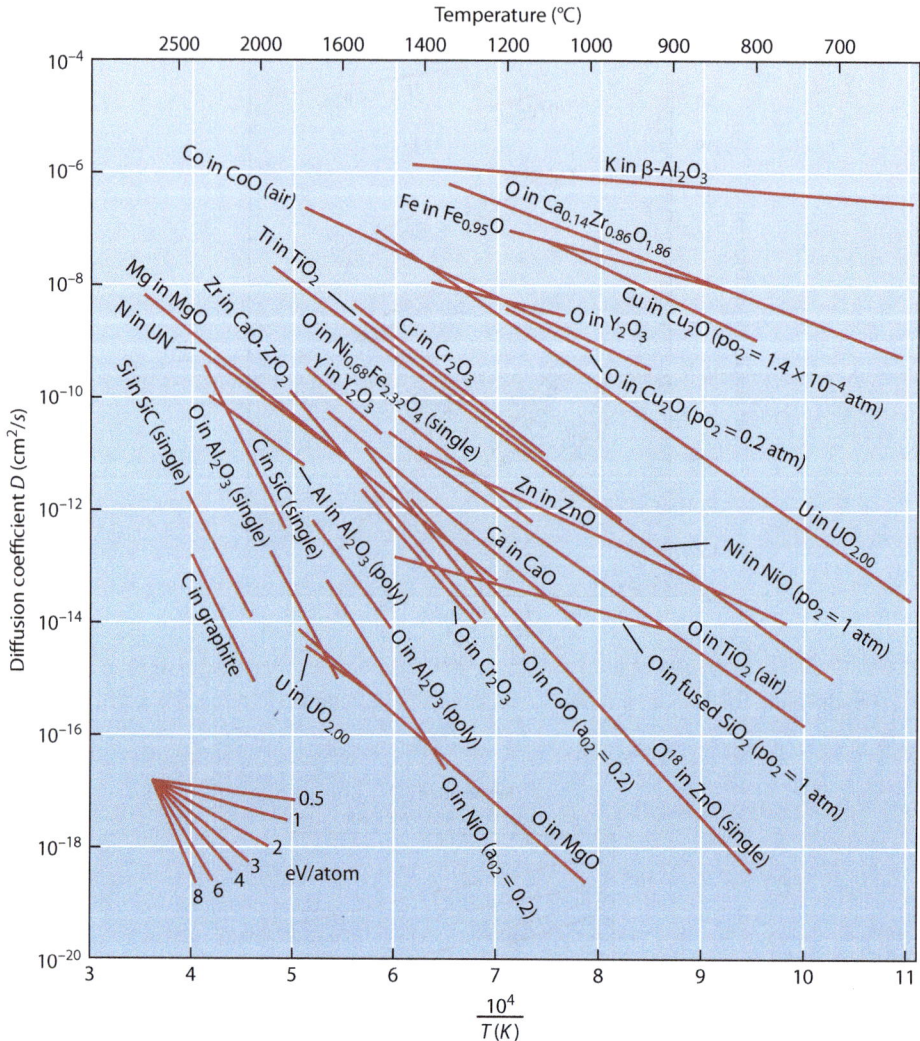

Figure 4.21 The diffusion coefficient for ions in various ceramics as a function of $10^4/T$. (*Based on Physical Ceramics: Principles for Ceramic Science and Engineering, by Y. M. Chiang, D. Birnie, and W. D. Kingery, Fig. 3-1, John Wiley, 1997.*)

Example Problem 4.11

Determine the activation enthalpy and preexponential constant for the diffusion of oxygen into silicon using the data presented in Figure 4.20.

Solution

The diffusion coefficient of oxygen diffusion into silicon (D_{OSi}) is plotted as a function of temperature in Figure 4.20. The diffusion coefficient as a function of temperature is presented in Equation 4.39. There are two unknowns in Equation 4.39: D_{0OSi} and ΔH_{DOSi}. By developing two equations with Equation 4.39, we can solve for the two unknowns. It is best to choose two values of D_{OSi} that are well separated and that are easy to read from the figure. For point 1 select $D_{OSi1} = 1 \times 10^{-12}\,cm^2/s = 1 \times 10^{-16}\,m^2/s$ and $1/T_1 = 0.00072\,K^{-1}$, or $T_1 = 1389\,K$. For point 2 select $D_{OSi2} = 1 \times 10^{-18}\,cm^2/s = 1 \times 10^{-22}\,m^2/s$ and $T_2 = 0.00118\,K^{-1}$, or $847\,K$. Taking the ratio of the two diffusion coefficients results in the elimination of one unknown (D_{0OSi}).

$$\frac{D_{OSi1}}{D_{OSi2}} = \frac{D_{0OSi}\exp-\left(\dfrac{\Delta H_{DOSi}}{kT_1}\right)}{D_{0OSi}\exp-\left(\dfrac{\Delta H_{DOSi}}{kT_2}\right)} = \exp-\left[\left(\frac{\Delta H_{DOSi}}{k}\right)\left(\frac{1}{T_1}-\frac{1}{T_2}\right)\right]$$

$$\frac{1\times10^{-16}\,m^2/s}{1\times10^{-22}\,m^2/s} = 1\times10^6 = \exp-\left[\left(\frac{\Delta H_{DOSi}}{k}\right)(0.00072-0.00118)\,K^{-1}\right]$$

Taking the natural log of each side of the ratio of the diffusion coefficients results in

$$13.82 = -\left(\frac{\Delta H_{DOSi}}{8.62\times10^{-5}\,eV/atom\cdot K}\right)(-0.00046\,K^{-1}) = -\Delta H_{DOSi}(-5.33)\frac{atom}{eV}$$

Solving for ΔH_{DOSi}

$$\Delta H_{DOSi} = \frac{13.82}{5.33}\frac{eV}{O\ atom} = 2.59\frac{eV}{O\ atom} = 2.5\times10^5\frac{J}{mole\ of\ O\ atoms}$$

The value of D_{0OSi} can be found by substituting ΔH_{DOSi} into any one of the data points, such as point 1.

$$D_{OSi1} = 1\times10^{-16}\frac{m^2}{s} = D_{0OSi}\exp-\left(\frac{\Delta H_{DOSi}}{kT_1}\right) = D_{0OSi}\exp-\left(\frac{2.59\,eV/atom}{(8.62\times10^{-5}\,eV/atom\cdot K)(1389\,K)}\right)$$

$$D_{OSi1} = 1\times10^{-16}\frac{m^2}{s} = D_{0OSi}\exp-\left(\frac{2.59\,eV/atom}{11.97\times10^{-2}\,eV/atom}\right) = D_{0OSi}\exp-(21.6) = D_{0OSi}(4.0\times10^{-10})$$

Solving for D_{0OSi}

$$D_{0OSi} = \frac{1\times10^{-16}\,m^2/s}{4.0\times10^{-10}} = 0.25\times10^{-6}\frac{m^2}{s}$$

Example Problem 4.12

Hydrogen gas is purified for use in fuel cells, as is shown schematically in Figure 4.18. The hydrogen purifier is a tube of wall thickness (t) with hydrogen at high partial pressure (P_h) on the inside of the tube and hydrogen at low partial pressure (P_l) on the outside of the tube. **Sievert's law** states that the concentration of a gas on a surface is proportional to the square root of the partial pressure of the gas. Therefore, the concentration gradient of hydrogen (dC_H/dx) through the metal tube wall is equal to

$$\frac{dC_H}{dx} = B\frac{P_h^{1/2}-P_l^{1/2}}{t}$$

where B is a constant dependent upon the gas and the metal. Experimentally it has been found by Ackerman and Koskinas that the flux of hydrogen ions through a tube of a palladium-25 weight percent silver alloy is given by

$$J_H = \left[4.86 \times 10^{-7} \left(\frac{P_h^{1/2} - P_l^{1/2}}{t} \right) \frac{\text{moles}}{\text{m}^2\text{s}} \right] \exp - \left(\frac{6.6 \times 10^3 \text{ J/mole}}{RT} \right)$$

where the preexponential constant contains terms from both the diffusivity of hydrogen and the solubility of hydrogen in palladium-25 weight percent silver. Calculate the flux of hydrogen ions through a tube of a palladium-25 weight percent silver alloy of wall thickness 1.24×10^{-4} m at a temperature of 651 K if the impure hydrogen is at a partial pressure of 0.69×10^6 Pa and the high-purity hydrogen is at a partial pressure of 3.45×10^4 Pa.

Solution

Substituting the given conditions into the equation for J_H,

$$J_H = \left[4.86 \times 10^{-7} \left(\frac{(0.69 \times 10^6 \text{ Pa})^{1/2} - (3.45 \times 10^4 \text{ Pa})^{1/2}}{1.24 \times 10^{-4} \text{ m}} \right) \frac{\text{moles}}{\text{m}^2\text{s}} \right] \exp - \left(\frac{6.6 \times 10^3 \dfrac{\text{J}}{\text{mole}}}{8.31 \dfrac{\text{J}}{\text{mole} \cdot \text{K}} 651 \text{ K}} \right)$$

$$J_H = \left[4.86 \times 10^{-7} (0.52 \times 10^7) \frac{\text{moles}}{\text{m}^2\text{s}} \right] \exp - (1.22)$$

$$J_H = 2.52 \frac{\text{moles}}{\text{m}^2\text{s}} (0.295) = 0.74 \frac{\text{moles}}{\text{m}^2\text{s}}$$

This is the flux of hydrogen ions; the moles of H_2 gas are half of this amount.

The steady-state diffusion Equation 4.38 predicts that atoms always move from high to low chemical concentration. Although this happens in many cases, there are cases where the atoms move from a low concentration to a high concentration, such as when metal atoms mix in a liquid but not in a solid. An alloy of 50% copper and 50% silver at room temperature is composed of nearly pure copper and pure silver, because the copper and silver do not mix in the solid. The copper and silver mix in the liquid; therefore, during solidification the copper and silver atoms must separate by diffusion. Equation 4.37 properly predicts this separation of copper and silver, because separation of the atoms into two phases reduces the Gibbs free energy. However, Equation 4.38 predicts that atoms move from a high concentration to a low concentration. The derivation of Equation 4.38 assumes that the atoms mix ideally, and copper and silver do not mix ideally; they separate. When atoms separate it is called uphill diffusion. It is not uphill diffusion when viewed from a Gibbs free energy viewpoint. Diffusion always decreases the Gibbs free energy. Nothing ever moves uphill spontaneously.

4.7.7   Time-Dependent Diffusion

The steady-state diffusion of Equation 4.38 applies where the chemical concentration gradient (dC/dx) is constant, as in the example of hydrogen diffusion through palladium metal shown in Figure 4.18. Equation 4.38 can also be applied where the concentration gradient is changing with time if the

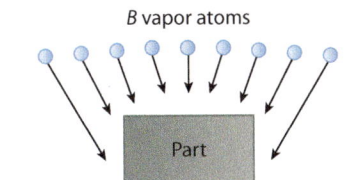

Figure 4.22 A part made of A-type atoms is exposed to a vapor of B-type atoms that are adsorbed onto the part, creating a concentration of B-type atoms of C_s on the surface of the part.

concentration gradient (dC/dx) is known at a particular time t and at a location x; then the flux of atoms $J(x)$ at position (x) and time (t) can be calculated. However, the steady-state diffusion equation has no predictive capability; it cannot predict the flux at some other position (x_2) and time (t_2) if the concentration gradient is not known at position (x_2) and time (t_2). To analyze problems where the concentration gradient changes with time and position, it is necessary to develop non-steady-state diffusion equations. An example of non-steady-state diffusion is the diffusion of copper and nickel, shown in Figure 4.15. At a time of 0 the nickel and copper atoms are completely separated, as shown in Figure 4.15b. After a time t the nickel atoms penetrate into the copper and the copper atoms into the nickel, as shown in Figure 4.15c. As the time increases, the amount of mixing at any particular site increases with time. Problems of this type are analyzed with Fick's second law, which is derived in Appendix 4E.

One classic solution to Fick's second law applies to processes, such as case hardening, where carbon atoms are diffused into a low-carbon steel surface. A schematic of the experimental configuration is shown in Figure 4.22, where the part made of A-type material is placed in an environment with a source of B-type material. This results in a surface concentration (C_s) of the B-type material. The B-type atoms could also be nitrogen gas adsorbed onto steel, a process called nitriding. The original impurity concentration of B-type atoms in the A-type material is C_0. Under these conditions a solution to Fick's second law gives the chemical concentration $[C_B(x, t)]$ of B-type atoms in the A-type material as a function of x and t, as shown in Equations 4.40 and 4.41.

$$\frac{C_B(x, t) - C_0}{C_s - C_0} = 1 - \text{erf}\left(\frac{x}{2(D_{BA}t)^{0.5}}\right) = 1 - \text{erf}(z) \qquad \textbf{4.40}$$

$$z = \frac{x}{2(D_{BA}t)^{0.5}} \qquad \textbf{4.41}$$

The erf(z) is the error function whose values are shown in Table 4.6. Values of the error function appear in many mathematical handbooks. The functional form of the erf(z) is shown in Equation 4.42:

$$\text{erf}(z) = 2\pi^{-0.5}\int_0^z \exp^{-y^2}dy \qquad \textbf{4.42}$$

In analyzing Equation 4.40, note that when time is equal to 0, z is equal to infinity for any finite value of x, and erf$(\infty) = 1$. At time equal to 0, it follows from Equation 4.40 that

$$\frac{C_B(x, t) - C_0}{C_s - C_0} = 1 - 1 = 0$$

or that $C_B(x,0) = C_0$ for any finite value of location (x). This means that the composition throughout the specimen, except at the surface $(x = 0)$, is equal to the original composition (C_0) at time (t) equal to 0.

Table 4.6 Error-Function Values

z	erf(z)	z	erf(z)	z	erf(z)
0	0	0.55	0.5633	1.3	0.9340
0.025	0.0282	0.60	0.6039	1.4	0.9523
0.05	0.0564	0.65	0.6420	1.5	0.9661
0.1	0.1125	0.70	0.6778	1.6	0.9763
0.15	0.1680	0.75	0.7112	1.7	0.9838
0.2	0.2227	0.80	0.7421	1.8	0.9891
0.25	0.2763	0.85	0.7707	1.9	0.9928
0.3	0.3286	0.90	0.7970	2.0	0.9953
0.35	0.3794	0.95	0.8209	2.2	0.9981
0.4	0.4284	1.0	0.8427	2.4	0.9993
0.45	0.4755	1.1	0.8802	2.6	0.9998
0.5	0.5205	1.2	0.9103	2.8	0.9999

http://en.wikipedia.org/wiki/Error_function

Also, if the x is greater than 0, and the time is very large, then z approaches 0 and the erf(z) approaches 0, resulting in

$$\frac{C_B(x, t) - C_0}{C_s - C_0} = 1 - 0 = 1$$

This result means that $C_B(x,t) = C_s$, or that the surface composition (C_s) penetrates into the bulk of the specimen. For any other positions and times, the chemical composition is determined with Equation 4.40. A schematic of compositions for different times is shown in Figure 4.23. Note that Equation 4.40 is dimensionless, because it is a ratio of chemical compositions. For this reason it is possible to utilize chemical concentration (atoms/m³) or chemical composition in atomic percent or weight percent in Equation 4.40.

Note that the chemical composition is a function of $x/(D_{BA}t)^{0.5}$. This observation alone allows for the solution of some diffusion problems. For example, assume that atoms of type B are diffused into type A for an hour. It is determined by either experiment or calculation that the chemical composition at a position 1 μm into the specimen is 1 atom percent. How much total time will it require to obtain the

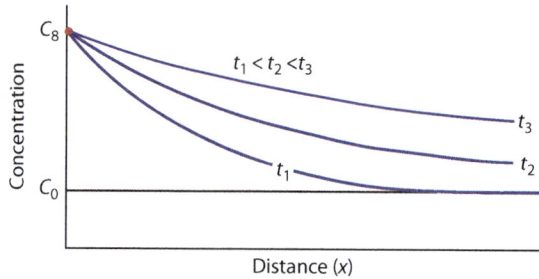

Figure 4.23 A concentration of atom type B into an A-type material, as a function of position x for non-steady-state diffusion at three different times t_1, t_2, and t_3.

composition of 1 atom percent at a depth of 2 μm at the same temperature? The left side of Equation 4.40 remains the same, because the chemical composition desired is still 1 atom percent, but the position of this composition and time are changed. Since the temperature is constant, the value of D_{BA} is constant, and the ratio of $x/t^{0.5}$ remains constant. If the location (x) is doubled, then the time must increase by a factor of 4 to keep the right side of Equation 4.40 constant, or it requires a total of four hours to obtain the same composition at double the depth.

Example Problem 4.13

Case hardening is the process of diffusing carbon into steel to produce a hard outer shell that resists abrasion and fatigue-crack initiation, and it leaves a softer core that resists fracture. Steel initially with 0.1 weight percent carbon is packed in carbon that maintains a surface concentration of 1 weight percent carbon at 1000°C. What is the time required to achieve 0.5 weight percent carbon at a depth of 1 millimeter into the steel? The iron-carbon phase diagram in Figure 5.18 shows that at 1000°C the iron phase is the FCC γ-phase iron for compositions from 0 to approximately 1.6 weight percent carbon. For this problem, the FCC γ phase is present.

Solution

The experiment described in this problem is the same as that described for Equation 4.40. From the problem we know that $C_C(0.001 \text{ m}, t) = 0.5$ weight percent, $C_0 = 0.1$ weight percent, $C_s = 1.0$ weight percent, and $x = 0.001$ m. D_{BA} and t must be determined. The information for determining D_{BA} is provided in Table 4.4. The value of the diffusion coefficient for the diffusion of carbon into γ-phase iron ($D_{C\gamma}$) is calculated from Equation 4.39.

$$D_{C\gamma} = D_{0C\gamma} \exp - \left(\frac{\Delta H_{C\gamma}}{kT} \right) = 2.3 \times 10^{-5} \frac{m^2}{s} \exp - \left(\frac{1.53 \text{ eV/atom}}{(8.62 \times 10^{-5} \text{ eV/atom} \cdot \text{K})(1273 \text{ K})} \right)$$

$$D_{C\gamma} = 2.3 \times 10^{-5} \frac{m^2}{s} \exp(-13.94) = 2.3 \times 10^{-5} \frac{m^2}{s} (0.9 \times 10^{-6}) = 2 \times 10^{-11} \frac{m^2}{s}$$

The time is found by first determining the value of z in the error function, and then inserting $D_{C\gamma}$ into Equation 4.40. Since all of the compositions in the diffusion equation are known, they can be substituted into the left side of Equation 4.40.

$$\frac{0.5 - 0.1}{1.0 - 0.1} = \frac{0.4}{0.9} = 0.444 = 1 - \text{erf}(z)$$

$$\text{erf}(z) = 0.556$$

Now we can find z in Table 4.6 from the value of erf(z). Since erf(z) = 0.556 does not appear in Table 4.6, it is necessary to interpolate between the error-function values of 0.55 and 0.50 to find z.

$$\frac{0.55 - 0.50}{0.5633 - 0.5205} = \frac{0.55 - z}{0.5633 - 0.556}$$

$$\frac{0.05}{0.0428} = \frac{0.55 - z}{0.0073}$$

$$0.55 - z = 0.0073(1.1682) = 0.0085$$

$$z = 0.55 - 0.0085 = 0.5415$$

Now that z is known, z is related to position (x), time (t), and $D_{C\gamma}$ in Equation 4.41.

$$z = 0.5415 = \frac{x}{2(D_{C\gamma}t)^{1/2}} = \frac{0.001 \text{ m}}{2\left(20 \times 10^{-12} \dfrac{\text{m}^2}{\text{s}} t\right)^{1/2}} = \frac{0.001 \text{ m}}{2\left(4.47 \times 10^{-6} \dfrac{\text{m}}{\text{s}^{1/2}}\right)t^{1/2}} = \frac{1.12 \times 10^2 \text{s}^{1/2}}{t^{1/2}}$$

Squaring z results in

$$z^2 = 0.293 = \frac{1.25 \times 10^4 \text{s}}{t}$$

Solving for time (t),

$$t = \frac{1.25 \times 10^4 \text{s}}{0.293} = 4.26 \times 10^4 \text{s} = 11.8 \text{ h}$$

4.8   MIXING, SOLUBILITY, DIFFUSION, AND PERMEABILITY IN POLYMERS

Much of the material presented above that concentrated on metals and ceramics has applications to polymers. For example, the Gibbs free energy, presented in Equation 4.13, applies to polymers in solvents and to the blending of polymers. Also, the diffusion of atoms or molecules through a thin polymer sheet is modeled with Fick's first law in Equations 4.37 and 4.38, and the diffusion of atoms or molecules into a solid polymer surface is modeled with the solution to Fick's second law in Equation 4.40. There are also significant differences between metals or ceramics and polymers, because the molecules of polymers do not significantly diffuse into each other if two different solid polymers are pressed against each other. This is because the positive activation enthalpy to move the LCMs is much larger than the negative temperature-entropy term. However, LCMs do diffuse when the polymer is a liquid.

4.8.1   Polymer Motion in Entangled Melts

The motion of polymer chains in liquids is divided into two segments. Each LCM is assumed to be contained inside a tube whose size is determined by the constraint provided by the surrounding polymer molecules. The polymer can move through this tube by diffusion in a manner similar to how a worm moves through a worm hole. This form of motion is called **reptation**. The second form of motion is that the constraining tube itself can move as a result of changes in the surrounding constraint provided by other LCMs. Based upon the reptation process alone, the diffusion coefficient (D) should be proportional to the inverse square of the molar mass (M) of the LCM. Experimentally it is observed that the diffusivity in polymer melts is proportional to the molar mass to the negative power 2.4, as shown in Equation 4.43.

$$D \propto M^{-2.4} \tag{4.43}$$

4.8.2 Mixing, Entropy, and Gibbs Free Energy for Polymer Melts

As in other liquids, some different polymers mix in the liquid state and others do not. When producing a polymer blend, it is important to know if the polymers mix in the liquid or if they separate, because the amount of mixing in the liquid is directly transferred to the solid. The polymers mix if the Gibbs free energy is reduced by the mixing process. The analysis of the mixing entropy (ΔS_m) for different polymers A and B has a form similar to that for substitutional B-type atoms in an A-type crystal in Equation 4.12. However, for polymers the mixing entropy is based upon volumes occupied by LCMs instead of atom fractions, as shown in Equation 4.44.

$$\Delta S_m = -kV\left[\frac{\phi}{v_A}\ln(\phi) + \frac{(1-\phi)}{v_B}\ln(1-\phi)\right] \qquad \textbf{4.44}$$

The volume fraction of the polymer occupied by the A type LCMs (ϕ) is given by Equation 4.45:

$$\phi = \frac{V_A}{V_A + V_B} \qquad \textbf{4.45}$$

where V is the total volume, V_A is the volume occupied by the A-type LCMs, and V_B is the volume occupied by the B-type LCMs. Similarly, for the B-type LCMs the volume fraction is given by Equation 4.46.

$$1 - \phi = \frac{V_B}{V_A + V_B} \qquad \textbf{4.46}$$

The volume occupied by one A-type molecule is v_A, and v_B is the volume occupied by a B-type molecule. Normally in this book the volume fraction occupied by the A-type material is indicated by v_A; however, in polymer science the volume fraction of the A-type material is usually indicated by the symbol ϕ. To be consistent with this literature, the symbol ϕ is used for volume fraction in the discussion of polymer interactions.

A single solvent molecule normally occupies a small volume relative to the total volume (V). For a solvent, the term V/v_A in Equation 4.44 is a very large number, and $\ln \phi$ is a negative number. A solvent can make a large positive contribution to the entropy of mixing, which is one reason why molecules dissolve in solvents. A large mixing entropy term when multiplied by the temperature can force the change in the Gibbs free energy to be negative.

For a large LCM with 10^7 monomers, the term V/v_A is reduced by a factor of approximately 10^7 relative to this term for a small solvent molecule, because the volume of the LCM (v_A) is increased by a factor of 10^7 relative to small solvent molecules. Thus the mixing-entropy contribution provided by an LCM is relatively small. If both A-type and B-type molecules are LCMs, then the mixing-entropy contribution from both terms in Equation 4.44 is small. Because of the small contribution provided by the mixing-entropy term, most LCMs with a high molecular mass do not mix even in liquids.

In polymers the enthalpy of mixing (ΔH_m) is given by Equation 4.47:

$$\Delta H_m = kTV\frac{\chi\phi(1-\phi)}{v} \qquad \textbf{4.47}$$

where χ is the interaction parameter between the two polymers, and v is a reference volume. The reference volume is approximately equal to the monomer volume. Values of the interaction parameter for a polymer blend or mixture are determined as a function of temperature in experiments such as neutron scattering.

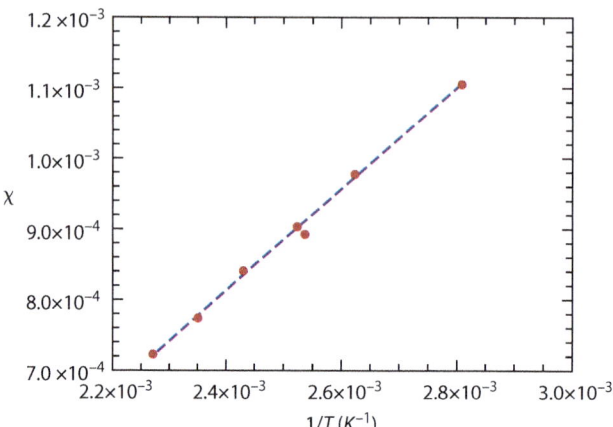

Figure 4.24 The interaction parameter (χ) for SPB 88 and SPB 78, as a function of the inverse of temperature in kelvin ($1/T$). *(Based on Balsara, N. P., Thermodynamics of Polymer Blends, in Physical Properties of Polymers Handbook, Mark, J.E. ed., AIP Press, Woodbury N.Y. (1996), p. 259.)*

The results of one set of experiments are shown in Figure 4.24. This is for a mixture of saturated polybutadiene (SPB) 88 and SPB 78. A polymer is saturated if there are no carbon-carbon double bonds on the LCM. The numbers "88" and "78" indicate the percentage of butadiene monomers on a molecule, and the remaining percentage is ethylene monomers. This interaction parameter is positive and increases with decreasing temperature, and the enthalpy of mixing is positive and increases with decreasing temperature. Mixing should not be expected at low temperatures, based upon the positive interaction parameter.

Example Problem 4.14

For a 50-volume-percent mixture of SPB 88 and SPB 78 with 2000 monomers per polymer molecule, determine the following; (a) the entropy of mixing; (b) the enthalpy of mixing at 393 K (120°C); (c) The Gibbs free energy of mixing at this temperature. In addition (d) should these two polymers mix at this temperature? Assume that the monomer volumes for both SPB 88 and SPB 78 are 0.116 (nm)^3, the reference volume is 0.100 (nm)^3, and the total volume is 1 m^3.

Solution

a) The entropy of mixing is calculated from Equation 4.44 with $V = 1$ m^3 and $\phi = 0.5$. The volume of each polymer molecule is equal to the monomer volume times the number of monomers:

$$v_A = v_B = (0.116 \times 10^{-27}\,\text{m}^3/\text{monomer})(2000\ \text{monomers/polymer molecule})$$

$$v_A = v_B = 0.232 \times 10^{-24}\,\text{m}^3/\text{molecule}$$

$$\Delta S_m = -1.38 \times 10^{-23}\ \frac{\text{J}}{\text{K} \cdot \text{molecule}}\ (1\ \text{m}^3)\left(2\frac{0.5\ \ln 0.5}{0.232 \times 10^{-24}\text{m}^3/\text{molecule}}\right)$$

$$\Delta S_m = -1.38 \times 10^{-23}\ \frac{\text{J}}{\text{K} \cdot \text{molecule}}\ (-3.0 \times 10^{24}\ \text{molecules}) = 41.2\frac{\text{J}}{\text{K}}$$

b) From Figure 4.24, the interaction parameter at 393 K is equal to 9.1×10^{-4}.
    The enthalpy of mixing is calculated from Equation 4.47, with $v = 0.100 \times 10^{-27} \, m^3$.

$$\Delta H_m = kTV \frac{\chi\phi(1 - \phi)}{v}$$

$$\Delta H_m = 1.38 \times 10^{-23} \frac{J}{K \cdot molecule} (393 \text{ K})(1 \text{ m}^3) \frac{(9.1 \times 10^{-4})(0.5)^2}{0.100 \times 10^{-27} m^3/molecule}$$

$$\Delta H_m = 542 \times 10^{-23} \frac{J}{molecule} (2.28 \times 10^{24} \text{ molecules}) = 12{,}338 \text{ J}$$

c) At 120°C the change in the Gibbs free energy is

$$\Delta G_m = \Delta H_m - T\Delta S_m = 12{,}338 \text{ J} - 393 \text{ K} \left(41.2 \frac{J}{K}\right) = 12{,}338 \text{ J} - 16{,}203 \text{ J} = -3864 \text{ J}$$

d) The Gibbs free energy change is negative, and these polymer molecules should mix at 393 K.

The mixing or separation of polymers depends upon the sign of the change in Gibbs free energy after mixing. For thermoplastic polymers, all of the molecular interactions are of the van der Waals type, and they are weak, resulting in a low enthalpy of mixing. If the enthalpy change upon mixing is negative or nearly zero, then the two different LCMs may mix in the liquid state. Some of the polymers that do mix include natural rubber and polybutadiene, polyamide (PA) 6 and PA 6,6, and polyphenylene and polystyrene (PS). A positive and relatively large enthalpy of mixing is likely to result in a positive change in the Gibbs free energy of mixing, and then the A-type and B-type LCMs may not mix even in the liquid state. Some of the polymers that do not mix include PS and PE, PA and PE, and PP and PE. As expected, as temperature is increased there is more tendency for mixing because of the increased contribution of the temperature-entropy change term in the Gibbs free energy. As a general rule, two LCMs do not mix even in the liquid state because of the small mixing entropy; however, there are exceptions to this when the enthalpy of mixing is negative, or very small and positive. For a more detailed discussion of this subject, see the book by Rubenstein and Colby or the article by Balsara referenced in Figure 4.24.

4.8.3   Polymer Solubility and Permeation

Solvents and polymers mix if the Gibbs free energy is reduced by the mixing. If the polymer and the solvent have a similar structure, there is a greater tendency toward mixing. A small amount of solvent mixing with a polymer can lead to polymer swelling, and a large amount of mixing can lead to the polymer dissolving in the solvent. Amorphous polymers are more subject to mixing with solvents than are crystalline polymers. Thermoset polymers with covalent cross-linking between LCMs are less subject to dissolving in solvents than are thermoplastic polymers.

The solubility parameter (δ) in Equation 4.48 is used to determine the tendency of a solvent to mix with a polymer.

$$\delta = \left(\frac{\Delta E}{V}\right)^{1/2}$$ **4.48**

ΔE is the energy in joules, or calories, per mole to evaporate either the polymer or the solvent at zero pressure, and V is the molar volume in cubic meters, or centimeters respectively, per mole. Values of the

Table 4.7 **Values of the Solubility Parameter (δ) of Selected Polymers and Solvents (1 Pa = 1 N/m^2)**

Polymer	δ (MPa)$^{0.5}$	Solvent	δ (MPa)$^{0.5}$
Epoxy resin	22.3	Acetone	20.2
Natural rubber	16.2	Benzene	18.8
Polyacrylonitrile	26.1	Carbon disulfide	20.4
Polyethylene	16.6	Ethanol	26.0
Polytetraforoethylene	12.7	Ethylene glycol	29.9
Polypropylene	18.8	Formic acid	24.7
Polystyrene	18.7	Methanol	29.6
Polyvinyl chloride	19.3	Methyl ethyl ketone	19.0
Polymethylmethacrylate	19.4[1]	Water	47.9
Polycarbonate	20.3[1]	Ammonia	33.3
Polyamide 6,6 (nylon)	27.8[1]	Xylene	18.0

Based on data from Du, Y., Xue, Y., and Frisch H. L., Solubility Parameters, in Physical Properties of Polymers Handbook, Mark, J. E. ed. AIP Press, Woodbury N.Y. (1996), p. 227 except where indicated.

1   Osswald, T.A., and Menges G., Materials Science of Polymers for Engineers, Hanser Publishers, Munich (2003), p. 231.

solubility parameter for polymers and solvents are given in Table 4.7. If the difference in the solubility parameters is less than 2 (MPa)$^{0.5}$ or less than 1 (cal/cm^3)$^{0.5}$, then in general the solvent will dissolve the polymer. For example, acetone has a solubility parameter of 20.2 (MPa)$^{0.5}$, and polyvinyl chloride has a solubility parameter of 19.8 (MPa)$^{0.5}$. The difference is only 0.4 (MPa)$^{0.5}$, and acetone should dissolve polyvinyl chloride. However, polyamide 6,6 (nylon) has a solubility parameter of 27.8 (MPa)$^{0.5}$, and the difference with acetone is 7.6 (MPa)$^{0.5}$. Acetone should not dissolve polyamide 6,6.

The penetration of atoms or molecules of gases or liquids into a polymer is called **permeation**. Permeation is composed of several processes. For gases to permeate a polymer, they must first be adsorbed onto the surface of the polymer. Then gases and liquids must be **absorbed** into the polymer surface and diffuse throughout the polymer. Absorption occurs when the gas atoms penetrate through the surface of the material and into the bulk, diffusing throughout the polymer. Permeation is the product of the solubility of the penetrating atoms or molecules in the host times the penetrating atoms diffusivity in the host.

If the concentration of foreign atoms or molecules is constant on both sides of a polymer film, the rate of diffusion of atoms and molecules through the polymer film follows Fick's first law, stated in Equation 4.38. The permeability (P) of a pure gas through a polymer membrane is determined with Equation 4.49 in an experimental configuration similar to that of Figure 4.18, with a polymer film replacing the palladium alloy film.

$$P = \frac{\dot{m}\Delta x}{A\rho(P_h - P_l)} \qquad\qquad \textbf{4.49}$$

A is the cross-sectional area, $\dot{m}$ is the mass of the atom or molecule permeating per unit time through the membrane, Δx is the membrane thickness, ρ is the density of the polymer, P_h is the high pressure, and P_l is the low pressure. The units of permeation from Equation 4.49 are (length)^2/pressure/seconds. Figure 4.25 shows data on the permeability of water into various polymers, as a function of the inverse of the absolute temperature. The activation enthalpy for permeation of water into these polymers is determined from

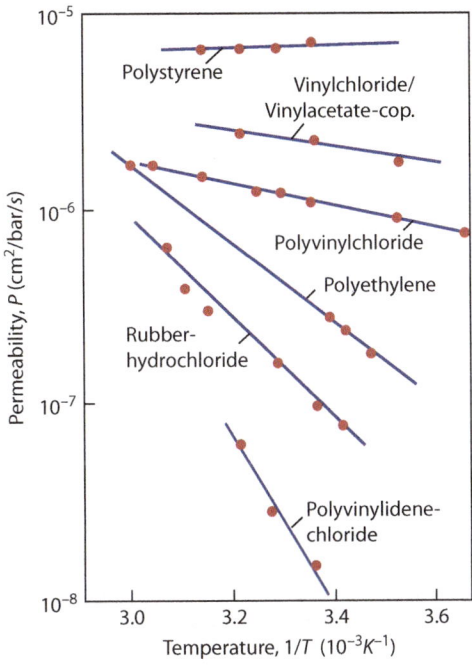

Figure 4.25 The permeability of water into various polymers, as a function of the inverse of the absolute temperature. 1 bar = 10⁵ N/m². *(Based on Osswald, T.A., and Menges G., Materials Science of Polymers for Engineers, Hanser Publishers, Munich (2003), p. 576.)*

the slopes of the plots. From these plots it is obvious that permeability (P) follows an Arrhenius-type equation with an activation enthalpy (ΔH_P), as shown in Equation 4.50.

$$P = P_0 \exp - \left(\frac{\Delta H_P}{kT} \right)$$ **4.50**

Since permeability is the product of solubility times diffusivity, atoms or molecules that are more soluble in a polymer tend to have a high permeability. Polymers that are more open, such as branched polyethylene (PE), have higher rates of diffusion than does high-density polyethylene (HDPE), which is more crystalline.

Polymer films are used in fuel cells to separate the electrolytes. The physical picture is very similar to Figure 4.18 for the purification of hydrogen. In the fuel cell, the polymer film allows the permeation of hydrogen, and it keeps different electrolyte molecules from permeating through the film. The permeation of foreign atoms or molecules into a bulk polymer follows equations similar to Equation 4.40 for case hardening of a metal.

4.9   WHAT IS A CATALYTIC CONVERTER?

The rate of change (R_c) of a process is increased by decreasing the activation enthalpy of the process, as shown in Equation 4.27. The science of changing the activation enthalpy of a process is called catalysis. Automobiles are equipped with catalytic converters, which remove harmful gases such as

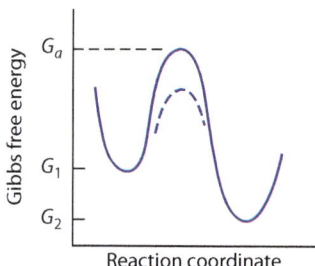

Figure 4.26 The Gibbs free energy as a function of reaction coordinate for a normal process (solid line) and a process with a catalyst (dashed line).

carbon monoxide (CO) from the automobile exhaust through a chemical reaction with oxygen (O_2), which is $2CO + O_2 \rightarrow 2CO_2$. In the exhaust gas the probability of molecules of CO and O_2 colliding and reacting is quite small, because of the low density of molecules in the exhaust gas. However, in the catalytic converter for a gasoline engine, the gases are adsorbed onto the surfaces of palladium (Pd) metal particles that are supported on a ceramic surface. The density of gas adsorbed onto the palladium metal surface is much higher than the exhaust gas density, and the probability of molecules of CO and O_2 reacting is increased. Also, the reaction of the molecules on the Pd metal surface has a smaller enthalpy of activation than for the reaction of the molecules in the gas. The result is a higher rate of formation of carbon dioxide (CO_2). The CO_2, which is less hazardous than carbon monoxide, then evaporates from the catalytic converter surface and enters the exhaust gas. A schematic of a reaction with a reduced activation enthalpy resulting from catalysis is shown in Figure 4.26.

Ice formation is another familiar catalytic effect that involves a surface. Ice initially forms on a surface, such as on the surface of an ice tray or along the edges of a pond, before it forms in the center of the water. One reason that the ice first forms on the tray surface is that the activation enthalpy to form ice on a surface is less than the activation enthalpy for ice to form in the water. The presence of a surface reduces the activation enthalpy required to form the new material. This is why clouds are seeded with small particles of salt or dust to produce rain. The particles provide a surface for the formation of water droplets in the clouds of water vapor. Catalytic effects during phase transformations, such as the solidification of a liquid on a surface, are discussed in Appendix 5B.3, Heterogeneous Nucleation, in Chapter 5.

Summary

- Heat (Q) is the amount of energy that is transferred from a hot system to a cold system because they are at different temperatures.
- In a solid when the temperature is increased, the amplitude of the vibrations of the atoms about their equilibrium atomic position increases.
- For a mole of material, the increment of heat (dQ) added to the material divided by the change in temperature (dT) is the heat capacity (C), and heat capacity is the amount of heat required to increase the temperature of a mole of material by one degree.
- Specific heat (c) is the heat capacity per unit mass, and for elements it is equal to the heat capacity divided by the molar mass (M).

- The heat capacity for solids at high temperature and at constant volume is equal to three times the universal gas constant ($3R$).
- The internal energy for a material is the sum of the kinetic energy and potential energy for all of the atoms.
- The first law of thermodynamics is that energy is conserved.
- The second law is that no heat engine can be 100% efficient.
- The third law is that a perfect crystal at 0 kelvin has 0 entropy.
- When the temperature of a rod is raised or cooled by an amount dT, the length of the rod increases or decreases in length by dL which is given by the original length (L_0) times the linear coefficient of thermal expansion (α_L) times the temperature change (dT).
- In general, strong chemical bonding results in low α_L, and weak chemical bonding results in high α_L. Thermal expansion results from the asymmetry of the interatomic potential as a function of interatomic separation.
- Heat (Q) always flows from an area of high temperature to one of low temperature that are different in temperature by dT over a distance dx. The heat flux (q) is the increment of heat (dQ) that crosses a unit area (A) per unit time (dt), and q is equal to the negative of the thermal conductivity (k_T) times the temperature gradient (dT/dx).
- Heat conducts through materials by exciting lattice vibrations and increasing their amplitude. Materials with strong bonding have a high thermal conductivity, as do metals with free electrons.
- A thermodynamic system consists of the atoms and molecules of whatever material or object is under study.
- Entropy is related to the disorder in a system and to the number of ways of arranging the system. If a system is completely ordered, and the atoms are at rest at zero kelvin, there is only one way to arrange the system, and the entropy is zero. If the system is disordered, there are many ways to arrange the system, and this system has a high entropy.
- Enthalpy is internal energy plus the pressure times the volume.
- The Gibbs free energy is the enthalpy minus the temperature times the entropy. The Gibbs free energy is a minimum when a system is at equilibrium at constant temperature and pressure.
- When atoms mix, this increases the disorder and the entropy.
- Point defects in a solid increase the disorder and the entropy. This results in an equilibrium number of point defects that minimizes the Gibbs free energy.
- The rate of change in a system that is thermally activated is given by the Arrhenius rate equation. The rate of change of a system at constant temperature and pressure is exponentially dependent upon the negative of the activation enthalpy divided by the absolute temperature.
- Atoms of similar size in solids mix by vacancy diffusion.
- Small atoms can mix into a solid of larger-size atoms by interstitial diffusion.
- Diffusion rates are higher along defects such as edge dislocations, surfaces, grain boundaries, and phase boundaries than through the bulk.
- Steady-state diffusion is present when the concentration gradient with respect to position of the diffusing species is constant. The most general steady-state diffusion rate equation predicts

that atoms move to decrease the Gibbs free energy. However, this equation is not of much experimental use, because the Gibbs free energy cannot be measured.

- Fick's first law of diffusion predicts the flux of atoms in a constant concentration gradient, and it predicts that atoms move in a direction to decrease the concentration gradient. Fick's second law allows the prediction of concentrations of atoms and molecules with changing concentrations.

- The entropy of mixing long-chain molecules (LCMs) is inversely proportional to the volume occupied by the molecules. Because the LCMs occupy a relatively large volume in comparison to that of an individual atom, there is very little contribution from the temperature entropy term in the Gibbs free energy to force the mixing of long-chain polymer molecules. As a result many LCMs do not mix even in the liquid state.

- The solubility parameter (δ) is equal to the energy (ΔE) in joules (or calories) per mole to evaporate either the polymer or the solvent at zero pressure, divided by the molar volume in cubic meters (or centimeters) per mole (V), all of this to the one-half power.

- A polymer is likely to mix with a solvent if their solubility parameters are similar.

- The penetration of atoms or molecules of gases or liquids into a solid polymer is called permeation. Permeation is composed of three processes: adsorption of atoms or molecules onto the surface of the polymer, absorption into the polymer surface, and diffusion through the polymer.

- The science of changing the activation enthalpy of a process is called catalysis.

Supplemental Reading Subjects and Authors

Full references are listed at the end of the book.

General:	*Askeland, Fulay, and Wright*
Thermodynamics:	*Callen; Kondepepudi and Prigogine; Ragone; Swalin; Gibbs*
Diffusion:	*Shewmon; Smithells*
Thermodynamics, diffusion, and permeation in polymers:	*Oswald and Menges; Mark; van Krevelen; Rubenstein and Colby; Balsara*

Homework

Concept Questions

1. Heat (Q) is the amount of _____ that is transferred from an area at high temperature to an area a at low temperature.

2. When the temperature of a material is increased, the _____ of the lattice vibrations increases.

3. For a mole of material, the increment of heat (dQ) added to the material divided by the change in temperature (dT) is the heat _____.

4. The heat capacity is also known as the _____ specific heat.

5. Specific heat (c) is the heat capacity per unit _____.

6. The primary reason for the high specific heat of polymers is their low monomer _____.

7. In semicrystalline polymers, there is a spike in the specific heat when the crystallites start to _____.

8. The sum of the kinetic and potential energy is equal to the _____ energy.

9. For a material that has no applied fields, the major contribution to the potential energy is the energy of chemical _____.

10. The _____ is the internal energy plus the pressure times the volume.

11. Internal energy is a _____ variable.

12. In a solid, the average kinetic energy is equal to the average _____ energy.

13. The high-temperature heat capacity at constant volume for a solid or liquid is equal to _____.

14. The change in length of a material per degree divided by the original length is the linear coefficient of thermal _____.

15. A(n) _____ material has the same properties in all directions.

16. For a polymer filament, the linear coefficient of thermal expansion would be expected to be (more or less) in the axial direction than in the transverse direction.

17. A material with an interatomic potential perfectly symmetric about the equilibrium interatomic separation would have a coefficient of thermal expansion equal to _____.

18. Copper melts at 1085°C and platinum melts at 1769°C. Which material should have the smaller coefficient of thermal expansion?

19. In heat transfer, the heat flux is directly related to the temperature _____.

20. Diamond has a notably high thermal conductivity because of its strong _____ _____.

21. What has the higher thermal conductivity, metallic aluminum or aluminum oxide? _____

22. Which should have the higher thermal conductivity, a single crystal gas-turbine blade or a polycrystal turbine blade? _____

23. HDPE is more crystalline than LDPE; of these two _____ should have a lower thermal conductivity.

24. The diffusion of atoms of similar size depends upon the presence of _____.

25. The entropy of a perfect crystal at absolute zero kelvin with the atoms at rest is equal to _____.

26. A mole of pure argon gas has (more or less) entropy than a mixture of a half mole of argon and a half mole of neon?

27. Which has the greater entropy, ice or water?_____

28. When the equilibrium concentration of vacancies is present in a crystal at constant temperature and pressure, the Gibbs free energy is at a(n) _____.

29. The probability that the energy to create a vacancy in a crystal is present in excess of the average energy in the crystal is given by the _____-_____ probability distribution.

30. If at constant temperature and pressure a change in a system produces a negative change in the Gibbs free energy, the change is _____.

31. The rate of change in a system depends upon the _____ enthalpy.

32. If the natural logarithm of the rate of a thermally activated process is plotted on the y axis as a function of the inverse of the absolute temperature on the x axis, the slope of this plot is equal to the _____ _____ divided by Boltzmann's constant.

33. The activation enthalpy for vacancy diffusion contains two contributions: the activation enthalpy for vacancy _____ and atom migration.

34. Which would have the higher activation energy for diffusion, carbon in iron or nickel in iron? The atomic radius of carbon is 0.077 nm, the atomic radius of nickel is 0.124 nm, and the atomic radius of iron is 0.124 nm. _____

35. The activation enthalpy for self-diffusion is directly related to the strength of chemical _____.

36. The activation enthalpy for vacancy diffusion along a grain boundary is (more or less) than that for diffusion through the bulk.

37. In the most general form of the first law of diffusion, the gradient of the Gibbs free energy with respect to the position of B-type atoms is equal to the chemical _____ on the B-type atoms.

38. Amorphous polymers are (more or less) subject to mixing with solvents than are crystalline polymers.

39. The solubility parameter of polycarbonate is 20.3 $(MPa)^{0.5}$ and for carbon disulfide it is 20.4 $(MPa)^{0.5}$. Should carbon disulfide dissolve polycarbonate? _____

40. The solubility parameter of polyethylene is 16.6 $(MPa)^{0.5}$ and for ethanol it is 26.0 $(MPa)^{0.5}$. Should ethanol dissolve polyethylene? _____

41. The solubility parameter of a polymer is related to the energy of _____ for the polymer.

42. The enthalpy of mixing for a pair of polymer molecules is related to the _____ parameter of the two molecules.

43. The science of changing the activation enthalpy of a process is called _____.

Engineer in Training–Style Questions

1. The internal energy of a material does not depend upon which of the following?
   (a) Energy of defects in the material
   (b) Kinetic energy of the vibrating atoms
   (c) Bond energy between atoms
   (d) Net energy resulting from a history of heating and cooling

2. The sequence from (a) to (d) in this question should be from high thermal conductivity to low thermal conductivity. Which material is out of sequence?
   (a) OUHMWPE fibers parallel to the fiber axis
   (b) PMMA
   (c) HDPE
   (d) LDPE

3. Which of the following does not result in high thermal conductivity?
    (a) Strong bonding
    (b) Free electrons
    (c) Grain boundaries
    (d) Small interatomic distances

4. Which of the following material states has the highest entropy?
    (a) Amorphous solid
    (b) Vapor
    (c) Liquid
    (d) Single crystal

5. Which of the following is not a state thermodynamic variable?
    (a) The Gibbs free energy
    (b) Entropy
    (c) Heat
    (d) Enthalpy

6. The mixing of which pair of the following atoms does not occur by vacancy diffusion?
    (a) Gold and silver
    (b) Copper and nickel
    (c) Platinum and palladium
    (d) Hydrogen and palladium

7. Which of the following polymer combinations is least likely to mix?
    (a) Liquids of two different long-chain molecules with a positive enthalpy of mixing.
    (b) A liquid solvent and solid polymer with solubility parameters of $19\,MPa^{0.5}$ and $20\,MPa^{0.5}$, respectively.
    (c) Liquids of two different long-chain molecules with a negative enthalpy of mixing.
    (d) Liquids of two different short molecules with a positive enthalpy of mixing.

Design Question

You are asked to select a material for a container on a ship for liquified natural gas that will be at $-163°C$ ($-260°F$). For the classes of metals, ceramics, polymers, and composites, give the advantages and disadvantages of the best possible material within each class of material for this design based upon subjects you have studied in Chapters 1 to 4 in this book. For example, you may decide that carbon fiber reinforced epoxy is the best possible composite material for this design. Give the advantages and disadvantages of carbon fiber reinforced epoxy for this design.

Problems

Problem 4.1   In a solar hot-water system, the hot-water tank is maintained at 130°C. It is desired to have the solar heater heat the feed water going into the tank to 180°C. Determine the amount of heat per kilogram of water required from the solar heater to heat the water from 130°C to 180°C. Assume that the hot-water heater runs at constant pressure.

Problem 4.2   Calculate the specific heat of copper, silver, and gold at constant volume using the high-temperature value of the heat capacity, and compare these values to their specific heats at constant pressure listed in Table 4.1.

Problem 4.3   The specific heat of copper is 385 J/kg · K, the melting temperature is 1356 K, and the latent heat of fusion is 2.05×10^5 J/kg. Compare the heat necessary to raise the temperature of 1 kg of copper from 300 K to 1356 K to the heat necessary to melt the copper at 1356 K.

Problem 4.4   Determine the average kinetic energy per atom due to thermal vibrations in a solid at 1000 K. You can assume that the heat capacity at constant volume for the solid is equal to the classical limit of $3R$ for all temperatures, and that the material is at constant volume during heating. Express your answer in eV/atom.

Problem 4.5   A piece of wire 15 m long is cooled from 40°C to −9°C and the change in length is −12.7 mm. (a) What is the linear coefficient of thermal expansion for this material? (b) Identify the material based upon the measured coefficient of thermal expansion.

Problem 4.6   A precision yellow brass casting is to be produced that should have a final length of 0.2000 m at room temperature (23°C). The casting is produced by pouring liquid brass into the cavity of a mold and allowing the liquid to solidify. The melting temperature of the brass is 800°C. What should be the mold cavity length?

Problem 4.7   A solid shaft of 1020 steel with an outside radius of 1.002 cm is to be inserted inside a hollow 1020 steel shaft with an inside radius of 1.000 cm. This is to be accomplished by heating the hollow shaft sufficiently so that the solid shaft can be inserted into the hollow shaft with a radial clearance of 0.002 cm. When the two shafts are brought to room temperature, they will be bonded together as a single unit. This process is called a shrink fit. To what temperature must the hollow shaft be heated for this process, assuming that the solid shaft is at 300 K?

Problem 4.8   Steel beams each 10 m in length at 25°C are used in a bridge design. The design temperature range is from −20°C to 30°C. The coefficient of linear expansion for the steel is 12.5×10^{-6} °C^{-1}. Calculate the total change of length possible for the beams in this design.

Problem 4.9   In the calculation of the centripetal forces in a nickel-based gas-turbine blade, the density of the blade at the operating temperature of 1000°C is needed. However, only the room-temperature density is available. Using the thermal expansion coefficient for nickel, estimate the density of nickel at 1000°C. You can assume that the expansion of the nickel in the turbine blade is isotropic.

Problem 4.10   What is the amount of heat transferred in an hour through a 1 m² window made of soda-lime glass that is 2 mm thick and has a thermal conductivity (k_T) of 1.7 watts/m · K if the outside temperature is 0°C and the inside temperature is 25°C?

Problem 4.11   In automobiles, brass radiators have been replaced with aluminum ones to reduce weight. Assume that the wall thickness in each radiator is 0.5 mm, the engine fluid is at a temperature of 400 K, and the ambient air temperature is 300 K.

(a)   Calculate the heat flux for each radiator material.

(b)   Estimate the relative area in percent and the relative weight of changing from a brass radiator to an aluminum radiator, assuming the same heat transfer rate (dQ/dt) for both designs and that the brass is 30 weight percent zinc and 70 weight percent copper with a density of 8.4 g/cm³.

Problem 4.12 Calculate the entropy of mixing for 2 moles of an alloy that has a composition of 30 atom percent copper and 70 atom percent nickel.

Problem 4.13 (a) Use a spreadsheet program such as Microsoft Excel or write a computer program to calculate and then plot the entropy change relative to the pure separated elements for mixing 1 mole total of A-type and B-type atoms that form an ideal solution. Make this calculation in increments of 0.1 atom fraction change. (b) What chemical composition has the largest entropy of mixing, and what is this value?

Problem 4.14 If the mixing of 10 atom percent copper into a silver crystal occurs at $T = 1000$ K (727°C), determine the change in the Gibbs free energy on a per-mole basis if the enthalpy of solution for copper into silver is 0.25 eV/atom. (b) Based upon your result, should it be possible to mix 10 atom percent copper into silver? Explain your answer, and discuss this result in comparison to Example Problem 4.8.

Problem 4.15 The enthalpy of formation of one vacancy in pure gold is 0.94 eV.

    (a) What is the fraction of atom sites that are vacant in gold at 1000 K?

    (b) Determine the contribution to the volume coefficient of expansion for gold that is due to the formation of vacancies. Use temperatures of 0 K and 1000 K.

    (c) Determine the fraction of the volume coefficient of expansion for gold that is due to the formation of vacancies.

Problem 4.16 Calculate the concentration of vacancies in silicon at the solid-liquid equilibrium temperature of 1683 K where single crystals of silicon are grown. The enthalpy of formation for vacancies in silicon is 2.3 eV/vacancy.

Problem 4.17 A thin-film gold conductor for an integrated circuit in a satellite application is deposited from a vapor, and the deposited gold thin film has a high resistivity. It is proposed that the high resistivity is due to a high nonequilibrium vacancy concentration, created when the vapor atoms rapidly condense into the thin film. To reduce the vacancy concentration, it is proposed to anneal (heat) the gold to a temperature that will result in only 1 vacancy for every 10^8 gold atoms, and then to cool the film slowly to room temperature. The enthalpy of formation for 1 vacancy in gold is equal to 0.94 eV. Specify the temperature for this anneal.

Problem 4.18 In a scientific presentation, an author stated that FCC metals melt when there is on the average one vacancy in the atom positions surrounding each atom.

    (a) Check the validity of this statement for copper by comparing the vacancy concentration when there is one vacancy in the atom positions surrounding each atom with the equilibrium vacancy concentration in copper at the melting temperature of copper (1356 K). The vacancy formation enthalpy in copper is 1.0 eV per vacancy.

    (b) Calculate the entropy of mixing the equilibrium number of vacancies into a copper crystal at the melting temperature, with a total number of crystal sites equal to 1 mole.

    (c) Calculate the change in the Gibbs free energy when the equilibrium number of vacancies forms relative to that of a perfect crystal with 1 mole of total sites.

Problem 4.19 The following data was recorded for the complete recrystallization of heavily cold-worked copper by B. F. Decker and D. Harker, *Trans. AIME* 188 (1950), p. 887. Cold work results in the formation of many dislocations and other defects in crystalline grains. Recrystallization is the growth of new low-defect grains from the highly defective grains.

Temperature (K)	Time (Seconds)
316	2.3×10^6
361	33,000
375	10,000
385	7,000
392	4,000
408	1,500

(a) Plot the logarithm of the rate of recrystallization versus the inverse of the absolute temperature $(1/T)$ to see if it is linear.

(b) Calculate the activation enthalpy for the recrystallization of copper.

(c) The activation enthalpy for vacancy diffusion in an annealed crystal of copper is 2.2 eV/atom. Discuss possible reasons for any difference in your result from part (b).

(d) Determine the temperature where copper completely recrystallizes in 1 hour. This is called the recrystallization temperature.

Problem 4.20 Determine the activation enthalpy and pre-exponential constant for the diffusion of oxygen (O^{18}) into ZnO using the data presented in Figure 4.21.

Problem 4.21 A laboratory-scale hydrogen purifier is made of a tube of palladium with 25 weight percent silver, with an inside radius of 5.11×10^{-4} m, an outside radius of 6.35×10^{-4} m, and a length of 5.81 m. The purifier is operated with an inside high pressure of 1×10^6 N/m^2 for low-purity hydrogen; an outside low pressure of 1×10^4 N/m^2, which is approximately 1 atmosphere, for the high-purity hydrogen; and a temperature of 673 K. Determine the number of moles per second of purified hydrogen gas that can be produced. Note that since the inside and outside area of the tube are not equal, the effective area (A) is given by the log normal area.

$$A = \frac{2\pi l \, (r_o - r_i)}{\ln\left(\dfrac{r_o}{r_i}\right)}$$

Problem 4.22 Nitrogen gas is used to harden iron by diffusing the nitrogen into BCC iron at 700°C. If the surface concentration of nitrogen is maintained at 0.1 weight percent, what is the expected concentration of nitrogen 1 mm from the surface after 10 hours? You can assume that at small values of z, erf(z) is approximately equal to z.

Problem 4.23 In case hardening, carbon is diffused into low-carbon iron to produce a hard surface layer that resists wear and deformation. At a temperature of 600°C it is observed that the hardened surface layer extends 0.001m into the surface of the iron after exposure to carbon for two hours. It is desired to have the hardened surface layer extend 0.002m into the iron. At 600°C how many hours of carburizing are required?

Problem 4.24 A steel alloy that has an initial carbon concentration of 0.1 weight percent carbon is to be case hardened by packing it in carbon that maintains a surface concentration of 1 weight percent carbon at 1000°C. After 11.8 hours the carbon concentration is 0.5 weight percent at a depth of 1 mm into the steel. What time is required to obtain a carbon concentration of 0.5 weight percent at a depth of 2 mm into the steel?

Problem 4.25 For a 25 volume percent mixture of SPB 88 and 75 volume percent SPB 78 with 2000 monomers per polymer molecule, determine the following: (a) The entropy of mixing. (b) The enthalpy of mixing at 100°C (373 K) using Figure 4.24. (c) The Gibbs free energy of mixing at 100°C (373 K). (d) Should these two polymers mix at this temperature? Assume that the monomer volumes for both SPB 88 and SPB 78 are 0.116 $(nm)^3$, the reference volume is 0.100 $(nm)^3$, and the total volume is 1 m^3.

A photograph of a blast furnace to produce iron from ore. The iron comes out of the furnace as a liquid. Why is iron a liquid when heated to a high temperature? The heat of fusion is needed to melt any material, and this inputs energy. Figure 4.7 shows a ball on a surface, and it always goes to the minimum internal energy state. Adding energy in the form of the heat of fusion to produce liquid iron is like raising a ball to a height (h). If we release the ball, it always drops down to a lower internal energy state. But the liquid iron is stable above the melting temperature in this higher internal energy state. To understand why materials melt and stay in this higher internal energy state, we must also consider the entropy change. In this chapter we will learn why materials melt, study mixtures of atoms and molecules, and develop phase diagrams.

© Prisma Bildagentur AG / Alamy

The goals of this chapter are to understand

- The energy and entropy changes that occur during a phase transformation
- The reasons why materials have phase transformations
- Nucleation and growth transformations
- The conditions for different phases to be in equilibrium
- Methods for reading a binary phase diagram
- The kinds of information obtainable from a phase diagram
- The reasons why different materials have different phase diagrams
- Binary phase diagrams for ceramic mixtures
- Martensitic transformations
- Time, temperature, transformation (TTT) diagrams
- The shape memory effect in metals and polymers
- TTT diagrams for amorphous to crystal transformations
- Phase diagrams for polymer blends

Chapter

5

Phase Transformations and Phase Diagrams

5.1   INTRODUCTION

You are designing an electromechanical device that has to operate at temperatures up to 200°C, and you want to solder two pieces of copper together. You call the stock room and locate solder that is a mixture of 60 weight percent lead and 40 weight percent tin. You check Appendix B and find that the melting temperature of lead is 327°C and that of tin is 232°C. You assume that there is a linear relationship between composition and melting temperature, and that the melting temperature of the solder is around 275°C. The assumption of linear relationships makes analysis easy. You solder the pieces and then test the part at 200°C. The solder melts and the two pieces of copper fall apart, and you wonder why. If you had checked the phase diagram of lead and tin in this chapter, you would have discovered that your solder melts at 183°C. This melting temperature is below that of either element in the solder and much lower than the melting temperature you calculated. This relationship is certainly not a linear one.

The melting of a material is a phase transformation. A **phase** is a homogeneous portion of a material that has uniform physical and chemical characteristics. A **phase transformation** is when a material changes from one phase to another. Phase transformations enter into many material processes, such as when parts are soldered or welded, or when metal is melted and cast into a shape that is then cooled to form a solid part. Phase transformations also occur in solids; for example, iron transforms upon heating from BCC to FCC at 912°C.

When the material being processed is made from two components, such as lead and tin, the phases present, their compositions, and the amount of each phase present are determined from binary phase diagrams that we will cover in this chapter. In Section 5.7 we will determine why our solder melted at 183°C and not at our calculated 275°C.

5.2   ENERGY AND ENTROPY CHANGES DURING PHASE TRANSFORMATIONS

Recall from Chapter 4 that for systems at constant temperature and pressure, the Gibbs free energy (GFE) is used to determine the equilibrium state of the system. When a phase transformation occurs in a single-phase material, such as H_2O, there is no change in chemical composition, but there is a change in structure from ice to water, and this changes the entropy. Ice is one phase and water is another phase. When there is a reversible heat transfer (ΔQ_R) at the absolute temperature T, the change in entropy (ΔS) is given by Equation 5.1.

$$\Delta S = \frac{\Delta Q_R}{T}$$

5.1

This equation was developed through the study of phase transformations and the efficiency of heat engines. Rudolph Clausius was the first to propose the concept of entropy, and the word itself is derived from the Greek word for transformation. The heat transfer is reversible if the system is always in equilibrium during the heat transfer. From a practical viewpoint, this means that the heat must be added or subtracted very slowly. With the entropy change calculated by Equation 5.1, Appendix 5A shows that the GFE is the energy available to do work by a system at constant temperature and pressure. The name Gibbs free energy (G) results from G being the energy that is free to do work at constant temperature and pressure.

As heat ΔQ is added to materials and the temperature increases, the entropy also increases. Additionally, when ice transforms to water, the latent heat of fusion (ΔQ_f), given in Table 5.1, must be added to the ice to form water. This addition increases the entropy, as shown by Equation 5.1. Also, from a statistical viewpoint water is more disordered than ice, and water has the larger entropy. The increase in entropy when ice melts to form water decreases the GFE, as shown by Equation 4.13, and this results in water at temperatures above 273 K and 1 atmosphere of pressure having a lower GFE than ice does.

If water is converted to ice at a temperature less than 273 K, there is a decrease in entropy. A form of the second law of thermodynamics is that the entropy of the universe must increase for any change in the universe. In addition there is a postulate of thermodynamics that a thermodynamic system always goes to a state of maximum entropy. The freezing of water to form ice would appear to violate the second law of thermodynamics and the postulate, because the entropy of the H_2O decreases. However, when disordered water changes to ordered ice at a temperature such as 270 K, the latent heat of fusion (ΔQ_f) is transferred to the environment resulting in an entropy change of $\Delta Q_f/270$ K for the environment around the ice. This change in entropy more than offsets the change in entropy due to water changing to ice that is equal to $-\Delta Q_f/273$, as shown in Example Problem 5.1. In any of the following discussions where atoms or molecules become more ordered, such as in solidification, and the entropy of the material decreases, it must be remembered that heat transfer to the environment occurs to always result in an increase in the entropy of the universe.

Table 5.1	The Melting Temperature (T_m), Heat (or Enthalpy) of Fusion (ΔQ_f), Vaporization Temperature (T_v), and Heat (or Enthalpy) of Vaporization (ΔQ_v) of Some Materials			
Material	(T_m °C)	(ΔQ_f kJ/mole)	(T_v °C)	(ΔQ_v kJ/mole)
Aluminum	660.32	10.71	2519	294
Argon	−189.36	1.18	−185.85	6.43
Boron	2075	50.2	4000	480
Germanium	938.25	36.94	2833	334
Gold	1064.18	12.55	2856	324
Lead	327.46	4.774	1749	179.5
MoO_3	802	48.7	1155	138
Water	0.0	6.01	99.97	40.65

Based on data from CRC Handbook of Chemistry and Physics, ed. D.R. Lide, 90th ed. 2009–2010, p. 6–130.

Example Problem 5.1

(a) Calculate the entropy change of H_2O when one mole of water transforms to ice at the equilibrium temperature of 273 K and 1 atmosphere of pressure (atm). (b) Calculate the entropy change to the thermodynamic system of H_2O and environment if one mole of water changes to ice at 270 K and one atm. (c) Calculate the entropy change to H_2O when one mole of water transforms to water vapor at the equilibrium temperature 373 K and at one atmosphere of pressure.

Solution

a) To calculate the entropy change to H_2O, use Equation 5.1. The heat of fusion of ice ΔQ_f from Table 5.1 is equal to 6.01 kJ/mole, and the equilibrium temperature is 273 K. When water changes to ice the heat is removed from the H_2O, and heat is negative when removed from the material.

$$\Delta S_1 = \frac{\Delta Q_1}{T} = \frac{-\Delta Q_f}{T} = \frac{-6.01 \text{ kJ/mole}}{273 \text{ K}} = -2.20 \times 10^{-2} \frac{\text{kJ}}{\text{mole} \cdot \text{K}}$$

b) Water will never transform completely to ice at 273 K, because ice and water are at equilibrium at 273 K. To transform ice to water it is necessary to super-cool the water to a temperature below 273 K to a temperature such as 270 K. If water is transformed to ice at 270 K the entropy change of the H_2O is still approximately given by ΔS_1 because it is assumed that the molecular arrangements of super-cooled water and ice at 270 K are essentially the same as at 273 K. However, when the ice is formed at 270 K the latent heat of fusion is transferred from the water to the environment at 270 K, and when considering the change in entropy of the environment, this is a positive heat transfer. It is assumed that the air around the water and ice is at 270 K. The entropy change due to this heat transfer (ΔS_2) is given by Equation 5.1.

$$\Delta S_2 = \frac{\Delta Q}{T} = \frac{6.01 \text{ kJ/mole}}{270 \text{ K}} = 2.23 \times 10^{-2} \frac{\text{kJ}}{\text{mole} \cdot \text{K}}$$

The total change in entropy ΔS is the sum of ΔS_1 and ΔS_2.

$$\Delta S = \Delta S_1 + \Delta S_2 = (-2.20 + 2.23) \times 10^{-2} \frac{\text{kJ}}{\text{mole} \cdot \text{K}} = 0.03 \times 10^{-2} \frac{\text{kJ}}{\text{mole} \cdot \text{K}}$$

For the freezing of water to form ice the net entropy change is positive, in agreement with the second law of thermodynamics.

c) The heat of vaporization of water (ΔQ_v) is 40.65 kJ/mole, and the equilibrium temperature of water and vapor is 373 K. The change in entropy of the H_2O during the vaporization process is equal to

$$\Delta S_v = \frac{\Delta Q_v}{T} = \frac{40.65 \text{ kJ/mole}}{373 \text{ K}} = 10.9 \times 10^{-2} \frac{\text{kJ}}{\text{mole} \cdot \text{K}}$$

In Example Problem 5.1, the entropy change upon transforming water to water vapor is positive, because the disorder increases in water vapor. Also, the entropy change in transforming water into vapor is nearly five times as large as the entropy change that would occur in transforming ice into water ($+2.20 \times 10^{-2}$ kJ/mole · K). This shows that there is a greater change in disorder when going from water to vapor that from ice to water.

5.3 PHASE TRANSFORMATIONS AND PHASE DIAGRAMS IN ONE-COMPONENT SYSTEMS

The melting of ice to form water is an example of a phase transformation. The components are the atoms or molecules in the phase. Molecules are components when the molecule is not altered during a phase transformation. Ice and water are two phases that are both made from the single component H_2O. The H_2O molecule is not altered when ice transforms to water; only the orientation of H_2O molecules relative to each other changes. In a pure material composed of one component at constant pressure, the melting temperature (T_m) is the temperature where the Gibbs free energy (GFE) per atom of the liquid and of the solid are equal, as shown in Figure 5.1. When two phases, such as ice and water, have the same GFE, they are in equilibrium with each other, and the H_2O molecules have no preference of one phase over the other. Figure 5.1 is a plot of the GFE per atom of a liquid (G^L) phase and solid (G^S) phase of some

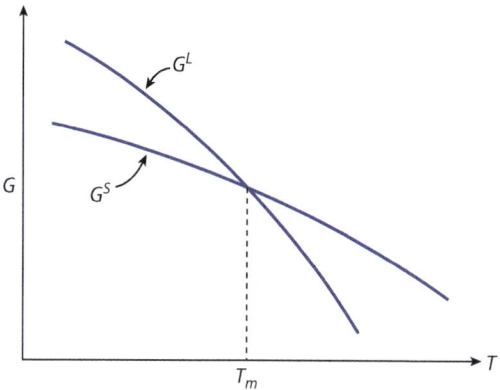

Figure 5.1 The GFE per atom for the liquid (G^L) phase and solid (G^S) phase of a pure element or compound, as a function of temperature (T). The melting temperature (T_m) is where G^L and G^S cross and are equal.

material, such as H_2O. In this book superscripts refer to phases. Although temperature is shown as a variable in Figure 5.1, it is assumed that for each point along the plot, the material is held at a constant temperature and pressure for a long time until the materials reach their appropriate energies. Therefore, the materials are at constant pressure and temperature for each point plotted in Figure 5.1. This section considers **reversible transformations**. If the transformation from the α phase to β phase is reversible, the α phase can be transformed to the β phase, and the β phase can be transformed back to the α phase without a change to the GFE for the total system.

Example Problem 5.2

An insulated container has both ice and water at 273 K. If 1 mole of water transforms to ice at 273 K at a constant pressure of 1 atmosphere, what is the change in the GFE of the water that transforms to ice?

Solution

From Example Problem 5.1, when 1 mole of water changes to ice, the entropy change is the negative of that calculated for ice transforming to water, or -2.2×10^{-2} kJ/mole $\cdot$ K. From Equation 4.14, the change in the GFE is given by

$$\Delta G = G - G_0 = \Delta E + P\Delta V - T\Delta S = \Delta H - T\Delta S$$

The heat transferred (ΔQ) is equal to the enthalpy change (ΔH). In Table 5.1 the heat of transformation is listed, but many tables present the numerically equal enthalpy of transformation. When a material freezes, the amount of heat $-\Delta Q_f$ is extracted from the material, and $-\Delta Q_f$ is equal to the enthalpy change (ΔH) for freezing. If water freezes at 273 K, the heat of 6.01 kJ/mole is extracted from the material, and this is negative because our convention is that heat input to the material is positive. Inserting these values into Equation 4.14 for the change in the GFE results in

$$\Delta G = \Delta H - T\Delta S = -6.01\frac{\text{kJ}}{\text{mole}} - 273 \text{ K}\left(-2.2 \times 10^{-2}\frac{\text{kJ}}{\text{mole} \cdot \text{K}}\right)$$

$$\Delta G = -6.01\frac{\text{kJ}}{\text{mole}} + 6.01\frac{\text{kJ}}{\text{mole}} = 0$$

The change in the GFE is equal to 0 if water transforms to ice when the materials are at the equilibrium temperature. At this temperature the transformation is reversible because there is no change in the GFE in the transformation. Below we will see that at the equilibrium temperature, the transformation takes an infinite amount of time to occur. Super-cooling the materials is necessary for the transformation to occur at a finite rate.

If a material can exist in multiple phases, the phase that has the lowest GFE per atom is the stable phase. Below the melting temperature the GFE per atom for the solid is less than for the liquid ($G^S < G^L$), and above the melting temperature the GFE per atom for the solid is more than for the liquid ($G^S > G^L$). The GFE of the solid is plotted for temperatures above the melting temperature, because it can be calculated even though the solid is not stable at these temperatures. A similar calculation can be made for liquid below the melting temperature. Appendix 5B.1 presents the theory of these calculations and the thermodynamics of phase transformations. The GFE per atom in Figure 5.1 is shown decreasing as temperature increases, because of the $-TS$ term in the GFE in Equation 4.13. In a material as temperature increases the internal energy increases and at the same time the $-TS$ term becomes more negative. The actual slope of the GFE versus the temperature plot must be calculated. The entropy of a liquid phase is greater than the entropy of a solid crystalline phase of the same material ($S^L > S^S$). Therefore, the curve for the liquid GFE per atom decreases more rapidly than for the solid as temperature is increased, as shown in Figure 5.1, because the term $-TS^L$ is more negative than the term $-TS^S$. This is why the G^S and G^L curves cross, and the material melts at the temperature where the two curves cross. The temperature dependence of the GFE of solids and liquids allows us to explain why a material melts.

A similar argument is made for vaporization. A vapor is more disordered than a liquid; therefore, the entropy of the vapor (S^V) is greater than the entropy of the liquid (S^L). Therefore, the vapor GFE per atom decreases more rapidly than for the liquid as the temperature increases. The GFE per atom of the vapor (G^V) and the liquid (G^L) are equal for one-component systems at the equilibrium vaporization temperature. Below the equilibrium vaporization temperature G^L is less than G^V, and above the equilibrium vaporization temperature G^L is greater than G^V. At some temperature, the vapor and liquid GFE per atom curves cross, and the material is a vapor at temperatures above the equilibrium vaporization temperature. Table 5.1 gives the equilibrium temperatures between some solids and liquids and the heat of fusion, and it also gives the equilibrium vaporization temperature and the heat of vaporization. Some solids sublime to a vapor rather than melt; in these materials the crossover of the solid and vapor Gibbs free energies per atom occurs at a lower temperature than does the crossover of the solid and liquid Gibbs free energies per atom.

The GFE of a material is a function of pressure in addition to temperature, as shown in Equation 4.13. The temperatures and pressures where phases are in equilibrium are shown in equilibrium one component **phase diagrams**, such as the phase diagram for H_2O in Figure 5.2. All of the phase diagrams in this book are equilibrium phase diagrams. In Figure 5.2, temperature in °C is on the x axis and pressure in atmospheres (atm) on a logarithmic scale is on the y axis. The phases present are labeled in the open areas. To determine the phase of H_2O present at a particular temperature and pressure, draw a vertical line at the temperature and a horizontal line at the pressure, and the point of intersection of these lines gives the phase present. For example, we know that under ambient conditions that H_2O is a liquid, and 23°C and 1 atmosphere of pressure (horizontal dashed line) lies in the region of liquid. Water freezes at 0°C and 1 atmosphere of pressure. Water and ice are in equilibrium under these conditions, and this point lies on the line between the solid and liquid phases. Water boils at 100°C and 1 atmosphere of pressure; therefore water and vapor are in equilibrium under these conditions, and this point lies on the line between vapor and liquid phases. Lines that separate two phases in the diagram give the conditions for equilibrium between the two phases. The line between liquid and vapor in Figure 5.2 gives the temperatures and

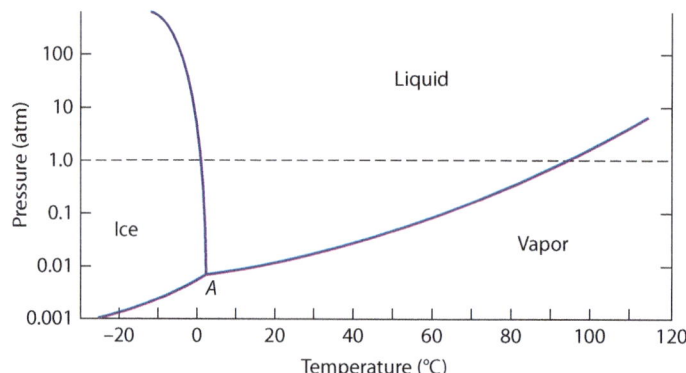

Figure 5.2 The phases of H_2O as a function of temperature in °C and pressure in atmospheres. (*Adapted from Chaing, Y.-M., Birnie, D. B., and Kingery, W. D., Physical Ceramics—Principles for Ceramic Science and Engineering, John Wiley & Sons, New York (1997), p. 269.*)

pressures where liquid and vapor are in equilibrium. There is a point (*A*) at 273.01 K (0.01°C) and 0.006 atmospheres of pressure where the phases of ice, water, and vapor all touch. The three phases are in equilibrium at this triple point, which also make it an invariant point because this is the only temperature and pressure where ice, water, and vapor are in equilibrium. In Section 5.4 we will discuss the Gibbs phase rule, which predicts that the triple point in H_2O is invariant. All elements and compounds have a phase diagram that shows the regions of phase stability and equilibrium among solids, liquid, and vapor.

5.3.1 Nucleation and Growth-Phase Transformations

One way to study a phase transformation is to record the temperature of the material as a function of time as a constant amount of heat is extracted from the material per unit of time. For example, pure iron transforms from the high temperature γ (FCC) phase to the low temperature α (BCC) phase below the equilibrium temperature of 912°C. If pure iron is heated to a temperature of 1000°C and heat is extracted at a constant rate, the γ phase cools at a constant rate for temperatures above 912°C, as shown schematically in Figure 5.3. Once the specimen is super-cooled to a temperature below 912°C by the amount ΔT, the temperature becomes constant with time while the transformation from the γ to the α structure occurs. Once the transformation starts, all of the extracted heat (ΔQ) comes from the heat of transformation at constant temperature. Phase transformations during cooling are exothermic, and heat is extracted from the material. Therefore ΔQ is negative, and the change in entropy is negative. If no super-cooling occurs ($\Delta T = 0$), the transformation is never completed, because 912°C is the equilibrium temperature where the γ and α phases of iron have the same GFE, and both phases can exist. The amount of time for the completion of the transformation decreases as the amount of super-cooling increases. Once the transformation is completed, the new α phase starts to cool as heat is extracted from it.

If we observe iron through a microscope while it is transforming at a temperature below 912°C, at the beginning of the transformation small nuclei of α iron form at grain-boundary intersections in the γ iron, as shown schematically in Figure 5.4a. The nuclei form by a few atoms coming together by diffusion in the new crystal structure. The theory of nuclei forming in the bulk of the material is presented in Appendix 5B.2 on homogenous nucleation. At the equilibrium temperature the GFE of the α iron and γ iron are equal, and there is no GFE change driving the transformation. As the amount of

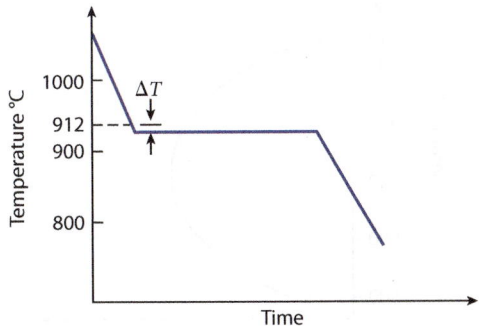

Figure 5.3 The temperature of an iron specimen as a function of time when heat is extracted at a constant rate from a temperature above 912°C through the γ-to-α phase transformation.

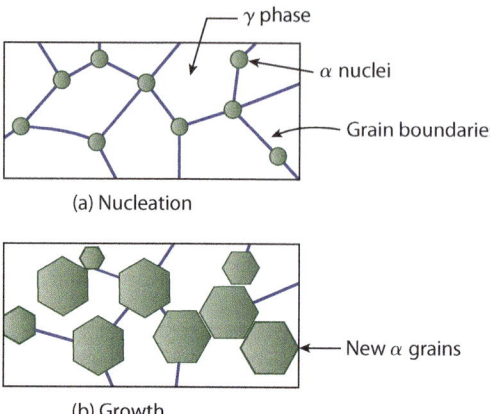

(a) Nucleation

(b) Growth

Figure 5.4 (a) Nuclei formation of the BCC α phase in grain boundaries of the FCC γ phase of iron. (b) Growth of the new α grains.

super-cooling increases, the reduction in the GFE increases for the transformation of γ iron to α iron, as shown in Figure 5.1. The α nuclei form preferentially at the grain boundaries or grain-boundary intersections, because the atoms in these areas are at a higher energy than are atoms in the bulk of the material. Transformation of γ iron to α iron at these locations results in a greater reduction in the GFE. The theory of nuclei forming at grain boundaries is presented in Appendix 5B.3. With significant super-cooling, the nucleus size of stable α is reduced so that only a few atoms have to come together to form in the α structure.

The nuclei of α iron grow as atoms diffuse across the γ-α interface from the γ phase to the α phase, as shown in Figure 5.4b. This diffusion-controlled transformation process is nucleation and growth. The growth rate is controlled by the rate of atoms diffusing across the γ-α interface. This is a complex function of time and temperature as shown in Figure 5.5, where a schematic of the growth rate of this transformation is plotted on the x axis and temperature is plotted on the y axis. At the γ-α equilibrium temperature of 912°C the growth rate is 0, because there is no chemical driving force ($\partial G_{Fe}/\partial x$) in Equation 4.37 for diffusion of iron atoms from the γ phase to the α phase. The atoms do not move preferentially from one phase to the other. As the temperature is decreased below 912°C, the chemical driving force for diffusion ($\partial G_{Fe}/\partial x$) increases because ΔG_{Fe} for the transformation becomes more negative with a

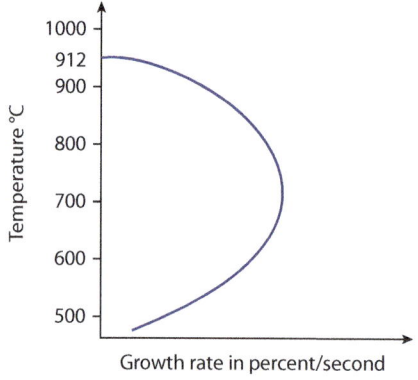

Growth rate in percent/second

Figure 5.5 A schematic of the growth rate of α-iron from γ-iron as a function of temperature.

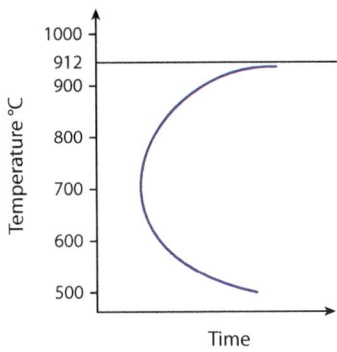

Figure 5.6 A schematic of the time for 50% transformation of γ to α-iron as a function of temperature, demonstrating the "C" curve.

decrease in temperature below the equilibrium temperature, as shown in Figure 5.1. As a result the rate of transformation from the γ phase to the α phase increases as the temperature decreases below 912°C. Figure 5.5 shows that as the temperature further decreases, the transformation rate eventually starts to decrease. The decrease in transformation rate is because the mobility of the iron atoms (B_{Fe}) decreases as the temperature decreases, as shown in Equation 5.2:

$$B_{Fe} = B_{Fe\,0}\exp -\left(\frac{\Delta H_{DFe}}{kT}\right) \qquad\qquad \textbf{5.2}$$

where B_{Fe0} is a constant and in this example ΔH_{DFe} is the activation enthalpy for diffusion of iron atoms. As the temperature is lowered toward absolute zero, the mobility of the atoms goes to 0; this causes the rate of the transformation to go to 0 as shown in Figure 5.5. A theory of phase-transformation rates is presented in Appendix 5B.4.

In the study of phase transformations, the transformation time is usually measured rather than the growth rate. The time for the transformation is inversely related to the rate of the transformation. The plot of the time for a given amount of transformation as a function of temperature is shaped like the letter "C," as shown in Figure 5.6.

Example Problem 5.3

Calculate the GFE change if 1 mole of water transforms to ice at 263 K. Assume that the entropy change and the enthalpy at 263 K are equal to those determined at the equilibrium temperature. This assumption is reasonable because these values depend upon the structure of the material, and it is a reasonable assumption that the structure of water and of ice at 263 K is essentially the same as that at 273 K. Both water and ice are present at 263 K during the transformation process; at the end of the transformation only ice is present.

Solution

The entropy and enthalpy changes are presented in Example Problem 5.2 to be

$$\Delta S = -2.2 \times 10^{-2}\frac{kJ}{mole \cdot K} \text{ and } \Delta H = -\Delta Q_f = -6.01\frac{kJ}{mole}$$

Inserting these values into Equation 4.14 results in

$$\Delta G = \Delta H - T\Delta S = -6.01\frac{kJ}{mole} - 263 \text{ K}\left(-2.2 \times 10^{-2}\frac{kJ}{mole \cdot K}\right)$$

$$\Delta G = -6.01\frac{kJ}{mole} + 5.79\frac{kJ}{mole} = -0.22\frac{kJ}{mole}$$

The change in the GFE is negative. This transformation is spontaneous and not reversible at 263 K.

5.4 THE GIBBS PHASE RULE

The observation in the H_2O phase diagram that the phases of ice, water, and vapor are in equilibrium at an invariant point, and that water and vapor are in equilibrium along a line in the phase diagram, can be expressed as an equation. The Gibbs phase rule allows for the calculation of the number of degrees of freedom (F) from the number of phases (P) and the number of components (C). If the material is made up of atoms that can bond with any of the other atoms in the different phases, then the components are the atoms in the material. If the material is made of molecules, such as H_2O, and the atoms of the molecule remain bonded to each other in all of the phases, then the component is molecules of H_2O. The possible degrees of freedom (thermodynamic variables) are temperature, chemical composition, and pressure, and then the Gibbs phase rule is given by Equation 5.3.

$$P + F = C + 2 \tag{5.3}$$

If pressure is fixed, such as at 1 atmosphere, and only temperature and composition can vary, then the Gibbs phase rule is given by Equation 5.4.

$$P + F = C + 1 \tag{5.4}$$

It helps to remember the Gibbs phase rule in Equation 5.4 as follows: The police (P) force (F) equals a captain (C) plus one.

Example Problem 5.4

Determine the number of degrees of freedom (F) for H_2O when pressure and temperature are both possible variables and the indicated phases (P) are present, and describe how this relates to the phase diagram in Figure 5.2, for the following:
(a) When only liquid is present
(b) When liquid and vapor phases are both present

Solution

a) H_2O is the only component and $C = 1$. When $C = 1$, composition cannot vary. Use Equation 5.3 because pressure is variable. If only liquid is present, $P = 1$ and $C = 1$.

$$P + F = C + 2$$

$$1 + F = 1 + 2$$

$$F = 2$$

The degrees of freedom are 2. Both temperature and pressure can be varied when only liquid is present.

b) When both liquid and vapor are present, $P = 2$ and $C = 1$.

$$2 + F = 1 + 2$$

$$F = 1$$

There is only one degree of freedom (F) if both liquid and vapor are present. If the pressure is selected as 1 atmosphere, then the temperature cannot also be selected. For liquid and vapor to be present at 1 atmosphere, the temperature can only be 100°C according to the phase diagram. Other pressures and temperatures where liquid and vapor are in equilibrium are given by the line between the liquid-phase and vapor-phase regions.

5.5 EQUILIBRIUM IN TWO-COMPONENT SYSTEMS

Many two-component materials have two phases in equilibrium that have different chemical compositions. Low-carbon steel has the two components iron and carbon, and at room temperature it has two phases. The first phase is α iron, which is BCC iron with a few carbon interstitial atoms, and the second phase is iron carbide (Fe_3C). Equilibrium between the two phases, α iron and Fe_3C, at constant temperature and pressure requires that an iron atom in the α phase does not change in GFE if it moves to the Fe_3C or if it moves back to the α phase, and this is also true for the carbon atoms in the α iron and Fe_3C phases. If this was not true, then the atoms would move to the phase that resulted in a decrease in the GFE, and the phases would not be in equilibrium. Equation 5.5 expresses this equilibrium:

$$\frac{\partial G^\alpha}{\partial N_{Fe}} = \frac{\partial G^{Fe_3C}}{\partial N_{Fe}} \qquad\qquad \textbf{5.5}$$

where ∂G^α is the change in the GFE of the α phase when the number of iron atoms changes by ∂N_{Fe}, and ∂G^{Fe_3C} is the change in the GFE of the iron-carbide phase when the number of iron atoms changes by ∂N_{Fe}. A change of one atom out of 10^{23} atoms can be considered an incremental change, and a differential equation is appropriate. A partial derivative is shown in Equation 5.5 because the GFE of the α phase is also a function of temperature and pressure, but these two variables are held constant. The partial derivatives in Equation 5.5 are called the chemical potential of the iron atoms in the α (μ_{Fe}^α) and Fe_3C ($\mu_{Fe}^{Fe_3C}$) phases. The chemical potential of a particular atom or molecule in a phase is the partial derivative of the GFE of the phase with respect to the number of atoms or molecules of that type. The chemical potential of an atom or molecule in a phase is also equal to the change in the GFE if one atom or molecule is added to the phase. Rewriting Equation 5.5 in terms of the chemical potentials of the iron atoms results in Equation 5.6.

$$\mu_{Fe}^\alpha = \frac{\partial G^\alpha}{\partial N_{Fe}} = \frac{\partial G^{Fe_3C}}{\partial N_{Fe}} = \mu_{Fe}^{Fe_3C} \qquad\qquad \textbf{5.6}$$

A similar equation applies to the carbon atoms in the α iron and iron-carbide phases when they are in equilibrium. Equation 5.6 can be generalized to apply to all of the different types of atoms ($i, j, k, \ldots$) in any number of phases ($\alpha, \beta, \gamma, \ldots$) that are in equilibrium, as shown in Equation 5.7.

$$\mu_i^\alpha = \mu_i^\beta = \mu_i^\gamma \qquad \qquad \textbf{5.7}$$

The units of chemical potential are energy per atom or molecule. Note that when two phases are in equilibrium, their Gibbs free energies are not necessarily equal; only the chemical potentials of the atoms or molecules in the phases are required to be equal. In this discussion superscripts refer to phases, and subscripts refer to atom types. Solid phases are usually represented by Greek symbols, and liquids are represented by L. In some cases other symbols are obvious; for example, with ice and water the symbols i and w are used.

When there is only one atom or molecular type, such as iron, the chemical potential is numerically equal to the GFE per iron atom. For one-component systems, when two phases are in equilibrium the chemical potentials are equal, as given in Equation 5.7, and the Gibbs free energies per atom are also equal as shown in Figure 5.1. However, when a phase contains two atom types, such as iron and carbon in Fe_3C, it is not possible to calculate the GFE per iron atom of this phase. This is because the bond energy is between the iron and carbon atoms, and it is impossible to determine how much of the energy goes to the iron atom and how much goes to the carbon atom. But a calculation that can be made is the GFE change when one atom of iron is added to a particle of Fe_3C; that GFE change is the chemical potential of the iron atoms in the iron carbide.

5.6   BINARY EQUILIBRIUM PHASE DIAGRAMS

To determine the melting temperature of a new high-temperature gas-turbine alloy, we use a phase diagram. When atoms of two different types (binary) are mixed together at a certain temperature, binary phase diagrams allow the determination of the phases present, the chemical composition of each phase, and the relative amount of each phase present at equilibrium. The thermodynamics we discussed above help us understand the appearance of the phase diagram; however, variables such as entropy and the GFE are not present in the phase diagram.

5.6.1   Continuous-Solid and Continuous-Liquid Solutions

Figure 5.7a shows a binary copper (Cu)-nickel (Ni) phase diagram. Cu-Ni alloys are utilized for heat exchangers and piping in corrosive environments such as seawater because of their high thermal conductivity and excellent resistance to corrosion. In this phase diagram, the chemical composition of the second component is plotted on the x axis in atom percent (bottom scale) or weight percent (top scale), and temperature is plotted on the y axis. We use atom percent in this discussion, because the number of atoms (N) is one of the fundamental thermodynamic variables. Also, when considering compounds, the atom percent corresponds to the stoichiometry of the compound; however, for

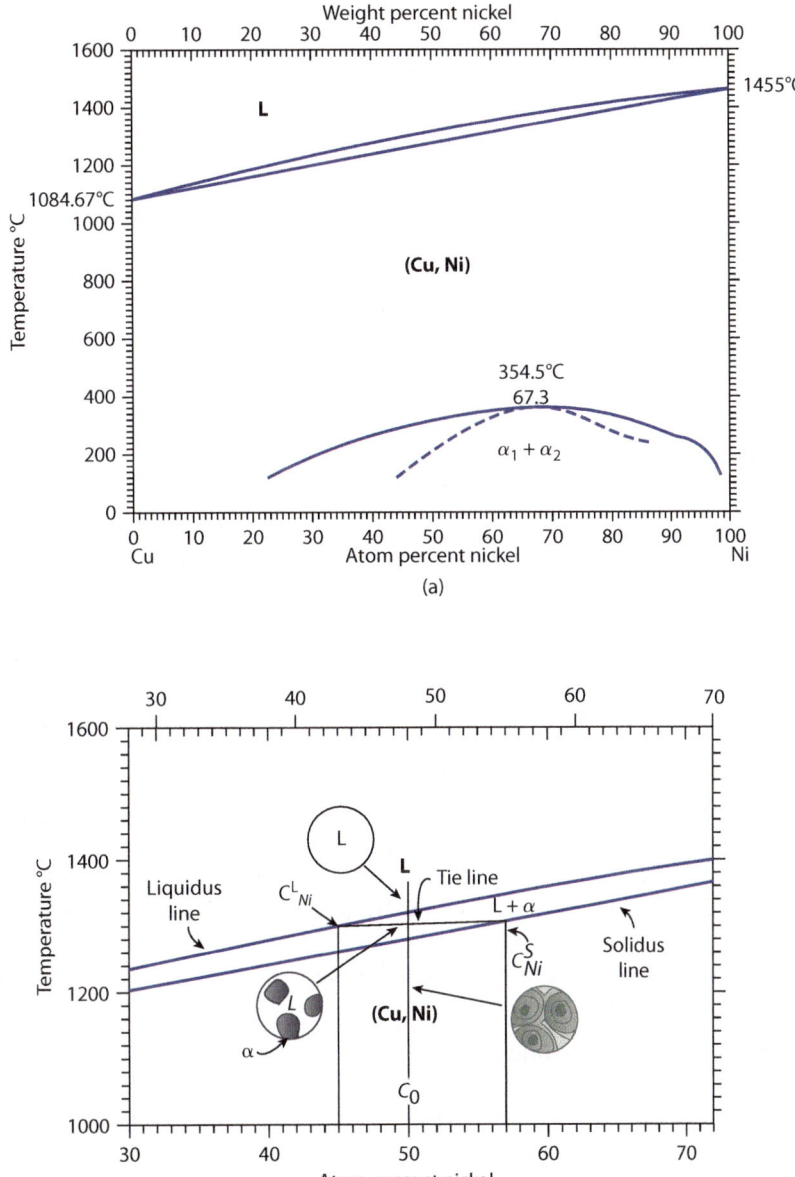

Figure 5.7 (a) The phase diagram of Cu and Ni. (b) A modified Cu-Ni phase diagram, with a vertical line at the original composition (C_0) of 50 atom percent Ni, a tie line at the temperature of 1300°C connecting the liquidus and solidus lines, and a vertical line that shows the liquid composition (C_{Ni}^L) of 45 atom percent Ni and the solid composition (C_{Ni}^S) of 57 atom percent Ni. Microstructure schematics are inserted for the original chemical composition of 50 atom percent Ni for temperatures above the liquidus line, for 1300°C, and for a temperature below the solidus line, such as 1200°C. (*Based on Binary Alloy Phase Diagrams 2nd ed. Editor Massalski,.T.B., ASM International (1990), p. 1444.*)

weight percent there is no obvious relationship with stoichiometry. Even so, weighing a material is much easier to do than to count atoms, and weight percent phase diagrams are usually used in practice. In binary phase diagrams of the type shown in Figure 5.7, the pressure is constant, and unless otherwise specified, 1 atmosphere is assumed. Pressure is not one of the variables plotted in binary phase diagrams, as it is for one-component phase diagrams. The phase that is present for an alloy of a particular chemical composition and temperature is determined by drawing a vertical line at the alloy chemical composition and a horizontal line at the alloy temperature. The intersection of these two lines is a point, and the phase or phases present at that point are labeled on the phase diagram. Some phase diagrams only label the single-phase regions, such as the solid solution of Cu and Ni (Cu, Ni) and liquid in Figure 5.7, and it is understood that the unlabeled region between two single-phase regions is a two-phase region composed of the two single-phase regions on either side. The regions of the phase diagram are also called phase fields. The solid that forms on the left side of the diagram at room temperature is normally designated the α phase. In this phase diagram the solid solution (Cu, Ni) is the α phase, and at temperatures above 354.5°C the α phase extends across the entire phase diagram.

In Figure 5.7a, at temperatures above 1455°C there is a continuous-liquid (L) region from 0 to 100 atom percent Ni. Any chemical-composition vertical line and temperature horizontal line above 1455°C results in a point in the liquid region, and there is only one continuous-liquid region. This indicates that Cu and Ni mix in the liquid for all chemical compositions. Not all atoms or molecules mix in the liquid phase; for example, liquid lithium and sodium do not mix at some compositions and temperatures, and many polymer molecules do not mix in the liquid. Atoms or molecules that are similar tend to mix in the liquid phase, and if atoms do not mix in the liquid phase, this indicates that the atoms are very different. Later in this chapter we will examine the phase diagrams of some metals where the liquids do not mix.

In Figure 5.7a, at temperatures below 1084°C and above 354.5°C, there is a continuous region (Cu, Ni) from 0 to 100 atom percent Ni. Any chemical-composition vertical line and temperature horizontal line in this region results in a point in the (Cu,Ni) phase region. The (Cu,Ni) phase has the FCC crystal structure of Cu with substitutional atoms of Ni; this is a substitutional-solid solution. In this case the (Cu,Ni) solid solution extends from pure Cu to pure Ni; this is continuous-solid solubility. On the right side of the phase diagram in Figure 5.7a, (Cu,Ni) is also FCC Ni with substitutional atoms of Cu. The Hume-Rothery rules define when continuous-solid solutions can occur between two metallic elements. Each of the metallic elements must have:

- Less than a 15% atomic radius difference

- The same crystal structure

- Similar electronegativities

- Similar valence

Cu and Ni both have the FCC crystal structure. Cu and Ni are adjacent to each other in the periodic table, and they have a similar valence. Their electronegativities are very similar at 1.9 and 1.8, respectively; their lattice parameters of 0.363 nm and 0.352 nm, respectively, yield a 3.1% difference. It is possible to have the (Cu,Ni) phase of any desired chemical composition at any temperature below the melting temperature of pure Cu (1084.87°C) and above 354.5°C. For example, a Cu-Ni alloy that has 50 atom percent Ni at 1000°C is in the (Cu,Ni) phase field, and only the (Cu,Ni) phase exists. This is determined by drawing a vertical line at 50 atom percent Ni and a horizontal line at 1000°C, and these two lines intersect in the (Cu,Ni) region. We will consider the region of the phase diagram below 354.5°C in Section 5.12, on miscibility gaps.

Example Problem 5.5

If the mixing of 50 atom percent Ni into a Cu crystal occurs at a temperature of 1000 K (727°C), determine the change in the GFE for 1 mole of total atoms if the enthalpy of solution for Ni into Cu is 0.03 eV/atom, according to data in Hultgren *et al.* (b) Based upon your result, should it be possible to mix 50 atom percent Ni into Cu at 1000 K? Explain your answer.

Solution

To determine the entropy of mixing the 50 percent Ni into Cu for a total of 1 mole of atoms, insert the chemical composition values into Equation 4.12. Since the problem is on a per-mole basis, R replaces k in Equation 4.12.

$$\Delta S_m = -RN(C_{Ni} \ln C_{Ni} + C_{Cu} \ln C_{Cu})$$

$$\Delta S_m = -8.31 \frac{J}{mole \cdot K}(1 \text{ mole})[0.5 \ln 0.5 + 0.5 \ln 0.5] = -8.31 \frac{J}{K}[0.5(-0.693) + 0.5(-0.693)]$$

$$\Delta S_m = -8.31 \frac{J}{K}[-0.693] = 5.67 \frac{J}{K}$$

The enthalpy of solution of Ni substituted into Cu is 0.03 eV per Ni atom. Converting 0.03 eV per Ni atom to joules per mole of Ni atoms results in

$$\frac{0.03 \text{ eV}}{Ni \text{ atom}} \times \frac{1.602 \times 10^{-19} \text{ J}}{1 \text{ eV}} \times \frac{6.023 \times 10^{23} \text{ Ni atoms}}{mole \text{ of Ni}} = \frac{0.29 \times 10^4 \text{ J}}{mole \text{ of Ni}}$$

However, in 1 mole of atoms only 50 atom percent are Ni. For 50 percent of 1 mole of Ni atoms, the enthalpy change is

$$C_{Ni}N\Delta H_{NiCu} = 0.50 \frac{mole \text{ of Ni}}{mole \text{ of atoms}}(1 \text{ mole of atoms})\left(\frac{0.29 \times 10^4 \text{ J}}{mole \text{ of Ni}}\right) = 0.145 \times 10^4 \text{ J}$$

The change in the GFE due to substituting 50 atom percent Ni atoms into Cu is then calculated with Equation 4.15.

$$\Delta G = G - G_0 = C_{Ni}N\Delta H_{NiCu} - T\Delta S_m = 0.145 \times 10^4 \text{ J} - (1000 \text{ K})5.67 \frac{J}{K}$$

$$\Delta G = 1450 \text{ J} - 5670 \text{ J} = -4220 \text{ J}$$

b) The change in the GFE is negative. The atoms should mix and form a solid solution at 1000°C.

5.6.2   Two-Phase Regions of Binary-Phase Diagrams

In the phase diagram of Figure 5.7a, two lines called the **liquidus** and **solidus** lines connect the melting temperature of Cu (1085°C) and of Ni (1455°C). These lines are labeled in a magnified version of the Cu-Ni phase diagram in Figure 5.7b. In between the liquidus and solidus lines is a region of α (Cu,Ni) plus liquid (L), as shown in Figure 5.7b. If a temperature-chemical composition point is above the liquidus line, this alloy is all liquid. For example, at the point 50 atom percent Ni and 1400°C, the alloy is entirely liquid with a chemical composition of 50 atom percent Ni, because this point is in the region labeled (L) for liquid. The single phase of liquid must have the same chemical composition as that of the original alloy. A schematic of the microstructure observable with a microscope at this temperature is indicated in Figure 5.7b, by the circle of liquid that is of uniform structure. If a temperature-chemical composition point is below the solidus line, this alloy is all solid. At the point 50 atom percent Ni and 1200°C, the alloy is entirely solid (Cu,Ni), as indicated by the single-phase region of (Cu,Ni) below the solidus line. The schematic of the microstructure shown at 1200°C is a polycrystalline solid. The darker shades indicate solid that formed early in the transformation process, and lighter shades indicate solid that formed later in the transformation, as discussed below.

If a temperature-chemical composition point is between the liquidus and the solidus lines, this alloy is a mixture of solid and liquid. In Figure 5.7b and in the following discussion, (Cu,Ni) and α are the same solid. For example, the point where the 50 atom percent Ni alloy is at 1300°C is in the L + α region between these two lines. At this point there is a mixture of liquid (L) and solid α, as is shown in the microstructure schematic. The equilibrium chemical compositions of the liquid and solid α phases are obtained by drawing a horizontal line (tie line) at 1300°C through the L + α phase field, as shown in Figure 5.7b. The **tie line** connects (ties) the two-phase boundaries of the two-phase field. Where the tie line intersects the liquidus line gives the chemical composition of the liquid phase (45 atom percent Ni), and the intersection of the tie line with the solidus line gives the chemical composition of the solid α phase (57 atom percent Ni).

When the original chemical composition of the alloy is 50 atom percent Ni, why is the chemical composition of the liquid phase 45 atom percent Ni and the solid α phase 57 atom percent Ni at 1300°C? Why are they not both 50 atom percent Ni? When the liquid and solid are at equilibrium, the chemical potentials of the Ni atoms in the liquid phase and the solid phase must be equal, as shown in Equation 5.7; otherwise, the Ni atoms would go to the phase that has the lower chemical potential. The same is true for the Cu atoms in the liquid and solid. When the liquid and solid are in equilibrium, there must be no change in the GFE of the combined liquid and solid if an atom or molecule goes from the liquid to the solid or in the reverse direction. The chemical potential depends upon the energy of the bonding. The strength of the bonding in the liquid and the solid is different, and the atom arrangements are different. There is no reason to expect at 1300°C for the chemical potential of the Cu (or Ni) atoms to be equal in both liquid and solid when the chemical composition of both the liquid and solid is equal to 50 atom percent. For a graphical demonstration of how the equality of chemical potentials of the Cu (or Ni) atoms results in the equilibrium chemical compositions of the liquid and solid in the Cu-Ni phase diagram, see Appendix 5C.

At 1300°C the same chemical compositions for the liquid phase (45 atom percent Ni) and solid α phase (57 atom percent Ni) results, no matter what original composition is selected as long as the intersection point of temperature and composition lies within the two-phase field of liquid and solid. To demonstrate this point, assume that Ni atoms are added to the 50 atom percent Ni alloy, so that the average Ni chemical composition is now 55 atom percent Ni. Is the liquid phase at 1300°C now more than 45 atom percent Ni after adding more Ni? Is the solid more than 57 atom percent Ni? The answer is no for both of these questions, because the intersection point of 55 atom percent Ni and 1300°C is still in the two-phase region of liquid and solid α. In this two-phase region, at 1300°C the liquid is always 45 atom percent Ni, and the solid is always 57 atom percent Ni, as given by the intersection of the tie line at 1300°C with the liquidus and solidus lines. What happened to the added 5% Ni? It seems as if the law of conservation of matter is being violated and the extra Ni has disappeared.

5.6.3 The Lever Rule

Application of the laws of conservation of matter show what is happening to the extra Ni atoms. To make this development as general as possible, consider the mixture to be made of A and B atoms, and the two-phase field has the two phases α and β; α and/or β could also be liquid. One known quantity is the original chemical composition (C_0), which is the percentage of B atoms added to produce the mixture of atoms. The equilibrium chemical compositions of the two phases (C^α and C^β) are known from the phase diagram at the temperature (T). What is not known is how much of the α and β phases is present. The amount of material in each phase is measured by the atom fractions of each phase (f^α and f^β). The atom fraction of α (f^α) is shown in Equation 5.8a as the total number of A and B atoms in the α phase (N^α) divided by the total number of atoms in the alloy (N).

$$f^\alpha = \frac{N^\alpha}{N} = \frac{N^\alpha_A + N^\alpha_B}{N} \qquad \textbf{5.8a}$$

$$f^\beta = \frac{N^\beta}{N} = \frac{N^\beta_A + N^\beta_B}{N} \qquad \textbf{5.8b}$$

C^α is the chemical composition of the α phase; as shown in Equation 5.9a, it is the number of B atoms in the α phase (N^α_B) divided by the total number of atoms in the α phase (N^α) multiplied by 100 to express the result in percent. The A atoms comprise the remainder of the α phase. Similarly, C^β is the chemical composition of the β phase; as shown in Equation 5.9b.

$$C^\alpha = 100\frac{N^\alpha_B}{N^\alpha} \qquad \textbf{5.9a}$$

$$C^\beta = 100\frac{N^\beta_B}{N^\beta} \qquad \textbf{5.9b}$$

C^α and f^α are often confused with each other, but they address completely different aspects of the material. C^α is the purity of the α phase; it says nothing about how much of the α phase is present. And f^α is the fraction of A and B atoms in the α phase; it says nothing about the purity of the α phase. To avoid confusion in the discussion of phase diagrams, atom percent is used for chemical composition (C^α), and atom fraction is reserved for f^α. However, when calculating the entropy of mixing in Equation 4.12, we must use the chemical composition in atom fraction and not percent.

Two equations account for the conservation of atoms. In a two-phase region the sum of the atom fractions in the two phases must be equal to 1.

$$f^\alpha + f^\beta = 1 \qquad \textbf{5.10}$$

From Equations 5.8a and 5.9a, the product $C^\alpha f^\alpha$ is the number of atoms of B in the α phase divided by the total number of atoms multiplied by 100, and $C^\beta f^\beta$ is the number of atoms of B in the β phase divided by the total number of atoms multiplied by 100.

$$C^\alpha f^\alpha = \left(\frac{N^\alpha_B}{N^\alpha}100\right)\left(\frac{N^\alpha}{N}\right) = \frac{N^\alpha_B}{N}100 \qquad \textbf{5.11a}$$

$$C^\beta f^\beta = \left(\frac{N^\beta_B}{N^\beta}100\right)\left(\frac{N^\beta}{N}\right) = \frac{N^\beta_B}{N}100 \qquad \textbf{5.11b}$$

Equation 5.12 shows the sum of Equations 5.11a and 5.11b is the total number of B atoms divided by the total number of atoms multiplied by 100, and this is equal to the original chemical composition (C_0) expressed in percent.

$$C^\alpha f^\alpha + C^\beta f^\beta = C_0 \qquad\qquad \textbf{5.12}$$

Solving Equations 5.10 and 5.12 simultaneously results in Equations 5.13a and 5.13b.

$$f^\alpha = \frac{C^\beta - C_0}{C^\beta - C^a} \qquad\qquad \textbf{5.13a}$$

$$f^\beta = \frac{C^a - C_0}{C^a - C^\beta} \qquad\qquad \textbf{5.13b}$$

Equations 5.13a and 5.13b apply to any two-phase field, and they are the **lever rules**. In the calculation of the atom fraction of α (f^α), the numerator is the distance in the phase diagram at temperature (T) from C_0 to C^β divided by the total length of the tie line from C^α to C^β. The larger the distance from C_0 to C^β, the higher the amount of α phase or the lower the amount of β phase. The equations are similar to calculating the equilibrium of forces in a lever system with a fulcrum at C_0. The derivation of the lever rules could also have been developed using weight percent for chemical composition (C) and weight fraction for f.

Example Problem 5.6

(a) For a Cu-Ni alloy with the original chemical composition of 50 atom percent Ni at 1300°C, calculate the atom fractions of the liquid and solid α phases. (b) For an alloy with the original chemical composition of 55 atom percent Ni at 1300°C, calculate the atom fractions of the liquid and solid α phases.

Solution

a) Apply the lever rule in Equations 5.13a and 5.13b to this problem. For the region containing solid α phase plus liquid, the β phase is replaced by liquid in these equations. In any other two-phase field, the α and β-phases are replaced by whatever two phases exist in the two-phase field. The equilibrium chemical compositions are $C^L = 45$ atom percent Ni, $C^\alpha = 57$ atom percent Ni, and $C_0 = 50$ atom percent Ni. From the lever rule in Equation 5.13, the atom fraction of liquid is

$$f^L = \frac{C^a - C_0}{C^a - C^L} = \frac{57 - 50}{57 - 45} = \frac{7}{12} = 0.58$$

and from Equation 5.10 the atom fraction of the α phase is

$$f^a = 1 - f^L = 1 - 0.58 = 0.42$$

Therefore, a fraction of 0.58 of the total number of atoms are in the liquid phase, and a fraction of 0.42 of the total atoms are in the solid α phase.

b) The Ni chemical composition is now increased to 55 atom percent. From the phase diagram at 1300°C, C^L = 45 atom percent Ni and C^α = 57 atom percent Ni, but now C_0 = 55 atom percent Ni. Now the atom fraction of the liquid phase is calculated from the lever rule in Equation 5.13.

$$f^L = \frac{C^\alpha - C_0}{C^\alpha - C^L} = \frac{57 - 55}{57 - 45} = \frac{2}{12} = 0.17 \text{ atom fraction liquid}$$

And the atom fraction of the α phase is calculated from Equation 5.10.

$$f^\alpha = 1 - f^L = 1 - 0.17 = 0.83 \text{ atom fraction } \alpha \text{ phase}$$

By adding more Ni in the two-phase field of liquid plus α, the chemical composition of the two phases does not change; the chemical composition is fixed by the requirement of the equality of the chemical potentials of the atoms in the two phases. What changes is the fraction of total atoms in the α and liquid phases. When more Ni is added, the fraction of the Ni-rich solid α phase increases to consume the extra Ni, and the law of conservation of matter is not violated. If the chemical compositions are in weight percents, then f^α and f^L are weight fractions.

5.6.4   Cooling through a Two-Phase Region

Assume that we melt a 50 atom percent Ni-Cu alloy by heating it to 1400°C, and then we cool the alloy. The phase diagram in Figure 5.7b shows this cooling process with a vertical line at 50 atom percent starting at 1400°C and going down in temperature. The first solid starts to form when this vertical line intersects the liquidus line at 1320°C, which is where the two-phase field of liquid plus α is first encountered. The chemical composition of the first solid to form is 60 atom percent Ni. This is obtained by the intersection of a tie line at 1320°C with the solidus line. The first solid is a small nucleus of a few atoms. As the alloy is cooled below 1320°C, more nuclei form and more solid grows onto the existing nuclei. This form of solidification is nucleation and growth. The nucleus is the darker area shown in the microscopic schematics in Figure 5.7b. The nuclei and the surrounding growth material grow together to form the grains of a polycrystalline solid. Using the lever rule, at 1320°C the atom fraction of α is 0. The alloy must be cooled to temperatures below 1320°C to obtain a significant amount of solid α. When the 50 atom percent Ni-Cu alloy is cooled to 1300°C, the atom fraction of the α phase is 0.42, as determined in Example Problem 5.6, and the solid α has a chemical composition of 57 atom percent Ni. For the chemical composition of the solid to adjust as it is cooled, the cooling rate must be extremely slow. When the alloy is cooled to approximately 1280°C, the solidus line is intersected, and for temperatures below 1280°C, the alloy is entirely solid α with an average chemical composition of 50 atom percent Ni.

Why doesn't the 50 atom percent Ni-Cu alloy solidify like water does at one temperature? Part of the reason is that water has only one component (H_2O), and the Cu-Ni alloy has the two components. Upon cooling a liquid 50 atom percent Cu-Ni alloy, the amount of solid increases with the decrease in temperature range from 1320°C to 1280°C. Through this temperature range the alloy mixture is like slush, with hardly any solid at 1320°C and hardly any liquid at 1280°C. At 1320°C a solid forms with 60 atom percent Ni that has the same chemical potential as the Ni atoms in the liquid of 50 atom percent Ni.

When the temperature is lowered, the chemical composition of the liquid and solid changes until the Ni atoms in the liquid have the same chemical potential as does a solid of 50 atom percent Ni. This occurs at 1280°C with a liquid composition of 40 atom percent Ni, and the remaining liquid transforms to a solid with 50 atom percent Ni. This discussion could be repeated by focusing on the Cu atoms in the Cu-Ni alloy.

The chemical compositions derived from the Ni-Cu phase diagram during cooling through the two-phase field of liquid-plus-α are the equilibrium chemical compositions, and they are obtained only for extremely slow cooling. A cooling rate in a foundry or in a manufacturing process that results in equilibrium chemical compositions may not be practical. If the alloy is more rapidly cooled through the liquid-plus-α two-phase region, the chemical compositions do not have sufficient time to adjust to the equilibrium value. For the 50 atom percent Ni chemical composition, the first nucleus of α that forms has the chemical composition of 60 atom percent Ni. Then as the alloy is cooled, more nuclei form and more metal atoms solidify onto the existing nuclei, like the concentric layers of an onion, as shown in Figure 5.7b. When solidification is complete, the average Ni chemical composition in the solid must be 50 atom percent. However, the initial solid is rich in Ni (60 atom percent), and for a rapid cooling rate most of the Ni atoms remain in place, because the rapid cooling rate does not allow the chemical composition to adjust to the equilibrium value by diffusion. As a result, the solid that forms at 1280°C must be less than 50 atom percent rich in Ni. The outer growth layers are lower in Ni chemical composition than the core of the grains, as indicated in the microstructure schematics in Figure 5.7b by the lighter shading of the solid α as the radii of the nuclei increase. The change in composition during solidification is called **coring**. Coring is eliminated by holding the 50 atom percent Ni-Cu alloy at a high temperature in the solid α region, such as 1200°C, and the diffusion of atoms results in a uniform chemical composition of 50 atom percent Ni with sufficient time. Since the phase diagram is an equilibrium diagram, the chemical compositions resulting from coring during a rapid cool do not appear in the phase diagram.

Example Problem 5.7

(a) Use the Gibbs phase rule to determine the number of degrees of freedom for a mixture of 50 atom percent Cu and 50 atom percent Ni at 1300°C and 1 atmosphere of pressure. (b) Determine the number of degrees of freedom for this same chemical composition at 1000°C and 1 atmosphere of pressure.

Solution

a) There are two components: Cu and Ni ($C = 2$), and at 1300°C there are two phases: α and liquid ($P = 2$). Using Equation 5.4 for 1 atmosphere of pressure,

$$P + F = C + 1$$

$$2 + F = 2 + 1$$

$$F = 1$$

There is one degree of freedom ($F = 1$). Once the temperature is selected as 1300°C, then one degree of freedom is utilized, and it is not possible to set the chemical composition of the phases in the two-phase field. The chemical compositions are controlled by the necessity of the atoms in the two phases having the same chemical potentials.

b) At 1000°C there is only the solid α phase ($P = 1$); there are two components, Cu and Ni ($C = 2$), and using Equation 5.4 for 1 atmosphere of pressure,

$$P + F = C + 1$$

$$1 + F = 2 + 1$$

$$F = 2$$

There are two degrees of freedom ($F = 2$). In the single-phase field of solid α, it is possible to set the temperature at 1000°C and the chemical composition of the solid α can be set at 50 atom percent Cu and 50 atom percent Ni, or at any other chemical composition that is desired within the single-phase field.

5.7 BINARY EUTECTIC PHASE DIAGRAMS

Binary eutectic phase diagrams are formed in two-component (binary) alloys, when the component atoms preferentially bond with atoms of the same type rather than the other type atoms. In alloys with eutectic phase diagrams the mixing of the metal components is limited, and atoms of the same type form in separate phases. Some metal alloys with eutectic phase diagrams have compositions that melt at temperatures significantly below the melting temperatures of the two components. Solders such as lead-tin and silver-copper are utilized in metal joining. Eutectic salt combinations are used to store solar thermal energy by melting the eutectic salt with solar energy. The energy is stored as the heat of fusion of the eutectic salt.

5.7.1 The Silver-Copper Eutectic-Phase Diagram

Silver solder is an alloy of silver (Ag) and copper (Cu). Solder is applied as a liquid that solidifies to join two solid pieces of metal without melting the two solid pieces. Ag-Cu alloys have the eutectic-phase diagram shown in Figure 5.8a.

The Ag-Cu eutectic phase diagram in Figure 5.8 is significantly different from the Cu-Ni phase diagram in Figure 5.7a. Figure 5.8 shows that at 200°C, and for all lower temperatures, there is a two-phase field of $\alpha + \beta$ for all chemical compositions. The α phase has the Ag crystal structure, and the β phase has the Cu crystal structure. Although both Cu and Ag have FCC crystal structures, they have lattice parameters that are 13% different: 0.409 nm for Ag and 0.362 nm for Cu. By convention the phase to the left of the phase diagram is designated the α phase, and the phase to the right of the diagram is designated the β phase. The α and β phases are terminal phases, because they appear at the ends of the phase diagram. The line at the left of the diagram that separates the single-phase field of α from the $\alpha + \beta$ two-phase field is the **solvus line**; it gives the maximum solubility of Cu in α as a function of temperature. Similarly, the solvus line at the right of the diagram separates the single-phase field of β from the $\alpha + \beta$ phase field, and it gives the maximum solubility of Ag in β as a function of temperature. The phase diagram shows that at temperatures below 200°C, Cu is essentially insoluble

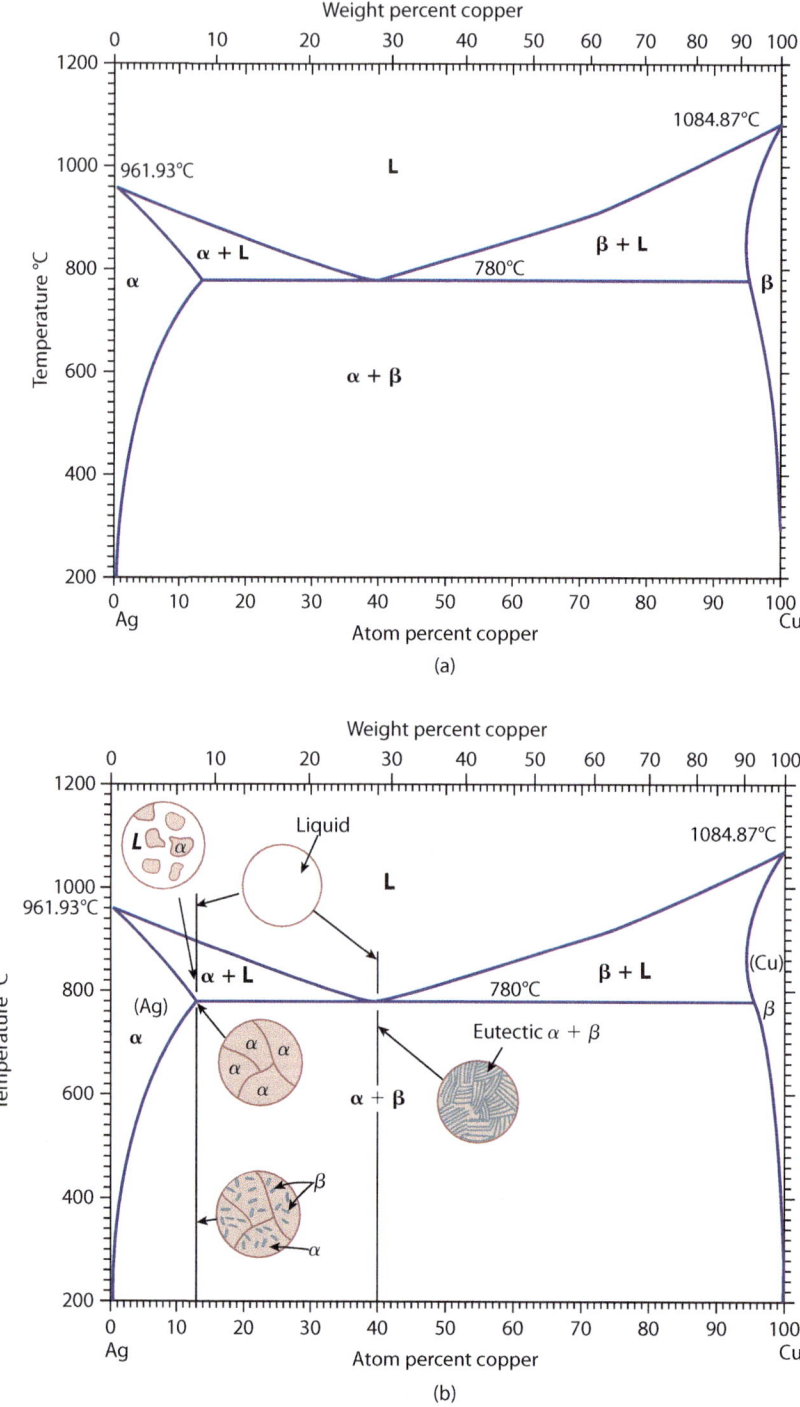

Figure 5.8 (a) The Ag-Cu eutectic phase diagram. (b) A modified Ag-Cu phase diagram with microstructure inserts for the eutectic composition of 40 atom percent Cu and for 13 atom percent Cu. (*Based on Binary Alloy Phase Diagrams 2nd ed. Editor Massalski, T.B., ASM International (1990), p. 29.*)

in the α phase and Ag is insoluble in the β phase. An Ag alloy with 50 atom percent Cu at equilibrium at 200°C, and for all lower temperatures, is a mixture of nearly pure Ag and nearly pure Cu with no mixing; this is segregation.

The segregation of Cu and Ag is explained by the GFE of the alloy. Cu and Ag do not bond well to each other, primarily due to the 13% atom size difference between Cu and Ag, which results in a relatively large positive enthalpy of mixing. The enthalpy of mixing Cu into Ag (α) (ΔH_{Cu}^{α}) is reported by Hultgren et al. to be 0.25 eV per atom, and the enthalpy of mixing Ag into Cu (β) is reported to be 0.39 eV per atom. The smaller enthalpy of mixing Cu into Ag crystals explains why the solubility of Cu into Ag is greater than the solubility of Ag into Cu at the same temperature. The GFE of an Ag crystal with a chemical composition of C_{Cu} Cu atoms relative to a pure Ag crystal is determined by Equation 4.15. We assume that the total entropy change of mixing the atoms (ΔS_T) is equal to the entropy of mixing (ΔS_m) calculated with Equation 4.12. If the GFE change (ΔG) given by Equation 4.15 is positive, the atoms do not mix, and this would be the case for Ag-Cu alloys at temperatures below 200°C. In Example Problem 4.8, we calculated that 20 atom percent Cu should not mix into an Ag crystal at 1000 K (727°C) because the GFE change is positive. However, in Homework Problem 4.14, we calculate that 10 atom percent Cu should substitute into an Ag crystal. The phase diagram confirms these calculations, for at 727°C the maximum solubility of Cu into Ag is approximately 11 atom percent.

Figure 5.8 shows that from 200°C to 780°C, the maximum amount of Cu that can be substituted into the α phase increases with temperature, and the amount of Ag that can be substituted into the β phase increases with temperature. The amount of Cu that minimizes the GFE of the α phase is given by Equation 4.16 using ΔH_{Cu}^{α}. According to Equation 4.16, as the temperature approaches absolute zero, the atom percent of Cu in Ag that minimizes the GFE approaches 0, and as temperature is increased this chemical composition increases. A similar argument can be made for the substitution of Ag into the Cu crystal structure (β phase). The temperature dependence of the solvus lines for the α and β phases is consistent with Equation 4.16; however, the equilibrium chemical compositions are not necessarily given by Equation 4.16. In between the α and β solvus lines, both the α and β phases are present and the α and β phases are in equilibrium with each other. At equilibrium the chemical potentials of the Cu atoms in the α and β phases are equal, and the chemical potential of the Ag atoms in the α and β phases are equal. However, the chemical potential of the Ag and Cu are not necessarily equal to each other. It can be shown that the condition stated in the previous sentence results in the minimum GFE for the combined α and β phases. To see a graphical demonstration of how the equilibrium chemical compositions of the α and β phases in an eutectic phase diagram results from Gibbs free energy plots, see Appendix 5D. In Appendix 5D it is graphically shown that the solvus line compositions for two solid phases that are in equilibrium are usually close to the composition that gives the minimum in the Gibbs free energy. The minimum Gibbs free energy for a single phase with two components is predicted by Equation 4.16.

At a temperature such as 700°C, there are three regions of the phase diagram from 0 to 100% Cu. For original chemical compositions from 0 to 9 atom percent Cu, there is the single phase of α. For original chemical compositions from 9 atom percent Cu to 96 atom percent Cu, there is a two-phase mixture of α and β. For any original chemical compositions within the two-phase α-plus-β region, the chemical composition of the α and β phases is obtained by drawing a tie line connecting the solvus lines of the α and β-phase regions at the temperature of 700°C. The intersections of this tie line at 700°C with the α and β solvus lines give the equilibrium chemical compositions of the α and β phases. The α-phase and β-phase chemical compositions are 9 atom percent Cu and 96 atom percent Cu, respectively. The atom fractions of the α and β phases in the two-phase field of α plus β are given by the lever rule in Equations 5.13a and 5.13b. For original chemical compositions from 96 to 100 atom percent Cu, there is the single-phase β.

Example Problem 5.8

At 700°C for an alloy with an original chemical composition of 50 atom percent Cu and 50 atom percent Ag, determine the following:
(a) The phases present
(b) The chemical composition of each phase
(c) The atom fraction of each phase

Solution

a) The intersection of a horizontal line at 700°C and a vertical line at 50 atom percent Cu is in the $\alpha + \beta$ phase field, and these two phases are present.
b) The 700°C tie line intersects the α solvus line at 9 atom percent Cu, and the β solvus line at 96 atom percent Cu. These are the chemical compositions of the α and β phases.
c) The atom fraction of the α phase is calculated with the lever rule in Equation 5.13.

$$f^{\alpha} = \frac{C^{\beta} - C_0}{C^{\beta} - C^{\alpha}} = \frac{96 - 50}{96 - 9} = \frac{46}{87} = 0.53$$

The atom fraction of the β phase is calculated with Equation 5.10.

$$f^{\beta} = 1 - f^{\alpha} = 1 - 0.53 = 0.47$$

The α phase has 53% of the total atoms, and 47% of the total atoms are in the β phase.

In the Ag-Cu phase diagram in Figure 5.8a, at 780°C there is a constant temperature line (isotherm) that connects the α phase at 13 atom percent Cu, and the β phase at 95 atom percent Cu. The liquid (L) phase also touches the 780°C isotherm at a chemical composition of 40 atom percent Cu. The constant temperature line at 780°C is the **eutectic isotherm**. Since the eutectic isotherm at 780°C intersects three phases (α, β, and liquid), these three phases are in equilibrium for any original chemical composition along the eutectic isotherm between 13 and 95 atom percent Cu. At 780°C the chemical potentials of the Cu in the α, liquid, and β are all equal at the chemical compositions of 13, 40, and 95 atom percent Cu, respectively; the same is true for Ag. The temperature where α, liquid, and β are in equilibrium is the **eutectic temperature**. Appendix 5D shows graphically how the equilibrium chemical compositions of α, β, and liquid phases result from the GFE energy plots of these three phases at 780°C.

What happens in this alloy at 780°C is best studied by producing an alloy that has a **eutectic composition** of 40 atom percent Cu. The eutectic composition is equal to the composition of the liquid that is in equilibrium with the α and β phases. If this composition alloy is heated to temperatures higher than 780°C, the alloy is entirely liquid with a chemical composition of 40 atom percent Cu, as shown in the phase diagram and in the microstructure schematic in Figure 5.8b. If this liquid alloy is cooled, the alloy remains liquid until 780°C is reached. The liquid, α, and β phases all are in equilibrium at 780°C, as shown in the eutectic reaction in Equation 5.14.

$$L = \alpha + \beta \qquad\qquad\qquad \textbf{5.14}$$

A binary eutectic phase diagram is a phase diagram of two components where the only reaction is a eutectic reaction.

When three phases are in equilibrium with two components at constant pressure, as in Equation 5.14, the Gibbs phase rule in Equation 5.4 results in 0 degrees of freedom. The eutectic reaction is an invariant reaction. There is no possibility to select the temperature of the equilibrium reaction or the composition of the participating phases in the reaction. If the alloy is cooled to a temperature just below 780°C, the

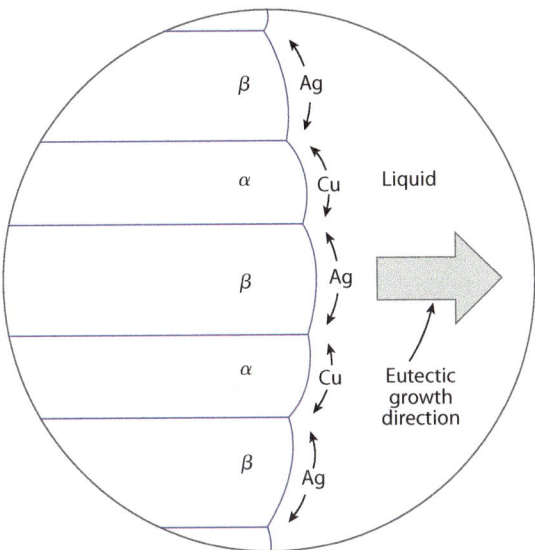

Figure 5.9 A schematic of the transformation of a liquid eutectic composition of copper (Cu) and silver (Ag) into layers of solid α and β phases by diffusion, forming the solid eutectic microstructure. (*Based on Callister, W.D., Materials Science and Engineering, An Introduction 6th ed. John Wiley &Sons NY (2003), p. 269.*)

liquid with 40 atom percent Cu transforms to a solid α phase with 13 atom percent Cu, and a β phase with 95 atom percent Cu. These compositions are determined from the intersection of a tie line just an increment of temperature below the eutectic isotherm at 780°C. In the Ag-Cu eutectic reaction, the Ag atoms must move from a random location in the liquid to the Ag-rich α phase, and the Cu atoms must move to the Cu-rich β-phase, as shown in Figure 5.9. So that the atoms of Cu and Ag in the liquid do not have to travel large distances to form the α and β phases, solid eutectic structures are alternating thin layers of α and β phases, as shown in Figure 5.9, and a photomicrograph of a eutectic microstructure is shown in Figure 5.10. Do not try to determine the atom fractions of phases at the invariant temperature where the three phases are in equilibrium (780°C), because we can only obtain the atom fractions of material when two phases are in equilibrium with the lever rule.

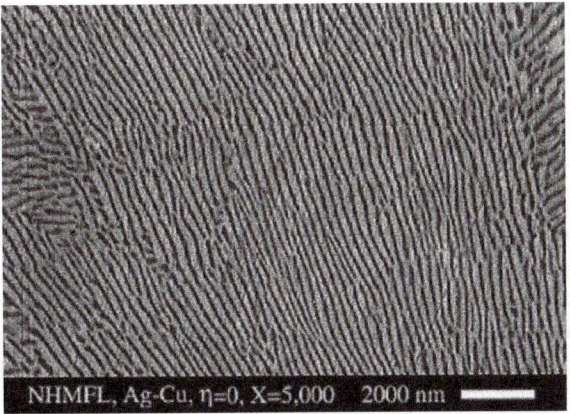

Figure 5.10 A micrograph of the microstructure of a eutectic composition silver-copper alloy showing alternating layers of light and dark α and β phases. (*Reprinted from Materials Science and Engineering: A, Volume 347, Issues 1–2, 25, Frank Heringhaus,Hans-Jörg Schneider-Muntau,Günter Gottstein, "Analytical modeling of the electrical conductivity of metal matrix composites: application to Ag–Cu and Cu–Nb," Pages No. 9–20, Copyright 2003, with permission from Elsevier.*)

Example Problem 5.9

A silver solder alloy with 40 atom percent Cu and 60 atom percent Ag, the eutectic composition, will be used to solder the heat-exchanger pipes for a super-heated steam turbine.
(a) What is the minimum temperature at which the silver solder will start to melt?
(b) If we heat the solder to 800°C to ensure it is all liquid with good flow characteristics, at what temperature will the solder start to solidify upon cooling?
(c) If this alloy is cooled to 779°C and held until equilibrium conditions are obtained, what phases are present, what are their compositions, and what is the atom fraction of each phase present?

Solution

a) The phase diagram shows that the solid α and β phases are in equilibrium with the liquid phase at the temperature of 780°C. It is necessary to heat the alloy above 780°C to ensure that the alloy is liquid, and the higher the temperature, the faster the alloy transforms to liquid. At 781°C the alloy will take a long time to melt, but it will eventually melt. The minimum temperature to start to melt the solder is 780°C, but from a practical viewpoint, it is necessary to heat the solder to temperatures above 780°C, such as 800°C.

b) Likewise the solder starts to solidify at temperatures below 780°C, but from a practical viewpoint it is necessary to cool the alloy to temperatures below 780°C to solidify the solder in a reasonable time.

c) At 779°C the equilibrium phases are $\alpha + \beta$. The α-phase chemical composition of 13 atom percent Cu is obtained by the intersection of a tie line between the α and β phases at 779°C with the α solvus line, and the β-phase chemical composition of 95 atom percent Cu is obtained by the intersection of this tie line with the β-phase solvus line. The atom fraction of the α phase is calculated with the lever rule in Equation 5.13.

$$f^\alpha = \frac{C^\beta - C_0}{C^\beta - C^\alpha} = \frac{95 - 40}{95 - 13} = \frac{55}{82} = 0.67$$

The atom fraction of the β phase is calculated with Equation 5.10.

$$f^\beta = 1 - f^\alpha = 1 - 0.67 = 0.33$$

At 779°C, a fraction of 0.67 of the total atoms in the alloy are in the α phase, and a fraction of 0.33 are in the β phase.

The Ag-Cu phase diagram shows that for increasing temperatures above 780°C, the solubility of Cu in the α phase and Ag in the β phase decreases, and the eutectic phase diagram has "ears." Equation 4.16 predicts that the chemical composition of substitutional Cu atoms in solid α that produces a minimum GFE should increase as temperature increases. Equation 4.16 and the shape of the phase diagram in these regions disagree. It is the requirement of equal chemical potentials for the atoms in the liquid and solid phases in equilibrium that forces the decrease in solubility of Cu in the α phase and of Ag in the β phase as temperature increases above the eutectic temperature of 780°C. Appendix 5D presents a graphical demonstration of the equilibrium chemical compositions in the "ears" of the phase diagram.

For the eutectic composition of 40 atom percent Cu, the equilibrium melting temperature is 780°C; this temperature is significantly lower than the melting temperature of either Ag (962°C) or Cu (1085°C). This alloy is therefore valuable in manufacturing processes as a solder that melts at a temperature less than either that of Ag or Cu. Silver solder is utilized in high-temperature applications where lead-tin solders would melt. Other elements, such as bismuth or zinc, are added to silver solder to reduce the melting temperature of the solder for lower-temperature applications, such as in domestic water systems where lead-tin solders are prohibited because of the potential for lead poisoning. The solidification of original alloy compositions that are not of the eutectic composition but are within the eutectic isotherm are covered in the solidification of lead-tin solders in Section 5.7.2.

Another important Ag-Cu alloy is sterling silver. Sterling silver is a silver alloy with approximately 13 atom percent Cu. This is the metal that silverware was originally made from. If the Cu is not added to the Ag, the silverware would be so soft that the fork tines would bend every time you tried to spear a piece of roast beef. Examining the phase diagram and the microstructures that develop during the cooling of liquid sterling silver explains why this composition is chosen and why the sterling silver is stronger than pure Ag. And this introduces the important strengthening mechanism called the precipitation hardening of metals. At temperatures above 900°C, an Ag alloy with 13 atom percent Cu is liquid, as indicated by the phase diagram in Figure 5.8b and the microstructure insert. Upon cooling the liquid alloy, the first solid starts to form at 900°C, where the liquidus line is intersected. At a temperature such as 820°C, the alloy is a mixture of liquid and α phases, as indicated in the phase diagram and the microstructure insert. At a temperature of 780°C, this alloy solidifies into the α phase, as indicated by the microstructure insert in Figure 5.8b, because the 13 atom percent Cu composition is in the α-phase region for a temperature of 780°C. At temperatures below 780°C the alloy is in the α-plus-β phase field. The β phase forms as small particles called precipitates, as indicated in the microstructure insert, and the precipitates strengthen the alloy. Although this metal alloy is not a particularly strong one, some of the strongest metal alloys are produced by precipitation hardening, as we discuss in Chapter 7, on strong solids.

5.7.2 The Lead-Tin Eutectic-Phase Diagram

Figure 5.11a shows the eutectic phase diagram of lead and tin. Lead-tin solder is an important material in manufacturing, frequently used for bonding metals to other metals, such as Cu electrical conductors or Cu tubing in heat exchangers. The melting and solidification behavior of lead-tin alloys is important for processing soldered joints. As shown in the lead-tin phase diagram in Figure 5.11a, a lead-tin alloy of 73.9 atom percent tin (the eutectic chemical composition) melts at 183°C, whereas pure lead melts at 327.5°C and pure tin melts at 232°C.

Figure 5.11a shows that the β phase of tin has the diamond cubic (DC) structure for temperatures below 13°C, and below 13°C tin is a semiconductor. The DC structure is the same crystal structure formed by the elements germanium, silicon, and diamond in Group IV of the periodic table. Above 13°C the β-phase structure is tetragonal, and this tin is a metal. In the following discussion, we assume that the temperature is always above 13°C, and that the metallic tetragonal phase of tin is present.

The solidification of the eutectic composition of lead-tin is similar to the solidification of the Ag-Cu eutectic composition that we covered above. The melting and solidification behavior of a 40 atom percent tin alloy of solder is significantly different than for the eutectic composition of 73.9 atom percent tin. Alloy compositions to the left of the eutectic composition are called **hypoeutectic** alloys; alloy compositions to the right of the eutectic composition are called **hypereutectic** alloys. If a hypoeutectic alloy with 40 atom percent tin is heated, it starts to melt at 183°C. However, it is not completely liquid until the temperature is above 265°C at point A in Figure 5.11b. If this alloy is heated to 300°C, it becomes completely liquid. If the alloy is then slowly cooled, solidification of the α phase starts at 265°C at point A in Figure 5.11b. The chemical composition of this α phase is 19 atom percent tin. This chemical composition is obtained by drawing a tie line at 265°C between the α and liquid phase fields, and this tie line intersects the α-phase boundary at 19 atom percent tin, marked as point B in Figure 5.11b. Applying the lever rule at 265°C, the atom fraction of α is 0, because the distance from the original chemical composition to the liquidus line $(C^L - C_0)$ is 0. To form a significant amount of solid α phase, the temperature must be reduced within the two-phase field of α-plus-liquid to a temperature such as 250°C.

At 250°C, the microstructure of the alloy of 40 atom percent tin and 60 atom percent lead has a matrix of liquid with **proeutectic** α-phase particles, as shown in the microstructure insert in Figure 5.11b. The proeutectic α phase is the α phase that forms before the eutectic reaction occurs. The proeutectic α-phase particles form by nucleation and growth within the liquid.

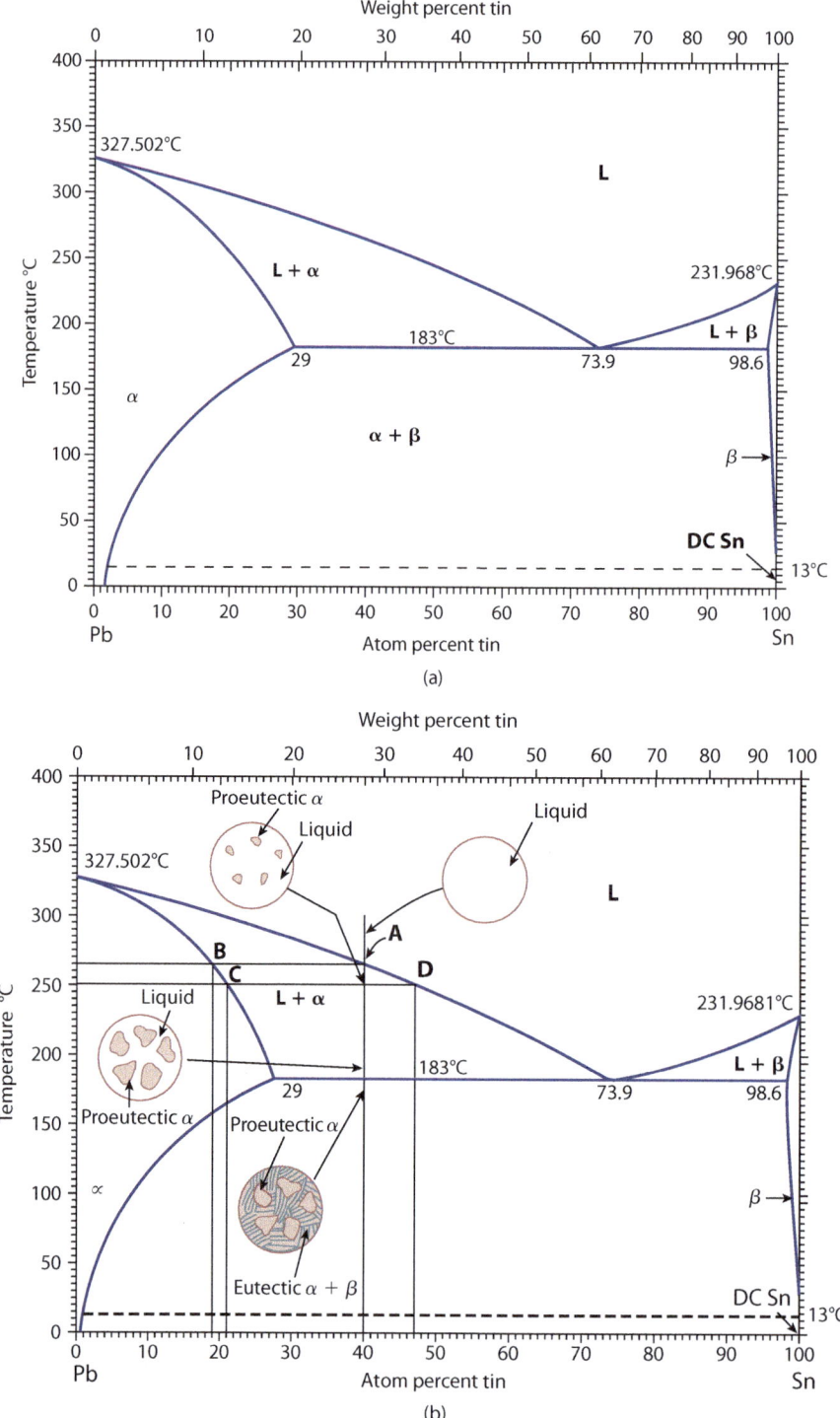

Figure 5.11 (a) The lead-tin (Pb-Sn) phase diagram. (b) A modified lead-tin phase diagram. Microstructure schematics, tie lines, and equilibrium chemical compositions are added for an original chemical composition of 40 atom percent tin. (*Based on Binary Alloy Phase Diagrams 2nd ed. Editor Massalski, T.B., ASM International (1990), p. 3016.*)

Example Problem 5.10

For an alloy of 40 atom percent tin and 60 atom percent lead, at 250°C:
(a) What phases are present?
(b) What are their chemical compositions?
(c) What is the atom fraction of each phase?

Solution

a) For this chemical composition and temperature, the phase diagram shows α and liquid phases present.
b) A tie line at 250°C connecting the α and liquid phases intersects the α-phase-field line at 21 atom percent tin (point C), and the liquid phase field at a chemical composition of 47 atom percent tin (point D).
c) The atom fraction of the α phase is determined with the lever rule in Equation 5.13.

$$f^\alpha = \frac{C^L - C_0}{C^L - C^\alpha} = \frac{47 - 40}{47 - 21} = \frac{7}{26} = 0.27 \text{ atom fraction of } \alpha \text{ phase}$$

The atom fraction of the liquid phase is calculated with Equation 5.10.

$$f^L = 1 - f^\alpha = 1 - 0.27 = 0.73 \text{ atom fraction of liquid}$$

As the alloy cools, the chemical composition of the remaining liquid tends toward the eutectic chemical composition of 73.9 atom percent tin, and the solid α phase's composition tends toward 29 atom percent tin. The change in microstructure from 250°C to 184°C is that the atom fraction of the proeutectic α phase increases, and the atom fraction of liquid is decreased. As shown in Figure 5.11b, at temperatures below 183°C the remaining liquid matrix transforms to a matrix of $\alpha + \beta$ solid eutectic structure surrounding proeutectic α-phase grains.

Figure 5.12 shows the microstructure of a hypoeutectic lead-tin alloy. The dark phase is the proeutectic α phase. The alternating light and dark layered phase is the eutectic microstructure that formed when the remaining liquid transformed to solid in the eutectic reaction. The dark phase in the eutectic microstructure is the α phase and the light phase is the β phase.

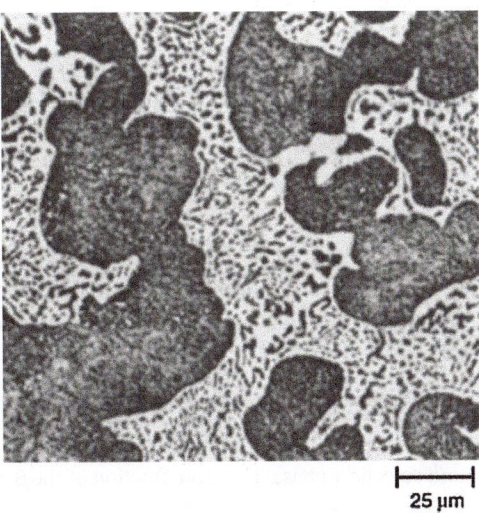

25 μm

Figure 5.12 The microstructure of a Pb-40 weight percent Sn alloy consisting of dark lead-rich proeutectic α-phase grains surrounded by the eutectic mixture. (*ASM Handbooks Online, http://www.asmmaterials.info, ASM International, 2004.*)

Example Problem 5.11

For an alloy of 40 atom percent tin and 60 atom percent lead,

(a) At 184°C, what phases are present, what are their chemical compositions, and what are their atom fractions?

(b) At 182°C, what phases are present, what are their chemical compositions, and what are their atom fractions?

(c) Determine the atom fraction of the $\alpha + \beta$ phase with the eutectic microstructure that formed during the eutectic reaction.

(d) What is the atom fraction of the α phase and of the β phase at room temperature?

Solution

a) At 184°C and at 40 atom percent tin, two phases are present: a proeutectic α phase and liquid. The proeutectic α phase's chemical composition is 29 atom percent tin, and the liquid is 73.9 atom percent tin. The atom fraction of the proeutectic α phase is calculated with the lever rule in Equation 5.13.

$$f^\alpha = \frac{C^L - C_0}{C^L - C^\alpha} = \frac{73.9 - 40}{73.9 - 29} = \frac{33.9}{44.9} = 0.76$$

The atom fraction of the liquid phase is calculated with Equation 5.10.

$$f^L = 1 - f^\alpha = 1 - 0.76 = 0.24$$

At 184°C the two-phase combination of liquid + proeutectic α is 0.76 atom fraction proeutectic α phase and 0.24 atom fraction liquid.

b) At 182°C the phases present are α and β. The α phase contains both proeutectic and eutectic α phases. The chemical composition of the α phase is 29 atom percent tin, and the β phase is 98.6 atom percent tin. The atom fraction of α (f^α) in the solid is calculated with the lever rule in Equation 5.13.

$$f^\alpha = \frac{C^\beta - C_0}{C^\beta - C^\alpha} = \frac{98.6 - 40}{98.6 - 29} = \frac{58.6}{69.6} = 0.84 \text{ atom fraction of } \alpha \text{ phase}$$

The atom fraction of the β-phase is calculated with Equation 5.10.

$$f^\beta = 1 - f^\alpha = 1 - 0.84 = 0.16 \text{ atom fraction of the } \beta \text{ phase}$$

c) The atom fraction of the liquid at 184°C is 0.24, and at 182°C the atom fraction of the solid with the eutectic microstructure that formed from the liquid is also 0.24, because the eutectic material came from the liquid phase in the eutectic reaction.

d) Slow cooling to room temperature causes both the α and β phases to reject solid-solution atoms. At room temperature there is only about 1 atom percent tin in the α phase and even less lead in the β phase. The β phase is essentially 100% tin. At room temperature, the equilibrium atom fraction of the α phase is calculated from the lever rule in Equation 5.13.

$$f^\alpha = \frac{C^\beta - C_0}{C^\beta - C^\alpha} = \frac{100 - 40}{100 - 1} = \frac{60}{99} = 0.61 \text{ atom fraction of } \alpha \text{ phase}$$

This is the sum of the proeutectic and eutectic α phase. The atom fraction of the β phase is given by Equation 5.10.

$$f^\beta = 1 - f^\alpha = 1 - 0.61 = 0.39 \text{ atom fraction of } \beta \text{ phase}$$

5.8   BINARY-PHASE DIAGRAMS WITH INVARIANT REACTIONS

According to the Gibbs phase rule, an invariant reaction occurs in a binary system ($C = 2$) at constant pressure whenever three phases are in equilibrium. The eutectic reaction is only one such possibility. Figure 5.13 presents some invariant reactions. Conventionally, in the reaction equations Greek symbols indicate solids, L represents a liquid, and the arrow indicates the reaction for cooling. For heating the reaction goes in the opposite direction. Figure 5.13 shows schematic phase diagrams for each of these reactions. A eutectoid reaction in a phase diagram appears similar to a eutectic reaction, except that solid γ phase is in equilibrium with the α and β phases rather than with the liquid phase. The γ is a solid phase with a crystal structure that is different from those of the α and β phases. For example, an extremely important eutectoid reaction in the iron-carbon phase diagram is the reaction $\gamma \rightarrow \alpha + Fe_3C$, where γ is FCC iron, α is BCC iron, and Fe_3C is iron carbide. The iron-carbon phase diagram with this reaction is presented in Figure 5.18. A peritectic reaction is present in the aluminum-nickel phase diagram, shown in Figure 5.15, when liquid and AlNi react to form Al_3Ni_2 at temperatures below 1133°C. Also, Figure 5.15 shows the peritectoid reaction that occurs when AlNi and $AlNi_3$ react at temperatures below 700°C to form Al_3Ni_5.

5.9   BINARY-PHASE DIAGRAMS WITH COMPOUNDS

Continuous-solid solutions occur when the atoms are very similar and the enthalpy of mixing is very small. In materials with eutectic phase diagrams the atoms segregate due to a large positive enthalpy of mixing. If there is a large negative enthalpy of mixing, compounds tend to form. If there is a negative enthalpy of mixing, the atoms bond with more strength to the other type of atom than to their own type. This often happens

Eutectic	$L \rightarrow \alpha + \beta$	α — $L + \alpha$ — L — $L + \beta$ — β / $\alpha + \beta$
Peritectic	$\alpha + L \rightarrow \beta$	α — $\alpha + L$ — L / $\alpha + \beta$ — β — $\beta + L$
Monotectic	$L \rightarrow L_1 + \alpha$	Miscibility gap — $L_1 + L_2$ — L — $L + \alpha$ — α / $L_1 + \alpha$
Eutectoid	$\gamma \rightarrow \alpha + \beta$	α — $\alpha + \gamma$ — γ — $\gamma + \beta$ — β / $\alpha + \beta$
Peritectoid	$\alpha + \beta \rightarrow \gamma$	α — $\alpha + \beta$ — β / $\alpha + \gamma$ — γ — $\gamma + \beta$

Figure 5.13   Five possible invariant reactions including the reaction name, the reaction, and a schematic of the phase diagram. *(Based on Askeland, D.R., Fulay, P.P., and Wright, W.J., The Science and Engineering of Materials, 6th ed. Cengage Learning, Stamford, CT. (2011), p. 418.)*

when the atoms have significantly different electronegativities. There are many important applications of materials with compounds. Compounds strengthen metal alloys. Carbon steels are strengthened with the compound Fe_3C; the compound that strengthens Al-Cu alloys is Al_2Cu, and the Ni alloys in high-temperature gas turbines are strengthened with Ni_3Al. To understand the processing procedures utilized to strengthen these alloys, we must understand their phase diagrams. Also, the fastest computers and the highest-efficiency solar cells are made from the semiconductor compound gallium arsenide (GaAs). The Ga-As phase diagram is used as an example because it is a simple-compound phase diagram. Ga is in Group IIIB with an electronegativity of 1.6, and As is in Group VB with an electronegativity of 2.0.

The gallium-arsenic (Ga-As) phase diagram is presented in Figure 5.14, and the compound GaAs corresponds to the vertical line at 50 atom percent As. This vertical line indicates that there is no variation in the chemical composition of GaAs; one Ga atom for one As atom is the only chemical composition of GaAs observed. When atoms, such as Ga and As, are strongly attracted to each other, the Ga atoms surround the As atoms, and the As atoms surround the Ga atoms. Ga-Ga and As-As bonds do not occur. Strong bonding between As and Ga produces ordering and compound formation.

At temperatures below 29.77°C, compositions of Ga with As percentages from 0 to 50 atom percent are a mixture of two phases the α and GaAs, and the amount of each phase in this two-phase region is given by the lever rule. The α phase is essentially pure Ga, and GaAs is the γ phase. At the chemical composition 50 atom percent As, the only phase is GaAs. Compositions of Ga with As percentages from 50 to 100 atom percent are a mixture of two phases β and GaAs, where the β phase is essentially pure As. Appendix 5E shows how the GFE of the α, GaAs(γ), β, and liquid phases results in the phases and compositions present at 25°C, 800°C, 1000°C, and 1238°C.

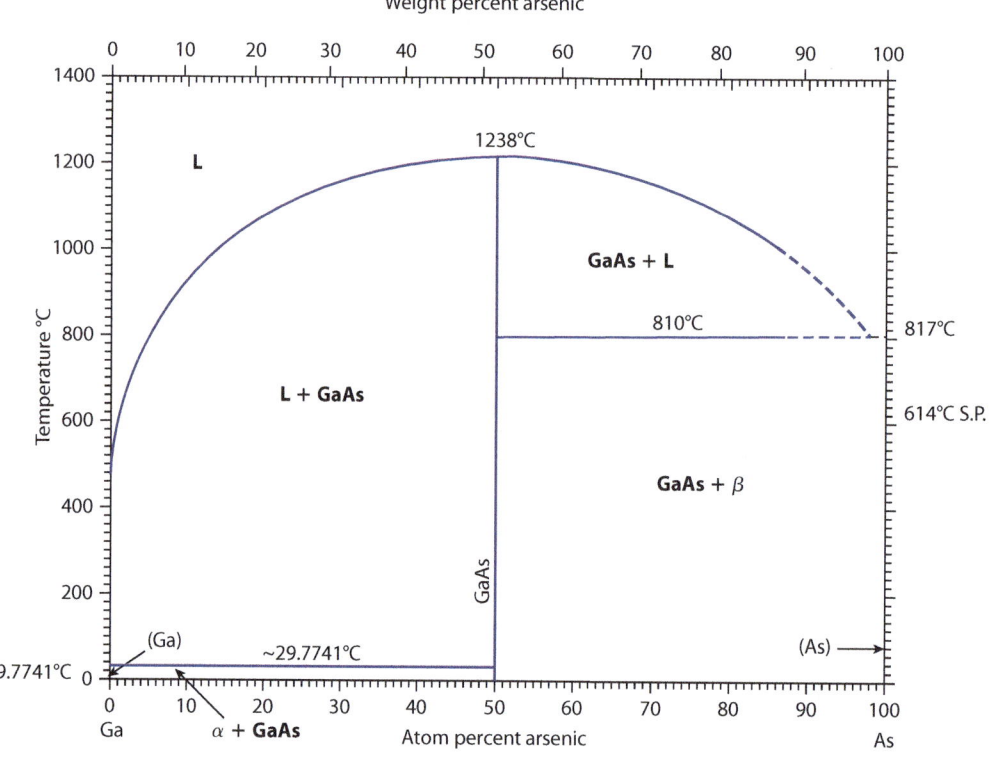

Figure 5.14   The gallium-arsenic (Ga-As) phase diagram. (*Based on Binary Alloy Phase Diagrams 2nd ed. Editor T.B. Massalski, ASM International (1990), p. 283.*)

Pure Ga melts at 29.77°C, and the α phase in the two phase region of α + GaAs also melts at 29.77°C, as shown by the isotherm in Figure 5.14 at 29.77°C extending from 0 to 50 atom percent As. Pure As sublimes at 614°C. For the chemical composition of 50 atom percent As, the compound GaAs has a melting temperature of 1238°C, which is much higher than for either Ga (29.77°C) or As (614°C sublimes). This high melting temperature for GaAs is an indication of the strong bonding of Ga and As. GaAs melts at 1238°C with no change in chemical composition, the liquid having the same chemical composition of 50 atom percent As. The point at which a solid phase melts to form a liquid of the same chemical composition is called a **congruent melting point**.

At 800°C for all original chemical compositions with less than 6 atom percent As, only liquid is present, and the chemical composition of this liquid is equal to the original chemical composition. At 800°C for all original chemical compositions with more than 6 atom percent As, but with less than 50 atom percent As, there is liquid in equilibrium with GaAs. The chemical composition for this liquid is 6 atom percent As and 94 atom percent Ga. For chemical compositions equal to 50 atom percent As, only solid GaAs is present. At 800°C for all original chemical compositions with As greater than 50 atom percent and less than 100 atom percent, GaAs is in equilibrium with the β phase, which is essentially pure As. However, for chemical compositions with As greater than 87 atom percent, the phase diagram is uncertain, as indicated by the dashed lines.

At 1000°C for chemical compositions from pure Ga to 14 atom percent As, only liquid exists, and the liquid chemical composition is equal to the chemical composition of the original material, because only one phase is present. At 1000°C for original chemical compositions above 14 atom percent As and less than 50 atom percent As, a mixture of liquid and GaAs is present, and the chemical composition for this liquid is 14 atom percent As. At 1000°C for chemical compositions equal to 50 atom percent As, only solid GaAs is present. At 1000°C for chemical compositions greater than 50 and less than 87 atom percent As, there is a mixture of GaAs with liquid that contains 87 atom percent As. For original chemical compositions with more than 87 atom percent As, the phase diagram is uncertain as indicated by the dashed lines.

If the chemical composition is 50 atom percent As, only GaAs is present and the congruent melting temperature of this compound is 1238°C. GaAs melts at one temperature just as H_2O melts at 0°C because GaAs is a compound just as H_2O is a compound. Combining Ga or As with GaAs reduces the melting temperature of the GaAs, as shown by the convex liquidus line at the top of the phase diagram, for above this line the alloy composition is all liquid.

Example Problem 5.12

An alloy of 30 atom percent As and 70 atom percent Ga is produced and then heated to 1250°C to thoroughly melt the mixture. The alloy is then slowly cooled so that it is always close to equilibrium. Answer the following about the alloy:

(a) At what temperature does a solid first form?
(b) What is the chemical composition of the first solid to form?
(c) Below what temperature is the alloy completely solid?
(d) At room temperature (23°C), what phases are present, what are their chemical compositions, and what is the atom fraction of each phase present?

Solution

a) The line for 30 atom percent As intersects the liquidus line at approximately 1180°C; therefore this is the temperature at which a solid first starts to form upon cooling.
b) The chemical composition of the first solid to form is found by drawing a tie line at 1180°C between the liquidus line and GaAs, and this intersects the GaAs compound line at 50 atom percent As. The first solid to form is GaAs, with the chemical composition of 50 atom percent As.

c) Below 29.77°C this alloy is an all-solid mixture of solid α phase and the compound GaAs.

d) At 23°C the α phase is pure Ga, and the γ phase has the chemical composition of 50 atom percent As and 50 atom percent Ga. This region is two phase, and the atom fraction of the α phase is given by the lever rule in Equation 5.13.

$$f^\alpha = \frac{C^\gamma - C_0}{C^\gamma - C^a} = \frac{50 - 30}{50 - 0} = \frac{20}{50} = 0.40$$

The atom fraction of the β phase is found from Equation 5.10.

$$f^\gamma = 1 - f^\alpha = 1 - 0.40 = 0.60$$

An intermediate compound, such as GaAs, is also an intermediate phase, but not all intermediate phases are intermediate compounds. The difference is demonstrated in the aluminum-nickel phase diagram shown in Figure 5.15. This phase diagram contains intermediate compounds such as $AlNi_3$ and AlNi. The $AlNi_3$ compound is used to strengthen high-temperature Ni-based alloys used in gas turbines. Intermediate compounds often have a very narrow range of chemical compositions where they are stable; they are nearly a vertical line in the phase diagram. AlNi is an example of an intermediate phase that contains an intermediate

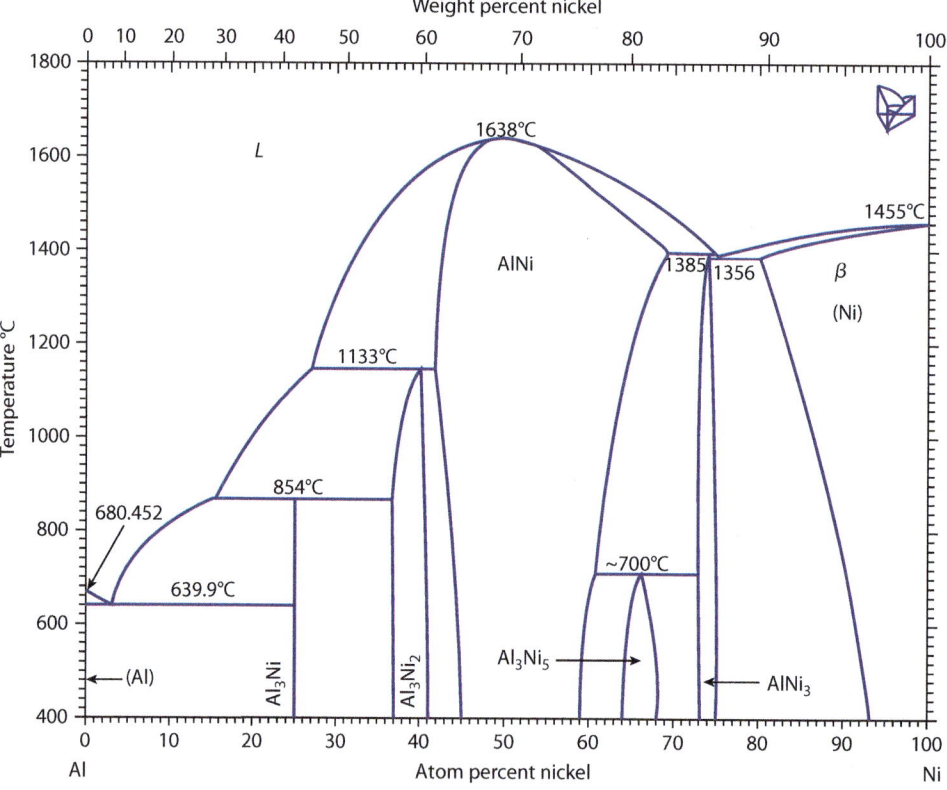

Figure 5.15 The aluminum-nickel phase diagram. (*Based on Binary Alloy Phase Diagrams 2nd ed. Editor Massalski, T.B., ASM International (1990), p. 183.*)

compound. At the stoichiometric composition AlNi the compound structure is simple cubic with Al atoms at 0,0,0 and Ni atoms at ½, ½, ½. AlNi has a large range of possible chemical compositions, so it is not necessary to have exactly one aluminum atom for every Ni atom in the AlNi phase. Note also the high melting temperature of AlNi of 1638°C, which is higher than those of Al and Ni. This higher melting temperature indicates that the Al-Ni bond is stronger than either the Al-Al or the Ni-Ni bond. Research continues on the use of AlNi as a high-temperature gas-turbine material, because of its high melting temperature. However, AlNi has been found to fracture too easily for use in aircraft gas turbines. Note that AlNi has a congruent melting point.

5.10 ANALYZING COMPLEX BINARY-PHASE DIAGRAMS

Some phase diagrams, such as that of aluminum-nickel (Al-Ni) in Figure 5.15, look overwhelming at first glance; however, they are no more difficult to use than a eutectic phase diagram. First determine the region of the phase diagram of interest; that is the region containing the intersection of the original chemical composition and the selected temperature. Only this region needs to be considered; it is not necessary to consider the entire phase diagram. The chosen point is usually in a single-phase region or a two-phase region. The single-phase regions are easy to analyze, because the single-phase chemical composition corresponds to the original chemical composition. In the Al-Ni phase diagram, only the single-phase regions are labeled. Areas between single-phase regions are two-phase regions consisting of the two single phases on either side of the two-phase region. In the two-phase regions, the chemical composition of the two phases is given by the compositions at the intersections of a tie line at the selected temperature, with the two-phase boundaries bordering the two-phase region. The atomic or weight fractions of the two phases are then obtained from the lever rule in Equations 5.13a and 5.13b. If the point formed by the intersection of the original chemical composition and temperature falls on an isothermal temperature line or at the invariant point, the phases participating in the invariant reaction are analyzed by moving an increment of temperature (ΔT) above and below the isothermal temperature. There are only one- or two-phase regions above or below the isothermal temperature, and the analysis for one- or two-phase regions is used. If the point formed by the original chemical composition and the temperature falls on a solvus line, liquidus line, or solidus line, assume that the atom or weight fraction of the phase corresponding to that line is 1.

Example Problem 5.13

For each of the original chemical compositions and temperatures at 1 atmosphere of pressure for the Al-Ni alloys listed below, give the following:
1. The phases present.
2. From the Gibbs phase rule, determine the degrees of freedom in the material and what thermodynamic variables apply to these degrees of freedom.
3. The chemical composition of each phase present.
4. The atom fraction of each phase.
   (a) 50 atom percent Ni at 1400°C.
   (b) 70 atom percent Ni at 1000°C.
   (c) 25 atom percent Ni at 854°C.
   (d) 25 atom percent Ni at 855°C.
   (e) 25 atom percent Ni at 853°C.

Solutions

a.1) The point 50 atom percent Ni and 1400°C is in the single-phase region AlNi.

a.2) $P = 1$, the Gibbs phase rule ($P + F = C + 1$) is $1 + F = 2 + 1$ and $F = 2$.
There are two degrees of freedom ($F = 2$). The temperature and chemical composition can be set in this single-phase region.

a.3) $C_{Ni}^{AlNi} = 50\%$ Ni atoms in AlNi. This is the stoichiometric chemical composition for AlNi.

a.4) $f^{AlNi} = 1$, all of the atoms are in AlNi.

b.1) The point 70 atom percent Ni and 1000°C is in the two-phase region of AlNi and AlNi$_3$.

b.2) $P = 2$; therefore the Gibbs phase rule is $2 + F = 2 + 1$ and $F = 1$. The one degree of freedom is used in setting the temperature to 1000°C. There is no possible choice in chemical composition.

b.3) From a tie line at 1000°C in the two-phase region of AlNi and AlNi$_3$ the chemical composition of the AlNi phase is $C_{Ni}^{AlNi} = 63$ atom percent Ni and $C_{Ni}^{AlNi_3} = 73$ atom percent Ni in AlNi$_3$. The stoichiometric chemical composition of AlNi is 50 atom percent Al and Ni. The 63 atom percent Ni is achieved by Ni atoms substituting for Al.

b.4) The atom fraction of AlNi is calculated from the lever rule in Equation 5.13.

$$f^{AlNi} = \frac{C^{AlNi_3} - C_0}{C^{AlNi_3} - C^{AlNi}} = \frac{73 - 70}{73 - 63} = \frac{3}{10} = 0.30 \text{ atom fraction AlNi}$$

The atom fraction of AlNi$_3$ is calculated from Equation 5.10.

$$f^{AlNi_3} = 1 - f^{AlNi} = 1 - 0.30 = 0.70 \text{ atom fraction of AlNi}_3$$

The fraction of the atoms in the AlNi phase is 0.30, and the fraction of the atoms in the AlNi$_3$ phase is 0.70.

c.1) At 25 atom percent Ni and 75 atom percent Al at 854°C, there are three phases in equilibrium: liquid, Al$_3$Ni$_2$ and Al$_3$Ni, and $P = 3$.

c.2) The Gibbs phase rule at 854°C is $3 + F = 2 + 1$, and $F = 0$. There is no possible freedom in selecting temperature or chemical composition when three phases are in equilibrium.

c.3) The chemical composition of the liquid, Al$_3$Ni$_2$, and Al$_3$Ni is obtained by the intersection of a tie line at 854°C with each of these phases.

$$C_{Ni}^{L} = 15 \text{ atom percent Ni, } C_{Ni}^{Al_3Ni_2} = 37 \text{ atom percent Ni, and}$$

$$C_{Ni}^{Al_3Ni} = 25 \text{ atom percent Ni.}$$

c.4) It is not possible to use the lever rule at 854°C, because there are three phases in equilibrium.

d.1) At 25 atom percent Ni and 75 atom percent Al at 855°C, liquid and Al$_3$Ni$_2$ are in equilibrium, and $P = 2$.

d.2) The Gibbs phase rule is $2 + F = 2 + 1$, and $F = 1$. There is one possible degree of freedom that was used in selecting the temperature of 855°C. There is no freedom in selecting the chemical compositions.

d.3) The chemical compositions of the liquid and of Al$_3$Ni$_2$ are obtained by the intersection of a tie line at 855°C with each of these phases. $C_{Ni}^{L} = 15$ atom percent Ni, and $C_{Ni}^{Al_3Ni_2} = 37$ atom percent Ni.

d.4) The atom fraction of liquid is found with the lever rule in Equation 5.13.

$$f^L = \frac{C^{Al_3Ni_2} - C_0}{C^{Al_3Ni_2} - C^L} = \frac{37 - 25}{37 - 15} = \frac{12}{22} = 0.55 \text{ atom fraction of liquid}$$

The atom fraction of Al$_3$Ni$_2$ is found from Equation 5.10.

$$f^{Al_3Ni_2} = 1 - f^L = 1 - 0.55 = 0.45 \text{ atom fraction of Al}_3\text{Ni}_2$$

e.1) At 25 atom percent Ni and 75 atom percent Al at 853°C, there is one phase Al$_3$Ni, and $P = 1$.
e.2) The Gibbs phase rule is $1 + F = 2 + 1$, and $F = 2$. There are two possible degrees of freedom that can be used in selecting the temperature of 853°C, and the chemical composition can be selected as long as it remains in the single-phase field of Al$_3$Ni. However, any variation of chemical composition from 25 atom percent Ni and 75 atom percent Al results in more phases than the single-phase Al$_3$Ni.
e.3) The chemical composition of the single phase is 25 atom percent Ni and 75 atom percent Al.
e.4) At 853°C: $f^{Al_3Ni} = 1.0$. All of the atoms are in the Al$_3$Ni phase.

5.11 BINARY-PHASE DIAGRAMS FOR CERAMICS

The phase diagrams for ceramic materials are similar in appearance to those of metal-metal systems, and they are interpreted in the same way. Figure 5.16 shows the phase diagram for Al$_2$O$_3$-Cr$_2$O$_3$. In this phase diagram, the compounds Al$_2$O$_3$ and Cr$_2$O$_3$ are the components, just as H$_2$O is the component in water, and just as Ni and Cu are components in the binary Cu-Ni phase diagram. The Al$_2$O$_3$-Cr$_2$O$_3$ phase diagram has the same continuous-solid solubility as we observed in the Figure 5.7a Cu-Ni phase diagram. The interpretation of this diagram is that at temperatures below the solidus line, all mixtures of Al$_2$O$_3$ and Cr$_2$O$_3$ are solid solutions, Cr$_2$O$_3$ substitutes into the Al$_2$O$_3$ crystal lattice, and Al$_2$O$_3$ substitutes into the Cr$_2$O$_3$ lattice. At temperatures above the liquidus line, all mixtures of Al$_2$O$_3$ and Cr$_2$O$_3$ are liquid, and the Al$_2$O$_3$ and Cr$_2$O$_3$ mix with each other. Between the liquidus and solidus lines, both liquid and solid

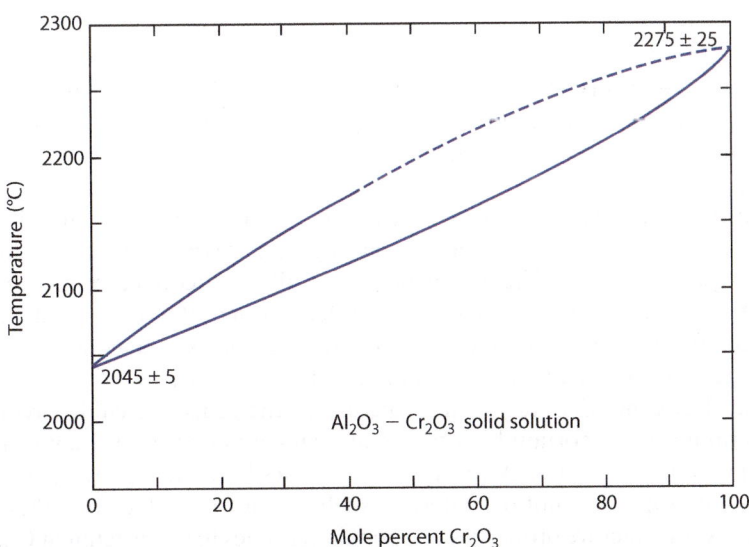

Figure 5.16 The Al$_2$O$_3$-Cr$_2$O$_3$ phase diagram. (*Based on Bunting, E.N., Nat. Bur. Stds. J, 6, 948 (1931).*)

structures exist. In the two-phase region of liquid-plus-solid solution, the same procedures developed for the Cu-Ni phase diagram are utilized to determine the chemical compositions and atom fractions of each phase present. In ceramic materials, the solidus line is important because heating the material above the solidus line results in an increased transport of atoms and chemical reactions during processes such as fusing particles of ceramics.

Example Problem 5.14

A ceramic mixture of 50 atom percent Al_2O_3 and 50 atom percent Cr_2O_3 is produced by heating the mixture to 2300°C. The mixture is then cooled to 2150°C. What phases are present at 2150°C, what are their chemical compositions, and what is the atom fraction of each phase?

Solution

At 2150°C this ceramic composition has two phases: liquid and solid. The liquid and solid chemical compositions are given by the intersection of a tie line at 2150°C with the liquidus and solidus lines. The liquid chemical composition is

$$C_{Cr_2O_3}^L = 32 \text{ molar percent } Cr_2O_3$$

The solid chemical composition is

$$C_{Cr_2O_3}^\alpha = 58 \text{ molar percent } Cr_2O_3$$

The molar fraction of liquid and solid α phase is given by the lever rule in Equation 5.13.

$$f^\alpha = \frac{C_0 - C^L}{C^\alpha - C^L} = \frac{50 - 32}{58 - 32} = \frac{18}{26} = 0.69 \text{ molar fraction of } \alpha \text{ phase}$$

The molar fraction of liquid is then found with Equation 5.10.

$$f^L = 1 - f^\alpha = 1 - 0.69 = 0.31 \text{ molar fraction of liquid}$$

Figure 5.17 shows the phase diagram for zirconia (ZrO_2) and up to 50 molecular percent calcia (CaO) that results in the compound $ZrCaO_3$. As is shown in Figure 5.17, eutectics and compounds can form in ceramic systems just as they can occur in metal systems. In Figure 5.17 the symbols C_{ss}, T_{ss}, and M_{ss} represent cubic, tetragonal, and monoclinic solid solutions (ss), respectively. In these solid solutions CaO substitutes for ZrO_2.

When CaO substitutes for ZrO_2, there is a deficiency of one oxygen atom, which creates oxygen vacancies in the zirconia. This is important technologically, because the electrical conductivity of zirconia depends upon the oxygen deficiency. The oxygen deficiency allows zirconia ceramics to be used as oxygen sensors in applications such as automobile exhaust systems, where the oxygen-deficient zirconia-calcia alloy absorbs oxygen from the exhaust and the conductivity changes. The conductivity of zirconia and applications as oxygen sensors are discussed in more detail in Section 16.9.

The cubic-phase of zirconia ZrO_2 (ss) is the structure of the gemstone cubic zirconia that is used in jewelry. Zirconia ceramics are important high-temperature materials because zirconia melts at a very high temperature of approximately 2700°C. Also, zirconia ceramics have the highest resistance to fracture of any ceramics when the proper amount of materials, such as CaO, are added and they are properly heat treated. We will discuss the fracture properties of zirconia ceramics in more detail in Chapter 11. Zirconia ceramics are also proposed for applications in high-temperature gasoline and diesel engines as cylinder liners, and valve coatings.

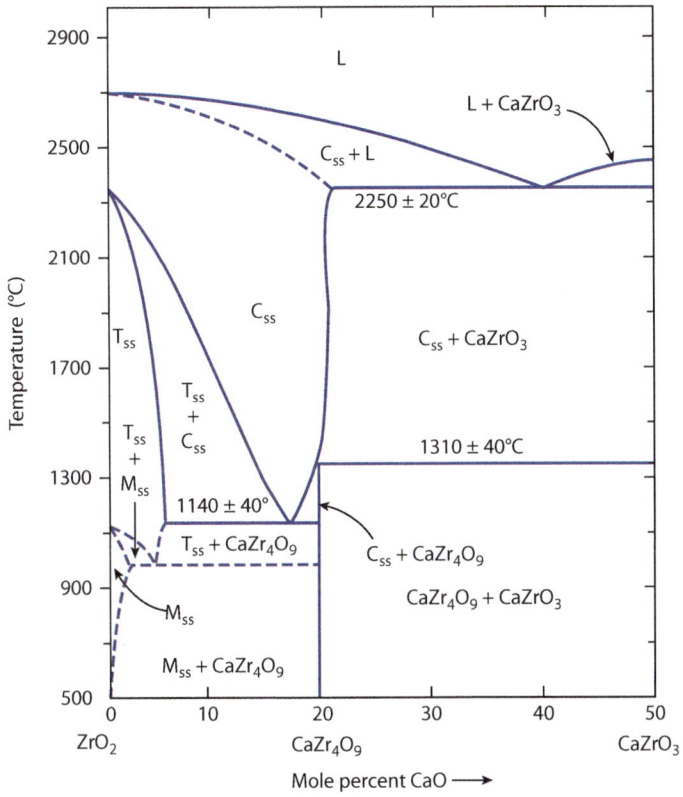

Figure 5.17 The ZrO$_2$-CaO phase diagram from ZrO$_2$ to ZrCaO$_3$. (*Based on Stubican, V.S. and Hellman, J.R., Advances in Ceramics 3 (1981), p. 25.*)

5.12 MISCIBILITY GAPS IN METAL AND CERAMIC BINARY-PHASE DIAGRAMS

In the Cu-Ni phase diagram in Figure 5.7a for temperatures below 354.5°C and for compositions centered on 67.3 atom percent Ni, the FCC α phase separates into two FCC phases α_1 and α_2, creating a miscibility gap. α_1 is rich in Cu, and α_2 is rich in Ni. A discussion of miscibility gaps is continued in Appendix 5F.

5.13 ORDERING IN METAL AND CERAMIC BINARY-PHASE DIAGRAMS

It is possible to have a metal or ceramic solid solution at elevated temperatures, and an ordered intermediate phase or compound at lower temperatures. A discussion of ordering reactions is continued in Appendix 5G.

5.14   THE IRON-CARBON PHASE DIAGRAM AND MARTENSITIC PHASE TRANSFORMATIONS

Carbon steel is made from iron and carbon. Steel is still one of the most widely used materials for engineering applications. We find in carbon steels a new type of transformation. A martensitic phase transformation occurs when the steel is rapidly cooled from an elevated temperature. A **martensitic** phase transformation occurs by deformation of the crystal rather than by diffusion of atoms as occurs during nucleation and growth phase transformations. Martensitic transformations also occur in other metals such as titanium alloys and gold-cadmium alloys; in ceramics, such as zirconia, which has potential applications as cylinder liners in diesel and gasoline engines; in the semiconductor indium thallium, and in titanium nickel (TiNi). The martensitic transformation in TiNi results in interesting applications, such as the coronary stents discussed in Chapter 1.

5.14.1   The Iron-Carbon Phase Diagram

Figure 5.18 shows the phase diagram for iron and carbon up to 6.67 weight percent carbon. This part of the phase diagram is of the most interest because it includes steels and cast irons. In Section 5.3.1 we

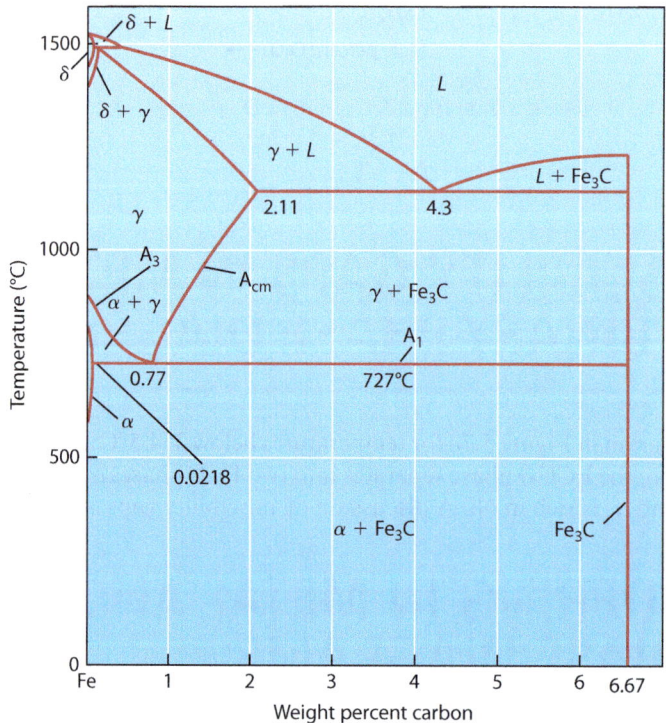

Figure 5.18   The iron-carbon phase diagram from 0 to 6.67 weight percent carbon (25 atom percent carbon) where the compound Fe$_3$C forms. (*Based on Askeland, D.R., Fulay, P.P., and Wright, W.J., The Science and Engineering of Materials, 6th ed. Cengage Learning, Stamford, CT. (2011), p. 472.*)

discussed the transformation of γ-phase iron into α-phase iron at 912°C. However, pure iron produced with normal processing procedures is a very low-strength metal. By adding carbon to iron to produce steel, the strength of the resulting material is significantly increased.

Plain-carbon steels have less than 1 weight percent carbon. Normally steel compositions are given in weight percent. For the remainder of the discussion of the iron-carbon system, "% C" refers to the weight percent carbon. In the iron-carbon phase diagram, there is a eutectoid reaction at 0.77% C. Steels that have less than 0.77% C are hypoeutectoid steels, and steels that have more than 0.77% C are hypereutectoid steels. Structural steels and the steels in auto bodies are low-carbon steels with typically 0.2 to 0.4% C. High amounts of carbon make the steel stronger, but it is more brittle than low-carbon steel. Railroad rails and wheels use steel with high-carbon contents, such as 0.70% C. Cast irons have more than 2.11% C. Chapter 8, Engineering Materials, covers steel alloys and cast iron for engineering applications in more detail. The purpose of this section is to study the phase transformations in plain-carbon steel.

To study the phase transformations in plain carbon steel, it is easiest to start with a steel of the eutectoid composition of 0.77% C, and to hold it above the eutectoid temperature of 727°C, such as 800°C. The structure at this temperature is 100% FCC γ phase with 0.77% C interstitial atoms of carbon in solid solution. If steel of the eutectoid composition of 0.77% C is cooled to 727°C, the γ phase is in equilibrium with the phases α + Fe_3C. The α phase is BCC with 0.0218% C interstitial atoms and Fe_3C is iron carbide. However, as discussed above, at the temperature of 727°C the transformation never completes. It is necessary to supercool the γ phase to a temperature such as 600°C to have it transform to the α + Fe_3C phase at a significant rate. **Pearlite** is the microstructure of the steel that results from the eutectoid reaction of $\gamma \rightarrow \alpha$ + Fe_3C. The name *pearlite* comes from the observation that the appearance of this material is layered like mother of pearl, as shown in Figure 5.19. Pearlite has plates of Fe_3C surrounded by α-phase layers. Figure 5.20 shows how the carbon atoms must move from the γ phase, where carbon has a uniform concentration of 0.77% C, to the Fe_3C phase with 6.7%C, and from the α phase, which has less than 0.0218% C. This transformation is dependent upon diffusion of carbon atoms, and the resulting microstructure depends upon the way that the metal is cooled.

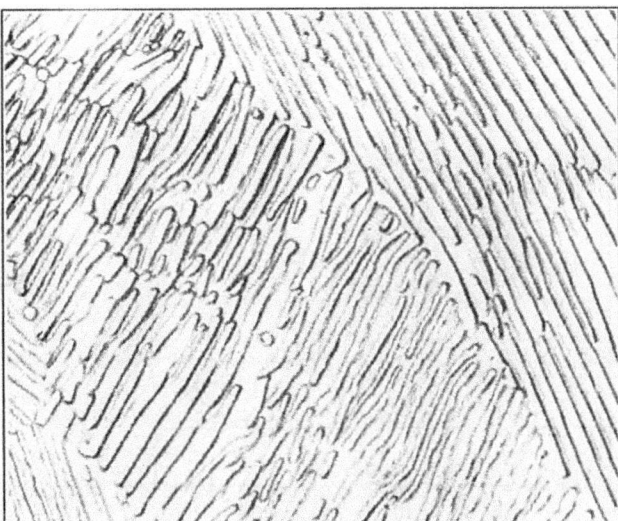

Figure 5.19 A photomicrograph of an eutectoid-composition iron-carbon steel, with pearlite microstructure consisting of plates of Fe_3C in a matrix of α (BCC) iron at a magnification of 2000. (*ASM Handbook Volume 7, Powder Metal Technologies and Applications, 1998, ASM International in ASM Handbooks Online, http://www.asmmaterials.info, ASM International, 2004.*)

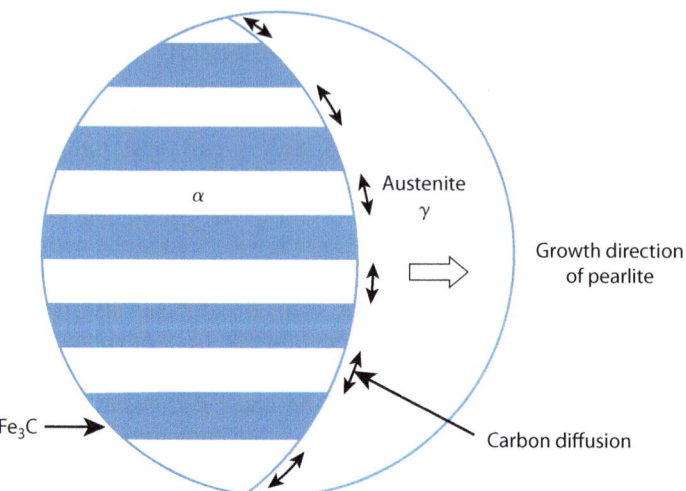

Figure 5.20 A schematic of the transformation of FCC austenite (γ) to BCC ferrite (α) and Fe$_3$C that requires the diffusion of carbon atoms.

5.14.2 Time-Temperature-Transformation Diagrams for Plain-Carbon Steel

The transformation of the γ phase to the α + Fe$_3$C phase is analyzed with time-temperature-transformation (TTT) diagrams. Figure 5.21 is the TTT diagram for the eutectoid composition of 0.77 % C in iron. Time is the x axis and temperature the y axis, and the transformation is indicated in the lines shown in the diagram. For nucleation and growth transformations, the TTT diagrams are developed by holding the material at a

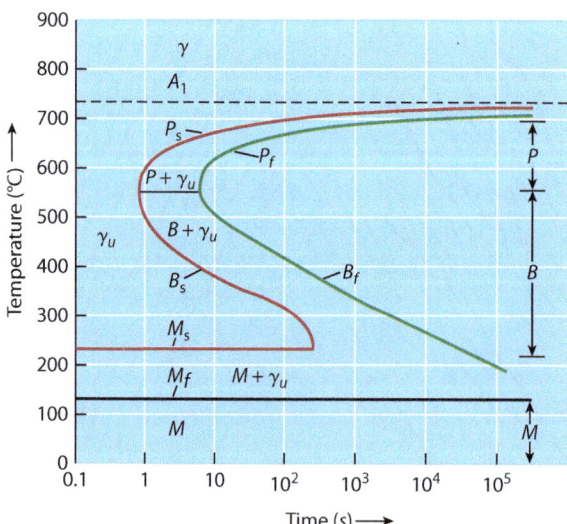

Figure 5.21 The time-temperature-transformation (TTT) diagram for the eutectoid composition iron-carbon steel. (*Based on Askeland, D.R., Fulay, P.P., and Wright, W.J., The Science and Engineering of Materials, 6th ed. Cengage Learning, Stamford, CT. (2011), p. 472.*)

fixed temperature (isothermal) and observing the percent transformation as a function of time. It is assumed that the specimen is held at the same temperature from beginning to end of the treatment. We continue with the example of the eutectoid composition of iron plus 0.77% C cooled rapidly from 800°C to 600°C and held at 600°C. The TTT diagram for this transformation in Figure 5.21 shows that no phase transformation is observed for approximately 1 second, which is the time it takes before the pearlite start line (P_s) is intersected. The **pearlite start line** (P_s) is the time required for a nucleus of pearlite to form and to grow to an observable size as a function of temperature. The A_1 in Figure 5.21 is for austenite or the γ (FCC) phase of iron. The γ phase is stable above 727°C and unstable below this temperature. The unstable γ is labeled γ_u. Note that the x axis is a $\log_{10}$ scale. One way to read the nonlinear $\log_{10}$ scale is to use the exponents of 10, because the exponents are linear on the log scale. For example, halfway between 1 and 10 on the logarithmic scale is $10^{0.5}$ seconds, or 3.16 seconds. Some TTT diagrams show a 50% transformation complete line, which is approximately 50% of the distance between the start and finish lines. The transformation at 600°C is finished when the 600°C line intersects the second solid line, the pearlite finish line (P_f), in $10^{0.8}$ seconds, or 6.3 seconds. The **pearlite finish line** gives the time when the transformation of unstable γ is fully transformed to pearlite. If this alloy is then cooled to room temperature, the pearlite structure remains, because pearlite is composed of the equilibrium phases of γ iron and Fe_3C. This is the case even though several lines are crossed upon cooling. These lines only pertain to the transformation of unstable austenite.

At isothermal temperatures below 550°C, the product of the eutectoid reaction is bainite (B) rather than pearlite. Bainite, shown in Figure 5.22, has a matrix of α phase with laths of Fe_3C. Laths are thin and narrow like a ribbon. The Fe_3C laths in bainite are smaller and more closely spaced than the plates of Fe_3C in pearlite, because at the lower temperature, the diffusion of the carbon atoms is slower. The carbon atoms cannot move over the distances required to produce the thicker plates seen in pearlite. At low temperatures, the chemical force driving the eutectoid reaction is large; however, the mobility of the atoms is low, as is demonstrated by Equation 5.2. As the isothermal temperature is reduced, the kinetics

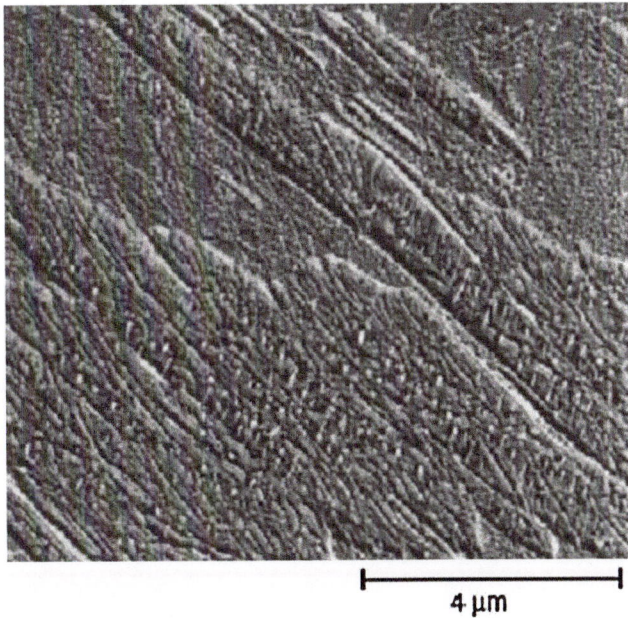

4 μm

Figure 5.22   A scanning electron micrograph of the bainite microstructure in an iron-carbon steel. The light particles oriented in one direction are the fine Fe_3C in a matrix of α (BCC) iron. (*ASM Handbook Volume 9, Metallography and Microstructures, 2004, ASM International, in ASM Handbooks Online, http://www.asmmaterials.info, ASM International, 2004.*)

favor a structure that requires a smaller diffusion distance, and the finer bainite structure forms at low temperature rather than the coarser pearlite structure.

If a eutectoid alloy is cooled rapidly (quenched) to 500°C and held at this temperature, bainite is first observed after 1 second, as shown in Figure 5.21 at the B_s line, and after 10 seconds the unstable austenite phase is fully transformed to bainite at the B_f line. If the alloy is then cooled to room temperature, the bainite structure remains, because the bainite finish line is crossed and the alloy is composed of the equilibrium phases of α iron and Fe_3C.

If the temperature of the iron-0.77%C alloy is quenched from a temperature above the eutectoid equilibrium temperature of 727°C to a low temperature such as 25°C, without the temperature-versus-time plot crossing the P_s or B_s start lines in Figure 5.21, then there is insufficient time for the pearlite or bainite structures to form. The horizontal M_s line is the start of a martensite transformation. In the martensite transformation, the FCC γ-phase iron transforms to a body-centered tetragonal unit cell (α') demonstrated in Figure 5.23 by distortion.

In the tetragonal unit cell, $a = b$, but c is not equal to a and b, and all unit cell corner angles are 90 degrees. The iron atoms are the large circles, and the small solid circles are carbon atoms. Two unit cells of FCC cubic iron are shown side by side in Figure 5.23. The FCC unit cell can also be indexed as a body-centered tetragonal (BCT) unit cell, as shown by the large solid atoms in Figure 5.23. However, if the a lattice parameter of the BCT unit cell is expanded and the c lattice parameter is compressed, the BCT lattice becomes BCC when $a = c$. The BCC structure is the equilibrium structure for iron-carbon alloys at temperatures below 727°C. If the alloy is quenched, there is insufficient time to transform the FCC structure to the BCC structure through the nucleation and growth process, which requires diffusion. The FCC structure transforms to the BCT structure that is close to the BCC structure by a diffusionless martensitic transformation. In the martensitic transformation, the lattice is distorted into a new lattice, and there is no long-range diffusion. The atoms move short distances of less than a lattice parameter. For the eutectoid-composition alloy, the FCC lattice parameter is $a_0 = 0.359$ nm. This FCC lattice parameter would produce a tetragonal unit cell of dimensions $a = b = 0.255$ nm and $c = 0.359$ nm, as it is drawn in Figure 5.23. If the carbon atoms go preferentially into the interstitial

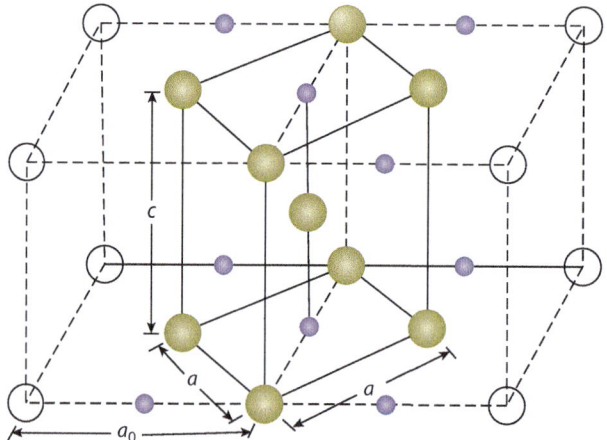

Figure 5.23   Two side-by-side FCC lattices (dashed lines). The large atoms (open circles and spheres) are iron atoms that are in the FCC structure at temperatures above 727°C, with carbon atoms (small solid spheres) in some interstitial atomic positions. When iron-carbon alloys are quenched to a low temperature, the BCT structure forms (large gold spheres and continuous lines). The BCT structure approaches the BCC structure if the carbon atoms occupy the interstitial positions shown. (*Adapted from Eisenstadt, M. M.,* Introduction to Mechanical Properties of Materials, *The Macmillan Co., New York (1971), p. 390.*)

positions shown in Figure 5.23, this expands the *a* lattice parameter and contracts the *c* lattice parameter producing a distortion of the lattice. This distortion of the unit cell results in lattice parameters of $a = b = 0.285$ nm and $c = 0.295$ nm, and it is called the **Bain distortion**. These unit-cell dimensions produce the body-centered tetragonal unit cell (α'), which is much closer to the BCC unit cell dimensions of 0.2866 nm. There is only 0.77% C, and only a few of the interstitial sites shown in Figure 5.23 are occupied. Figure 5.24 shows that the microstructure of martensite in low-carbon steel is acicular, or needle like.

The small amount of motion of the carbon atoms in the martensite transformation occurs more rapidly than does the longer-range diffusion that is required to form pearlite or bainite. The interface between the austenite (γ) and martensite (α') phases can move at a velocity of approximately one-third the velocity of sound during the transformation. Because the distances moved by the carbon atoms and the iron atoms in the martensitic transformation are less than a lattice parameter, the martensitic transformation can occur very rapidly. The percent of material transformed from the FCC γ phase to martensite depends only upon the temperature; it does not depend upon time, as did the transformation of FCC γ iron to pearlite or bainite. The martensite transformation in plain carbon steel is **athermal**, meaning that the amount of transformation depends upon the temperature but not upon the time spent at the temperature. An athermal martensitic transformation does not have the isothermal temperature region shown in Figure 5.3. There are some metals that have time-dependent martensitic transformations; however, most martensitic transformations are athermal.

As shown in Figure 5.21, the iron + 0.077% C alloy is 100% martensite if the alloy is cooled from 800°C to room temperature in fewer than 10 seconds. To demonstrate this, draw a line from the time 0.1s and 800°C to 10s and 0°C. This line does not cross the P_s or B_s start lines; therefore, no pearlite or bainite are present. It only passes through the martensite start (M_s) and finish (M_f) lines. Therefore, 100% of the material is martensite. The martensite start line (M_s) is the temperature where any remaining unstable γ phase starts to transform to α', and the martensite finish line (M_f) is the highest temperature where all of the γ-phase has transformed to α' through a martensitic transformation.

Iron-carbon martensite is a very hard, brittle material that is seldom used in engineering applications in the quenched condition. The metastable iron-carbon martensite (α') is tempered or heated to an elevated

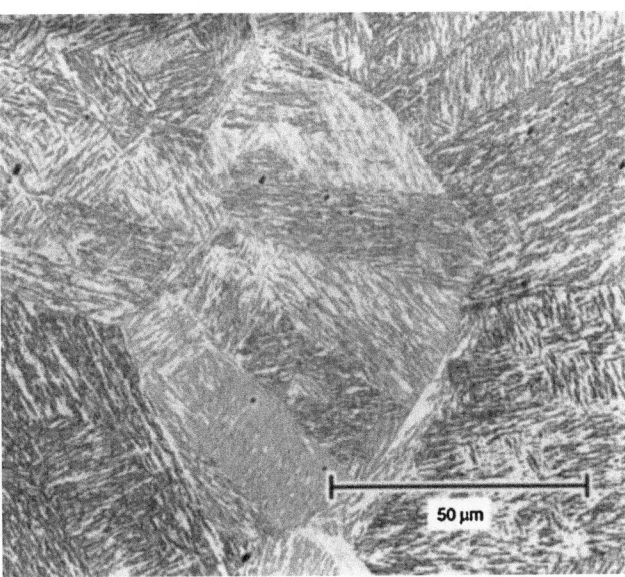

Figure 5.24 A photomicrograph of the acicular martensite microstructure of a water-quenched low-carbon steel. (*ASM Handbook Volume 9, Metallography and Microstructures, 2004, ASM International, in ASM Handbooks Online, http://www. asmmaterials.info, ASM International, 2004.*)

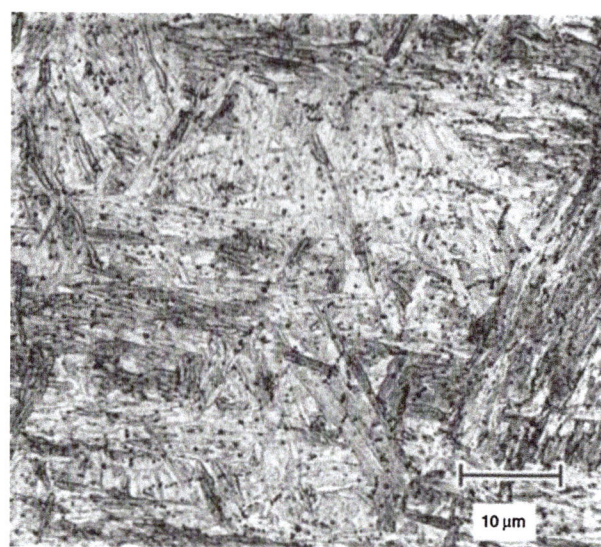

Figure 5.25 A photomicrograph of tempered martensite, showing dark particles of Fe$_3$C in a matrix of α-phase iron. (*ASM Handbook Volume 9, Metallography and Microstructures, 2004, ASM International, in ASM Handbooks Online, http://www.asmmaterials.info, ASM International, 2004.*)

temperature, such as 600°C, where the martensite transforms to BCC α-phase iron and fine particles of Fe$_3$C, shown in Figure 5.25. The quenching and tempering process results in the Fe$_3$C precipitate being nucleated in many sites throughout the bulk of the alloy. The quenched and tempered iron-carbon alloys are high strength with good ductility. Quenched and tempered steels are utilized for such applications as cutting tools, crank shafts, springs, gears, and pistons.

Example Problem 5.15

Determine the transformation products for the following cooling procedures of a eutectoid steel ($C_c = 0.77\%$ C) that is initially equilibrated at 730°C.
(a) Quench to 600°C, then hold for 10 seconds, then slow cool to room temperature.
(b) Quench to 600°C, then hold for 3.2 ($10^{0.5}$) seconds, then quench to room temperature.
(c) Quench to 600°C, then hold for 1 second, then quench to room temperature.
(d) Quench to 500°C, then hold for 1 second, then quench to room temperature.
(e) Quench to 500°C, then hold for 3.2 ($10^{0.5}$) seconds, then quench to room temperature.
(f) Quench to 500°C, then hold for 10 seconds, then slow cool to room temperature.
(g) Quench to room temperature in 1 second, then heat to 500°C and hold for 1000 seconds.

Solution

The TTT diagram of Figure 5.21 is for the eutectoid-composition steel. Also, from the iron-carbon phase diagram in Figure 5.18, we observe that at 730°C the phase present is 100% austenite (γ), also designated as A$_1$ in Figure 5.21.

a) After quenching to 600°C, there is still 100% unstable austenite (γ_u). It is assumed that the temperature falls to 600°C instantly. After holding at 600°C for 10 seconds, the austenite fully transforms to pearlite, because the time has crossed the pearlite finish line (P_f). After the cool to room temperature, the pearlite (α + Fe$_3$C) is still present, because it is stable once it forms at temperatures below 727°C, as shown in Figure 5.18. The result is 100% pearlite.

b) After holding at 600°C for 3.2 ($10^{0.5}$) seconds, only about 50% of the unstable austenite (γ_u) is transformed to pearlite. After the quench to room temperature, the 50% pearlite (α + Fe$_3$C) is still present, but the unstable austenite (γ_u) is transformed to martensite because the M_s and M_f lines are crossed during the quench. The result is 50% pearlite and 50% martensite.

c) After holding at 600°C for 1 second, 100% unstable austenite (γ_u) is still present. After the quench to room temperature, the unstable austenite (γ_u) is transformed to martensite because the M_s and M_f lines are crossed during the quench. The result is 100% martensite.

d) After holding at 500°C for 1 second, 100% unstable austenite (γ_u) is still present. After the quench to room temperature, the unstable austenite (γ_u) is transformed to martensite because the M_s and M_f lines are crossed during the quench. The result is 100% martensite.

e) After holding at 500°C for 3.2 ($10^{0.5}$) seconds, only about 50% of the unstable austenite (γ_u) is transformed to bainite. After the quench to room temperature, the 50% bainite (α + Fe$_3$C) is still present, but the unstable austenite (γ_u) is transformed to martensite because the M_s and M_f lines are crossed during the quench. The result is 50% bainite and 50% martensite.

f) After quenching to 500°C, there is still 100% unstable austenite (γ_u). After holding at 500°C for 10 seconds, the unstable austenite fully transforms to bainite, because the bainite finish (B_f) is crossed. After the cool to room temperature, the bainite (α + Fe$_3$C) is still present. The result is 100% bainite.

g) After quenching to room temperature in 1 second, 100% martensite is present because only the M_s and M_f lines are crossed during the quench. After heating to 500°C for 1000 seconds, tempered martensite is present.

5.15 REVERSIBLE MARTENSITIC PHASE TRANSFORMATIONS

The martensitic phase transformation in iron-carbon alloys is not reversible. This is due to the formation of dislocations that accommodate the large deformation that occurs during the Bain distortion. The formation of dislocations allows the material to be continuous during the Bain distortion of the martensitic transformation. The formation of dislocations in a material is not a reversible process. If iron-carbon martensite is heated to above 727°C, where austenite is the equilibrium structure, the new austenite phase forms by a diffusion-controlled nucleation and growth process. The austenite does not form by a reversal of the martensitic transformation.

Reversible martensitic transformations occur in materials where dislocations do not form as a result of the martensitic transformation. In reversible martensitic transformations, the distortion is accommodated by the stretching and bending of the atomic bonds, which is reversible. Also, in a reversible martensitic transformation the change in the GFE is zero. A negative change in the GFE occurs due to the change in structure that is balanced by a positive change in GFE due to the bond stretching and bending. In nonreversible martensitic phase transformations, such as in iron-carbon alloys, the GFE change is negative.

Reversible martensitic phase transformations have been observed in a number of metal alloys, such as AuCd, CuAlNi, and CuZnAl; however, the alloy TiNi has the most engineering applications. This discussion focuses on the TiNi alloy. At temperatures above 640°C, TiNi is a random mixture of Ti and Ni atoms in a BCC structure, and below this temperature the Ti and Ni atoms order into the CsCl structure, shown in Figure 2.7. In analogy with iron-carbon alloys, the CsCl structure of TiNi is the austenitic phase. The martensite start temperature (M_s) for TiNi is approximately 70°C, the temperature of hot water. Wang has described the martensitic transformation as a shear distortion, as indicated for

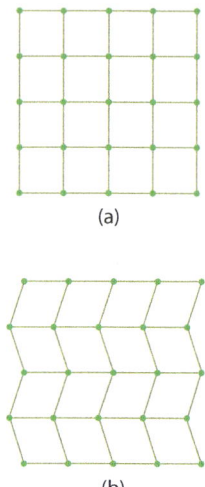

(a)

(b)

Figure 5.26   (a) A two-dimensional schematic of the cubic austenite phase. (b) The distorted martensite lattice formed by reversible displacements.

only two dimensions in Figure 5.26. This figure shows only two of the possible distortion variants of the TiNi austenitic phase in the transformation to martensite. Each cube lattice parameter in TiNi ($a = b = c$) can be distorted in positive and negative x, y, and z directions. One of the characteristics of the reversible martensitic transformation in TiNi is that the overall shape of the specimen does not change during the martensitic transformation, even though the crystal lattice changes shape, as shown in Figures 5.26a and 5.26b. The martensite finish temperature (M_f), where there is 100% martensite in TiNi, is typically 30 to 40°C below the martensite start temperature. In the TiNi martensitic transformation the amount of martensite depends only upon the temperature and not upon time.

If the martensitic phase in Figure 5.26b is heated, it transforms back to the cubic austenite phase shown in Figure 5.26a by the reversal of the bond stretching and bending. Upon heating, the transformation from martensite to austenite begins at the **austenite start temperature** (A_s), shown in Figure 5.27, and ends at the **austenite finish temperature** (A_f), where 100% of the martensite has transformed to austenite. In TiNi it is possible to go through a cycle of cooling austenite to form martensite and heating the martensite to form austenite, as shown in Figure 5.27.

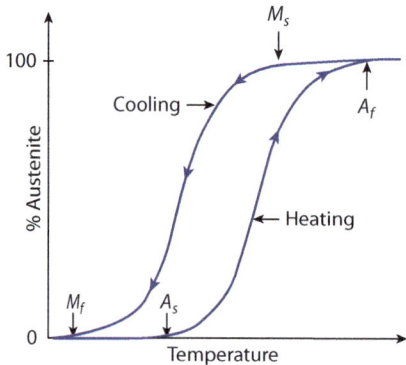

Figure 5.27   The % austenite as a function of temperature, as a result of cooling and heating TiNi.

5.15.1 The Shape-Memory Effect and Training in Metal Alloys

The shape-memory effect is a result of the reversible martensitic transformation in TiNi. Other metal alloys with a reversible martensitic transformation also have a shape-memory effect. To demonstrate the shape-memory effect, obtain a wire of TiNi that was formed straight at a temperature well above the A_f temperature but below 640°C, where the structure is cubic as shown in Figure 5.26a. If the M_f temperature of the wire is approximately 30°C, then the wire is 100% martensite at room temperature, and it has a structure like Figure 5.26b. Bend the wire into any desired shape as long as the bends are not too severe. When the wire is bent, the different variants of the distortion shown in Figure 5.26 allow the metal to bend. Then when the wire is heated above the A_f temperature, it reverts back to the cubic structure shown in Figure 5.26a and the wire returns to its initial straight shape. If the wire is bent into any shape, such as an S, at temperatures above the A_f temperature, the wire returns to the S shape when heated to temperatures above A_f. That is why it is necessary to obtain a wire that was formed straight above the A_f temperature if you want it to return to a straight shape.

Coronary stents utilize the shape memory effect, as we discussed in Chapter 1. However, it is necessary to reduce the temperature of the martensitic transformation so that human blood temperature of 37°C is above A_f. The temperatures M_s, M_f, A_s, and A_f can be adjusted by changing the TiNi alloy composition. Figure 5.28 shows how the M_s temperature is changed by altering the Ni-to-Ti ratio. Also, replacing Ni atoms with Co atoms reduces the M_s temperature; for example, the alloy $TiNi_{0.8}Co_{0.2}$ has an M_s of approximately −13°C according to Wang.

It is also possible to train TiNi to assume a certain shape when cooled below the M_f temperature, by bending TiNi into a certain shape at temperatures below M_f and heating above A_f, and repeating the cycle of bending to a certain shape below M_f and heating above A_f. Eventually the TiNi will assume the desired shape when cooled to temperatures below the M_f temperature.

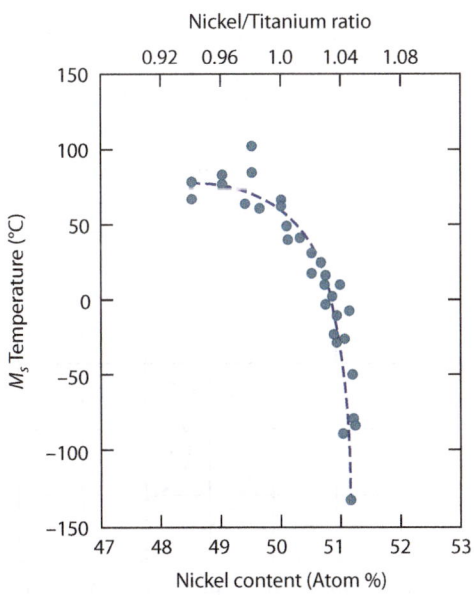

Figure 5.28 A plot of the martensite start temperature (M_s) as a function of the nickel content for TiNi. The data points are from several sources. (*Based on figure Nitinol Ms vs Ni content by Tomduerig*)

Superelasticity is another valuable property of metals with reversible martensitic transformations. This property has led to the use of TiNi to straighten the alignment of teeth. Superelasticity is covered in Section 6.8.

5.16   SHAPE-MEMORY POLYMERS

A shape-memory effect has been developed in polymer composites. However, there is not any phase transformation in shape-memory polymers (SMPs). SMPs are discussed in Appendix 5H.

5.17   LIQUID-TO-CRYSTAL AND LIQUID-TO-AMORPHOUS MATERIAL TRANSITIONS

If a material with a complex crystal structure is cooled very slowly from the liquid state to a temperature below the melting temperature, a crystalline material forms. Examples of such materials include silica (SiO_2), which has tetrahedra of silicon and oxygen atoms, as shown in Figure 2.28a; and polyethylene, which is shown in Figure 2.17. The transformation from a liquid to a crystalline solid occurs by nucleation and growth, as discussed above. If the cooling rate is rapid, and there is not sufficient time for the silica tetrahedra to form a crystalline nucleus, then a crystal does not form, and the structure of the liquid remains at temperatures below the melting temperature of the crystalline material, as discussed in Section 2.10.4.

TTT diagrams similar to Figure 5.21 can be constructed for materials that have nucleation and growth transformations to a crystalline structure, or alternatively to an amorphous structure, such as the schematic for glass shown in Figure 5.29. In Figure 5.29, the area labeled G contains only amorphous glass, the $G + C$ area contains glass and crystalline material, and the C area contains only crystalline material. The line on the left (start line) gives the time when crystalline material is first observed at a constant temperature, and the line on the right (finish line) gives the time when all of the material has transformed to a crystal. If glass is cooled so rapidly that the temperature as a function of time misses the start line in the TTT diagram where crystalline material is first observed in Figure 5.29, nucleation of crystalline material does not occur. Instead this material forms amorphous solid glass. For example,

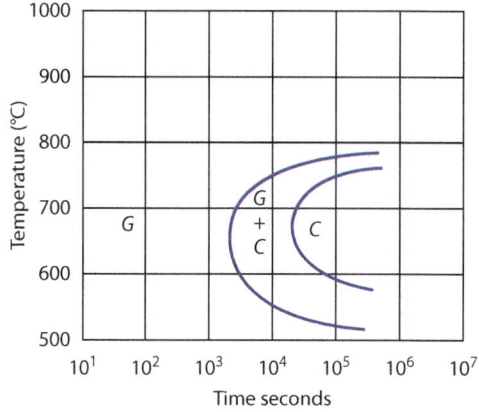

Figure 5.29   A TTT diagram for a material that can be either glass (G), crystalline (C), or a combination of both ($G + C$), depending upon the temperature and the isothermal hold time.

if this glass is cooled from 800°C to 500°C in 10^3 seconds, only glass is observed. However, if glass is quenched from 800°C to 600°C and then held at 600°C, crystals are first observed after $10^{3.5}$ seconds, and after 10^5 seconds the material is fully crystalline.

Glass-ceramics are materials where some crystallization has occurred within glass. Often nucleating agents such as titanium dioxide are added to glass to promote the crystallization. If the crystallization is only partial, the material can be processed using glass processing procedures. These materials have found application because of their thermal shock resistance in cookware.

TTT diagrams similar to Figure 5.29 can also be constructed for polymers such as polyethylene. In polymers the mode of crystal nucleation depends upon the molecule. Flexible long-chain molecules such as polyethylene form crystalline nuclei by the folding of individual molecules into small (nano or micro) lamellae, as shown in Figure 2.21b. Rigid long-chain molecules must form crystalline nuclei of different parallel molecules. The growth of polymer crystalline nuclei depends upon the movement of segments of long-chain polymers to the crystalline nucleus.

It is difficult to produce amorphous pure metals because of the extremely high quench rates required. For example, the estimated cooling rate required to produce amorphous silver is 10^{10} K/s. However, it is possible to make amorphous metal alloys, known as metallic glass, by mixing metal atoms with atoms of very different size. One amorphous metal alloy that has applications as a permanent magnet is iron with 15 to 25 atom percent boron. Iron has an atomic radius of 0.124 nm, and boron has an atomic radius of 0.046 nm.

5.18 PHASE DIAGRAMS FOR LIQUID POLYMER BLENDS

A liquid polymer blend is created by combining two or more different liquid polymers. Phase diagrams for liquid polymer blends are calculated from the temperature, entropy, and enthalpy of mixing. Equation 4.44 presents the entropy of mixing polymers. The enthalpy of mixing is calculated from the interaction parameter in Equation 4.47. Figure 4.24 presents the interaction parameter for a mixture of saturated polybutadiene (SPB) 88 and SPB 78, and Figure 5.30 presents a phase diagram based upon entropy and enthalpy calculations for these two mixtures.

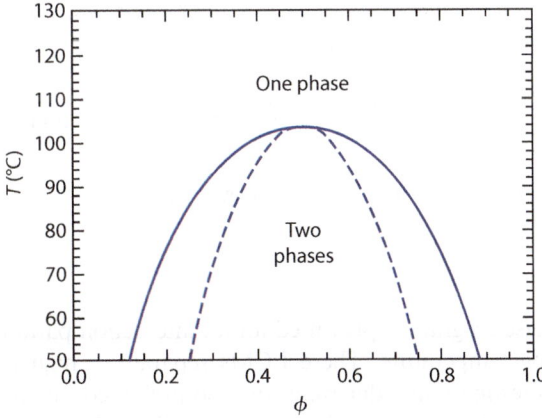

Figure 5.30 The calculated phase diagram for a mixture of saturated polybutadiene (SPB) 88 and SPB 78 with 2000 monomers per molecule. The *x* axis is volume fraction of the SPB 88, and the *y* axis is temperature. (*Based on Balsara, N. P., Thermodynamics of Polymer Blends, in Physical Properties of Polymers Handbook, Mark, J.E. ed., AIP Press, Woodbury N.Y. (1996), p. 259.*)

In Figure 5.30 the x axis is the volume fraction of the first listed component; otherwise it is similar to phase diagrams for metals and ceramics. The phase diagram shows that at temperatures above 102°C, these two LCMs form a homogenous mixture for all compositions. For temperatures below 102°C and for compositions inside the solid line, the two polymers separate into two phases. For any blends inside the solid lines, the composition based upon volume fractions of each molecular type of the two phases is given by the intersection of a tie line at the temperature of interest with the solid line. The dashed lines in Figure 5.30 are the spinodal lines. This phase diagram has a miscibility gap and spinodal line similar to that observed in the Cu-Ni phase diagram in Figure 5.7. Miscibility gaps and spinodal lines are discussed in Appendix 5F. As with metals and ceramics at constant temperature and pressure, two polymer molecules mix if the GFE of mixing is negative, and if the GFE of mixing is positive the LCMs separate into two phases. In Example Problem 4.14, we calculated that the GFE of mixing is negative at 120°C for a 50% mixture of SPB 88 and SPB 78 with 2000 monomers per molecule; therefore the molecules should mix at this temperature. In Homework Problem 4.25, we calculated that the GFE of mixing is also negative for a 25% mixture of SPB 88 and 75% SPB 78 with 2000 monomers per molecule at 373 K (100°C), and the molecules should also mix at this temperature. These two calculations are in agreement with the phase diagram in Figure 5.30.

For polymers the terms *alloy* and *blend* have slightly different meanings than for metals. For polymers an alloy refers to a combination of two or more polymers that form a homogenous single phase. A blend is a combination of polymers that form two or more phases. In metals the term *alloy* refers to a mixture of a metal with another element no matter how many phases are present.

Example Problem 5.16

For a polymer blend of 40 volume % (vol %) SPB 88 and 60 vol % SPB 78, at 50°C:
(a) How many phases are present?
(b) What are their compositions?
(c) What volume fraction is in each phase?

Solution

a) A polymer blend of 40 vol % SPB 88 and 60 vol % SPB 78 at 50°C is in the two-phase region. By convention, we will call the phases α and β.
b) A tie line at 50°C intersects the solid line, resulting in the α-phase composition of 12 vol % SPB 88 and 88 vol % SPB 78, and the β-phase composition is 88 vol % SPB 88 and 12 vol % SPB 78.
c) The volume fractions of each phase are given by the lever rule.

$$f^a = \frac{\phi^\beta - \phi_0}{\phi^\beta - \phi^a} = \frac{0.88 - 0.40}{0.88 - 0.12} = \frac{0.48}{0.76} = 0.63 \text{ volume fraction } \alpha$$

$$f^\beta = 1 - f^a = 1 - 0.63 = 0.37 \text{ volume fraction } \beta$$

In the article by Balsara, a phase diagram is presented for an interaction parameter that is negative and decreases linearly with decreasing temperature. These LCMs separate at high temperatures and mix at low temperatures. Some very interesting phase diagrams are also presented for cases where the interaction parameter is nonlinear. A difference between the phase diagrams for polymers and those for metals and ceramics is that the phase diagrams for polymers are calculated from the GFE based upon experimental values of the interaction parameter, whereas the phase diagrams shown in this book for metals and ceramics are completely experimental.

Phase diagrams are also used when polymers are mixed with solvents, for the polymer and the solvent can be in equilibrium. We discussed the solubility of polymers in solvents in Section 4.8.3. In copolymers, different polymers are mixed on the same LCM by chemical reactions; however, the polymerization reaction is not an equilibrium process, and this process cannot be represented on an equilibrium phase diagram.

5.19 TERNARY PHASE DIAGRAMS

Ternary phase diagrams have been developed for three components, and they are technologically important for alloys of metals and ceramics with three components. However, learning how to use a ternary phase diagram does not teach us anything new about how materials interact with each other to mix, segregate, or form compounds. For this reason, we will not be covering this subject in this book. Books on phase diagrams discuss ternary phase diagrams.

Summary

- When there is a reversible heat transfer (ΔQ_R) at the absolute temperature T, the change in entropy (ΔS) is given by $\Delta S = \dfrac{\Delta Q_R}{T}$.

- When all entropy changes are considered, the entropy will increase in any material change, and it is postulated that the state with the maximum entropy is the most probable state.

- At the melting temperature (T_m) of a solid, the liquid and solid Gibbs free energies are equal, and at the vaporization temperature (T_v) of a liquid, the liquid and vapor Gibbs free energies are equal.

- Phase diagrams of one-component systems show the phases present as a function of temperature and pressure.

- In the nucleation and growth phase transformation, nuclei of the new phase form by a few atoms coming together in the new crystal structure by diffusion. The nucleus then grows by atoms diffusing from the old phase to the new phase.

- At the equilibrium temperature for liquid and solid phases of the same material, the phase transformation rate is 0. For phase transformations to occur at a finite rate during cooling, some amount of super-cooling is necessary. At low temperatures the transformation rate decreases because the diffusion rate decreases.

- The Gibbs phase rule calculates the number of degrees of freedom (F) from the number of phases (P) and the number of components (C). When the possible degrees of freedom (thermodynamic variables) are temperature, chemical composition, and pressure, then the Gibbs phase rule is $P + F = C + 2$. If pressure is fixed, such as at 1 atmosphere, and temperature and composition can vary, then the Gibbs phase rule is $P + F = C + 1$.

- The chemical potential of a particular atom or molecule in a phase is the partial derivative of the Gibbs free energy of the phase, with respect to the number of atoms or molecules of that type. The chemical potential of an atom or molecule in a phase is also equal to the change in the Gibbs free energy if one atom or molecule is added to the phase.

- When two phases are in equilibrium, the chemical potentials of the atoms or molecules of a particular type in the two phases are equal.

- Binary phase diagrams are a plot of the temperature versus composition for two components, and they show the phases present and the chemical composition of each phase.

- The Hume-Rothery rules define when continuous-solid solutions can occur between two metallic elements: The elements should differ by less than 15% in atomic radius; they must have the same crystal structure, similar electronegativities, and a similar valence.

- The lever rules are used to calculate the atom or weight fraction of each phase present in a two-phase region of a binary-phase diagram.

- When two different types of atoms, molecules, or compounds form a eutectic-phase diagram, the different types bond more strongly with the same type than with the second type. At the eutectic composition and eutectic temperature, liquid and two solid phases are in equilibrium. The eutectic composition melts at a lower temperature than either component alone. Lead-tin and silver-copper solders have eutectic phase diagrams, as do eutectic salts that are used for energy storage.

- If there are two components, and three phases are in equilibrium, this is an invariant point according to the Gibbs phase rule. Eutectic, eutectoid, peritectic, peritectoid, and monotectic phase diagrams have invariant reactions.

- When compounds form with two different types of atoms or molecules, the different types bond more strongly with the other type than with themselves. Many compounds have high melting temperatures. Some important compounds in binary-phase diagrams include Ni_3Al, which strengthens superalloys for high-temperature gas turbines; TiNi, which is used in the shape-memory alloys; Fe_3C, which is used to strengthen carbon steels; Al_2Cu, which strengthens some aircraft aluminum alloys; and Ti_3Al, which strengthens some titanium alloys for aircraft structures and turbine fan blades.

- In phase diagrams for ceramics, such as Al_2O_3 and Cr_2O_3, the compounds Al_2O_3 and Cr_2O_3 are the components. Otherwise ceramic phase diagrams are similar to metal phase diagrams.

- A martensitic phase transformation occurs when carbon steel is rapidly cooled from an elevated temperature. A martensitic phase transformation occurs by deformation of the crystals rather than by nucleation and growth by diffusion.

- Time-temperature-transformation (TTT) diagrams are plots of phases present as a function of temperature and time, and they are used to follow the progress of isothermal nucleation and growth transformations and martensitic transformations.

- Time in the TTT diagrams only applies to the amount of time for the isothermal nucleation and growth transformations.

- The martensitic phase transformation in iron-carbon alloys is not reversible due to the formation of dislocations that accommodate the large deformation that occurs during the Bain distortion. The formation of dislocations in a material is not a reversible process.

- Reversible martensitic transformations occur in materials, such as TiNi, where dislocations do not form as a result of the martensitic transformation. In reversible martensitic transformations, the distortion is accommodated by the reversible stretching and bending of the atomic bonds. The reversible martensitic transformation in TiNi leads to the shape-memory effect in metals used for biomedical devices, such as stents to increase blood flow to the heart.

- Shape-memory polymers are made from two different polymers. One polymer is elastic. The other polymer responds to a stimulus that allows it to change reversibly from compliant to rigid and back to compliant.
- TTT diagrams can be constructed for materials that have a nucleation and growth transformation to a crystalline structure from an amorphous structure.
- Phase diagrams for liquid polymer blends are calculated from the temperature and from the entropy and enthalpy of mixing.

Supplemental Reading Subjects and Authors

Full references are listed at the end of the book.

General:	*Askeland, Fulay, and Wright*
Thermodynamics of phase transformations:	*Porter and Easterling; Ragone; Swalin; Volmer and Weber*
Thermodynamics and phase diagrams in polymers:	*Mark; Rubenstein and Colby; Balsara*
Martensitic transformations:	*Reed-Hill and Abbaschian*
Martensitic transformation in TiNi:	*Wang*

Homework

Concept Questions

1. If a transformation from solid to liquid occurs such that the solid and liquid are always in equilibrium, then the transformation is _____.

2. A transformation occurs in a material, and the Gibbs free energy change in the material is negative. The transformation is _____.

3. When α-phase iron is in equilibrium with iron carbide, the iron atoms in the two phases have equal _____ _____.

4. Nickel and silver have lattice parameters of 0.352 nm and 0.408 nm, respectively; an electronegativity of 1.8 for each, an FCC structure, and a similar valence. Should they be able to form a continuous-solid solution?

5. In a phase diagram, the _____ line gives the temperature where the first solid forms upon cooling an alloy of given composition C_0 from the liquid phase.

6. In a phase diagram, the _____ line gives the temperature where the first liquid forms upon heating an alloy of given composition C_0 from the solid phase.

7. The lattice parameter of nickel is 0.352 nm, and for copper it is 0.363 nm. The electronegativities are 1.8 and 1.9, respectively, and they are both FCC metals. It is expected that there is _____ solid solubility of nickel in solid copper at elevated temperature.

8. When a liquid copper-nickel alloy is rapidly cooled, the center of the grains has a higher concentration of nickel atoms than the average concentration. This is _____.

9. The purity of an alloy is given by the _____ _____.

10. The weight _____ gives the weight of material present in each phase of a two-phase region, but this says nothing about the phase purity.

11. For a binary phase diagram in a two-phase region, use the _____ rule to find the atom or weight fraction of the phases present.

12. The α _____ line in a phase diagram separates the α-phase field from the $\alpha + \beta$-phase field, and it gives the maximum solubility of the B-type atoms in the A-type crystal.

13. Solders are made from metals with a(n) _____ phase diagram, because alloys of the metals melt at a temperature lower than that of either of the two components individually.

14. When two elements form a eutectic phase diagram, in the eutectic reaction the atoms of the solid elements _____.

15. At a pressure of 1 atmosphere, a mixture of two metals (A and B) has the three phases α, β, and γ in equilibrium. Can the temperature be changed and still have the α, β, and γ phases in equilibrium?

16. At a pressure of 1 atmosphere, a mixture of two metals (A and B) has the two phases α and β equilibrium at the selected temperature (T). Can the chemical composition of the α phase be changed by adding more A-type atoms at temperature (T)?

17. In a(n) _____reaction, liquid transforms to two solids α and β.

18. In a(n) _____reaction, one solid γ phase transforms to two solids σ and β.

19. In a(n) _____reaction, one solid α phase and a liquid phase react to form one solid phase β.

20. In a(n) _____reaction, two solid phases α and γ react to form one solid phase β.

21. The microstructure of the grains of solid α and β phases that form in the lead-tin eutectic reaction is _____ of the α and β phases.

22. If elements A and B chemically bond to each other more strongly than a random mixture of A and B elements chemically bond, it is expected that a _____ of A and B should form.

23. The Ga-As phase diagram has a compound GaAs (γ phase) that is a single vertical line at 50 atom percent As. Can a γ phase be produced that has 55 atom percent As?

24. An intermediate _____ can have a wider region of possible compositions than does an intermediate compound.

25. When an intermediate phase melts with no change in chemical composition, this is a(n) _____ melting point.

26. In the aluminum-nickel phase diagram in Figure 5.15, what is the reaction type that leads to the formation of Al_3Ni at 854°C?

27. A(n) _____ reaction is when one liquid transforms to another liquid and a solid upon cooling.

28. In polymer blends, the enthalpy of interaction is determined from the experimentally determined _____ parameter.

29. If an iron-0.77 % C steel is very slowly cooled from 800°C to room temperature, the expected microstructure is described as _____.

30. If an iron-0.77% C steel is very rapidly cooled from 800°C to room temperature, the expected microstructure is described as _____.

31. If as-quenched martensite is heated to 500°C for one hour, it is then_____ martensite.

32. The distortion of the FCC unit cell to a tetragonal unit cell after quenching a carbon steel from 800°C to room temperature is the _____ distortion.

33. For a martensitic transformation to be reversible, there cannot be the production of _____ to accommodate strain.

34. In the shape-memory effect in metals, the martensitic transformation must be _____.

35. In a reversible martensitic transformation, the Gibbs free energy of the transformation is equal to _____.

36. For a blend of long-chain molecules, if the enthalpy of mixing is positive, it is expected that the polymers should mix in the liquid at all temperatures above the melting temperature. True or false?

Engineer in Training–Style Questions

1. At the equilibrium melting temperature of a solid, if an atom of solid changes to liquid, the change in the Gibbs free energy is equal to:
    (a) The heat of fusion
    (b) Zero
    (c) A negative value
    (d) A positive value

2. At a temperature below the melting temperature of the solid, if a mole of liquid changes to solid, the change in the Gibbs free energy is equal to:
    (a) The negative of the heat of fusion
    (b) Zero
    (c) A negative value
    (d) A positive value

3. For a given element, which of the following has the greatest entropy?
    (a) A perfect single crystal
    (b) A polycrystal
    (c) A liquid
    (d) A vapor

4. Which of the following is not true about a nucleation and growth transformation from a liquid to a solid?
    (a) The transformation rate continuously increases with the amount of super-cooling.
    (b) The transformation rate is equal to 0 at the equilibrium temperature.
    (c) The transformation rate goes to 0 at temperatures approaching 0 kelvin.
    (d) The transformation rate reaches a maximum at a temperature between the equilibrium temperature and 0 kelvin.

5. For a mixture of two different elements A and B, at 1 atmosphere of pressure three phases are in equilibrium (α, β, and liquid), and it is desired to keep all three phases in equilibrium. Which of the following statements is true?
   (a) If the temperature is changed, the composition of the α, β, and liquid phases changes.
   (b) The composition of the α phase is changed by adding more A-type atoms.
   (c) The composition of the liquid phase is changed by adding more A-type atoms.
   (d) Nothing can be changed.

6. Two phases α and β are in equilibrium, and they are made of A-type and B-type atoms. Which of the following statements is true?
   (a) The Gibbs free energies of the α and β phases are equal.
   (b) The chemical potentials of the A-type atoms in the α and β phases are equal.
   (c) The Gibbs free energies of the α and β phases are equal to 0.
   (d) The chemical potentials of the A-type and B-type atoms in the α phase are equal.

7. Which of the following is not necessary for a continuous-solid solution?
   (a) Similar atomic numbers
   (b) Atomic radii differ by less than 15%
   (c) The same crystal structure
   (d) A similar electronegativity

8. Which of the following best describes the microstructure resulting from a eutectic reaction?
   (a) Alternating grains of α and β phases
   (b) Grains of the α phase with precipitate particles of the β phase
   (c) A volume with only α-phase grains and another volume with only β-phase grains
   (d) The grains contain alternating plates or rods of the α and β phases

9. Which of the following is a eutectoid reaction?
   (a) $L \rightarrow \alpha + \beta$
   (b) $\gamma \rightarrow \alpha + \beta$
   (c) $L + \alpha \rightarrow \beta$
   (d) $\gamma + \alpha \rightarrow \beta$

10. Iron with 0.20% carbon is a(n)
   (a) Cast iron
   (b) Hypereutectoid steel
   (c) Hypoeutectoid steel
   (d) Eutectoid steel

11. Bainite is formed by the following procedure with a eutectoid composition carbon steel starting at 800°C:
   (a) Quench to 500°C, then hold for 100 seconds, then cool to room temperature.
   (b) Quench to 600°C, then hold for 100 seconds, then cool to room temperature.
   (c) Quench to 700°C, then hold for 100 seconds, then cool to room temperature.
   (d) Quench to room temperature in 10 seconds.

12. The highest-strength carbon steels with some ductility have a microstructure described as:
   (a) Martensite
   (b) Pearlite
   (c) Bainite
   (d) Tempered martensite

Design-Related Questions

1. You are asked to design a metal alloy for use in salt water. The most important design parameter is corrosion resistance followed by other mechanical properties such as yield strength. It is known that dissimilar metals in contact in salt water can result in corrosion of one of the materials. What type of metal alloy phase diagrams should you consider first for this application?
   (a) Continuous solid solution
   (b) Eutectic
   (c) Compound formation
   (d) Peritectic

2. You are asked to design a metal alloy for a high-temperature application. What type of metal alloy phase diagrams should you consider first for this application?
   (a) Continuous solid solution
   (b) Eutectic
   (c) Compound formation
   (d) Peritectic

3. You are asked to design a metal alloy for a part with a room-temperature application. The part is to be made by casting the metal into shape. When casting the part, liquid metal is poured into a mold that has a cavity with the shape of the part. To reduce cost by conserving energy and to utilize an inexpensive furnace that does not have a high-temperature capability, it is desired that the metal alloy not have a high melting temperature. What type of metal alloy phase diagrams should you consider first for this part?
   (e) Continuous solid solution
   (f) Eutectic
   (g) Compound formation
   (h) Peritectic

Problems

Problem 5.1 (a) Gold has a melting temperature of 1063°C and a latent heat of fusion of 12,700 J/mole. Calculate the entropy change in the gold when 1 mole of gold freezes at 1063°C. (b) What is the sign of the change in entropy of the gold, and how does this correlate with the change in order in the material?

Problem 5.2 Iron transforms from the FCC to BCC α-phase iron upon cooling at an equilibrium transformation temperature of 1185 K (912°C), and with a latent heat of this transformation upon cooling of -900 J/mole of iron atoms. Determine the change in the GFE of the iron if 1 mole of α-phase iron transforms to BCC α-phase iron at 1085 K (812°C) and at 1 atmosphere of pressure during cooling. (b) Is this transformation spontaneous or is it reversible?

Problem 5.3 A Cu-Ni alloy is produced that is 43 atom percent Ni. At a temperature of 1260°C:

(a) What phases are present?

(b) What are their chemical compositions?

(c) What is the atom fraction of each phase present?

Problem 5.4   An alloy is produced that is 40 atom percent Ni and 60 atom percent Cu. The alloy is initially heated to 1300°C to melt the entire alloy. All of the following questions relate to this chemical-composition alloy.

    (a)   The alloy is then slowly cooled. At what temperature does the first solid start to form, and what is the first solid chemical composition?

    (b)   At 1250°C what phases are present, what are their chemical compositions, and what is the atom fraction of each phase present?

    (c)   At what temperature does the entire alloy become solid for very slow cooling, and what is the chemical composition of this solid?

Problem 5.5   Use the Gibbs phase rule and Figure 5.7 to determine the degrees of freedom for an alloy of 40 atom percent Ni and 60 atom percent Cu at the temperatures of 1300°C and 1250°C, assuming that the pressure is 1 atmosphere.

Problem 5.6   A silver solder alloy with 40 atom percent Cu and 60 atom percent Ag is at 600°C. At this temperature:

    (a)   What phases are present?

    (b)   What is the chemical composition of each phase?

    (c)   What is the atom fraction of each phase?

Problem 5.7   A silver conductor is to be in service at 400°C, and it is desired to strengthen this conductor by adding copper. However, to keep the maximum conductivity at 400°C, it is desired that there be no β phase present to scatter electrons. What is the maximum amount of copper that can be added to silver and not have any β phase form at the operating temperature?

Problem 5.8   Silver solder is produced that is 60 atom percent copper and 40 atom percent silver, and this alloy is initially heated to 1000°C to melt the alloy. The alloy is then slowly cooled.

    (a)   At what temperature does the first solid form, and what is the chemical composition of this first solid to form?

    (b)   At what temperature does the alloy completely transform to a solid; what phases are present after the alloy is completely solid, and what are their chemical compositions at the transformation temperature?

    (c)   At room temperature, what is the atom fraction of each phase present?

Problem 5.9   You want to produce a casting of sterling silver that is silver plus 13 atom percent copper. Pure silver is too soft for many applications, and adding 13 atom percent copper strengthens the silver without significantly changing the color. To produce the sterling silver casting, you are going to melt commercially pure silver and copper in a furnace, and then you will pour the liquid metal mixture into a mold. The following questions relate to this alloy.

    (a)   In selecting a furnace, what must be the minimum value of the high-temperature capability of your selected furnace?

    (b)   In transferring the liquid metal from the furnace to the mold, what is the minimum temperature that the metal can reach so that none of the metal has solidified?

(c) After the casting has cooled to room temperature, what phases are present, what is the equilibrium chemical composition (C_i) of each phase present, and what is the atom fraction (f^i) of each phase present?

Problem 5.10 An alloy of 73.9 atom percent tin (Sn) and 26.1 atom percent lead (Pb) is mixed and heated to 250°C. See the lead-tin phase diagram in Figure 5.11a.

(a) At 250°C, what phases are present, what are their chemical compositions, and what is the atom fraction of each phase?

(b) This Pb-Sn mixture is then cooled to room temperature (25°C). What phases are present, what are their chemical compositions, and what is the atom fraction of each phase present?

Problem 5.11 An alloy of 80 atom percent arsenic and 20 atom percent gallium is produced, and the alloy is heated to 1250°C to thoroughly melt the mixture. The alloy is then slowly cooled so that it is always close to equilibrium. Answer the following about the alloy:

(a) At what temperature does a solid first form?

(b) What is the chemical composition of the first solid to form?

(c) Below what temperature is the alloy completely solid?

(d) At room temperature (23°C), what phases are present, what are their chemical compositions, and what is the atom fraction of each phase present?

Problem 5.12 The ordered structure AlNi has been investigated as a high-temperature gas-turbine material; however, brittle fracture is a problem with this alloy. Theoretical calculations have determined that the enthalpy of the ordered AlNi structure is 0.15 eV per atom less than for a disordered structure. The AlNi ordered structure is a cube with Ni at 0, 0, 0 and Al at $\frac{1}{2}, \frac{1}{2}, \frac{1}{2}$.

(a) What is the Bravais lattice type for AlNi?

(b) What is the GFE change if this alloy disorders at 1900 K?

(c) Based upon your result for (b), predict if AlNi is ordered or disordered at 1900 K.

(d) From the Al-Ni phase diagram in Figure 5.15, to what temperature does the AlNi phase remain ordered?

Problem 5.13 For each of the original chemical compositions and temperatures at 1 atmosphere of pressure for the aluminum (Al)–nickel (Ni) alloys (see Figure 5.15) listed below, give the following:

1. The phases present.

2. From the Gibbs phase rule, determine the degrees of freedom in the material. Can the composition of the phase or phases present be adjusted by changing the original chemical composition?

3. The chemical composition of each phase.

4. The atom fraction of each phase.

   (a) 95 atom percent Ni and 1200°C.

   (b) 80 atom percent Ni and 1000°C.

(c) 40 atom percent Ni and 1134°C.

(d) 40 atom percent Ni and 1133°C.

(e) 40 atom percent Ni and 1132°C.

Problem 5.14   Design a nickel (Ni)–aluminum (Al) alloy for a gas-turbine blade that is strengthened with $AlNi_3$ that will have an atom fraction of 0.1 of the $AlNi_3$ strengthening compound in a matrix of β-phase Ni at an operating temperature of 1000°C. Specify the atom percent of Ni and aluminum in this alloy.

Problem 5.15   A ceramic mixture of 20 atom percent Cr_2O_3 and 80 atom percent Al_2O_3 is produced by heating the mixture to 2300°C (see Figure 5.16). The mixture is then cooled to room temperature.

(a) What is the minimum temperature at which this ceramic mixture is entirely liquid?

(b) What is the maximum temperature at which this ceramic mixture is entirely solid?

Problem 5.16   A eutectoid-composition carbon steel is heated to 800°C, equilibrated, cooled to 500°C, held for 10 seconds, and then cooled rapidly to room temperature.

(a) What is the name of the microstructure in this alloy?

(b) What is the weight fraction of α-phase iron and of Fe_3C, assuming equilibrium amounts at the transformation temperature?

Problem 5.17   A 0.30% C carbon steel is heated to 1000°C and equilibrated; then it is slowly cooled to 600°C, where it is held until an equilibrium structure develops, and then it is quenched to room temperature.

(a) After this procedure, what phases are present, and what are their weight fractions?

(b) After this procedure, approximately what weight fraction of the alloy is in the proeutectoid α phase, and what weight fraction is in the pearlite phase?

Problem 5.18   Determine the transformation products and the approximate percent after each step for the following three cooling procedures, for steel with the eutectoid composition that is initially equilibrated at 730°C.

1.(a) Quench to 650°C and hold for 100 seconds.

1.(b) Then cool to room temperature.

2.(a) Quench to 650°C and hold for 2 seconds ($2 = 10^{0.3}$).

2.(b) Then quench to room temperature.

3.(a) Quench to 650°C and hold for 10 seconds.

3.(b) Then quench to room temperature.

4.(a) Quench to 400°C and hold for 3.16 seconds ($3.16 = 10^{0.5}$).

4.(b) Then quench to room temperature.

5.(a) Quench to 400°C and hold for 25 seconds ($25 = 10^{1.4}$).

5.(b) Then quench to room temperature.

6.(a) Quench to 400°C and hold for 200 seconds ($200 = 10^{2.3}$).

6.(b) Slow cool to room temperature.

7.(a) Quench to 0°C in 10 seconds.

7.(b) Heat to 600°C and hold for 1000 seconds.

Problem 5.19 A polymer blend of 60 vol % SPB 88 and 40 vol % SPB 78 is heated to 130°C and then cooled.

(a) Upon slow cooling, at what temperature does the single phase start to separate into two phases?

(b) How many phases are present at 70°C?

(c) What is the composition of each phase at 70°C?

(d) What is the volume fraction of each phase present at 70°C?

The top photo shows a strength test of a Boeing 787 wing. The bottom photo is a close-up of the wing test.

Copyright © Boeing

The goals of this chapter are to understand

- The ways to determine the mechanical properties of metals, ceramics, and polymers that can be measured on smooth specimens. These engineering properties include elastic strain, plastic strain, stress, elastic modulus, yield strength, strain hardening, ultimate tensile strength, toughness, ductility, and hardness.

- True stress and strain.

- The major types of loading, including tension, compression, shear, and hydrostatic pressure, and how to calculate stress, strain, and elastic modulus for each type of loading.

- The origin of elastic strain and plastic strain in metals, ceramics, and polymers.

- The way the elastic modulus relates to the chemical bonding between atoms.

- The relationship between yield strength and ductility for a given material.

Introduction to Mechanical Properties

6.1 INTRODUCTION

Engineers design automobiles whose passengers survive crashes, buildings that survive earthquakes, gas turbines that operate at 1000°C, and ships that survive impacts with icebergs. For these designs, the engineer must be able to determine the mechanical response of the structure to applied forces. When a force is applied to a material, it deforms; this means that the material's dimensions change. Elastic deformation occurs when the material returns back to its original dimensions when the force is removed. Plastic deformation is permanent deformation. Most mechanical parts and structures are designed to experience only elastic deformation, and plastic deformation is avoided. However, in the design for extreme forces, it is normally desirable to have plastic deformation rather than fracture. Therefore, plastic deformation must be understood by designers if they are to incorporate it into a design that avoids catastrophic failure. The energy absorption of an automobile is an example. Automobiles are designed to absorb energy of the impact by using "crumple zones," which crush in an accident, as shown in Figure 1.3. If the automobile is made so that it does not crumple, more of the shock of the impact is transmitted to the passengers, who could be killed by the shock of the impact. There are designs where fracture is desired rather than plastic deformation, such as in breakaway highway signs.

6.2   STRESS AND STRAIN

6.2.1   Tension and Compression

Figure 6.1 shows the uniaxial forces and the resulting deformation that can be applied to a material including tension, compression (the negative of tension), and shear. When we apply a **tensile** force (F) as schematically shown in Figure 6.1a, the specimen elongates from length l_0 to length l. When we **compress** a specimen with a force (F) as demonstrated in Figure 6.1b, the specimen contracts from length l_0 to length l. In both tension and compression, the length is changed by (Δl), as shown by Equation 6.1.

$$\Delta l = l - l_0 \qquad\qquad\qquad \textbf{6.1}$$

The magnitude of the deformation (Δl) in response to an applied force (F) depends upon the size of the specimen.

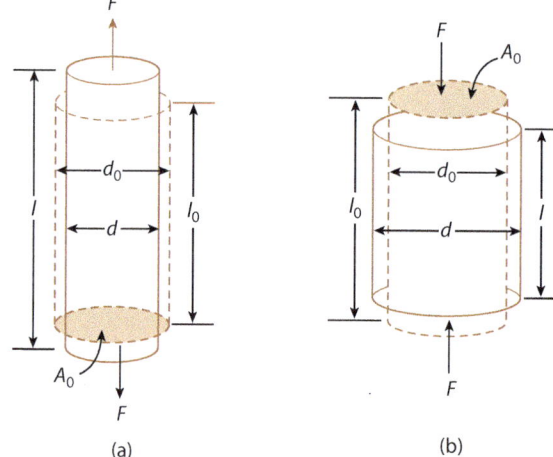

(a)   (b)

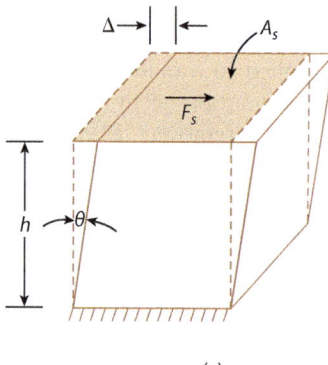

(c)

Figure 6.1   (a) A schematic of the application of a tensile force (F) to a cylindrical specimen of original area (A_0) and length (l_0). The dashed lines indicate the original specimen size, and the solid lines are the final size. (b) Application of a compressive force. (c) Application of a shear force (F_s) to a specimen with an area of (A_s) that shears the specimen by an angle (θ). (*Adapted from Callister, W. D., Materials Science and Engineering: An Introduction, 6th ed., John Wiley & Sons, New York (2003), p. 114.*)

The applied force and the deformation are normalized to eliminate the size dependence of the calculated specimen response. The uniaxial tensile or compressive force (F) is normalized by dividing by the specimen's original cross-sectional area (A_0) normal to the force, as shown in Figures 6.1a and 6.1b, the result is the normal stress (σ), which is calculated with Equation 6.2.

$$\sigma = \frac{F}{A_0} \qquad \qquad \textbf{6.2}$$

The units of stress are force per unit area or newtons per square meter (N/m^2) in the SI system of units. One newton per square meter is also one pascal (Pa). The abbreviations for 10^6 Pa and 10^9 Pa are MPa and GPa, respectively. The deformation (Δl) produced by a force in the x direction (F_x) is normalized by dividing by the original length of the specimen (l_0) resulting in the dimensionless **strain** (ε_x), which is calculated with Equation 6.3.

$$\varepsilon_x = \frac{\Delta l}{l_0} \qquad \qquad \textbf{6.3}$$

If we take a rubber band and stretch it extensively, we notice that while the length increases, the cross-sectional area decreases, as shown in Figure 6.1a. The negative of the ratio of strain in the direction of the loading axis (ε_x) to the strain in a direction perpendicular to the loading axis (ε_y) is **Poisson's ratio** (ν), which is shown in Equation 6.4.

$$\nu = -\frac{\varepsilon_y}{\varepsilon_x} \qquad \qquad \textbf{6.4}$$

The negative sign results in positive values of Poisson's ratio, because ε_y is always opposite in sign to ε_x. The value of ε_y in a tension or compression test is determined by substituting the initial and final diameter of a round specimen or the width of a square or rectangular specimen into Equation 6.3. Table 6.1 shows values of Poisson's ratio for some metals, ceramics, and polymers. Poisson's ratio ranges from 0.082 for SiO_2 to 0.49 for rubber. Many materials have a Poisson's ratio of approximately 0.3.

Table 6.1 Young's Modulus (E), Shear Modulus (G), and Poisson's Ratio (ν) for Some Selected Metals, Ceramics, and Polymers

Metals [1] 20°C	$\dfrac{E}{GPa}$	$\dfrac{G}{GPa}$	ν
Aluminum	70.3	26.1	0.345
Bismuth	31.9	12.0	0.330
Cadmium	49.9	19.2	0.300
Chromium	279.1	115.4	0.210
Copper	129.8	48.3	0.343
Gold	78.0	27.0	0.44
Iron	211.4	81.6	0.293
Iron (cast)	152.3	60.0	0.27
Lead	16.1	5.59	0.44
Magnesium	44.7	17.3	0.291
Nickel	199.5	76.0	0.312
Niobium	104.9	37.5	0.397

(Continued)

Table 6.1 *Continued*

Metals [1] 20°C	$\dfrac{E}{GPa}$	$\dfrac{G}{GPa}$	v
Platinum	168.0	61.0	0.377
Silver	82.7	30.3	0.367
Tantalum	185.7	69.2	0.342
Tin	49.9	18.4	0.357
Titanium	115.7	43.8	0.321
Tungsten	411.0	160.6	0.280
Vanadium	127.6	46.7	0.365
Zinc	108.4	43.4	0.249
Ceramics [2]			
Diamond	1022	468	0.092
SiC	402	170	0.181
Al_2O_3	402	163	0.233
MgO	300	128	0.175
ZrO_2-12%Y_2O_3	233	89	0.31
SiO_2	95	44	0.082
Polymers [3]			
Polystyrene	3.2	1.2	0.33
PMMA	4.15	1.55	0.33
Nylon 6,6	2.35	0.85	0.33
LDPE	1.0	0.35	0.45
Rubber	0.002	0.0007	0.49

1   Kay, G.W.C. and Laby, T.H., Tables of Physical and Chemical Constants, 14th ed. Longman Group Ltd., London, (1973), p. 31.

2   Carter, C.B. and Norton M.G. Ceramic Materials: Science and Engineering, Springer, N.Y. (2007), p. 293.

3   van Krevelen, D. W., Properties of Polymers, Correlations with Chemical Structure, Elsevier, Amsterdam (1972), p. 150.

Example Problem 6.1

A steel rod 0.010 m in radius and 2.000 m long is subject to an axial tensile force of 0.100 mega newtons (MN). It is observed that the rod elongates 0.308×10^{-2} m. Calculate the tensile stress and the axial strain in the rod.

Solution

The tensile stress is given by Equation 6.2.

$$\sigma = \frac{F}{A_0} = \frac{0.100 \times 10^6 \text{ N}}{\pi (0.010 \text{ m})^2} = 3.183 \times 10^8 \frac{\text{N}}{\text{m}^2}$$

The axial strain is given by Equation 6.3.

$$\varepsilon_x = \frac{\Delta l}{l_0} = \frac{0.308 \times 10^{-2} \text{ m}}{2.00 \text{ m}} = 0.154 \times 10^{-2}$$

In most engineering applications, the stress and strain defined in Equations 6.2 and 6.3 provide an adequate description of the response of the material. However, if the cross-sectional area and length of the specimen changes significantly during a tension or compression test, the stress and strain in the material is not accurately defined by Equations 6.2 and 6.3, because in these equations the area and length are constant fixed values. The **true tensile or compressive stress** (σ_t), shown in Equation 6.5, is the applied force (F) divided by the cross-sectional area (A_i) measured at the time that the force is measured.

$$\sigma_t = \frac{F}{A_i} \qquad \qquad \textbf{6.5}$$

For most stress-strain tests, only the original area (A_0), the force (F), and the deformation (Δl) are measured. It is difficult to measure A_i during a tensile test; thus A_i is not normally measured. Also, in most engineering applications the change in area is very small; under these circumstances, assuming that A_i is equal to A_0 is a reasonable approximation. However, if the change in cross-sectional area is large, it is more accurate to utilize the true stress.

An increment of true strain ($d\varepsilon_t$) is defined as an increment in length change (dl) divided by the actual length (l) at the instant of measurement, as shown in Equation 6.6.

$$d\varepsilon_t = \frac{dl}{l} \qquad \qquad \textbf{6.6}$$

The total **true strain** of a specimen that experiences a length change from l_0 to l is then obtained with Equation 6.7, by integrating the increment of true strain ($d\varepsilon_t$) in Equation 6.6 from l_0 to l.

$$\varepsilon_t = \int_{l_0}^{l} \frac{dl}{l} = \ln \frac{l}{l_0} = \ln \frac{(l_0 + \Delta l)}{l_0} = \ln(1 + \varepsilon) \qquad \qquad \textbf{6.7}$$

With Equation 6.7, it is possible to convert between engineering strain and true strain. The engineering stress in Equation 6.2 when multiplied by the ratio A_0/A_i results in the true stress given by Equation 6.5. In the remainder of this book, tensile stress and strain refer to the engineering definitions in Equations 6.2 and 6.3; only if true stress and true strain are specified do Equations 6.5 and 6.7 apply.

Example Problem 6.2

A force of 1200 N is applied axially to a latex rubber sheet that is initially 0.250 m long, 0.020 m wide, and 0.00100 m thick. The new dimensions of the sheet when subject to the load are 1.750 m long, 0.009 m wide, and 0.00046 m thick. The thickness is measured with a micrometer, resulting in a more accurate measurement than for the length. Calculate the following for this material:
(a) Engineering strain and true strain
(b) Engineering stress and true stress
(c) True Poisson's ratio and engineering Poisson's ratio

Solution

a) The engineering strain in the axial direction (ε_x) is given by Equation 6.3.

$$\varepsilon_x = \frac{\Delta l}{l_0} = \frac{1.750 \text{ m} - 0.250 \text{ m}}{0.250 \text{ m}} = \frac{1.500}{0.250} = 6.000$$

The true axial strain is given by Equation 6.7.

$$\varepsilon_{tx} = \ln \frac{l}{l_0} = \ln \frac{1.750 \text{ m}}{0.250 \text{ m}} = \ln 7.000 = 1.950$$

b) Equation 6.2 is used to calculate the engineering stress.

$$\sigma = \frac{F}{A_0} = \frac{1200 \text{ N}}{(0.020 \text{ m})(0.00100 \text{ m})} = \frac{1200 \text{ N}}{2.000 \times 10^{-5} \text{ m}^2} = 600 \times 10^5 \frac{\text{N}}{\text{m}^2} = 60 \text{ MPa}$$

The cross-sectional area with the load applied is

$$A = wt = (0.009 \text{ m})(0.00046 \text{ m}) = 4.1 \times 10^{-6} \text{ m}^2$$

The true stress with the applied load is

$$\sigma_t = \frac{F}{A_i} = \frac{1200 \text{ N}}{4.1 \times 10^{-6} \text{ m}^2} = 293 \times 10^6 \frac{\text{N}}{\text{m}^2} = 293 \text{ MPa}$$

The true stress is nearly five times as large as the engineering stress.

c) Poisson's ratio is given by

$$\varepsilon_y = -\nu\varepsilon_x$$

The engineering axial strain (ε_x) is calculated in part (a) to be 6.000, and the transverse strain (ε_y) is calculated from the change in width.

$$\varepsilon_y = \frac{\Delta w}{w_0} = \frac{0.009 \text{ m} - 0.020 \text{ m}}{0.020 \text{ m}} = \frac{-0.011}{0.020} = -0.550$$

The engineering Poisson's ratio is

$$\nu = -\frac{\varepsilon_y}{\varepsilon_x} = -\frac{-0.550}{6.000} = 0.092$$

Using true values, the strain in the y direction is

$$\varepsilon_{ty} = \ln \frac{w}{w_0} = \ln \frac{0.009 \text{ m}}{0.020 \text{ m}} = \ln 0.450 = -0.800$$

The true Poisson's ratio is then

$$\nu_t = -\frac{\varepsilon_{ty}}{\varepsilon_{tx}} = -\frac{-0.800}{1.950} = 0.410$$

The calculated true value of Poisson's ratio is closer to the value of 0.49 for Poisson's ratio listed in Table 6.1.

In some applications, such as for parts used in deep-ocean exploration, equal tensile or compressive stresses are applied in all directions resulting in a **hydrostatic stress** ($\sigma_x = \sigma_y = \sigma_z$). As shown in Equation 6.8, compressive hydrostatic stress is a pressure (P) that is equal to

$$P = \sigma_x = \sigma_y = \sigma_z \qquad\qquad \textbf{6.8}$$

Hydrostatic stresses produce a change in volume (ΔV) to a body with an original volume (V_0). The **volumetric strain** (ε_V) is given in Equation 6.9, by the change in volume (ΔV) divided by the original volume (V_0).

$$\varepsilon_V = \frac{\Delta V}{V_0} \qquad\qquad \textbf{6.9}$$

For an isotropic material, one with equal properties in all directions, subject to a hydrostatic stress, the volumetric strain (ε_V) is three times the strain along the x axis (ε_x), as shown in Equation 6.10:

$$\varepsilon_V = 3\varepsilon_x \qquad\qquad \textbf{6.10}$$

6.2.2 Shear

Shear deformation occurs when we apply a force (F_s) parallel to the top surface of a specimen, as shown in Figure 6.1c, and the top of the material is displaced relative to the bottom by Δ. As shown in Equation 6.11, the **shear strain** (γ) of a block is defined as the displacement of the top of the block (Δ) divided by the height of the block (h).

$$\gamma = \frac{\Delta}{h} = \tan\theta \qquad\qquad \textbf{6.11}$$

θ is the change in angle caused by the shear force. As shown in Equation 6.12, **shear stress** (τ) is defined as the shear force (F_s), shown in Figure 6.1c, divided by the area (A_s) over which the shear force acts.

$$\tau = \frac{F_s}{A_s} \qquad\qquad \textbf{6.12}$$

It is not necessary to define a true stress or strain for shear, because the area of the specimen in shear does not change. Also, it is assumed in a first approximation that the height of a material subject to a shear stress, as shown in Figure 6.1c, does not change.

6.3 ELASTIC MODULI

When a stress is applied to a material, there is a corresponding strain. Once the stress is removed from the material, if the material then returns to its original dimensions, the strain in the material is elastic. **Elastic strain** is recovered strain. When we stretch and then release a rubber band, the rubber band returns to its original dimensions. This is elastic strain. Rubber is called an elastomer because of its extensive elastic strain.

   If the tensile stress is plotted versus the axial strain, many materials exhibit a linear relationship between stress and strain, as shown in Figure 6.2. Stress-strain plots similar to Figure 6.2 are determined in test machines that apply a force using either mechanical or hydraulic systems and the elongation of the specimen is measured by mechanical or electronic systems. Tensile tests are covered in more detail in Section 6.7.1.

   For materials with a linear relationship between stress and strain, the ratio of the tensile stress to the axial elastic strain is the **elastic modulus** (E), as shown in Equation 6.13.

$$E = \frac{\sigma}{\varepsilon_e} \qquad\qquad \textbf{6.13}$$

The tensile elastic modulus is also known as **Young's modulus**, and the linear relationship between stress and strain expressed in Equation 6.13 is **Hooke's law**. The tensile elastic modulus for some metals, ceramics, and polymers is listed in Table 6.1.

   Some materials do not have a linear relationship between stress and elastic strain. Figure 6.3 is a schematic of a nonlinear elastic strain versus stress plot. For such a material the elastic modulus depends

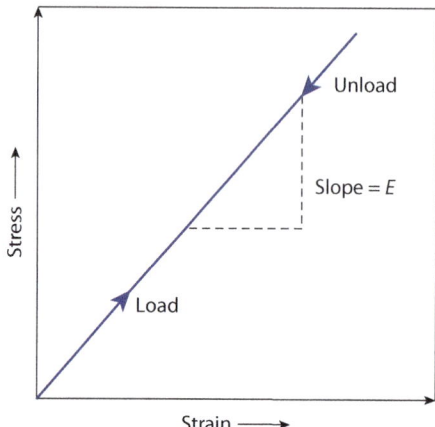

Figure 6.2   Schematic of a stress-strain diagram showing a linear relationship during loading and unloading. Loading and unloading follow the same path. (*Based on Askeland, D.R., Fulay, P.P., and Wright, W.J., The Science and Engineering of Materials, 6th ed., Cengage Learning, Stamford, CT. (2011), p. 199.*)

upon the value of the strain where the elastic modulus is defined. There are several ways to define the elastic modulus. The **secant modulus** $[E_{sec}(\varepsilon_1)]$ at a stress σ_1 and corresponding strain ε_1 is the stress divided by the strain, as shown by Equation 6.14, and it is the slope of the dashed line in Figure 6.3.

$$E_{sec}(\varepsilon_1) = \frac{\sigma_1}{\varepsilon_1} \qquad \textbf{6.14}$$

The secant modulus is the average modulus between zero strain and the strain where the stress is measured.

Another possible calculation is the **tangent elastic modulus** $[E_{tan}(\varepsilon_1)]$ measured at the strain ε_1, as shown in Equation 6.15 and in Figure 6.3.

$$E_{tan}(\varepsilon_1) = \frac{\Delta\sigma_1}{\Delta\varepsilon_1} \qquad \textbf{6.15}$$

The tangent elastic modulus is the slope of the stress-strain diagram at the value of strain specified, as shown in Figure 6.3. Often the tangent modulus is specified at zero strain. However, for materials such

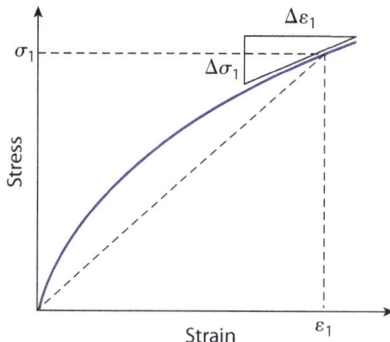

Figure 6.3   A schematic of a nonlinear stress-strain diagram. The secant elastic modulus at ε_1 (dashed line) is σ_1/ε_1 and the tangent elastic modulus is $\Delta\sigma_1/\Delta\varepsilon_1$ at ε_1.

as rubber where the strains can be large, it is possible to specify other magnitudes of strain where the modulus is defined.

Example Problem 6.3

Calculate the secant elastic modulus of the material in Example Problem 6.1 at a strain of 0.154×10^{-2}, assuming that all of the deformation is elastic.

Solution

From Example Problem 6.1,

$$\sigma_1 = 3.183 \times 10^8 \text{ N/m}^2 \text{ and } \varepsilon_1 = 0.154 \times 10^{-2}$$

Applying Equation 6.14 for the secant modulus,

$$E_{sec}(\varepsilon_1) = \frac{\sigma_1}{\varepsilon_1} = \frac{3.183 \times 10^8 \text{ N/m}^2}{0.154 \times 10^{-2}} = 20.67 \times 10^{10} \frac{\text{N}}{\text{m}^2}$$

The ratio of the hydrostatic stress or pressure (P) to volumetric strain (ε_v) is the **bulk modulus** (B), as shown in Equation 6.16.

$$B = \frac{P}{\varepsilon_V} \qquad\qquad \textbf{6.16}$$

The ratio of the shear stress (τ) to the elastic shear strain (γ) is the **shear modulus** (G), as shown in Equation 6.17.

$$G = \frac{\tau}{\gamma} \qquad\qquad \textbf{6.17}$$

In isotropic materials, Equations 6.18 and 6.19 show the following relationships between the elastic moduli:

$$E = 2G(1 + \nu) \qquad\qquad \textbf{6.18}$$

$$E = 3B(1 - 2\nu) \qquad\qquad \textbf{6.19}$$

Polycrystalline and amorphous materials that are processed so they do not induce directional properties should be isotropic. A block of annealed polycrystalline aluminum that has randomly oriented grains, or a block of polyethylene with randomly oriented long-chain molecules (LCMs), should be isotropic. However, if the polycrystalline aluminum is rolled into foil, it has a different elastic modulus in the direction of rolling than it does perpendicular to the sheet. In the polycrystalline aluminum, the close-packed directions are more easily extended in the rolling direction, resulting in a nonrandom orientation of the grains. If polyethylene is drawn into a fine filament at a temperature just above the glass transition temperature, the long-chain molecules are extended in the direction of the drawing. Then the strong covalent bonds are aligned along the filament axis, and there are weak van der Waals bonds perpendicular to the filament axis holding the long-chain molecules together. A fine polyethylene filament has a high elastic modulus in the axial direction and a lower one perpendicular to the fiber axis.

A shock absorber is produced by sandwiching natural rubber 0.01 m thick between two thin steel plates. The area of the sandwich is 10^{-3} m². The bottom steel plate is fixed, and a shear force of 3000 N is applied parallel to the surface of the top steel plate. The shear modulus of this rubber is 3.4 MPa (1 MPa = 10^6 Pa). Calculate (a) the shear stress in the rubber, (b) the shear angle in the rubber, and (c) the linear displacement of the top steel plate relative to the bottom steel plate. You can assume that the strain in the steel plates is insignificant in comparison to that in the rubber.

Solution

The shear stress (τ) is calculated from Equation 6.12.

$$\tau = \frac{F_s}{A_s} = \frac{3000 \text{ N}}{1 \times 10^{-3} \text{ m}^2} = 3000 \times 10^3 \frac{\text{N}}{\text{m}^2} = 3.000 \text{ MPa}$$

The shear strain (γ) is calculated from Equation 6.17.

$$\gamma = \frac{\tau}{G} = \frac{3.000 \text{ MPa}}{3.4 \text{ MPa}} = 0.880$$

From Equation 6.11 the shear angle (θ) is calculated from the shear strain (γ).

$$\tan\theta = \gamma = 0.880$$

$$\theta = \tan^{-1}(0.880) = 41°$$

The displacement (Δ) of the upper steel plate relative to the fixed lower plate is calculated from Equation 6.11.

$$\Delta = \gamma h = (0.88)(0.01 \text{ m}) = 0.880 \times 10^{-2} \text{ m} = 0.880 \text{ cm}$$

6.3.1   Anelastic Materials

In **anelastic** materials the unloading stress-strain plot does not follow the loading stress-strain plot; however, the strain does return to 0 with sufficient time when the stress is 0, as shown schematically in Figure 6.4. Carbon steels are anelastic, because when the body-centered cubic (BCC) crystal structure of steel is stressed in tension along a cube axis, the carbon atoms preferentially align in interstitial sites along the tensile cube axis, and the lattice parameter elongates in the direction of the tensile force. When

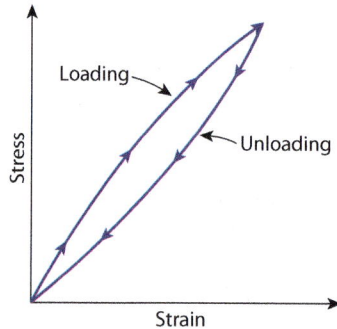

Figure 6.4   A schematic of a stress-strain diagram for an anelastic material.

the stress is removed, the carbon atoms become randomly oriented along the various cube unit cell axes, and the unit cell returns to BCC. Carbon steels at low applied stresses are normally considered elastic because most of the strain in steel is recovered immediately upon unloading. The anelastic response of carbon steel has important applications where vibrational energy must be absorbed, such as in the bed of rotating machine tools and in drive shafts. Rubber is another material that is anelastic.

6.3.2 Interatomic Potentials and Elastic Moduli

At the atomic level, tensile or compressive elastic strain corresponds to the movement of atoms away from their equilibrium atomic positions, and when the stress is released, the atoms return to their equilibrium atomic positions. The interatomic potential introduced in Chapter 2 is utilized to evaluate the elastic modulus of a crystal. We will demonstrate this concept by deriving the tangent elastic modulus for a pair of atoms. Assume that the interatomic potential between two atoms is given by an equation similar to Equation 2.9. A reference atom is fixed at the position $r = 0$, and a second atom is placed at the equilibrium interatomic separation of $r = r_0$. If a positive force F_A is applied to the atom at r_0, its position is increased to $r_0 + dr$. The applied force necessary to move the second atom from r_0 to r is also the gradient of the interatomic potential energy at r, as shown in Equation 6.20.

$$F_A = \frac{dV_P}{dr} \qquad \textbf{6.20}$$

The cross-sectional area of the atom pair is A, and the stress is F_A/A. The strain (ε) is $(r - r_0)/r_0$, and we assume that the cross-sectional area of the atom pair does not change. Inserting these values into Equation 6.15 for the tangent modulus at zero strain results in Equation 6.21.

$$E_{tan}(\varepsilon = 0) = \frac{d\sigma}{d\varepsilon} = \frac{d\left(\dfrac{F_A}{A}\right)}{d\left(\dfrac{r - r_0}{r_0}\right)} = \frac{r_0}{A}\frac{d^2V_P}{dr^2} \qquad \textbf{6.21}$$

The elastic modulus for this one-dimensional pair of atoms is proportional to the curvature of the interatomic potential. A high curvature to the interatomic potential due to strong chemical bonding indicates that it takes a large force to change the separation between the atoms, resulting in a high elastic modulus. All of the elastic moduli for three-dimensional solids can be derived from the interatomic potential. The derivation is more complicated in three dimensions, but the physics is the same.

There is a strong correlation between melting temperature and elastic modulus. For example, tungsten has a high elastic modulus of 411 GPa (1 GPa = 10^9 Pa) and a melting temperature of 3683 K; quartz (SiO_2) has an elastic modulus of 95 GPa and a melting temperature of 2983 K; and LDPE has an elastic modulus of approximately 1.2 GPa and a melting temperature of approximately 400 K. A high melting temperature indicates strong chemical bonding, and strong chemical bonding also results in a high elastic modulus.

All of the elastic moduli for the metals in Table 6.1 are for elements. When these metals are alloyed, the elastic modulus does not change significantly; however, small changes can be observed. For example, the elastic modulus of aluminum is 70.3 GPa. In one of the strongest alloys of aluminum, there are additions of approximately 5.5 weight percent zinc, 1.6 weight percent copper, 2.5 weight percent magnesium, and some other elements. The elastic modulus of this alloy is 72 GPa. The greater the amount of the alloy additions, the greater is the possible change in the elastic modulus.

6.3.3   Elastic and Anelastic Strain in Rubber

The mechanism of elastic strain in rubber is significantly different than that in crystalline solids. The strain as a function of stress in rubber, as shown in Figure 6.5, is understood by observing the coiled and cross-linked structure of rubber, as shown in Figure 2.24. In vulcanized rubber, the coiled long-chain molecules are cross-linked with sulfur atoms every several hundred mer units. When rubber is strained, the coiled long-chain molecules of rubber uncoil, and the uncoiling results in the low elastic modulus observed in rubber for strains up to 5, as shown in Figure 6.5. As the long-chain molecules uncoil, van der Waals bonds on different chains are broken and re-formed with different neighbors. The cross-linked long-chain molecules move in a viscous-like medium created by the van der Waals bonds between the long-chain molecules. The cross-linking of the long-chain molecules restricts their sliding relative to each other. When the strain on rubber is greater than 6, the long-chain molecules with covalent bonds between carbon atoms become straightened and stretched in the direction parallel to the applied tensile stress. This results in the large increase in the elastic modulus of rubber at high strain, and the elastic modulus of rubber is a function of the magnitude of the strain. When the applied stress on rubber is reduced to 0, the stretched long-chain molecules return to the coiled structure, and the cross-linked molecules return approximately to their original positions relative to each other. However, energy is absorbed by the movement of the long-chain molecules in the viscous-like medium. This results in the anelastic behavior in rubber, and it makes rubber a good energy-absorbing material. For many applications rubber is considered to be elastic. However, it requires time for the complete recovery of all strain, because of the viscous-like medium in which the long-chain molecules move. The anelastic property is what gives rubber its energy-absorbing capability, as we will discuss in the next section. The mechanical properties of rubber and synthetic rubbers are treated in Section 8.3.5.

Because rubber changes from a structure with coiled molecules at random orientations relative to each other to a structure with molecules stretched parallel to the load direction, an entropy change in the molecular structure takes place. This entropy change is used to calculate the elastic modulus of rubber rather than using Equations 6.20 and 6.21. The thermodynamics of this derivation are beyond the scope of this text; however, the result is that the relationship between stress (σ) and the ratio of final length to initial length ($L/L_0 = \lambda$) is given by Equation 6.22:

$$\sigma = \frac{\rho RT}{M_c}\left(\lambda - \frac{1}{\lambda^2}\right) \qquad \textbf{6.22}$$

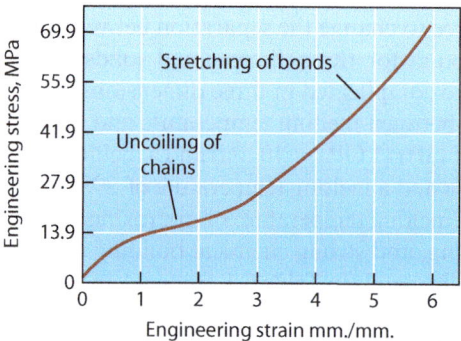

Figure 6.5   A stress-strain diagram for rubber. (*Based on Askeland, D.R., Fulay, P.P., and Wright, W.J., The Science and Engineering of Materials, 6th ed., Cengage Learning, Stamford, CT. (2011), p. 632.*)

where ρ is the rubber density, M_c is the average molecular mass of the rubber between cross-link locations, R is the universal gas constant, and T is the temperature in kelvin. The term before the parentheses in Equation 6.22 is proportional to the rubber elastic modulus, and the term in the parentheses is proportional to the elastic strain. Equation 6.22 shows the stress required for a given value of λ increases with temperature. This observation is the opposite to that observed in crystalline solids where the elastic modulus decreases with an increase in temperature. The stress required for a given value of λ also increases for a smaller value of M_c. This observation is expected because a smaller M_c results from more cross-links, and more cross-links produce a harder and stiffer rubber. A more advanced treatment of rubber elasticity is found in textbooks such as that by Rubenstein and Colby.

6.4 STRAIN ENERGY

The area under a tensile stress-strain diagram represents the work per unit volume necessary to strain the material to ε. Consider first the area under the linear stress-strain diagram in Figure 6.2 from the origin to a stress σ and strain ε. The units of stress (F/A) are N/m^2, and the units of strain ($\Delta l/l$) are m/m. As shown in Equation 6.23a, the area under the linear stress-strain diagram is equal to

$$\text{Area} = \frac{1}{2}\sigma\varepsilon = \frac{F\Delta l}{2Al} = \frac{W}{V} = SED \qquad \textbf{6.23a}$$

where W is the work necessary to elongate the specimen by Δl, V is the volume of the specimen, and SED is the strain-energy density or work per unit volume. The factor of $\frac{1}{2}$ is necessary because the force starts at 0 and ends at F. For a linear elastic material, the strain energy per unit volume needed to strain the material uniaxially to ε is given by Equation 6.23b.

$$SED = \frac{1}{2}\sigma\varepsilon = \frac{E\varepsilon^2}{2} = \frac{\sigma^2}{2E} \qquad \textbf{6.23b}$$

Similar equations, such as Equation 6.24, can be developed for calculating shear-strain energy density (SSED),

$$SSED = \frac{1}{2}\tau\gamma = \frac{G\gamma^2}{2} = \frac{\tau^2}{2G} \qquad \textbf{6.24}$$

and for calculating hydrostatic-strain energy density (HSED), as shown in Equation 6.25.

$$HSED = \frac{1}{2}P\varepsilon_V = \frac{B\varepsilon_V^2}{2} = \frac{P^2}{2B} \qquad \textbf{6.25}$$

Equations 6.23 through 6.25 do not apply if the material response is nonlinear. For nonlinear elastic materials, such as shown in Figure 6.3, we must integrate the area under the stress-strain diagram to obtain the elastic-strain energy density.

Example Problem 6.5

Calculate the elastic-strain energy density absorbed in the specimen in Example Problem 6.1. Assume that the material is linearly elastic to this strain.

Solution

The uniaxial strain energy density is given by Equation 6.23a.

$$SED = \frac{1}{2}\sigma\varepsilon = \frac{(3.183 \times 10^8 \text{ N/m}^2)(0.154 \times 10^{-2} \text{ m/m})}{2} = 0.245 \times 10^6 \frac{\text{N} \cdot \text{m}}{\text{m}^3} = 0.245 \times 10^6 \frac{\text{J}}{\text{m}^3}$$

Figure 6.4 demonstrates the loading and unloading characteristics of an anelastic material. The area inside the loop resulting from loading and unloading is the hysteresis loop, and it represents the strain-energy density absorbed by the material in a loading and unloading cycle. More strain energy is required to initially strain the material (loading) than is recovered during unloading, and the difference in these two energies is absorbed by the material. Absorption of energy without producing permanent damage in the material is important for damping mechanical vibrations in machines or for damping sound vibrations. The ability to absorb energy is why rubber is utilized for motor mounts and other vibration-isolation applications.

6.5   PLASTIC STRAIN

Plastic strain is permanent strain that remains in a material after the applied stress is reduced to 0. There is plastic strain in the crumple zones of an automobile involved in a front end collision, as shown in Figure 1.3. Figure 6.6 shows the response of a material such as a metal alloy that is stressed sufficiently beyond the linear elastic stress-strain region in a tensile test to a region where the stress-strain plot is

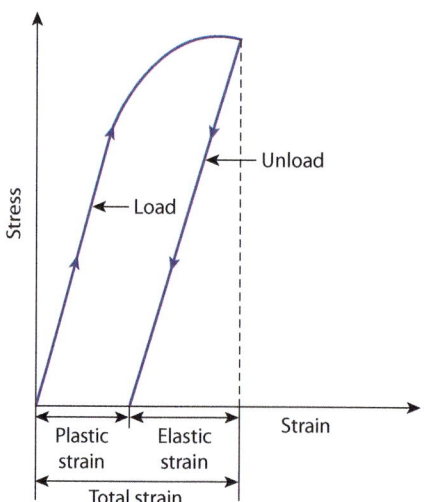

Figure 6.6   A schematic of a stress-strain diagram for a material demonstrating elastic (recovered) and plastic (permanent) strain upon unloading.

nonlinear. For many materials, such as metal alloys, the unloading has the same slope as the elastic modulus has, and the strain that is recovered is the elastic strain (ε_e). The elastic-strain recovery during unloading is due to the atoms returning to their original positions. The strain that permanently remains after the stress is reduced to 0 is the plastic strain (ε_p), as shown in Figure 6.6.

Example Problem 6.6

A low-carbon steel is stressed in tension to 21.0×10^7 pascal (Pa) with a total strain of 0.10; then the specimen is unloaded. Determine the expected elastic strain and the plastic strain, assuming that the elastic modulus is 211 GPa.

Solution

The relationship among applied stress, elastic modulus, and elastic strain is given by Equation 6.13. The stress is equal to 21.0×10^7 Pa, and the elastic modulus (E) is equal to 211 GPa. Solving for the elastic strain,

$$\varepsilon_e = \frac{\sigma}{E} = \frac{21.0 \times 10^7 \text{ Pa}}{211 \times 10^9 \text{ Pa}} = 0.10 \times 10^{-2}$$

The plastic strain is then the total strain minus the elastic strain.

$$\varepsilon_p = \varepsilon - \varepsilon_e = 0.10 - 0.001 = 0.099$$

6.5.1   Plastic Strain in Polymers

The origin of plastic strain depends upon the type of material. In thermoplastic amorphous polymers without cross-linking, plastic strain occurs when the van der Waals bonds between the LCMs are broken by an applied stress, and the LCMs slide past each other, as shown in Figure 6.7. The sliding of the LCMs past each other is due to shear stresses on the molecules, which produces a shear strain, just as the sliding of a sheet of paper over another sheet of paper produces shear strain. Once the LCMs slide relative to each other the they rebond, and there is no restoring force to bring the chains back to their original positions. Plastic strain orients the LCMs in the direction of tensile stress, just as it does if a ball of randomly oriented strings is pulled. In Section 6.6 we will discuss how a shear stress results from an applied tensile force.

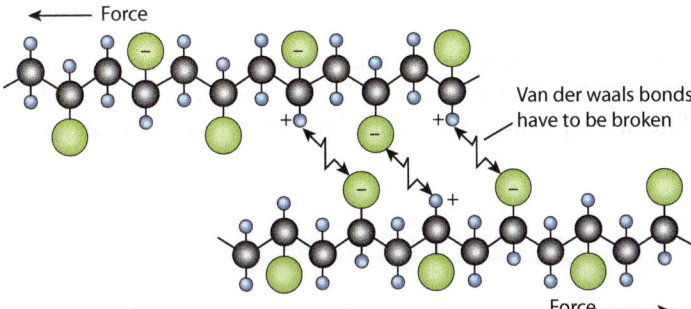

Figure 6.7   A schematic of an applied force breaking the van der Waals bonds between hydrogen ($+$) and chlorine ($-$) atoms in PVC and allowing the LCMs to slide past each other producing permanent shear strain. (Based on Askeland, D.R., Fulay, P.P., and Wright, W.J., The Science and Engineering of Materials, 6th ed., Cengage Learning, Stamford, CT. (2011), p. 40.)

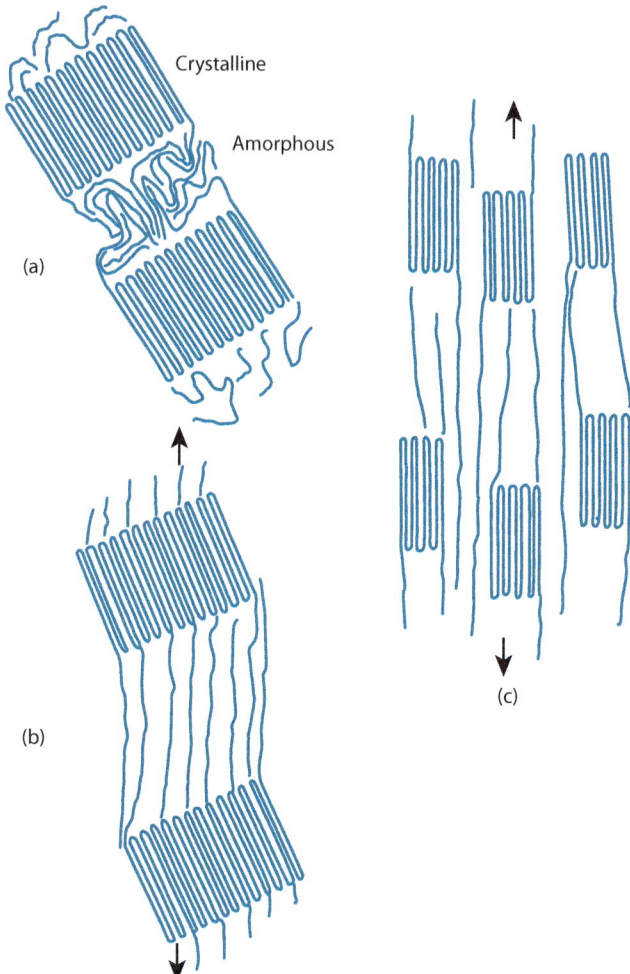

Figure 6.8   The deformation of a semicrystalline polymer: (a) Two lamellae and surrounding amorphous material. (b) The amorphous chains are deformed first. (c) At high strain the lamellae are rotated and break apart into blocks.

In semicrystalline polymers, shown in Figure 6.8a, the amorphous areas between the crystalline areas are deformed first and the LCMs are then oriented in the tensile direction, as shown in Figure 6.8b. At high strain the crystallites are separated and rotated into the tensile direction, as shown in Figure 6.8c.

6.5.2   Plastic Strain in Glass

In silica glass the silicon atoms are surrounded by four oxygen atoms in a tetrahedral arrangement, as shown in Figure 2.28, and the oxygen atoms act as bridges between the tetrahedra, as shown in Figure 3.3. At temperatures above the glass transition temperature, the bonds between the silicon atoms and the bridging oxygen atoms are continuously broken and re-formed. If a stress is applied to silica glass at a high temperature when the silicon-oxygen atom bonds are momentarily broken, the bridging oxygen atom can rebond with a different tetrahedron of atoms, resulting in plastic strain. The higher the temperature, the higher the frequency of bond breaking, and the greater the amount of plastic strain for a given applied stress.

6.5.3 Plastic Strain in Crystalline Metals and Ceramics

It is possible to calculate the shear stress necessary to slide the planes of atoms of a perfect crystal of a metal, such as copper, over each other simultaneously, as shown in Figure 6.9, to produce plastic deformation. The theoretical shear strength is approximately the shear modulus divided by 30 (G/30). For copper, the theoretical shear strength is approximately 1.4×10^9 Pa. However, in a real crystal of high-purity annealed copper, the experimental stress where plastic strain begins is approximately 0.6×10^6 Pa. The results from the theoretical calculation and the experiment differ from each other because the real crystal is not perfect. A real crystal contains defects such as dislocations. Figure 6.10 shows a transmission electron micrograph of dislocations in molybdenum metal. The dislocations are the dark lines. All metals have dislocations that form during solidification. Crystals of silicon with covalent bonding have been produced that are dislocation free, but most silicon crystals have a few dislocations.

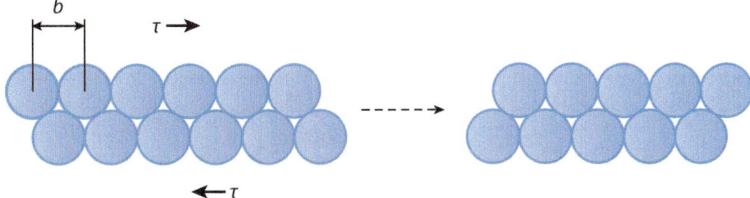

Figure 6.9 The shear deformation of a perfect crystal created by sliding the top plane of atoms over the lower plane.

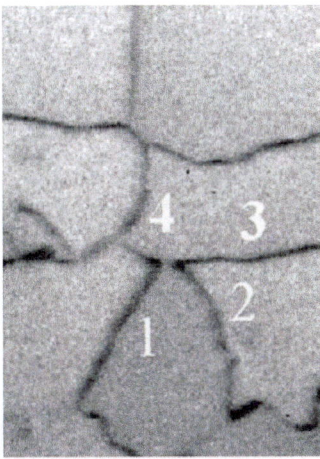

Figure 6.10 A tranmission electron micrograph illustrating dislocations (dark lines) in molybdenum. As an indication of magnification the distance between the top of the 1 and the intersection of the horizontal and vertical parts of the 4 is 200 nm. The numbers have no meaning. (*Reprinted by permission from Macmillan Publishers Ltd: NATURE, Vol 440, Vasily V. Bulatov,Luke L. Hsiung,Meijie Tang,Athanasios Arsenlis,Maria C. Bartelt et al., "Dislocation multi-junctions and strain hardening," copyright 2006.*)

In crystalline metals at low temperatures relative to their melting temperatures, plastic strain results primarily from the motion of dislocations. In most high-purity crystalline metals, dislocation motion is easy to produce at room temperature. The relationship between dislocation motion and plastic strain in crystalline materials is shown in Figure 6.11 for a crystal of dimension $L \times L \times L$.

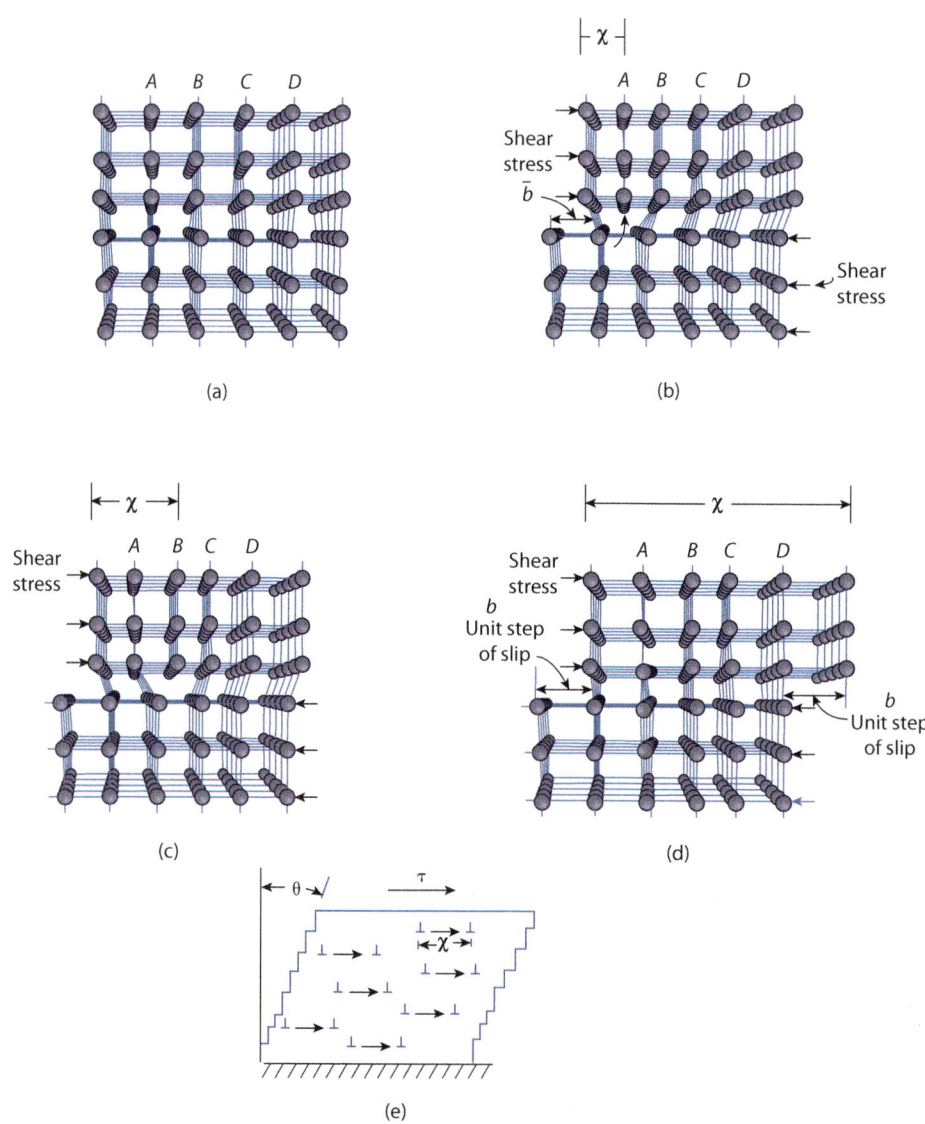

Figure 6.11 (a) A dislocation-free crystal. (b) A shear stress is applied, and a dislocation forms on the left side of the crystal and moves to position A. (c) The shear stress is increased further, and the dislocation moves to position B. (d) The dislocation moves out of the crystal, and the upper part of crystal is displaced to the right by a magnitude b relative to the lower part of the crystal. (e) A schematic of a crystal with N dislocations represented by the symbol ($\perp$), producing a plastic shear strain γ_p and a change of shape by the angle θ. In Figure 6.11e the arrows indicate the direction of positive dislocation motion over the average distance x. ((b)–(d) Based on A. G. Guy, *Essentials of Materials Science, McGraw-Hill, 1976*.)

In Figure 6.11b, a shear stress (τ) is applied, and an edge dislocation has formed at the left side of the crystal at position A, causing a step on the left side of the crystal of magnitude equal to the Burger's vector (b). Dislocations have been observed to form at surfaces and interfaces, and dislocations are naturally present in metal crystals. For this discussion, it is not important how the dislocation is formed. Figures 6.11b and c are schematics of the motion of a dislocation from one position to another. All that is necessary to move the dislocation from position A to position B is to shuffle atoms around the dislocation. The atoms do not have to move over each other, as shown in Figure 6.9. Also, the configuration of atoms around the dislocation is the same at positions A and B if there are no other defects in the vicinity of the dislocation and if the surface of the crystal is a large distance from the dislocation. Then the energy of the dislocation is the same at positions A and B. However, a small amount of energy in crystals is required to shuffle the atoms around the dislocation; therefore, an energy loss of internal friction is associated with moving the dislocation. As a result the applied shear stress required to move the dislocation is much less than the theoretical shear strength. If the dislocation moves all the way across the crystal, it creates a step of magnitude equal to the Burger's vector (b) on the right side of the crystal, as shown in Figure 6.11d; Equation 6.26 shows that the crystal has a plastic shear strain (γ_p) of magnitude

$$\gamma_P = \frac{b}{L}$$ 6.26

If the dislocation moves only part of the distance across the crystal (x), then the fraction of the distance moved across the crystal is x/L. This is also the fraction of the plastic strain resulting from dislocation motion by the distance (x); therefore, the plastic strain is shown in Equation 6.27 as

$$\gamma_P = \frac{bx}{L^2}$$ 6.27

If there are N dislocations, indicated by the symbols $\perp$ moving an average distance x, as shown schematically in Figure 6.11e, the plastic shear strain is shown in Equation 6.28 as

$$\gamma_P = \frac{Nbx}{L^2}$$ 6.28

If we assume that each dislocation extends across the length of the crystal, then each dislocation line has a length L, and the total length of dislocation lines is NL. If we multiply both the numerator and denominator in Equation 6.28 by L, the resulting Equation 6.29 calculates the shear plastic strain as

$$\gamma_P = \frac{NLbx}{L^3}$$ 6.29

The term NL/L^3 is the total length of mobile dislocations per unit volume or the **mobile dislocation density** (ρ). Equation 6.30 shows the plastic shear strain as

$$\gamma_P = \rho bx$$ 6.30

The plastic shear strain is a function of the mobile dislocation density (ρ), the average extent of movement of the dislocations (x), and the magnitude of the Burger's vector (b). Equation 6.30 relates the macroscopic engineering property of plastic strain to atomic-level defects (dislocations) and their motion. In this derivation of the relationship between plastic strain and dislocation motion, we assume that the dislocations are of length L and that they move the length of the crystal. In real crystals the dislocations do not have this configuration. The transmission electron micrograph in Figure 6.10 of the dislocations in molybdenum shows that the dislocations are not in straight lines. However, the average displacement (x) of the mobile dislocation density (ρ) of dislocations with a magnitude of the Burger's vector (b) produces a plastic shear strain (γ_p). In the derivation above, we also only considered edge dislocations; however, screw dislocations also produce shear plastic strain.

In Chapter 3 we observed that the Burger's vector of a dislocation connected two adjacent lattice points in a crystal. Also, the shortest Burger's vector is in the most closely packed direction in the most closely packed plane. For example, in face-centered cubic (FCC) metals, shown in Figure 6.12, the close-packed planes are of the {111} family of planes, and the Burger's vector is one of the <110> type directions that lie in the close-packed plane. A possible Burger's vector on the (111) plane is $\frac{a}{2}[1\bar{1}0]$. In the cubic system the [111] direction is normal to the (111) plane. The dot product of $[1\bar{1}0] \cdot [11\bar{1}]$ is 0; thus the $[1\bar{1}0]$ direction lies in the (111) plane. An edge dislocation is confined to move by the mechanism depicted in Figure 6.11, called **slip**, on the plane defined by the Burger's vector and the tangent vector, and this plane is called the **slip plane**. The **slip directions** on the slip plane are the most closely packed directions on the slip plane. The slip plane and the slip direction form the **slip system** for a crystal. For screw dislocations the Burger's vector is parallel to the tangent vector and no slip plane is defined. Screw dislocations can move on any plane. However, dislocations are the most mobile on the most closely packed planes.

The body-centered cubic (BCC) crystal, shown in Figure 2.5, is not close packed. The most densely packed planes are the {110} family of planes, and the most densely packed direction on the {110} family of planes is in <111> type directions. This is the most common slip system for BCC metals; however, other slip systems have been observed. On the (110) plane a possible Burger's vector is $\frac{a}{2}[1\bar{1}1]$. The dot product $[1\bar{1}1] \cdot [110]$ is 0; thus the Burger's vector $\frac{a}{2}[1\bar{1}1]$ lies in the (110) plane.

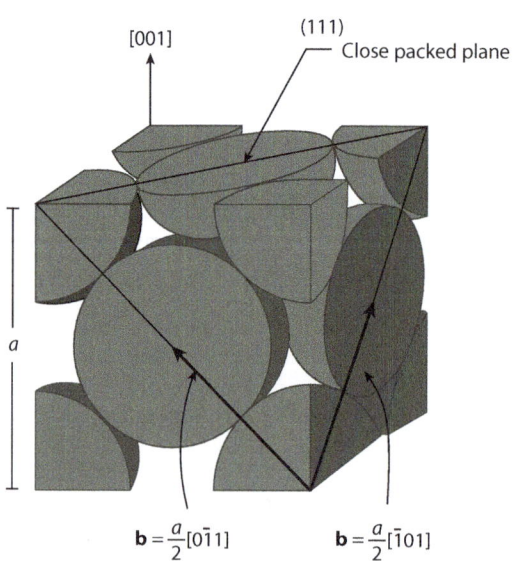

$$\mathbf{b} = \frac{a}{2}[0\bar{1}1] \qquad\qquad \mathbf{b} = \frac{a}{2}[\bar{1}01]$$

Figure 6.12 A face-centered cubic (FCC) crystal with the [001] direction vertical, showing the (111) close-packed plane and possible Burger's vectors of $\mathbf{b} = \frac{a}{2}[0\bar{1}1]$ and $\mathbf{b} = \frac{a}{2}[\bar{1}01]$. *(Adapted from Callister, W. D., Materials Science and Engineering: An Introduction, 6th ed., John Wiley & Sons, New York (2003), p. 33.)*

(a) Determine the dislocation density of the thin film shown in Figure 6.10, assuming that the thickness (t) of the thin film is 1.00×10^{-7} m, and that all of the dislocations are parallel to the surface of the thin film. The actual thickness of a thin film can be determined in a transmission electron microscope by tilting a flat plane, such as a twin plane, by a known angle and observing the apparent change in projected area of the plane.

(b) Calculate the average distance moved by the dislocations in Figure 6.10 as a result of 3% plastic strain ($\gamma_p = 0.03$). Assume that the dislocation density does not change during the strain of the specimen, and that the Burger's vector (**b**) is $\frac{a}{2}[111]$. The lattice parameter of the BCC molybdenum is 0.315 nm.

Solution

a) A total dislocation length of $NL = 19$ cm is measured in Figure 6.10. This can be done with a device such as that used to measure distance on a map. The magnification is given as 1.3 cm = 200 nm. Correcting the measured dislocation length on the photograph for the magnification results in the actual dislocation length (NL) in the material.

$$NL = 19 \text{ cm} \left[\frac{200 \text{ nm}}{1.3 \text{ cm}} \right] = 2923 \text{ nm} = 2.9 \times 10^{-6} \text{ m}$$

The area of the image is 4.3 cm by 5.8 cm or 25 cm². Correcting this area for the magnification results in an area of

$$A = 25 \text{ cm}^2 \left[\frac{200 \text{ nm}}{1.3 \text{ cm}} \right]^2 = 59 \times 10^4 \text{ nm}^2$$

The assumed thickness is 100 nm. The volume (V) of this thin film is then

$$V = At = (0.59 \times 10^6 \text{ nm}^2)(100 \text{ nm}) = 0.59 \times 10^8 \text{ nm}^3 \left[\frac{1 \times 10^{-27} \text{ m}^3}{1 \text{ nm}^3} \right] = 0.59 \times 10^{-19} \text{ m}^3$$

The dislocation density (ρ) is then

$$\rho = \frac{NL}{V} = \frac{2.9 \times 10^{-6} \text{ m}}{0.59 \times 10^{-19} \text{ m}^3} = 4.9 \times 10^{13} \frac{\text{m}}{\text{m}^3} \text{ dislocation length per unit volume}$$

b) The distance moved by the dislocations during the strain of 3% is calculated from Equation 6.30.

$$\gamma_P = \rho b x = 0.03$$

The Burger's vector for the BCC Mo is calculated from

$$\mathbf{b} = \frac{a}{2}[111] = \frac{0.315 \times 10^{-9} \text{ m}}{2} (1^2 + 1^2 + 1^2)^{1/2} = 0.273 \times 10^{-9} \text{ m}$$

The only unknown is the distance moved (x). Solving for x,

$$x = \frac{\gamma_P}{\rho b} = \frac{0.03}{(4.9 \times 10^{13} \text{ m/m}^3)(0.273 \times 10^{-9} \text{ m})} = 0.22 \times 10^{-5} \text{ m} = 2200 \text{ nm}$$

The width of the image in Figure 6.10 is 4.3 cm and with the magnification of 1.3 cm = 200 nm the real image width is 662 nm. For 3% strain the dislocations move more than 3 times the distance across the image in Figure 6.10.

In crystalline materials with ionic and covalent bonding, much more energy is required to move the dislocation than in high-purity metals. One reason for this difference in required energies is that the Burger's vector is larger in these crystalline materials than in simple metals. The Burger's vector connects adjacent lattice points, and lattice points are identical to each other. Thus all the lattice points must have the same type of atoms. In an ionic crystal the Burger's vector connects ions of the same charge, and in an ionic crystal such as MgO, shown in Figure 3.10, the same-charge ions are at double the spacing of the opposite-charge ions. The shear stress necessary to move a unit length of dislocation is directly proportional to the magnitude of the Burger's vector. Because of the large Burger's vector in ionic crystals, the critical stress necessary to move a dislocation is greater than in a metal. Also, in covalently bonded materials, there is resistance to changes in the angle of bonds. For example, in diamond the angle of the bond between atoms is 109.5 degrees, and it requires a significant force to change this angle. Shuffling of atoms around a dislocation in a covalently bonded material requires changes in the angles between atoms, and this strain in covalently bonded materials requires more energy than it does in metals. Other than the difficulty of dislocation motion in crystalline ionic and covalently bonded materials, plastic strain is related to dislocation motion in these materials by Equation 6.30, just as it is in metals. Ionic and covalently bonded crystalline materials plastically deform at elevated temperatures by dislocation motion, where the thermal vibrations help to activate the dislocation motion.

6.6 TENSILE STRESS-STRAIN DIAGRAMS FOR SINGLE CRYSTALS AND POLYCRYSTALS

If plastic deformation is due to shear stress producing a shear strain, then how does a shear stress develop during a compression or tension test to produce shear strain? Figure 6.13 shows a specimen subject to a tensile force (F). Assume that this is a single crystal metal at room temperature, and plastic deformation is due to dislocation motion on the close-packed planes. Dislocations move most easily on a close-packed plane and in a close-packed direction that are inclined at angles of ϕ and λ, respectively, to the tensile axial force (F), as shown in Figure 6.13.

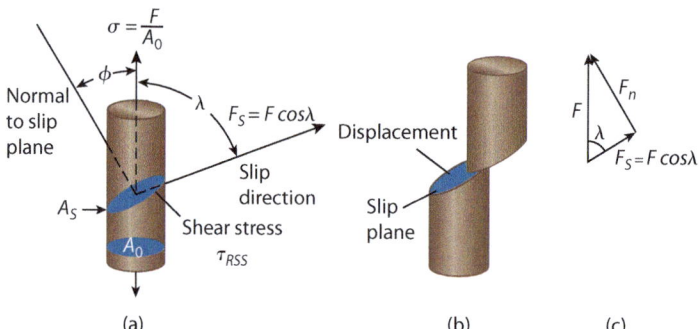

Figure 6.13 (a) A cylindrical single-crystal specimen with cross-sectional area A_0 that is subject to an axial force (F), with a slip plane of area (A_s), whose normal is oriented at an angle ϕ relative to the axial direction, and with the slip direction at an angle of λ relative to the axial direction. (b) The displacement along the slip plane. (c) The resolution of the axial force (F) into a force normal to the slip plane (F_n) and a shear force (F_s) parallel to the slip direction. (*Based on Askeland, D.R., Fulay, P.P., and Wright, W.J., The Science and Engineering of Materials, 6th ed., Cengage Learning, Stamford, CT. (2011), p. 132.*)

The force (F) resolved into the slip direction (F_s) is given by Equation 6.31.

$$F_S = F \cos \lambda \qquad \textbf{6.31}$$

The force (F_s) lies in the slip plane A_s; therefore, F_s is a shear force. The resolved shear stress (τ_{RSS}) in the direction of the dislocation motion is given by Equation 6.32, which is Schmid's equation.

$$\tau_{RSS} = \frac{F_S}{A_S} = \frac{F \cos \lambda}{\dfrac{A_0}{\cos \phi}} = \frac{F}{A_0} \cos \lambda \; \cos \phi = \sigma \cos \lambda \; \cos \phi \qquad \textbf{6.32}$$

In crystals, dislocation motion on a slip plane begins at a constant **critical resolved shear stress** (τ_{CRSS}). When the resolved shear stress is equal to the critical value (τ_{CRSS}), the axial tensile stress is equal to the **proportional limit** (σ_{PL}), which is a function of ϕ and λ for a single crystal, as shown in Equation 6.33.

$$\tau_{CRSS} = \sigma_{PL} \cos \lambda \; \cos \phi \qquad \textbf{6.33}$$

At the proportional limit, plastic strain begins and the linear relationship between stress and strain ends, as shown in the stress-strain diagrams in Figure 6.14a. Table 6.2 gives the critical resolved shear stress (τ_{CRSS}) for some FCC single crystals. The proportional limit in single crystals is not a constant; the proportional limit stress depends upon the orientation of the crystal planes and the slip direction. For example, Figure 6.14a shows stress-strain diagrams of anthracene single crystals with differently oriented slip planes and directions. What is constant is the τ_{CRSS}. Figure 6.14b shows the proportional limit in

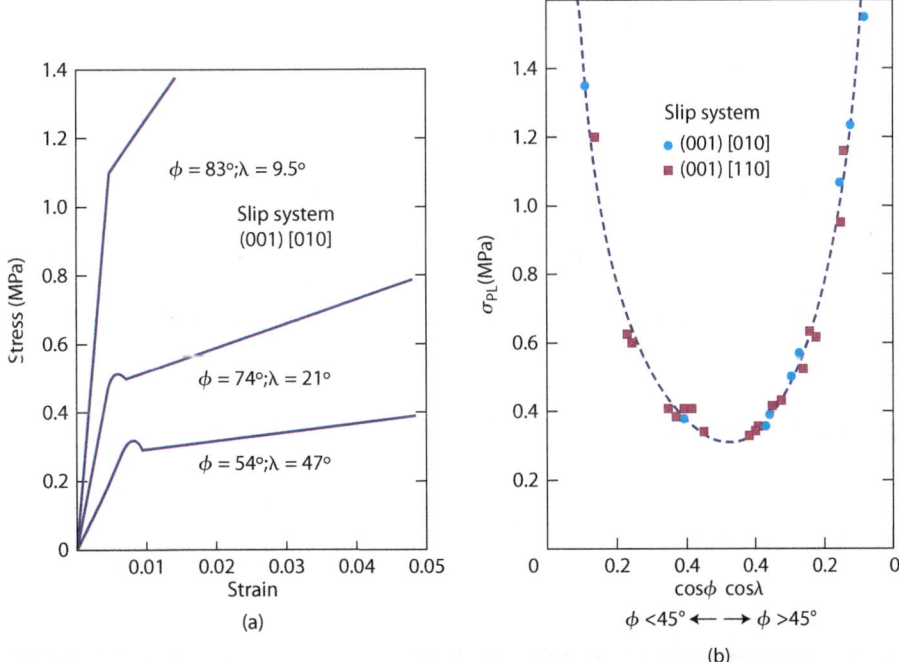

Figure 6.14 (a) Tensile stress-strain diagrams for anthracene single crystals with different slip-system orientations of ϕ and λ. (b) The axial proportional limit stress of anthracene single crystals as a function of $\cos \phi \cos \lambda$ for values of ϕ less than and greater than 45°. The dashed line is a plot of Equation 6.33 with $\tau_{CRSS} = 0.137$ MPa. *(MPa. Original data from Robinson, P.M., and Scott, H.G., Acta Met. 15 (1967) 1581. Based on Hertzberg, R.W., Deformation and Fracture Mechanics of Engineering Materials, 3rd ed. John Wiley & Sons N.Y. (1989), p. 85.)*

Table 6.2   The Critical Resolved Shear Stress for Some FCC Metals

Metal	Purity	Slip System	Critical Resolved Shear Stress MPa
Cu*	99.999	{111} <110>	0.63
Ag†	99.999	{111} <110>	0.37
Au‡	99.99	{111} <110>	0.91
Al§	99.996	{111} <110>	1.02

Based on data from Reed-Hill, R.E., and Abbaschian, R., *Physical Metallurgy Principles*, 3rd. ed. PWS, Boston (1994), p. 141.

*Rosi, F. D., *Trans. AIME*, 200,1009 (1954).

†daC. Andradc, E. N., and Henderson, C., *Trans. Roy. Soc.* (London), 244,177 (1951).

‡Sachs, G., and Weerts, J., *Zeitschrift fur Physik*, 62,473(1930).

§Rosi, F. D., and Mathewson, C. W., *Trans. AIME*, 188,1159 (1950).

anthracene single crystals as a function of $\cos \lambda \cos \phi$. The dashed line is a plot of Equation 6.33 with a critical resolved shear stress of 0.137 MPa. The function $\cos \lambda \cos \phi$ is a maximum of 0.5 at 45°, and then it decreases for values of λ and ϕ either larger or smaller than 45°. The maximum value of $\cos \lambda \cos \phi$ results in the minimum value of the proportional limit, because τ_{CRSS} is a constant. This can be seen in Figure 6.14b, where the minimum tensile strength for anthracene crystals occurs at $\phi = \lambda = 45°$. The proportional limit stress (σ_{PL}) increases as $\cos \lambda \cos \phi$ decreases from 0.5 and ϕ increases or decreases from 45°.

Polycrystalline materials have many grains of many different orientations. If the grains are randomly oriented, some grains will be oriented with both λ and ϕ oriented at 45°, and plastic strain is expected to occur in these grains first. In a polycrystalline material with a critical resolved shear stress of τ_{CRSS}, the proportional limit stress (σ_{PL}) might be expected to be approximately twice the critical resolved shear stress (τ_{CRSS}) when 45° is substituted into Equation 6.33. However, grains that are not oriented with their close-packed planes and directions at 45° constrain the plastic strain of those that are oriented at 45°, and as a result the proportional limit for a polycrystalline material is given by Equation 6.34:

$$\sigma_{PL} \approx m\tau_{CRSS} \qquad\qquad 6.34$$

where m is approximately equal to 3.1 for FCC metals and 2.75 for BCC metals.

Example Problem 6.8

An FCC crystal of aluminum with the critical resolved shear stress given in Table 6.2 is pulled in tension along the [001] direction. What is the predicted proportional limit stress of this crystal?

Solution

For an FCC crystal the close-packed planes are {111}, and the close-packed directions are <110>. Figure 6.12 shows the (111) close-packed plane and two possible Burger's vectors $b = \frac{a}{2}[0\bar{1}1]$ and $b = \frac{a}{2}[\bar{1}01]$ that lie in the (111) plane. The normal direction to the (111) plane is the [111] direction, and the [111] direction has an angle ϕ relative to the [001] direction whose cosine is equal to

$$cos\ \phi = \frac{1}{\sqrt{3}}$$

The <110> type directions are at an angle of 45° relative to the [001] direction, and the cosine of λ is equal to

$$cos\ \lambda = \frac{1}{\sqrt{2}}$$

The proportional limit stress is calculated with Equation 6.33, with τ_{CRSS} equal to 1.02 MPa from Table 6.2.

$$\tau_{CRSS} = \sigma_{PL}\ cos\ \lambda\ cos\ \phi$$

$$1.02 \times 10^6\,\mathrm{Pa} = \sigma_{PL}\frac{1}{\sqrt{2}}\frac{1}{\sqrt{3}}$$

Solving for the proportional limit stress of the crystal,

$$\sigma_{PL} = \sqrt{2}\sqrt{3}(1.02 \times 10^6\,\mathrm{Pa}) = 2.45(1.02 \times 10^6\,\mathrm{Pa}) = 2.50 \times 10^6\,\mathrm{Pa}$$

6.7 TENSILE STRESS-STRAIN DIAGRAMS AND MECHANICAL PROPERTIES FOR ENGINEERING MATERIALS

6.7.1 Introduction to Tensile Tests

A tensile test to fracture a smooth specimen provides some very important information about an engineering material's response to stress. From a tensile test we can determine the tensile elastic modulus, Poisson's ratio, yield stress, ultimate tensile strength, ductility, and toughness. All of these parameters are important to the selection of the proper material for an application.

A drawing of a standard $\frac{1}{2}$-inch-diameter tensile-test specimen is shown in Figure 6.15a. The American Society for Testing Materials (ASTM) publishes standard procedures for testing materials. A few of the publications covering tensile or compression tests for materials covered in this book include the following:

E8—Standard Test Methods for Tension Testing of Metallic Materials

C39—Standard Test Method for Compressive Strength of Cylindrical Concrete Specimens

C1273—Standard Test Method for Tensile Strength of Monolithic Advanced Ceramics at Ambient Temperatures

D638—Standard Test Method for Tensile Properties of Plastics

D3039—Standard Test Method for Tensile Properties of Polymer Matrix Composite Materials

For example ASTM E8 covers the tensile testing of flat plate specimens from 6.35 to 38.1 mm width and the testing of round tensile test specimens 2.87 to 12.7 mm diameter.

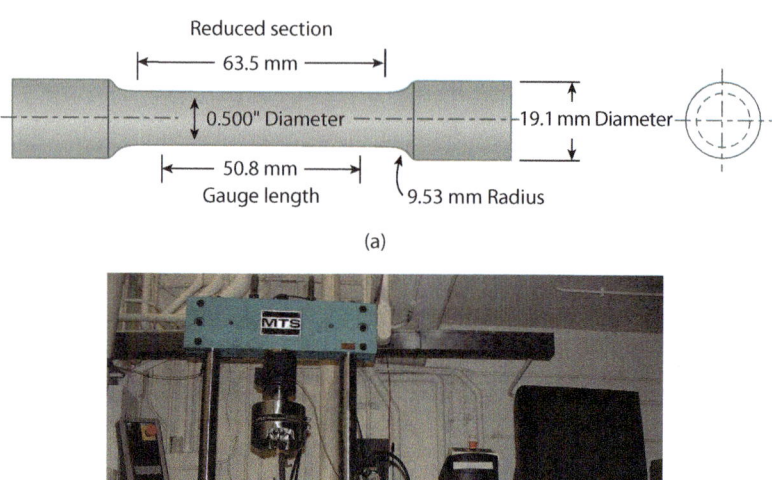

Figure 6.15   (a) A drawing of a standard $\frac{1}{2}$-inch-diameter tensile-test specimen. (b) A picture of a computer-controlled servo-hydraulic test facility. (*Photo by C.M. Gilmore*)

A servo-hydraulic system that can perform tensile and compressive tests is shown in Figure 6.15b. For a tensile test, this machine's upper jaw is fixed, and the lower jaw is pulled down by a hydraulic system that stretches the specimen between the jaws. A compression test is conducted by replacing the jaws with flat plates with the specimen placed between the plates. The lower plate is then raised and pressed against the specimen by the hydraulic system to create a compressive stress.

6.7.2   Stress-Strain Diagrams and Material Types

Figure 6.16 shows room-temperature stress-strain diagrams for sapphire (Al_2O_3), low-carbon steel, natural rubber, and polymethyl-methacrylate (PMMA) tested at 122°C. This chapter will explain the significant differences in these stress-strain diagrams. The stress-strain diagram for sapphire is nearly linear until fracture; fracture is indicated by the x in Figure 6.16a. The stress-strain diagram for natural rubber is nonlinear, as shown in Figure 6.16c, and the slope increases rapidly at high stress. The stress and strain scales for rubber are on the right and on the top, respectively. For low-carbon steel and the PMMA in Figures 6.16b and 6.16d, respectively, the stress-strain diagrams are linear for low stresses; however, at higher stresses they are nonlinear, and the slope becomes negative after each stress reaches a maximum.

6.7.3   Yield Stress

The stress-strain diagram for cold-worked (CW) 1020 steel (0.2 weight percent C) is shown in Figures 6.17a and 6.17b. For materials that have an initial linear elastic region, the proportional limit (σ_{PL}) is the stress where the linear relation between stress and strain ends and where plastic strain begins.

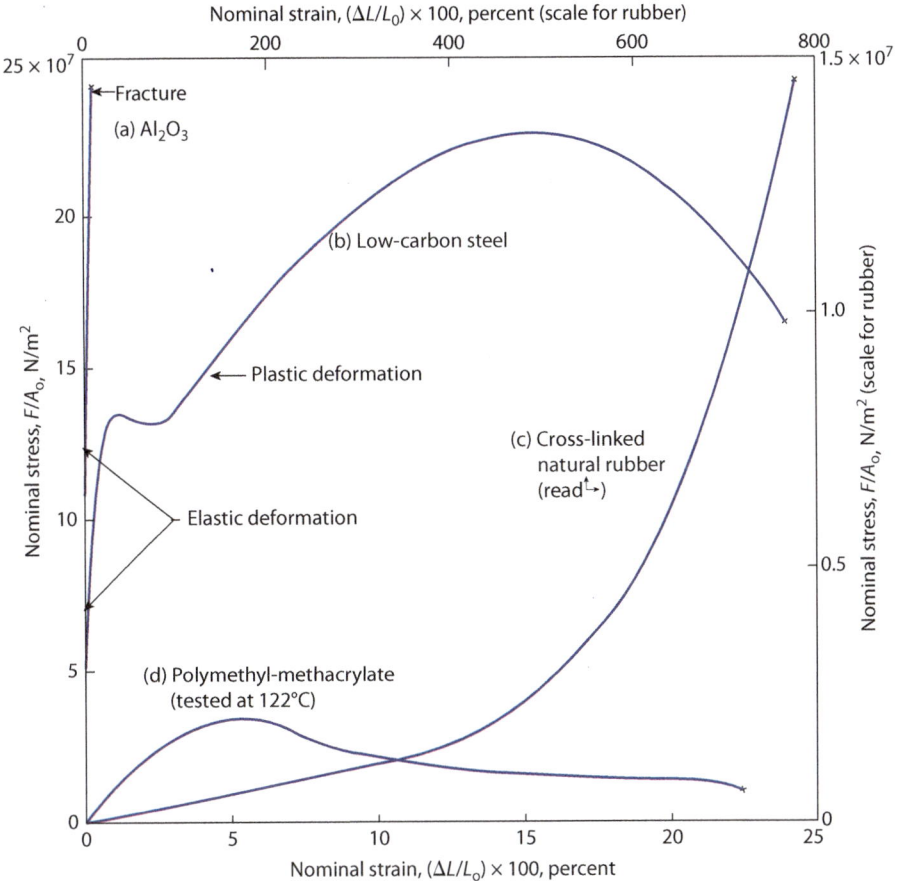

Figure 6.16 Tensile stress-strain diagrams for four different types of materials: (a) sapphire (Al_2O_3), (b) low-carbon steel, (c) natural rubber, all tested at room temperature, and (d) polymethyl-methacrylate (PMMA) tested at 122°C. The rubber scales are on the right and top. (*Based on Guy, A.G. Introduction to Materials Science, McGraw-Hill, N.Y. (1972), p. 400.*)

However, the value of the proportional limit depends upon the precision of the measurement. A very precise measurement system would detect plastic strain at a lower stress than would a less precise measurement. Therefore, for most engineering applications the **yield stress** (σ_y) is utilized rather than the proportional limit. To determine the yield stress, draw a line parallel to the initial slope of the stress-strain diagram, with a strain offset of 0.002, as shown in Figure 6.17a. The stress where this line intersects the uniaxial tensile stress-strain diagram is the yield stress. For this material the value is 455 MPa.

The tensile yield strength of some commercially pure (usually 99% pure) polycrystalline metals and plain-carbon steels in the annealed condition is shown in Table 6.3. In the literature the terms *yield stress* and *yield strength* are both found. The strength is the resistance provided by the material to the applied stress. When the applied stress is equal to the yield strength, the material yields. The applied stress when the material yields is also called the yield stress. Note that the yield stress listed in Table 6.3 for annealed 1020 steel is only 295 MPa in comparison to 455 MPa for the cold worked 1020 steel shown in Figure 6.17a. Cold work increases the yield strength of annealed steel. We will cover annealing and cold work effects on yield strength in Chapter 7. In Table 6.3 plain-carbon steels are designated 10XX. The last two numbers give the weight percent carbon in the steel with a decimal point before the number represented by the first X.

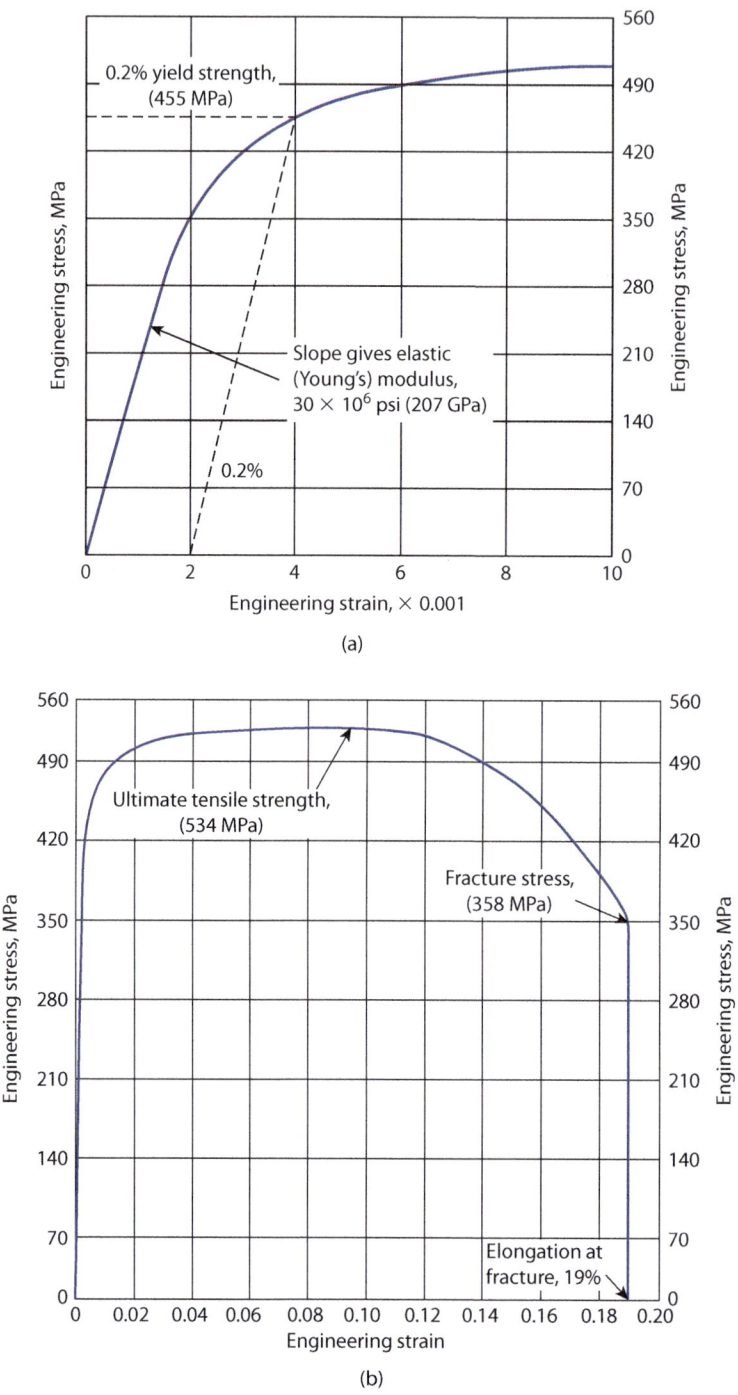

Figure 6.17 (a) The low-strain region of the stress-strain diagram for CW 1020 steel, showing the construction to determine the yield strength. (b) The full stress-strain diagram of CW 1020 steel to fracture. (*Brooks, C.R., Heat Treatment, Structure, and Properties of Nonferrous Alloys, American Society for Metals, (1982), p. 2. Based on Atlas of Stress-Strain Curves 2nd ed. ASM International Metals Park, (2002), p. 74.*)

Material	σ_y (MPa)	σ_u (MPa)	ε_f (%)
Al (FCC)	34.5	89.6	45
Cu (FCC)	33.3	210	60
Fe (BCC)	50	540	—
Ti (HCP)	140	220	54
Mg (HCP)	90	160	15
1020 steel	295	395	36.5
1040 steel	350	515	30
1060 steel	370	625	22
1080 steel	380	615	25

Table 6.3 Yield Stress (σ_y), Ultimate Tensile Strength (σ_u), and Engineering Strain at Fracture (ε_f) of Some Commercially Pure Metals and Plain-Carbon Steels

1 Based on data primarily from http://www.matweb.com.

The 1020 steel has 0.20 weight percent carbon as the primary alloy element. All 10XX steels are BCC iron plus iron carbide. The numbering systems and the properties of other metal alloys are discussed in Chapter 8. We will describe the ultimate tensile strength, strain to fracture, and reduction in area in later sections of this chapter.

Example Problem 6.9

The stress-strain diagram for CW 1020 steel is shown in Figure 6.17. Determine the tangent elastic modulus at zero strain.

Solution

The tangent elastic modulus at zero strain is obtained by drawing a tangent to the stress-strain diagram at zero strain and then determining the slope of this line. This tangent line intersects the 0.414 GPa horizontal line at a strain of approximately 0.002. From Equation 6.15

$$E_{tan}(\varepsilon = 0) = \frac{\Delta\sigma_0}{\Delta\varepsilon_0} = \frac{(0.414 - 0)\ \text{GPa}}{2.0 \times 10^{-3} - 0} = \frac{0.414\ \text{GPa}}{2.0 \times 10^{-3}} = 207\ \text{GPa}$$

The value of the elastic modulus listed for iron in Table 6.1 is 211 GPa. Our calculation result is very close. Metal alloys have an elastic modulus that is usually close to the pure metal if the concentration of alloy elements is low. The elastic moduli of many alloys are presented in Chapter 8.

The yield strengths of the metals in Table 6.3 are related to their crystal structures (we discussed crystal structures in Chapter 2), their defects (we discussed defects in Chapter 3), and their phases present (we discussed phases in Chapter 5). Because the FCC metals have multiple {111} close-packed planes for dislocation motion, they have a low yield strength and a high plastic strain at fracture. In each grain there are {111} planes that are favorably oriented for dislocation motion according to Equation 6.33. The BCC metals are next to lowest in yield strength. The BCC metals have no close-packed planes, but there are many planes that have a relatively high packing density where dislocation motion can occur, such as the {110} planes. The HCP metals have the highest yield strength of the commercially pure metals, because dislocation motion occurs primarily on the basal plane in HCP metals, and in each grain of the polycrystalline metal there is only one orientation of the basal plane; this orientation may not be favorably oriented for dislocation motion according to Equation 6.33. For a polycrystalline metal to yield, plastic strain must occur in most of the grains; otherwise the elastic grains constrain the plastic grains from extensive plastic strain.

In plain-carbon steels the yield strength is considerably increased relative to that of pure iron, because the iron carbide precipitates (Fe_3C) that form in plain-carbon steels impede dislocation motion. The greater the amount the carbon added, the higher the yield strength. The ultimate tensile strength, which we will discuss in Section 6.7.6, also increases with the amount of carbon; however, the ultimate tensile strength reaches a plateau at 0.60 weight percent carbon. The phase diagram for iron and carbon in Figure 5.18 shows that 0.77 weight percent carbon corresponds to the eutectoid composition. Also, the strain at fracture of the plain-carbon steels decreases as the amount of carbon is increased from 0.20 to 0.60 weight percent: As the amount of carbon is increased, there is more of the brittle iron carbide phase (Fe_3C) that reduces the amount of plastic strain at fracture. The plastic strain before fracture is relatively constant for 0.60 and 0.80 weight percent carbon steel.

Observe that in the low-carbon steel stress-strain diagram of Figure 6.16b, we can see an initial sharp yield stress, called the upper yield stress; and then the stress drops to a lower level, called the lower yield stress. Upper and lower yield stresses are present for many carbon steels. In Figure 6.16b the upper yield stress is at approximately 135 MPa and the lower yield stress is at approximately 131 MPa. The **upper yield stress** corresponds to the stress necessary to move the dislocations away from carbon atoms in steel. Once the dislocations are away from the carbon atoms, the **lower yield stress** is the stress required to move the dislocations through the crystal. We will discuss the effect of the carbon atoms on the yield stress in Chapter 7, on strong solids. In materials with an upper and lower yield stress, the upper yield stress is reported as the yield stress. Pure metals do not have an upper and lower yield stress.

In metals and polymers that are not brittle, the yield stress in compression is normally equal to the yield stress in tension. Ceramics do not yield at room temperature. Ceramics normally fracture before they yield, or they fracture when they yield.

6.7.4   Resilience

The resilience of a material is related to its ability to return stored strain energy when unloaded. Engineers use two measurements of resilience: the modulus of resilience and the impact resilience. The **modulus of resilience** is important in applications where the material is to be used as a spring, and **impact resilience** is important if the material is to be used as a shock absorber.

The modulus of resilience is the maximum strain energy per unit volume that a material can absorb and then fully return when unloaded. Springs are usually made from metals that are linear elastic up to a sharp yield stress that corresponds to the initiation of plastic strain. If the yield stress is assumed to be the maximum stress where the strain is all elastic, the elastic-strain energy density that is returned by the material at the yield stress (R) is determined using Equation 6.23b. Equation 6.35 gives the modulus of resilience in a material where plastic deformation begins at the yield stress.

$$R = \frac{1}{2E}\sigma_y^2 \qquad\qquad \textbf{6.35}$$

The units of resilience are energy per unit volume or joules per cubic meter. The resilience is the strain energy density recovered by unloading the specimen from the yield stress.

What is calculated in Equation 6.35 can be understood by calculating the resilience of a material such as the 1020 steel shown in Figure 6.17a, where there is plastic strain at the yield stress. During unloading

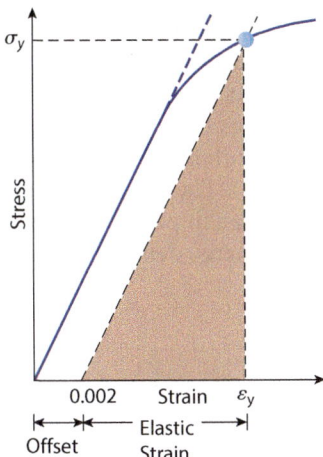

Figure 6.18 Resilience (shaded area) determination from the elastic-strain energy recovered at the yield stress.

from the yield stress for a linear elastic material, the magnitude of the elastic strain recovered (ε_{y-el}) is given by Equation 6.36.

$$\varepsilon_{y-el} = \frac{\sigma_y}{E} \qquad\qquad \textbf{6.36}$$

Figure 6.18 shows the yield stress (σ_y), the total yield strain (ε_y), the elastic strain at yield (ε_{y-el}), the plastic strain at yield ($\varepsilon_{y-pl} = 0.002$), and the resilience is the shaded area. Note that in calculating the resilience with Equation 6.35 only the elastic strain given by Equation 6.36 is used. The total strain at the yield stress is not used, because the total strain at the yield stress contains a plastic strain of 0.002, and the plastic strain is not recovered. If a material is unloaded from the yield stress, the stress-strain diagram follows the dashed line in Figure 6.18 that is parallel to the elastic modulus line down to a permanent strain of 0.002 at zero stress. Brittle materials such as ceramics, glass, and brittle polymers fracture before they yield. For these materials, the resilience is equal to the elastic-strain energy density to fracture the material.

From Equation 6.35 we see that materials can have a high modulus of resilience by having a high yield stress or a small elastic modulus. Strong materials, such as the steel used in springs, have a high yield stress and a high modulus of resilience.

Example Problem 6.10

Calculate the resilience of the 1045 steel in Figure 6.22. You can assume that the elastic modulus of 1045 steel is 211 GPa, as indicated by Table 6.1 for iron.

Solution

From Figure 6.22, the yield stress of 1045 steel is 600 MPa. Using Equation 6.35,

$$R = \frac{\sigma_y^2}{2E} = \frac{(600 \times 10^6 \text{ Pa})^2}{2(211 \times 10^9 \text{ Pa})} = \frac{36 \times 10^{16} \text{ (Pa)}^2}{0.422 \times 10^{12} \text{ Pa}} = 85.3 \times 10^4 \frac{\text{J}}{\text{m}^3}$$

Impact Resilience of Polymers

Rubber is a good material for absorbing large amounts of strain energy if large strains are allowable. For rubber it is difficult to calculate a modulus of resilience, because in rubber the stress-strain plot is nonlinear and a function of strain rate. Because of the nonlinear stress-strain plot the impact resilience test is used on rubber and other polymeric materials. In this test, a steel ball is dropped from an initial height (h_1) onto a flat specimen of the material; the steel ball bounces off the material to a height (h_2). The impact resilience is the ratio h_2/h_1. An impact resilience of 1 indicates a purely elastic material, and an impact resilience of 0 is a completely plastic material. This test does not determine how much strain energy density is absorbed or released.

6.7.5   Strain Hardening

Strain hardening is the increased material strength resulting from plastic strain; strain hardening is also called work hardening. Strain hardening is an important strengthening process in low-strength metals. Copper weapons and tools were strain hardened by hammering before the development of bronze and carbon steel. The yield stress of the low-carbon steel sheet metal in automobile bodies is increased by a factor of approximately ten by strain hardening, from approximately 0.7×10^8 Pa to 0.7×10^9 Pa. In Chapter 7 we will discuss strengthening processes, including strain hardening.

The 1020 steel in Figure 6.17 demonstrates extensive strain hardening after the yield stress. To observe strain hardening, it is easiest to analyze a material that has plastic strain at low stress, as shown in Figure 6.19. For the material in Figure 6.19 it is not necessary to consider the transition from elastic to plastic strain. In a material that strain hardens, the stress necessary to obtain additional plastic strain increases. The resulting stress-strain diagram is not linear, because there is a combination of elastic and plastic strain. Elastic strain occurs whenever the material supports a stress, because the atomic bonds resist the stress, and the interatomic spacing changes whenever the applied stress changes.

In Figure 6.19, upon initial loading the yield strength is σ_y. The specimen is stressed well beyond the initial yield stress. Then the specimen is unloaded and reloaded as indicated by the directional arrows. In unloading, the stress-strain diagram has the same slope as the elastic modulus of the material, because in unloading only the elastic strain is recovered. The recovered elastic strain is due to the atoms returning to their equilibrium atomic positions, and the magnitude of the elastic strain is given by Equation 6.13.

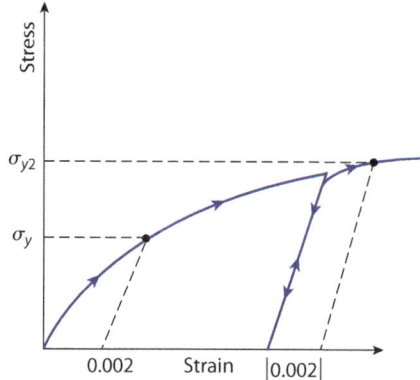

Figure 6.19   A schematic of a stress-strain diagram for a material exhibiting plastic strain at low stress, with an initial yield stress of (σ_y). Unloading of the material demonstrates recovery of the elastic strain and the permanent plastic strain, and reloading demonstrates a second yield stress of σ_{y2} and strain hardening.

After unloading, the residual strain equals the plastic strain. Upon reloading in the original loading direction, the stress-strain diagram in most materials follows the same path as during the unloading with a slope of the elastic modulus. Most materials yield upon reloading at a stress (σ_{y2}) approximately equal to the previous maximum stress before unloading. The increment in the yield stress ($\sigma_{y2} - \sigma_y = \Delta\sigma$) is the strain hardening or strengthening. The increment in strain ($\Delta\varepsilon$) associated with the increase in strength ($\Delta\sigma$) is composed of both elastic and plastic strain. The elastic strain increment ($\Delta\varepsilon_e$) for a linear elastic material is given by Equation 6.37.

$$\Delta\varepsilon_e = \frac{\Delta\sigma}{E} \qquad \textbf{6.37}$$

Equation 6.38 shows that the plastic strain increment ($\Delta\varepsilon_p$) is the remaining strain increment.

$$\Delta\varepsilon_p = \Delta\varepsilon - \Delta\varepsilon_e \qquad \textbf{6.38}$$

Since the magnitude of the total strain in a material can be large when there is plastic strain, the nonlinear analytical relation between stress and strain for a strain-hardening material is written in terms of true stress and strain (σ_t, ε_t), as in Equation 6.39:

$$\sigma_\tau = K\varepsilon_t^n \qquad \textbf{6.39}$$

where the strength coefficient K is a constant with the units of stress, and n is the strain-hardening exponent. Values of n and K are presented in Table 6.4 for some metals. We will discuss alloys such as 4340 steel, 304 stainless steel, and 2024 aluminum in Chapter 8. The magnitude of the strain-hardening exponent (n) is obtained from the slope of the log of the true stress versus the log of the true strain diagram.

Polymer scientists and engineers use viscoelastic equations to model the increase in stress that occurs with strain. We will cover viscoelastic models in Chapter 9. Strain hardening of polymers occurs during the processing by extrusion of parts and drawing filaments. We will cover these processes in Chapter 13.

Table 6.4	Tabulation of the Strain-Hardening Exponent (n) and the Strength Coefficient (K) Values for Several Metal Alloys, in Units of MPa	
Material (structure)	**n**	**K MPa**
Titanium (HCP)[2]	0.05	1208
Low-carbon steel, annealed (BCC)[1]	0.26	530
4340 steel, annealed (BCC)[1]	0.15	640
304 stainless steel (FCC)[1]	0.45	1275
Aluminum 1100-O (FCC)[1]	0.20	180
Aluminum alloy 2024, heat treated (FCC)[1]	0.16	690
Copper, annealed (FCC)[1]	0.54	315
Copper plus 30 weight percent zinc, annealed (FCC)[1]	0.49	895
Molybdenum (BCC)[2]	0.13	725

1 Kalpakjian, S., Manufacturing Processes for Engineering Materials, Addison-Wesley Publ. Co. (1984), p, 38.
2 Askeland, D.R., Fulay, P.P., and Wright, W.J., The Science and Engineering of Materials, 6th ed., Cengage Learning, Stamford, CT. (2011), p. 295.

6.7.6 The Ultimate Tensile Strength

Figure 6.20 is a schematic of an engineering stress-strain diagram to failure in a metal that is initially linear elastic, yields, and then fractures after significant plastic strain. The engineering stress reaches a maximum, and then it decreases until fracture is reached. The maximum engineering strength in tension is the **ultimate tensile strength** (σ_u). The maximum load-carrying capacity of the material is the ultimate tensile strength times the original area. When the strain exceeds the value at the ultimate tensile strength, the load-carrying capacity of the specimen decreases.

   The strain in a tensile-test specimen is uniform if the strain is less than that at the ultimate tensile strength, as shown by the specimen insert in Figure 6.20. However, for strains greater than that at the ultimate tensile strength, the strain in the specimen is not uniform, and a neck develops in the specimen, as shown by the specimen insert in Figure 6.20. The condition for the transition from uniform strain to necking is obtained by relating the applied force (F) to the true stress (σ_t) and the true area (A) as shown in Equation 6.40.

$$F = A\sigma_t \qquad\qquad\qquad \textbf{6.40}$$

At the ultimate tensile strength, the applied force (F) is a maximum. Thus the total differential of applied force (dF) is equal to 0, and true stress and area are variables, as shown in Equation 6.41.

$$dF = A\,d\sigma_t + \sigma_t\,dA = 0 \qquad\qquad \textbf{6.41}$$

Rearranging variables results in Equation 6.42.

$$\frac{d\sigma_\tau}{\sigma_\tau} = -\frac{dA}{A} \qquad\qquad\qquad \textbf{6.42}$$

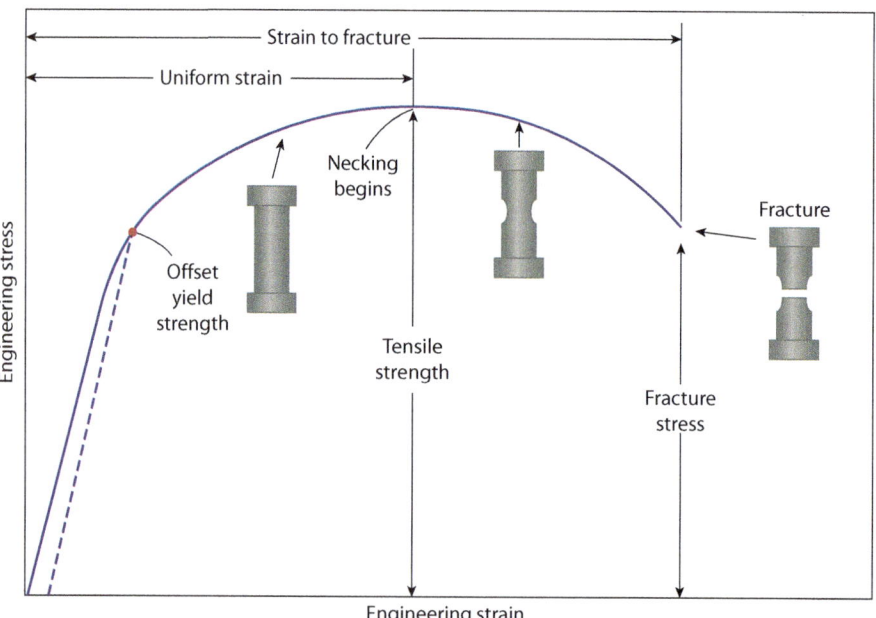

Figure 6.20 A hypothetical engineering stress-strain diagram showing the 0.002 offset yield strength, the ultimate tensile strength, necking, strain to fracture, and the fracture stress. (*Adapted from from Moosbrugger, C., ed., Atlas of Stress-Strain Curves, 2nd ed, ASM International, Metals Park, OH (2002), p. 1.*)

The engineering ultimate tensile strength (σ_u) corresponds to the condition where the increment of true stress (strength) increase divided by the true stress equals the negative of the increment of cross-sectional area change, divided by the area.

Figure 6.21 compares an engineering stress-strain diagram to a true stress-strain diagram. Point a is the ultimate tensile strength on the engineering stress-strain diagram, and a' corresponds to this point on the true stress-strain diagram. The true stress is determined at the minimum cross section, which is at the neck; and the true stress continues to increase even at stresses above the ultimate tensile strength, as shown schematically in Figure 6.21, where both engineering and true stress-strain diagrams are plotted. The continuous increase in true stress occurs because in the calculation of true stress the applied force is divided by the actual area at the neck. The decrease in the engineering stress for strains greater than the value at the ultimate tensile strength is because in the calculation of engineering stress, the applied force is divided by the original area, and the decrease in area at the neck is not considered.

Figure 6.22 is a true stress-strain diagram for 1045 steel, which is is a medium-strength carbon steel. Observe the difference between the true stress-strain diagram of 1045 steel in Figure 6.22 and the engineering stress-strain diagram of the low-carbon steel in Figure 6.16. Figure 6.16 shows a maximum engineering stress at the ultimate tensile strength, and then the stress decreases. In contrast, Figure 6.22 shows there is never a decrease in the true stress. However, the load-carrying capacity of the specimen does decrease for strains greater than that at the ultimate tensile strength, but this is not observed in the true stress-strain diagram of Figure 6.22.

In Figure 6.21, note that at the ultimate tensile strength, the true strain is less than the engineering strain as indicated by points a' and a. This is because up to the ultimate tensile strength, the true strain is measured relative to the actual length of the specimen, as demonstrated in Equations 6.6 and 6.7. The actual length of the specimen increases with increasing strain, and the length is in the denominator of the equation for true strain; in contrast, engineering strain is measured relative to the original length, as shown in Equation 6.3.

Observe also in Figure 6.21 that the true strain to failure is greater than the engineering strain to failure as indicated by points b' and b. Why have the relative magnitudes of true and engineering strain reversed? To understand this reversal we have to develop a new measurement of true strain for strains greater than that at the ultimate tensile strength. For a material that has significant plastic strain at the ultimate tensile strength, we assume that the plastic strain is much larger than the elastic strain, as demonstrated in

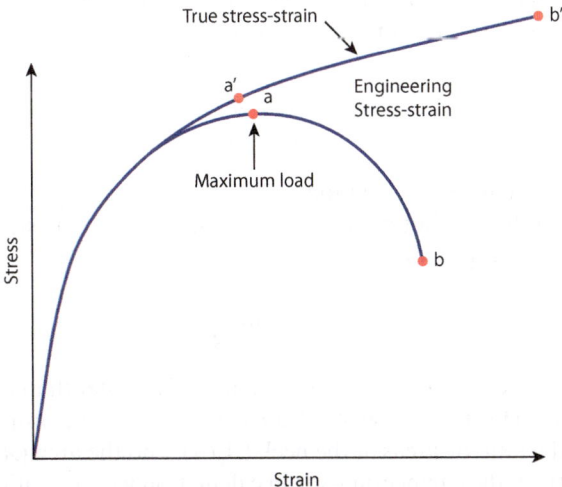

Figure 6.21   Engineering and true stress-strain diagrams indicating the ultimate tensile strength (a) and failure (b).

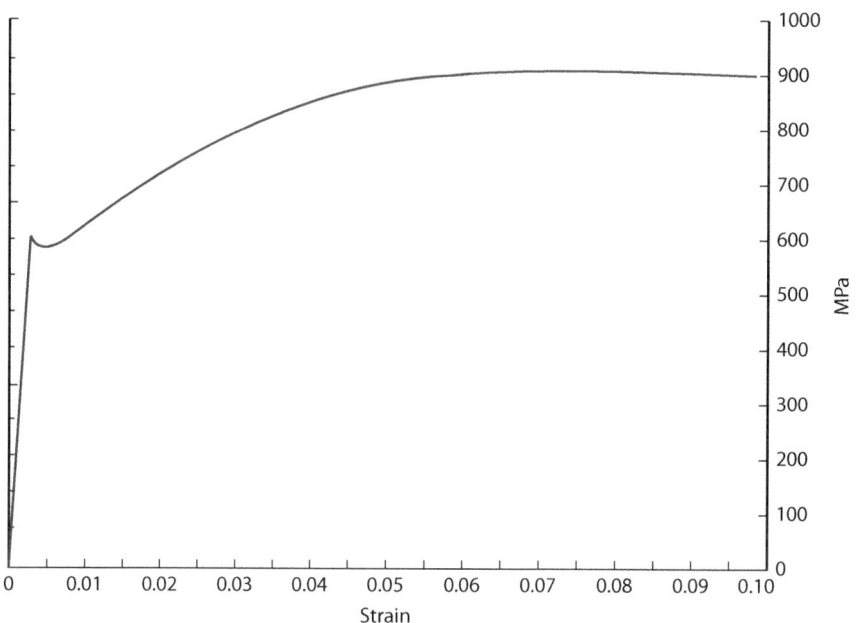

Figure 6.22 A true stress-strain diagram for 1045 steel in mega pascals MPa. *(Based on Boyer, H.E., (Editor), Atlas of Stress-Strain Curves, ASM International, Metals Park, OH (1987), p. 193.)*

Example Problem 6.11. Plastic strain conserves volume, because the motion of a dislocation in a crystal does not change the volume, nor does the sliding of long-chain molecules past each other in a polymer change the volume. Assuming that volume is conserved results in Equation 6.43:

$$V = V_0 = Al = A_0 l_0 \qquad\qquad \textbf{6.43}$$

where V, A, and l are, respectively, the instantaneous volume, area, and length of a volume of material at the neck of the specimen, and V_0, A_0, and l_0 are the original values. Solving Equation 6.43 for l/l_0 results in Equation 6.44.

$$\frac{l}{l_0} = \frac{A_0}{A} \qquad\qquad \textbf{6.44}$$

The true uniform strain is defined in Equation 6.7 as the natural log of the ratio of the final length, divided by the original length (l/l_0), but by using Equation 6.44 the true strain at the neck (ε_t) is defined in terms of areas, as shown in Equation 6.45.

$$\varepsilon_t = \ln \frac{A_0}{A} \qquad\qquad \textbf{6.45}$$

As shown in Figure 6.21, the true fracture strain (point b') is greater than the engineering fracture strain (point b) in a ductile metal. This is because for strains greater than the ultimate tensile strength, the true strain is calculated from the ratio of areas at the neck (A_0/A), and the area (A) becomes a small number in a ductile material. In contrast, the engineering-strain calculation with Equation 6.3 assumes that the strain is uniform over the length of the specimen, even though the strain is concentrated at the neck.

Example Problem 6.11

Estimate the elastic and plastic strain at the ultimate tensile strength in the 1020 steel specimen in Figure 6.17b.

Solution

At the ultimate tensile strength, the total strain is approximately 0.09. The elastic strain at the ultimate tensile strength is determined from the strain recovered during unloading, assuming that the unloading stress-strain plot follows the slope of the elastic modulus. The elastic modulus from of the 1020 steel is equal to 207×10^9 Pa. The ultimate tensile strength for 1020 steel in Figure 6.17b is 534×10^6 Pa. The elastic strain at the ultimate tensile strength can be calculated from

$$\varepsilon_{e-u} = \frac{\sigma_u}{E} = \frac{534 \times 10^6 \text{ Pa}}{207 \times 10^9 \text{ Pa}} = 2.58 \times 10^{-3}$$

The plastic strain (ε_p) is the total strain (ε) minus the elastic strain at the ultimate tensile strength.

$$\varepsilon_p = \varepsilon - \varepsilon_{e-u} = 90 \times 10^{-3} - 2.58 \times 10^{-3} = 87.4 \times 10^{-3}$$

For the case of 1020 steel at the ultimate tensile strength, the elastic strain is much smaller than the plastic strain, as assumed in Equation 6.43 above.

Example Problem 6.12

For a specimen made from the low-carbon steel shown in Figure 6.16, calculate the maximum load-carrying capacity for a standard tensile-test specimen as shown in Figure 6.15a.

Solution

The maximum load-carrying capacity is given by Equation 6.2, with the stress set equal to the ultimate tensile strength. The ultimate tensile strength for the low-carbon steel in Figure 6.16 is 225 MPa. The diameter of the tensile specimen is 1.27 cm = 1.27×10^{-2} m. The cross-sectional area (A_0) of the tensile specimen is then

$$A_0 = \frac{\pi d^2}{4} = \frac{\pi (1.27 \times 10^{-2} \text{m})^2}{4} = \frac{\pi (1.61 \times 10^{-4} \text{m}^2)}{4} = 1.27 \times 10^{-4} \text{m}^2$$

$$F_{max} = \sigma_{uts} A_0 = \left(225 \times 10^6 \frac{\text{N}}{\text{m}^2}\right)(1.27 \times 10^{-4} \text{m}^2) = 286 \times 10^2 \text{N}$$

6.7.7 Toughness

The **toughness** of a material is the strain energy per unit volume required to fracture the specimen in a tensile test. The toughness is determined by integrating the area under the stress-strain diagram to fracture. The combination of high strength and high strain to fracture results in a high toughness. Materials with a very small strain to fracture are brittle materials, and they have a low toughness. If a numerical model of the stress-strain plot is available, then the toughness can be determined with numerical integration. If a stress-strain plot is available only in printed form, it is possible to integrate the area with a planimeter designed to determine the area of a plane surface, or the area can be determined by the method of trapezoids.

Example Problem 6.13

Determine the toughness of the 1020 steel specimen in Figure 6.17b.

Solution

The area under the stress-strain diagram is the strain energy per unit volume to fracture the specimen, or the toughness. One way to estimate this area is to break the stress-strain diagram into trapezoids and then sum the area of each trapezoid. Use increments of 0.02 strain and number the areas starting from 0 strain, as shown in Figure 6.17c.

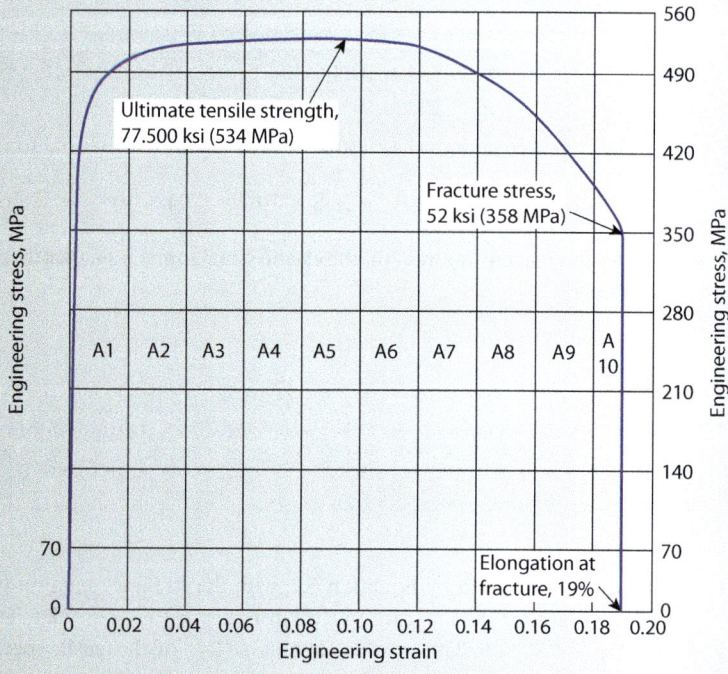

Figure 6.17 (c) The full stress-strain diagram of CW 1020 steel to fracture with areas for the toughness calculation indicated. (*Based on Atlas of Stress-Strain Curves 2nd ed. ASM International Metals Park, (2002), p. 74.*)

Area number one (A1) is 0.02 wide, and the area is approximated by a starting stress of 350 MPa and an ending stress of 500 MPa. The strain energy per unit volume represented by area A1 is then

$$A1 = \frac{1}{2}\left[0.02\,\frac{m}{m}\left(350 \times 10^6\,\frac{N}{m^2} + 500 \times 10^6\,\frac{N}{m^2}\right)\right] = 8.5 \times 10^6\,\frac{N \cdot m}{m^3} = 8.5 \times 10^6\,\frac{J}{m^3}$$

Calculating the areas of A2 to A10,

$$A2 = \frac{1}{2}\left[0.02\,\frac{m}{m}\left(500 \times 10^6\,\frac{N}{m^2} + 520 \times 10^6\,\frac{N}{m^2}\right)\right] = 10.2 \times 10^6\,\frac{J}{m^3}$$

$$A3 = \frac{1}{2}\left[0.02\,\frac{m}{m}\left(520 \times 10^6\,\frac{N}{m^2} + 524 \times 10^6\,\frac{N}{m^2}\right)\right] = 10.4 \times 10^6\,\frac{J}{m^3}$$

$$A4 = \frac{1}{2}\left[0.02\,\frac{m}{m}\left(524 \times 10^6\,\frac{N}{m^2} + 526 \times 10^6\,\frac{N}{m^2}\right)\right] = 10.5 \times 10^6\,\frac{J}{m^3}$$

$$A5 = \frac{1}{2}\left[0.02\,\frac{\text{m}}{\text{m}}\left(526 \times 10^6\,\frac{\text{N}}{\text{m}^2} + 524 \times 10^6\,\frac{\text{N}}{\text{m}^2}\right)\right] = 10.5 \times 10^6\,\frac{\text{J}}{\text{m}^3}$$

$$A6 = \frac{1}{2}\left[0.02\,\frac{\text{m}}{\text{m}}\left(524 \times 10^6\,\frac{\text{N}}{\text{m}^2} + 520 \times 10^6\,\frac{\text{N}}{\text{m}^2}\right)\right] = 10.4 \times 10^6\,\frac{\text{J}}{\text{m}^3}$$

$$A7 = \frac{1}{2}\left[0.02\,\frac{\text{m}}{\text{m}}\left(520 \times 10^6\,\frac{\text{N}}{\text{m}^2} + 488 \times 10^6\,\frac{\text{N}}{\text{m}^2}\right)\right] = 10.1 \times 10^6\,\frac{\text{J}}{\text{m}^3}$$

$$A8 = \frac{1}{2}\left[0.02\,\frac{\text{m}}{\text{m}}\left(488 \times 10^6\,\frac{\text{N}}{\text{m}^2} + 455 \times 10^6\,\frac{\text{N}}{\text{m}^2}\right)\right] = 94.3 \times 10^6\,\frac{\text{J}}{\text{m}^3}$$

$$A9 = \frac{1}{2}\left[0.02\,\frac{\text{m}}{\text{m}}\left(455 \times 10^6\,\frac{\text{N}}{\text{m}^2} + 390 \times 10^6\,\frac{\text{N}}{\text{m}^2}\right)\right] = 8.45 \times 10^6\,\frac{\text{J}}{\text{m}^3}$$

The area $A10$ starts at a strain of 0.18 and ends at 0.19 for a strain width of 0.01. The stress in $A10$ starts at 390 MPa and fracture occurs at 350 MPa.

$$A10 = \frac{1}{2}\left[0.01\,\frac{\text{m}}{\text{m}}\left(390 \times 10^6\,\frac{\text{N}}{\text{m}^2} + 350 \times 10^6\,\frac{\text{N}}{\text{m}^2}\right)\right] = 3.70 \times 10^6\,\frac{\text{J}}{\text{m}^3}$$

The toughness is the sum of all of the above areas, or the strain energy per unit volume represented by the area under the stress-strain plot.

$$\text{Toughness} = A1 + A2 + A3 + A4 + A5 + A6 + A7 + A8 + A9 + A10$$

$$\text{Toughness} = (8.5 + 10.2 + 10.4 + 10.5 + 10.5 + 10.4 + 10.1 + 9.43 + 8.45 + 3.70) \times 10^6\,\frac{\text{J}}{\text{m}^3}$$

$$\text{Toughness} = 92.2 \times 10^6\,\frac{\text{J}}{\text{m}^3}$$

6.7.8 Stress-Strain Diagrams for Polymers

A schematic of a stress-strain diagram for a polymer with an initial linear elastic region, followed by a yield point and plastic strain, is shown in Figure 6.23, and the change in shape of a polymer specimen is indicated by the tensile specimens related to points (a) to (e) on the stress-strain diagram. The stress-strain diagram is similar to that of a metal; however, there is also a significant difference. As with metals, for many polymers there is an initial linear elastic stress-strain region where the strain is uniform along the specimen. After the polymer yields, plastic strain concentrates with the formation of a neck. The strain hardening of the polymer in the neck region results in the strain transferring to regions adjacent to the neck, and in this way the neck propagates along the specimen. The propagation of the neck in a polymer is different than the behavior of a metal. In a metal once the neck forms, the strain concentrates in the neck, and the neck does not propagate. The equations related to the stress-strain diagram developed in this chapter apply to metals, ceramics, and polymers, except that Equation 6.39 is not used by polymer scientists or engineers.

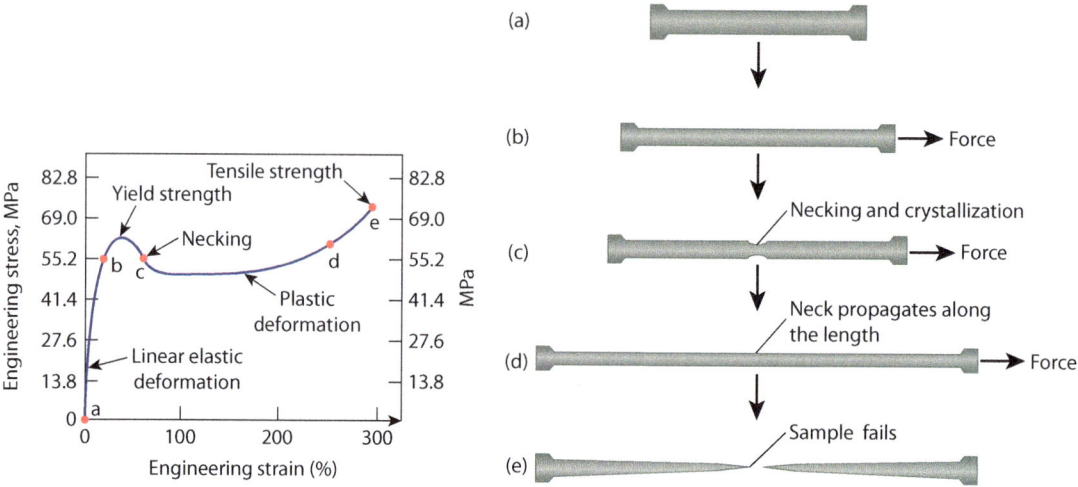

Figure 6.23 A schematic of a stress-strain diagram for a polymer with a yield stress, subsequent plastic strain, and illustrations of specimen deformation at points (a) to (e) on the stress-strain diagram. (*Based on Askeland, D.R., Fulay, P.P., and Wright, W.J., The Science and Engineering of Materials, 6th ed., Cengage Learning, Stamford, CT. (2011), pp. 625-626.*)

Different polymers have variations of the stress-strain plot in Figure 6.23, as shown in Figure 6.24. Table 6.5 shows that polymers, such as PMMA and polystyrene (PS) tested at room temperature have a relatively high ultimate tensile strength for polymers, but a very small elongation to fracture of only 3%. These are hard and brittle materials. In these materials, the elastic portion of the stress-strain diagram is present, but there is not significant plastic strain. These materials fracture before yield, and no yield strength is indicated. The toughness of these materials is relatively low for polymers, because of the small strain to fracture the polymer. Both PMMA and PS have steric hindrance. Steric hindrance is the resistance to the sliding of LCMs provided by bulky side group molecules.

LDPE and PTFE have a low yield strength and a very large elongation to failure of 400 to 600%. In these materials, the stress-strain plot is dominated by plastic strain. These materials are soft and tough. Their toughness is relatively large for a polymer, because of the extensive plastic strain; however, the low

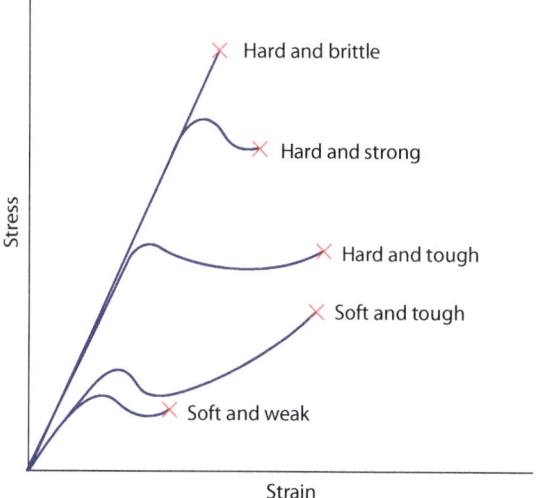

Figure 6.24 A schematic of stress (*y* axis)-strain (*x* axis) diagrams for various types of polymers.

Table 6.5 Yield Stress (σ_y), Ultimate Tensile Strength (σ_u), Strain at Yield (ε_y), and Engineering Strain to Fracture (ε_f) of some Polymers. These are typical values, but the actual values for a particular sample can vary significantly from these values.

Material	σ_y (MPa)	ε_y (%)	σ_u (MPa)	ε_f (%)
LDPE	8.5	20	13	600
HDPE	30	9	27	450
PP	32	12	33	650
PS	Does not yield		50	3
PVC	50	3	55	30
PTFE	13	62.5	23	400
PMMA	Does not yield		70	3
Nylon 6,6	57	25	80	170
Phenolformaldehyde	Does not yield		55	1
Polyester resin	Does not yield		45	1

Based on van Krevelen, D. W., Properties of Polymers, Correlations with Chemical Structure, Elsevier, Amsterdam (1972), p. 181.

yield and fracture strength limit the magnitude of their toughness. The LCMs in these two materials do not have steric hindrance because these are LCMs with only hydrogen atoms (LDPE) or fluorine atoms (PTFE) attached to the carbon atoms.

Polymers that have an elastic region, a relatively high yield stress, and a significant amount of plastic elongation to failure (such as 100 to 600%) are hard and tough. Polymers that have this form of stress-strain plot include HDPE, PP, and nylon-6,6. PVC is an example of a hard and strong polymer. It has a relatively high yield strength, but a small strain to fracture in comparison to that of HDPE.

Figure 6.16 shows that PMMA tested at 122°C is moderately ductile; however, if this same PMMA is tested at room temperature, it has only 3% elongation at fracture, as shown in Table 6.5. It is virtually brittle. We will cover the effect of temperature upon polymer properties in more detail in Chapter 9. In polymers such as PMMA at room temperature, the yield stress is higher than the fracture stress, because plastic strain by the sliding of LCMs is impeded by the large methyl and acrylic side groups on the LCM, as shown in Figure 2.19. To slide past each other at low temperatures, the LCMs require a larger stress than at high temperatures.

Phenolformaldehyde is a three-dimensional network polymer, and polyester resin has cross-links that form a three-dimensional network. In these polymers the long-chain molecules cannot slide over each other and yield, because covalent bonds connect the molecules in three dimensions. These are hard and brittle polymers.

6.7.9 Fracture and Ductility of Smooth Tensile Specimens

Figure 6.25 shows the types of failure observed in tensile tests of smooth metal specimens. The ductility of a fractured tensile specimen is determined by comparing the initial cross-sectional area (A_0) of the specimen to the final cross-sectional area at the location of fracture (A_f). One measure of ductility is

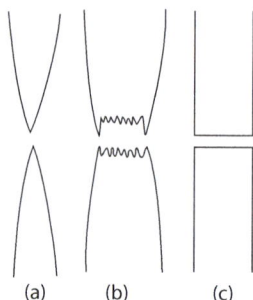

(a)   (b)   (c)

Figure 6.25   Smooth metal specimens may produce any of three fracture shapes during tensile tests: (a) from a ductile material, (b) from a moderately ductile material, and (c) from a brittle material.

the fractional-cross sectional area change at fracture (RA) expressed in percent (%RA), as shown in Equations 6.46a and 6.46b, respectively.

$$RA = \frac{A_0 - A_f}{A_0}$$

6.46a

$$\%RA = 100 \left(\frac{A_0 - A_f}{A_0} \right)\%$$

6.46b

Also, Equation 6.47 shows the true plastic strain to fracture (ε_{tf}), determined with Equation 6.45, which is a measure of the ductility to fracture.

$$\varepsilon_{tf} = \ln \frac{A_0}{A_f}$$

6.47

In ductile materials, as shown in Figure 6.25a, the reduction in cross-sectional area can approach 100%; the elongation can approach 700% as shown for low-density polyethylene in Table 6.5. In these materials, plastic strain usually begins at very low stresses with virtually no initial purely elastic region. Materials of this type include annealed high-purity soft metals such as lead, copper, aluminum, iron, and gold tested at room temperature. The onset of plastic strain in these pure annealed metals begins at a very low stress because dislocations move easily in pure well-annealed metals. Simple polymers such as polyethylene and polytetraflouroethylene (PTFE, Teflon™) tested at room temperature are nearly 100% ductile, because LCM sliding happens easily. An extremely ductile material fails when the cross-sectional area of the specimen goes to zero. We can observe this mode of failure by stretching a piece of chewed bubble gum.

Materials that are moderately ductile fail as shown in Figure 6.25b. These materials usually have initial elastic strain followed by plastic strain, as shown by the Figure 6.16 examples of low-carbon steel and the PMMA tested at 122°C. Other examples of moderately ductile materials at room temperature include aluminum, copper, and titanium alloys, as well as many polymers such as nylon and polycarbonate. A material with a 25% RA at fracture is considered a moderately ductile material.

Brittle materials have no significant permanent reduction in cross-sectional area after fracture, as shown in Figure 6.25c. You can break a piece of chalk and observe this type of fracture. The sapphire in Figure 6.16a is a brittle material. Brittle materials have insignificant plastic strain at fracture. Materials are brittle if the fracture stress is lower than the yield stress. In ionically bonded crystalline materials, such as NaCl, MgO, Al_2O_3, and ZrO_2, dislocation motion is difficult at room temperature, because of the large Burger's vector, and as a result they are brittle materials. In covalently bonded crystalline materials, such as silicon, diamond, and germanium, dislocation motion is difficult at room temperature because covalent bonds do not easily bend. The absence of dislocation motion in crystalline materials makes them brittle. Some metals, such as cast iron, are brittle at room temperature because they have a high volume fraction of a brittle phase, such as iron carbide, in the alloy. Other materials with brittle fracture include glass, chalk, concrete, magnesia, and silicon.

6.7.10 The Relationship between Yield Strength and Ductility

In general, the higher the strength of a given type of material, such as carbon steel, the lower its ductility. In Table 6.3, 1080 steel has a yield strength of 380 MPa and 25% *RA*, and 1020 steel has a yield strength of 295 MPa and 36.5% *RA*. Table 6.5 shows that PMMA does not yield, but its fracture strength is typically 70 MPa and its elongation to failure is 3%; in contrast, LDPE has a yield strength of only 8.5 MPa and an elongation to failure of 600%.

In Chapter 8, on engineering materials, we will discuss high-strength, low-alloy steels that develop their high strength by the addition of alloy elements to low-carbon steel, without significantly reducing the ductility of the low-carbon steel. Likewise, some polymers have increased strength while maintaining significant ductility, such as nylon 6,6, which has a typical yield strength of 57 MPa and an elongation at fracture of 170%.

6.7.11 Tensile and Compressive Strength of Brittle Ceramic Materials

The tensile and compressive fracture strength of some brittle ceramic materials is presented in Table 6.6. These materials do not yield, and there is no yield strength listed. Note in Table 6.6 that a ceramic material's compressive fracture strength can be much higher than its tensile fracture strength, by a factor

Table 6.6   The Compressive Strength and Tensile Strength of Some Selected Ceramics

Material	Compressive Strength MPa	Tensile Strength MPa
Alumina (85% dense)[1]	1620	125
Alumina (99.8% dense)[1]	2760	205
SiC (sintered)[2]	3864	173
Si_3N_4 (hot pressed)[2]	3450	552
Si_3N_4 (reaction bonded)[2]	1035	138
ZrO_2 (partially stabilized 9% MgO)[1]	1860	449
ZrO_2 (transformation toughened)[1]	1760	350

1   Guide to Engineering Materials Vol. 1(1) ASM, Metals Park, OH (1986), pp. 16, 64, and 65.

2   Askeland, D.R., Fulay, P.P., and Wright, W.J., The Science and Engineering of Materials, 6th ed., Cengage Learning, Stamford, CT. (2011), p. 575.

of 5 to 10. This is because in brittle ceramic materials, tensile fracture usually begins at some defect that forms a small crack; the fracture occurs when the small crack propagates through the material to form a large crack. For example, a smooth pane of glass is difficult to fracture by bending the glass, but if the glass is scribed with a cutter that makes a small scratch on the glass surface, then the glass is easily broken along the scribe line. Uniaxial compressive failure of brittle materials is either due to the propagation of shear cracks at approximately 45° to the compression axis, or the failure is due to tensile strains perpendicular to the compression axis resulting from Poisson's ratio. Brittle materials can withstand a very high hydrostatic compressive stress.

We will cover the processing treatments, such as hot pressing, for ceramic materials in Chapter 13, on materials processing, and we will discuss partially stabilized and transformation-toughened ZrO_2 in Chapter 11, on fracture and fatigue.

6.7.12   Strength Tests for Brittle Materials and Composite Materials

Compression tests can be conducted on ceramics, brittle polymers, composite materials, and wood. However, it is difficult conducting a tension test using a standard specimen like the one shown in Figure 6.15, because the specimens tend to break at the grip. If a brittle specimen is threaded, it will often break at the thread, and this yields an invalid test result. The specimen must break in the uniform area to give a valid test result. Other gripping designs for tension tests have been developed that do not include threads, but brittle specimens still tend to break at any gripping point because of stress concentrations caused by the grips. The dog-bone grip shown in Figure 6.26 is designed primarily for brittle specimens, such as concrete. The grip and specimen are designed so that fracture should occur at the minimum

Figure 6.26   Dog-bone grips for tensile testing of brittle materials with a concrete specimen. (*Photo by C.M. Gilmore*)

cross-sectional area; however, even with this design fracture often occurs at the point where the specimen first contacts the grip. Only the tensile fracture strength is determined with this test. The fracture load and minimum cross-sectional area are measured and used in Equation 6.2 to determine the fracture strength. However, because the cross-sectional area of the specimen is not uniform, the stresses and strains in the specimen are not uniform. Therefore, many of the properties discussed in this chapter cannot be determined with this test. For example, the elastic modulus cannot be determined, because the specimen strain cannot be determined.

The three-point bend test shown in Figure 6.27 avoids the problem of gripping brittle ceramic materials. In a bend test, the bottom surface of a rectangular cross-sectional beam has the maximum tensile stress (σ_b), given by Equation 6.48, where F is the load, L is the spacing between the supports, w is the width of the specimen, and h is the specimen thickness.

$$\sigma_b = \frac{3FL}{2wh^2} \qquad\qquad \textbf{6.48}$$

The top surface of the specimen has the maximum compressive stress, and this stress is of equal magnitude but opposite sign to the tensile stress on the bottom surface. In a bend test, a brittle material always fractures in tension at the bottom surface, because of the much lower tensile strength than compressive strength, as shown in Table 6.6. The fracture strength resulting from a three-point bend test is the flexural strength, which is calculated by inserting the load at fracture into Equation 6.48. The elastic modulus determined from the three-point bend test is the flexural modulus (E_b), given by Equation 6.49:

$$E_b = \frac{FL}{4wh^3\delta} \qquad\qquad \textbf{6.49}$$

where δ is the deflection at the center of the beam. Equations 6.48 and 6.49 are developed in strength or mechanics of materials courses and are not derived here. Equations for a beam with a circular cross section can also be developed. There is also a four-point bend test where the load is applied at two locations equidistant from the two support points.

In uniaxial composite materials and wood, flexural strength and modulus are often determined rather than a normal tensile strength and modulus. If a uniaxial composite material such as fiberglass or wood is tested in a standard tension test with the fibers parallel to the tensile axis, the specimens fracture at the grip along the interface between the fibers and the matrix, and the strong fibers pull out of the grip. Chapter 12, on composite materials, provides a more complete discussion of the fracture of uniaxial composite materials and wood.

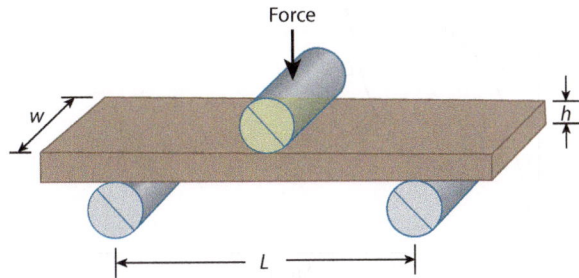

Figure 6.27   A three-point bend-test configuration. (*Based on Askeland, D.R., Fulay, P.P., and Wright, W.J., The Science and Engineering of Materials, 6th ed., Cengage Learning, Stamford, CT. (2011), p. 219.*)

6.8 SUPERELASTICITY

In Chapter 5, we discussed materials such as TiNi that have a reversible martensitic transformation. These materials exhibit the shape-memory effect, and in addition they are **superelastic** when deformed above the martensite start (M_s) temperature. *Superelastic* means that very large strains such as 10% are recovered. Superelasticity is also called pseudoelasticity because it is not conventional elasticity due to the atoms returning to their equilibrium interatomic position. In materials with a reversible martensitic transformation, such as TiNi, the superelasticity is due to the recovery of the distortion of the crystal lattice that occurs during the martensitic transformation. In a conventional high-strength material, the recovered strain is just fractions of a percent. The superelastic effect is what makes these materials ideal for orthodontic applications.

A schematic of a stress-strain diagram for a superelastic alloy deformed above the M_s temperature where the alloy is in the austenitic phase is shown in Figure 6.28. In the initial part of the diagram, up to point A, the austenitic phase exhibits normal linear elastic behavior. In materials that undergo a martensitic transformation, the high-temperature phase is often called austenite in analogy with the designation in the iron-carbon system. At higher stresses the stress-strain diagram deviates from linearity. However, this is not a yield point; the applied stress induces the austenite to transform to martensite through the distortion shown in Figure 5.26b. Since the temperature is above the M_s temperature, the Gibbs free energy change alone is not sufficient to drive the transformation into martensite. However, if the applied force assists the distortions shown in Figure 5.26b, then the martensitic transformation can be induced above the M_s temperature. The extent of the initial normal elastic region depends upon the ambient temperature relative to the M_s temperature. The higher the ambient temperature above the M_s temperature, the greater the extent of the austenite elastic region, and the greater is the stress necessary to induce the transformation to martensite. The region of the stress-strain diagram labeled B is where the stress-induced transformation from austenite to martensite occurs, and the extent of the strain in this

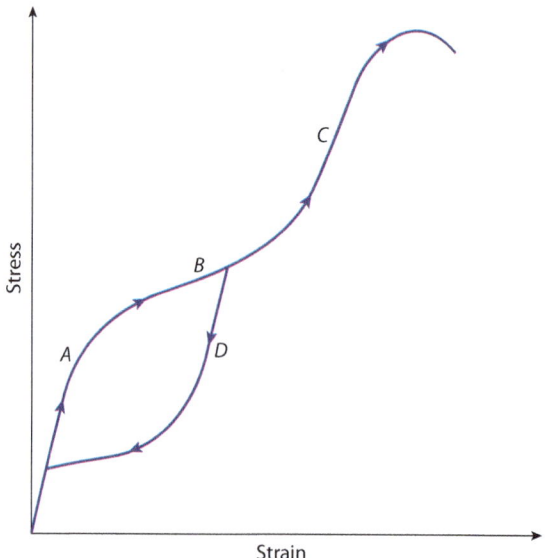

Figure 6.28 The stress-strain diagram for a superelastic material at a temperature above the martensite start temperature.

region is typically 0.1. If the material is unloaded in region B, the martensite transforms back to austenite along the path D, and the distortion in the martensite is recovered. All of the strain is recovered as long as the temperature is above the M_s and as long as the martensite is not plastically deformed. The strain recovery is superelastic.

At the end of the superelastic region (B) the austenite is all transformed to martensite, and the distortions are all in the most favorable orientation possible that permits continuity of the material. At higher stresses than in the stress-induced martensite region (B) there is region (C), where the martensitic phase is elastically deformed. At even higher stresses than region C the stress-strain curve is nonlinear where the martensite phase is plastically deformed.

Superelastic materials, such as TiNi, are ideal for orthodontic applications. The wires are installed and stretched at temperatures above M_s so that the wire is strained in the austenitic phase. When the teeth start to move in response to the stress in the wires, the superelastic strain recovery allows the stress to continue to pull the teeth into their ultimate positions. In a normal elastic wire, such as stainless-steel wire, a small movement of the teeth results in the stress being reduced quickly toward zero.

6.9 HARDNESS, MICROHARDNESS, AND NANOHARDNESS

A hardness test measures the material resistance to penetration by an indenter, as shown in Figure 2.12. Hardness tests are the most frequently performed tests on materials. Hardness tests are relatively simple and easy to run, the part is not destroyed, only a small indentation is made, and the hardness value gives some indication of the mechanical properties of the material. Hardness tests are used in quality control to assure that the material used to produce a part conforms to specifications. Table 6.7 presents some of the different hardness measurement systems, the indenter type, the load (F), and typical applications.

Table 6.7 Various Hardness Tests Showing the Indenter, the Load, and Typical Applications

Test	Indenter	Load	Application
Brinell	10-mm ball	3000 kg	Cast iron and steel
Brinell	10-mm ball	500 kg	Nonferrous alloys
Rockwell A	Brale	60 kg	Very hard materials
Rockwell B	1.59 mm ball	100 kg	Brass, low-strength steel
Rockwell C	Brale	150 kg	High-strength steel
Rockwell D	Brale	100 kg	High-strength steel
Rockwell E	3.18 mm ball	100 kg	Very soft materials
Rockwell F	1.59 mm ball	60 kg	Aluminum, soft materials
Vickers	Diamond square pyramid	10 kg	All materials
Knoop	Diamond elongated pyramid	500 g	All materials

Based on data from Askeland, D.R., Fulay, P.P., and Wright, W.J., The Science and Engineering of Materials, 6th ed., Cengage Learning, Stamford, CT. (2011), p. 222.

The indenter is either a small hard steel or tungsten carbide ball, or a pointed diamond. The Rockwell brale and the Vickers and Knoop indenters are all pointed diamonds. The Rockwell balls are steel, and the Brinell balls are steel for soft materials and tungsten carbide for hard materials.

In the Brinell (*HB*), Vickers (*HV*) , and Knoop (*HK*) hardness tests after withdrawl of the indenter, the dimensions of the indentation are measured with a microscope using length scales printed on the microscope lens. The shape of the indentation and the dimension measured are shown in Table 6.8 for the *HB*, *HV*, and *HK*.

Equations 6.50 to 6.52 give *HB, HV,* and *HK* in terms of the dimensions shown in Table 6.8, the diameter of the Brinell ball (*D*), and the applied load (*F*).

$$HB = \frac{2F}{\pi D[D - (D^2 - d^2)^{0.5}]} \qquad \textbf{6.50}$$

$$HV = \frac{1.854F}{[0.5(d_1 + d_2)]^2} \qquad \textbf{6.51}$$

$$HK = \frac{14.2F}{l^2} \qquad \textbf{6.52}$$

The dimensions of *HB*, *HK*, and *HV* are force per unit area. Originally the hardness units utilized were kilograms force per square millimeter. However, the recent trend is to use SI units of either newtons per square meter or newtons per square millimeter. The units in any quote of hardness values should be clearly stated.

In the Rockwell tests, the amount of penetration of the indenter into the material with a fixed load is automatically measured by the machine, and the hardness is obtained by reading a dial indicator. The simplicity of reading a dial indicator makes the Rockwell test appropriate for production testing. Table 6.7 shows some of the different Rockwell hardness scales. The different Rockwell hardness scales are necessary because the hardness of different materials spans a large range. For a soft material such as polyethylene, a large indenter such as a $\frac{1}{8}$ inch steel ball is utilized in the Rockwell E scale. For a quenched and tempered high-strength steel, the diamond brale indenter with a 150-kg force is utilized for the Rockwell C scale.

Table 6.8   The Shape and Dimension Designations of Hardness Indentations for *HB, HV,* and *HK*

Test	Shape of Indentation Top View
Brinell	
Vickers microhardness	
Knoop microhardness	

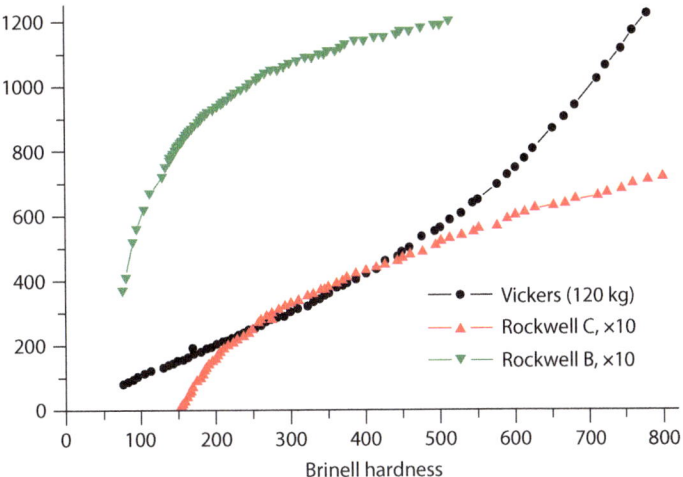

Figure 6.29   A comparison of various hardness scales using the Brinell hardness as a standard. (*Based on http://en.wikipedia.org/wiki/Hardness_comparison*)

Figure 6.29 compares some of the different hardness scales using Brinell as a standard. The approximate Brinell hardness of some different materials are diamond—6000, tool steels—800, high-strength steels—200, structural steel—150, copper and aluminum alloys—60, and plastics—20.

For some materials, the hardness and the ultimate tensile strength (σ_u) are approximately correlated. For example, the approximate ultimate tensile strength in MPa of steels is related to the Rockwell B hardness (R_B) of a material through Equation 6.53:

$$\sigma_u \cong 3.45\, R_B\, \text{MPa} \qquad\qquad \textbf{6.53}$$

Similar equations exist for the other types of hardness tests. A hardness test gives some indication of the expected ultimate tensile strength of a material when a tensile test cannot be run.

A microhardness test utilizes a micron-size (10^{-6} m) indentation. Microscopes are necessary to observe these small indentations. Examples of microhardness tests are the Vickers and Knoop tests when low loads (F) such as 2 N or less are utilized. Microhardness tests can be utilized to determine the hardness of different areas on the surface of a specimen, such as different grains or phases in a specimen if each indentation can be completely contained within that area.

With the development of scanning tunneling microscopes and atomic-force microscopes, it is now possible to observe indentations of nanometer dimensions. Nanometer-scale indenters have been created to produce extremely small but observable indentations. The determination of nanohardness is an area of research at present, but it is not used for the standard evaluation of materials.

Summary

- The main types of force that can be applied to a material are tension, compression (which is the negative of tension), and shear. A tensile force applied to a block elongates the block of material in the direction of the tensile force. A shear force applied to a surface of a block changes the angles at the corners of the block.

- If the uniaxial tensile or compressive force is divided by the specimen's original cross-sectional area normal to the force, the result is the normal stress.
- The axial strain is the deformation produced by an axial force in the axial direction divided by the original length of the specimen.
- Poisson's ratio is the negative of the strain in a direction perpendicular to the loading axis divided by the strain in the direction of the loading.
- The true stress is the applied force divided by the cross-sectional area measured at the instant that the force is measured.
- An increment of true strain is defined as an increment in length change divided by the length at the instant of measurement. The total true strain of a specimen that experiences a length change from l_0 to l is then obtained by integrating the increment of true strain from l_0 to l.
- A hydrostatic state of stress is when the stresses in all directions are all equal. A compressive hydrostatic stress is a pressure.
- Hydrostatic stresses produce a volumetric strain equal to the change in volume divided by the original volume.
- Shear deformation occurs when we apply a force parallel to the top surface of a specimen, and the top of the specimen is displaced relative to the bottom. The shear strain is defined as the relative displacement of the top of the specimen divided by the height.
- Shear stress is defined as the shear force divided by the area over which the shear force acts.
- When a stress is applied to a material, a corresponding strain is created. If the stress is reduced to 0 and the material returns to its original dimensions, then the strain was elastic.
- In anelastic materials, the unloading stress-strain plot does not follow the loading stress-strain plot, and upon unloading the strain does not go to 0 instantly. However, the strain does return to 0 with sufficient time, with 0 stress.
- Elastic strain in materials, except for elastomers, is due to the stretching of interatomic bonds. The elastic modulus for these materials, except for elastomers, is proportional to the curvature of the interatomic potential.
- The elastic strain in elastomers, such as rubber, is due to the uncoiling of long-chain molecules. The elastic modulus of rubber is a function of the change in entropy resulting from the uncoiling of these long-chain molecules.
- The area under the stress-strain diagram is the strain energy per unit volume of material needed produce the strain the material.
- Plastic strain is permanent strain that remains in a material after the applied stress is reduced to 0.
- Plastic strain in thermoplastic amorphous polymers without cross-linking occurs when the van der Waals bonds between the long-chain molecules are broken by an applied stress, freeing the long-chain molecules to slide past each other.
- In plastic strain of a semicrystalline thermoplastic polymer, the amorphous areas between the crystalline areas are deformed first and the long-chain molecules are oriented in the tensile direction. Then the crystallites are rotated into the tensile direction. At higher plastic strain the crystallites are separated.
- Plastic strain of silica glass at temperatures above the glass transition temperature occurs because the bonds between the silicon atoms and the oxygen atoms are continuously broken, and if there is an applied stress the atoms can move and then bond with different atoms.

- In crystalline metals at low temperatures relative to their melting temperatures, plastic strain results primarily from the motion of dislocations.
- The shear stress to move a unit length of dislocation is directly proportional to the magnitude of the Burger's vector. In ionic crystals, the critical stress to move a dislocation is greater than that observed in metals, because the Burger's vector in an ionic material is greater than in a metal.
- In covalently bonded crystalline solids, there is resistance to change in the angle of bonds. Shuffling atoms around a dislocation in a covalently bonded crystalline solid requires changes in the angles between atoms, and this shuffling requires more strain energy in covalently bonded solids than it does in metals.
- Ionic and covalently bonded materials plastically deform by dislocation motion at elevated temperatures, where the thermal vibrations help to activate the dislocation motion.
- In crystals, dislocation motion on a particular close-packed plane begins at the constant critical resolved shear stress. A constant critical resolved shear stress for dislocation motion results in a tensile stress where dislocation motion initiates (the proportional limit) for a single crystal that is a function of the orientation of the close-packed plane and direction.
- The proportional limit in a polycrystalline metal is equal to approximately 3.1 times the critical resolved shear stress for FCC metals and 2.75 times the critical resolved shear stress for BCC metals.
- The yield stress is the stress where a line drawn parallel to the initial slope of the stress-strain diagram with a strain offset of 0.002 intersects the uniaxial tensile stress-strain diagram.
- The modulus of resilience is the maximum strain energy that a material can absorb and then fully return when unloaded.
- In the impact resilience test for rubber and polymers, a steel ball is dropped from an initial height h_1 onto a flat specimen of the material. The steel ball bounces off the material to a height h_2. The impact resilience is the ratio h_2/h_1.
- The maximum engineering strength in tension of a material is the ultimate tensile strength.
- The toughness of a material is the strain energy per unit volume required to fracture the specimen in a tensile test. The toughness is determined by integrating the area under the stress-strain diagram to fracture.
- One measure of ductility is the percent cross-sectional area change at fracture.
- Another measure of ductility is the true plastic strain to fracture determined as the log of the initial cross-sectional area divided by the final cross-sectional area at the fracture location.
- In very ductile materials, the change in cross-sectional area approaches 100%. For moderately ductile materials, the change in sectional area is approximately 25%. In brittle materials, the change in cross-sectional area is approximately 0.
- In general, the higher the strength of a given material the lower its ductility.
- In brittle ceramic materials, the compressive fracture strength is five to ten times higher than the tensile fracture strength.
- Tensile mechanical properties are obtained for brittle and composite materials with dog-bone-shaped specimens and three- or four-point bend tests.
- Superelasticity in materials with a reversible martensitic transformation, such as TiNi, is due to the recovery of the distortion of the crystal lattice that occurs during the martensitic transformation.

- The hardness of a material is the resistance of the material to penetration by an indenter forced into the material with a fixed load.
- For some materials, there is an approximate correlation between hardness and the ultimate tensile strength.
- A microhardness test is performed by a micron-size (10^{-6} m) indenter that creates a micron-size indentation.
- With the development of scanning tunneling microscopes and atomic-force microscopes, we can now observe indentations of nanometer dimensions (10^{-9} m), and indenters have been developed to this very small scale. The determination of nanohardness is an area of research at present, but it is not used for the standard valuation of materials.

Supplemental Reading Subjects and Authors

Full references are listed at the end of the book.

General:	*Askeland, Fulay, and Wright*
Mechanical behavior of materials:	*Courtney; Dieter; Hertzberg*
Mechanical behavior of ceramics:	*Carter and Norton*
Mechanical behavior of polymers:	*Osswald and Menges; Kinney; McCrum, Buckley, and Bucknall; Rubenstein and Colby; Schultz; van Krevelen; Winding and Hiatt*
Superelastic TiNi:	*Wang*

Homework

Concept Questions

1. The tensile force divided by the original cross-sectional area of a specimen is the engineering normal _____

2. The change in length resulting from an applied force on a specimen divided by the original length is the engineering _____.

3. The negative of the ratio of the strain transverse to an applied force divided by the strain parallel to the applied force is equal to the _____ ratio.

4. The tensile force divided by the actual cross-sectional area of a specimen is the _____ stress.

5. If an applied stress is equal in three perpendicular directions, the stress is _____.

6. A(n) _____ material has properties that are equal in three perpendicular directions.

7. In a tensile specimen with uniform strain, the true strain is equal to the natural log of 1 plus the _____ _____.

8. Recovered strain is _____ strain.

9. Permanent strain is _____ strain.

10. In an isotropic material subject to a hydrostatic stress, the volumetric strain is equal to _____ times the strain in the x direction.

11. A shear strain causes changes in _____ in a specimen.

12. The linear relationship between stress and elastic strain is known as _____ law.

13. _____ modulus is equal to the ratio of the applied normal stress to the elastic strain.

14. In a nonlinear elastic material, the ratio of the applied stress to the total strain is the _____ modulus.

15. In a nonlinear elastic material, the ratio of the increment in applied stress to the increment in strain measured at the total strain is the _____ modulus.

16. The ratio of the hydrostatic stress to the elastic volumetric strain is equal to the _____ modulus.

17. A material that requires time to fully recover strain after the stress is reduced to 0 is _____.

18. In crystalline metals, plastic strain at low temperature is due to _____ motion.

19. In thermoplastic polymers, plastic strain is due to the sliding of _____ _____ _____.

20. In glass, plastic strain is due to the _____ and reformation of atomic bonds.

21. In a crystalline metal at low temperature, the magnitude of the plastic strain is equal to the mobile dislocation density times the Burger's vector times the average _____ moved by the dislocations.

22. In the tensile test of single crystals, the proportional limit stress depends upon the angle of the slip plane and the slip _____ relative to the tensile axis.

23. In the tensile test of single crystals, the proportional limit stress is not constant; the _____ _____ _____ _____ is constant.

24. The elastic modulus of a material, except for elastomers, is proportional to the _____ of the interatomic potential.

25. The tensile stress where plastic strain begins is equal to the _____ limit.

26. The tensile stress where plastic strain is equal to 0.002 is equal to the _____ stress.

27. The modulus of _____ is the maximum strain energy that a material can absorb and then fully return when unloaded.

28. At the ultimate tensile strength, the slope of the engineering stress-strain diagram is equal to _____.

29. The area under the stress-strain curve to fracture a material is equal to the _____.

30. The increase in specimen strength with plastic strain is _____ hardening.

31. The ratio of the change in the cross-sectional area at the location of fracture to the original cross-sectional area is equal to the _____.

32. The logarithm of the ratio of the initial cross-sectional area to the area at the location of fracture of a specimen is equal to the _____ _____ to fracture.

33. Thermosetting polymers are more brittle than thermoplastic polymers, because the long-chain molecules are _____.

34. For the low-carbon steel in Figure 6.16, at what strain would a neck start to form in the specimen?

35. The tensile yield stress in ductile metals and polymers is approximately _____% of the compressive yield strength.

36. In crystalline metals and ceramics, elastic strain is due to the stretching or compression of _____ bonds resulting from changes in the lattice parameter.

37. The _____ strength of a brittle or composite material is the maximum tensile stress determined from a three-point bend test.

38. Hardness is a measure of the resistance of a material to _____ by an indenter.

Engineer in Training–Style Questions

1. During loading in a stress-strain test of a material, we observe that the plot is not linear and has no discontinuities. We can assume that:
    (a)   The material is nonlinear elastic.
    (b)   The material is anelastic.
    (c)   There is elastic and plastic strain.
    (d)   None of the above can be assumed.

2. Which of the following materials would you expect to have the lowest Young's modulus?
    (a)   Aluminum metals
    (b)   Alumina ceramic
    (c)   Polyvinylchloride
    (d)   Rubber

3. At the yield stress in a metal,
    (a)   Dislocation motion starts.
    (b)   The plastic strain is equal to 0.002.
    (c)   Plastic strain begins.
    (d)   The specimen starts to neck.

4. Which of the following pure annealed materials would you expect to have the highest yield stress?
    (a)   FCC copper
    (b)   BCC iron
    (c)   HCP magnesium
    (d)   LDPE

5. Which of the following materials would you expect to be isotropic?
    (a)   A casting of aluminum
    (b)   Aluminum foil
    (c)   Aluminum wire
    (d)   A polyethylene sheet 0.01 cm thick

6. One reason that dislocation motion is difficult in covalently bonded materials such as diamond is that:
    (a)   The interatomic distance is large.
    (b)   Covalent bonds resist changes in angle.
    (c)   The curvature of the interatomic potential is low.
    (d)   The interatomic potential is shallow.

7. Plastic strain in silica glass at a temperature just above the glass transition temperature is due to:
    (a)   Dislocation motion
    (b)   The breaking and reforming of silicon-oxygen bonds with different neighbors
    (c)   The sliding of different silicone long-chain molecules
    (d)   Glass is always a liquid and flows like a liquid at any temperature.

8. What is the weight percent carbon in a 1040 steel?
    (a)   10%
    (b)   4%
    (c)   0.40%
    (d)   1%

9. Which of the following would not contribute to a high modulus of resilience?
    (a)   A high ductility
    (b)   A low elastic modulus
    (c)   A high yield stress
    (d)   A high yield strain

10. At the ultimate tensile strength,
    (a)   The true stress is at its maximum.
    (b)   The specimen always fractures.
    (c)   The maximum load-carrying capacity is experienced.
    (d)   The material yields.

11. The toughness of a material is:
    (a)   The strain energy per unit volume needed to fracture the material
    (b)   The maximum engineering stress
    (c)   The true stress at fracture
    (d)   The ductility times the stress at fracture

12. The ratio of the tensile fracture strength of ceramics to the compressive fracture strength is typically:
    (a)   0.1 to 0.2
    (b)   5 to 10
    (c)   1
    (d)   0.5

13. Which of the following tests would probably not be successful for a brittle ceramic material?
    (a)   Three-point bend test
    (b)   Tensile test with a standard tensile-test specimen, as shown in Figure 6.15a
    (c)   Tensile test with a dog-bone specimen, as shown in Figure 6.26
    (d)   Compression test with a cylinder of ceramic

14. If TiNi is deformed 5% at a temperature above the martensite start temperature, and then the strain is fully recovered upon reduction of the stress to 0, this material is
    (a)   Super or pseudoelastic
    (b)   Anelastic
    (c)   Elastic
    (d)   Plastic

15. If TiNi is deformed 5% at a temperature below the martensite finish temperature, and then the force is reduced to 0, how would you expect the material to respond?
    (a)   Superelastically with all strain recovered
    (b)   With a large elastic strain and a small amount of plastic strain
    (c)   With a small elastic strain and a large plastic strain
    (d)   With all plastic strain

16. In the Brinell hardness test, the indenter is steel or tungsten, and the shape is a:
    (a)   sphere with a 10-mm diameter
    (b)   diamond pyramid
    (c)   diamond cone
    (d)   steel cone

Problems

Problem 6.1: A circular steel rod 0.005 m in radius and 4.000 m long is subject to an axial force of 0.250×10^5 N. We observe that the rod elongates 0.616×10^{-2} m.

Calculate the tensile stress and the axial strain in the rod.

Problem 6.2: For the material in Problem 6.1:

(a) Calculate the change in radius of the specimen while subject to the stress. Assume that the deformation of this specimen is elastic and that Poisson's ratio is equal to 0.29.

(b) Is there any significant difference between the true stress and the engineering stress calculated in Problem 6.1?

(c) Calculate the true strain for this application and compare it to the engineering strain calculated in Problem 6.1.

Problem 6.3: Compare the engineering and true secant elastic moduli for the natural rubber in Example Problem 6.2 at an engineering strain of 6.0. Assume that the deformation is all elastic.

Problem 6.4: A stress of 210 MPa is applied to a low-carbon steel with an elastic modulus of 211 GPa. After the stress is applied and then removed, we note that the permanent strain at 0 stress is 0.1. What was the total strain when the stress was equal to 210 MPa?

Problem 6.5: Compare the true and engineering strain energy density absorbed in the natural rubber in Example Problem 6.2 at an engineering strain of 6.00. Also compare the strain energy absorbed in the natural rubber to that absorbed in the steel specimen in Example Problem 6.5. Note that this stress is a very high one for the steel. You should find that the rubber absorbs much more strain energy per unit volume than the steel does. Explain why rubber can absorb so much energy. For this calculation assume that the natural rubber stress-strain diagram is linear elastic up to the engineering strain of 6.00.

Problem 6.6: An iron specimen is plastically deformed in shear by 1%, and it has a dislocation density of 1×10^{14} m/m³. Assume that the dislocation density did not change in the 1% strain of this specimen, the Burger's vector (**b**) is $\frac{a}{2}[1\bar{1}1]$, the slip plane is (110), the shear stress is applied to the (110) plane, and the lattice parameter of the BCC iron is 0.286 nm.

(a) Calculate the magnitude of the Burger's vector for these dislocations in iron.

(b) Calculate the average distance moved by the mobile dislocations as a result of the 1% shear strain.

Problem 6.7: A copper specimen is stressed in tension to 76×10^6 Pa and a total strain of 0.01 m/m; then the specimen is unloaded. We observe after the test that the deformation was uniform. Determine the expected recovered elastic strain and the permanent plastic strain.

Problem 6.8: Sketch possible interatomic potentials as a function of interatomic separation for two materials A and B. Material A has a high melting temperature and vaporization temperature and a high elastic modulus. Material B has a low melting and vaporization temperature and a low elastic modulus.

Problem 6.9: One type of strain gauge consists of a long thin wire or foil that is folded multiple times, as shown in Figure 6.30. The strain gauge is embedded in an adhesive tape so that the gauge can be attached to a specimen. The wire strains the same amount as the specimen does. Determine the relationship between the change in resistance ΔR and the strain (ε). Assume that the resistivity (ρ) of the wire is constant over this strain range and that the material has a Poisson's ratio of v.

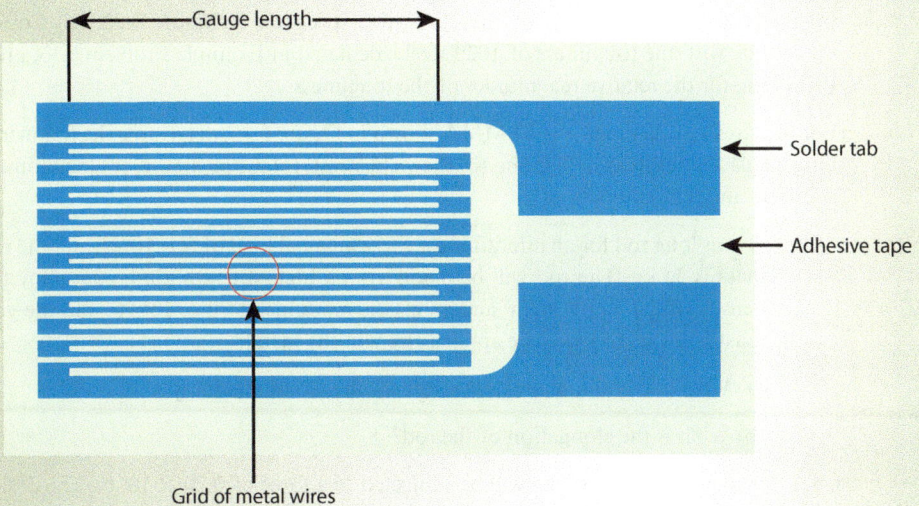

Figure 6.30 A schematic of an SR-4 strain gauge, which consists of a metal wire grid bonded to an adhesive tape that attaches to a specimen. There are solder tabs at the ends of the wire for connection to an Ohm meter. This device is named after its inventors: Simmons, Ruge, and two assistants. *(H.Pollok, „Umwertung der Skalen" ("Conversion of Scales"), Qualität und Zuverlässigkeit, Ausgabe 4/2008)*

Problem 6.10: The critical resolved shear stress for high-purity FCC copper is 0.63 MPa. A polycrystalline tensile test specimen is made from this copper material. What is the predicted tensile proportional limit stress for this specimen?

Problem 6.11: A single crystal of BCC niobium is oriented in a tensile test so that the [001] crystal direction is parallel to the load axis. The critical resolved shear stress for niobium is 20×10^6 Pa for dislocation motion on the {011} type planes and in the <111> type directions at 295 K. What is the tensile proportional limit stress for this single crystal at 295 K? Assume that the slip occurs on the {011} type planes and in the <111> type directions.

Problem 6.12: Calculate the resilience of 1040 steel listed in Table 6.3. Assume that the elastic modulus of this steel is the same as that of iron.

Problem 6.13: For the 1045 steel shown in Figure 6.22, determine the elastic modulus, the upper yield point, and the lower yield point.

Problem 6.14: For a specimen of 1020 steel with the stress-strain diagram shown in Figure 6.17, calculate the maximum load-carrying capacity for a standard tensile test specimen as shown in Figure 6.15a.

Problem 6.15: Estimate the elastic and plastic strain at the ultimate tensile strength in the low-carbon steel specimen in Figure 6.16.

Problem 6.16: A 1040 steel specimen has a yield stress of 350 MPa, an ultimate tensile strength of 515 MPa, and an elastic modulus of 211 GPa. The specimen is stressed to 400 MPa. What is the elastic strain energy density in this specimen?

Problem 6.17: A steel rod with an elastic modulus of 211×10^9 Pa that is loaded in tension must be able to absorb and return an energy density of 1×10^6 J/m^3 without yielding. What is the minimum possible yield strength for this steel rod?

Problem 6.18: (a) Determine the toughness of the 1045 steel specimen in Figure 6.22. (b) Compare this magnitude with the toughness of 1020 steel calculated in Example Problem 6.13, and explain the reason for the relative magnitudes of the toughness.

Problem 6.19: For the stress-strain diagram of PMMA shown in Figure 6.16, determine the following: (a) the tangent elastic modulus at 0 strain, (b) the yield stress, (c) the resilience, (d) the ultimate tensile strength, and (e) the toughness.

Design Problem 6.1: A 1-m-long rod for an aircraft must carry an axial load of 5×10^6 N, and the rod design mass is 54 kg. The rod will be made from a high-strength aluminum alloy that has a density of 2.7×10^3 kg/m^3 and an elastic modulus of 70×10^9 Pa, and the yield stress must be equal to twice the applied stress in the metal.

(a) What is the minimum yield stress for the aluminum alloy?

(b) What is the elongation of the rod?

Design Problem 6.2: A steel rod 0.50 m long will be subjected to a force of 0.20×10^6 N. The design of the rod dictates that the rod cannot elongate more than 0.001 m, must have the minimum weight possible, and must behave elastically for this load. The elastic modulus of steel is 211×10^9 Pa.

(a) What is the maximum allowable engineering strain in the rod?

(b) What is the minimum possible proportional limit for this steel?

(c) What is the required cross-sectional area for this rod?

Photographs of a Damascus-style steel knife. Damascus steel was produced in the Middle East from about 1100 to 1700 CE. Damascus steel is resilient, strong, and tough, and a sword made of it holds a sharp edge. The technique for making Damascus steel originated in India in approximately 500 BCE, where it was called wootz steel. The Indians started with wrought iron produced from blooms that were very low in carbon, as discussed in Section 1.1. The wrought iron was sealed in a small ceramic crucible about the size of a rice bowl, along with specific organic materials such as wood and leaves that provided carbon. The sealed crucible was placed in a furnace capable of melting the iron and carbon. The result was a high-quality uniform carbon steel. The iron makers in Damascus adopted this technique to produce Damascus steel. However, the carbon-source additions to the crucible were kept secret by the iron makers. Unfortunately, the process to produce Damascus steel has been lost, and modern attempts to reproduce it have failed. The steel has a distinctive pattern of banding that is similar to the process of pattern welding, where a steel billet is forged into thin strips that are folded back on each other and welded together by heat and more forging. Pattern welding creates many thin layers of steel and a pattern due to different exposed layers of steel. Interestingly, carbon nanotubes are found in Damascus steel swords.

Kondor83 / Shutterstock.com

The goals of this chapter are to understand

- The way metals are strengthened
- The way the ductility of strain-hardened metals is increased by annealing
- The way yield strength in metals relates to dislocation density, grain size, dispersion content, and precipitates
- Strengthening brittle ceramics
- Whiskers and their strength
- Strengthening polymers by chain molecular mass, crystallinity, and chain orientation
- Strengthening polymers by modifying polymer chemistry
- Three-dimensional polymers
- Strengthening polymers by cross-linking
- Thermosetting and thermoplastic elastomers

Chapter 7

Making Strong Materials

7.1 INTRODUCTION

The yield stress of high-purity annealed bulk metals, such as copper, aluminum, and iron, or of the polymer polyethylene (PE) is too low for high-performance structural applications. Processing and chemical modifications can significantly increase the yield stress of these materials. The tensile yield strength of annealed commercial-purity aluminum is only 34.5 MPa; however, aircraft-grade aluminum alloys have a yield strength of over 500 MPa. Bulk low-density PE, the material that is in milk bottles, has an ultimate tensile strength of approximately 13 MPa; however, PE can also be processed into a fiber with an ultimate tensile strength of 3300 MPa. This is stronger than some high-strength steels. In this chapter we will learn the techniques that are utilized to produce these strong solids.

7.2 STRENGTHENING METALS

In crystalline metals, plastic deformation at temperatures below half of the melting temperature is due to dislocation motion. Dislocation motion is easy in high-purity single crystals of metals such as copper, aluminum, and iron that have a low dislocation density, because there are few obstacles to dislocation motion. Easy dislocation motion results in a metal with a low yield strength. All bulk metals have dislocations that are introduced during the solidification process. To produce a bulk metal with a high yield strength, it is necessary to introduce obstacles to the motion of the dislocations present in the metal.

7.2.1   Strain Hardening Metals

Figures 6.16 and 6.17 show that metals strain-harden when plastically strained. Strain hardening is the increase in the yield strength of a metal when it is plastically strained. When a metal is plastically strained, it is observed that the dislocation density in the metal increases, as shown in Figure 7.1; the dislocations are the dark lines. During plastic strain, dislocations are created at defects in the crystal, such

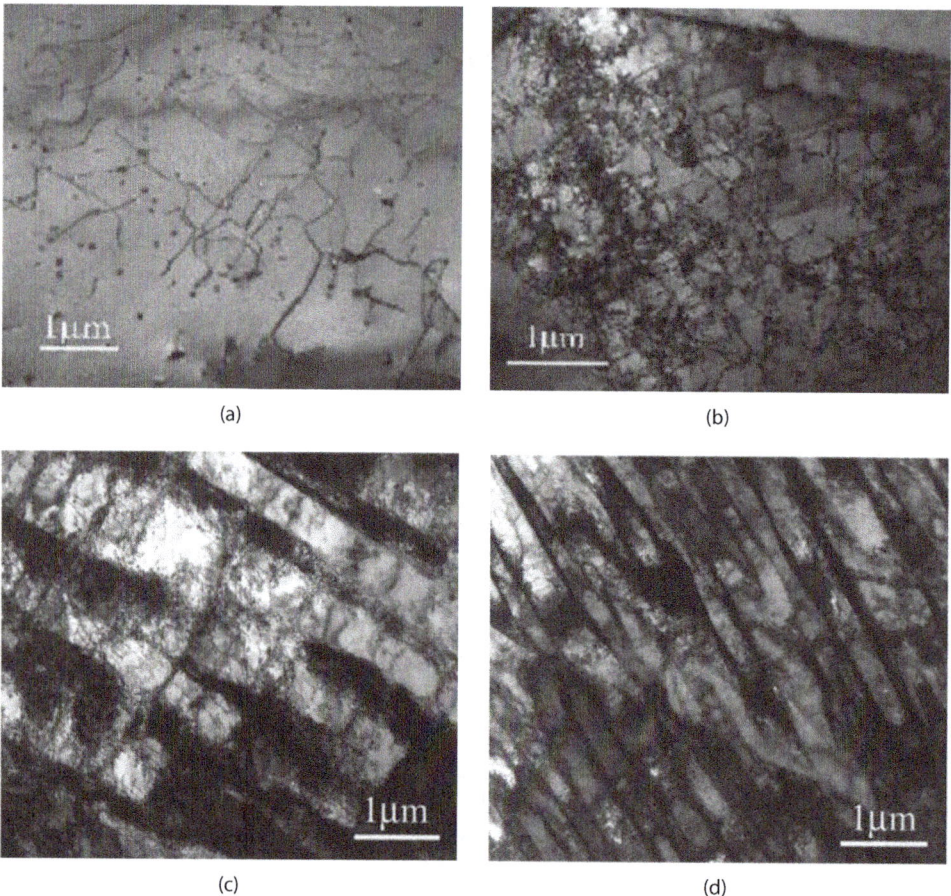

(a)   (b)

(c)   (d)

Figure 7.1   Transmission electron micrographs of dislocations in a carbon (0.15 weight percent) silicon (0.15−0.35 weight percent) steel annealed for 1 hour at 900°C. (a) 0% cold work with a dislocation density of 5×10^{13} m^{-2}. (b) 5% cold work by rolling with a dislocation density of 2×10^{14} m^{-2}. (c) 10% cold work by rolling with a dislocation density of 6×10^{14} to 10×10^{14} m^{-2}. (d) 20% cold work by rolling. The dense cellular structure of dislocations in (c) and (d) prevents accurate determination of the dislocation density. (*© 2009 IEEE. Reprinted, with permission, from Kikuchi, H., et al., "Effect of Microstructure Changes on Barkhausen Noise Properties and Hysteresis Loop in Cold Rolled Low Carbon Steel," IEEE Trans, Magnetics, 45 (6) 2744–2747.*)

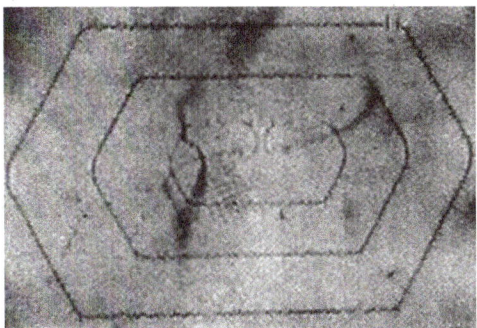

Figure 7.2 Transmission electron microscope image of a Frank-Read dislocation source in silicon, with two completed dislocation loops and one loop being created. (*Professor Dinker B. Sirdeshmukh, Lalitha Sirdeshmukh, K. G. Subhadra, "Defects in Crystals II: Dislocations," Atomistic Properties of Solids. Springer Series in Materials Science Volume 147, 2011, pp 511-559, Figure 36, with kind permission from Springer Science+Business Media B.V.*)

as at surfaces, grain boundaries, and crack tips. Dislocations in certain configurations can produce new dislocations by a mechanism shown in Figure 7.2 called a **Frank-Read source**. At the origin of the Frank-Read source is a pinned dislocation segment. When a critical value of resolved shear stress is applied to the Frank-Read source, multiple dislocation loops are created, as shown in Figure 7.2. The Frank-Read source can continuously produce dislocation loops.

Figure 7.3 shows that the resolved shear stress at yield in copper is a linear function of the square root of the dislocation density, and that the resolved shear stress increases by nearly 100 when the dislocation density increases from 10^{10} to 10^{15} m/m³. The tensile yield strength (σ_y) and the shear yield strength (τ_y) are related to the dislocation density (ρ) through the relationships shown in Equations 7.1a and 7.1b:

$$\sigma_y = \sigma_0 + k_{dt}\, G\rho^{\frac{1}{2}} \qquad\qquad \textbf{7.1a}$$

$$\tau_y = \tau_0 + k_{ds}\, G\rho^{\frac{1}{2}} \qquad\qquad \textbf{7.1b}$$

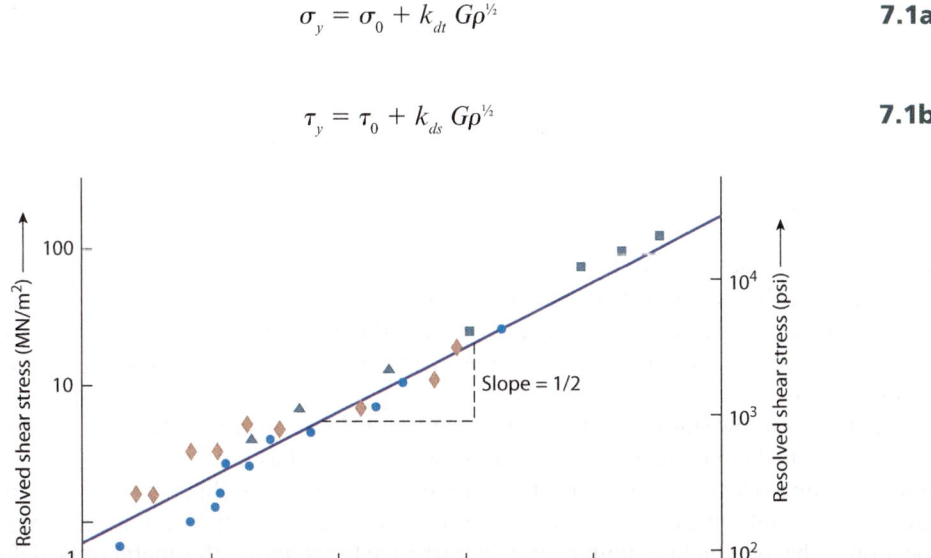

Figure 7.3 The resolved shear stress at yield in various forms of copper tested at room temperature, as a function of the square root of the dislocation density. □—polycrystal, o—single crystal with one slip system, ◊—single crystal with two slip systems, △—single crystal with six slip systems. (*Based on Courtney, T.H., Mechanical Behavior of Materials, 2nd ed. McGraw-Hill, N.Y. (2000), p. 180.*)

where G is the shear modulus, σ_0 and τ_0 are, respectively, the tensile and shear yield strengths of a crystal with a dislocation density approaching 0, and k_{dt} and k_{ds} are constants. The values of τ_0 and k_{ds} for shear are approximately one-third of the values of σ_0 and k_{dt} for tension.

Why does the yield strength increase as the dislocation density increases when dislocations are responsible for plastic deformation? Immobile dislocations act as obstacles to the motion of other dislocations. If the slip plane for a dislocation is not one of the most closely packed planes for that crystal structure, the edge dislocation is immobile on that plane. Immobile dislocations can be created by dislocation reactions. At high stress, several close-packed planes in metals can have sufficiently high resolved shear stress that dislocations are mobile on these different planes, and the dislocations on these different close-packed planes can react where the planes intersect. When two dislocations react, they form a new, third dislocation whose Burger's vector is the vector sum of the two initial dislocations. It is possible when dislocations intersect from different close-packed planes that the resultant dislocation will not be on a close-packed plane. For example, in an FCC crystal two dislocations on close-packed planes $\mathbf{b}_1$ and $\mathbf{b}_2$ in Equation 7.2 react to form the dislocation $\mathbf{b}_3$:

$$\mathbf{b}_1 + \mathbf{b}_2 = \frac{a}{2}[110] + \frac{a}{2}[1\bar{1}0] = a[100] = \mathbf{b}_3 \qquad \qquad \textbf{7.2}$$

The dislocation $\mathbf{b}_3$ does not lie on a {111} close-packed plane and is not a mobile dislocation in a FCC metal crystal. The higher the dislocation density, the higher the density of immobile dislocations. Consider the following analogy. The automobile is great for moving people from one location to another, particularly on uncrowded superhighways. However, during rush hour in a busy city, if the automobile is trapped in an intersection after the light changes, the trapped automobile becomes an obstacle to the mobility of other automobiles.

An example of strain hardening is when the yield strength of steel is increased from approximately 7×10^7 Pa by cold-rolling a plate into a sheet with a yield strength of approximately 7×10^8 Pa. In cold-rolling, the thickness of a plate of metal is reduced at a temperature where strain hardening occurs. Other processes, such as cold-forging, stamping, and drawing, also strain harden the material, as discussed in Chapter 13, on material processing. Cold work is a general term applied to all of the processes that produce strain hardening at temperatures below the recrystallization temperature. Strain hardening is probably the least expensive process for strengthening metals, because no expensive alloy elements are added and it is only necessary to strain the metal with standard metal deformation processes.

The ability of a metal to be strain hardened depends upon the stacking-fault energy. If the stacking-fault energy is high, the dislocations remain as full dislocations. Pure screw dislocations can move from one close-packed plane to an intersecting close-packed plane; this is called cross-slip. Cross-slip of screw dislocations results in the possibility of more dislocation reactions as presented in Equation 7.2, and more strain hardening.

If a metal has a low stacking-fault energy, the full dislocations separate into two partial dislocations separated by a stacking fault, as discussed in Appendix 3A. It has been observed that partial dislocations do not cross-slip as easily as full dislocations can. If there is no cross-slip, the metal does not strain harden as much as a metal with cross-slip would because there are not dislocation reactions by intersecting dislocations. The low stacking-fault energy of cartridge brass allows this metal to be deformed into thin-walled tubes without extensive strain hardening.

(a) Evaluate the terms τ_0 and k_{ds} in Equation 7.1b for the copper shown in Figure 7.3. (b) Estimate the shear yield strength of copper with a dislocation density of 1×10^{16} m^{-2}.

Solution

a) From Table 6.1, the shear modulus of copper is 48.3 GPa. In Figure 7.3, τ_0 is the shear yield stress when the dislocation density is extrapolated to 0. Since it is not possible to extrapolate this figure to a 0 dislocation density, we can set up two equations to solve for the two unknowns of τ_0 and k_{ds}. At a dislocation density of 10^{11} m^{-2}, the shear yield strength is $10^{0.33}$ MPa, or 2.1 MPa. At a dislocation density of 10^{15} m^{-2}, the shear yield strength is $10^{2.21}$ MPa, or 162 MPa. Setting up two equations based upon Equation 7.1b,

$$\tau_y = \tau_0 + k_{ds} G\rho^{1/2}$$

1. 2.1×10^6 Pa $= \tau_0 + k_{ds}(48.3 \times 10^9$ Pa$)(10^{11}$ m$^{-2})^{1/2} = \tau_0 + k_{ds}(48.3 \times 10^9$ Pa$)(10 \times 10^{10}$ m$^{-2})^{1/2}$
2. 162×10^6 Pa $= \tau_0 + k_{ds}(48.3 \times 10^9$ Pa$)(10^{15}$ m$^{-2})^{1/2} = \tau_0 + k_{ds}(48.3 \times 10^9$ Pa$)(10 \times 10^{14}$ m$^{-2})^{1/2}$

Multiplying out the terms and rewriting Equations 1 and 2,

1. 2.1×10^6 Pa $= \tau_0 + k_{ds}(153 \times 10^{14}$ Pa $\cdot$ m$^{-1})$
2. 162×10^6 Pa $= \tau_0 + k_{ds}(153 \times 10^{16}$ Pa $\cdot$ m$^{-1})$

Subtracting Equation 1 from 2 eliminates τ_0 and results in

$$160 \times 10^6 \text{ Pa} = k_{ds}(151 \times 10^{16} \text{ Pa} \cdot \text{m}^{-1})$$

Solving for $k_{ds} = \dfrac{160 \times 10^6 \text{ Pa}}{151 \times 10^{16} \text{ Pa} \cdot \text{m}^{-1}} = 1.06 \times 10^{-10}$ m

Substituting the value of k_{ds} into Equation 2 to solve for τ_0,

$$162 \times 10^6 \text{ Pa} = \tau_0 + 1.06 \times 10^{-10} \text{ m } (153 \times 10^{16} \text{ Pa} \cdot \text{m}^{-1})$$

$$\tau_0 = 162 \times 10^6 \text{ Pa} - 162 \times 10^6 \text{ Pa}) = 0$$

For copper, the value of τ_0 is insignificant in this problem. This result confirms that copper with a very low dislocation density has a very low yield strength.

b) For a dislocation density of 10^{16} m^{-2},

$$\tau_y - k_{ds}G\rho^{1/2} = (1.06 \times 10^{-10} \text{ m}^{-1})(48.3 \times 10^9 \text{ Pa})(1 \times 10^{16} \text{ m}^{-2})^{1/2}$$

$$\tau_y = (51.2 \times 10^{-1} \text{ Pa} \cdot \text{m}^{-1})(1 \times 10^8 \text{ m}^{-1}) = 51.2 \times 10^7 \text{ Pa}$$

If the dislocation density is increased to 1×10^{16} m^{-2}, the shear yield strength is predicted to be 512 MPa. If the tensile yield strength of a polycrystalline specimen is two to three times the shear yield strength, the tensile yield strength of copper should be at least 1000 MPa. This strength is stronger than those of the carbon steels in Table 6.3.

7.2.2 Annealing Metals

Annealing produces an effect opposite of strain hardening. When a strain hardened metal is annealed, it is heated to an elevated temperature for a period of time to reduce the metal's strength. It is assumed no reactions such as precipitation hardening occur during the heating. Figure 7.4 shows the fraction of residual strain hardening in iron that was prestrained by 5 % at 0°C as a function of annealing time

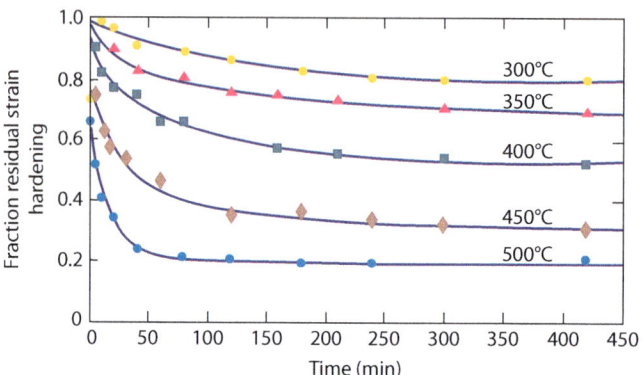

Figure 7.4 The fraction of residual strain hardening after the indicated annealing temperature and time in iron that was prestrained by 5% at 0°C. (*Based on Leslie, W.C., Michalak, J.T., and Aul, F.W., Annealing of Cold-Worked Iron in Iron and its Dilute Solid Solutions, eds. Spencer, C.W., and Werner, F.E., Interscience Publ. NY (1963), p. 128.*)

and temperature. Higher temperatures and longer times give a lower residual strain hardening. Heating a metal increases the mobility of dislocations, and dislocations of opposite signs attract each other. If two dislocations of opposite signs come together on the same plane, they form a continuous plane of atoms, and the two dislocations are eliminated from the crystal, as shown in Figure 7.5a and 7.5b. The elimination of the two dislocations reduces the dislocation density and reduces the yield strength of the metal. In a **recovery anneal**, some of the ductility is recovered, and the yield strength is reduced.

To obtain the maximum ductility after extensive strain hardening, the metal must be **recrystallized**. Recrystallization is the formation and growth of new grains. A metal recrystallizes if it is extensively cold-worked and then heated to an appropriate temperature. Cold-worked brass is shown in Figure 7.6a, note the dark straight lines in the grains where dislocations have come to the surface and created steps on the surface, as schematically shown in Figure 6.11. The new recrystallized grains shown in Figure 7.6b are of the same crystal structure as the prior grains, and the recrystallized grains are of a very low dislocation density. A fully recrystallized metal has zero residual strain hardening.

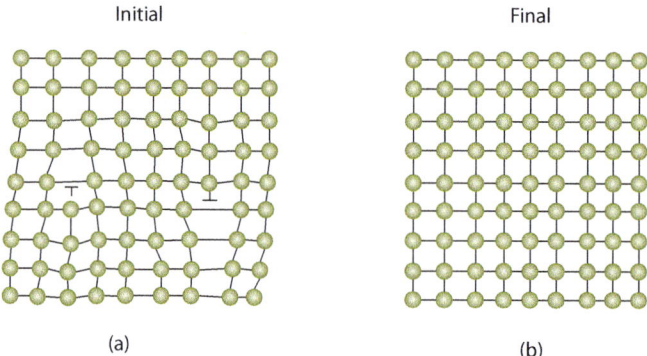

Figure 7.5 (a) A crystal with two edge dislocations of opposite signs on the same slip plane. The two dislocations attract each other. (b) The dislocations come together and form a continuous plane of atoms, and the two dislocations are annihilated. (*Adapted from Eisenstadt, M. M., Introduction to Mechanical Properties of Materials, The Macmillan Co., N.ew Y.ork (1971), p. 240.*)

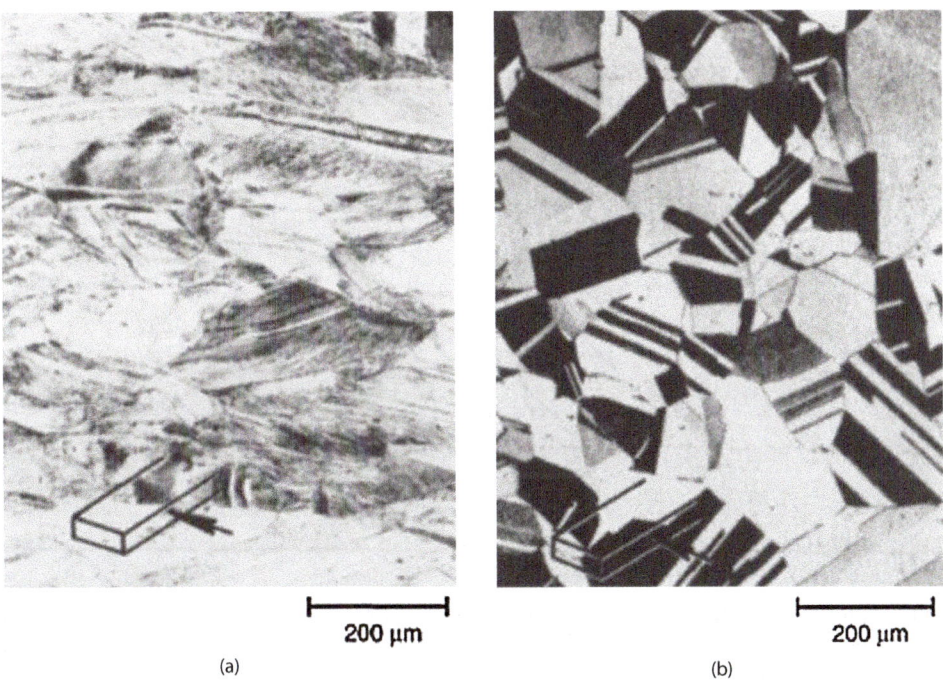

200 μm   200 μm

(a)   (b)

Figure 7.6 (a) The microstructure of cartridge brass cold worked 37%. The rolling direction in (a) and (b) is indicated by the arrow pointing to the metal slab in the lower left side of the figure. (b) The microstructure of cartridge brass cold worked and recrystallized to a grain size of 120 μm. (*ASM Handbooks Online, http://www.asmmaterials.info, ASM International, 2004.*)

7.2.3   Grain-Boundary Strengthening of Metals

Grain boundaries in a polycrystalline material are the interface between two crystals, or grains, that are of different orientations, and the close-packed planes are of different orientations in each grain, as shown in Figure 7.7. The edge dislocation with Burger's vector $\mathbf{b}_1$ is mobile in grain 1, because it lies in the close-packed planes of grain 1 that are schematically shown by the straight lines in grain 1. If the dislocations with Burger's vector $\mathbf{b}_1$ were to move across the grain boundary to grain 2, the Burger's vector is conserved, and $\mathbf{b}_1$ does not lie in the close-packed planes indicated by the straight lines of grain 2. The edge dislocations of Burger's vector $\mathbf{b}_1$ would not be mobile in grain 2. Instead, the edge dislocations in grain 1 pile up at the grain boundary, as shown in Figure 7.7. Grain boundaries are obstacles to dislocation motion, and they increase the yield stress. Figure 7.8 shows dislocation pileups at a twin boundary in stainless steel.

Figure 7.9 shows that the yield strength, or flow stress, of annealed commercially pure titanium (A70) with a yield strength of 483 MPa is linear when plotted as a function of grain diameter to the power of negative one-half. The symbol $\sigma(0.08)$ associated with the data in Figure 7.9 indicates that 0.08 true pre-strain was applied prior to grain growth, and pre-strain hardens the metal. The tensile yield strength (σ_y) and the shear yield strength (τ_y) of metals follow the Hall-Petch relationships shown in Equations 7.3a and 7.3b:

$$\sigma_y = \sigma_0 + k_{gt}d^{-\frac{1}{2}} \qquad\qquad \textbf{7.3a}$$

$$\tau_y = \tau_0 + k_{gs}d^{-\frac{1}{2}} \qquad\qquad \textbf{7.3b}$$

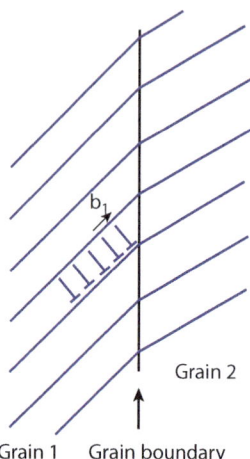

Figure 7.7   A schematic of dislocations with Burger's vector $\mathbf{b}_1$ in grain 1 in a pileup at a grain boundary. Grain 2 has close-packed planes indicated by the straight lines at a different orientation than the close-packed planes in grain 1.

In Equation 7.3, d is the average grain diameter, σ_0 and τ_0 are, respectively, the tensile and shear yield strength of a specimen when the grain size is infinite or when $d^{-\frac{1}{2}}$ is 0, and k_{gt} and k_{gs} are constants. Equation 7.3 shows that metals with a smaller grain size have a higher yield strength. With a smaller grain size, there are more grain boundaries per unit volume of material.

It has been known for a long time that small grain sizes give superior properties to steels. Engineering drawings often specify a fine-grained steel. Researchers are presently developing techniques to produce materials with grain sizes as small as 100 nm or possibly even smaller. Grain-boundary strengthening is a

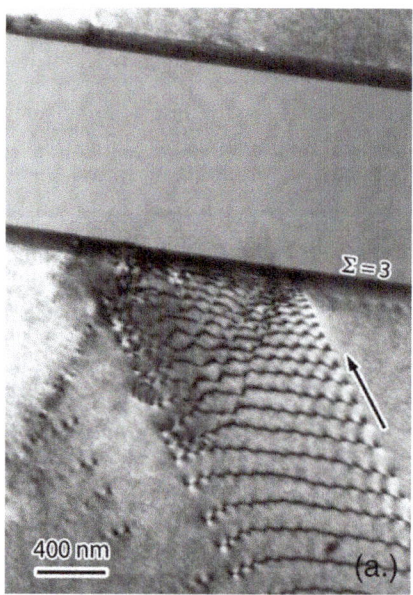

Figure 7.8   A transmission electron micrograph of dislocations (dark wavy lines along arrow) piling up at a twin boundary (straight dark line with $\Sigma = 3$) in stainless steel. (*Reprinted from Acta Materialia, Volume 59, Issue 1, Michael D. Sangid,Tawhid Ezaz,Huseyin Sehitoglu,Ian M. Robertson, "Energy of slip transmission and nucleation at grain boundaries," Pages No. 283–296, Copyright 2011, with permission from Elsevier.*)

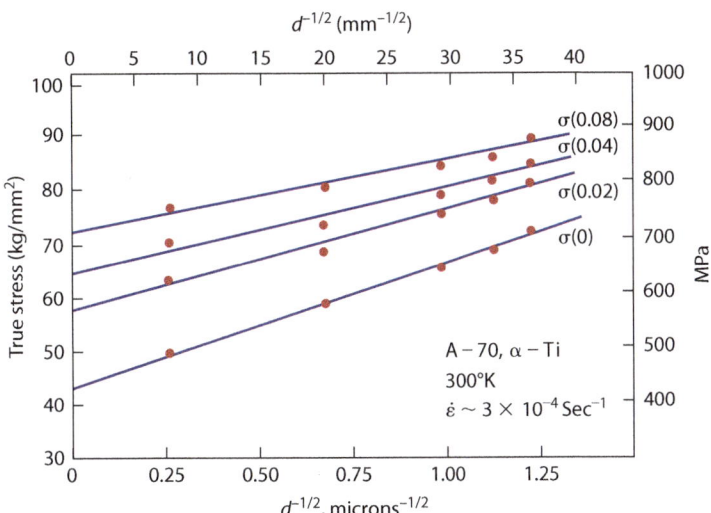

Figure 7.9   The yield stress as a function of the grain size to the negative power of one-half for A70 titanium specimens that were prestrained by 0, 2, 4, and 8%. *(Based on Jones, R.L., and Conrad, H.,TMS-AIME 245 (1969) 779.)*

relatively inexpensive technique of strengthening metals, because it is not necessary to add any expensive alloy elements, and standard metal processing procedures can be utilized. One way to produce small grains is to rapidly cool a metal or ceramic from the liquid phase to room temperature.

Example Problem 7.2

Evaluate the terms σ_0 and k_{gt} in Equation 7.3, using the SI system of units for the titanium shown in Figure 7.9, with a pre-strain of 2%.

Solution

In Figure 7.9, σ_0 is 58 kg of force per square millimeter. σ_0 is the stress when the grain size becomes infinite or when $d^{-\frac{1}{2}}$ becomes 0. The SI unit for stress is newtons per square meter, and the conversion factor from kilograms of force per square millimeter is

$$1\frac{kg_f}{mm^2} = 9.806 \text{ MPa}$$

The value of σ_0 is

$$\sigma_0 = 58\,\frac{kg_f}{mm^2} = 569 \text{ MPa}$$

On the $d^{-\frac{1}{2}}$ axis of Figure 7.9, if the grain size is equal to 1 micron (1×10^{-6} m), then on the $d^{-\frac{1}{2}}$ axis the value is $1.0 \times 10^3\,\text{m}^{-\frac{1}{2}}$. Analyze Equation 7.3 at $d^{-\frac{1}{2}} = 1.25\,(\text{microns})^{-\frac{1}{2}} = 1.25 \times 10^3\,\text{m}^{-\frac{1}{2}}$, where the yield strength is

$$\sigma_y = 83\,\frac{kg_f}{mm^2} = 814 \text{ MPa} = \sigma_0 + k_{gt}d^{-\frac{1}{2}} = 569 \text{ MPa} + k_{gt}(1.25 \times 10^3\,\text{m}^{-\frac{1}{2}})$$

Substituting values of σ_y, σ_0, and $d^{-\frac{1}{2}}$ into Equation 7.3a and solving for k_{gt} results in

$$k_{gt} = \frac{(814 - 569) \times 10^6\,\text{Pa}}{1.25 \times 10^3\,\text{m}^{-1/2}} = \frac{245 \times 10^6\,\text{Pa}}{1.25 \times 10^3\,\text{m}^{-1/2}} = 196 \times 10^3\,\text{Pa} \cdot \text{m}^{1/2}$$

High-strength metals are being developed by reducing the grain size to nanometer dimensions. Predict the yield strength of the titanium in Figure 7.9 with 2% pre-strain if the grain size is reduced to 100 nm, using the results of Example Problem 7.2.

Solution

$$d = 100 \times 10^{-9} \text{ m} = 10 \times 10^{-8} \text{ m}$$
$$d^{1/2} = 3.16 \times 10^{-4} \text{ m}^{1/2}$$
$$d^{-1/2} = 0.316 \times 10^{4} \text{ m}^{-1/2}$$

From Example Problem 7.2, σ_0 and k_{gt} are equal to

$$\sigma_0 = 58 \frac{\text{kg}_f}{\text{mm}^2} = 569 \text{ MPa}$$

$$k_{gt} = 196 \times 10^3 \text{ Pa} \cdot \text{m}^{1/2}$$

Putting the values σ_0, k_{gt}, and $d^{-1/2}$ into Equation 7.3a results in

$$\sigma_y = \sigma_0 + k_{gt} d^{-1/2} = 569 \times 10^6 \text{ Pa} + (196 \times 10^3 \text{ Pa} \cdot \text{m}^{1/2})(0.316 \times 10^4 \text{ m}^{-1/2})$$
$$\sigma_y = 569 \times 10^6 \text{ Pa} + 62 \times 10^7 \text{ Pa} = 1189 \text{ MPa}$$

The yield strength is more than doubled to 1189 MPa by this reduction in grain size to 100 nm, relative to titanium with a large grain size and a yield strength of 569 MPa.

7.2.4 Dispersion Strengthening

A metal is **dispersion strengthened** by the addition of particles of a hard material. Examples of dispersion strengthening are nickel-thoria (ThO_2), shown in Figure 7.10, that is used in high-temperature gas turbines; and cobalt-tungsten carbide, shown in Figure 3.22, that is used in carbide cutting tools and in dentist drills. In the example of cobalt-tungsten carbide, the ductile metal cobalt is the continuous matrix material that holds the material together, and the hard dispersion particle is tungsten carbide. Materials such as nickel-thoria and cobalt-tungsten carbide are composite materials, and the equations for dispersion strengthening are the same as for composite materials. Table 7.1 shows additional dispersion-strengthened systems developed for other applications.

The tensile strength of dispersion-strengthened materials is modeled with the rule of mixtures. In the rule of mixtures, the contribution to the tensile strength from each phase present is in proportion to the volume fraction of the phase present, as shown in Equation 7.4.

$$\sigma_u = v_d \sigma_d + v_m \sigma_m \qquad \qquad \textbf{7.4}$$

The tensile strength of the composite material is σ_u, v_d is the volume fraction of the dispersion, σ_d is the tensile strength of the dispersion, v_m is the volume fraction of the matrix material, and σ_m is the tensile strength of the matrix material. The volume fraction of the dispersion is the total volume of the dispersion particles (V_d) divided by the total volume of the material (V). An equation similar to 7.4 can also be written for shear stress, and also for the elastic modulus of the composite. The mechanism of dispersion strengthening a ductile matrix is that the hard dispersion particles stop dislocation motion

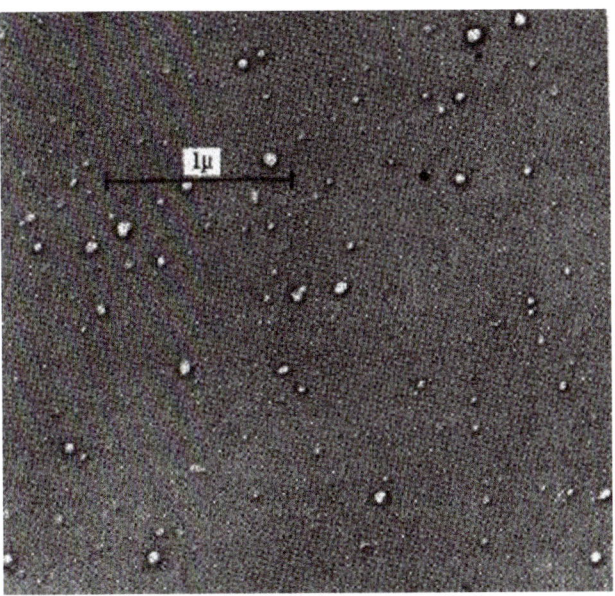

Figure 7.10   A scanning electron micrograph of nickel-thoria. The nickel is the continuous gray material and the thoria particles are white. (*Cremens, W. S.; Norris, L. F.; Weeton, J. W. "Overlay copy technique to provide high-contrast electron micrographs for automatic metallographic analysis," NASA Technical Report, Glenn Research Center, 1966.*)

in the ductile matrix when the dislocations come to the hard phase boundary, and the hard dispersion particles constrain the ductile metal particles from deforming.

Dispersion strengthening in some materials can create advantages: For nickel-thoria at high temperatures, the thoria particles stop nickel grain growth and the ceramic thoria particles are stable and strong, thereby preserving strength at high temperatures.

Conversely, dispersion-strengthened nickel-thoria has several disadvantages. For example, the nickel and thoria are not in equilibrium with each other, and interdiffusion and reactions between the nickel and thoria particles occurs. Because the matrix and the dispersion often melt at significantly different temperatures, it is usually necessary to use powder metallurgy techniques, which we will discuss in Chapter 13, to produce these metal alloys. Small voids at the particle-matrix interfaces may be present

Table 7.1   Applications of Some Dispersion-Strengthened Metal-Ceramic Systems

System	Applications
Ag-CdO	Eiectrical contact materiais
Al-Al$_2$O$_3$	Possible use in nuclear reactors
Be-BeO	Aerospace and nuclear reactors
Co-ThO$_2$Y$_2$O$_3$	Possible creep-resistant magnetic material
Ni-20% Cr-ThO$_2$	Turbine engine components
Pb-PbO	Battery grid
Pt-ThO$_2$	Filaments, electrical components
W-ThO$_2$, ZrO$_2$	Filaments, heaters

1   Based on data from Askeland, D.R., Fulay, P.P., and Wright, W.J., The Science and Engineering of Materials, 6th ed., Cengage Learning, Stamford, CT. (2011), p. 654.

that can act as crack initiation sites in powder processed metals. In Chapter 13 we discuss how the voids form during powder processing. The requirement to use powder metallurgy techniques makes dispersion strengthening expensive in comparison to other metal-strengthening techniques. Dispersion hardening can be combined with other strengthening processes for the matrix, such as the solid-solution and precipitation-strengthening techniques that we will discuss in Sections 7.2.5 and 7.2.6.

Example Problem 7.4

A dispersion-strengthened aluminum alloy is to be made with a 75 vol% commercial purity aluminum matrix with 25% aluminum oxide particles. Table 6.1 shows an elastic modulus of 70.3 GPa for aluminum, and Table 6.3 shows a yield strength of 34.5 MPa. Assume that the alumina has an elastic modulus of 386 GPa and a fracture strength of 400 MPa. Estimate the yield strength and elastic modulus of this alloy.

Solution

From Equation 7.4 the yield strength is estimated as

$$\sigma_c = v_d \sigma_d + v_m \sigma_m = 0.25(400 \text{ MPa}) + 0.75(34.5 \text{ MPa}) = 100 \text{ MPa} + 25.9 \text{ MPa} = 125.9 \text{ MPa}$$

The elastic modulus is estimated as

$$E_c = v_d E_d + v_m E_m = 0.25(386 \text{ GPa}) + 0.75(70.3 \text{GPa}) = 96.5 \text{ GPa} + 52.7 \text{ GPa} = 149 \text{ GPa}$$

In these calculations, the estimation of the elastic modulus is probably more reliable than that of the yield strength. This is because in the elastic-modulus calculation, the summation is for similar processes; in each material the strain is elastic. In the yield-strength estimation, the aluminum yield strength corresponds to a plastic strain of 0.002; however, the alumina is always elastic. We are summing the stresses relating to two different modes of strain, and the result may not be accurate. However, the calculation gives an estimate of the result that would be tested by experiment.

7.2.5   Solid-Solution Strengthening

A solid solution is formed by adding either substitutional atoms or interstitial atoms to a host material. Nickel and chromium are substitutional atoms added to iron to produce stainless steel, magnesium substitutional atoms strengthen aluminum, and Cr_2O_3 is substitutional in Al_2O_3. Also, very small amounts of carbon (up to 0.022 weight percent at 727°C) are interstitial in carbon steel. **Solid-solution strengthening** occurs when foreign solid-solution atoms interact with the strain field of edge dislocations. At the edge dislocation shown in Figure 7.11, the atoms above the slip plane are in compression, and the atoms below

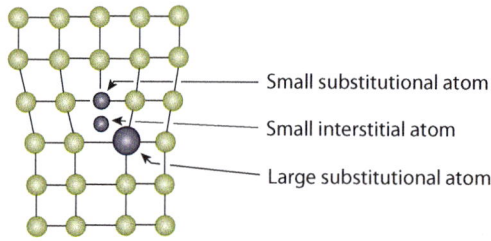

Figure 7.11   An edge dislocation in a crystal with solid-solution atoms in low-energy positions.

the slip plane are in tension. Small substitutional atoms are attracted to the area of compressive strain above the slip plane, because the small substitutional atoms allow some of the compressive strain to be relaxed. The relaxation of strain is analogous to when a large person gets off a crowded subway and is replaced by a small person. This reduces the compressive strain in the subway car. In Figure 7.11, large substitutional atoms are attracted to the area of tensile strain of the edge dislocation, because the large atom relaxes some of the tensile strain around the edge dislocation. Interstitial atoms are also attracted to the large interstitial spaces below the slip plane. Since atoms around the dislocation can move by diffusion, they can move to the area around the dislocation that minimizes the strain energy of the solid-solution atoms and the edge dislocation. Pure screw dislocations do not have any tensile- or compressive-strain fields; therefore, they are not attracted to solid-solution atoms. However, most dislocations are of mixed character, and solid-solution atoms are attracted to the edge component of mixed dislocations.

If a shear stress is applied to a solid-solution strengthened metal, it is more difficult to move the dislocation away from the solid-solution atoms in the metal that have lowered the energy of the dislocation than in the pure metal. The tensile yield strength (σ_y) and shear yield strength (τ_y) increase with the solute chemical composition in atom fraction (c_s) according to the relationships shown in Equations 7.5a and 7.5b:

$$\sigma_y = \sigma_0 + k_{st}c_s^{\frac{1}{2}} \qquad\qquad \textbf{7.5a}$$

$$\tau_y = \tau_0 + k_{ss}c_s^{\frac{1}{2}} \qquad\qquad \textbf{7.5b}$$

where σ_0 and τ_0 are the tensile and shear yield stress at a zero concentration of solid-solution atoms, and k_{st} and k_{ss} are constants. Figure 7.12 shows that the yield strength of iron is linear with respect to the atomic fraction of solid-solution carbon atoms to the one-half power.

An upper yield point and a lower yield point are present in low-carbon steel in Figure 6.16, and in 1045 steel in Figure 6.22. The upper yield point is the stress necessary to force the dislocations away from the carbon solid-solution atoms. The concentration of solid-solution atoms at a stationary dislocation is greater than the average concentration of solid-solution atoms in the solid, because the solid-solution atoms are attracted to the dislocation. Once the dislocation is forced away from the solid-solution atoms, a lower stress equal to the lower yield point is required to move the dislocations through the metal with the average distribution of solid-solution atoms.

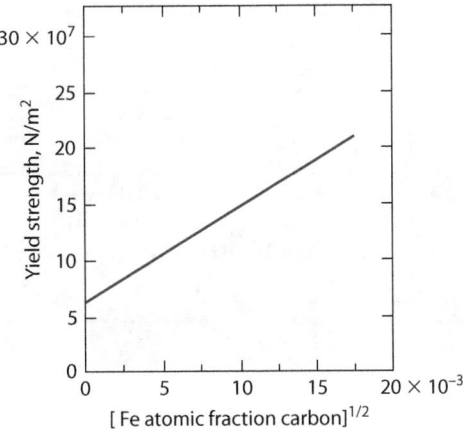

Figure 7.12 The yield strength of iron as a function of the atom fraction of carbon to the one-half power. (*Based on Guy, A.G. Introduction to Materials Science, McGraw-Hill, N.Y. (1972), p. 226.*)

Solid-solution strengthening can be used in conjunction with other types of strengthening such as strain hardening, grain-boundary strengthening, and dispersion strengthening. Solid-solution strengthening is the strengthening method of choice for metals that are to be welded or heated during processing, because heating during processing does not affect their subsequent strength during use. The solid-solution atoms are still present after the processing in the same concentration as before the heating, unless impurities have been added by processes such as welding.

If the concentration of the solid-solution atoms is too large, then a precipitate is formed. For example, at 727°C the maximum amount of solid-solution carbon in iron is 0.022 weight percent. If more than this amount of carbon is added, then a precipitate of iron carbide forms.

7.2.6 Precipitation Hardening

In **precipitation-hardened** metals, a new second phase ideally forms as small particles in the host metal that increase the yield strength. Precipitation hardening results in some of the strongest metals. For example, if aluminum is alloyed with 4 atom percent copper, the yield strength of the precipitation-hardened alloy is increased by a factor of nearly 10 times relative to that of the annealed commercial-purity aluminum. Precipitation-hardened aluminum alloys are utilized in aircraft because of their high strength and low density. The term *precipitation* is a familiar one from weather reports when the forecast is for rain or snow. A precipitate is a new phase that forms, and in the case of weather the precipitate may be rain or snow. The Fe_3C particles that form in carbon steels are a precipitate. Figure 7.13 shows the precipitate particles in an aluminum-lithium alloy.

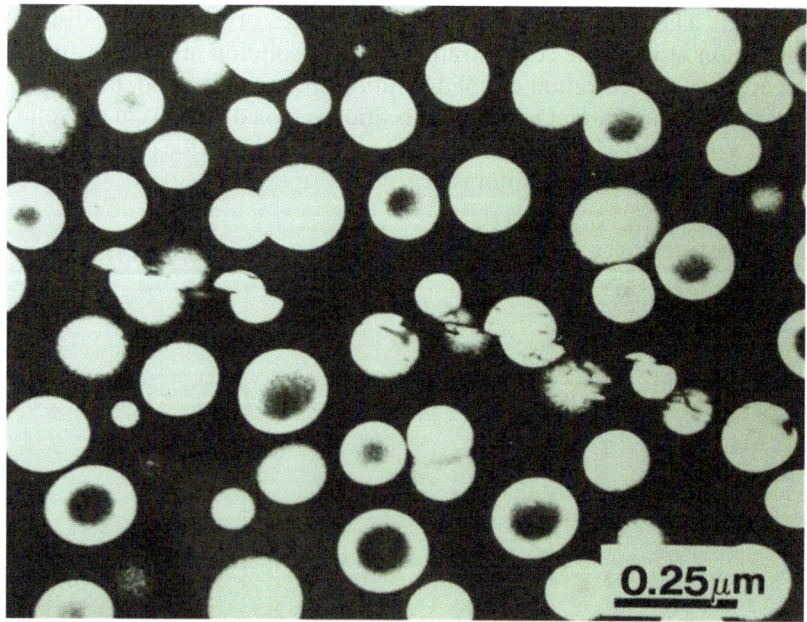

Figure 7.13 A transmission electron micrograph of coherent Al_3Li precipitate particles in an Al–12.9 atom percent Li alloy over-aged for 50 h at 260°C. The Al_3Li particles are the light-colored phase, because this is an image taken from the electron beam diffracted by the Al_3Li (dark field image). (*Photo courtesy of Dr. S. Baumann, Department of Materials Science and Engineering, Lehigh University, presently with Alcoa Mill Products.*)

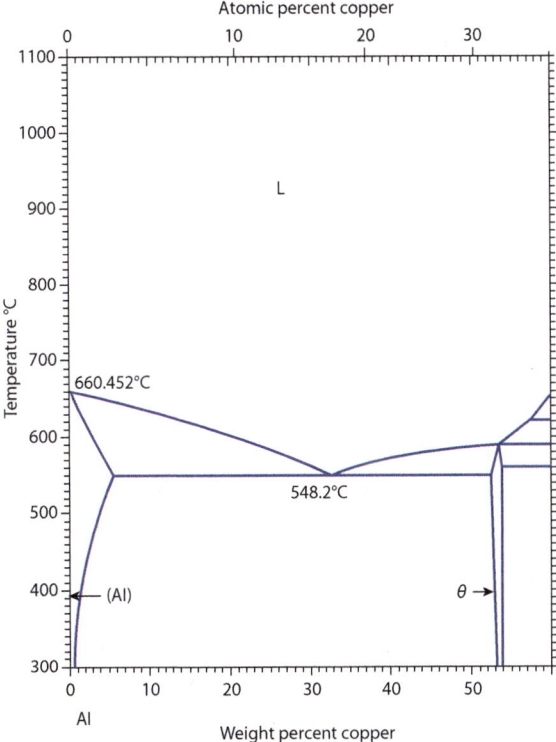

Figure 7.14   The aluminum-copper phase diagram from 0 to 60 weight percent copper. (*Based on American Society for Metals Handbook Vol. 3, ASM International, Metals Park, OH (1992).*)

Aluminum-copper alloys provide an excellent example of precipitation hardening. In aluminum plus 4 weight percent Cu at temperatures above 450°C, the Cu atoms are in solid solution, as shown in the phase diagram of Figure 7.14 and schematically in Figure 7.15a. However, at temperatures below 450°C, precipitate particles of Al_2Cu form in the aluminum matrix as shown in the phase diagram of Figure 7.14 and schematically in Figures 7.15b and 7.15c.

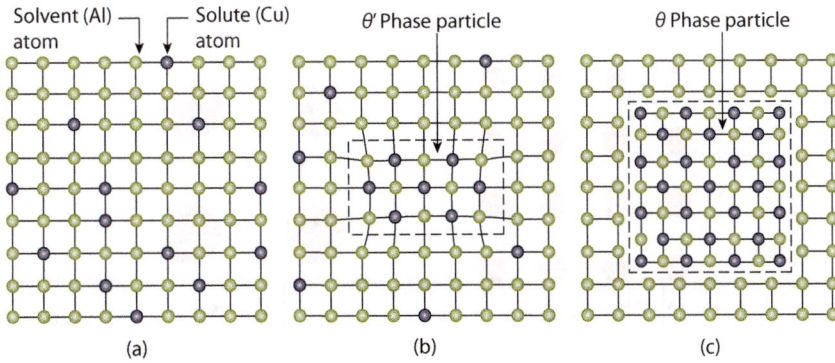

Figure 7.15   Schematics of the stages of formation of the precipitate Al_2Cu (θ). (a) Solid solution is present at the solution temperature above 450°C. (b) A coherent precipitate (θ') forms during the initial stage of aging at temperatures below 450°C. (c) An incoherent precipitate (θ) forms after longer aging times.

The precipitate particles of Al_2Cu block the movement of dislocations in the aluminum matrix. This blockage decreases the amount of plastic deformation of the aluminum, and increases the yield strength. The tensile yield strength (σ_y) and shear yield strength (τ_y) of a precipitation-hardened alloy are given by Equations 7.6a and 7.6b:

$$\sigma_y = \sigma_0 + \alpha\,\frac{Gb}{L} \qquad\qquad \textbf{7.6a}$$

$$\tau_y = \tau_0 + \alpha\,\frac{Gb}{L} \qquad\qquad \textbf{7.6b}$$

where σ_0 and τ_0 are the tensile and shear yield strength, respectively, of the matrix metal without the precipitate. G is the metal matrix shear modulus, b is the magnitude of the Burger's vector of the metal matrix, α is a constant equal to 2 for the tensile yield stress, α is equal to 1 if calculating the shear yield stress, and L is the open space between the precipitate particles. The open spacing is shown schematically in Figures 7.16a and 7.16b. The increase in yield strength is inversely proportional to the spacing between the precipitate particles, and the yield strength is maximized if the open space between the particles (L) is minimized.

It is necessary to follow certain heat-treating procedures to obtain the desired precipitation hardening. Using aluminum plus 4 weight percent copper as an example, the first step of the heat-treating procedure is the solution treatment to obtain a uniform distribution of the copper atoms in solid solution, as shown schematically in Figure 7.15a. This distribution is achieved by holding the alloy at a temperature above 450°C for a period of time. This composition and temperature is in the α-phase solid-solution region of the phase diagram in Figure 7.14. At this temperature with sufficient time, any previously formed precipitate dissolves to form a solid solution. Then the alloy is quenched to a low temperature, by placing it in a medium such as water, oil, a salt solution, or even liquid nitrogen. The quenching medium can be at various temperatures to obtain the desired quench rate and final temperature. The purpose of the quench is to cool the metal rapidly to start the precipitation process without allowing diffusion that would alter the distribution of the copper atoms. During the quench and in the initial stages of aging, very small precipitate particles (θ') nucleate, such as those shown in Figure 7.15b. When nuclei first form, they are usually coherent with the matrix. In a **coherent precipitate** the atoms of the precipitate align themselves with the atoms of the host. If there is a lattice parameter difference between the matrix and a coherent precipitate, this results in a strain in the precipitate and the matrix that can extend over large distances in the material. Internal stresses are calculated from the lattice strains with the elastic modulus of each material. Internal stresses are a result of strains that are not due to applied external forces. As the quenched alloy is aged further at a temperature such as 250°C, the existing nuclei grow into larger incoherent precipitate

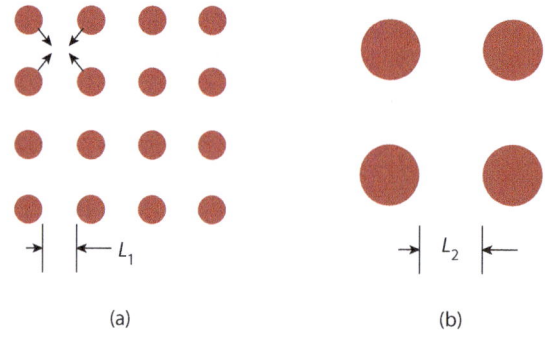

(a)   (b)

Figure 7.16   (a) Small precipitate particles with a small spacing of L_1. (b) Aged precipitate particles have coalesced into larger particles with a larger spacing of L_2.

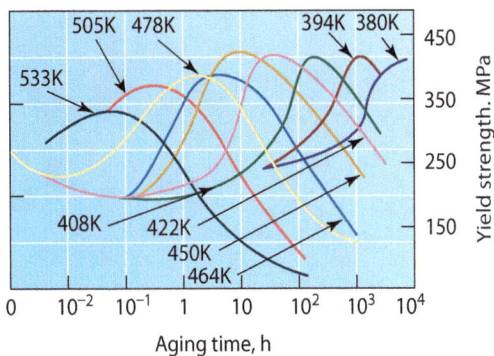

Figure 7.17 Isothermal aging curves at temperatures of (107°C), (121°C), (135°C), (149°C), (191°C), (232°C), and (260°C) for the 2014 aluminum alloy. (*Based on Hatch, J.E. ed. Aluminum Properties and Physical Metallurgy, ASM, Metals Park, OH (1984), p. 178.*)

particles (θ), such as shown in Figure 7.15c. For an **incoherent precipitate**, there is an interface between the matrix and the particle, as shown in Figure 7.15c, and the atoms of the precipitate particle are not aligned with those of the matrix. The aging is usually conducted at an elevated temperature, such as 250°C for Al-Cu alloys, or it may be naturally aged at room temperature. **Aging** is holding a metal at a specified temperature and time for the purpose of forming a precipitate by nucleation and growth.

The yield strength of the aged alloy depends upon the time and temperature of the aging, as shown in Figure 7.17. There is a peak in the yield strength as a function of aging time for the aluminum-copper alloy 2014 that contains 4.4 weight percent copper. In the early stages of aging, where the yield strength is increasing rapidly with time, more precipitate particles are forming, and they are also growing in size. As a result, the particle spacing (L) decreases and the yield strength (σ_y) increases. Internal stresses resulting from coherent precipitates also significantly harden an alloy, because the dislocations interact with the internal stresses. If the precipitate changes from coherent to incoherent the strength decreases with further aging. An **over-aged alloy** is one aged beyond its peak strength. At some point in the aging, the equilibrium amount of precipitate is formed, and past this point any additional aging will not form additional precipitate. As an alloy is further over-aged, the spacing between the precipitate particles increases, because all of the Al_2Cu that can form is now precipitated; however, the average size of the precipitate particles continues to increase. The way for this to happen is for the number of precipitate particles to decrease as their size increases. Overaging causes growth of the largest precipitate particles by consuming the smaller particles. This consumption increases the spacing (L) between the particles, and it decreases the yield strength according to Equation 7.6. Also, the presence of an interface between the incoherent precipitate and the matrix eliminates the long-range internal stresses, which contributes to a decrease in the yield strength of the alloy. In some applications, such as in corrosive environments and cyclic loading, over-aged alloys can have a longer life than alloys aged to peak strength.

Figure 7.17 shows that at low temperatures, such as 107°C, the time to achieve peak strength is nearly 10^4 hours, because the kinetics of diffusion to form the precipitate is slow. The diffusion rate decreases exponentially as temperature decreases. Increasing the temperature increases the rate at which diffusion occurs and the precipitate particles form. Observe that the peak strength occurs at shorter times for increasing temperatures. At 260°C, the alloy is aged to peak strength in less than 10^{-1} hour or in 6 minutes.

Why is it necessary to go through the complicated procedure of solution treating, quenching, and aging to form the precipitate? Why not just cool an alloy from the solution-treatment temperature to the aging temperature? The quench forces the precipitates to nucleate in a uniform distribution throughout the grain, as shown in Figure 7.18a, because the solid-solution atoms cannot diffuse very far at the quench

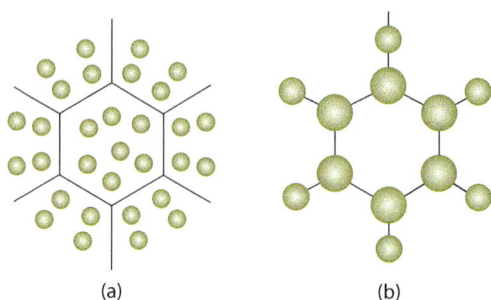

Figure 7.18   (a) A schematic of evenly distributed precipitate particles formed by solution treating, quenching, and aging. (b) Precipitate particles preferentially form in grain boundaries if the alloy is cooled from the solution-treating temperature to the aging temperature.

medium temperature. An alloy simply cooled to the aging temperature preferentially forms precipitate particles at its grain boundaries, as shown schematically in Figure 7.18b. At the aging temperature, such as 250°C, the atoms of copper would diffuse to the grain boundary, where they would form a precipitate within the grain boundary. This would result in an alloy of low strength within the grains of the material, and one that would have brittle grain boundaries. Solution treating, quenching, and aging produces high-strength metal alloys that also have ductility.

One advantage of precipitation hardening over dispersion hardening is that the precipitate particle is in equilibrium with the matrix. No reactions occur at elevated temperatures between the precipitate and the matrix, whereas high-temperature reactions can take place with dispersion hardening. The equilibrium with the matrix is an advantage for nickel that is strengthened with the Ni_3Al precipitate for use in high-temperature gas turbines. However, if an aluminum–4 weight percent copper precipitation-hardened alloy is heated to 450°C, the precipitate is dissolved in the matrix, as shown in the phase diagram in Figure 7.14. The temperature limit for precipitation hardening depends upon the stability of the precipitate as the temperature increases.

Example Problem 7.5

Predict the average spacing between precipitate particles in the 2014 aluminum alloy shown in Figure 7.17 that was aged at 149°C to peak strength. Assume that the as-quenched yield strength of 138 MPa is the value of σ_0. This is the value of the yield strength of the alloy aged for the shortest time at 0°F. We can assume that no precipitate has formed in this alloy. We can also assume that the lattice parameter of FCC aluminum is 0.405 nm.

Solution

The tensile yield strength of a precipitation-hardened alloy is given by Equation 7.6:

$$\sigma_y = \sigma_0 + \alpha \frac{Gb}{L} \text{ with } \alpha = 2$$

For aluminum, assume that the alloy shear modulus (G) is equal to 26.1 GPa, which is the pure aluminum shear modulus found in Table 6.1. Alloying changes the shear modulus only by small amounts.
For FCC metals, the magnitude of the Burger's vector (**b**) is a vector of the type

$$\mathbf{b} = \frac{a}{2}[110] = \frac{0.405 \times 10^{-9}\,\text{m}}{2}[1^2 + 1^2 + 0]^{1/2} = 0.2025 \times 10^{-9}\,\text{m}\,[2^{1/2}] = 0.286 \times 10^{-9}\,\text{m}$$

From Figure 7.17, the maximum yield strength at 149°C is approximately 430 MPa. The only unknown is the open space between the particles (L).

$$\sigma_y = \sigma_0 + \alpha\frac{Gb}{L} = 430 \times 10^6\,Pa = 138 \times 10^6\,Pa + 2\left[\frac{(26.1 \times 10^9\,Pa)(0.286 \times 10^{-9}\,m)}{L}\right]$$

$$\sigma_y = 430 \times 10^6\,Pa = 138 \times 10^6\,Pa + \frac{14.9\,Pa \cdot m}{L}$$

Solving for L,

$$L = \frac{14.9\,Pa \cdot m}{292 \times 10^6\,Pa} = 0.051 \times 10^{-6}\,m = 51\,nm$$

The spacing of 51 nm is comparable to some of the open spaces observed between particles in Figure 7.13.

7.3   STRENGTHENING CERAMICS

Most ceramics are brittle at room temperature, although at high temperatures ceramics are more ductile. If glass is heated over a flame, it is easily plastically deformed. A glass blower plastically deforms a lump of glass into a beautiful vase by heating the glass and then blowing it into shape with a blow pipe. Most ceramics are brittle at room temperature because the fracture stress is lower than the yield stress. The focus in strengthening ceramics is on increasing the fracture stress, and this is primarily achieved by eliminating defects from the ceramic. However, some single crystals with ionic bonding, or with a combination of ionic and covalent bonding, do yield at room temperature, and for these materials some of the strengthening techniques for metals apply.

7.3.1   Strengthening Single Crystal Ceramics

Single crystals of NaCl and MgO both yield and exhibit plastic strain at room temperature. It has been found that solid-solution strengthening procedures discussed above for metals also apply to ductile single crystals of NaCl and MgO. For example, it is observed that the compressive yield strength of MgO is more than doubled by solid-solution strengthening with 6.2 molar percent NiO. Also, the increase in the yield strength produced by calcium substitutional atoms in NaCl single crystals follows Equation 7.5.

Single crystals of ceramics with more covalent bonding and with more complex crystal structures are not ductile at room temperature. For example, single crystals of alumina (Al_2O_3) are brittle up to temperatures of approximately 900°C.

7.3.2   Strengthening Polycrystalline Ceramics

Why is polycrystalline MgO brittle up to temperatures of 1000°C, whereas a single crystal of MgO can be plastically deformed at room temperature? The reason for this is that during plastic deformation of polycrystalline materials, it is necessary for the material to remain continuous. The grains must to be able to change shape without opening spaces in the material. For polycrystalline material to remain

2 μm

Figure 7.19 A photomicrograph of sintered ceramic barium magnesium tantalite, showing the grain structure with voids between the grains. (*Heather Shirey*)

continuous, each grain must have at least five active slip systems. In single crystals, extensive plastic strain is possible with only one active slip system, because the surface of the single crystal can deform freely. For materials like polycrystalline MgO to have five active slip systems requires such a high critical resolved shear stress that it is not possible to activate all five slip systems before brittle fracture occurs. In a polycrystalline ceramic with only a few active slip systems, a crack will develop where dislocations in the active slip systems intersect a grain boundary. We can assume that all polycrystalline ceramics are brittle at room temperature.

To increase the strength of a brittle polycrystalline ceramic, the fracture strength must be increased. We will cover the fracture of materials in Chapter 11, and in that chapter we will find that the most effective way to increase the fracture strength of a brittle material is to reduce the size of defects, such as cracks and pores.

Because of the high melting temperatures of most ceramics, melting is not a technique used to shape a ceramic into a product. Most ceramic parts are shaped as powders, such as in hot isostatic pressing (HIP). In HIP, powders are first pressed into the shape of the part. Then the part is subjected to a high pressure and temperature, which results in the diffusion bonding, or sintering, of the powder particles but does not melt them. HIP results in some pores or voids between the powder particles, as shown in Figure 7.19.

The fracture strength and elastic modulus of a ceramic are both increased by minimizing the porosity. The porosity is minimized by using higher temperatures and pressures, along with longer times to allow diffusion to eliminate the voids. Table 7.2 gives the density, flexural strength, and Young's elastic modulus of various alumina products. The theoretical density (ρ_t) of alumina is 3.97 g/cm^3. If each product is pure alumina, the fraction of porosity (P) is calculated from Equation 7.7:

$$P = \frac{\rho_t - \rho_o}{\rho_t} = 1 - \frac{\rho_o}{\rho_t} \qquad\qquad \textbf{7.7}$$

where ρ_o is the observed density.

Various models have been proposed that relate the strength and Young's modulus to porosity. One equation for flexural strength (σ_f) is given by Equation 7.8:

$$\sigma_f = \sigma_0 \exp(-nP) \qquad\qquad \textbf{7.8}$$

Table 7.2	The Density (g/cm³), Flexural Strength in MPa, and Young's Modulus in GPa of Sintered Alumina. The data should not be used for design purposes.

Density	Flexural Strength	Young's Modulus
3.42	296	221
3.60	338	276
3.70	352	303
3.90	379	370
3.92	400	386

Based on data from (http://www.coorstek.com/materials/ceramics/alumina.asp)

where σ_0 is the strength with 0 porosity, n is a number from 4 to 7, and the porosity is in volume fraction. A similar equation has been proposed for Young's modulus (E), as shown in Equation 7.9.

$$E = E_0 \exp(-mP)$$ **7.9**

It has been found that the strength of some ceramics, such as Si_3N_4 and ZrO_2, is increased through microstructure control. In Si_3N_4, acicular (elongated) grains, shown in Figure 7.20, are produced by adding other ceramics such as MgO, Al_2O_3, CaO, and Y_2O_3. During sintering of the Si_3N_4, these additional ceramics form a liquid that promotes the growth of acicular grains of the Si_3N_4. The acicular microstructure of the Si_3N_4 resists fracture, in a way similar to the fracture resistance found in a composite material. Increased energy is required to pull the acicular grains from the matrix. Si_3N_4 is produced with flexural strengths up to 1100 MPa and a Young's modulus up to 320 GPa.

5 µm

Figure 7.20   A scanning electron microscope image of the microstructure of sintered Si_3N_4, showing the acicular morphology of the grains. (*Photo courtesy of Dr. J. Wallace of the National Institute of Standards and Technology and presently with the Nuclear Regulatory Commission.*)

The fracture strength of ZrO_2 is increased by controlling the martensitic phase transformation that occurs in this ceramic. By adding other ceramics, such as MgO, CaO, MgO, Y_2O_3, and Ce_2O_3, the stability of the ZrO_2 is modified. The added ceramics lower the temperature of the martensitic transformation of the tetragonal phase to the monoclinic phase for ZrO_2 to the range of the service temperature of the part resulting in partially stabilized (PS) zirconia. The ZrO_2–CaO phase diagram in Figure 5.17 shows these phases. The ZrO_2 mixed with approximately 10 weight percent CaO is heat treated to produce the tetragonal phase mixed with the cubic phase at the service temperature. Under high stress, the tetragonal phase at cracks and pores transforms to the monoclinic phase with a volume expansion that produces compressive stresses at the tips of cracks. The compressive stress at the cracks impedes their propagation, which increases the resistance to fracture. We will cover this subject in more detail in Chapter 11. Modified ZrO_2 ceramics are proposed for high-temperature applications in gasoline and diesel engines.

It is also possible to make composite materials that will increase the ultimate tensile strength of a brittle material. Steel-reinforced concrete is an example of such a material. If a crack develops in brittle concrete, the crack propagation is stopped by the ductile steel reinforcing bars. Similarly, the brittle carbon-carbon tiles on the space shuttle have their strength increased by the graphite fibers that reinforce the carbon matrix. We will cover these composite materials in more detail in Chapter 12.

7.3.3   Strengthening Amorphous Ceramics

The most common amorphous ceramic is glass. The strength of bulk glass is increased through the process of tempering. In tempering, the glass is heated above the glass transition temperature but below the melting temperature. Then the surface of the glass is cooled, usually with jets of air. The surface temperature of the glass is reduced below the glass transition temperature, and it is rigid. The center of the glass is above the glass transition temperature, and it is a super-cooled liquid. As the center of the glass cools; it thermally contracts, and simultaneously it forces the rigid surface to contract. The forced contraction of the rigid surface results in surface compressive strains and stresses. The rigid surface also restricts the center of the glass from thermal contraction as it cools; thus in the center the atoms are stretched beyond the equilibrium interatomic positions, resulting in tensile strains and stresses. The result of this tempering process is compressive residual stress in the outer surface and tensile stress in the center of the glass. Since the applied stress is 0, the sum of all the internal stresses in the material must equal 0. Fracture of glass normally starts at the surface at a scratch or a crack, and the surface compressive stress inhibits the fracture of tempered glass and increases the fracture strength.

In the process of chemical tempering, ions larger than Si, such as K and Na, are substituted for the Si in SiO_2. The larger ions are introduced in the surface by diffusion at temperatures below the annealing temperature of glass, and the diffusion rates are accelerated by applied electric fields. Chemically tempered glass is produced with tensile strengths of 350 MPa. This strength is comparable to the strength of structural steel.

If glass fibers are drawn from a melt without damaging the surface of the fibers, and if they are immediately tested in uniaxial tension, the glass fibers have an incredibly high fracture stress (σ_f) of approximately 0.7×10^4 MPa. The glass fibers in fiberglass have a fracture stress of 0.5×10^4 MPa. If we test ordinary window glass, we would see that the tensile fracture stress drops to only 69 MPa. The difference is due to cracks and other defects in the ordinary glass. The freshly drawn glass fibers are defect free. Glass is made incredibly strong by eliminating the surface cracks and other defects. One practical application of high-strength glass fibers is its use in fiberglass.

7.4 WHISKERS: NEARLY PERFECT SINGLE CRYSTALS

Thermodynamically, it should be possible to make crystals that are free of dislocations and cracks, because there is no equilibrium concentration of dislocations or cracks in crystals. With no dislocations or cracks present, the shear yield strength of a crystal is theoretically calculated to be approximately the shear modulus divided by 30 (G/30). For iron, G/30 is 2720 MPa. This shear yield strength is the stress necessary for simultaneously moving all of the atoms in a plane of atoms over another plane of atoms, as shown in Figure 6.9. Experimentally, the shear yield strength of a well-annealed iron crystal with a low dislocation density is 14 MPa. This difference of nearly 200 times is due to the presence of dislocations in the iron crystal, and due to the ease of moving dislocations through a crystal with few barriers to dislocation motion.

It is possible to produce dislocation-free silicon single crystals. However, it is not possible at this time to grow bulk metal dislocation-free single crystals, because of the ease of dislocation formation in metals relative to that in silicon. The dislocations form in metals during the solidification process.

We can grow **whiskers** of metals, sapphire, silicon, silicon carbide, silicon nitride, and graphite in a vacuum that have strengths that approach the theoretical strength. The whiskers are typically a fraction of a micron in diameter and millimeters to a centimeter in length. A whisker has a single screw dislocation down its center, as schematically shown in Figure 7.21. The whisker grows by a screw-dislocation rotation mechanism that is explained with Figure 7.21. As new atoms are deposited on the end of the whisker, they diffuse to the step where the screw dislocation emerges from the end. The atoms attach to the whisker at the step, because here the deposited atom can form more bonds than it could on the flat surface. When

Figure 7.21 The atom arrangement around a screw dislocation, and the configuration of atoms in a whisker. *(Adapted from Barrett, C. R., Nix, W. D., and Tetelman, A. S., The Principles of Engineering Materials, Prentice-Hall, Englewood Cliffs, NJ (1973), p. 82.)*

Table 7.3	Tensile Strength (TS) and Elastic Modulus (E) of Some Whiskers	
Material	TS in GPa	E in GPa
Al_2O_3	20.7	428
Cr	8.9	242
Graphite	20.7	703
SiC	20.7	483
Si_3N_4	13.8	380

Based on data from Askeland, D.R., Fulay, P.P., and Wright, W.J., The Science and Engineering of Materials, 6th ed., Cengage Learning, Stamford, CT. (2011), p. 669.

atoms attach to the step, this causes the step to move, and the whisker length increases as the step spirals around the axis of the whisker.

The single screw dislocation up the whisker center has no resolved shear stress given by Equation 6.32. Both λ and ϕ are 90° if the whisker is loaded in tension along the whisker axis. There is no resolved shear stress forcing the screw dislocation to move. Also, the whisker is free of cracks. The tensile strength of whiskers is approximately that of the theoretical calculation. For example, the tensile yield strength of an iron whisker in the (111) orientation has been measured as nearly 12 GPa by Brenner. The theoretical shear yield strength is $G/30$, and the tensile theoretical yield strength is approximately $G/10$. The experimental result for iron is even more than the $G/10$ approximation to the theoretical strength of 8.16 GPa.

The tensile strength and the Young's elastic modulus for some whiskers are presented in Table 7.3. For an isotropic material with a Poisson's ratio of 0.3, the shear modulus is calculated with Equation 6.18 to be Young's modulus divided by 2.6.

Whiskers are incorporated into structural composite materials; however, the cost of whiskers is at present very high. Whiskers can only be considered for applications where high cost is not a concern.

7.5   CARBON NANOTUBES, GRAPHENE, AND NANOWIRES

Carbon nanotube and graphene schematics are shown in Figures 2.16e and 2.16f. A carbon nanotube consists of graphite wrapped into a cylinder of approximately 1 nm in diameter. The nanotube can be a single-walled sheet or a multiwalled set of sheets. Graphene is a single sheet of carbon in the graphite structure. At the time of this writing, carbon nanotubes have been grown up to 0.18 m in length. At present carbon nanotubes and graphene are the strongest and stiffest materials ever tested. Measured tensile strengths of carbon nanotubes range from 11 to 150 GPa, and Young's elastic modulus values range from 270 to 5000 GPa. The range of nanotube properties is due to different orientations of the graphite sheets, whether the nanotube is single-walled or multiwalled, the presence of vacancies, and the difficulty in testing a tube only 1 nm in diameter. The elastic modulus of graphene is determined to be 1000 GPa, and the tensile strength is 130 GPa based upon a nanoindentation experiment. Also, carbon nanotubes and graphene have an extremely high thermal conductivity of 3800 $Wm^{-1}K^{-1}$ and approximately 5000 $Wm^{-1}K^{-1}$, respectively.

Because of their small size, most of the mechanical applications of carbon nanotubes are in composite materials. Figure 1.11 shows a bundle of oriented bulk carbon nanotubes that could be used in composite materials. Bulk carbon nanotubes are carbon nanotubes that are mass produced and not individually selected for ideal properties. Present carbon nanotube composite applications include parts for wind turbines and high-performance sporting equipment, such as bicycles, skis, arrows, surfboards, and hockey sticks.

An interesting experimental biomedical application is to insert carbon nanotubes around cancerous cells and then to expose the nanotubes to radiofrequency waves. The radiofrequency waves heat the nanotubes in a manner somewhat analogous to microwave heating. However, the actual heating mechanism is still under study. The localized heat in the carbon nanotubes kills the cancer cells.

Future possible mechanical applications of carbon nanotubes include body armor and most any mechanical application where composite materials are applied. Interestingly, Damascus steel swords contain carbon nanotubes. If the process for making Damascus steel swords could be rediscovered, it might be possible to develop a process to strengthen steel with carbon nanotubes.

Mechanical applications of graphene have not yet been developed. One reason is that techniques for isolating graphene were developed only recently, in 2005. Also, sheet-type structures are not easily incorporated into composite materials. However, with all of the aerospace and biomedical applications that require a strong, lightweight skin structure, applications will likely develop soon for this material.

There are potential electronic applications for carbon nanotubes and graphene, because they are good conductors of electricity, and they can be made into semiconductors. Carbon nanotubes and graphene could replace conductors and semiconductors in electronic integrated circuits.

Research is being conducted on nanowires of metals, such as gold, and on ceramics, such as SiC, BN, and Si_3N_4. Nanometer-diameter wires of these materials are also very strong. However, research on these materials is not as advanced as that for carbon nanotubes.

7.6 STRENGTHENING THERMOPLASTIC POLYMERS

A polymer such as LDPE typically has a tensile strength of 13 MPa. However, stronger polymers are produced in a number of ways: by increasing the molecular mass of the LCMs, eliminating side chains, increasing the crystallinity, orienting the molecular chains in the direction of loading, modifying the structure of the LCMs, and cross-linking the LCMs. For example, PE is made into a fiber with a tensile strength of 3300 MPa by orienting the polymer chains along the fiber axis. In the following sections, we will first discuss the increases in strength possible with a given polymer. Then we will cover increases in strength by chemical modifications to the mer structure.

7.6.1 Strengthening a Specific Thermoplastic Polymer

Strengthening Thermoplastic Polymers by Increased Molecular Mass
One way to increase the strength of a polymer is to increase the molecular mass of the LCMs, by increasing the number of mer units in each chain. Increasing the average length of the chains increases the number of molecules from different chains that are bonded to each other, and this increases the strength.

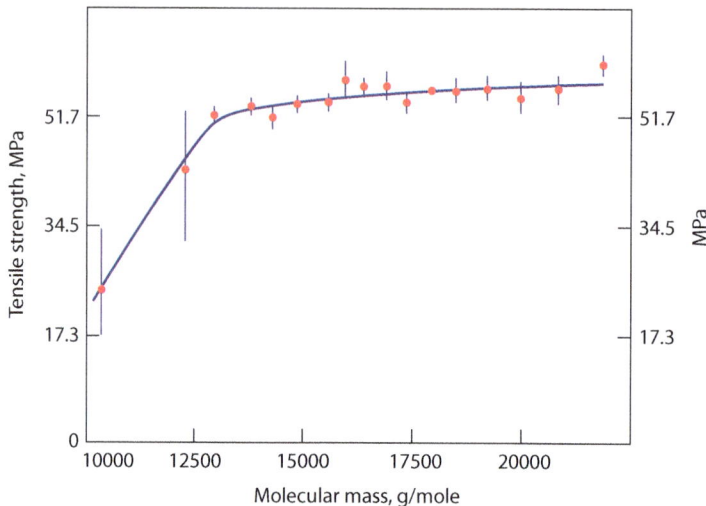

Figure 7.22   The tensile strength of polycarbonate as a function of molecular mass in atomic mass units. (*Based on Golden, J.H., Hammant, B.L., and Hazell, E.A. J. Polym. Sci. 2A, (1964) 4787.*)

It requires less force to shear two short molecules by each other than two longer molecules that are in contact along a longer distance. Polymers with low molecular mass are often liquids, and as the molecular mass increases the strength increases, as shown for polycarbonate in Figure 7.22. This figure shows that at a low molecular mass, the strength of polycarbonate increases rapidly as the chain molecular mass increases, but after a critical value of approximately 12,500 g/mole, there is no further increase in tensile strength. There is a limit to the strengthening effect of molecular mass.

Strengthening Thermoplastic Polymers by Reducing Side Branches

It is possible to change the strength of a given polymer by controlling the side branching. Side branches are shown in Figure 3.5. Extensive side branching results in low density and strength, because the LCMs cannot come close together to bond. Weak bonding between LCMs results in a low tensile strength of 7 to 17 MPa for LDPE, as shown in Table 7.4. HDPE has longer LCMs than does LDPE, and also less side branching. HDPE has a tensile strength from 20 to 40 MPa, as shown in Table 7.4.

Strengthening Thermoplastic Polymers with Crystallinity

The degree of crystallinity affects the tensile strength of polymers. Figure 7.23 shows the tensile strength of PE from different commercial sources as a function of degree of crystallinity. The tensile strength is more than doubled when the crystallinity increases from 65% to 95%. If the polymer is crystalline, the molecules are located in positions that result in the strongest bonding and the highest strength. Crystallinity is increased by slow cooling from the melt, because then the LCMs have more time to orient into the strongest bonding positions and into a crystalline structure. Crystallinity is increased if side branching is minimized and the LCM is uniform, as in PE and PTFE.

Strengthening Thermoplastic Polymers by Orienting LCMs

The LCMs of a polymer are oriented in the direction of material flow by deforming the polymer rapidly at a temperature just above the glass transition temperature. If a polymer is deformed below the glass transition temperature, it is brittle and fractures, and if it is deformed well above the glass transition temperature, the polymer flows and deformation cannot orient the LCMs. In an unstrained amorphous polymer, the LCMs

Table 7.4	Values for the Elastic Modulus (E), Tensile Strength (TS), and Elongation to Failure (ε_f) of Selected Polymers		
Polymer	E (GPa)	TS (MPa)	ε_f (%)
LDPE[1]	0.14–0.3	7–17	200–900
HDPE[1]	0.7–1.4	20–40	100–1000
OUHMWPE[2]	172	3300	—
PP[1]	1.1–2	30–40	100–600
PVC (rigid)[1]	1–3.5	40–75	30–80
PS[1]	2.4–3.2	30–60	1–4
PC[1]	2.1–2.4	70–90	100–120
PEEK[1]	3.6	90–200	50
PMMA[1]	2.5–3.3	55–75	3–5
ABS[1]	2–2.8	30–50	5–70
Kevlar 29[3]	59	3500	—
Kevlar 49[3]	124	3600	2.3
PF[4]	5.6–12	25	0.4–0.8
Epoxy[1]	3–5	30–90	1–2

1 Brostow, W., et al., Mechanical Properties, Physical Properties of Polymers Handbook, Mark, J. E. ed., AIP Press, Woodbury, N.Y. (1996), p. 331

2 Askeland, D.R., Fulay, P.P., and Wright, W.J., The Science and Engineering of Materials, 6th ed., Cengage Learning, Stamford, CT. (2011), p. 693.

3 Wen, J., Some Mechanical Properties of Typical, Polymer-Based Composites, in Physical Properties of Polymers Handbook, Mark, J. E. ed., AIP Press, Woodbury, N.Y. (1996), p. 372.

4 Osswald, T.A., and Menges G., Materials Science of Polymers for Engineers, Hanser Publishers, Munich (2003), Table I.

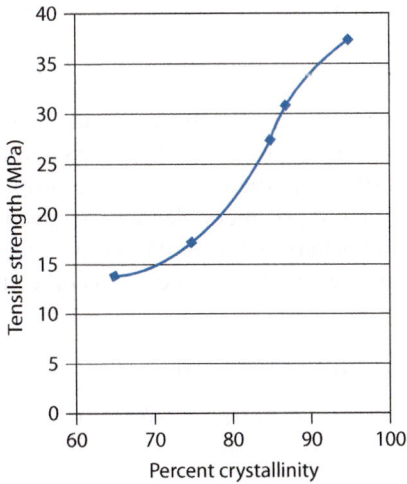

Figure 7.23 The tensile strength of commercial PE from different sources as a function of the percent crystallinity. (*Based on Boening, H.V., Polyolefins: Structure and Properties, Elsevier Press, Lausanne, (1966), p. 57.*)

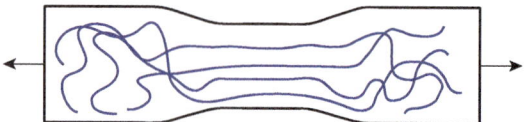

Figure 7.24   A schematic of the orientation of LCMs in a polymer deformed rapidly at a temperature just above the glass transition temperature. In the wide-grip area, the molecules have the original random orientation, because the polymer is not deformed there. In the reduced-gauge section, the polymer is deformed and the LCMs are oriented in the direction of deformation.

are randomly oriented. In ductile polymers subject to a tensile force, a neck forms at the initial peak stress after the elastic portion of the tensile stress-strain diagram, as shown in Figure 6.23. With continued strain, the neck propagates. In the necked region, the LCMs become oriented in the direction of material flow, as shown in Figure 7.24. Then the strong carbon-carbon bonds along the LCM resist the stress, rather than the weak dipole bonds between different LCMs. Orienting the LCMs strengthens the polymer.

We can take advantage of how the LCMs orient during plastic deformation to make strong polymers. For example, when PE is drawn into a fiber of OUHMWPE, the ultimate tensile strength is increased to approximately 3300 MPa; and the ultimate tensile strength of the oriented aramid polymer fiber Kevlar is 3600 MPa. The molecular structure of an aramid polymer is shown in Figure 7.25.

The production of fibers such as OUHMWPE and Kevlar is by the process of **solution spinning**. In solution spinning, the fiber material is initially dissolved in a solvent, such as paraffin oil, where the fiber's LCMs maintain a local crystalline arrangement. The solution and crystalline LCMs are extruded through a small orifice, and the solvent evaporates, leaving the crystalline LCMs in a fiber. The fiber is then drawn at a temperature above the glass transition temperature and below the melting temperature, in a process similar to wire drawing, to orient the LCMs. This elongates the LCMs in the direction of the fiber axis. In this way a polymer fiber is produced that has a tensile strength comparable to that of high-strength steel wires.

These polymers also have a high elastic modulus in the direction of the fiber axis, because the covalent bonds of the molecules are aligned along the fiber axis. Figure 7.26 shows the increase in elastic modulus that occurs in OUHMWPE as a function of draw ratio. Draw ratio is the amount of strain in the axial direction. The orienting of the polymer molecules is different than that resulting from strain hardening a metal by drawing it into a wire. If the oriented polymer is heated, the chain orientations do not change, and the strength is retained. If a drawn metal wire is heated, the strength decreases because of annealing.

Strengthening and Weakening of Thermoplastic Polymers with Blends and Alloys

In a **polymer blend**, two or more polymers are combined, but the polymers are separate in two or more different phases. PS and PE do not mix, and they will form a blend if combined. Adding PS to PE strengthens the PE. The properties of the blended polymers are predicted with volumetric averages of the polymers blended, as given in Equation 7.4. Polymer blends are normally produced by first mixing pellets of the individual polymers, and then producing the desired shape by processes such as extrusion. We will discuss the extrusion of polymers in Section 13.4.1.

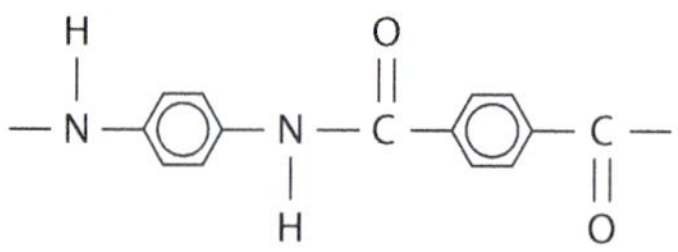

Figure 7.25   The mer structure of an aramid polymer. (*Based on Askeland, D.R., Fulay, P.P., and Wright, W.J., The Science and Engineering of Materials, 6th ed., Cengage Learning, Stamford, CT. (2011), p. 670.*)

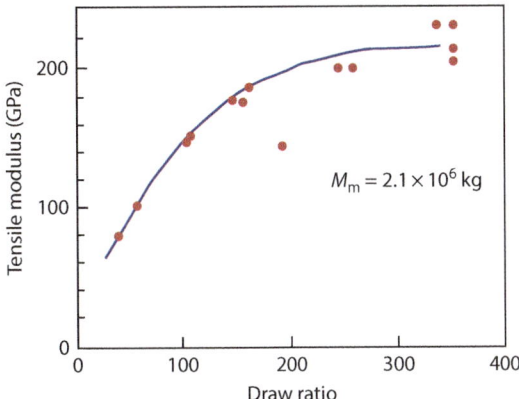

Figure 7.26 The Young's modulus of UHMWPE as a function of draw ratio to produce OUHMWPE at a high draw ratio. (*Based on Osswald, T.A., and Menges G., Materials Science of Polymers for Engineers, Hanser Publishers, Munich (2003), p. 417.*)

In a **polymer alloy**, two or more polymers are combined, and they mix to form a homogenous single phase. For example, polyphenelene ether (PPE) and PS mix, and when combined they form an alloy. The properties of an alloy can significantly differ from those of any of the components, and also significantly differ from an average of the components. Alloys are produced by heating pellets of the polymers above the melting temperature of the component polymers.

There are many possible blends and alloys of polymers. Producing a blend or alloy to obtain desired properties is much less expensive than developing new polymers. However, the production of blends and alloys of polymers impedes the recycling of polymers.

A **plasticizer** is a solvent that lowers the strength and elastic modulus of the polymer with which it is mixed. Pure PVC is a brittle polymer at room temperature, but with the addition of the plasticizer dioctylphthalate, the glass transition temperature of PVC is reduced to below room temperature. PVC that has a glass transition temperature below room temperature can be applied where a ductile polymer is required. Plasticizers are added to polymers so that they can be deformed into shape at lower temperature. Figure 7.27 shows the effect of a 50% addition of plasticizer on the shear modulus of PVC. The figure shows that the shear modulus in pure PVC is much higher at room temperature (23°C) than in the mixture of PVC and 50% plasticizer.

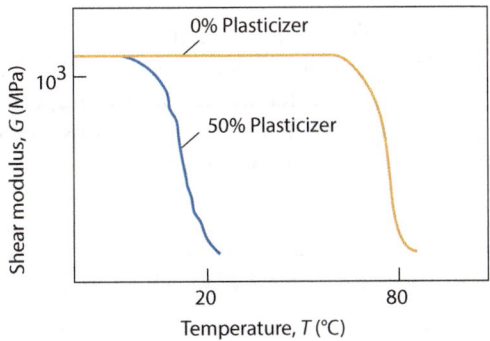

Figure 7.27 The shear modulus as a function of temperature for PVC and for PVC with 50% plasticizer. (*Based on Osswald, T.A., and Menges G., Materials Science of Polymers for Engineers, Hanser Publishers, Munich (2003), p. 410.*)

7.6.2   Strengthening Thermoplastic Polymers by Mer Modification

A way to strengthen polymers is to replace the hydrogen atoms on a simple mer, such as PE, with different atoms or different groups of atoms. In PVC, one of the hydrogen atoms on the PE mer is replaced by a chlorine atom. The bonding between LCMs in PE is primarily by fluctuating-dipole bonds. PVC also has fluctuating-dipole bonds, and the chlorine adds permanent-dipole bonds between the LCMs, as shown in Figure 2.22. The permanent-dipole bonds increase the tensile strength and the elastic modulus of PVC, as shown in Table 7.4. Figures 2.19 and 2.20 show some of the common polymer molecular structures, such as PE, PVC, and PMMA.

In PMMA some of the hydrogen atoms in PE are replaced by much larger methyl and acrylic groups. Steric hindrance is the increased difficulty of the LCMs of PMMA with large methyl and acrylic side groups to slide past each other relative to the LCMs of PE with uniform hydrogen atoms. This results in PMMA having a higher room-temperature yield strength of 55 to 75 MPa relative to that of HDPE with a yield strength of approximately 20 to 40 MPa, as shown in Table 7.4.

It is also possible to have atoms and molecules other than carbon such as oxygen, nitrogen, or benzene as the backbone of the LCM. See, for example, in Figure 2.20 polycarbonate (PC) and polyetheretherketone (PEEK) and in Figure 7.25 the aramid Kevlar. When these atoms and molecules replace carbon it can result in a stiffer LCM, stronger permanent-dipole bonding between LCMs, and steric hindrance. PEEK has one of the highest tensile strengths for nonoriented polymers of 90 to 200 MPa and a relatively high elastic modulus of 3.6 GPa, as shown in Table 7.4. PEEK has one of the highest melting temperatures (334°C) of all thermoplastic polymers; thus it is used in high-temperature applications and in high-temperature composites for aerospace applications. PEEK and other polymers such as polypenylene ether (PPE), polyetherketone (PEK), polysulfone (PSU), and polysthersulfone (PES), are high-performance thermoplastics.

7.6.3   Strength of Thermoplastic Copolymers

In copolymers, different polymers are combined into a single LCM, as shown in Figure 3.6. Impact-resistant polystyrene (IPS) is produced by forming a copolymer of PS (PS ≈ 30%) with butadiene rubber (BR ≈ 70%). The impact resistance of IPS is as much as seven times greater than that of PS. Styrene acrylonitrile (SAN) is a random amorphous copolymer of PS (PS ≈ 70%) and polyacrylonitrile (PAN ≈ 30%). PAN is used as a fiber (Orlon™). The impact resistance of SAN is greater than that of PS, but not as high as that of IPS.

To create a polymer of greater impact resistance than what SAN offers, BR is added to SAN to produce the copolymer acrylonitrile butadiene rubber styrene (ABS). Depending upon how much of each component is present in the copolymer, different mechanical properties result. The properties listed in Table 7.4 are representative of the most widely used compositions. ABS has become one of the most commonly utilized room-temperature engineering plastics.

7.7   STRENGTH OF THERMOSET POLYMERS

There are two types of thermoset polymers: network polymers, such as phenolformaldehyde (PF), which is commercially known as Bakelite™, and cross-linked polymers, such as rubber, epoxy, and polyester.

7.7.1 Strength of Network Thermoset Polymers

Figure 7.28a shows that formaldehyde (CH_2O) reacts with two phenols (the ring structures) to form PF with the by-product of H_2O (water). The attachment of formaldehyde to phenols can occur at the three locations noted by the three circled hydrogen atoms in Figure 7.28a. Figure 7.28b shows attachments at multiple locations in a planar arrangement; however, the orientation of the ring structures can be out of the plane. This results in a three-dimensional network of covalent bonds. The resulting polymer is of relatively high strength, as shown in Table 7.4, and the elastic modulus is higher than any of the nonoriented thermoplastic polymers, because PF has covalent bonds in three dimensions. Also, the plastic-strain mechanism of LCM sliding cannot occur in this solid with three-dimensional covalent bonding, and the strain to fracture is less than 1%, as shown in Table 7.4. The resistance to fracture of PF is increased with fillers, such as chopped glass fibers or cellulose, resulting in a composite material. Other network polymers include urea formaldehyde and melamine formaldehyde.

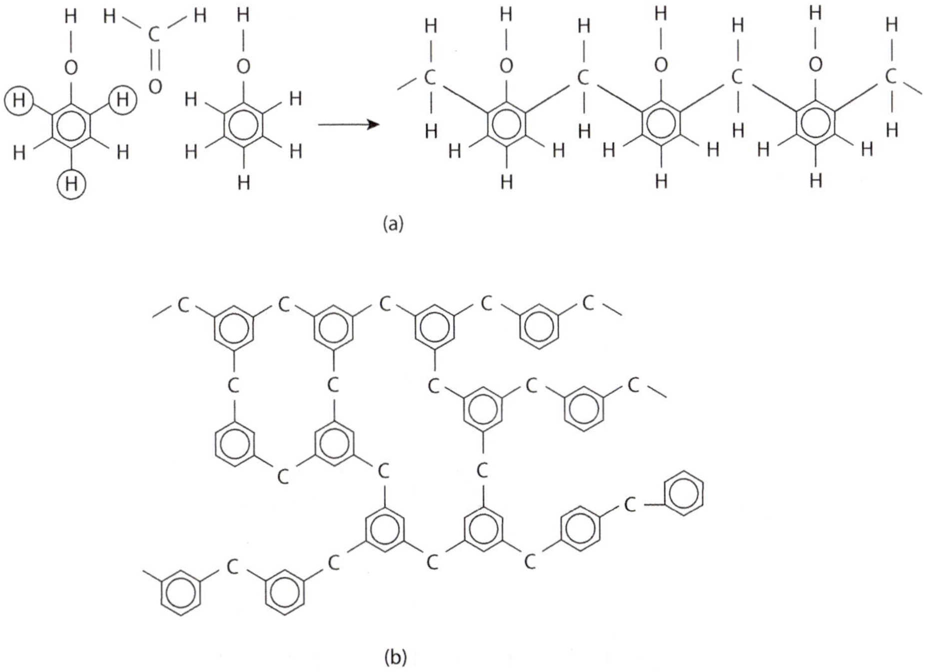

(a)

(b)

Figure 7.28 (a) Formaldehyde reacts with two phenol molecules, which are the ring structures, to form phenolformaldehyde (PF). The formaldehyde can react with any one of the three circled hydrogen atoms on the phenol on the far left. (b) A three-dimensional network of covalent bonds results from the reaction of formaldehyde at the three possible locations on the phenol molecule. (*Based on Askeland, D.R., and Phule, P.P., The Science and Engineering of Materials, 4th ed. Thomson-Brooks/Cole, Pacific Grove, CA (2003), p. 705.*)

7.7.2   Strength of Cross-Linked Thermoset Polymers

Cross-linking of the LCMs in polymers results in a significant increase in strength and elastic modulus, as well as a reduction in ductility. The cross-linking of LCMs is schematically shown in Figure 2.23. Rubber is one of the most common cross-linked polymers. In rubber liquid, latex is cross-linked with sulfur in a process called vulcanization to form solid rubber, as shown in Figure 2.24. Natural rubber that has not been vulcanized is thermoplastic. Unvulcanized rubber has a very low elastic modulus, as shown in Figure 7.29, and plastic deformation occurs by the relative sliding of the LCMs. The sulfur atoms that are added to natural rubber during the vulcanization process cross-link the natural-rubber LCMs. This cross-linking reduces the LCM sliding, and it produces a thermosetting elastomer with a significantly higher elastic modulus and strength, as shown in Figure 7.29. When stretched rubber is unloaded, the LCMs return to positions that are close to their original coiled positions, because of the sulfur cross-link bonds. Vulcanized natural rubber can be elongated over 600% without fracture, and it will return to its original length.

A thermoplastic liquid polymer epoxy resin is shown in Figure 7.30a along with an ethylene diamine hardener molecule. The epoxy is in the circled groups at the ends of a polymer LCM that usually contains phenol groups, such as bisphenol A. In ethylene diamine the two ends of the molecule are amines (NH_2), and the remainder of the molecule is ethylene molecules (C_2H_4) repeated n times. The thermoplastic liquid polymer epoxy resin is converted to a rigid solid thermosetting epoxy through a chemical reaction with the hardener that cross-links the epoxy resin LCMs, as shown in Figure 7.30b. In the chemical reaction, the bond between the oxygen and CH_2 on the epoxy group is broken, and the CH_2 bonds with NH at the end of the ethylene diamine hardener. The oxygen on the epoxy group bonds with a hydrogen atom that is broken from the end of the ethylene diamine. The amine at the other end of the ethylene diamine molecule reacts with another epoxy resin molecule to cross-link two LCMs, as shown in Figure 7.30b.

The mechanical properties of epoxy are shown in Table 7.4. Note that the tensile strength of epoxy is from 30 to 90 MPa, in comparison to the tensile strength of LDPE of 7 to 17 MPa. Also, the elastic modulus of the epoxy is 3 to 5 GPa; this is an order of magnitude higher than the value for LDPE of

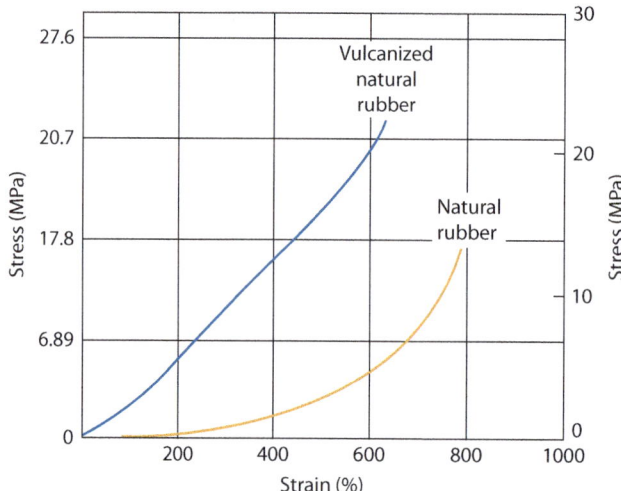

Figure 7.29   Stress-strain diagrams for natural rubber and vulcanized natural rubber. (*Based on Eisenstadt, M.M., Introduction to Mechanical Properties of Materials, The Macmillan Co. N.Y. (1971), p. 89.*)

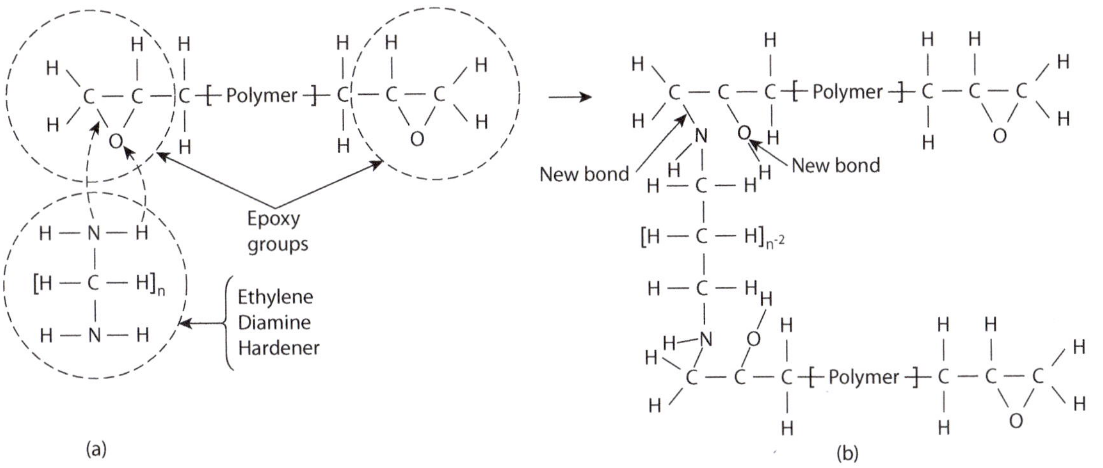

Figure 7.30 (a) A single linear epoxy resin polymer and an ethylene diamine hardener. (b) The reaction of two linear thermoplastic epoxy resin polymers with an ethylene diamine hardener to form a cross-linked thermosetting epoxy.

0.14 to 0.3 GPa. The reason for the higher value of elastic modulus for epoxy resin is because the cross-linked epoxy has covalent bonds between the LCMs. When a stress is applied to epoxy, covalent bonds resist the applied stress. The cross-link covalent chemical bonds prevent the epoxy LCMs from sliding past each other. This increases the yield strength of the epoxy, but it also results in a brittle material that has only a 1 to 2% elongation to failure.

Once the liquid epoxy resin and liquid hardener are mixed, the mixture remains liquid for sufficient time to be cast into a shape, or infiltrated into reinforcing fibers or fabrics to produce composite materials. This makes epoxy ideal for use with composite materials, because it is not necessary to heat the epoxy resin and hardener mixture to obtain a liquid. Thermoplastic composite matrix materials, such as PEEK, must be heated above the melting temperature to have a liquid for infiltration into the composite fabric. Other cross-linked polymers similar to epoxy include polyesters, vinyl esters, and polyurethanes.

Thermoset polymers do not melt when heated, because the LCMs are connected by covalent bonds. In the presence of oxygen, thermoset polymers oxidize and deteriorate. The absence of a melting temperature makes it difficult to recycle thermoset polymers. Thermoset polymers and composite materials made with a thermoset polymer matrix are usually ground into powder and can be used as filler in other polymers or in asphalt.

Summary

- When a metal is plastically strained and strain hardens, the dislocation density in the metal increases. The yield strength of the metal increases proportional to the square root of the dislocation density.
- When a strain-hardened metal is annealed, it is heated for a period of time to reduce the yield strength and the dislocation density. In a recovery anneal, some of the ductility is recovered, and the yield strength is reduced. To obtain the maximum ductility, the metal must be recrystallized. Recrystallization is the formation and growth of new grains.

- Grain boundaries are obstacles to dislocation motion, and they increase the yield strength of metals. The yield strength of metals is proportional to the negative one-half power of the average grain diameter. This is the Hall-Petch relationship.

- A metal is dispersion strengthened by the addition of particles of a hard material. The tensile strength of dispersion strengthened materials is modeled with the rule of mixtures using volume fraction. The mechanism of dispersion strengthening a ductile matrix is that the hard dispersion particles stop dislocation motion in the ductile matrix when the dislocations come to the hard-phase boundary, and the hard-dispersion particles constrain the ductile metal particles from deforming.

- Solid-solution strengthening in a metal occurs when foreign solid-solution atoms interact with the strain field of edge dislocations.

- In precipitation-hardened metals, a new second phase forms as small particles in the host metal, which increase the yield strength. The increase in yield strength is inversely proportional to the spacing between the precipitate particles, and the yield strength is maximized if the open space between the particles (L) is minimized. "Aging" is to hold a metal at a specified temperature and time for the purpose of forming a precipitate. The process of formation of the precipitate is by nucleation and growth. In the early stages of aging, the yield strength increases rapidly with time as coherent precipitate particles are formed and grow in size. This decreases the particle spacing (L) and increases the yield strength (σ_y). If the precipitate changes from coherent to incoherent this corresponds to peak strength. An over-aged alloy is one aged beyond its peak strength. As an alloy is over-aged, the spacing between the precipitate particles increases.

- Most ceramics are brittle at room temperature, because the fracture stress is lower than the yield stress. The focus in strengthening ceramics is on increasing the fracture stress, and this is primarily achieved by reducing defects in the ceramic. The tensile strength and elastic modulus are proportional to the negative exponential of the porosity.

- The strength of bulk glass is increased by the process of tempering. The result of the tempering process is compressive residual stress in the outer surface, and this increases the tensile strength. In chemical tempering of glass, a compressive stress in the surface results from the substitution of larger atoms for silicon atoms.

- If glass fibers are drawn from a melt without damaging the surface of the fibers, and they are immediately tested in uniaxial tension, the glass fibers have an incredibly high fracture stress of approximately 0.7×10^4 MPa. The freshly drawn glass fibers are defect free, whereas normal window glass has defects.

- It is possible to grow whiskers that have strengths that approach the theoretical strength of the material. Whiskers are typically a fraction of a micron in diameter and millimeters to a centimeter in length. A whisker has a single screw dislocation down its center that results from the whisker growth mechanism.

- A carbon nanotube is a sheet (single walled) or sheets (multiwalled) of graphite wrapped into a cylinder of approximately 1 nm in diameter. Graphene is a single sheet of carbon in the graphite structure. Carbon nanotubes have been grown up to 0.18 m in length. At present, carbon nanotubes and graphene are the strongest and stiffest materials ever tested.

- The strength of a polymer is increased if the molecular mass of the LCMs is increased by increasing the number of mer units in each chain.

- Side branching in polymers reduces their density and strength, because the LCMs cannot come close together to bond.

- If a polymer is crystalline, the molecules are located in positions that result in the strongest bonding and the highest strength.
- The LCMs of a polymer are oriented in the direction of deformation by deforming the polymer rapidly at a temperature just above the glass transition temperature. In this way a polymer fiber is produced that has an ultimate tensile strength comparable to that of high-strength steel wires.
- In a polymer blend, two or more polymers are combined, but the polymers separate into two or more different phases. The properties of a blend are normally an average of the properties of the individual polymers.
- In a polymer alloy, two or more polymers are combined, and they mix into a homogenous single phase. The properties of an alloy can be significantly different from those of any of the components, and significantly different from those of an average of the components.
- A way to strengthen polymers and to increase the elastic modulus is to replace the hydrogen atoms on a simple mer, such as for PE, with different atoms or molecules. This can result in both fluctuating- and permanent-dipole bonds between the LCMs. If hydrogen atoms are replaced by large molecules, this provides steric hindrance that resists sliding between LCMs.
- Plasticizers are solvents that mix with polymers and lower their strength and elastic modulus.
- In copolymers, different polymers are combined into a single LCM. Depending upon how much of each component is present in the copolymer, different mechanical properties result. The copolymer ABS has become one of the most commonly utilized room-temperature engineering plastics.
- There are two types of thermoset polymers: network and cross-linked. Network polymers are three-dimensional polymers, such as phenolformaldehyde (PF). PF has a relatively high tensile strength and elastic modulus, but a strain to fracture of only 1%. In cross-linked polymers, such as rubber, epoxy, and polyester, a chemical reaction links LCMs together for three-dimensional covalent bonding. Cross-linking of the LCMs in polymers results in an increase in strength and elastic modulus and a reduction in strain to fracture.

Supplemental Reading Subjects and Authors

Full references are listed at the end of the book.

General:	*Askelund, Fulay, and Wright*
Strengthening metals:	*Cahn; Courtney; Dieter; Hertzberg; Reed-Hill; Reed-Hill and Abbaschian; Brenner*
Strengthening ceramics:	*Carter and Norton; Chaing, Birnie, and Kingery*
Strengthening polymers:	*Hearle; Kinney; McCrum, Buckley, and Bucknall; Schultz; Osswald and Menges; Strong; van Krevelen; Winding and Hiatt*

Homework

Concept Questions

1. In metals, at temperatures below half of the melting temperature, plastic strain is due to _____ motion.

2. To increase the room-temperature yield strength of a metal containing dislocations, it is necessary to introduce _____ to dislocation motion.

3. Strain hardening results from _____ strain.

4. Edge dislocations whose tangent and Burger's vector are not on a most closely packed plane are _____.

5. The increase in tensile and shear yield strength due to the strain hardening of metals is proportional to the dislocation density to the _____ power.

6. A(n) _____ source produces new dislocations when a segment of a dislocation is pinned at two ends on a slip plane and the resolved shear stress reaches a critical value.

7. A recovery _____ is when a strain-hardened metal is heated to an elevated temperature and the ductility is increased, but no new grains form.

8. If a strain-hardened metal is heated to an elevated temperature and new grains form, this process is _____.

9. If a crystal contains only two edge dislocations with Burger's vectors of opposite signs and they come together on the same slip plane, the resulting number of dislocations is equal to _____.

10. As a result of an anneal, the dislocation density is _____.

11. The yield strength of metals is proportional to the negative _____ power of the grain diameter.

12. When edge dislocations on a close-packed plane in a grain cannot slip in an adjacent grain with differently oriented close-packed planes, dislocations form a(n) _____ at a grain boundary.

13. A steel with a small grain size has a (<u>higher or lower</u>) yield strength than the same steel with a larger grain size.

14. _____ strengthening is when small hard particles are added to a ductile metal.

15. In dispersion-strengthened metals, the strength and elastic modulus of the alloy are proportional to the _____ fraction of the dispersion.

16. One disadvantage of dispersion-strengthened metals at high temperatures is that the dispersion and the matrix are not in _____ with each other, and as a result the matrix and dispersion may react with each other.

17. Metal alloys dispersion strengthened with oxide particles are usually produced by _____ metallurgy techniques.

18. In solid-solution strengthened metals, substitutional and interstitial solid-solution atoms interact with _____ dislocations in the metal.

19. A large substitutional atom is attracted to the region of an edge dislocation that has _____ strains.

20. The increase in strength due to solid-solution strengthening is proportional to the _____ power of the atom fraction of solute atoms.

21. The concentration of solid-solution atoms at a stationary edge dislocation is greater than the average concentration of solid-solution atoms in the solid, and this results in the _____ yield point in low-carbon steels.

22. Strengthening by cooling a solid solution and forming small particles of a new phase is called _____ hardening.

23. The increase in yield strength of a precipitation-hardened alloy is _____ proportional to the spacing between the particles.

24. The first step in the precipitation-hardening process is the _____ treatment to produce a uniform distribution of solute atoms in the grains.

25. The second step in the precipitation-hardening process is to _____ the alloy to maintain a uniform distribution of atoms in the grains at low temperature.

26. The third step in the precipitation-hardening process is to _____ the alloy to form the precipitate.

27. During precipitation-hardening, at peak strength the precipitate changes from coherent to _____.

28. _____ stresses result from strains that are not due to applied external forces.

29. Whiskers have one _____ dislocation up the center of the whisker.

30. In brittle ceramics, the fracture strength is lower than the _____ strength.

31. Polycrystalline MgO is brittle at room temperature because to maintain continuity with plastic strain in a polycrystalline material, there must be at least _____ active slip systems.

32. In brittle materials, the easiest way to increase the _____ strength is to decrease the size of cracks and pores.

33. In partially stabilized zirconia (ZrO_2), the addition of materials such as MgO or CaO lowers the temperature of the _____ transformation of the tetragonal to the monoclinic phase.

34. To temper glass, the glass is initially heated to a temperature above the glass transition temperature but below the _____ temperature.

35. In tempered glass, the stress in outer surface is in a state of _____.

36. In _____ tempering of glass, ions larger than Si, such as K and Na, are introduced into the surface.

37. One method of strengthening the ceramic Si_3N_4 is by producing a(n) _____ microstructure.

38. The theoretical and observed tensile yield strength of whiskers is approximately equal to _____ of the elastic modulus.

39. A carbon _____ is one or more sheets of graphite wrapped into a cylinder terminated at the ends.

40. A sheet of graphite that is one atom layer thick is called _____.

41. Side branches on a polymer (<u>increase or decrease</u>) the density, crystallinity, yield strength, and elastic modulus.

42. An increase in crystallinity in a polymer (<u>increases or decreases</u>) the yield strength.

43. To orient the LCMs of a polymer, it must be rapidly deformed just above the _____ transition temperature and below the melting temperature.

44. If two polymers mix in the liquid state, this is called a(n) _____.

45. A(n) _____ is a material added to a polymer to lower the glass transition temperature.

46. The chlorine atoms on PVC polymer LCMs increase the amount of _____-dipole bonding between LCMs.

47. The LCMs of PMMA cannot easily slide past each other at room temperature, because of _____ hindrance.

48. PEEK, along with other polymers, such as PPE, PEK, PSU, and PES, is a high-performance _____ polymer.

49. LCMs of ABS have mers of PAN, BR, and PS on one polymer chain, resulting in a(n) _____.

50. Thermoset polymers (<u>do or do not</u>) melt when heated to a high temperature in the presence of oxygen.

51. The epoxy resin LCMs are _____ with covalent chemical bonds after mixing with a hardener.

52. In a network thermoset polymer, there is _____-dimensional covalent bonding.

Design-Related Questions

1. _____hardening is the least expensive process for strengthening metals, because no expensive alloy elements are added, and it is only necessary to deform the metal with standard deformation processes.

2. Solid-solution strengthening is (<u>compatible or incompatible</u>) with designs that require the part to be heated during fabrication by processes such as welding or hot-working.

3. If a precipitation-hardened alloy that is aged to peak strength is heated during processing to produce the part, the strength may be (<u>increased or decreased</u>).

4. The requirement to use powder metallurgy techniques makes oxide dispersion-strengthened metals (<u>expensive or inexpensive</u>) in comparison to other metal-strengthening techniques.

5. In developing a precipitation-hardened aluminum-copper alloy, what is the maximum amount of copper in weight percent that can be completely in solution in a solid aluminum-copper alloy at any temperature?

6. In selecting an unalloyed alumina material for a design, it is expected that the one with the _____ density would have the highest tensile strength.

7. The use of polymer blends and alloys in a design (<u>improves or impedes</u>) the ability to recycle the material.

8. Thermoset polymers (<u>can or cannot</u>) be melted for processing or recycling.

Engineer in Training–Style Questions

1. Strain hardening in metals is due to
    (a) A reduction in dislocation density due to dislocations running out of the crystals
    (b) The creation of immobile dislocations by dislocation reactions
    (c) A reduction in dislocation density by dislocation annihilation
    (d) Foreign atoms pinning dislocations

2. Which of the following strengthening processes in metals does not require the addition of foreign atoms or compounds to a pure metal?
    (a) Grain boundary
    (b) Dispersion

    (c)   Precipitation

    (d)   Solid solution

3.   Which of the following is not expected to provide solid-solution strengthening?

    (a)   A large substitutional atom in the tensile strain region of an edge dislocation

    (b)   A small substitutional atom in the compressive strain region of an edge dislocation

    (c)   A small interstitial atom in the compressive strain region of an edge dislocation

    (d)   A small interstitial atom in the tensile region of an edge dislocation

4.   The peak strength in a precipitation-hardened alloy occurs when

    (a)   The alloy is quenched.

    (b)   The precipitate is coherent.

    (c)   The precipitate changes from coherent to incoherent.

    (d)   The precipitate is incoherent.

5.   Which of the following aging treatments of a precipitation-hardened alloy would be expected to result in the most stable precipitate particles?

    (a)   As-quenched

    (b)   Under-aged

    (c)   Aged to peak strength

    (d)   Over-aged

6.   Which of the following procedures would not increase the fracture strength of a polycrystalline ceramic?

    (a)   Decreasing the porosity

    (b)   Solid-solution alloy additions to increase the yield strength

    (c)   Creating an acicular microstructure

    (d)   Mixing ductile steel wires with the ceramic

7.   Which of the following procedures would not be expected to increase the yield strength of LDPE?

    (a)   Quenching the LDPE to room temperature from the liquid state

    (b)   Increasing the crystallinity

    (c)   Increasing the molecular mass

    (d)   Decreasing the side branching

8.   Which of the following materials is not cross-linked?

    (a)   Rubber

    (b)   Epoxy

    (c)   UHMWPE

    (d)   Polyester

9.   Which of the following processes would you expect to produce small-diameter PE fibers that have a very high tensile strength?

    (a)   Draw fibers from a liquid

    (b)   Draw fibers at a temperature just below the glass transition temperature

    (c)   Draw fibers at a temperature just above the glass transition temperature

    (d)   Draw fibers at room temperature

Problems

Problem 7.1: Predict the dislocation density present if polycrystalline copper is work hardened to a shear yield strength of 500 MPa, using the data in Figure 7.3 and the results of Example Problem 7.1.

Problem 7.2: Annealed polycrystalline low-carbon iron has a tensile yield strength of 60×10^6 Pa, and a dislocation density of 1×10^{13} m^{-2}. This iron is cold-rolled to a tensile yield strength of 600×10^6 Pa.

   (a) Estimate the dislocation density of the cold-rolled iron.

   (b) You must make some assumptions to solve this problem. Justify your assumptions.

Problem 7.3: In Figure 7.4, the time for the fraction of residual strain in iron to drop to 0.6 at 400°C is 137 minutes, and at 450°C it is 15 minutes.

   (a) Calculate the activation enthalpy for the recovery of cold work in iron.

   (b) Comment on the magnitude of the activation enthalpy for recovery in comparison to the activation enthalpy for vacancy diffusion in iron, and justify your result.

Problem 7.4: Predict the yield strength of the titanium in Figure 7.9 with 4% pre-strain if the grain size is reduced to 100 nm.

Problem 7.5: For the titanium with a 2% pre-strain in Figure 7.9, what grain size would result in a yield stress of 1000 MPa? You can use the results of Example Problem 7.2 to determine your answer.

Problem 7.6: The room-temperature yield strength of very-low-carbon steel as a function of the square root of the interstitial solid-solution carbon concentration is shown in Figure 7.12.

   (a) Evaluate all of the constants in the equation for the tensile yield strength as a function of solid-solution chemical composition.

   (b) The maximum equilibrium composition of interstitial carbon possible in ferritic iron is 0.0011 atom fraction. If an iron-carbon alloy of this composition could be produced at room temperature, what should the yield strength be? You must solve this problem analytically; you can check your result graphically.

   (c) If the composition of 0.0011 atom fraction of carbon is exceeded, the extra carbon forms iron carbide. If a specimen is produced with 0.0015 atom fraction of carbon, do you expect the strength to be given by an extension of the line in the figure? Briefly explain your answer.

Problem 7.7: A dispersion-strengthened cutting tool will be made of a 80 vol% commercial purity nickel matrix and 20 vol% synthetic diamond. Nickel has an elastic modulus, given in Table 6.1, of 200 GPa and a tensile fracture strength of 400 MPa, and synthetic diamond has an elastic modulus of 900 GPa and a tensile fracture strength of 1000 MPa. Estimate the tensile strength and elastic modulus of this alloy.

Problem 7.8: In the precipitation-hardened aluminum-lithium alloy in Figure 7.13, the average spacing measured between precipitate particles is 0.10×10^{-6} m. The shear modulus of the alloy is 28×10^9 Pa. The tensile yield strength of the alloy as quenched without any precipitates is 100×10^6 Pa. Aluminum is FCC with a lattice parameter of 0.404 nm. Predict the tensile yield strength of this alloy after formation of the precipitates.

Problem 7.9: Determine the increase in the open spacing between precipitate particles that occurs when the 2014 aluminum alloy shown in Figure 7.17 is aged for 1000 hours at 149°C, relative to the value at peak strength that is calculated in Example Problem 7.5.

Problem 7.10: Assume that the as-quenched yield strength of 100 MPa is the value of σ_0, that the lattice parameter of FCC aluminum is 0.404 nm, and that the shear modulus of this alloy is equal to that of aluminum given in Table 6.1.

Problem 7.11: An aluminum alloy with 4 weight percent copper is to be precipitation hardened. The alloy is solution treated and quenched, and after the quench the shear yield strength is 38×10^6 Pa. Assume that the copper atoms remain in solid solution after the quench. The alloy is then artificially aged to a shear yield strength of 180×10^6 Pa. Determine the spacing between the precipitate particles that could produce this increase in yield strength. The shear modulus of this alloy is 27×10^9 Pa, the lattice parameter (a) of the aluminum alloy is 0.404×10^{-9} m, and aluminum is FCC with an atom at each lattice site.

Problem 7.12: Isothermal aging curves in Figure 7.17 show that peak yield strength at aging temperatures of (232°C) and (260°C) occurs in times of $10^{0.6}$ hours (4 hours) and $10^{-0.6}$ hours (0.25 hours), respectively. Calculate the activation enthalpy for precipitate formation in this alloy.

Problem 7.13: (a) Calculate the porosity of the alumina materials in Table 7.2, assuming that the alumina is pure and that any change in density from the theoretical value of 3.97 g/cc is due to porosity. (b) Plot the natural log of the flexural strength and Young's modulus as a function of porosity to see if the plots agree with Equations 7.7 and 7.8. (c) Evaluate the terms σ_0, E_0, m, and n. (d) Test your results by comparing calculated values of σ_f and E for a density of 3.42 g/cc to the measured values.

Problem 7.14: A polymer blend will be 75% PVC and 25% volume fraction LDPE, to provide additional ductility to the PVC. Estimate the elastic modulus, tensile strength, and elongation to failure of the proposed blend if the LDPE and PVC have the following properties:

LDPE: $E = 0.2$ GPa, $\sigma_u = 10$ MPa, and $\varepsilon_f = 500\%$

PVC: $E = 3.0$ GPa, $\sigma_u = 50$ MPa, and $\varepsilon_f = 30\%$

A photograph of the first flight of the Wright flyer. The Wright brothers were the first to fly a heavier-than-air aircraft. On December 17, 1903 the aircraft traveled 36.6 m (120 ft) in 12 seconds at Kitty Hawk, North Carolina. The frame of the plane was built of spruce and ash wood covered with muslin cloth. The 12-horsepower (hp) internal-combustion gasoline engine was primarily made of cast iron and steel; however, to lighten the engine, the crankcase was made of aluminum with 8% copper. Aluminum had recently become available commercially with the development of the Hall-Heroult electrolytic refining process in 1886.

The materials in aircraft have changed significantly since the Wright flyer. The frames of planes are now made from high-strength aluminum or titanium alloys. The skin of commercial aircraft is now high-strength aluminum sheet and graphite-fiber-reinforced epoxy in the Boeing 787. Supersonic aircraft have skins of titanium alloys, and the highest-temperature locations of the outer surface of the space shuttle are covered with graphite-fiber-reinforced carbon. The engines in jet aircraft are turbines that can develop up to 35,000 hp with titanium alloys in the compressor and superalloys in the highest-temperature locations.

Library of Congress, Prints & Photographs Division, LC-DIG-ppprs-00626;
© Rhea Eason/Alamy

The goals of this chapter are to understand

- What metals, ceramics, and polymers are available for engineering designs
- The mechanical properties of each of these materials, including, density, tensile strength, yield strength, elastic modulus, and strain to failure
- The designation systems for metals
- The processing procedures that are indicated with material designations
- The ways to relate the strength of these materials to the strengthening mechanisms and processing procedures that affect strength
- Some of the present sources of information for determining mechanical properties
- Some typical applications of materials

Chapter 8

Engineering Materials

8.1 METAL ALLOYS

The primary commercial structural metal alloys include the face-centered cubic (FCC) alloys of aluminum, copper, nickel, and austenitic stainless steel; body-centered cubic (BCC) iron; and hexagonal close-packed (HCP) magnesium and titanium. The tables of this chapter use several numbering systems, including that of the Aluminum Association (AA), the American Iron and Steel Institute (AISI), the American Society for Testing Materials (ASTM), the Society of Automotive Engineers (SAE), and the unified numbering system (UNS).

In the UNS, the designation of each metal alloy consists of a letter followed by five digits. The UNS digits often incorporate numbers from other systems; however, this is not always true. The letter indicates the type of metal; for example, stainless steels start with the letter S, as shown in the following list:

Axxxxx—Aluminum Alloys
Cxxxxx—Copper Alloys, including Brass and Bronze
Fxxxxx—Iron, including Ductile Irons and Cast Irons
Gxxxxx—Carbon and Alloy Steels
Hxxxxx—Steels—AISI H Steels
Jxxxxx—Steels—Cast
Kxxxxx—Steels, including Maraging, HSLA, Iron-Based Superalloys

L5xxxx—Lead Alloys, including Babbit Alloys and Solders
M1xxxx—Magnesium Alloys
Nxxxxx—Nickel Alloys
Rxxxxx—Refractory Alloys
  R03xxx—Molybdenum Alloys
  R04xxx—Niobium (Columbium) Alloys
  R05xxx—Tantalum Alloys
  R3xxxx—Cobalt Alloys
  R5xxxx—Titanium Alloys
  R6xxxx—Zirconium Alloys
Sxxxxx—Stainless Steels and Iron-Based Superalloys
Txxxxx—Tool Steels
Zxxxxx—Zinc Alloys.

8.1.1   Treatment of Metal Alloys

Metals can be purchased in many different conditions. The two main conditions for metals are **wrought** and **cast**. Wrought alloys are those produced by mechanical deformation processes, such as rolling, extruding, and drawing. The conditions for wrought alloys include recrystallized, annealed, stress relieved, hot rolled, cold rolled, and quenched and tempered. Recrystallization produces the lowest-strength and highest-ductility metal. The strength increases and ductility decreases for metals in this order: annealed, stress relieved, hot rolled, cold rolled, and quenched and aged or tempered. Not all metal alloys can be strengthened by quenching and tempering; there must be a precipitate that forms in sufficient concentration and appropriate distribution for the alloy to be heat treatable. In the tables below, data is presented for an annealed material and a high-strength material when it is available. In cast alloys, liquid metal is poured into a mold that has a cavity of the desired solid shape. Cast alloys can be heat treated to increase the strength.

The tables of strength below show that the strongest alloys are produced by the heat treatment of quenching and aging or tempering. Several quenching media are used for metal alloys, and in order of decreasing cooling rate, the most frequently used quenching media are saltwater brine, water, oil, and air. Agitating the metal can also increase the cooling rate by moving vapor bubbles away from the cooling metal. Quenching can produce deformation of the metal and residual stresses. Residual stresses are relieved from steel during tempering and from other metal alloys by aging. Tempering of steel and aging of appropriate composition alloys of aluminum, magnesium, and copper forms precipitate particles that provide strength.

8.1.2   Aluminum Alloys

Aluminum is approximately ten times as expensive as carbon steel; it has a lower elastic modulus than does carbon steel, and carbon steels are available with a higher yield strength and tensile strength than those of any aluminum alloys. However, aluminum has properties that make it the metal of choice for some designs. For example, aluminum alloys have a density of approximately 2.71 g/cc, and for carbon steels the density is approximately 7.87 g/cc. Aluminum alloys are therefore one-third the density of steel, which makes aluminum preferable where weight is an important factor, such as in aircraft structures, fuel-efficient automobiles, trucks, and trains and in objects that must be lifted by hand, such as cookware and hand tools. The **specific strength**, which is the strength divided by the density, and the specific elastic modulus are both high for aluminum, as we will discuss in Section 8.1.8. Aluminum is corrosion resistant in air but not in seawater. In air Al_2O_3 forms on the surface of aluminum alloys and provides a protective oxide from further oxidation. Aluminum oxidation resistance results in applications such as aluminum window and door frames. Aluminum has good thermal conductivity and electrical conductivity, resulting in applications in heat exchangers and as electrical conductors. Since aluminum is FCC, it is quite easily deformed, and it is available in wrought forms of sheet, plate, bars, wire, rod, and extrusions. Aluminum is readily cast into shape because it has a relatively low melting temperature of 660°C. Pure aluminum is ductile to very low temperatures, resulting in cryogenic applications.

The AA has the following numbering system for wrought aluminum alloys:

Alloy Series	Major Alloying Element
1xxx	99.00% minimum aluminum
2xxx	Copper
3xxx	Manganese
4xxx	Silicon
5xxx	Magnesium
6xxx	Magnesium and silicon
7xxx	Zinc
8xxx	Other elements

Cast aluminum alloys have the following AA numbering system.

Alloy Series	Major Alloying Element
1xx.x	99.00% minimum aluminum
2xx.x	Copper
3xx.x	Silicon plus copper and/or magnesium
4xx.x	Silicon
5xx.x	Magnesium
6xx.x	Unused series
7xx.x	Zinc
8xx.x	Tin
9xx.x	Other elements

The UNS for aluminum has A as the first letter, followed either by a "9" for wrought alloys, or by a digit from 0 to 6 for cast alloys. The final four digits in the number usually correspond to the AA number. Table 8.1 presents some of the commercial aluminum alloys with both the UNS and AA number.

Aluminum alloys have the following designations indicating the treatment:

F—as fabricated
Hxy— strain hardened
O—annealed and recrystallized
T—solution heat treatment plus age hardening.

For Hxy, the first digit x designates heat treatment after the strain hardening as follows:

x = 1—strain hardened
x = 2—strain hardened and partially annealed
x = 3—strain hardened and stabilized

The second digit y designates the relative amount of strain hardening.

y = 2—quarter-hard
y = 4—half-hard
y = 8—full-hard
y = 9—extra-hard

For example, "H32" indicates the aluminum was strain hardened to quarter-hard and then stabilized. The solution-treatment and age-hardening designations (T) are as follows:

T1—Naturally aged (aged at ambient temperature)
T2—Annealed (applies to cast products only)
T3—Solution heat treated, cold worked, and then naturally aged
T4—Solution heat treated and then naturally aged
T5—Artificially aged
T6—Solution heat treated and then artificially aged
T7—Solution heat treated and then artificially over-aged
T8—Solution heat treated, cold worked, and then artificially aged
T9—Solution heat treated, artificially aged, and then cold worked
T10—Artificially aged and then cold worked

The Website *www.keytometals.com* has an excellent presentation of the temper designations of metal alloys.

Table 8.1 presents the AA designation, the UNS designation, the treatment as thermal (T) or mechanical (H_{xy}), the chemical composition of the primary alloy elements (CC), the room-temperature density (ρ), the ultimate tensile strength (σ_u), the yield strength (σ_y), the elongation to failure (ε_f), and the elastic modulus (E). Density is only included for the first listing of an alloy type because heat treatment does not significantly change the density of an alloy. Data are presented for one of the most commonly used alloys for each designation area as an example; however, there are many more alloys in addition to those listed. Students can search other alloys on *www.matweb.com.* Similar data are available from the American Society for Metals Handbook, but *www.matweb.com* is available to all students. Table 8.1 lists the typical value for each alloy; however, if a range of mechanical properties is given on *www.matweb.com*, then the lowest value is listed here. If a range of chemical compositions is given, then a number in the middle of the range is presented, and only the major alloy elements are presented. In this chapter, chemical composition in percentages always refers to weight percent. Data are presented for alloys in the annealed condition and in a high-strength condition appropriate for that alloy if it is available. From the annealed and high-strength data students can observe the effects of strengthening processes covered in Chapter 7. Also, when data are available for different thicknesses, the result for 12.7 mm is reported. In Table 8.1 there is a listing for (a) wrought aluminum alloys and for (b) cast alloys.

Table 8.1 The AA and UNS Designations, Treatment (T), Room-Temperature Density (ρ), Ultimate Tensile Strength (σ_u), Yield Strength (σ_y), Strain to Failure (ε_f), and Elastic Modulus (E)

AA (UNS)—T	ρ g/cc	σ_u MPa	σ_y MPa	ε_f %	E GPa
(a) Wrought Aluminum Alloys					
1100 (A91100)-O	2.71	89.6	34.5	45	68.9
1100 (A91100)-H18		165	152	15	68.9
2024 (A92024)-O	2.78	179	75.8	20	73.1

Table 8.1 *Continued*

AA (UNS)—T	ρ g/cc	σ_u MPa	σ_y MPa	$\varepsilon_f\%$	E GPa
2024 (A92024)-T62		415	340	5	72.4
2090 (A92090)-O	2.59	210	190	11	76.0
2090 (A92090)-T86		550	520	6	76.0
3003 (A93003)-O	2.73	110	41.4	40	68.9
3003 (A93003)-H18		200	186	10	68.9
4032 (A94032)-T6	2.68	379	317	9	78.6
5052 (A95052)-O	2.68	193	89.6	30	70.3
5052 (A95052)-H38		290	255	8	70.3
6061 (A96061)-O	2.70	124	55.2	30	68.9
6061 (A96061)-T6		310	276	7	68.9
7075 (A97075)-O	2.81	228	103	17	71.7
7075 (A97075)-T6		572	503	11	71.7
8090 (A98090)-T651	2.54	515	450	6	77.0
(b) Cast Alloys (all are sand cast)					
295 (A02950)-T62	2.81	249	193	2	70.0
356 (A03560)-T6	2.68	207	138	3	72.4
443 (A04430)-O-F	2.69	117	48.3	3	71.0

The data are primarily from www.matweb.com. The data should not be used for design purposes.

Below are listed the chemical compositions of alloy elements in weight percent, and applications of the alloys listed in Table 8.1.

(a) Wrought Alloys

1100 (not heat treatable)—Si + Fe < 1—Food-handling and chemical-handling equipment, storage equipment, heat exchangers

2024 (heat treatable)—Cu = 4.4, Mg = 1.5—Aircraft structures, gears and shafts, bolts, computer parts, hydraulic valve parts, missile parts, pistons, fastening devices, gears, truck wheels, machine parts

2090 (heat treatable)—Cu = 2.7, Li = 2.3—Low-density aircraft structure parts, cryogenic tanks

3003 (not heat treatable)—Mn = 1.25, Si<0.6—Cooking utensils, pressure vessels, piping

4032 (heat treatable)—Si = 12.3, Cu = 0.9, Mg = 1.05, Ni = 0.9—Master-brake cylinders, transmission valves, bushings for rack-and-pinion steering systems, bearings, hydraulic applications, forged pistons

5052 (not heat treatable)—Mg = 2.6, Cr = 0.25—Aircraft fuel and oil lines, aircraft fuel tanks, appliance parts, rivets, wire

6061 (heat treatable)—Mg = 1.0, Si = 0.6—Aircraft fittings, marine fittings and hardware, electrical fittings and connectors, brake pistons, valves and valve parts, truck bodies, canoes, railroad cars, furniture, pipelines

7075 (heat treatable)—Zn = 5.6, Mg = 2.5, Cu = 1.6—Highly stressed aircraft structures

8090 (heat treatable)—Li = 2.5, Cu = 1.3, Mg = 0.9, Zr = 0.1—Low-density aircraft structures

(b) Cast Alloys

> 295 (heat-treatable castings)—Cu = 4.5, Si = 1.1—Flywheel and rear axle housings, bus and aircraft wheels, crankcases
>
> 356 (heat treatable castings)—Si = 7.0, Mg = 0.3—Aircraft pump parts, automobile transmission cases, water-cooled cylinder blocks
>
> 443—Si = 5.3—Cooking utensils, food-handling equipment, marine fittings, thin-section castings, die castings

The commercially pure 1100-series alloy can only be work hardened. The 3000- and 5000-series alloys are solid solution hardened, and in addition they can be work hardened; however, they cannot be precipitation hardened. The 2000-, 6000-, 7000-, and 8000-series alloys are precipitation hardened. The 7000-series alloys are the highest-strength aluminum alloys. One of the strongest of the 7000-series alloys is 7075, with a yield strength of 572 MPa.

Aluminum alloys are normally welded in an inert atmosphere; welding in oxygen results in extensive oxidation of the metal. The commercially pure aluminum (1100) and the solid-solution-strengthened alloys (3xxx and 5xxx) can be welded. Some of the precipitation-hardened alloys such as 6061 can be welded. Other precipitation-hardened alloys, such as 7075 and 2024, are not recommended for welding, because welding changes the precipitate distribution and the strength in the alloy. If these alloys are welded, they can fail in the weld zone, because the weld zone would not have the strength of the base metal.

8.1.3   Copper Alloys

Copper is an FCC metal, and it is very ductile and of low strength in the pure state. Because of its high ductility, pure copper is easily mechanically worked into wrought forms of wire, sheet, rod, pipe, plate, and extruded shapes. Wrought copper can be in the ductile annealed (O) condition, or it can be in the hard (H) cold-worked condition. Since copper has the FCC crystal structure, it can be extensively deformed and strain hardened. Most copper alloys are solid solution hardened, and they can be strain hardened and grain size strengthened. The yield strength of many wrought copper alloys can be increased by a factor of approximately 4 by strain hardening. Some alloys, such as cartridge brass, can be plastically deformed extensively into thin-walled tubes suitable for ammunition casings. Copper-beryllium alloys are precipitation hardened. The yield strength of heat-treated copper-beryllium (1.9 weight precent) is increased by a factor of 13 relative to annealed pure copper, and the elastic modulus is increased by 14%. These alloys are the highest-strength copper alloys, and they are wear- and corrosion resistant. Thus they have found application as bearing materials in high-performance applications. Copper also has a relatively low melting temperature of 1085°C, and it is quite easily cast into shape.

Copper and most of its alloys are corrosion resistant in air and in saltwater; these properties result in applications, such as heat exchangers and marine parts. Copper and copper alloys also have high thermal and electrical conductivity, and these properties result in many applications as thermal and electrical conductors. Copper is also ductile at very low temperatures because of its FCC crystal structure, and it has cryogenic applications.

Copper alloys are known by names, such as brass. **Brass** alloys are copper plus zinc, and they are utilized in many saltwater applications because of their corrosion resistance. Brass is also utilized in applications that require extensive plastic deformation without significant work hardening, such as in the formation of bullet cartridge cases. **Bronze** alloys are copper plus any element other than zinc. The brass and bronze alloys are solid solution strengthened, and they can be cold worked, as shown in Table 8.2. Copper-nickel alloys are highly corrosion resistant, and they are utilized in applications such as seawater heat exchangers, although their thermal conductivity is more than an order of magnitude less than that of pure copper.

Nickel-silvers are copper alloys with nickel and zinc as the primary alloying elements not necessarily silver. Copper alloys with 1 to 6 weight precent lead are free machining grades that are used for machined parts.

Listed below is the UNS system for copper alloys. Copper alloys all start with the letter C followed by five digits. C10100 to C79900 are for wrought alloys, and C80100 to C99750 are for cast alloys.

(a) Wrought Alloys

C10100–C15760—Pure and low-alloy copper greater than 99% Cu
C16200–C16500—Cadmium copper
C17000–C17700—Copper-beryllium (beryllium bronzes)
C18000–C19900—High-alloy copper greater than 96% Cu
C20500–C29800—Copper-zinc brasses
C31000–C35600—Leaded brasses (Cu + Zn + Pb)
C40400–C49080—Tin brasses (Cu + Zn + Sn + Pb)
C50100–C52900—Phosphor bronzes (Cu + Sn + P)
C53200–C54800—Leaded phosphor bronzes (Cu + Sn + P + Pb)
C55180–C56000—Cu-Ag-P and Cu-P brazing filler metal
C60600–C64400—Aluminum bronzes
C64700–C66100—Silicon bronzes
C66200–C66420—Copper alloys
C66700–C67820—Manganese bronzes
C68000–C69950—Silicon brasses and other Cu-Zn alloys
C70100–C72950—Copper-nickel alloys
C73150–C79900—Nickel silvers and leaded nickel silvers (Cu + Ni + Zn + Pb) not Ag

(b) Cast Alloys

C80100–C81200—Greater than 99% Cu
C81300–C82800—Chromium copper and beryllium copper (>96% Cu)
C83300–C85800—Red, yellow, and leaded brasses
C86100–C86800—Manganese bronzes and leaded manganese bronzes
C87300–C87900—Silicon brasses and bronzes
C89320–C89940—Cu-Sn-Bi-(Se, Zn, or Ni) alloys
C90200–C94500—Tin bronzes and leaded tin bronzes
C94700–C94900—Nickel-tin bronzes
C95200–C95810—Aluminum bronzes
C96200–C96800—Copper-nickel alloys
C97300–C97800—Nickel silver alloys (Cu + Ni + Zn) not Ag
C98200–C98840—Leaded copper alloys
C99300–C99750—Other copper alloys

The treatment designations for copper alloys are as follows:

- Annealed—O
- Annealed to a grain-size requirement of 0.xxx mm—OSxxx
- As manufactured—M
- Cold worked—H
- Heat treated—T

Within each category above are many subcategories that are explained on *www.keytometals.com.* Some examples of tempers of alloys presented in Table 8.2 are as follows:

- O61—annealed.
- OS050—annealed to a grain size of 0.050 mm
- H00—cold worked to one-eighth hard
- H01—cold worked to one-fourth hard
- H02—cold worked to one-half hard
- H04—cold worked to full hard
- H10—cold worked to extra-spring hard
- M01—as sand cast
- TF00—cold worked to one-eighth hard and then precipitation hardened
- TQ50—quench hardened and temper annealed

Table 8.2(a) lists some common wrought copper alloys, and Table 8.2(b) lists some cast alloys.

| Table 8.2 | The Common Name, Treatment (T), Room-Temperature Density (ρ), Ultimate Tensile Strength (σ_u), Yield Strength (σ_y), Strain to Failure (ε_f), and Elastic Modulus (*E*) for Copper Alloys |

Name (UNS)	T	ρ g/cc	σ_u MPa	σ_y MPa	ε_f %	*E* GPa
(a) Wrought Copper Alloys						
Electrolytic Tough pitch	H00	8.89	250	65	40	115
Beryllium Copper	TF00	8.36	1130	890	3–10	131
Cartridge	H00	8.53	380	275	48	110
Brass	H04		525	435	8	110
High	H01	8.83	325	250	30	110
Conductivity Bronze	H10		540	525	3	110
Phosphor	OSO50	8.86	325	130	64	110
Bronze	H04		560	515	10	110
Copper	O61	8.94	380	125	36	150
Nickel	H02		515	485	15	150
(b) Cast Alloys						
Leaded Yellow Brass	M01	8.45	235	83	35	76
Aluminum	M01	7.47	515	205	12	110
Bronze	TQ50		620	310	6	110

Based on data primarily from www.matweb.com.

Below are listed the common names, UNS designations, chemical compositions of alloy elements in weight percent, and applications of the copper alloys in Table 8.2.

(a) Wrought Alloys

Electrolytic tough pitch, C11000—all<1—Gutters and downspouts, roofing, gaskets, automobile radiators, electrical buss bars, nails, printing rolls, rivets, electrical wire, electrical conductors

Beryllium copper, C17200, Be = 1.9—Springs, bellows, firing pins, bushings, valves, diaphragms

Cartridge brass, C26000, Zn = 30—Radiator cores, flashlight shells, lamp fixtures, fasteners, locks, hinges, ammunition casings, plumbing accessories, rivets

High-conductivity bronze, C40500, Zn = 4, Sn = 1—Meter clips, terminals, fuse clips, contact and relay springs, washers

Phosphor-bronze, C51000, Sn = 5.0, P = 0.35—Bellows, diaphragms, bourdon tubing, clutch discs, fasteners, lock washers, wire brushes, chemical hardware, textile machinery, welding rods

Copper-nickel, C71500, Ni = 31, Fe = 0.6—Condensers, condenser plates, distiller tubing, evaporator and heat-exchanger tubing, ferrules, saltwater piping.

(b) Cast Alloys

Leaded yellow brass, C85400, Zn = 28, Pb = 2.5, Sn = 1—Furniture hardware, ornamental castings, radiator fittings, ship trimmings, battery clamps, valves, fittings

Aluminum-bronze, C95400, Al = 11, Fe = 4, Ni < 2.5—Pump impellers, bearings, gears, worm gears, bushings, valve seats and guides, nonsparking hardware

8.1.4   Iron Alloys

Plain-Carbon and Alloy Steels

Plain-carbon and low-alloy steels are the primary materials for building structures, automobiles, bridges, and so on. **Steel** is made from iron, carbon, and other alloy elements. **Plain-carbon steels** have carbon as the main strengthening element and at least 0.45% but less than 1.5% manganese, and no other significant alloy elements. Manganese is added to all steels to deoxidize the steel and to provide solid solution strengthening. When the steel is deoxidized or denitrided, the elements oxygen and nitrogen are removed from solid solution and form oxide or nitride particles. In steel the carbon content is typically below 1%. **Low-carbon steels** contain less than 0.25% carbon. These steels are not quenched and tempered, and the microstructure is ferrite plus pearlite. These alloys are relatively low-strength but ductile, and they are utilized in applications such as automobile bodies, structural beams, and piping. **Medium-carbon steels** contain between 0.25% and 0.6% carbon. These steels can be quenched and tempered for high strength. **High-carbon steels** have more than 0.6% carbon, and they also can be quenched and tempered. The designation of steels is based upon the alloy elements in the steel. In the AISI/SAE system, each number contains four digits, as shown in Table 8.3. The first digit indicates the alloy elements other than carbon. For example, plain-carbon steels with less than 1.5% manganese (Mn) are given the designation 10XX, where 0.XX is the carbon content in weight percent. 1020 steel has 0.20 weight percent carbon. The 1.5% maximum for manganese is noted for the 10xx steels, and this applies to all steel types listed in Table 8.3, except for the 13xx steels, which have 1.6% to 1.9% manganese.

Table 8.3	Major Classifications of Steel in the AISI/SAE System

10xx—Mn = <1.5

11xx—Free machining S = 0.08–0.33

12xx—Free machining S = 0.10–0.35, P = 0.04–0.12

13xx—Mn = 1.6–1.9

2xxx—Ni = 3.5–5

3xxx—Ni = 1.25–3.5, Cr = 0.65–1.57

40xx—Mo = 0.20–0.30

41xx—Mo = 0.08–0.35, Cr = 0.4–1.1

43xx—Mo = 0.2–0.3, Ni = 1.65–2.00, Cr = 0.4–0.9

46xx—Mo = 0.15–0.30, Ni = 0.70–2.00

5xxx—Cr = 0.2–1.15

61xx—Cr = 0.5–1.6, V = >0.15

7xxx—W

8xxx—Ni = 0.2–0.7, Cr = 0.4–0.6, Mo = 0.08–0.4

9xxx—Si = 1.2–2.2, Ni = 0–3.5, Cr = 0–1.4

The major alloy elements are listed in weight percent. Manganese of less than 1.5% but more than 0.45% is present in all alloys, unless otherwise listed.

Plain-carbon steels are the cheapest steels, but they have limitations. As the strength is increased by adding carbon and by quenching and tempering, the ductility is significantly decreased. Plain-carbon steels become brittle at low temperatures; we will cover this subject in more detail in Chapters 9 and 11. Plain-carbon steels are not corrosion resistant. For all of these reasons, alloy steels are developed to improve the properties of carbon steels.

Low-alloy steels typically contain a total of less than 5% of alloy elements. In alloy steels, other first numbers are applied as shown in Table 8.3. For example, the alloy steel 4340 typically has 1.82% Ni, 0.80% Cr, 0.25% Mo, 0.7% Mn, and 0.40% C. Mo significantly strengthens steel. Nickel, copper, and silicon provide solid solution strengthening to steel. Silicon and aluminum are added to steel to combine with oxygen and nitrogen, to form oxide and nitride particles that deoxidize and denitride the steel.

Table 8.4 presents the AISI/SAE designation, UNS designation, major alloying elements, density, tensile strength, yield strength, elongation to failure, and elastic modulus of some carbon and low-alloy steels. The UNS designation system for plain-carbon and alloy steels starts with the letter G, followed by five digits, the first four of which are the AISI/SAE designation given in Table 8.3. In Table 8.4 under treatment, steel that is annealed is designated by A, oil-quenched and tempered by OQT, as-rolled by AR, and cold-rolled by CR. The amount of carbon is not listed for some alloy steels, to leave room for other alloy elements; the amount of carbon can be deduced from the AISI/SAE designation. The chemical composition (CC) for a given alloy can occupy two lines in the table. The density is not included, because it is generally equal to 7.85 g/cc, and Young's elastic modulus is approximately equal to 205 GPa for all of these alloys.

Table 8.4	The AISI/SAE Designation (UNS designation), Treatment (T), Chemical Composition of Alloy Elements (CC), Ultimate Tensile Strength (σ_u), Yield Strength (σ_y), and Strain to Failure (ε_f) for Wrought Plain-Carbon and Low-Alloy Steels

AISI/SAE (UNS)	T	CC wt%	σ_u MP	σ_y MPa	ε_f %
1020 (G10200)	A	C = 0.20, Mn = 0.45	395	295	37
1020 (G10200)	CR		420	350	15
1040 (G10400)	A	C = 0.40, Mn = 0.75	515	350	30
1040 (G10400)	OQT		722	500	27
1060 (G10600)	A	C = 0.60, Mn = 0.75	625	370	22
1060 (G10600)	OQT		965	634	20
1080 (G10800)	AR	C = 0.80, Mn = 0.75	965	585	12
1080 (G10800)	OQT		1165	841	15
4140 (G41400)	A	C = 0.40, Mo = 0.20,	655	415	26
4140 (G41400)	OQT	Cr = 0.90, Mn = 0.88	965	800	18
4340 (G43400)	A	Mo = 0.25, Ni = 1.8	745	470	22
4340 (G43400)	OQT	Cr = 0.8, Mn = 0.7	1255	1165	14
6150 (G61500)	A	Cr = 0.95, V = >0.15	670	415	23
6150 (G61500)	OQT	Mn = 0.8, Si = 0.2	1240	1225	15

Based on data primarily from www.matweb.com.

Below are applications of the plain-carbon and low-alloy steels listed in Table 8.4.

1020 Shafts, lightly stressed gears, hard-wearing surfaces, pins, chains, case-hardened parts where core strength is not critical

1040 Machine, plow, and carriage bolts; tie wire, cylinder-head studs, machined parts, U-bolts, concrete reinforcing rods, forgings, noncritical springs

1060 Spring wire, forging dies, railroad wheels

1080 Music wire, springs, chisels, hammers, axe heads, forging die blocks

4140 Gears for aircraft gas-turbine engines and transmissions

4340 Piston pins, bearings, gears, dies, pressure vessels

6150 Automotive leaf springs, valve springs, piston rods, pump parts, spline shafts

Some of the things to note in Table 8.4 are that increasing the amount of carbon in plain-carbon steels from 0.20 to 0.80% nearly doubles the yield strength, but it also decreases the elongation to failure by a factor of 4. Adding less than 1% of the carbide formers Cr, V, and Mo to a 0.40% carbon steel can double the yield strength. Cr, V, and Mo react with carbon more readily than does iron at certain temperatures, and they form small carbide particles rather than the large iron carbide plates in pearlite. Alloy steels with Cr, V, Mo, and Nb are normally quenched and then tempered at a temperature that maximizes the reaction of these alloy elements with carbon relative to the reaction of iron with carbon. As a result, these alloy elements can produce a large increase in strength due to precipitation hardening if the spacing (L) between the particles is small, where the yield strength is given by Equation 7.6. The alloys 4140, 4340, and 6150 in the quenched-and-tempered condition are high-strength steels, and they are utilized in applications where the additional strength of the steel results in a weight reduction of the parts. For

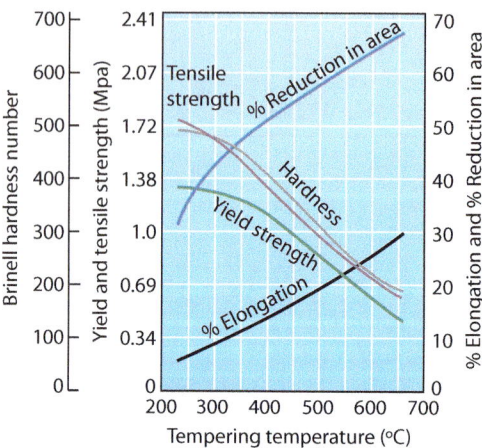

Figure 8.1 The effect of tempering temperature on the mechanical properties of 1050 steel. (*Based on Askeland, D. R., Fulay, P. P., and Wright, W. J., The Science and Engineering of Materials, 6th ed., Cengage Learning, Stamford, CT (2011), p. 504.*)

example, in automobiles trucks, trains, and other transportation vehicles, the weight reduction results in fuel savings that pay for the cost of the alloy additions.

Table 8.4 shows that the highest-strength steels with reasonable ductility are produced by quenching and tempering. Figure 8.1 shows the effect of tempering temperature on the mechanical properties of a 1050 steel. In general, as the tempering temperature increases, the yield strength and the tensile strength decrease, and the ductility increases.

The hardenability of steels is measured by the Jominy test. **Hardenability** is the ability of steel to be hardened to a specified depth. For the Jominy test, a steel rod 4 inches long and 1 inch in diameter is heated to a temperature and for a time to produce 100% austenite (FCC). A very-low-carbon steel would have to be heated to above 910°C, and eutectoid steel would have to be heated to above 727°C to produce 100% austenite. Then one end of the steel rod is sprayed with a jet of water. The sprayed end cools very rapidly, and the other end cools more slowly, almost air-cooling. After cooling, flats are ground on two opposite sides of the round. Then the hardness of the steel is measured as a function of length in $\frac{1}{16}$-inch intervals along the Jominy specimen. The results of this test are shown in Figure 8.2 for various steel alloys. Figure 8.2 shows that 4340 steel has a high hardenability, and that the hardenability of 1050 steel is low. Alloy elements that delay the formation of ferrite and iron carbide during the quench and produce instead martensite result in a high hardenability.

In high-strength low-alloy (HSLA) steels, very small amounts of elements, such as Cr, V, Mo, and Nb, are added to form carbides other than Fe_3C, and this results in an increase in the strength of quenched-and-tempered steel. The alloy elements are quoted in the SAE and AISI standards with a range that allows the producers to achieve specified mechanical properties; however, the midrange of strength is listed in Table 8.5. Alloy additions to HSLA steels also result in much better corrosion resistance than that of plain-carbon steel. The alloy amounts are very small; therefore, the density and elastic modulus do not vary significantly from those of iron, and they are assumed to be 7.87 g/cc and 205 GPa respectively. The ASTM designates the HSLA alloys with the letter A, followed by three digits and a letter grade that indicates the strength. The further the letter from the start of the alphabet, the higher the strength of the alloy. The UNS system for the HSLA alloys starts with the letter K. Table 8.5 gives the properties of some HSLA steels.

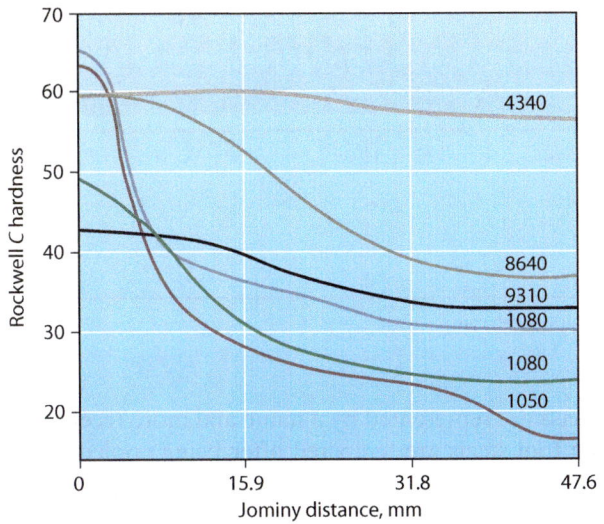

Figure 8.2 The Rockwell C hardness of a variety of steels as a function of distance along a Jominy end quench test specimen in mm. *(Based on Askeland, D.R., Fulay, P.P., and Wright, W.J., The Science and Engineering of Materials, 6th ed., Cengage Learning, Stamford, CT. (2011), p. 512.)*

Table 8.5	The ASTM Designation and Grade (UNS designation), Ultimate Tensile Strength (σ_u), Yield Strength (σ_y), and Strain to Failure (ε_f) for Some HSLA Steels		
ASTM (UNS)	σ_u **MPa**	σ_y **MPa**	ε_f **%**
A588-B (K12043)	≥483	≥345	≥21
A633-A (K01802)	434–572	≥290	≥23
A633-E (K12202)	552–689	≥414	≥23
A690 (K12249)	≥483	≥345	≥18

Based on data primarily from www.matweb.com.

The ASTM designation, alloy chemical composition in weight percent, and applications for the HSLA steels in Table 8.5 are listed below.

A588-B: C = <0.2, Cr = 0.55, Cu = 0.3, Mn = 1.0, Ni = 0.38, Si = 0.23, V = 0.05—Welded, bolted, and riveted structures, bridges and buildings

A633-A: C = <0.18, Mn = 1.18, Nb = <0.05, Si = 0.23—Welded, bolted, and riveted structures, bridges and buildings above −45°C

A633-E: C = <0.22, Mn = 1.33, Nb = 0.030, Si = 0.33, V = 0.076—Welded, bolted, and riveted structures, bridges and buildings above −45°C

A690: C = <0.22, Cu = 0.5, Mn = 0.75, Ni = 0.58—Corrosion resistant in saltwater splash zones, used in docks, sea walls, bulkheads, and related applications

Tool Steels

Tool steels are high-carbon steels that are alloyed with carbide-forming elements such as W, Mo, Cr, and V, as shown in Table 8.6. These steels are normally quenched and tempered to a high strength. These steels are used when a sharp edge or limited wear is required, as in cutting tools,

Table 8.6	Name (UNS designation), Temper Designation (T), Room Temperature Density (ρ), Ultimate Tensile Strength (σ_u), Yield Strength (σ_y), Strain to Failure (ε_f), and Elastic Modulus (E) for High-Alloy Tool Steels

Name (UNS)	T	ρ g/cc	σ_u MPa	σ_y MPa	ε_f %	E GPa
A9 (T30109)	AQT	7.78	2200	1780	5	210
H13 (T20813)	OQT	7.80	1990	1650	9	210
W1 (T72301)	WQT	7.83	1680	1500	3.5	205

Based on data primarily from www.matweb.com.

knives, dies, and drills. Tool steels are represented by a name and more recently by the UNS designation starting with the letter T. The tool steels are tempered after being cooled in air (AQT), oil-quenched (OQT), or water-quenched (WQT), as shown in Table 8.6 under the temper designation (T).

The chemical compositions of alloy additions to iron and applications of these tool steels are listed below.

A9—C = 0.5, Cr = 5.1, Mo = 1.55, Ni = 1.5, Si = 1.0, V = 1.1, Mn = <0.5—Cold-forming, blanking, and bending dies; forming rolls, drill bushings, knurling tools, master gauges and dies
H13—C = 0.36, Cr = 5.20, Mo = 1.36, S = 1.0, V = 1.0—Pressure die-casting tools, extrusion tools, forging dies, hot shear blades, stamping dies, plastic molds, aluminum die-casting tools
W1—C = 0.60 to 1.4, Mn = 0.25, Si = 0.25—Blacksmith tools, wood-working tools

Stainless Steel

Stainless steel is iron alloyed with at least 11% Cr. The primary reason for the development of stainless steels, listed in Table 8.7, is for improved corrosion resistance relative to that of

Table 8.7	The AISI Designation, UNS Designation, Room-Temperature Density (ρ), Ultimate Tensile Strength (σ_u), Yield Strength (σ_y), Strain to Failure (ε_f), and Elastic Modulus (E) for Stainless Steels

AISI (UNS)	ρ g/cc	σ_u MPa	σ_y MPa	ε_f %	E GPa
Annealed Ferrite					
403 (S40300)	7.80	515	275	35	200
409 (S40930)	7.76	434	241	33	200
Quenched-and-Tempered Martensite-Ferrite					
440 (S44002)	7.80	1750	1280	4	200
Precipitation-Hardened Martensite					
17-4 (S17400)	7.80	1158	1117	16	197
17-7 (S17700)	7.81	1517	1310	5	200
Annealed Austenite					
301 (S30111)	8.03	515	205	40	212
304 (S30400)	8.00	505	215	70	197
316LS (S31603)	7.81	586	434	88	193

Based on data primarily from www.matweb.com.

carbon and low-alloy steels. Iron reacts with OH^- in water to form rust, and rust is not a protective oxide on iron. If at least 11 weight percent Cr is added to iron, then the oxide that forms on iron is predominantly chromium oxide and not iron oxide. Chromium oxide is protective on iron alloyed with at least 11% Cr.

The chemical composition of alloy additions to iron, and applications of the stainless steels in Table 8.7, are listed below.

403—C = <0.15, Cr = 12.3—Compressor blades, turbine parts

409—C = 0.01, Cr = 11, Mn = 0.30, Ni = 0.25, Nb = 0.15, Si = 0.41, Ti = 0.17—Exhaust manifolds, exhaust pipes, catalytic converters, mufflers, tail pipes, fuel filter housings, electrical transformer housings, heat-exchanger tubing, farm equipment components, blades and vanes in power-generation turbines

440—C = 0.60 to 0.75, Cr = 17—Cutlery, surgical tools, bearings

17-4—C = <0.07, Cr = 17, Ni = 4, Cu = 4, Nb + Ta = 0.30—Oil-field valve parts, chemical process equipment, aircraft fittings, fasteners, pump shafts, nuclear reactor components, gears, paper-mill equipment, missile fittings, jet engine parts

17-7—C = 0.09, Cr = 17, Ni = 7, Al = 1.1—High-strength springs, bellows, aerospace applications, knives, pressure vessels

301—C = <0.15, Cr = 17, Ni = 7, Mn = <2—Automobile molding and trim, wheel covers, conveyor belts, kitchen equipment, roof-draining systems, hose clamps, springs, truck and trailer bodies, railway and subway cars

304—C = <0.08, Cr = 19, Ni = 9.3—Beer kegs, bellows, chemical equipment, coal-hopper linings, cooking equipment, cooling coils, cryogenic vessels, dairy equipment, evaporators, flatware utensils, feedwater tubing, flexible metal hose, food processing equipment, hospital surgical equipment, hypodermic needles, kitchen sinks, marine equipment and fasteners, nuclear vessels, refrigeration equipment, pressure vessels, sanitary fittings, valves, shipping drums, textile processing equipment, tubing

316LS—(LS = low sulfur, S = <0.01)—C = <0.30, Cr = 18, Ni = 14, Fe = 62, Mn = <2, Mo = 2.25 to 3.5—Medical implants

Another element that is added to stainless steel is nickel. Nickel is FCC, whereas at room temperature iron is in the BCC ferrite phase. However, at temperatures above 912°C iron transforms to the FCC austenite phase. Adding FCC nickel to BCC iron reduces the temperature where ferrite and austenite are in equilibrium, or where the nickel stabilizes the FCC austenite phase of iron. Approximately 9 weight percent nickel is sufficient to stabilize the austenitic phase of stainless steel at room temperature. The addition of nickel in the austenitic stainless steels also results in their being more corrosion resistant in a saltwater environment than are the BCC ferritic stainless steels. Austenitic stainless steels with the FCC crystal structure are more ductile than are BCC ferritic stainless steels at low temperatures. Austenitic stainless steels are used in the annealed condition, and they have a relatively low yield strength. Austenitic stainless steel is also not ferromagnetic, whereas ferritic stainless steel is ferromagnetic. You can check to see if a stainless steel is austenitic with a magnet. If the steel is not attracted to a magnet, it is austenitic. If it is attracted to the magnet it is ferritic. Ferritic stainless steels with a lower carbon content are used in the annealed condition, while those with a higher carbon content, such as 440, are quenched and tempered to form tempered martensite. The stainless steel 17-7PH is precipitation hardened with the intermetallic compound Ni_3Al, and after quenching and tempering it has a tempered martensitic structure.

In the AISI designation system, austenitic stainless steels have a three-digit number that begins with a 2 or 3, and the ferritic stainless steels have a three-digit number that begins with a 4. In the UNS designation system, all stainless steels begin with the letter S, followed by the three AISI digits. In the precipitation-hardened stainless steels 17-4 and 17-7, the amount of chromium in each is 17%, and the amount of nickel is 4 and 7%, respectively.

Cast Iron
Cast iron is iron with a carbon content from 2 to 4.5 weight percent. The phase diagram of iron and carbon in Figure 5.18 shows that at the chemical composition 4.26 weight percent carbon,

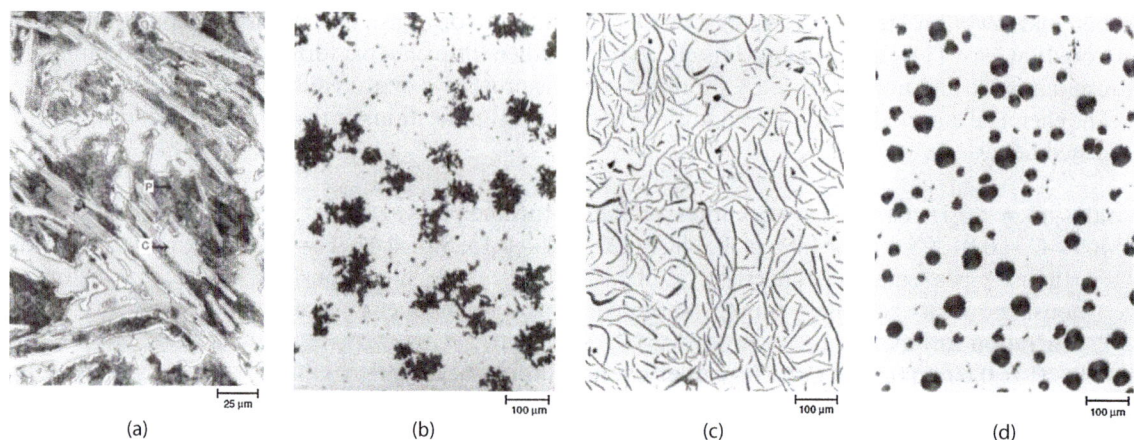

(a)   (b)   (c)   (d)

Figure 8.3   (a) White cast iron with a matrix of pearlite (P) and particles of cementite (C). (b) Malleable cast iron with a ferrite matrix and graphite particles. (c) Gray cast iron with a ferrite matrix and flakes of graphite. (d) Annealed ductile iron with a matrix of ferrite and graphite nodules. (*ASM Handbooks Online, http://www.asmmaterials.info, ASM International, 2004.*)

there is a eutectic reaction of liquid transforming to γ-phase iron plus Fe_3C at a temperature of 1153°C. This temperature is much lower than the 1538°C melting temperature of pure iron. Mixtures of iron and carbon of approximately 4% are more easily melted and cast than are low-carbon steels. If iron with a carbon composition close to eutectic is cooled to room temperature, it forms white cast iron, and its microstructure, shown in Figure 8.3a, has a matrix of pearlite and particles of cementite (Fe_3C). The name white cast iron comes from the white appearance of the fracture surface. White cast iron is extremely hard and brittle, and it cannot be machined. This iron is used in some applications because of its resistance to wear and abrasion. However, white cast iron is often changed into other forms of cast iron.

Through heat treatment it is possible to convert iron carbide to graphite and ferrite. At high concentrations of carbon and at high temperatures, graphite and iron is more stable than Fe_3C is, and Fe_3C decomposes to iron and graphite. If white cast iron is held at temperatures around 850°C for extended periods of time, the iron carbide transforms to clusters of graphite in a matrix of ferrite iron to form malleable iron, with a ductility of 6 to 10%, as is shown in Table 8.8 with the microstructure shown in Figure 8.3b.

Heat treating cast iron alloyed with silicon decomposes the iron carbide into iron plus flakes of graphite to produce gray cast iron, as shown in Figure 8.3c. Gray cast iron is brittle, because the flake form of graphite forms a sharp crack tip in the iron matrix. When gray cast iron fractures, it breaks preferentially along the graphite flakes, producing a gray fracture surface; this results in the name "gray cast iron."

By adding magnesium to the cast iron, nodular graphite forms, as shown in Figure 8.3d. In the as-cast condition, this composition of iron has a matrix of pearlite with carbon nodules. However, by heat treating at 700°C for several hours, the matrix is converted to more-ductile ferrite with the carbon diffusing to the carbon nodules; this is the microstructure shown in Figure 8.3d, and it results in the high ductility of 18%, as shown in Table 8.8. If the nodular iron is quenched from above the eutectoid temperature of 727°C, the resulting matrix is martensite and can be tempered.

In the UNS designation system, cast irons all begin with the letter F. For gray irons the number begins with a 1, malleable iron begins with 2, and ductile iron begins with a 3. Previous to the UNS system, cast irons were specified by the SAE and/or ASTM systems. For ductile cast irons, the ASTM designation is the tensile strength and yield strength in ksi, and the elongation in percent. The treatments for ductile cast iron are annealed (A) and oil-quenched-and-tempered (OQT).

SAE-ASTM (UNS)	T	σ_u MPa	σ_y MPa	ε_f %
Gray				
J431-G1800 (F10004)		118		
J431-G3000 (F10006)		>207		
J431-G4000 (F10006)		>276		
Malleable				
32510 (F22200)		345	224	10
50005 (F23530)		483	345	5
Ductile				
60-40-18 (F32800)	A	414	276	18
120-90-02 (F36200)	OQT	827	621	2

Table 8.8 The SAE or ASTM Designation (UNS designation), Treatment, Ultimate Tensile Strength (σ_u), Yield Strength (σ_y), and Strain to Failure (ε_f) for Cast Irons

Based on data primarily from www.matweb.com.

The chemical compositions of alloy additions to cast iron and applications of the cast irons in Table 8.8 are listed below.

J431-G1800—C = 3.55, Mn = 0.65, Si = 2.55—Miscellaneous iron castings where strength is not a primary consideration

J431-G3000—C = 3.25, Mn = 0.75, Si = 2.10—Gasoline and diesel cylinder blocks, cylinder heads, flywheels, differential casings, pistons, medium-duty brake drums and clutch plates

J431-G4000—C = 3.15, Mn = 0.83, Si = 1.95—Diesel engine castings, liners, cylinders, pistons

32510—C = 2.5, Mn = 0.40, Si = 1.4—For general engineering service at room and elevated temperature, good machinability, and good thermal shock resistance

50005—C = 2.35, Mn = 0.75, Si = 1.38—For general engineering service at room and elevated temperature

60-40-18—C = >3.0, Si = <2.5—Pressure-containing parts for use at elevated temperature, valves and fittings for steam-plant and chemical-plant equipment

120-90-02—C = 3.7, Si = 2.3, Cu = 0.5, Mn = 0.5—Pinions, gears, rollers, slides

8.1.5 Magnesium Alloys

The main advantage of magnesium as a structural metal is a low density of $1.7 \times 10^3 \text{kg/m}^3$, and a low melting temperature of 651°C allows it to be cast at low temperature. Magnesium has a relatively low elastic modulus (45 GPa), strength, ductility, melting temperature, and corrosion resistance relative to other structural metals. However, because of the low density, the specific strength is still relatively high, and the specific strength results in applications where weight is critical, such as in hand tools and aerospace applications. The low ductility is a result of the limited number of slip planes in the hexagonal close-packed crystal structure. The only close-packed plane is the basal plane. Magnesium can be solid solution strengthened with aluminum, and precipitation hardened with alloy additions of manganese, zinc, zirconium, and thorium. The wrought alloys can be strain hardened; however, the small ductility limits this strengthening mechanism. The cast alloys have very low ductility for a metal. Magnesium alloys are applied where low weight is more of a requirement than high strength and high ductility, and where corrosion resistance is not critical. Table 8.9 presents the mechanical properties of some magnesium alloys.

The magnesium alloy designation starts with two letters related to the two major alloy elements. The letters for the alloy elements in Table 8.9 are aluminum (A), thorium (H), zirconium (K), silicon (S), and zinc (Z). A listing of additional element designations is presented at *www.keytometals.com.* The two alloy designations are followed by two digits that give the weight percentage of the two alloy elements, in the same order as the element designation, rounded off to the nearest integer. These two composition digits are followed by a letter. The first alloy with this composition is given the letter A following the percentages, the second alloy with these percentage compositions is assigned a B, and so on. This is followed by a dash and the treatment designation. The treatment designations are:

F—As fabricated
O—Annealed
H10, H11—Slightly strain hardened
H23, H24, H26—Strain hardened and partially annealed
T4—Solution heat treated
T5—Artifically aged only
T6—Solution heat treated, cold worked, and artificially aged

The UNS designation for magnesium alloys starts with the letter M, followed by a 1 and four digits. Table 8.9 presents the name, temper designation, UNS designation, density, tensile strength, yield strength, elongation to failure, and elastic modulus of some magnesium alloys.

| Table 8.9 | Name-Temper Designation (T) (UNS designation), Room-Temperature Density (ρ), Ultimate Tensile Strength (σ_u), Yield Strength (σ_y), Elongation to Failure (ε_f), and Elastic Modulus (E) for Some Wrought (a) and Cast (b) Magnesium Alloys |

Name—T (UNS)	ρ g/cc	σ_u MPa	σ_y MPa	ε_f %	E GPa
(a) Wrought Alloys					
AZ31B-O (M11311)	1.77	255	150	21	45
AZ31B-H26 (M11311)		275	190	10	45
AZ80A-F (M11800)	1.80	330	230	11	45
AZ80A-T6 (M11800)		340	250	5	45
HK31A-O (M13310)	1.80	200	125	30	45
HK31A-H24 (M13310)		250	180	15	45
(b) Cast Alloys					
AS41A-F (M10410)	1.78	214	138	6–15	45
AZ91D-F (M11916)	1.81	230	150	3	44.8
ZK61A-T6 (M16611)	1.83	310	195	10	45

Based on data primarily from www.matweb.com.

The chemical compositions of alloy additions to magnesium and applications of the magnesium alloys in Table 8.9 are listed below.

AZ31B—Al = 3.0, Zn = 1.0, Mn = >0.2—Forgings and extruded bar, rod, shapes, and tubing with moderate mechanical properties

AZ80A—Al = 8.0, Zn = 0.50, Mn = >0.12—Extruded products and press forgings. This alloy can be heat treated

HK31A—Th = 3.25, Zr = 0.70—Sheet and plate with excellent weldability and formability

AS41A—Al = 4.25, Mn = 0.35, Si = 1.0 Used as cast for automotive structural die-casting parts up to 177°C

AZ91D-F—Al = 9.0, Mn = >0.13, Zn = 0.67—Used for die castings, excellent corrosion resistance for a Mg alloy

ZK61A—Zn = 6.0, Zr = 0.80—Simple highly stressed aerospace and military castings of uniform cross section. Intricate castings are subject to microporosity and cracking due to shrinkage. Not readily welded.

8.1.6 Superalloys

A **superalloy** is a metal alloy based upon nickel, cobalt, or iron that is designed for high-temperature use. Nickel is the base metal for most of the superalloys that are utilized in high-temperature gas turbines. Some superalloys are listed in Table 8.10; compositions and properties of many more superalloys are available in the *ASM Handbook* and in Chapter 9. In the nickel-based superalloys, aluminum and titanium are added to form the precipitates Ni_3Al or Ni_3Ti for high-temperature strength. As shown in the

Table 8.10 The Name, Room-Temperature (RT) Density (ρ), Ultimate Tensile Strength (σ_u), Yield Strength (σ_y), and Strain to Failure (ε_f), for Some Superalloys, Followed on the Next Line by High-Temperature Mechanical Properties

Name—UNS	ρ g/cc	σ_u MPa	σ_y MPa	ε_f %
Inconel 718 (RT)	8.19	1375	1100	25
At 650°C		1100	980	18
Inconel 6000 (RT)	8.1	1295	1285	4
At 538°C		1155	1010	6
Nimonic 901 (RT)	8.14	1200	875	15
At 550°C		1030	800	10
Pyromet CTX-3 (RT)	8.28	1180	830	14
At 538°C		1040	673	16
Rene 95 (RT)	8.2	1560	1140	8.6
At 760°C		1170	1100	15
Waspalloy (RT)	8.19	1276	897	23
At 650°C		1115	690	34
MM-0011-PC (RT)	8.5	1035	825	8
At 870°C		825	550	6
MM-0011-DS (RT)	8.5	1085	855	12
At 870°C		950	710	13
PWA 1480-SC (RT)	8.7	1140	895	4
At 980°C		685	495	20

Based on data primarily from www.matweb.com and from Sims, C., et al. Superalloys II: High Temperature Materials for Aerospace and Industrial Power, John Wiley, N.Y. (1987).
1. This ultimate strength is for 650°C (1200°F).

aluminum-nickel phase diagram in Figure 5.15, the Ni_3Al precipitate is stable at temperatures up to 1385°C, and this precipitate provides strength to temperatures in excess of 1000°C. Superalloys are also alloyed with high-melting-temperature metals such as chromium, niobium, molybdenum, tungsten, and tantalum. All of these elements solid solution strengthen superalloys, and they are strong carbide formers. A number of different carbides form in superalloys, and the carbides provide high-temperature strength. Both chromium and aluminum provide superalloys with high-temperature corrosion and oxidation resistance, because of the formation of protective oxides of Cr_2O_3 and Al_2O_3 on the surface rather than oxides of nickel, cobalt, and iron. Some superalloys, such as Inconel 6000, are strengthened with dispersion particles, such as thoria (ThO_2) or yttria (Y_2O_3). Although refractory metals such as niobium, molybdenum, tungsten, and tantalum have higher melting temperatures than do the superalloys, the superalloys have a combination of relatively high melting temperature, corrosion resistance, ductility, and moderate density that makes them the metal of choice for high-temperature gas-turbine materials. The UNS designation system for superalloys has a variety of different designations. Nickel-based alloys start with the letter N, cobalt-based alloys start with R3, and iron-based alloys start with either a K or an S.

Even though superalloys have a relatively high melting temperature, some parts, such as turbine blades, are cast into shape. One reason for casting is that superalloys are very strong at the high temperature where deformation processing into shape is performed. Superalloys are cast as polycrystals (PCs), directionally solidified (DS), and single crystals (SCs). We will cover the technique for producing DS and SC turbine blades by investment-casting in Chapter 13. Figure 9.13 shows turbine blades made with each of these techniques.

In DS the casting is cooled from one end. This forces the interface between solid and liquid to propagate from the cooler end to the hot end, and the grains grow parallel to the direction of cooling. At high temperature the grains are weaker and melt at a lower temperature than crystalline material, because the grain boundary is more disordered and in a higher energy state, and the impurities concentrate at the grain boundary. In a turbine blade, grain boundaries parallel to the blade axis have zero tensile stress on the grain boundaries resulting from the rotation of the turbine. This is shown by using Equation 6.32, and assuming that the slip plane and slip direction are both along the grain boundary parallel to the blade axis, then both φ and λ are equal to 0 and the resolved shear stress is equal to 0. For this reason, DS turbine blades can operate at higher temperatures than PC blades can. However, grain boundaries are more subject to high-temperature corrosion than crystalline materials are, and for this reason SC turbine blades are superior to DS blades for the highest-temperature applications. Table 8.10 lists MM-0011-DS and MM-0011-PC; observe the higher strength of the DS alloy relative to that of the PC alloy.

The alloy PWA 1480-SC is the only SC alloy in Table 8.10, and its properties are for room temperature and for 980°C. The elastic moduli of superalloy types at room temperature and at 980°C are presented in Table 8.11. SC turbine blades have no grain boundaries, and they allow for the highest operating temperatures.

The chemical composition of the superalloys in Table 8.10 and some applications are listed below.

Inconel 718—Ni = 53, Al = 0.5, Cr = 19, Fe = 17, Mo = 3.0, Nb = 5.3, Ti = 0.9—Material of choice for applications below 650°C, gas turbines, rocket motors, space craft, space shuttle, nuclear reactors, pumps, pump seals, and tooling. Precipitation-hardened with outstanding weldability, and excellent creep strength to 700°C.

Inconel 6000—Cr = 15, Al = 4.5, W = 4, Ti = 2.5, Co = 2.0, Mo = 2, Ta = 2, Y_2O_3 = 1.1—Oxide-dispersion-strengthened alloy available as powder metallurgy hot isostatic pressed (HIP) preforms.

Nimonic 901—Ni = 43, Cr = 12.5, Fe = 33, Mo = 5.75, Ti = 2.95—Gas-turbine discs and shafts, high creep resistance to 600°C.

Pyromet CTX3—Fe = 39, Ni = 38, Co = 14, Nb + Ta = 5, Ti = 1.5—Precipitation hardenable, low coefficient of thermal expansion over a broad temperature range, good thermal fatigue resistance.

Rene 95—Ni = 62, Al = 3.5, Cr = 13, Co = 8.0, Mo = 3.5, Nb = 3.5, Ti = 2.5, W = 3.5—One of the highest-strength alloys available for service at temperatures from 425°C to 650°C, high-pressure turbine discs and turbine blades. Available as powder metallurgy (HIP) preforms or extruded billet.

Waspalloy—Ni = 57, Cr = 19.5, Co = 13, Mo = 4.3, Ti = 3, Al = 1.4, Fe = 1, Zr = 0.7—Turbine hardware, space shuttle turbo pump seals.

MM-0011—Ni = 60, Cr = 8.3, Co = 10, Mo = 0.7, W = 10, Ta = 3, Al = 5.5, Ti = 1, Hf = 1.5—One of the highest-temperature DS and SC cast alloys for applications up to 1100°C, used for turbine blades and nozzles.

PWA 1480—Ni = 63, Cr = 10, Co = 5.0, W = 4.0, Ta = 12, Al = 5, Ti = 1.5—One of the highest-temperature DS and SC cast alloys for applications up to 1100°C, used for turbine blades.

Some superalloys are shaped by powder metallurgy (PM) processes, and the dispersion-strengthened alloys must be processed by PM techniques. Superalloys in Table 8.10 that are formed with PM techniques include Rene 95 and Inconel 6000. Inconel 6000 is oxide dispersion strengthened.

Table 8.11	The Elastic Modulus (E) in GPa for Different Superalloy Types at Room Temperature (RT) and at 980°C	
Superalloy Type	E at RT	E at 980°C
Nickel-based PC	199	145
Nickel-based DS & SC	128	87
Cobalt-based PC	225	137
Iron-based PC	202	151

Based on data from Superalloys II, p. 565.

1. At 650°C, because iron-based superalloys are not capable of achieving the high temperatures of nickel- and cobalt-based alloys.

8.1.7 Titanium Alloys

Titanium (Ti) alloys have several advantages over other metal alloys. They are one of the most corrosion-resistant metals in severe environments, such as ocean water or chemical processing plants; they are less dense than steel; they have a relatively high melting temperature; and they have strength comparable to that of steel, as shown in Table 8.12. The excellent corrosion resistance of Ti is due to the formation of a surface layer of TiO_2 that protects the metal at moderate temperatures. Bare Ti metal is highly reactive with oxygen. For example, when machining Ti, care must be taken that the metal chips do not catch fire because of rapid oxidation. The protective layer of oxide also makes Ti corrosion resistant when used as medical implants, and it is not rejected by the body. Medical-implant alloys usually have extra-low interstitials (ELIs). In titanium the primary interstitial is oxygen and for the ELI grade the maximum is 0.2 weight percent. Ti ELI alloys are also used in mechanical applications that require high resistance to fracture. Because of its strength at high temperature and its corrosion resistance, Ti is used for the skin material of high-performance aircraft, such as supersonic jets, and in jet-engine compressors. Unfortunately, the attraction of Ti for oxygen at high temperature limits the use of Ti to temperatures of approximately 550°C, for above this temperature the Ti adsorbs extensive amounts of oxygen and other interstitial atoms and becomes brittle. The attraction of Ti for oxygen is so effective that ultra-high-vacuum systems use hot Ti metal to remove oxygen from the system.

Table 8.12 The Alloy Name, Heat Treatment (T), Ultimate Tensile Strength (σ_u), Yield Strength (σ_y), Strain to Failure (ε_f), and Elastic Modulus (E) for Some Wrought Ti Alloys

Name	T	σ_u MPa	σ_y MPa	ε_f %	E GPa
α-Alloys					
CP Ti—Grade 1	A	330	240	30	100
CP Ti—Grade 4	A	660	590	20	105
5Al-2.5Sn—ELI	A	775	720	15	110
Near-α-Alloys					
8Al-1Mo-1V	DA	950	941	18	124
8Al-1Mo-1V	STA	1180	1070	17	120
6Al-2Sn-4Zr-2Mo	DA	940	860	15	113.8
6Al-2Sn-4Zr-2Mo	STA	1035	966	15	113.8
$\alpha + \beta$ Alloys					
6Al-4V	A	950	880	14	113.8
6Al-4V	STA	1170	1100	10	114
6Al-4V-ELI	A	860	790	15	113.8
β-Alloys					
10V-2Fe-3Al	A	970	900	9	110
10V-2Fe-3Al	STA	1240	1200	6	106
13V-11Cr-3Al	A	950	860	18	99
13V-11Cr-3Al	STA	1210	1140	7	110

Based on data primarily from www.matweb.com.

The nominal chemical compositions of the Ti alloys in Table 8.12, the UNS designation, the density in g/cc (lb/in³), and some applications are listed below.

CP Ti (R50250 & R50700)—4.51 (0.163)—Airframe components, cryogenic vessels, heat exchangers, condenser tubing, and pickling baskets

Ti-5Al-2.5Sn (R54521)—4.48 (0.162)—Gas-turbine engine casings and rings, chemical processing equipment

Ti-8Al-1Mo-1V (R54810)—4.37 (0.158)—Fan and compressor blades, discs, spacers, seals, and rings

Ti-2Sn-4Zr-2Mo (R54620)—4.54 (0.164)—High temperature jet engine parts, compressor blades, discs, spacers, high-performance automotive valves, afterburner structures, and hot airframe skin applications

Ti-6Al-4V (R56400)—4.43 (0.160)—Fan and compressor blades, discs, rings, airframe, fasteners, components, hubs, and forgings

Ti-10V-2Fe-3Al (R54610)—4.65 (0.168)—High-strength airframe components, landing gear structures, and helicopter rotors

Ti13V-11Cr-3Al (R58010)—4.82 (0.174)— High-strength airframe components up to 315°C (600°F), and used primarily for forgings

Pure Ti is hexagonal close packed (HCP) α phase at room temperature. However, at temperatures above approximately 890°C, Ti is in the BCC β phase up to the melting temperature of 1668°C. The **β-transus** is the minimum temperature where a Ti alloy is 100% BCC. Some alloy elements, such as aluminum and oxygen, raise the temperature of the β-transus and stabilize the α-phase structure. Other elements such as BCC metals iron, vanadium, and chromium reduce the β-transus temperature and stabilize the BCC β phase. This is similar to the stabilization of the FCC γ phase in stainless steel with FCC nickel. Tin is another common alloy element in Ti alloys; its room-temperature structure is tetragonal, and it is neutral in its effect on the β-transus. The β-transus for Ti alloys is given on the Web site *www.matweb.com* and in the book by Lutjering and Williams. Table 8.12 presents the four categories of commercial Ti alloys: α, near-α, $\alpha + \beta$, and β.

Alpha alloys include commercial purity (CP) Ti and alloys of Ti with aluminum and only small amounts of β-phase stabilizing elements. CP Ti is solid solution strengthened with interstitial oxygen in a manner similar to the carbon strengthening of steel. The higher the amount of oxygen, the higher the strength. However, CP Ti is not heat treatable like iron-carbon alloys are. Alloys of Ti solid solution hardened with aluminum (<5%) and tin, such as Ti-5Al-2.5Sn, are also α-phase Ti alloys that are not heat treatable. The α-phase Ti alloys do not significantly strain harden; therefore, the α-phase alloys are normally used in the annealed condition. Grain-size strengthening is an effective strengthening mechanism in α-phase Ti alloys. Since the HCP structure of Ti is not isotropic, α-phase Ti alloys that are significantly plastically deformed by processes such as rolling or drawing are textured (not isotropic). The single-phase α Ti alloys are the most corrosion-resistant Ti alloys. Solid solution strengthened α-phase alloys that do not respond to heat treatment are suitable for welding. Alloys with greater than 5% aluminum are heat treatable. At concentrations of aluminum greater than 5 %, the precipitate Ti_3Al forms in Ti alloys. The formation of Ti_3Al improves high-temperature creep strength, but it decreases ductility and fracture resistance. Titanium alloys strengthened by the formation of the precipitate Ti_3Al are not suitable for welding.

If sufficient amounts of β-phase stabilizing elements, such and V, Fe, or Mo, are added to Ti, the alloy is in a two-phase field of $\alpha + \beta$. In near-α Ti alloys, there is primarily α-phase material with small amounts of β-phase material, and in the $\alpha + \beta$ Ti alloys, there are comparable amounts of the α and β phases. In two-phase $\alpha + \beta$ alloys, such as Ti-6Al-4V, the α-phase stabilizing element (Al) concentrates in the α phase, and the β-phase stabilizing element (V) concentrates in the β phase. The $\alpha + \beta$ alloys are available in the annealed condition, or in the solution-treated and aged condition.

There are different microstructures available for wrought $\alpha + \beta$ Ti alloys in the annealed condition. In the processing of wrought $\alpha + \beta$ alloys in the annealed condition, there are four steps: homogenization; deformation, such as rolling, forging, or drawing; recrystallization; and annealing. The final microstructure depends upon the temperatures of these processes. All $\alpha + \beta$ alloys are first homogenized at a temperature above the β-phase transus. After deformation the recrystallization is conducted at temperatures above the β-transus for β-annealed alloys. The alloy is then cooled to a temperature in the $\alpha + \beta$-phase field and annealed. The process of β-annealing produces lamellar microstructures. In lamellar $\alpha + \beta$ microstructures, the β-phase grains are elongated and the α-phase forms in the β-phase grain boundaries. The cooling rate from the β-phase recrystallization temperature to the $\alpha + \beta$ annealing temperature determines the size of the lamellar β-phase grains. Fast cooling produces smaller grains. Grain-size strengthening is an effective strengthening mechanism in annealed $\alpha + \beta$ alloys.

If after deformation the recrystallization process of $\alpha + \beta$ Ti alloys is carried out at a temperature below the β-transus in the $\alpha + \beta$-phase field, two different microstructures are possible. At higher recrystallization temperatures in the $\alpha + \beta$-phase field, a lamellar $\alpha + \beta$ matrix with relatively equiaxed grains of α-phase results, called the bimodal or duplex-annealed. The other possible microstructure is equiaxed grains of α and β phases that results from lower recrystallization temperatures in the $\alpha + \beta$-phase field. In a condition called "mill annealed," the alloy is not recrystallized after deformation-processing; it is only annealed at a temperature well below the β-transus.

Some $\alpha + \beta$ Ti alloys are reported as solution treated and aged. This refers to alloys that are first recrystallized (solution treated) at a temperature close to the β-transus, and then aged at a temperature where the intermetallic compound Ti_3Al precipitates, resulting in higher-strength alloys. For example, the solution-treated and aged (STA) Ti-6Al-4V shown in Table 8.12 is solution treated at 900°C to 955°C and aged at 540°C. The β-transus for this alloy is 995°C. At temperatures above 550°C the Ti_3Al goes into solution; therefore, holding at 540°C is aging, and holding at 600°C is annealing, not aging, because no precipitate forms at 600°C.

Near-α-alloys are annealed or solution treated and aged in a manner similar to that of the $\alpha + \beta$ alloys discussed above. Some near α-alloys intended for high temperature applications have additions of silicon to form the precipitate Ti_5Si_3, which is stable at temperatures up to 1040°C. The STA near α-alloys have the highest temperature-creep resistance of the Ti alloys, and they are used for higher-temperature applications, such as gas-turbine fan blades.

The β-alloys are alloyed with BCC β-phase stabilizing elements, such as vanadium, iron, chromium, and molybdenum. When these alloys are quenched from above the β-transus, the single phase of β is metastable at room temperature. No commercial Ti alloys have such high amounts of β-phase stabilizing elements that the single β-phase structure is stable at room temperature. The equilibrium phases are $\beta + \alpha$ with more β than α, and the β-transus for β-alloys is typically from 700°C to 900°C. Most β-alloys are processed by β-annealing, as discussed above. At room temperature they have a lamellar structure with β-phase grains and α-phase at the grain boundaries. In annealed β-alloys, the final annealing temperature is close to the β-transus, where the α phase does not precipitate. If the final heat treating is at a temperature well below the β-transus, this results in the precipitation of the α phase in the β-phase matrix, and this strengthens the alloy. This final heat treatment is an aging step, because a precipitate is formed; it is not annealing. These alloys are listed as STA. Advantages of β-phase alloys relative to α-phase and $\alpha + \beta$ alloys are that they can be deformed in the more-ductile BCC structure at lower temperatures, and they have high strength in the STA condition. The main applications of β-alloys are in highly stressed aircraft structures, such as landing gears. In Table 8.12, under treatment (T) the letter A represents annealed; DA means duplex-annealed, and STA is solution treated and aged.

Ti alloys are normally designated by their composition, and as an abbreviation they are known as "Ti" with the weight-percent numerals of each major alloy element. For example, a Ti alloy with additions of 6% Al and 4% V is known as Ti-64. The ASTM has developed a system that refers to the different alloys as grades. For example, Grades 1 through 4 are all commercially pure Ti (CP Ti) where the interstitial content, which is primarily oxygen, and the strength increase with the grade number. Grade 5 is the alloy Ti—6% Al—4% V. Grade 6 is the alloy Ti—5% Al 2.5% Sn, and so on until reaching Grade 38, which is Ti—4% Al—2.5% V—1.5% Fe. In the UNS system, Ti alloys start with the letter R, followed by a 5, with the remaining digits usually containing the alloy percentage numbers; for example, Ti-6Al-4V is R56400. The alloys in Table 8.12 are in the wrought condition. The same Ti alloys can be cast into shape by using graphite molds, and investment casting; or Ti alloy powder can be hot isostatic pressed.

8.1.8   The Specific Strength and Modulus of Structural Metals

The different metal alloys discussed above have different strengths, elastic moduli, and densities. In many applications, such as aircraft and high-efficiency or high-performance automobiles, it is important to have the highest strength per unit of density. The **specific yield strength** or the **specific modulus** are the yield strength or the elastic modulus divided by the **specific gravity** (*SG*). The *SG* is equal to the density of a material divided by the density of water. The specific gravity is dimensionless. The density of water in the SI system of units is 1×10^3 kg/m³ at 4°C. For example, the density of the aluminum alloy 7075 is 2.81×10^3 kg/m³, and the density of water is 1×10^3 kg/m³. The specific gravity of 7075 aluminum is 2.81. The yield strength of 7075 is 503 MPa, and the specific yield strength is 503 MPa divided by 2.81, or 179 MPa. In the English system of units, the specific gravity is equal to the value in the SI system of units. Table 8.13 gives the *SG*, specific yield strength, and specific elastic modulus of the strongest materials that have at least 10% elongation to failure that are listed in the tables above.

In Table 8.13, the aluminum alloy 7075-T6 has the highest specific yield strength, and 4340 steel has the highest specific elastic modulus. Thus if weight is critical and yield strength is the most critical design

| Table 8.13 | The Alloy Name, Treatment (T), Room-Temperature Specific Gravity (SG), Specific Yield Strength ($S\sigma_y$), and Specific Elastic Modulus (SE) for the Strongest Alloys in the Tables above |

Alloy Name	T	SG	$S\sigma_y$ MPa	SE GPa
Al-7075	T6	2.81	179	25.5
Beryllium copper	TF00	8.36	106	15.7
4340 steel	OQT	7.85	148	26.1
Mg-ZK61A	T6	1.83	106	24.6
Inconel 718	STA	8.19	134	24.4
Ti-8Al-1Mo-1V	STA	4.37	130	14.6

The data should not be used for design purposes.

factor, then high-strength aluminum alloys provide the best design solution of the materials in Table 8.13. If stiffness or elastic deflection is the most critical design factor, then steel alloys provide the best design solution. Example Problem 8.1 demonstrates one approach if both specific yield strength and specific elastic modulus are important. Other considerations must also be taken into account for any design, such as cost, corrosion resistance, fracture, and fatigue. In Chapter 14 we will consider these other factors in the choice of a material for a design.

Example Problem 8.1

In a design where weight is a critical factor, such as in a space shuttle frame, and both specific yield strength and specific elastic modulus are of equal importance, which of the materials 7075-T6 or 4340 OQT in Table 8.13 is the best material choice?

Solution

To sum the influence of both specific yield strength and elastic modulus on a design, one solution is to normalize each property by dividing the individual values for each property by the maximum value of that property. For specific yield strength the maximum value is 179 MPa and for specific elastic modulus the maximum value is 26.1 GPa. Make a table with the normalized values and the sum of the normalized values.

Material	$S\sigma_y$/179 MPa	SE/26.1 GPa	Sum
7075-T6	1.0	0.98	1.98
4340 OQT	0.83	1.0	1.83

From this analysis the 7075-T6 has the highest sum if specific yield strength and specific elastic modulus are of equal importance. If stiffness is twice as important as strength, then a multiplying factor of 2 is applied to the specific elastic modulus values.

8.1.9 Noble Metals

The **noble metals** are silver, gold, platinum, palladium, rhodium, ruthenium, iridium, and osmium. The noble metals are either of the FCC or the hexagonal close-packed crystal structure. They tend to be corrosion resistant. Gold is the only metal that does not oxidize under ambient conditions. Also, the noble metals are good thermal and electrical conductors. Some of the noble metals have high melting temperatures, such as platinum (1769°C), palladium (1552°C), rhodium (1966°C), ruthenium (2310°C), and iridium (2454°C), and they can be applied at high temperature. Silver (961°C) and gold (1063°C) have relatively low melting temperatures. Low melting temperatures allow gold and silver to be cast into shape.

Pure noble metals are ductile and of low strength, and they must be strengthened for structural applications. However, these metals are too expensive for standard structural applications. Most applications of noble metals are of a specialized nature where cost is not a significant factor, such as in jewelry, medical devices, and outer-space hardware. However, platinum-conductor spark plugs for gasoline engines are manufactured today. Palladium is used in gasoline engine catalytic converters, and platinum is used in diesel engine catalytic converters

8.1.10   Refractory Metals

The **refractory** metals are metals with a very high melting temperature. The refractory metals and their melting temperatures are niobium (2468°C), molybdenum (2610°C), tungsten (3410°C), and tantalum (2996°C). These are all BCC transition-series metals, and they have strong bonding that is a combination of metallic and covalent. As a result, refractory metals have high elastic moduli and high yield strengths. These metals are applied where temperatures are extremely high; however, all of these metals are subject to high-temperature oxidation. Tungsten is a filament material in light bulbs, X-ray tubes, and cathode-ray tubes; however, the tungsten in these devices must be in a vacuum or in an inert-gas atmosphere. Tungsten is also used in nonconsumable welding electrodes. Molybdenum is used as an extrusion-die material because of its high strength and hardness. Tantalum is very corrosion resistant at temperatures below 150°C; as a result, it has found applications as a medical prosthesis material among other applications. All of these metals are also used in small amounts in high-strength steel alloys to provide solid solution strengthening, and they are strong carbide formers.

8.1.11   Other Metals

Zirconium is a hexagonal close-packed metal that is similar to Ti; however, its density is higher than Ti. Zirconium is used as a cladding (coating) material for uranium fuel rods, because it is relatively transparent to thermal neutrons. Zirconium, like Ti, is corrosion resistant; thus it has applications in corrosive environments.

Zinc is a low-strength metal that is subject to corrosion. One application of zinc is as a sacrificial anode to protect other materials from corrosion. An example is galvanized steel which is steel coated with zinc. We will cover the corrosion of metals and galvanizing in Chapter 10.

8.2   CERAMICS, GLASS, AND GLASS-CERAMICS

The most widely used structural ceramic is silica (SiO_2), in the form of glass. The strength of glass depends upon the form of the glass and the treatment. Freshly drawn silica glass filaments have an extremely high strength of 24 GPa, which approaches the theoretical strength of the atomic bonds.

The glass filaments in fiberglass also have a high strength of approximately 3GPa. Glass strength is significantly reduced by scratches and abrasion. Bansal and Doremus report that the room-temperature flexural strength of abraded soda-lime silicate glass and borosilicate glass can be as low as 103 MPa; however, bulk soda-lime glass exposed to water has a strength of approximately 35 MPa. The design strength of glass depends upon the extent of environmental exposure and surface condition. Because of all the factors that can reduce the strength of glass, glass strength is not frequently reported, and it is not listed for the ceramics in Table 8.14. The design strength of glass should be taken as a low number, such as 20 MPa or less. *The Handbook of Glass Properties* by Bansal and Doremus in the list of references at the end of the book contains an extensive compilation of the elastic moduli and viscosity of various glasses.

Glass-ceramics are materials that start as glass, but then during heat treatment crystalline material is nucleated within the glass, as shown in Figure 5.29. Depending upon the additions to the silica, many different engineering properties are possible. Additions of Al_2O_3 and Li_2O result in glass-ceramics that have a low coefficient of thermal expansion and good thermal shock resistance. These glass-ceramics are used in cookware, such as Corningware™ and Pyroceram™. Other glass-ceramic applications of similar

Table 8.14	The Identification (ID) of Ceramic Materials, Density (ρ), Flexural Strength (*FS*) (MPa), Young's Modulus (*E*), Poisson's Ratio (υ), and Melting Temperature (T_m) for Crystalline Materials or Softening Temperature for Glass

ID	ρ (g/cm³)	*FS* MPa	*E* GPa	υ	T_m (°C)[1]
Glass[2]					
Soda-lime glass	2.5	—	72	0.25	700[2]
Pyrex borosilicate	2.23	—	64	0.20	820[2]
Potash-soda-lead	2.86	—	61	0.21	630[2]
Lime-MgO-Al₂O₃	2.52	—	88	0.25	915[2]
Sodium borosilicate	2.24	—	60	0.22	710[2]
E-glass fibers	2.61	—	72	0.22	840[2]
Alumina—The data are from *http://www.coorstek.com/materials/ceramics.php*					
85% Al₂O₃	3.42	296	221	0.22	2054
90% Al₂O₃	3.60	338	276	0.22	2054
94% Al₂O₃	3.70	352	303	0.21	2054
96% Al₂O₃	3.72	358	303	0.21	2054
99.5%Al₂O₃	3.90	379	370	0.22	2054
99.9%Al₂O₃	3.92	400	386	0.22	2054
Silicon carbide—The data are from *http://www.coorstek.com/materials/ceramics.php*					
Sintered	3.15	480	410	0.21	2837
Reaction-bonded	3.10	462	393	0.20	2837
CVD-99.9995%	3.21	517	434	0.21	2837
Silicon nitride[3]					
Hot-pressed	3.31	906	311	0.27	2151[4]
Reaction-bonded	2.5	338	179	0.23	2151[4]

(Continued)

Table 8.14	*Continued*				
ID	ρ (g/cm^3)	*FS* MPa	*E* GPa	υ	T_m (°C)[1]
Zirconia—The data are from *http://www.coorstek.com/materials/ceramics.php*					
MgO PS	4.01	450	200	0.30	2677[5]
Y$_2$O$_3$ PS sintered	6.02	1240	210	0.23	2677[5]
Y$_2$O$_3$ PS HIPed	6.07	1720	210	0.23	2677[5]

The data should not be used for design purposes.
1. Melting temperatures for crystalline ceramics from Carter, C. B., and Norton, M.G., *Ceramic Materials: Science and Engineering*, Springer, NY (2007), p. 622.
2. Glass data and softening temperatures from Bansal, N. P., and Doremus, R. H., *Handbook of Glass Properties*, Academic Press, Orlando, Fl (1986), p. 34.
3. Silicon nitride data is from Schwartz, M., *Handbook of Structural Ceramics*, McGraw-Hill, NY (1992), p. 3.15.
4. This material sublimes.
5. The melting temperature is for pure zirconia.

compositions include aircraft nosecones and radomes. Machineable glass-ceramics contain uniformly distributed small crystalline grains with planes of easy cleavage that allow small cracks to form during machining, rather than the large cracks that would occur in glass or a ceramic. One advantage of glass-ceramics is that they are processed in the same manner as glass, and as a result they are pore free.

Of the crystalline ceramics, the use of alumina (Al$_2$O$_3$) exceeds that of all other ceramics. Other ceramics with structural applications include silicon carbide (SiC), silicon nitride (Si$_3$N$_4$), and zirconia (ZrO$_2$). The density, flexural strength, Young's elastic modulus, and melting temperature of these ceramics are presented in Table 8.14. The physical properties of ceramics result from their strong ionic and covalent bonding. Polycrystalline ceramics are brittle at room temperature, they are used as insulators or semiconductors, and they have high melting temperatures. Because of the physical properties of ceramics, their primary applications involve high temperature and high electrical resistivity. Some commercial ceramics are listed in Table 8.14, and the materials in the table are then discussed after the table. The percentages listed in the ID column are the percentage of the listed compound. In most cases the remainder is porosity. Table 8.14 shows that for the crystalline materials as the density increases, the strength and elastic modulus also increases. As density increases for the crystalline materials, the porosity decreases.

Silica is SiO$_2$. Crystalline silica is quartz, and in the amorphous condition it is silica glass. Quartz remains structurally rigid to a higher temperature than glass does; therefore, it is used in higher-temperature applications. The cristobalite form of quartz melts at 1717°C, and, for example, E-glass softens at 840°C. Quartz is also piezoelectric, and it is used as an actuator for clocks and watches. To alter the properties and lower the viscosity, silica is mixed with a variety of oxides, such as soda (Na$_2$O), lime (CaO), potash (K$_2$O), magnesia (MgO), lead oxide (PbO), boron oxide (B$_2$O$_3$), and alumina (Al$_2$O$_3$). Silica in the form of glass is used in many domestic and industrial applications and as fiber in fiberglass, in fiber-optic cable, and as both fiber and foamed thermal insulation. Typical glass compositions used in various applications are shown in Table 8.15.

Alumina (Al$_2$O$_3$) is primarily found in the hexagonal (corundum) crystal structure. The quality of alumina is primarily dependent upon the percent alumina in the material, as shown in Table 8.14. In the lower-percent alumina, the difference is due primarily to porosity. Polycrystalline alumina products are sintered under high pressure and temperature. A higher-percentage alumina requires a higher temperature, pressure, time, and cost. Alumina is primarily utilized in refractory applications and as an electrical insulator. Alumina is the primary electrical insulator material in gasoline-engine spark plugs,

| Table 8.15 | The Applications of a Variety of Glass Chemical Compositions |

Glass	SiO_2	Al_2O_3	CaO	Na_2O	B_2O_3	MgO	PbO	Others
Fused silica	99							
Vycor™	96				4			
Pyrex™	81	2		4	12			
Glass jars	74	1	5	15		4		
Window glass	72	1	10	14		2		
Plate glass/Float glass	73	1	13	13				
Light bulbs	74	1	5	16		4		
Fibers	54	14	16		10	4		
Thermometer	73	6		10	10			
Lead glass	67			6			17	10% K_2O
Optical flint	50			1			19	13% BaO, 8% K_2O, ZnO
Optical crown	70			8		10		2% BaO, 8% K_2O
E-glass fibers	55	15	20		10			
S-glass fibers	65	25				10		

Based on data from Askeland, D.R., Fulay, P.P., and Wright, W.J., The Science and Engineering of Materials, 6th ed., Cengage Learning, Stamford, CT. (2011), p. 587.

as shown in Figure 1.2. Alumina is also used as a medical implant material and in dental reconstruction. Other applications include abrasives and cutting tools. Alumina in single crystal form is white sapphire if it is pure, or if it has impurity atoms of iron and titanium it is blue sapphire, and it is ruby if it has impurity atoms of chromium. Ruby is used as a laser material and as a precious gemstone.

Silicon carbide (SiC) forms in the zincblende (ZnS) crystal structure, which is a derivative of the diamond structure. SiC is FCC with an atom of silicon (Si) at 0, 0, 0 and an atom of carbon (C) at $\frac{1}{4}, \frac{1}{4}, \frac{1}{4}$. In diamond there are carbon atoms at both of these atom positions. SiC is very hard, it has good high-temperature strength, and it is a semiconductor. It also has good high-temperature oxidation resistance, because it forms a protective oxide of SiO_2. In Table 8.14, "sintered" refers to sintering of silicon-carbide powder at temperatures of 2000°C or higher. Silicon-carbide powder can be made by reacting silica with carbon at temperatures of approximately 2000°C. Reaction-bonded SiC is a result of combining SiC powder with carbon and molten silicon. The reaction of the molten silicon with carbon results in SiC that bonds the SiC particles together. In the chemical vapor deposition (CVD) of silicon carbide, gases that contain silicon and carbon are adsorbed onto a surface, where they react to form solid silicon carbide. Applications of SiC include abrasive grinding wheels, particulate and fiber reinforcement, a coating for carbon fibers and carbon used at high temperature to prevent oxidation, and high-temperature furnace heating elements.

Silicon nitride (Si_3N_4) crystals form in a tetrahedral structure. In Table 8.14, "hot pressed" refers to silicon nitride powder that is bonded under high temperature and pressure, resulting in a part with nearly full density. Silicon nitride powder is made by reacting silicon powder with nitrogen. In reaction-bonded silicon nitride, a part is made out of pure silicon and then reacted with nitrogen at high temperature. The resulting material has a porosity of 15 to 20%, as shown by the low density in Table 8.14, and the mechanical properties are significantly reduced. However, there is essentially zero size change in the reaction-bonded Si_3N_4 parts, and this is an aid in processing parts. Silicon nitride maintains good strength up to 1000°C; however, the strength decreases rapidly above 1200°C. Attempts to increase the high-temperature strength are focused on reinforcing silicon nitride with particles and filaments by producing

Figure 8.4 A photograph of a variety of silicon nitride bearings. *David W. Richerson and Douglas W. Freitag; Oak Ridge National Laboratory*

composite materials. Silicon nitride is used as rotors in automotive turbochargers, cutting tools, and bearings. Figure 8.4 shows a variety of silicon nitride bearings. It is expected that silicon nitride will have more automotive engine applications, including valves, valve guides, piston rings, piston pins, and cam follower pads. Potential gas-turbine applications include vanes, seals, and ducts.

Zirconia (ZrO_2) is used to make oxygen sensors for exhaust gas in gasoline engines, because the ionic conductivity depends upon the zirconia oxygen concentration. Partially stabilized (PS) zirconia is proposed for structural applications because of its relatively high resistance to fracture. PS zirconia is discussed in Section 7.3.2., and Chapter 11, on fracture, explains the stabilization of zirconia in more detail. Zirconia is stabilized by a number of oxides, including MgO and Y_2O_3, that are included in Table 8.14. HIPed in Table 8.14 is hot isostatic pressed. HIP is a process of heating powders of the material to a high temperature under pressure to bond powder particles into a solid piece. HIP is discussed in more detail in Chapter 13, on processing materials. Applications of PS zirconia include metal extrusion dies, cutting tools, and thermal barrier coatings for high-temperature gas-turbine blades. Future applications could include coatings for valves, pistons, and cylinder liners for high-temperature diesel and gas engines. Single crystal cubic zirconia is a semiprecious gemstone.

Barium titanate ($BaTiO_3$) is a capacitor material, and it is also piezoelectric. Capacitors and piezoelectric materials are discussed in Chapter 16.

Boron carbide (B_4C) is the third-hardest material known. As a result, it is applied as bullet-proof armor and where abrasion resistance is required.

Cordierite ($2MgO\text{-}2Al_2O_3\text{-}5SiO_2$) has electronic applications, and it is a support material for Pt, Pd, and Rh catalytic converters for gasoline and diesel engines in automobiles and trucks.

Diamond (C) is the hardest natural material. Industrial-grade diamonds are used as abrasives and as cutting tools. Coatings of diamond-like material (noncrystalline carbon with sp^3 bonding) can be produced by vapor deposition, and the coatings have properties similar to those of diamond. Diamond cannot be used in high-temperature oxidizing atmospheres because it converts to carbon dioxide. Single crystal diamond is a precious gemstone.

Lead zirconium titanate ($Pb\text{-}ZrTiO_3$) is a piezoelectric material, and it has applications in sound and ultrasound creation and detection systems.

Sialon is silicon aluminum oxynitride with various amounts of each component; one possible composition is $Si_3Al_3O_3N_4$. Applications include cutting tools and conditions that combine high temperature and wear.

Titanium dioxide (TiO_2) or titania is a semiprecious gemstone called rutile when in the single crystal form. Titania is also a white pigment in paints and in sunblock lotion.

Titanium diboride (TiB_2) is a good conductor of both electricity and heat, and it has a high toughness for a ceramic. One application is high-temperature heating elements.

Tungsten carbide (WC) is used in cutting tools.

8.3   POLYMERS

8.3.1   Commodity Thermoplastic Polymers

Table 8.16 presents mechanical properties of more commonly used, or commodity, thermoplastic polymers, and the table includes one composite material: PP 40% glass fiber filled under grade. The heading Grade in Table 8.16 includes any fillers or modifications to the polymer. Composite materials are covered in Chapter 12. IS in Table 8.16 is the Izod impact strength, which is a measure of the resistance

Table 8.16	The Mechanical Properties of Some Common (commodity) Polymers, Including the Polymer Name, Abbreviation, Grade If Any, Density (ρ), Young's Elastic Modulus (E), Tensile Strength (σ_u), Strain to Fracture (ε_f), Izod Impact Strength (IS), and Structure. Under impact strength, NB stands for did not break, and under structure, A stands for amorphous and C represents crystalline.

Polymer	Grade	ρ g/cm³	E GPa	σ_u MPa	ε_f %	IS J/m	Structure
PE LD, LDPE Polyethylene, low density		0.915–0.93	0.14–0.3	7–17	200–900	NB	C
PE HD, HDPE Polyethylene, high density		0.94–0.97	0.7–1.4	20–40	100–1000	30–200	C
PE UHMW, UHMW PE Polyethylene, ultra-high molecular weight		0.93–0.94	0.1–0.7	20–40	200–500	NB	C
PP polypropylene	Homopolymer	0.90–0.91	1.1–2	30–40	100–600	20–75	C
PP polypropylene	−40% glass fiber filled	1.22–1.23	6.8–7.2	60–110	1.5–4	75–110	
PP polypropylene	Copolymer	0.89–0.905	0.9–1.2	28–40	200–500	60–750	C
PVC Polyvinyl chloride	Rigid (RPVC)	1.32–1.58	1–3.5	40–75	30–80	20–1000	A

(Continued)

Table 8.16 *Continued*

Polymer	Grade	ρ g/cm³	E GPa	σ_u MPa	ε_f %	IS J/m	Structure
PVC	Flexible (FPVC, plasticized)	1.16–1.70	0.05–0.15	6–25	150–400	—	A
PS Polystyrene		1.04–1.05	2.4–3.2	30–60	1–4	13–25	A
SB Styrene-butadiene	Rubber-modified PS High-impact PS, HIPS	0.98–1.10	1.5–2.5	15–40	15–60	50–400	A
ABS Acrylonitrile-butadiene-styrene	Medium IS	1.03–1.06	2–2.8	30–50	15–30	130–320	A
ABS Acrylonitrile-butadiene-styrene	High IS	1.01–1.04	1.6–2.5	30–40	5–70	350–600	A
SAN Styrene-acrylonitrile		1.07–1.09	3.4–3.7	55–75	2–5	15–30	A
ASA Acrylate-styrene-acrylonitrile		1.05–1.07	2.2–2.4	30–50	20–40	450–600	A

Based on data from Brostow, W., Kubat, J, and Kubat, M., Mechanical Properties, Physical Properties of Polymers Handbook, Mark, J. E. ed., AIP Press, Woodbury, N.Y. (1996), p. 331.

to fracture, and tests of this type are covered in Section 11.2.7. Some of the commodity polymers in Table 8.16 are discussed below. The discussion below covers only the properties of the pure polymer and not fillers or composite materials.

Polyethylene (PE) is the most extensively utilized thermoplastic polymer. Table 8.16 includes LDPE, HDPE, and UHMWPE. Polyethylene has a high toughness because of the high elongation to failure of up to 1000%. It is corrosion resistant, and it has excellent insulating properties. Some of the applications of polyethylene include containers, bottles, chemical tubing, and electrical insulation. Polyethylene is also made into films for packaging and into sheets for use in retaining ponds for water and waste liquids. OUHMWPE is made into high-strength fibers whose mechanical properties are presented in Table 7.4. Spectra and Dyneema™ are two commercial forms of OUHMWPE.

Polyvinyl chloride (PVC) is the second most extensively utilized thermoplastic polymer. PVC has a higher tensile strength (40 to 75 MPa) than does HDPE. PVC has high corrosion resistance resulting in many of its applications. PVC is used for water and sewer pipe, building siding, and electrical conduits. PVC is often blended with other polymers to modify its properties. Table 8.16 shows the properties of plasticized PVC that is modified to improve the ductility to between 150 and 400%.

Polypropylene (PP) is the third most extensively utilized thermoplastic polymer. The tensile strength for PP of 30 to 40 MPa is comparable to that of HDPE. PP maintains its strength to higher temperatures than does PE; this results in some higher-temperature polymer applications for PP. Some of the applications for PP include containers such as bottles, laboratory ware, appliance parts, automobile fenders, automobile fan shrouds, and heater ducts. PP is also made into film wrap material.

Polystyrene (PS) is the fourth most extensively utilized thermoplastic polymer. The phenylene ring in each mer unit results in PS being amorphous and of a very high transparency. The phenylene ring also provides steric hindrance that prevents long-chain molecules from sliding past each other; thus PS has a high tensile strength of 30 to 60 MPa, as shown in Table 8.16, and a low elongation to failure of 1 to 4%. PS is subject to environmental degradation, and it is attacked by chemicals such as solvents and oils. Applications of PS include automobile interior parts, appliance parts, and domestic products. PS is also foamed to form insulated containers. PS is often mixed with other polymers to improve its ductility.

PS copolymers (SBR, ABS, and **SAN)** can be tailored to achieve desired engineering properties. PS is made more impact resistant by making a copolymer with butadiene rubber (BR) to form styrene butadiene rubber (SBR), which is also called high-impact polystyrene (HIPS). Applications of SBR include automobile interior parts, appliance housings, dials and knobs, bicycle helmets, and household items.

The amorphous copolymer ABS is composed of polyacrylonitrile (PAN), BR, and PS. The mer structures of PS and PAN are shown in Figure 2.19, and the mer structure of BR is presented in Table 8.19 below with the discussion of rubbers. Because the amounts of each polymer can be varied in ABS, it is possible to obtain a wide range of mechanical properties. PAN provides environmental resistance and high-temperature strength, PS provides rigidity and transparency, and BR provides toughness and elongation. Because ABS is the most widely used polymer for low-cost engineering plastics applications, ABS is considered a commodity plastic. ABS is used in the housing for small tools such as drills and saws, safety helmets, golf cart fenders, snowmobile covers, housings for electronics such as televisions and radios, and luggage shells. The housing for the leaf blower shown in Figure 8.5 is made of ABS plastic.

SAN is a random amorphous copolymer of PS and PAN, typically with about 65 to 70% PS. The mechanical properties of PS and SAN are similar; however, SAN is somewhat stronger and has higher heat resistance, by about 10°C. Some SAN applications include dishwasher-safe containers; oil-resistant

Figure 8.5 A leaf blower with a housing made of ABS plastic. (*Photo by C. M. Gilmore*)

containers; packaging for food, pharmaceuticals, and cosmetics; automotive instrument panel lenses; appliance knobs; blender and mixer bowls; medical syringes; construction safety glazing; and plastic mugs. PAN itself is used as a fiber, and one trade name is Orlon™.

8.3.2   Engineering Thermoplastic Polymers

Table 8.17 presents the mechanical properties of some thermoplastic engineering plastics. Most engineering plastics have the following properties:

- A relatively high strength and elastic modulus
- Retention of mechanical properties over a wide range of usage temperatures
- A relatively high toughness
- Dimensional stability over the usage temperature range
- Environmental resistance

Table 8.17   The Mechanical Properties of Some Thermoplastic Engineering Polymers, Including the Polymer Name, Abbreviation, Grade If Any, Density (ρ), Young's Elastic Modulus (E), Tensile Strength (σ_u), Strain to Fracture (ε_f), Izod Impact Strength (IS), and Structure (A) Amorphous or (C) Crystalline

Polymer	Grade	ρ g/cm^3	E GPa	σ_u MPa	ε_f %	IS J/m	Structure
PA 6[a] Polyamide 6 (Polycaprolactam)		1.13	3	80	50–120	30–120	C
PA 6[b] Polyamide 6 (Polycaprolactam)			1.5	50	160–200	160	
PA 6[a] Polyamide 6 (Polycaprolactam)	30–35% glass fiber	1.35–1.42	8–10	170–180	2–4	50	C
PA 6[b] Polyamide 6 (Polycaprolactam)	30–35% glass fiber		5.5	110		95	
PA 6,6[a] Polyamide 6,6 [Poly(hexamethyleneadipamide)]		1.14	3.4	75–90	20	30–55	C
PA 6,6[b] Polyamide 6,6 [Poly(hexamethyleneadipamide)]			1.7–2	50	80	50–110	C
PA 11[a] Polyamide 11 [Poly(11-aminoundecanoic acid)]		1.04	1.5	45–50	400–500	100–NB	C

(Continued)

Table 8.17   *Continued*

Polymer	Grade	ρ g/cm^3	E GPa	σ_u MPa	ε_f %	IS J/m	Structure
POM Polyacetal Polyoxymethylene	Homopolymer	1.42	3.1	65–70	25–75	60–120	C
POM Polyacetal Polyoxymethylene	Copolymer	1.41	2.8	65–72	40–75	50–80	C
PET Poly(ethylene terephthalate)		1.29–1.40	3	50	50–300	12–40	C
PBT Poly(butylene terephthalate)		1.31	2.3–2.5	50–60	120–200	40–55	C
PBT Poly(butylene terephthalate)	+30% glass fiber	1.52	10	100–140	2–4	80–130	C
PC Polycarbonate		1.2	2.1–2.4	70–90	100–120	650–1000^c	A
CA Cellulose acetate		1.27–1.32	1.5–2.5	25–45	10–70	100–450	A
CAB Cellulose acetate butyrate		1.18	1.4–1.8	30–35	30–100	50–500	A
PMMA Poly(methyl methacrylate)		1.17–1.20	2.5–3.3	55–75	3–5	10–20	A
PTFE Polytetrafluoroethylene		2.15–2.20	0.41	7–30	200–400	150	C
PSU (PSO) Polysulfone		1.25	2.5–2.6	70	50–100	65–70	A
PES Polyethersulfone		1.37	2.5	80–90	40–80	75–120	A
PPS Poly(phenylene sulfide)		1.35	3.6	65–75	1–2	70	C
PPO (PPE) Poly(phenylene oxide) or -ether	Modified with PS	1.06–1.08	2.2–2.7	50–60	200–350	200–370	A
PEEK Polyetheretherketone		1.32	3.6	90–200	50	80	C
PEEK Polyetheretherketone	30% glass fiber	1.49	10	100	2	100	C

Based on data from Brostow, W., Kubat, J, and Kubat, M., Mechanical Properties, Physical Properties of Polymers Handbook, Mark, J. E. ed., AIP Press, Woodbury, N.Y. (1996), p. 332.
^{a}dry as molded.
^{b}at 50% relative humidity.
^{c}thickness 3.2 mm.

Polyamide (PA) 6,6 listed in Table 8.17 has a repeating amide group, as shown for polyamide (nylon) in Figure 2.20. The 6,6 after polyamide means that there are 6 plus 6 carbon atoms in the repeating mer unit. Despite the complexity of the mer unit, nylons are highly crystalline; thus nylon is not transparent. There are permanent dipoles from the N—H bonds and the C—O bonds. The

N and the O ends of these dipoles are negative. Nylon's tensile strength of 75 to 90 MPa and its ductility of 20 to 80% are relatively high, and its crystallinity also adds to its strength. Nylon also has a low coefficient of friction. Nylon is sensitive to environmental conditions; for example, it absorbs polar molecules, such as water, because of the permanent-dipole bonds. Applications for nylon include gears, bearings, roller wheels, impellers, and zippers; and nylon is drawn into fiber to form fabrics, carpets, and rope.

Polycarbonate (PC) has important engineering applications because of its combination of high tensile strength of 70 to 90 MPa, its high elongation to failure of 100 to 120%, and its relatively high resistance to fracture. The high tensile strength results from the steric hindrance provided by the complex mer structure shown in Figure 2.20. The complex mer unit also results in an amorphous structure, and PC is transparent. PC has good elevated-temperature properties for a polymer, and it has high electrical resistivity. PC is resistant to many chemicals, but it is attacked by chlorinated and polar solvents and alkali solutions. Applications include mechanical parts such as gears and cams, CDs, transparent parts (such as windows, headlight and taillight housings, safety-glass lenses, safety shields, and bulletproof windows), and housings for tools, appliances, and computers. One trade name for PC is Lexan™.

Cellulose acetate (CA) is based upon cellulose derived from materials such as wood and cotton. The raw cellulose is chopped and chemically treated with acetic acid, acetic anhydride, and sulfuric acid to form CA. CA is used for tool and brush handles, eyeglass frames, pen barrels, and cellophane packaging film. Movie film is made from CA; CA is spun into a fiber known as acetate. Cellulose acetate butyrate (CAB) is tougher than cellulose acetate and is used for data keyboards, light covers, steering wheels, car trim parts, and packaging. CAB is produced by adding butyric acid to the mix of acids used to form CA. The butyric acid results in the formation of butyryl groups on the polymer.

Polymethylmethacrylate (PMMA) is an amorphous polymer with very high transparency. PMMA is amorphous because of the large methyl and acrylate groups attached to the LCM, shown in Figure 2.19. PMMA sheets are often utilized as a substitute for glass where fracture resistance is required. PMMA is also known commercially as Plexiglass™ and Lucite™. In Table 8.17 note the relatively high tensile strength of 55 to 75 and low elongation to failure of 2.0 to 5.5% for PMMA. The strength of PMMA is discussed in Section 7.6.2. PMMA has good chemical and environmental resistance. Applications of PMMA include glazing materials for boats, aircraft, and skylights; automobile light covers; safety shields; and eyeglass lenses.

Polytetrafluoroethylene (PTFE) is partially crystalline; it has a relatively low tensile strength of 7 to 30 MPa and a high ductility of 200 to 400%. PTFE has a high specific gravity of 2.14 to 2.20. In PTFE all of the hydrogen atoms on the polyethylene LCM are replaced with fluorine, as shown in Figure 2.19. PTFE has a low coefficient of friction, which gives it self-lubricating properties. PTFE has a low surface energy, as shown in Table 3.2. The low surface energy results in other materials not easily bonding or sticking to it. PTFE has useful mechanical properties up to 260°C, which result in many cookware applications. PTFE has high electrical resistivity because of the strong carbon-fluorine bond, and it has strong resistance to chemical and environmental attack. Applications of PTFE include pipe and pump parts for chemicals, high-temperature electrical-cable insulation, molded electrical components, electrical tape, coatings for fry pans, gaskets, O-rings, and seals. One commercial name of PTFE is Teflon™.

Polyester (PET) is an engineering thermoplastic polymer whose properties are not listed in Table 8.17 that has extensive usage. The polyester mer is listed in Figure 2.20. The mer backbone has a combination of carbon oxygen and benzene rings that give the polymer strength and stiffness. Polyester has extensive use in thin sections because it can be strengthened by orienting the LCMs and by partially crystallizing the polymer. However, the amount of crystallization is not sufficient to impede the transparency of the polymer. Thermoplastic PET is the primary polymer in beverage containers, and other thin section applications include photographic film backing, magnetic tape backing, electrical insulation, and decorative films. Other engineering applications include pump housings, windshield wiper arms, sunroof frames, and light-duty gears. It should be noted that there is also a thermoset polyester that is discussed below.

8.3.3 High-Performance Thermoplastic Polymers

In Table 8.17 the polymers PSU, PES, PPS, PPE, and PEEK are considered high-performance thermoplastic polymers. These polymers have a backbone composed primarily of benzene rings. Figure 2.20 shows the mer units for PES and PEEK. When the benzene rings are linked by an oxygen atom it is an ether linkage, and when the linkage is a by a C=O group it is a ketone linkage. The PEEK mer has two oxygen atom linkages and one C=O linkage; thus the name polyetheretherketone. The backbone of benzene rings produces polymers of high strength, high stiffness, and high impact resistance; good high-temperature performance and solvent resistance; and low flammability. These polymers are often combined with reinforcing fibers in high-performance composite materials.

Polysulfones (PSU and PES) have mer units with benzene rings joined with an SO_2 group, as shown in Figure 2.20. The SO_2 group adds stiffness and strength to the molecule, and also resistance to high-temperature deformation. For example, PES has a high tensile strength of 80 to 90 MPa and an excellent strain to fracture of 40 to 80%. PES is amorphous and can be clear. Applications of PES include hot-water pipes, circuit-breaker parts, circuit boards, automobile parts near the engine, and dishwasher parts.

Polyetheretherketone (PEEK) has a mer unit with benzene rings, as shown in Figure 2.20. PEEK is one of the highest-tensile-strength polymers (90 to 200 MPa), with a strain to fracture of 50%. PEEK also has the highest melting temperature of any thermoplastic polymer of 334°C and good resistance to chemical attack. The main application of PEEK is as a matrix material for carbon-fiber-reinforced composites in the aerospace industry.

8.3.4 Thermosetting Polymers

The mechanical properties of some thermosetting polymers are presented in Table 8.18. Thermosetting polymers have three-dimensional covalent bonds that result in their relatively high hardness, tensile strength, and elastic modulus, but also account for their low ductility.

Table 8.18	The Polymer Name, Room-Temperature Density (ρ), Tensile Strength (σ_u), Strain to Failure (ε_f), and Elastic Modulus (E) for Some Thermosetting Polymers			
Name	ρ g/cm^3	σ_u MPa	ε_f %	E GPa
Epoxy	1.2–1.3	30–90	1–4	3–5
Polyester	1–1.4	30–40	1.5–2.5	2–4.5
Phenolformaldehyde	1.4	25	0.4–0.8	5.6–12

Based on data from Brostow, W., Kubat, J, and Kubat, M., Mechanical Properties, Physical Properties of Polymers Handbook, Mark, J. E. ed., AIP Press, Woodbury, N.Y. (1996), p. 333, and from Osswald, T.A., and Menges G., Materials Science of Polymers for Engineers, Hanser Publishers, Munich (2003), Table I.

Epoxy is widely utilized because the reaction that cross-links the LCMs is accomplished by mixing two liquids, as we discussed in Section 7.7.2. Epoxy is utilized as the matrix in the production of composite materials, because after the liquid epoxy resin and liquid hardener are mixed, they remain liquid for approximately an hour before the mixture starts to solidify. In this time the mixed liquid can be worked into the fiber material to produce the composite. Epoxy has low cure shrinkage; thus the final rigid solid is approximately the same shape as the liquid was. The tensile strength of epoxy as shown in Table 8.18 is relatively high at 30 to 90 MPa, and its elastic modulus is 3 to 5 GPa. The strain to fracture for all thermoset polymers is small relative to that of most thermoplastic polymers. Epoxy has good chemical and environmental resistance and a high electrical resistivity. Epoxy is applied as a coating to protect containers, appliances, and automotive parts. The high electrical resistivity results in applications as high-voltage insulators, electrical-wire coatings, and as an encapsulator for integrated circuits.

Polyester is cross-linked by using unsaturated polyester (PET) to form a thermoset polymer. The PET mer shown in Figure 2.20 is saturated and is the form used to make LCMs of thermoplastic PET. To produce unsaturated PET, two hydrogen atoms are removed from the region of the PET mer that has an ethylene structure and replaced by a double bond between the two carbon atoms. The unsaturated PET is made from short molecules so that it is a liquid. To produce solid thermoset PET, the double bond between the carbon atoms is opened by an initiator such as peroxide that allows for cross-linking between polyester molecules. The tensile strength of thermoset PET as shown in Table 8.18 is from 30 to 40 MPa, and the elastic modulus is from 1.5 to 2.5 GPa. Epoxy is stronger than thermoset PET, but thermoset PET is utilized in many of the same applications as epoxy because the components of the polyester are less toxic than the hardener in epoxy.

Phenolics are three-dimensional network solids, as shown in Figure 7.28. Phenolics, such as PF, have a lower tensile strength and elongation to failure than do other thermosets, but they have a higher elastic modulus, as shown in Table 8.18. One commercial phenolic is Bakelite™. Phenolics have a high stiffness, a high electrical and thermal resistivity, and a high chemical and environmental resistance. Applications of phenolics include automobile-transmission parts, housings for electrical hand tools, electrical resistive parts, electrical insulators, adhesives for plywood and particleboard, and appliance parts and panels.

8.3.5   Thermosetting Elastomers

Vulcanized natural rubber (NR) with cross-links of sulfur, as shown in Figure 2.24, is an example of a thermosetting elastomer. The polymer in NR is *cis*-polyisoprene. The mechanical properties of vulcanized NR with a few percent sulfur results in properties typical of those listed in Table 8.19. The most important property of elastomers is the very high elongation to failure of up to 900% for NR. If an elastomer does not fracture, and the stress is removed, most of the strain is recovered immediately. The strain is elastic, and the name *elastomer* (from "elastic" and "polymer") follows. The name *rubber* comes from one of the earliest applications of this material, as an eraser that rubs out pencil markings.

The amount of sulfur in the rubber determines its strength and flexibility. Flexible rubber gloves only have 2 to 3% sulfur. Automobile tires have from 3 to 4 weight percent sulfur, and carbon black is added to the rubber in tires to increase the strength and the tear and abrasion resistance. Carbon black changes the color of NR from light tan to black. Vulcanized NR is utilized in applications where elasticity, flexibility, and energy absorption are required. Applications include rubber gloves, rubberized fabric, automobile tires, vibration isolation mountings, and various restraining devices such as elastic bands. Higher amounts of sulfur produce more-rigid rubber for applications such as battery cases.

Synthetic rubbers were developed during World War II because the primary supply of NR was interrupted from South East Asia. The interruption resulted in the development of a number of synthetic

Table 8.19 The Polymer Abbreviation, Polymer Name, Grade, Density (ρ), Elastic Modulus (E), Tensile Strength (σ_u), and Elongation to Failure (ε_f) for Various Rubbers

Polymer	Grade	ρ g/cm^3	E MPa	σ_u MPa	ε_f
NR (Natural rubber) *Cis*-polyisoprene	Unfilled vulcanisate	0.93	1–2	17–30	650–900
SBR Styrene-butadiene rubber	Unfilled, vulc. (23–25% styrene)	0.93–1.0	1–2	1.4–2.8	450–600
IIR (Butyl rubber) Isobutylene-isoprene rubber	Unfilled, vulc.	0.91–0.98	—	17–21	750–950
NBR (nitrile rubber) Acrylonitrile-butadiene rubber	Unfilled, vulc. (AN content 26–27%)	0.92		4–7	350–800
CR (Chloroprene rubber) Poly(2-chloro-1, 3-butadiene)	Unfilled, vulc.	1.2–1.25	1–3	13–22	800–1000
EPDM Ethylene-propylene rubber	Unfilled, vulc.	0.85–0.87	—	1.2	400

Based on data from Brostow, W., Kubat, J, and Kubat, M., Mechanical Properties, Physical Properties of Polymers Handbook, Mark, J. E. ed., AIP Press, Woodbury, N.Y. (1996), p. 333.

rubbers, including butadiene, butyl, neoprene (also called chloroprene), and ethylene propylene. The molecular structures of NR, butadiene, butyl, and chloroprene are shown in Table 8.20. All of these mers have a carbon-carbon double bond that allows for cross-linking with sulfur, as shown in Figure 2.24, except for butyl rubber (polyisobutylene). Synthetic rubber with approximately 90% of the polymer *cis*-polyisoprene is now produced. However, natural rubber is still preferred because of its lower cost and better properties.

The difference between the mer of butadiene rubber (BR) and that of NR is that BR has a hydrogen atom located where NR has a methyl group, as shown in Table 8.20. The methyl group provides steric hindrance to NR, resulting in a higher strength and stiffness than seen in BR. As a result of this low strength, BR is primarily used as a copolymer with other polymers such as PS and acrylonitrile (AN) to form SBR and acrylonitrile rubber (NBR), formerly known as buna-n, as well as the ABS copolymers we discuss in Section 8.3.1. The properties of vulcanized SBR with approximately 24% styrene are presented in Table 8.19. SBR has a much lower tensile strength than NR does, but a similar stiffness and good elongation to failure. SBR has poor oil and oxidation resistance, and it is sensitive to UV radiation. Applications of SBR include footware, wire insulation, adhesives, gaskets, and seals. Vulcanized NBR with 26 to 27% AN has a higher tensile strength than does SBR, as shown in Table 8.19, and good elongation to failure; however, the properties are still not equal to those of NR. NBR has improved oil and oxidation resistance relative to that of SBR, but NBR is more expensive than SBR. Applications of NBR include oil and fuel lines, gaskets, seals, conveyor belts, and coatings for printer rolls.

Butyl rubber, or **polyisobutylene,** has two methyl groups attached to a carbon and two hydrogen atoms attached to a second carbon, as shown in Table 8.20. Note that butyl rubber and butadiene rubber are different rubbers. Butyl rubber does not have any of the carbon-carbon double bonds that are necessary for vulcanization; thus butyl rubber is usually copolymerized with approximately 2% isoprene to give some

Table 8.20   The Polymer Name, Mer Unit, and Some Applications of Elastomers

Polymer	Repeat Unit	Applications
Polyisoprene		Tires, golf balls, shoe soles
Polybutadiene (or butadiene rubber or Buna-S)		Industrial tires, toughening other elastomers, inner tubes of tires, weatherstripping, steam hoses
Polyisobutylene (or butyl rubber)		Adhesives, ball bladders, tire inner tubes, high temperature hoses, vibration dampers, sound dampers, weather stripping
Polychloroprene (Neoprene)		Hoses, cable sheathing
Butadiene-styrene (BS or SBR rubber)		Tires
Butadiene-acrylonitrile (Buna-N)		Gaskets, fuel hoses
Silicones		Gaskets, seals

Based on data from Askeland, D.R., Fulay, P.P., and Wright, W.J., The Science and Engineering of Materials, 6th ed., Cengage Learning, Stamford, CT. (2011), p. 633.

carbon-carbon double bonds to allow for vulcanization. Table 8.19 shows that vulcanized butyl rubber with isoprene has strength and elongation to failure comparable to that of NR, and it has good oxidation and weathering resistance. The principal applications of butyl rubber include vibration dampers, inner tubes for tires, high-temperature hoses, weather stripping, and sound dampers.

In **chloroprene (CR),** also known as **neoprene rubber,** the methyl group in *cis*-polyisoprene NR is replaced by a chlorine atom, as shown in Table 8.20. The chlorine atom provides steric hindrance to the molecule and permanent-dipole bonding. This results in properties similar to those of NR; however, the strength of CR is slightly lower than that of NR. CR has good oil resistance. Applications of CR include fuel hoses, boots, shoe soles, fire-resistant mattress foam, wet suits, and coatings for dry suits.

Ethylene propylene diene monomer (EPDM) rubbers are made by producing a copolymer of ethylene and propylene diene. Propylene diene has two methyl groups in the mer replacing two hydrogen atoms on the ethylene mer. This copolymer does not have any carbon-carbon double bonds for vulcanization. Instead, the polymer is heated in the presence of peroxide, and the peroxide removes hydrogen from some locations on the polymers, which allows different LCMs to become cross-linked to each other. EPDM rubber is stiffer than NR but not as high in strength, as shown in Table 8.19. Applications of EPDM rubbers include seals for automobile windows, doors, and trunks; roofing membranes, hoses, tubing, wire and cable insulation, and tire sidewalls.

8.3.6   Thermoplastic Elastomers

Thermosetting elastomers, such as rubber, cannot be processed into shape once they are cross-linked and solidify, because heating does not soften the polymer. They must be shaped during the cross-linking process. **Thermoplastic elastomers** are shaped by heating and deformation. Thermoplastic elastomers use areas of strong bonding that are usually polar bonds to anchor the polymer chains together, as shown in Figure 8.6. These areas of strong bonds provide the elastic recovery. Areas of weak bonds where the LCMs are more distant from each other easily deform and provide the large deformation of an elastomer. A number of different thermoplastic elastomers have been developed that use this strong and weak bonding combination that are based upon PS, polyurethane, polyester, and polyamides. Applications of these thermoplastic elastomers include footware, molded and extruded goods, and elastic fibers. Spandex™ and Lycra™ are thermoplastic elastomer fibers.

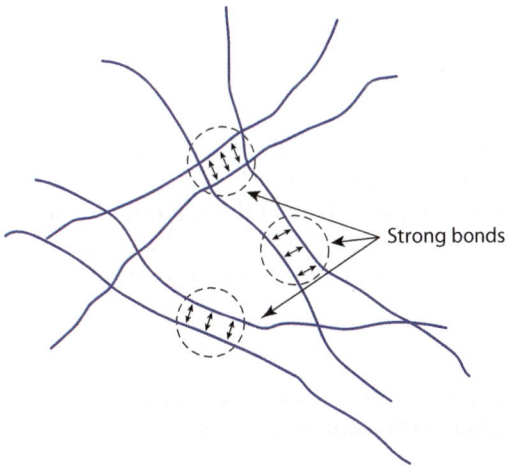

Figure 8.6   A schematic of areas with strong bonding and areas with weak bonding in a thermoplastic elastomer.

8.3.7 Fluoroelastomers

Fluoroelastomers are based upon fluorinated hydrocarbons, particularly vinyldiene fluoride. Vinyldiene fluoride has two fluorine atoms replacing two hydrogen atoms on the ethylene mer. Fluoroelastomers offer excellent resistance to chemical attack up to temperatures of 200°C. Applications of fluoroelastomers include O-rings, gaskets, oil seals, lubricants, hoses, and tubing. One commercial name of a fluoroelastomer is Viton™, which is used in O-rings.

8.3.8 Silicones

Silicones are polymers based upon silicon rather than carbon. Silicon is in Group IV in the periodic table, just below carbon, and there are some chemical similarities between the two elements; however, there are also significant differences. Silicon does not form double bonds as does carbon; therefore, the polymerization processes used for carbon-based mers do not apply to silicon, and the cross-linking processes used for carbon-based polymers also do not apply to silicon. In silicone polymers, silicon alternates with oxygen along the backbone of the LCM, as shown in Table 8.20, to produce polysiloxanes, or silicones. Low-molecular-mass silicones are oils, and higher-molecular-mass silicones are solids. The mer in Table 8.20 has methyl groups attached to the silicon atoms; however, the methyl groups can be replaced by a benzene ring forming a phenyl group, or other atoms or groups of atoms to develop certain properties.

Silicone elastomers are produced by adding active side groups to silicone polymers that can react with molecules, such as water and carbon dioxide to vulcanize the silicone. Both water vapor and carbon dioxide are absorbed into silicone at room temperature resulting in room temperature vulcanization (RTV) of silicone. Silicone caulk and silicone waterproofing treatment for fabrics use RTV silicone elastomers. There is also high temperature vulcanization (HTV) based upon the addition of molecules such as peroxide to silicone polymers to produce elastomers. Special liquid hardeners are available that act like the hardener in epoxy. Some advantages of silicones are low surface tension, solvent resistance, thermal stability, oxidation resistance, low freezing temperature, low volatility at high temperature, low flammability, and chemical inertness. Some of the applications for silicones include sealants, caulks, waterproofing, adhesives, and implant materials.

8.3.9 Polyurethane

A basic **urethane** reaction is shown in Figure 8.7, where a molecule with an hydroxyl group (O—H) reacts with a molecule containing an isocyanate group (N=C=O) to form a urethane. R^1 and R^2 can be different atoms or molecules that do not participate in the reaction. R^1 and R^2 can change the physical properties of the resulting urethane. The hydroxyl group can be in the form of an alcohol containing (C—O—H). **Polyurethane (PUR)** is a result of the chemical reaction of a polyol (multiple alcohol or

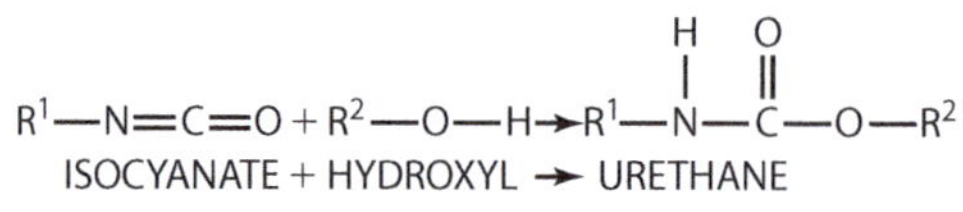

$$R^1—N≡C≡O + R^2—O—H → R^1—N—C—O—R^2$$
ISOCYANATE + HYDROXYL → URETHANE

Figure 8.7 The reaction of molecules containing isocyanate and hydroxyl groups to form a urethane molecule.

multiple O—H groups) and a polyisocyanate that has multiple N=C=O groups. PURs are like epoxy in that the polyol is considered to be the resin, and the polyisocyanate the hardener. Both the polyol and polyisocyanate are usually liquids that react to form a solid.

Depending upon the molecular mass of the polyol and polyisocyanate and the R^1 and R^2, the resulting PUR can be thermoplastic, thermosetting, a thermoset elastomer, a thermoplastic elastomer, a rigid solid, a soft solid, or a liquid. There are double bonds on the PUR molecule that can be opened to allow for cross-linking and the formation of thermoset polymers.

Thermoset PUR is used for automobile bumpers and in shoe soles. PUR thermoplastic elastomer fibers include Spandex and Lycra. PUR coatings are used on wood, metal, and rubber. PUR is made into foams and fibers. PUR foams are used for flexible bedding, seat cushions, and carpet padding. Rigid PUR foam is used in construction for insulation and for packaging.

Example Problem 8.2

A fastener bolt for the airframe of an aircraft has a cross-sectional area of 2.0 square cm and a length of 5.0 cm. The bolt must be able withstand a tensile force of 5.0×10^4 N, and as a factor of safety this tensile force can produce a stress of no more than 50% of the tensile yield stress in the bolt. For damage tolerance the bolt must have at least 10% strain to fracture. Because weight is critical for this aircraft design, the maximum weight possible for this bolt is 30 g. The airframe fastener is not subject to high or low temperatures. Of the materials covered in this chapter, what would be the best material to select? Because this is an aircraft part, it is not necessary to choose the cheapest material, but excessive cost should be avoided.

Solution

Ceramics are eliminated from consideration because they should not be used in a design with any significant tensile forces. The weight requirement could limit the selection of materials. The maximum density (ρ) for the part is calculated from the maximum weight divided by the volume of the part.

$$\rho = \frac{W}{V} = \frac{W}{Al} = \frac{30g}{2.0 \times 10^{-4}\,m^2(5.0 \times 10^{-2}\,m)} = \frac{30 \times 10^{-3}\,kg}{10 \times 10^{-6}\,m^3} = 3.0 \times 10^3\,\frac{kg}{m^3} = 3.0\,\frac{g}{cm^3}$$

Of the materials listed in this chapter the metal alloys of aluminum and magnesium all satisfy the weight requirement as do all polymers. The stress requirement may eliminate other materials from consideration. The applied stress (σ) is the force divided by the cross-sectional area.

$$\sigma = \frac{F}{A} = \frac{5.0 \times 10^4\,N}{2.0 \times 10^{-4}\,m^2} = 2.5 \times 10^8\,\frac{N}{m^2} = 250 \times 10^6\,\frac{N}{m^2}$$

The yield stress of the material must be at least twice the applied stress.

$$\sigma_y = 500 \times 10^6\,\frac{N}{m^2}$$

There are no magnesium alloys that have this yield strength. Of the aluminum alloys, the alloys 2090-T86 and 7075-T6 have sufficient yield strength.

The strain to failure for the 2090-T86 is only 6%. It is therefore not sufficiently ductile. The alloy 7075-T6 has 11% strain to failure and satisfies all of the requirements. There are no polymers or polymers reinforced with glass listed in these tables that meet the strength requirement.

Summary

- The primary commercial structural metal alloys include the FCC alloys of aluminum, copper, nickel, and austenitic stainless steel; BCC iron, steel, and ferritic stainless steels; and hexagonal close-packed magnesium and titanium.

- Commercial metals are referred to using several numbering systems, including that of the Aluminum Association (AA), the American Iron and Steel Institute (AISI), the American Society for Testing Materials (ASTM), the Society of Automotive Engineers (SAE), and the unified numbering system (UNS).

- The two main conditions for metals are wrought and cast. Wrought alloys are those that are produced by mechanical deformation processes, such as rolling, extruding, and drawing. The conditions for wrought alloys include recrystallized, annealed, stress-relieved, hot-rolled, cold-rolled, solution treated quenched (STQ) and aged, and STQ and tempered. In cast alloys, liquid metal is poured into a mold that forms the desired solid shape. Cast alloys can be heat treated to increase the strength.

- The properties of aluminum that make it the metal of choice for designs are a low density of approximately 2.71 g/cc, a high specific yield strength, good corrosion resistance in air, good thermal and electrical conductivity, good ductility, a relatively low melting temperature of 660°C that allows for casting, and ductility to very low temperatures. Some aluminum alloys are strain hardened, and the highest strength aluminum alloys are precipitation hardened.

- Copper is very ductile and of low strength in the pure state. Commercially pure copper and its alloys are known by names, such as cartridge brass. Because of its high ductility, pure copper and most of its alloys are mechanically worked into wrought forms. Some alloys, such as cartridge brass, can be plastically deformed extensively into thin-walled tubes. Copper has a relatively low melting temperature of 1085°C, and it can be cast into shape. Most copper alloys are solid solution hardened, and they can be strain hardened. The copper-beryllium alloys are precipitation hardened. Copper and most of its alloys are corrosion resistant in air and in saltwater. The high thermal and electrical conductivity of copper and copper alloys results in many applications. Copper is also ductile at very low temperatures because of its FCC crystal structure, and it has cryogenic applications.

- Plain-carbon and low-alloy steels are the primary materials for building structures, automobiles, bridges, and so on. Steel is made from iron, carbon (<1%), and other alloy elements. Manganese, usually less than 1%, is added to all steels to deoxidize the steel and to provide solid-solution strengthening. Plain low-carbon steels contain less than 0.25% carbon, and they are of low strength. These steels are utilized in applications such as automobile bodies, structural beams, and piping. Medium-carbon steels contain between 0.25 and 0.6% carbon. These steels can be quenched and tempered for high strength. High-carbon steels have more than 0.6% carbon, and they also can be quenched and tempered. Plain-carbon steels are the cheapest steels, but as the strength is increased by adding carbon and quenching and tempering, the ductility is significantly decreased. Plain-carbon steels become brittle at low temperatures, and they are not corrosion resistant. Alloy steels are developed to improve the properties of carbon steels. Low-alloy steels typically contain a total of less than 5% of alloy elements Mn, Al, Cu, Ni, Si, V, and Mo.

- The hardenability of steels is measured by the Jominy test. Hardenability is the ability of steel to be hardened to a specified depth.

- In high-strength low-alloy (HSLA) steels, very small amounts of elements, such as Cr, V, Mo, and Nb, are added to form carbides other than Fe_3C, and these carbides result in an increase in the strength of quenched-and-tempered steel. The alloy elements are designated with a range that is designed to achieve specified mechanical properties.

- Tool steels are high-carbon steels that are alloyed with carbide-forming elements such as W, Mo, Cr, and V. These steels are normally solution treated quenched and tempered to high strength. Applications of these steels are where a sharp edge or limited wear is required, as in cutting tools, knives, dies, and drills.

- Stainless steels were developed for improved corrosion resistance relative to that of plain carbon and low-alloy steels. At least 11 weight percent Cr is added to iron to form protective chromium oxide. Stainless steels alloyed primarily with Cr have the BCC ferrite crystal structure. If approximately 9 weight percent nickel is added to stainless steel, it is sufficient to stabilize the austenitic FCC phase at room temperature. The austenitic stainless steels have better corrosion resistance in a saltwater environment than ferritic BCC stainless steels. Austenitic stainless steels are used in the annealed condition, and they have a relatively low yield strength. Austenitic stainless steel is not ferromagnetic, whereas ferritic stainless steel is ferromagnetic. Lower-carbon-content ferritic stainless steels are used in the annealed condition. Higher-carbon-content ferritic stainless steels are quenched and tempered to form tempered martensite. The stainless steel 17-7PH is precipitation hardened with the intermetallic compound Ni_3Al, and after quenching and tempering it has a tempered martensite structure.

- Cast iron has a carbon content from 2 to 4.5 weight percent. At the chemical composition of iron plus 4.26 weight percent carbon, there is a eutectic reaction of liquid transforming to γ-phase iron plus Fe_3C at a temperature of 1153°C. If liquid iron with a carbon composition close to eutectic is cooled to room temperature, it forms white cast iron. White cast iron is extremely hard and brittle, and it cannot be machined. White cast iron is used in some applications because of its resistance to wear and abrasion. If white cast iron is held at temperatures of 850°C for extended periods of time, the iron carbide transforms to clusters of graphite in a matrix of ferrite iron to form malleable iron with a ductility of 6 to 10%. Cast iron alloyed with silicon and heat treated decomposes the iron carbide into iron plus flakes of graphite to produce gray cast iron. Gray cast iron is brittle, because the flake form of graphite forms a sharp crack tip in the iron matrix. Adding magnesium to cast iron transforms Fe_3C into nodular graphite and iron, resulting in nodular or ductile cast iron with a high ductility of 18%.

- The main advantage of magnesium as a structural metal is a low density of 1.7×10^3 kg/m³, and a low melting temperature of 651°C allows it to be cast at low temperature. Magnesium has a relatively low elastic modulus of 45 GPa; as well as a relatively low strength, ductility, and corrosion resistance; relative to other structural metals. Magnesium can be solid solution strengthened with aluminum, and precipitation hardened with alloy additions of manganese, zinc, zirconium, and thorium. The wrought alloys can be strain hardened; however, the small ductility limits this strengthening mechanism. The cast alloys have a very low ductility for a metal. Magnesium alloys are applied where low weight is more important than high strength and high ductility, and where corrosion resistance is not critical.

- A superalloy is a metal alloy based upon nickel, cobalt, or iron that is designed for high-temperature use. Nickel is the base metal for most of the superalloys that are utilized in high-temperature gas turbines. In nickel-based superalloys, Al and Ti are added to form the precipitate Ni_3Al or Ni_3Ti, respectively, for high-temperature strength. Superalloys are also alloyed with high-melting-temperature metals such as Cr, Nb, Mo, W, and Ta. All of these elements solid solution strengthen superalloys, and they are strong carbide formers. Both Cr and Al provide superalloys with high-temperature corrosion and oxidation resistance because of the formation of protective oxides of Cr_2O_3 and Al_2O_3, respectively, on the surface rather than base-metal oxides. Some superalloys are strengthened with dispersion particles such as thoria (ThO_2) or yttria (Y_2O_3), and these alloys and some others are processed by powder-metallurgy techniques. Some superalloy parts, such as turbine blades, are cast into shape. Superalloys are cast as polycrystals (PC), directionally solidified (DS), and single crystals (SC). SC turbine blades are superior to PC and DS for the highest-temperature applications.

- Titanium alloys are one of the most corrosion-resistant metals in severe environments; they are less dense than steel; they have a relatively high melting temperature; and they have strengths comparable to that of steel. The excellent corrosion resistance of titanium is due to the formation of a surface layer of TiO_2 that protects the metal at moderate temperatures, and the TiO_2 coating results in Ti metal medical implants not being rejected by the body. Because of its strength at high temperature and its corrosion resistance, titanium is used for the skin material of high-performance aircraft such as supersonic jets and in jet-engine compressors. The attraction of titanium for oxygen at high temperature limits the use of titanium to temperatures of approximately 550°C. Pure titanium is hexagonal close-packed α phase at room temperature; however, at temperatures above approximately 890°C, which is the β-transus, titanium is BCC β phase up to the melting temperature. Al and oxygen raise the β-transus and stabilize the α-phase structure. BCC metals Fe, V, Mo, and Cr reduce the β-transus temperature and stabilize the BCC β phase. There are four categories of commercial titanium alloys: α, near α, $\alpha + \beta$, and β. Alpha-phase alloys include commercial purity (CP) titanium and alloys of titanium with aluminum and only small amounts of β-phase stabilizing elements. Alloys with greater than 5% aluminum are heat treatable, and they are not suitable for welding. If sufficient amounts of β-phase stabilizing elements are added to titanium, the alloy is $\alpha + \beta$. The $\alpha + \beta$ alloys are available in the annealed condition or in the solution-treated and aged condition (STA). The β alloys are alloyed with β-phase stabilizing elements. When these alloys are quenched from above the β-transus, the single phase of β is metastable at room temperature. Advantages of β alloys are that they can be deformed in the more-ductile BCC structure at lower temperatures, and they have high strength in the STA condition. The main applications of β-titanium alloys are in highly stressed aircraft structures, such as landing gears.

- In many applications, such as aircraft and high-efficiency or high-performance automobiles, it is important to have a high specific yield strength and a high specific modulus. The specific strength and specific modulus are the strength and elastic modulus divided by the specific gravity. High-strength aluminum alloys have a high specific yield strength, and steel has a high specific elastic modulus.

- The most widely used structural ceramic is silica (SiO_2), in the form of glass. The design strength of glass should be taken as a low number, such as 20 MPa or less. Of the crystalline ceramics, the use of alumina (Al_2O_3) exceeds that of all other ceramics. Other ceramics with structural applications include silicon carbide (SiC), silicon nitride (Si_3N_4), and zirconia (ZrO_2). The physical properties of ceramics result from their strong ionic and covalent bonding. Polycrystalline ceramics are brittle at room temperature, they are insulators or semiconductors, and they have high melting temperatures. Because of the physical properties of ceramics, their primary applications involve a high temperature and a high resistivity.

- The most commonly used commodity thermoplastic polymers, in order of use, are PE, PVC, PP, PS, and copolymers such as ABS. ABS is the most widely used polymer for low-cost engineering plastics applications. Bulk commodity polymers have a much lower tensile yield strength and elastic modulus than do metals and much lower melting temperatures; however, the density is much less than for metals. Commodity thermoplastic polymers are applied where the stresses and temperatures are low, and where low weight is important. Relative to commodity thermoplastics, engineering thermoplastic polymers have a higher strength and elastic modulus, retention of mechanical properties over a wide range of use temperatures, a relatively high toughness, dimensional stability over the use temperature range, and environmental resistance. Engineering thermoplastic polymers include PA, PC, PMMA, CA, PMMA, and PTFE. In high-performance thermoplastic polymers, the backbone is composed primarily of benzene rings. This produces polymers of high strength, stiffness, and impact resistance; good high-temperature performance; low flammability; and good solvent resistance. High-performance thermoplastic polymers include PSU, PES, PPS, PPE, and PEEK. The 334°C melting temperature of PEEK is the highest of any thermoplastic polymer, and PEEK has good resistance to chemical attack. The main application of PEEK is as a matrix material for carbon-fiber-reinforced composites in the aerospace industry.

- Thermosetting polymers have three-dimensional covalent bonds that result in a relatively high hardness, tensile strength, and elastic modulus but low ductility. Epoxy and polyester are cross-linked by mixing two liquids. Epoxy and polyester are utilized as the matrix in the production of composite materials and as cast parts. Phenolics are three-dimensional network solids. Phenolics have high stiffness, high electrical and thermal resistivity, and high chemical and environmental resistance.

- Thermosetting elastomers include vulcanized natural rubber (NR) with cross-links of sulfur and synthetic rubbers. The amount of sulfur added to NR determines the strength and flexibility of the rubber. The most important property of elastomers is the very high elongation to failure of up to 900%. If an elastomer does not fracture, and the stress is removed, most of the strain is recovered immediately. Synthetic rubbers include butadiene, butyl, neoprene (also called chloroprene), and ethylene propylene.

- Thermoplastic elastomers are shaped by heating and deformation. Thermoplastic polymers use areas of strong bonding that are usually polar bonds to anchor the polymer chains together and provide the elastic recovery. Areas of weak bonds easily deform and provide the large deformation of an elastomer. Applications of these thermoplastic elastomers include footwear, molded and extruded goods, and elastic fibers.

- Silicones are polymers based upon silicon rather than carbon. In silicone polymers, silicon alternates with oxygen along the backbone of the LCM. Low-molecular-mass silicones are oils, and higher-molecular-mass silicones are solids. Elastomers are produced by cross-linking silicone polymers with water, carbon dioxide, or special liquid hardeners. Some advantages of silicones are low surface tension, solvent resistance, thermal stability, oxidation resistance, low freezing temperature, low volatility at high temperature, low flammability, and chemical inertness. Some of the applications for silicones include sealants, caulks, waterproofing, adhesives, and implant materials.

- Fluoroelastomers are based upon fluorinated hydrocarbons, particularly vinyldiene fluoride. Fluoroelastomers offer excellent resistance to chemical attack up to temperatures of 200°C (392°F). Applications of fluoroelastomers include O-rings, gaskets, oil seals, lubricants, hoses, and tubing.

- Polyurethane (PUR) results from the chemical reaction of a polyol (multiple alcohol or multiple O—H groups) and an isocyanate that has multiple nitrogen-carbon-oxygen (NCO) groups of atoms. Polyurethanes are like epoxy in that the polyol is considered to be the resin and isocyanate the hardener. Both the polyol and isocyanate are usually liquids. There are double bonds on the PUR molecule that can be opened up for cross-linking to form a thermoset polymer. PUR can be thermoplastic, thermosetting, a thermoset elastomer, a thermoplastic elastomer, a rigid solid, a soft solid, or a liquid, depending upon the atoms or molecules on the mer and the molecular mass. PUR is also made into foams, coatings, and fibers. Thermoset PUR is used for automobile bumpers and in shoe soles.

Supplemental Reading Subjects and Authors

Full references are listed at the end of the book.

General:	*Askeland, Fulay, and Wright*
Mechanical properties of metals:	*ASM Handbook*
Mechanical properties of superalloys:	*Sims, Stoloff, and Hagel; Donachie and Donachie*
Mechanical properties of titanium:	*Lutjering and Williams*
Mechanical properties of ceramics:	*Carter and Norton; Chaing, Birnie, and Kingery*
Mechanical properties of glass:	*Bansal and Doremus*
Mechanical properties of polymers:	*Brostow, Kubat and Kubat; van Krevelen*

Homework

Concept Questions

1. In the UNS system of material designations, if an alloy designation starts with a G, it is a plain-carbon or low-alloy _____.

2. If an alloy form is produced by mechanical deformation, such as rolling, it is _____.

3. The alloy treatment that provides the lowest-strength and highest-ductility alloys is the _____ condition.

4. Quenching into a saltwater brine produces a (faster or slower) cooling rate than quenching into oil.

5. If the AA designation of an alloy begins with a 7, the major alloy element is _____.

6. If the AA designation of an alloy has three digits, it is a(n) _____ alloy.

7. Aluminum alloys such as 2024 and 7050 (are or are not) recommended for welding.

8. If an AA alloy's numerical designation is followed by the letter F, this indicates that the alloy is as _____.

9. If an aluminum alloy's numerical designation is followed by H18, it is strain hardened to _____ hardness.

10. If the AA numerical designation of an alloy is followed by T7, the alloy is solution heat treated and artificially _____.

11. The highest-strength aluminum alloys, such as 7075-T6, are strengthened by _____ hardening.

12. Aluminum alloys must be welded in an inert atmosphere because of the formation of aluminum _____ if welding is performed in air.

13. Aluminum-lithium alloys were developed for aircraft applications because of their relatively high yield strength and low _____.

14. Aluminum alloys are readily deformed into shape because they have the _____centered-cubic crystal structure.

15. Cartridge brass can be extensively deformed into thin-walled sections because it does not _____ harden.

16. _____ is the major alloy element in the highest-strength copper alloys.

17. _____ is an alloy of zinc and copper.

18. Free _____ grades of copper have 1 to 6 weight percent lead.

19. In nickel-silver copper alloys, the major alloy elements are nickel and _____.

20. If a copper alloy is in the (O) condition, it is _____.

21. If a copper alloy is in the OS050 condition, it is annealed with a grain size of _____mm.

22. Many of the mechanical applications of copper alloys are due to high thermal conductivity and excellent _____ resistance.

23. 1020 steel contains _____ weight percent carbon.

24. In 1020 steel, the only significant element other than iron and carbon is _____.

25. The ability of a steel to be hardened to a specific depth is the _____.

26. A way to measure the hardenability is the _____ end quench test.

27. In general, as the yield strength of a given alloy increases, the ductility _____.

28. In the UNS designation system, high-strength, low-alloy steels start with the letter ___.

29. In the ASTM designation system of HSLA steels, the letter grade A following the three digits in the designation system indicates that this alloy is of the _____ strength.

30. An alloy in the UNS designation system starting with the letter T is a _____ steel.

31. The primary purpose of adding chromium to stainless steel is to improve the resistance to _____.

32. Austenitic stainless steel has the _____-centered-cubic crystal structure at room temperature.

33. Austenitic stainless steel (is or is not) ferromagnetic.

34. In stainless steel, _____ is added to stabilize the austenitic phase.

35. The LS designation of stainless steel used for medical implants indicates low _____.

36. The stainless steel 17-7PH is precipitation hardened with the intermetallic compound _____.

37. The main reason for use of stainless steel is resistance to _____.

38. If 15 weight percent Cr is added to iron, the oxide that forms on the surface of this alloy is _____ oxide.

39. Austenitic stainless steel is (more or less) corrosion resistant than ferritic stainless steel.

40. _____ is the stainless steel used in beer kegs.

41. If the AISI designation of a stainless steel begins with a 4, it has the _____-centered-cubic crystal structure.

42. At the chemical composition of iron and 4.26 weight percent carbon and a temperature of 1153°C, there is a(n) _____ reaction.

43. Iron with 2 to 4.5 weight percent carbon is _____ iron.

44. If iron with 4.26 weight percent carbon is cooled to room temperature in a normal manner from the melting temperature, it is called _____ cast iron.

45. In ductile cast iron, the carbon is in the form of _____ rather than as iron carbide.

46. If a cast iron has the designation 60-40-18, the yield strength is equal to _____ psi.

47. If a cast iron has the UNS designation F2XXXX, it is _____ cast iron.

48. Automobile engine blocks are made with _____ cast iron.

49. A magnesium alloy that begins with the letters AK has major alloy elements of aluminum and _____.

50. A magnesium alloy that has the designation AZ31 has _____ percent aluminum.

51. Magnesium alloys have low ductility because of their _____ close-packed crystal structure.

52. The low ductility and low melting temperature of magnesium alloys results in many parts being _____ into shape.

53. In nickel-based superalloys, aluminum and titanium are added to provide _____ hardening.

54. Aluminum and _____ are added to superalloys to provide corrosion and oxidation resistance.

55. Superalloys that are dispersion-strengthened with an oxide must be produced by _____ metallurgy techniques.

56. The corrosion resistance of titanium alloys is due to the formation of a protective layer of titanium _____ on the surface.

57. At room temperature a crystal of pure titanium is _____ close-packed.

58. The minimum temperature where titanium alloys are 100% BCC is the β _____.

59. Adding BCC elements such as vanadium to titanium (<u>raises or lowers</u>) the β-phase transus.

60. There are four categories of titanium alloys: α, near α, _____, and β.

61. CP titanium alloys are primarily solid solution strengthened by the addition of the element _____.

62. Single-phase alpha titanium alloys (<u>can or cannot</u>) be welded.

63. Alpha titanium alloys with concentrations of aluminum greater than 5% can be precipitation hardened with the precipitate _____.

64. In titanium alloys called "mill annealed," the alloy is not _____, it is only annealed after deformation.

65. The most corrosion-resistant titanium alloys are the _____ alloys.

66. The highest-strength titanium alloys are the STA _____ alloys.

67. The β alloys of titanium are precipitation hardened by the precipitation of the _____ phase by heat treating at a temperature well below the β-transus.

68. The specific yield strength of the alloy Ti-6Al-4V in Table 8.12 is equal to _____ MPa.

69. The specific elastic modulus of the magnesium alloy AZ31B is equal to _____GPa.

70. The only metal that does not oxidize is _____.

71. The noble metals are good electrical and _____ conductors.

72. A common property of _____metals is their high melting temperature.

73. A primary application of _____is as a cladding for nuclear fuel rods.

74. Galvanized steel is coated with the metal _____.

75. The most widely used ceramic is silica in the form of _____.

76. The primary types of chemical bonding in ceramics are ionic and _____.

77. For a particular type of ceramic, the strength and elastic modulus both increase as the _____ increases.

78. The ceramic that is used as a spark-plug insulator in gasoline engines is _____.

79. One application of _____ is as an exhaust gas oxygen sensor.

80. The primary use for tungsten carbide is for _____ tools.

81. The most extensively utilized commodity thermoplastic polymer is _____.

82. PVC that is _____ has an increased elongation to failure relative to that of rigid PVC.

83. Polypropylene has (higher or lower) strength at elevated temperatures than PE does.

84. Polystyrene has a (high or low) strain to fracture at room temperature relative to that of most other thermoplastic polymers.

85. ABS is a(n) _____ of PAN, BR, and PS.

86. High-impact polystyrene is a copolymer of polystyrene and _____ rubber.

87. _____ is the most widely used low-cost commodity plastic for engineering applications.

88. Nylon derives some of its strength from dipole bonds, but this also leads to it _____ polar molecules such as water.

89. The structure of polycarbonate is _____. As a result, it is transparent, and this leads to applications such as safety-glass lenses.

90. A polymer that is made from wood or cotton is _____ _____.

91. PMMA is used as a substitute for _____where fracture resistance is required.

92. PTFE has low _____ energy. This results in other materials not sticking to it and applications in cookware.

93. The thermoplastic polymer with the highest melting point is _____.

94. If two benzene rings are linked by an oxygen atom it is an _____ linkage.

95. The main application of PEEK is as a(n) _____ material for carbon-fiber-reinforced composites in aircraft.

96. An advantage of epoxy over thermoplastic polymers as the matrix in composite materials is that the epoxy resin and hardener are _____ when mixed and can be impregnated into the fabric of the composite material.

97. The components of polyester are (more or less) toxic than epoxy.

98. The most important property of elastomers is the high elastic _____.

99. Natural rubber is made from the polymer _____.

100. Butadiene rubber has a very low tensile strength; therefore, it is primarily used as a(n)_____.

101. The mer of butyl rubber has no carbon-carbon double bonds; therefore, its molecules cannot be _____ with sulfur.

102. Thermoset polyester (PET) is produced by cross-linking a(n) _____form of PET.

103. In a thermoplastic elastomer there are areas of _____ bonding that allow for large strains, and areas of _____ bonding that anchor the LCMs together.

104. In silicones the backbone polymer is composed of alternating silicon atoms and _____ atoms.

105. Polyurethanes result from the reaction of a(n) _____ and an isocyanate.

Engineer in Training–Style Questions

1. Which of the following is not a physical property of pure aluminum?
    (a) Low density
    (b) High specific yield strength
    (c) Good ductility at low temperatures
    (d) Oxidation resistant in air

2. Which of the following aluminum alloys is strain hardened and stabilized to full hardness?
    (a) 5052-H38
    (b) 2024-T62
    (c) 3003-H18
    (d) 2090-T86

3. Which of the following aluminum alloys should you consider the primary candidate for a design where specific yield strength is the most critical design factor?
    (a) 2024-T62
    (b) 7075-T6
    (c) 2090-T86
    (d) 8090-T651

4. Which of the following aluminum alloys is not suitable for welding?
    (a) 1100
    (b) 3003
    (c) 5052
    (d) 7075-T6

5. Which of the following copper alloys can be precipitation hardened to a relatively high yield strength?
    (a) Beryllium-copper
    (b) Cartridge brass
    (c) Copper-nickel
    (d) Aluminum-bronze

6. Which of the following steels is not quenched and tempered to increase the strength?
    (a) 4340
    (b) 1040
    (c) 1010
    (d) 1060

7. What is the weight percent carbon in 1020 steel?
    (a) 2
    (b) 0.2
    (c) 0.1
    (d) 1

8. Which of the following steels has the highest hardenability?
   (a) 1050
   (b) 1080
   (c) 4320
   (d) 4340

9. Which of the following is not a property of an austenitic stainless steel?
   (a) Ferromagnetism
   (b) Resistant to corrosion in saltwater
   (c) High ductility
   (d) FCC crystal structure

10. If iron with 4.3 weight percent carbon is heated to 850°C for an extended period of time, what is the resulting form of cast iron?
    (a) White
    (b) Malleable
    (c) Gray
    (d) Nodular

11. What is the primary advantage of magnesium alloys?
    (a) High corrosion resistance
    (b) High strength
    (c) Low density
    (d) High ductility

12. What is the microstructure of superalloys that are used for the highest-temperature gas-turbine blades?
    (a) Small-grain-size polycrystals
    (b) Single crystals
    (c) Metals with ceramic dispersions
    (d) Directionally solidified polycrystals

13. What is the base metal for superalloys that are used for the highest-temperature gas turbine applications?
    (a) Nickel
    (b) Iron
    (c) Aluminum
    (d) Titanium

14. Which of the following applications is not suitable for a titanium alloy?
    (a) Human body implants
    (b) Parts in saltwater, such as a ship propeller
    (c) High-strength, low-weight parts
    (d) A gas-turbine part at 700°C

15. Which of the following materials has the highest specific yield strength at room temperature?
    (a) Ti-8Al-1Mo-1V-STA
    (b) 4340 steel OQT
    (c) Al-7075-T6
    (d) Inconel 718 superalloy

16. Which of the following is not a suitable application for a noble metal?
    (a) Large parts in seawater, such as a ship hull
    (b) Catalytic converters
    (c) Human body implants
    (d) Satellite parts

17. Which of the following is not a property of refractory metals?
    (a) High melting temperature
    (b) Good oxidation resistance at high temperature
    (c) High strength
    (d) High elastic modulus

18. The best way to increase the tensile strength of a polycrystalline ceramic is to
    (a) Decrease the grain size
    (b) Add alloy elements
    (c) Decrease porosity
    (d) Solution treat, quench, and age

19. An architect wants to use glass in a highly tensile-stressed critical application on the exterior of a building, and you are asked to select a suitable material. The best solution for this request is to:
    (a) Use tempered glass
    (b) Flame treat the outer surface of the glass to eliminate all scratches and cracks
    (c) Use the highest-strength glass available
    (d) Suggest a redesign with another material

20. The design of a part that operates at room temperature requires a material that is of low density, corrosion-resistant, and low cost. The stresses on the part are very low. Which of the following materials might be appropriate?
    (a) PE
    (b) A magnesium alloy
    (c) Alumina ceramic
    (d) PEEK

21. The design of a part that operates at temperatures up to 150°C (302°F) requires a material that is of low density and corrosion resistant. The stresses on the part are up to 50 MPa (7250 psi), and it must be possible to recycle the material. Which of the following materials might be appropriate?
    (a) PE
    (b) A magnesium alloy
    (c) Epoxy
    (d) PEEK

Problems

Problem 8.1   A fastener bolt for the airframe of a glider has a cross-sectional area of 1.0 cm² and a length of 4.0 cm. The part must be able withstand a tensile force of 1.0×10^4 N, and as a factor of safety this tensile force can produce a stress of no more than 50% of the tensile yield stress in the bolt. For damage tolerance the bolt must have at least 10% strain to fracture. Because weight is critical for this glider design, the maximum weight possible for this bolt is 8.0 g. The bolt is not subject to high or low temperatures. Of the materials covered in this chapter, what would be the best material to select? Because this is an aircraft part, it not necessary to choose the cheapest material, but excessive cost should be avoided.

Problem 8.2   A fastener bolt for the engine on a supersonic jet aircraft has a cross-sectional area of $2.0\ cm^2$ and a length of 4.0 cm. The bolt part must be able withstand a tensile force of 10×10^4 N, and as a factor of safety this tensile force can produce a stress of no more than 50% of the tensile yield stress in the bolt. For sufficient resistance to fracture this critical bolt must have at least 15% strain to fracture. Because weight is critical for this jet aircraft design the maximum weight possible for this bolt is 40 g. The bolt is subject to temperatures up to 400°C. To minimize the possibility of high temperature plastic strain, the selected material should not be subject to more than one-third of its melting temperature. Of the materials covered in this chapter what would be the best material to select? Because this is a supersonic jet aircraft, high-priced materials can be utilized.

Problem 8.3   To reduce the weight of an automobile it is decided to redesign the arm that controls the steering of the front wheels. The present control arm is made of oil quenched and tempered 1040 steel, and the maximum design force creates a stress that is 50% of the metal yield stress. It proposed to change to a low alloy steel of higher strength, and the new design is to be 50% of the yield stress. It is also determined that the new steel must have at least 15% ductility to provide sufficient resistance to fracture. The present control arm is 1.0 cm. in radius and 30 cm long. The maximum tensile force expected on the control arm is 7.85×10^4 N. Of the low-alloy steels presented in this chapter, what steel that meets the design requirements and provides the greatest weight reduction.

     (a) What is the new arm radius?

     (b) What weight reduction is provided by this new steel?

Problem 8.4   In an effort to increase the miles per gallon in automobiles, many steel parts are being replaced with polymer parts where the strength of steel is not required. The blades of the fan that cool the radiator experience a centripetal stress of less than 1×10^5 Pa when the fan blade material density is reduced to approximately $1 \times 10^3\ kg/m^3$. Vibration of the fan blades is the greatest problem, and vibration amplitude is minimized by maximizing the elastic modulus (stiffness) of the blade material. In addition, it is determined that the Izod impact strength must be a minimum of 100 for sufficient resistance to fracture and fatigue. To control cost it is necessary to select a commodity polymer. Polymers with glass or other material reinforcement are too expensive for this design. What material discussed in this chapter would best suit the design requirements?

Problem 8.5   The frame of a space shuttle type vehicle must have a high yield strength and high stiffness, and the most important design factor is weight. Of all the materials presented in this chapter, what material might be the most suitable for the frame of a space shuttle? Assume that there will be both tensile and compressive stresses. For a space shuttle, cost is not a limiting factor.

     (a) You can eliminate entire classes of materials from consideration with a brief statement about their unsuitability.

     (b) What material has the highest specific yield strength? Give the yield strength, specific gravity, specific yield strength, elastic modulus, and specific elastic modulus for this material.

     (c) What material has the highest specific elastic modulus? Give the yield strength, specific gravity, specific yield strength, elastic modulus, and specific elastic modulus for this material.

     (d) Compare the materials with the highest specific yield strength and highest specific elastic modulus for suitability in the space shuttle frame.

     (e) Discuss the suitability of the top-rated material for this design from the viewpoint of the ability to produce a frame.

The high thrust of the turbo fan engines from the F-22 Raptor allows for the high performance of this aircraft. The surfaces of the turbine blades in these engines reach temperatures of up to 1150°C. Designing turbine blades for reliable performance at these temperatures requires an understanding of the effect of time, temperature, and stress on the turbine blade materials.

U.S. Air Force photo

The goals of this chapter are to understand

- The effect of temperature on the elastic modulus, yield strength, and ductility of materials
- The relationship among plastic strain rate, stress, temperature, and viscosity in liquids
- The models of time and temperature dependence of creep strain in metals and ceramics
- The use of time-to-rupture parameters in design
- The mechanisms of creep strain in metals, ceramics, glass, and polymers
- Viscoelastic models of strain rate and creep in polymers and glasses
- The mechanism and model for stress relaxation

Chapter 9

Time, Temperature, and Mechanical Properties

9.1   INTRODUCTION

When jet-propelled aircraft were first introduced during World War II, the operational time of a turbine blade with a surface temperature of 650°C was about 10 to 25 hours. After that, chunks of metal would start flying out of the back of the engine. Modern turbine blade materials allow the turbine blade to run at surface temperatures of up to 1150°C in high-performance engines and for times of 20,000 hours in commercial aircraft before they are inspected. It is the blade material that limits the turbine operating temperature. Turbine designers would like to run the turbine with blade temperatures of 2000°C; however, that is not possible with present materials. By understanding the mechanisms of high-temperature plastic deformation, materials scientists and engineers have been able to develop new turbine alloys that have significantly increased the turbine operating temperature relative to the first jet aircraft gas turbines. In this chapter the mechanisms of plastic deformation in materials at elevated temperature are presented along with models and equations to predict plastic strain and rupture time.

9.2 ELASTIC MODULUS AND TEMPERATURE

The elastic strain in crystalline materials corresponds to the stretching of primary atomic bonds. Elastic strain at a particular stress is time independent, because the atoms respond so quickly to changes in stress that the elastic strain is instantaneous. However, the elastic modulus of crystalline materials is temperature dependent. The elastic modulus (E) decreases as temperature increases for most materials and it starts to decrease rapidly as the melting temperature is approached. For example, E for aluminum at room temperature is 70.3 GPa, and at 327°C E is approximately 50 GPa. For steel at room temperature E is 211 GPa, and at 527°C E is approximately 120 GPa. The reason the elastic modulus decreases when the temperature increases is that at high temperature, the atoms are on the average farther apart than at low temperature; therefore, the bonding is weaker. Figure 9.1 shows that the elastic modulus of glass drops from 57 GPa, at 0°C to 45.5 GPa at 300°C. One example of the elastic modulus increasing with increasing temperature is in TiNi when there is a phase transformation upon heating.

9.3 PLASTIC STRAIN IN LIQUIDS AND AMORPHOUS SOLIDS

Viscosity (η) relates the plastic shear strain rate to shear stress in liquids and supercooled liquids, such as glasses and polymers. Figure 9.2 shows a liquid between two solid plates. The bottom plate is fixed and the top plate of area (A) is subject to a shear force (F), producing a shear stress (τ), a velocity (v) of the liquid, and a plastic shear strain (γ_p) in the liquid. This configuration is similar to a lubricating fluid

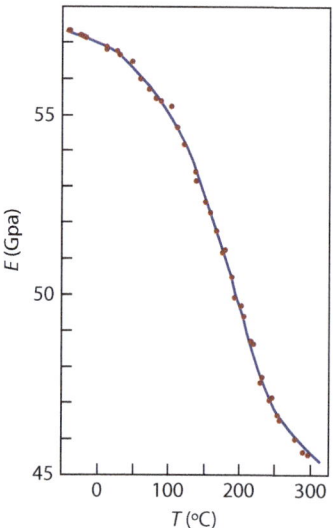

Figure 9.1 The tensile elastic modulus (E) as a function of temperature (T) for a soda-lime glass. (*Based on Bansal, N.P., and Doremus, R.H., Handbook of Glass Properties, Academic Press, Orlando, Fl. (1986), p. 319.*)

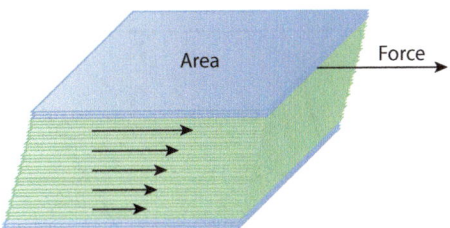

Figure 9.2 The viscous flow of a liquid (green) between a fixed bottom plate (blue) and a movable top plate with an applied shear force. The arrows indicate the liquid or material velocity (v). (*Based on http://www.insula.com .au/physics/1250images/Image448.gif*)

between a fixed bearing and a rotating shaft. The applied shear force results in a continuous increase of the plastic strain of the liquid with time, and the plastic shear strain rate is $d\gamma_p/dt$.

In a **Newtonian liquid** the plastic shear strain rate ($d\gamma_p/dt$) is linearly related to the applied shear stress (τ) and to the viscosity (η), as shown in Equation 9.1.

$$\frac{d\gamma_p}{dt} = \frac{\tau}{\eta} \qquad \textbf{9.1}$$

The SI unit for viscosity is Pa · s. An older unit for viscosity is the poise (10 poise = 1 Pa · s). In non-Newtonian liquids, the relationship between plastic shear strain rate and applied shear stress is nonlinear, as shown in Equation 9.2:

$$\left(\frac{d\gamma_p}{dt}\right)^n = \frac{\tau}{\eta} \qquad \textbf{9.2}$$

where n is a number that can be less than or more than 1 but is not equal to 1. The temperature dependence of the viscosity is given by Equation 9.3, where η_0 is a pre-exponential constant, ΔH_p is the enthalpy of activation for plastic strain, and k is Boltzmann's constant.

$$\eta = \eta_0 \exp\left(\frac{\Delta H_p}{kT}\right) \qquad \textbf{9.3}$$

The temperature dependence of the log of viscosity of silica glass, soda-lime glass, and pure boron oxide (B_2O_3) as a function of $1000/T$ is shown in Figure 9.3. As temperature is increased, resulting in smaller values of $1000/T$, the viscosity decreases. We discussed the plastic strain in silica (SiO_2) glass in Section 6.5.2. Briefly, at higher temperatures the silicon-oxygen bonds are momentarily broken and can re-form with different neighbors, allowing the glass to flow. The reason that soda-lime glass has a smaller viscosity than silica glass (SiO_2) does at the same temperature is that the soda (Na_2O) and lime (CaO) ions break up the bond between silicon and oxygen in the glass, as shown schematically in Figure 3.3. The sodium and calcium atoms become interstitial atoms, and the extra oxygen atoms allow silicon-oxygen tetrahedra to be terminated with oxygen without connecting to another silicon atom.

You probably have experienced the change in viscosity of liquids like honey or maple syrup. When the syrup is removed from the refrigerator, it does not easily flow onto our pancakes. However, if we heat the syrup, it flows more easily. By increasing the temperature of the syrup, we have decreased the viscosity, as predicted by Equation 9.3. A decrease in viscosity results in a greater shear plastic strain rate according

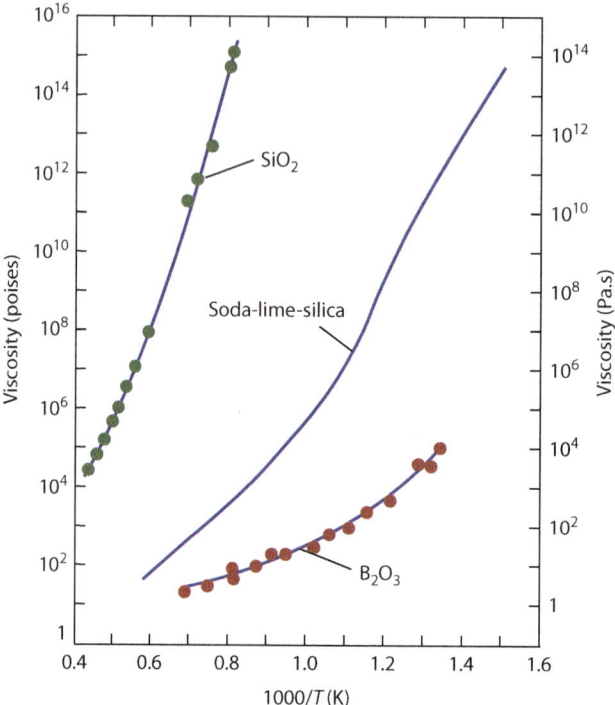

Figure 9.3 The temperature dependence of the viscosity of silica glass (SiO_2), soda-lime glass, and boron oxide (B_2O_3). *(Based on Kingery, W.D., John Wiley & Sons, N.Y. (1960), p. 574.)*

to Equation 9.1. Substituting Equation 9.3 into Equation 9.1 for a Newtonian fluid, the result is a form of the Arrhenius rate equation, as shown in Equation 9.4.

$$\frac{d\gamma_p}{dt} = \frac{\tau}{\eta_0} \exp\left(-\left(\frac{\Delta H_p}{kT}\right)\right) \qquad \textbf{9.4}$$

This equation shows that as the shear stress or temperature is increased, the plastic shear strain rate of the liquid is increased.

If a tensile stress (σ) is applied to an amorphous material at temperatures above the glass transition temperature, the relationship between viscosity and the tensile strain rate analogous to Equation 9.1 is as shown in Equation 9.5.

$$\frac{d\varepsilon_p}{dt} = \frac{\sigma}{3\eta} \qquad \textbf{9.5}$$

The temperature dependence of the tensile strain rate for this material is given by Equation 9.6.

$$\frac{d\varepsilon_p}{dt} = \frac{\sigma}{3\eta_0} \exp\left(-\left(\frac{\Delta H_p}{kT}\right)\right) \qquad \textbf{9.6}$$

Liquids do not resist a tensile stress. Tensile stresses are resisted by amorphous and semicrystalline super-cooled glasses and polymers.

9.4 CREEP STRAIN AND STRESS RUPTURE IN SOLIDS

Creep strain is time-dependent plastic strain. If a constant stress, which can be more or less than the yield stress, is applied to a material, plastic strain can accumulate with time. Creep strain is particularly important if the temperature is at least half the absolute melting temperature. The strain measured as a function of time appears as schematically shown in Figure 9.4. Equation 9.7 shows the initial instantaneous elastic strain (ε_e) given by Hooke's law.

$$\varepsilon_e = \frac{\sigma}{E} \qquad\qquad \textbf{9.7}$$

The elastic strain is followed by **stage I creep** strain. In stage I creep, there is a nonlinear transient plastic strain rate ($d\varepsilon_I/dt$) that is inversely related to time (t) and directly related to stress and temperature through the term $B(\sigma, T)$, as shown in Equation 9.8.

$$\frac{d\varepsilon_I(\sigma, t)}{dt} = \frac{B(\sigma,T)}{t} \qquad\qquad \textbf{9.8}$$

Equation 9.8 predicts that the transient plastic strain rate decreases with time, as is observed in Figure 9.4. Integrating Equation 9.8 with respect to time results in Equation 9.9:

$$\varepsilon_I(\sigma, t) = B(\sigma,T)\ln t + C \qquad\qquad \textbf{9.9}$$

where C is a constant of integration.

During the transient stage I creep, all materials strain harden due to the plastic strain. In metals and ceramics there is a simultaneous softening resulting from recovery anneal caused by the thermal vibrations at the elevated temperature. The plastic strain rate decreases with increased strain in stage I creep, because the strain-hardening rate exceeds the softening rate.

The transient stage I creep changes to **stage II creep**, or steady-state creep, in metals and ceramics when the stain-hardening rate is equal to the rate of softening due to annealing. The higher the temperature, the higher the annealing rate and the higher the stage II creep strain rate. In crystalline metals and

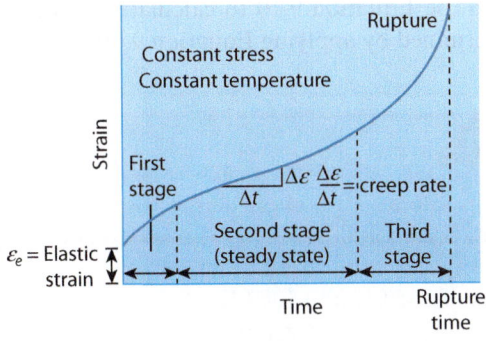

Figure 9.4 A schematic of the strain in a specimen resulting from the application of a constant stress and constant elevated temperature that are maintained for an extended time. (*Based on Askeland, D.R., Fulay, P.P., and Wright, W.J., The Science and Engineering of Materials, 6th ed., Cengage Learning, Stamford, CT. (2011), p. 277.*)

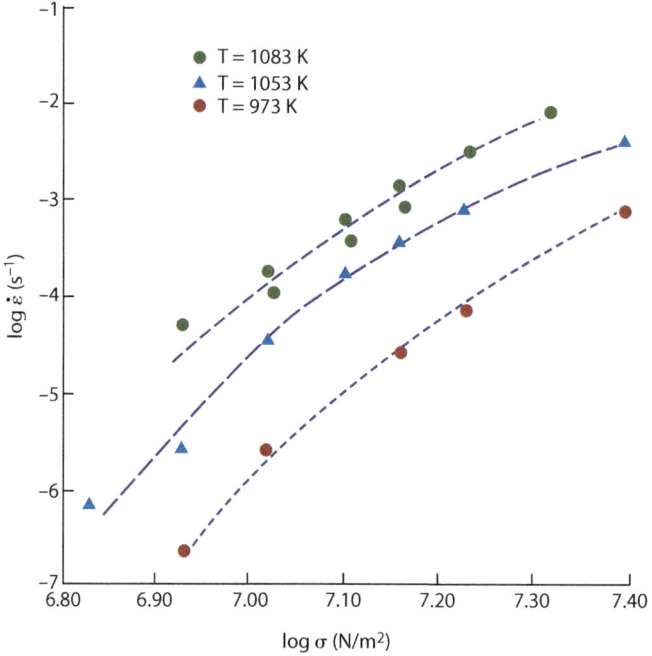

Figure 9.5   The log of the stage II creep strain rate of zirconium as function of the log of stress, at three temperatures. *(Based on Courtney, T.H., Mechanical Behavior of Materials, McGraw-Hill, N.Y. (1990), p. 284.)*

ceramics the tensile stage II creep rate ($d\varepsilon_{II}/dt$) is given by Equation 9.10, which is similar to Equation 9.6 but is not linear in stress.

$$\frac{d\varepsilon_{II}}{dt} = A\sigma^n \exp - \left(\frac{\Delta H_p}{kT}\right) \qquad\qquad \textbf{9.10}$$

   The logarithm of the stage II creep rate data for zirconium metal is plotted as a function of the logarithm of stress at three different temperatures in Figure 9.5. Both higher stresses and higher temperatures produce higher stage II creep rates. The exponent of the stress in Equation 9.10 is obtained from the slope of the lines in Figure 9.5. If the creep strain rate is required at a temperature not presented in Figure 9.5, it is necessary to use Equation 9.10 to calculate this rate. The value of the enthalpy of activation for creep can be determined by applying Equation 9.10 at two different temperatures.

Example Problem 9.1

The log of the stage II creep rate as a function of the log of stress is plotted at three different temperatures for zirconium in Figure 9.5.
(a) Determine the enthalpy of activation in eV/atom for creep in zirconium at $\log \sigma = 7.10$ and between the temperatures of 1053 K and 1083 K.
(b) Determine the exponent of stress in the creep rate equation for stage II creep at 1083 K. Use data between stresses of $\log \sigma = 7.00$ and $\log \sigma = 7.20$.
(c) Determine the preexponential constant A at 1083 K and at $\log \sigma = 7.16$.

Solution

a) There are three unknowns to be determined in (a), (b), and (c). If the ratios of the strain rates at two different temperatures and the same stress are taken, then the stress to the power n and the constant A cancel out of the equations. Two of the unknowns are eliminated from the equations, and the enthalpy of activation can be determined. For the first point take $T_1 = 1083$ K and $\log \sigma_1 = 7.10$; then $\log (d\varepsilon_1/dt) = -3.35$. Taking the antilog of -3.35 gives the strain rate of

$$\frac{d\varepsilon_1}{dt} = 4.467 \times 10^{-4}\,\text{s}^{-1}$$

$$\frac{d\varepsilon_1}{dt} = 4.467 \times 10^{-4}\,\text{s}^{-1} = A\sigma^n \exp -\left(\frac{\Delta H_p}{kT_1}\right) = A\sigma^n \exp -\left(\frac{\Delta H_p}{k(1083\text{ K})}\right)$$

For the second point take $T_2 = 1053$ K and the same stress of $\log \sigma_2 = 7.10$. Then $\log (d\varepsilon_2/dt) = -3.81$

$$\frac{d\varepsilon_2}{dt} = 1.55 \times 10^{-4}\,\text{s}^{-1} = A\sigma^n \exp -\left(\frac{\Delta H_p}{kT_2}\right) = A\sigma^n \exp -\left(\frac{\Delta H_p}{k(1053\text{ K})}\right)$$

Now take the ratio of the strain rates at the two temperatures.

$$\frac{\dfrac{d\varepsilon_1}{dt}}{\dfrac{d\varepsilon_2}{dt}} = \frac{4.467 \times 10^{-4}\,\text{s}^{-1}}{1.55 \times 10^{-4}\,\text{s}^{-1}} = \frac{A\sigma^n \exp -\left(\dfrac{\Delta H_p}{k(1083\text{ K})}\right)}{A\sigma^n \exp -\left(\dfrac{\Delta H_p}{k(1053\text{ K})}\right)} = \frac{\exp -\left(\dfrac{\Delta H_p}{k(1083\text{ K})}\right)}{\exp -\left(\dfrac{\Delta H_p}{k(1053\text{ K})}\right)}$$

$$2.884 = \exp -\frac{\Delta H_p}{k}\left(\frac{1}{1083\text{ K}} - \frac{1}{1053\text{ K}}\right) = \exp -\frac{\Delta H_p}{k}(9.234 \times 10^{-4}\,\text{K}^{-1} - 9.497 \times 10^{-4}\,\text{K}^{-1})$$

$$2.884 = \exp -\frac{\Delta H_p}{k}(-2.63 \times 10^{-5}\,\text{K}^{-1})$$

Take the natural log of each side of the equation and insert the value of Boltzmann's constant.

$$1.059 = -\frac{\Delta H_p}{8.62 \times 10^{-5}\,\dfrac{\text{eV}}{\text{atom} \cdot \text{K}}}(-2.63 \times 10^{-5}\,\text{K}^{-1}) = \Delta H_p\left(0.305\,\frac{\text{eV}}{\text{atom}}\right)$$

Solve for the enthalpy of activation for creep strain.

$$\Delta H_p = 3.47\frac{\text{eV}}{\text{atom}}$$

b) To solve for the exponent of the stress, take the ratio of the strain rates at two different stresses at the same temperature. This eliminates A from the equations and the temperature-dependent terms. Use a temperature of 1083 K and a stress of $\log \sigma_1 = 7.02$ or $\sigma_1 = 1.05 \times 10^7$ Pa.

$$\log\left(\frac{d\varepsilon_1}{dt}\right) = -4$$

$$\frac{d\varepsilon_1}{dt} = 1.0 \times 10^{-4}\,\text{s}^{-1}$$

At 1083 K use data at $\log \sigma_2 = 7.16$ or $\sigma_2 = 1.45 \times 10^7$ Pa.

$$\log\left(\frac{d\varepsilon_2}{dt}\right) = -3$$

$$\frac{d\varepsilon_2}{dt} = 1.0 \times 10^{-3}\ \text{s}^{-1}$$

Take the ratio of the two strain rates and all of the terms cancel except for the stress term.

$$\frac{\dfrac{d\varepsilon_1}{dt}}{\dfrac{d\varepsilon_2}{dt}} = \frac{1.0 \times 10^{-4}\text{s}^{-1}}{1.0 \times 10^{-3}\text{s}^{-1}} = 1.0 \times 10^{-1} = \frac{A\sigma_1^n \exp-\left(\dfrac{\Delta H_p}{k(1083\ \text{K})}\right)}{A\sigma_2^n \exp-\left(\dfrac{\Delta H_p}{k(1083\ \text{K})}\right)} = \frac{\sigma_1^n}{\sigma_2^n} = \left(\frac{\sigma_1}{\sigma_2}\right)^n$$

Take the log of the above equation.

$$-1 = n \log\left(\frac{\sigma_1}{\sigma_2}\right) = n \log\frac{1.05 \times 10^7\ \text{Pa}}{1.45 \times 10^7\ \text{Pa}} = n \log(0.72) = n(-0.14)$$

$$n = \frac{1.0}{0.14} = 7.0$$

To evaluate A, insert the values of n and the enthalpy of activation of creep strain into the equation of one data point. For example, use $T = 1083$ K and a stress of $\log \sigma_1 = 7.16$ or $\sigma_1 = 1.45 \times 10^7$ Pa. In Figure 9.5 at this stress the log of the strain rate is -3.

$$\log\frac{d\varepsilon_1}{dt} = -3$$

$$\frac{d\varepsilon_1}{dt} = 1.0 \times 10^{-3}\ \text{s}^{-1}$$

Put this strain rate into the strain-rate equation.

$$\frac{d\varepsilon_1}{dt} = 1.0 \times 10^{-3}\ \text{s}^{-1} = A\sigma_1^n \exp-\left(\frac{\Delta H_p}{kT_1}\right) = A(1.45 \times 10^7\ \text{Pa})^7 \exp-\left(\frac{3.47\dfrac{\text{eV}}{\text{atom}}}{\left(8.62 \times 10^{-5}\dfrac{\text{eV}}{\text{atom}\cdot\text{K}}\right)(1083\ \text{K})}\right)$$

$$1.0 \times 10^{-3}\text{s}^{-1} = A(13.5 \times 10^{49}) \exp\left(\frac{-3.47\dfrac{\text{eV}}{\text{atom}}}{9.34 \times 10^{-2}\dfrac{\text{eV}}{\text{atom}}}\right) = A(13.5 \times 10^{49}) \exp(-37.1)$$

$$1.0 \times 10^{-3}\text{s}^{-1} = A(13.5 \times 10^{49})(7.72 \times 10^{-17}) = A(104 \times 10^{32})$$

Solve for A in units of Pa.

$$A = \frac{1.0 \times 10^{-3}\,s^{-1}}{1.04 \times 10^{34}} = 0.96 \times 10^{-37}$$

With these constants, the value of the strain rate can be determined at any temperature and stress between the conditions used in this problem by substituting into the strain rate equation.

In **stage III creep**, shown in Figure 9.4, the material is progressing toward failure. In most applications, failure by creep only occurs at temperatures more than 50% of the melting temperature in degrees kelvin. Some design applications require only the time to rupture, because the accumulation of plastic strain is not critical. In a jet engine, the spacing between the turbine blade and the inner surface of the engine is a critical dimension, because the turbine blade will hit the engine if it elongates too much. However, a pipeline carrying high-pressure hot gases in a power plant may be able to deform a significant amount without failure of the power plant, but it must not deform to the point of rupture. **Rupture** is failure of a specimen or part, and rupture is often used to describe failure when temperature and time are associated with the failure. The time to rupture a material as a function of both stress and temperature is important engineering design information. Rupture is the final result of the third stage of creep.

Stress-rupture test data for a temperature-compensated rupture time for aluminum is shown in Figure 9.6 as a function of stress. Figure 9.6 shows that if the time to rupture (t_r) is multiplied by the same exponential term as appears in the plastic-strain-rate Equations 9.4, 9.6, and 9.10, all of the data fall on a single line that is a function of stress. We can write an equation for a function $\theta_r(\sigma)$ that is only a function of stress and is equal to the values on the x axis of Figure 9.6, as shown in Equation 9.11a.

$$\theta_r(\sigma) = t_r \exp - \left(\frac{\Delta H_p}{kT} \right) \qquad \textbf{9.11a}$$

The inverse of the time to rupture $(1/t_r)$ is the rate of approaching rupture. Solving Equation 9.11a for $1/t_r$ results in Equation 9.11b, which is similar in form to Equations 9.4, 9.6, and 9.10.

$$\frac{1}{t_r} = \frac{1}{\theta_r(\sigma)} \exp - \left(\frac{\Delta H_p}{kT} \right) \qquad \textbf{9.11b}$$

In engineering applications where calculations of creep rupture time are required, the most commonly utilized creep relationship is the empirical **Larson-Miller parameter** (P_{LM}). We assume that the rate of approaching rupture is proportional to the rate of accumulation of stage II plastic strain. P_{LM} can be developed from the stage II creep-strain-rate equation by first taking the natural log of Equation 9.10, as shown in Equation 9.12.

$$\ln \frac{d\varepsilon_{II}}{dt} = -\frac{\Delta H_p}{kT} + g(\sigma) = \ln \dot{\varepsilon}_{II} = \ln \frac{c}{t_r} \qquad \textbf{9.12}$$

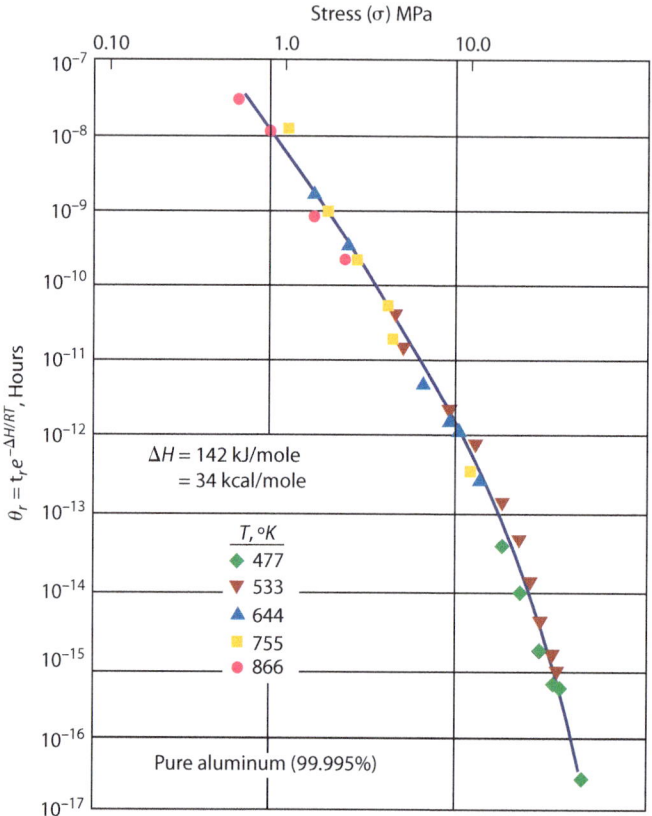

Figure 9.6 Temperature-compensated rupture time as a function of stress, for five different temperatures. (*Based on Dorn, J.E., Creep and Recovery, ASM, Metals Park, OH (1957), p. 267.*)

The term $g(\sigma)$ contains the stress term and the constant A from Equation 9.10. It is assumed that the time to rupture is inversely proportional to the stage II strain rate, as shown by the similarity between Equation 9.10 and Equation 9.11b. Rewriting Equation 9.12 results in Equation 9.13.

$$T\left[\ln t_r - \ln c + g(\sigma)\right] = \frac{\Delta H_p}{k} \qquad\qquad \textbf{9.13}$$

P_{LM} is the absolute temperature times the first two terms in the brackets in Equation 9.13. P_{LM} uses the logarithm to the base 10, as shown in Equation 9.14.

$$P_{LM} = T\left[\log t_r + C\right] \qquad\qquad \textbf{9.14}$$

P_{LM} is a function of stress, because when the term $T\left[\ln t_r - \ln c\right]$ in Equation 9.13 is added to $Tg(\sigma)$, the sum is equal to the term on the right-hand side of Equation 9.13, and this is a constant over significant temperature ranges. The constant (C) is typically a number approximately equal to 20. P_{LM} is plotted as a function of stress for a variety of wrought superalloys in Figure 9.7. Figure 9.7 is utilized in design when two of the three parameters of temperature, stress, and time to failure are known; then the third parameter is found from Figure 9.7.

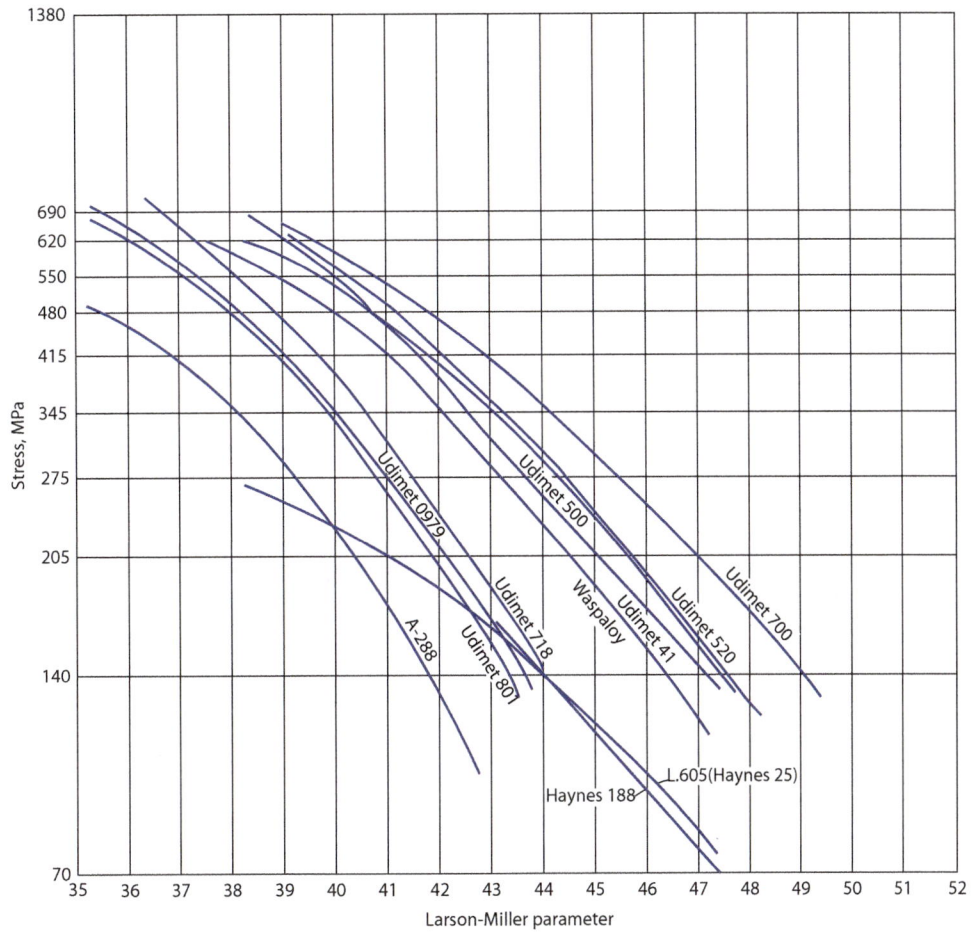

Figure 9.7 The Larson-Miller parameter (P_{LM}) times 10^{-3} as a function of stress for a variety of wrought superalloys. For this plot C = 20, the temperature is in degrees rankine (R), and time is in hours (h). The chemical compositions of wrought superalloys in this figure are presented in Table 9A.1 in Appendix 9A. (*Based on Donachie, M.J. and Donachie, S.J, Superalloys: A Technical Guide, American Society for Metals International, Metals Park, OH, (2002), p. 248.*)

Example Problem 9.2

A part made of Udimet 41 is subject to an applied stress of 205 MPa at a temperature of 1000 K. What is the expected time to failure for this part?

Solution

From Figure 9.7, P_{LM} for an applied stress of 205 MPa is 45×10^3. The equation for P_{LM} from this figure is

$$45 \times 10^3 = T[20 + \log t_r] = 1800\, R[20 + \log t_r]$$

Solve for the time to rupture (t_r).

$$\log t_r = \frac{45 \times 10^3}{1.8 \times 10^3} - 20 = 25 - 20 = 5$$

$$t_r = 10^5\, \text{h} = 100{,}000\, \text{h}$$

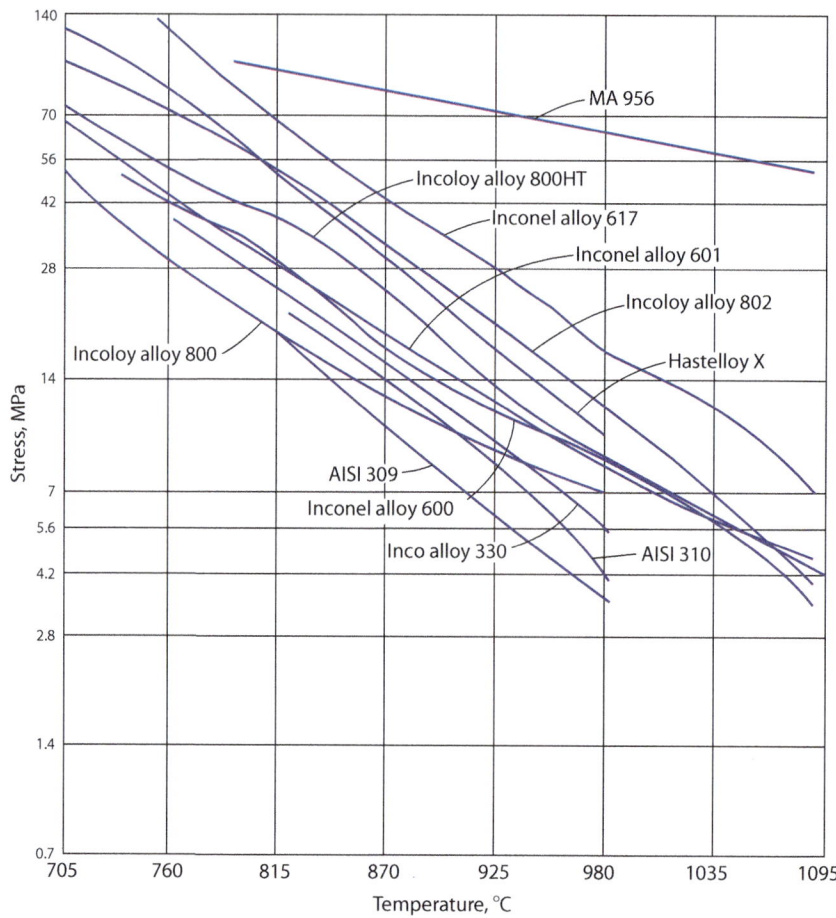

Figure 9.8 The 10,000-hour creep rupture strength of some wrought solid solution strengthened superalloys, nickel-based superalloys, and stainless steels as a function of temperature. The alloy MA 936 is iron based and oxide dispersion strengthened. (*Based on Donachie, M.J. and Donachie, S.J, Superalloys: A Technical Guide, American Society for Materials International, Materials Park, OH, (2002), p. 244.*)

Figure 9.8 presents the 10,000-hour rupture strength of wrought solid solution strengthened superalloys at temperatures from 705°C to 1095°C. The mechanical properties of some of these alloys are presented in Section 8.1.5, and the chemistry of all of these alloys is presented in Appendix 9A.

9.4.1   The Mechanisms of Stage II Creep in Crystalline Metals

At temperatures below half of the melting temperature in crystalline metals, the primary mechanism of creep strain is dislocation glide. Dislocation glide occurs when dislocations move across a crystal, as demonstrated in Figure 6.11. For glide to occur in high-yield-strength materials, dislocations must

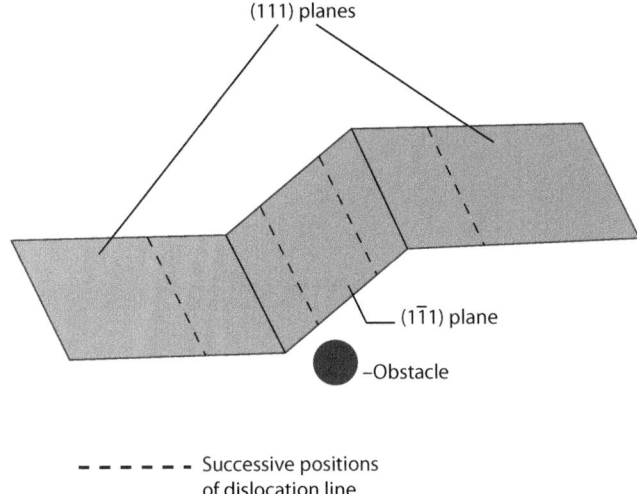

(111) planes

(1$\bar{1}$1) plane

–Obstacle

— — — — — Successive positions
of dislocation line

Figure 9.9 A schematic of a screw dislocation (dashed line) cross-slipping in an FCC metal lattice from a (111) slip plane to a (1$\bar{1}$1) slip plane, and back to a (111) slip plane. This process can occur when there is an obstacle to continuous glide on the (111) slip plane. (*Adapted from Courtney, T. H., Mechanical Behavior of Materials, 2nd ed., McGraw-Hill, New York (2000), p. 100.*)

circumvent the strengthening mechanisms that produce a high-yield stress. Thermal vibrations resulting from high temperatures assist dislocations in circumventing obstacles. At low temperatures, screw dislocations can cross-slip around an obstacle, as shown in Figure 9.9. For a screw dislocation, there is no unique slip plane, because the Burger's vector and the tangent vector to the dislocation line are parallel to each other, and the slip plane is defined by the plane of the Burger's vector and the tangent vector. A screw dislocation that is blocked by an obstacle can **cross-slip** from one close-packed plane to another close-packed plane. In FCC crystals, there are many possible close-packed slip planes and cross-slip planes. This contributes to the high ductility of FCC metals.

At temperatures of approximately half of the melting point or higher, the climb of edge dislocations over obstacles, shown schematically in Figure 9.10, makes a significant contribution to creep strain in metals. If a vacancy diffuses to the center of an edge dislocation, that segment of the edge dislocation climbs up one lattice distance. Vacancies must diffuse to the center of the dislocation along the entire length that is blocked by the obstacle for the entire dislocation to climb. In this way an edge dislocation climbs over obstacles. At higher temperatures there is a higher concentration of vacancies, and they diffuse faster. The edge dislocation climb mechanism is also called dislocation creep.

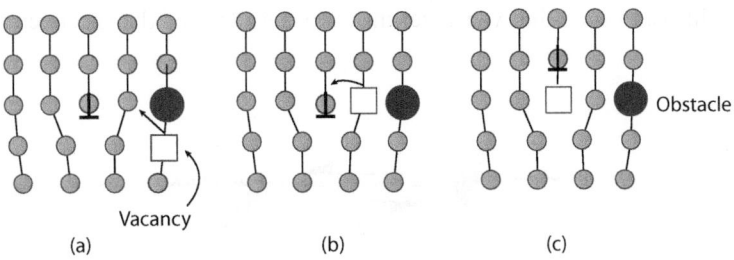

Vacancy

(a)   (b)   (c)

Obstacle

Figure 9.10 (a) A dislocation is blocked from glide by an obstacle on the slip plane. (b) A vacancy diffuses toward the dislocation. (c) The dislocation moves up to another slip plane by dislocation climb when vacancies go to the center of the dislocation. The dislocation can now glide on the new slip plane above the obstacle.

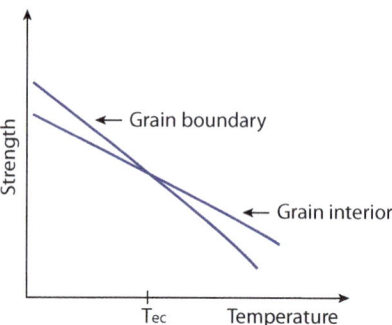

Figure 9.11   A schematic of the strength of the interior crystalline part of a grain and the grain boundary region, as a function of temperature. The equi-cohesive temperature (T_{ec}) is where the grain boundary and the grain interior strength are equal.

As the temperature of a polycrystalline metal is increased, the grain boundaries weaken more rapidly than the crystalline material does, as shown schematically in Figure 9.11, because the grain boundaries are more disordered and super-cooled liquid-like in structure than the crystalline material is. At temperatures below the equi-cohesive temperature (T_{ec}), the grain boundaries are stronger than the grains. Grain boundaries strengthen metals at low temperatures. At temperatures higher than the equi-cohesive temperature, the grain boundaries are weaker than the crystalline material and grain boundary sliding contributes to creep strain and the formation of grain boundary cracks, as shown in Figure 9.12. Also at elevated temperatures, diffusion of atoms occurs along the grain boundaries. The transport of atoms along the grain boundaries results in a change of shape, called **Coble creep**.

Because of grain boundary sliding and the general weakness of grain boundaries at elevated temperature, some high-performance gas-turbine blades and stator blades are made from directionally solidified metals, which are shown in Figure 9.13b, resulting in columnar grains. Grain boundaries that are parallel to the turbine blade axis have no resolved shear stress resulting from the tensile forces due to rotation of the turbine.

Single crystal turbine blades have a longer high temperature rupture life than polycrystals or directionally solidified superalloys. Single crystal turbine blades have no grain boundaries. They are not weakened by grain-boundary sliding, and single crystal turbine blades are more resistant to high-temperature corrosion. Grain boundaries are a location of higher energy than the crystal. As a result of this higher energy, the atoms at the grain boundary are more reactive than the atoms in the bulk of the crystal. Also, impurity atoms tend to concentrate in the grain boundaries, because there is more space for foreign atoms in the disordered grain boundary than in the ordered crystal. The impurities in the grain boundary can be more reactive than the bulk of the crystal. We will discuss the procedures to produce single crystals in Chapter 13.

At temperatures above half the melting temperature in a metal subject to tension, there is transport of material by vacancy diffusion such that vacancies move to the edges of the specimen and atoms move in

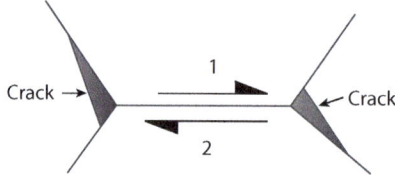

Figure 9.12   A schematic of grain boundary sliding between two grains 1 and 2. The grain boundaries are the dark lines. As a result of a resolved shear stress, displacement occurs in the directions indicated by the two arrows on either side of the grain boundary and cracks form at the grain boundary intersections with neighboring grains.

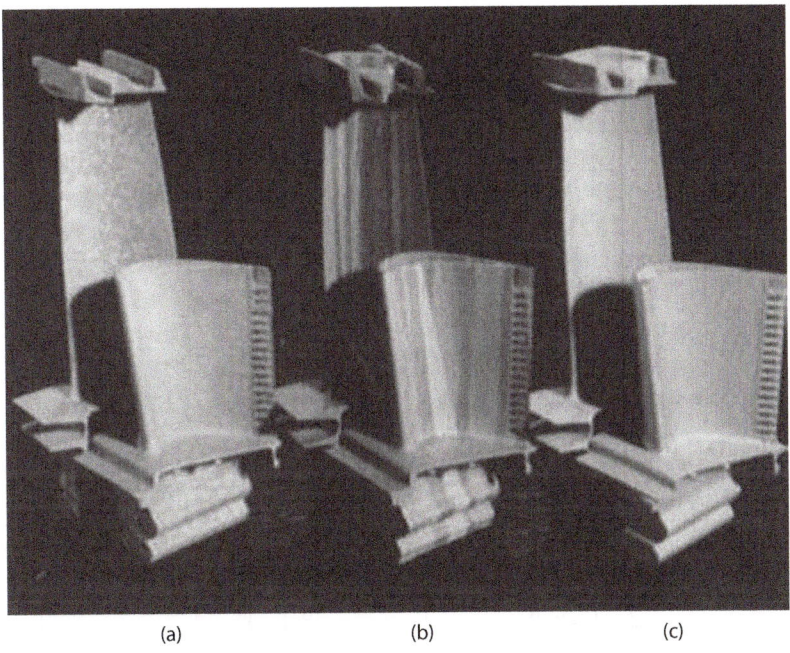

Figure 9.13 (a) A polycrystalline gas-turbine blade (front) and stator vane (behind) showing the grain structure. (b) A directionally solidified blade and vane. (c) A single crystal turbine blade and vane. (*Reprinted from Donachie, M.J. and Donachie, S.J, Superalloys: A Technical Guide, 2nd Edition, 2002, ASM International.*)

the direction of the tension to elongate the specimen, as shown schematically in Figure 9.14. At the atomic level, the applied tensile stress is a force that acts on each atom. As each atom moves by diffusion in the crystal, the tensile stress provides a force that pulls each atom in the direction of the applied tensile stress. This produces a net flow of atoms in the direction of the tensile stress. This flow of atoms causes the

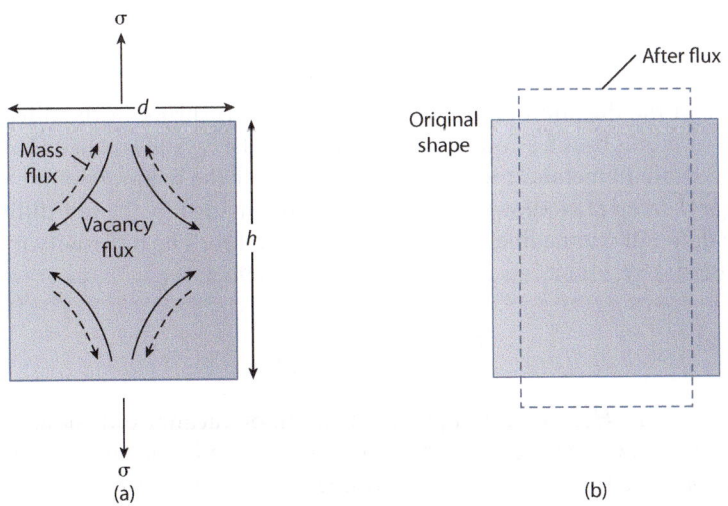

Figure 9.14 (a) High-temperature mass transport by diffusion in a metal or ceramic material as a result of a tensile stress. Atoms move in the direction of the stress, elongating the specimen, and vacancies move in the opposite direction, decreasing the width. (b) The change in dimensions resulting from mass transport. (*Adapted from Courtney, T. H., Mechanical Behavior of Materials, 2nd ed., McGraw-Hill, New York (2000), p. 301.*)

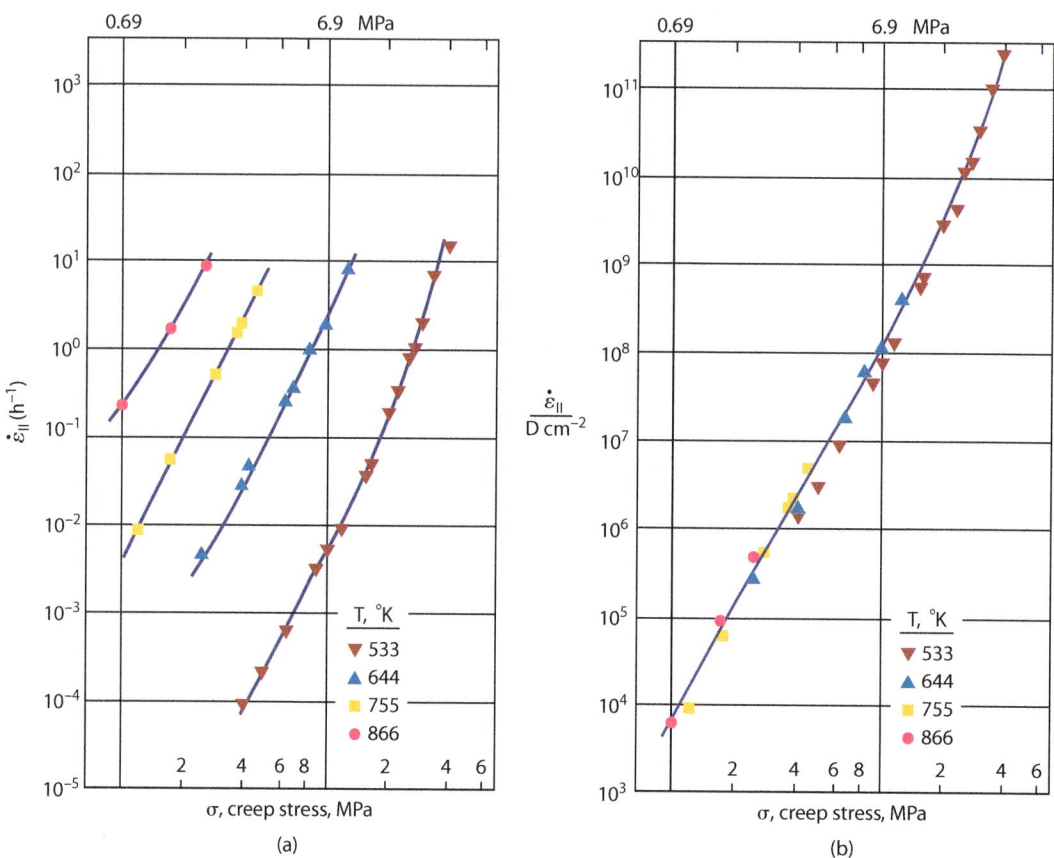

Figure 9.15 (a) The log of the stage II creep rate in aluminum as a function of the log of stress, at four temperatures. (b) The log of the stage II creep rate divided by the diffusivity plotted as a function of the log of stress, at four temperatures, on one line. (*Based on Sherby, O.D. and Burke, P.M., Prog. Mater. Sci.,13 (1968), p. 331.*)

specimen to elongate in the direction of the tensile stress and to shrink in the direction normal to the tensile axis. The flow of atoms by diffusion through the bulk of the crystal is called **Herring-Nabarro creep**.

Most of the creep strain in metals at temperatures above half the melting point is a result of vacancy diffusion. If the stage II creep rates shown in Figure 9.15a are divided by the self-diffusivity of the metal, all of the creep-rate data fall on one line, as shown in Figure 9.15b. The diffusivity for vacancy diffusion (D) is given in Equation 4.39, and it has the form of Equation 9.15:

$$D = D_0 \exp - \left(\frac{\Delta H_v}{kT} \right) \qquad \textbf{9.15}$$

where D_0 is a constant, and ΔH_v is the enthalpy of activation of vacancy diffusion.

The discussion above shows that the mechanism for creep strain in metals changes as the temperature increases. Plastic deformation processes such as dislocation glide, which require a low enthalpy of activation, are activated at lower temperatures; and plastic deformation processes such as vacancy diffusion, which require a higher enthalpy of activation, are activated at higher temperatures. Diffusion is required for dislocation climb, Herring-Nabarro creep, and Coble creep. The enthalpy of activation required for creep in metals increases with temperature, as shown for aluminum in Figure 9.16. In Figure 9.16, below $T/T_m = 0.25$ corresponds to the glide of dislocations and cross-slip. From T/T_m values of 0.25 to 0.5

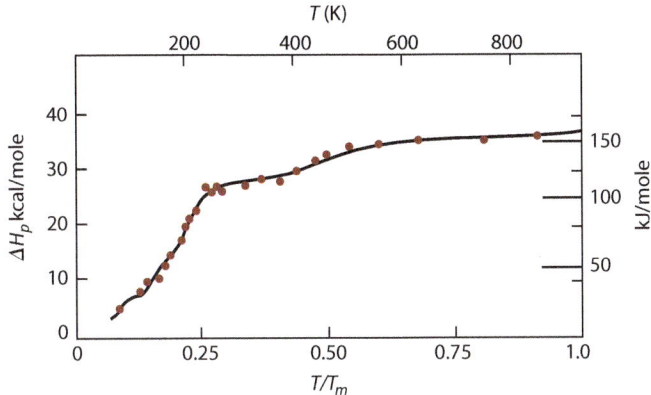

Figure 9.16 The variation of the enthalpy of activation for creep in aluminum as a function of temperature. *(Based on Sherby, O.D. and Burke, P.M., Prog. Mater. Sci.,13 (1968), p. 328.)*

corresponds to dislocation climb around obstacles, and above $T/T_m = 0.5$ is diffusional flow including Herring-Nabarro creep and Coble creep. The enthalpy of activation for creep in aluminum reaches a plateau equal to the activation enthalpy for diffusion, given in Table 4.4 of 135 kJ/mole at approximately 466 K; 466 K is half of the melting point of 933 K for aluminum.

9.4.2 The Mechanisms of Stage II Creep in Crystalline Ceramics

Because dislocation motion is difficult at low temperature in covalently bonded materials, it is necessary to increase the temperature in ceramics to approximately 50% of the absolute melting temperature before dislocation glide occurs. At lower temperatures, brittle fracture is likely to occur before dislocation glide. At high temperatures, the mechanisms of stage II creep in crystalline ceramics are similar to those in metals and the stage II creep rate obeys Equation 9.10.

The observed stage II creep mechanisms in both metals and ceramics are a function of grain size. For example, in MgO with a 32-μm grain size, Coble creep is not observed to be a significant creep mechanism. However, for MgO that has a grain size of 10 μm, Coble creep is the dominant stage II creep mechanism at low stress and temperatures at 50% of the melting temperature. Herring-Nabarro creep in ceramics only occurs at the highest of temperatures. This is because activation enthalpy for vacancy diffusion in crystalline ceramics is very large.

9.4.3 Deformation-Mechanism Maps of Stage II Creep in Crystalline Metals and Ceramics

The various creep mechanisms for metals and ceramics discussed above are displayed on deformation-mechanism maps, as shown for silver in Figure 9.17. The homologous temperature, or the temperature divided by the melting temperature (T/T_m) both in absolute degrees, is the horizontal axis on the bottom, and the temperature in degrees Celsius is on the top. The normalized stress, or the tensile stress divided by the shear modulus, is the vertical axis on the left, and the actual tensile stress is the vertical scale on the

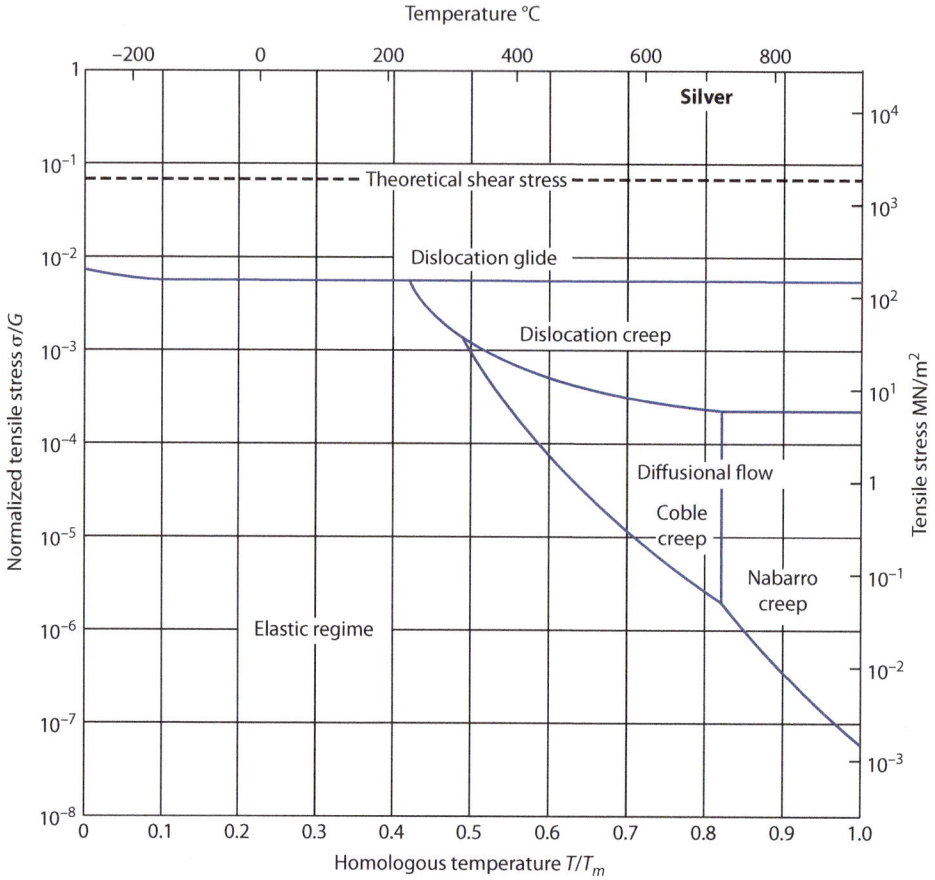

Figure 9.17   A deformation-mechanism map for pure silver with a grain size of 32 μm and a strain rate of 10^{-8} s^{-1}. (*Based on Crossman, F.W., and Ashby, M.F., Acta Met. 23 (1975), p. 887.*)

right. The strain rate of various deformation mechanisms is determined by computation with deformation models, and the mechanism with the highest strain rate is the mechanism shown in the map. At low temperature and low stress, corresponding to the bottom left side of the figure, the deformation is elastic. If the stress is increased at low temperature, the dislocation glide boundary line is encountered at a normalized stress of a little less than 10^{-2}. The boundary between two deformation mechanisms is where computations yield equal strain rates for the two deformation mechanisms. Likewise, at points where three mechanisms come together, all three deformation mechanisms result in the same strain rate. At the top of Figure 9.17, the dotted line is the theoretical shear yield strength, which is the stress required to shear a perfect crystal with no dislocations. The theoretical shear yield strength is approximately equal to one-tenth the shear modulus. The theoretical shear strength is included as a reference; no metal, except a whisker, has this yield strength. Dislocation glide is the dominant creep mechanism for any normalized stress above $10^{-2.2}$ for all temperatures up to the melting temperature. At temperatures approaching the melting temperature ($T/T_m = 1.0$), and at normalized stress less than $10^{-2.2}$ but more than $10^{-3.6}$, the dominant creep mechanism is dislocation creep; this is the dislocation climb described above. At temperatures above $T/T_m = 0.82$ and at normalized stress less than $10^{-3.6}$, the deformation mechanism is primarily Herring-Nabarro creep. Herring-Nabarro creep results from atom flux through the bulk of the metal. Coble creep dominates at homologous temperatures between 0.82 and 0.50 and for normalized stress below where dislocation creep occurs.

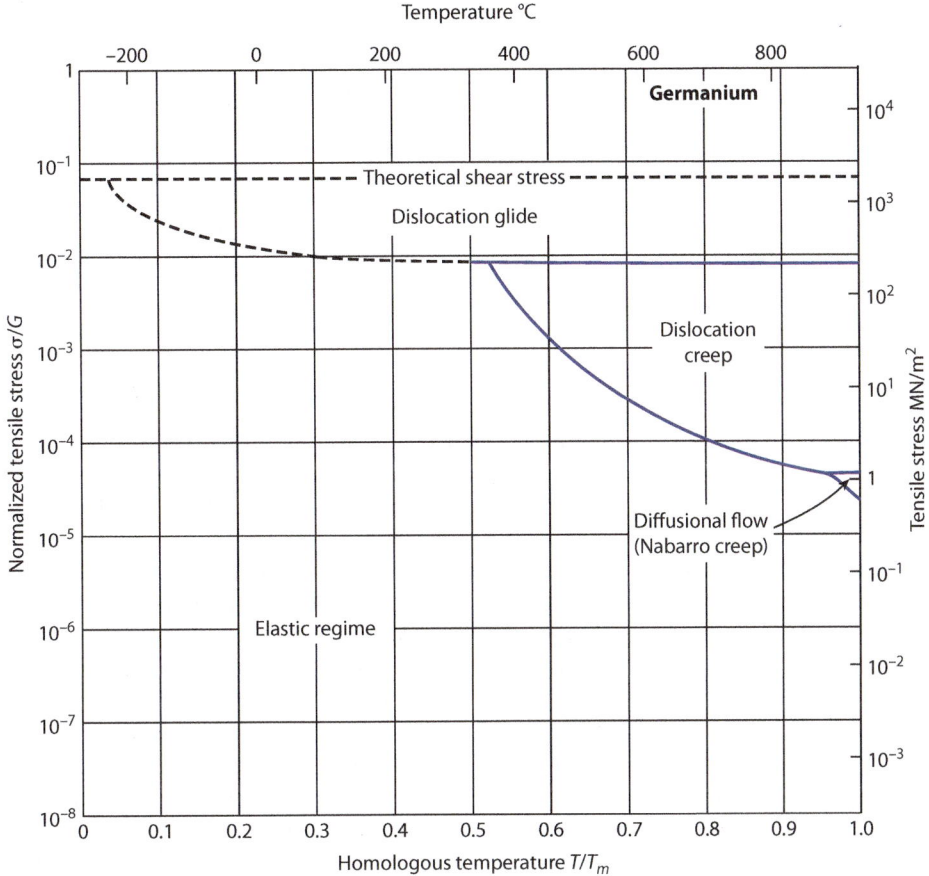

Figure 9.18 A deformation-mechanism map for germanium with a grain size of 32 μm and a strain rate of 10^{-8} s^{-1}. (*Based on Crossman, F.W., and Ashby, M.F., Acta Met. 23 (1975), p. 887.*)

The mechanisms for creep in crystalline ceramics are similar to those for crystalline metals, as is shown in the deformation-mechanism map of Figure 9.18 for covalently bonded germanium. Because dislocation motion is difficult at low temperatures in covalently bonded materials, it is necessary to increase the temperature in germanium and other ceramics relative to metals before dislocation glide occurs. The dislocation glide line is dashed at low temperatures, because brittle fracture is likely to occur at these stresses before dislocation glide can take place. Herring-Nabarro creep, which is diffusional flow, occurs only at the highest of temperatures. This is because the activation enthalpy for vacancy diffusion in germanium is large relative to metals, and diffusional flow requires vacancy diffusion.

9.4.4 Superplastic Strain in Metals and Ceramics

Superplasticity occurs when materials demonstrate extremely high plastic strains, from hundreds to thousands of percent, as shown in Figure 9.19. Normally superplasticity occurs at temperatures above 50% of the absolute melting temperature. Superplasticity is a result of high-temperature

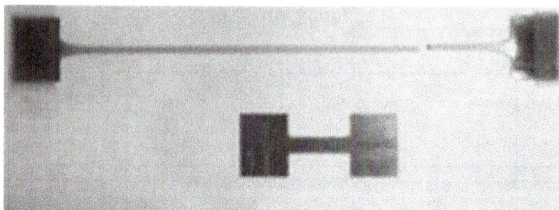

Figure 9.19 A photograph of superplastic strain (top image) of an Al-Mn-Mg-Sc alloy. The original dimensions of the specimen are shown in the lower image. (*PNNL*)

grain boundary sliding and the diffusional flow of atoms. During the grain boundary sliding, the grains maintain their shape, but they slide relative to each other. Superplasticity requires equiaxed grains normally less than 10 μm in diameter. The strain rates in superplastic deformation are typically on the order of $10^{-3}\,\mathrm{s}^{-1}$; these strain rates are higher than those normally associated with creep strain.

Materials that are superplastic are also **strain-rate sensitive**. In a material that is strain-rate sensitive, the true stress is a function of the strain rate to a power that approaches 1. In superplastic materials this power is greater than 0.8 and less than 1. When a neck starts to form during plastic deformation, plastic strain localizes at the neck. In a highly strain-rate sensitive material, the increased strain rate at the neck significantly increases the required stress for further strain. Further strain is transferred to a less strained region of the specimen that can plastically strain at a lower stress. The high strains observed in superplasticity require that a neck not form as a result of plastic strain. Metals that form a neck are not highly strain-rate sensitive, and they are not superplastic. All metals and ceramics creep at high temperatures, but not all metals and ceramics are superplastic at high temperatures. Some titanium, aluminum, and iron alloys and ceramics that are superplastic are listed in Table 9.1. We will discuss superplastic forming of metals and ceramics in Chapter 13.

Table 9.1 Some Superplastic Metal Alloys and Ceramics, along with the Superplastic Temperature in °C, the Percent Elongation, and the Strain Rate

Alloy	Superplastic Temperature	% Elongation (At Strain Rate Listed Here)	Strain Rate (1/S)
Ti-6% Al-4% V	927	1000−2000	2×10^{-4}
Al-4.5% Cu-0.5% Zn (Supral 100)	450	600−1000	10^{-3}
Al-4.5% Zn-4.5% Ca	565	500	10^{-3}
Fe-26% Cr-6.5% Ni	900	1000	5×10^{-5}
3mol. % Yttria-97% Zirconia	1450	>160	10^{-4}
Al_2O_3 - 0.05% MgO - 0.05% Yttria	1550	65	10^{-4}
80wt.% of (3mol.%Yttria, 97% Zirconia) - 20wt.% Al_2O_3	1650	500	10^{-3}

Reprinted from Askeland, D.R., and Phule, P.P., The Science and Engineering of Materials, 4th ed. Thomson-Brooks/Cole, Pacific Grove, Ca. (2003), p. 343.

9.4.5 Plastic Strain and Creep in Noncrystalline Ceramics

Dislocation motion does not apply to amorphous materials, because dislocations can only be identified in an ordered material. Creep strain in glass requires that the bonds between different SiO_2 tetrahedra be broken, so that tetrahedra of silicon and oxygen can rotate or move and then rebond. The breaking of the bonds is a thermally activated process, and higher temperatures result in more bonds being broken. Also, the addition of compounds such as soda (NaO) and lime (CaO) results in nonbridging oxygen atoms as shown in Figure 3.3. The addition of soda and lime increase the creep rate and decrease the viscosity of the glass, as is shown in Figure 9.3. The plastic strain rate of a noncrystalline ceramic, such as glass, at temperatures above the glass transition temperature is modeled with Equations 9.1 through 9.6.

In glass the temperatures are named where the viscosity has certain values. The working point is the temperature where the viscosity is equal to 10^3 Pa · s (10^4 P). In the study of viscosity, the poise unit (P) is also used. The SI unit of viscosity is Pa · s, and 10 poise is equal to 1 Pa · s. At the working point, glass is easily formed. The softening point is the temperature where the viscosity is equal to $10^{6.6}$ Pa · s ($10^{7.6}$ P). At the softening point, a rod of glass 24 cm long and 0.7 mm in diameter elongates 1 mm/minute under its own weight. The annealing point is the temperature where the viscosity is equal to $10^{12.4}$ Pa · s ($10^{13.4}$ P). At the annealing point, glass that has large internal strains becomes strain free in 15 minutes.

9.4.6 Plastic Strain, Creep, and Stress Rupture in Polymers

The plastic strain rate in polymers is modeled with Equations 9.1 through 9.6, where the plastic strain rate is related to the viscosity, stress, and temperature. The viscosity of polymers depends upon the molecular mass of the polymer. For low molecular mass, the viscosity linearly increases with molecular mass. However, above a critical molecular mass, the viscosity depends upon the molecular mass to the power of approximately 3.4. The viscosity of polymers depends upon temperature in an Arrhenius-type relationship, as given by Equation 9.3.

Most polymer liquids are not Newtonian. Water is an example of a Newtonian fluid. If the exponent (n) in Equation 9.2 for non-Newtonian liquids is less than 1, then as the shear strain rate increases, the slope of a plot of applied shear stress versus strain rate decreases relative to the value at lower strain rates, as shown in Figure 9.20. These polymers are called shear-thinning polymers, and most polymers are

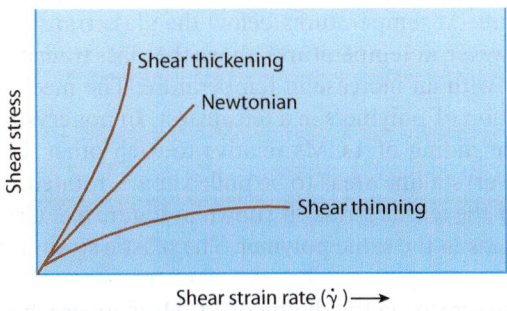

Figure 9.20   A schematic of the applied shear stress as a function of applied shear strain rate for Newtonian, shear-thinning, and shear-thickening polymers. (*Based on Askeland, D.R., Fulay, P.P., and Wright, W.J., The Science and Engineering of Materials, 6th ed., Cengage Learning, Stamford, CT. (2011), p. 202.*)

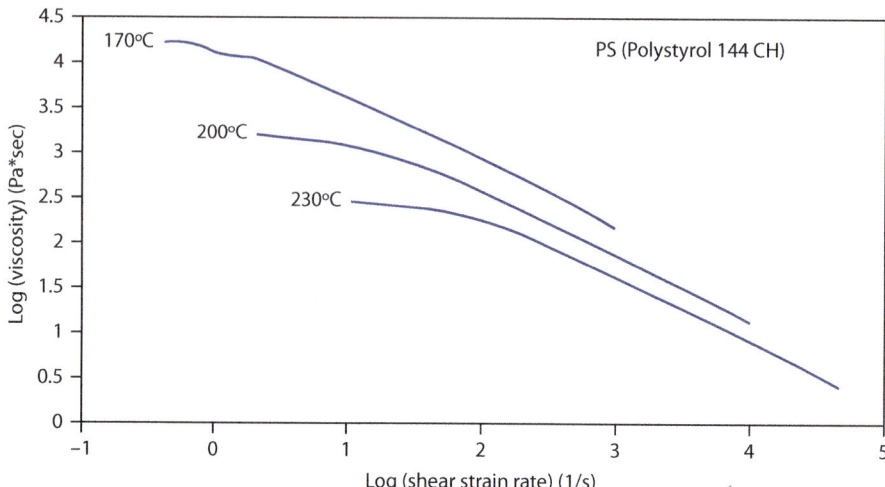

Figure 9.21 A plot of the logarithm of the viscosity of Polystyrol as a function of the logarithm of the shear strain rate. (*Based on Osswald, T.A., and Menges G., Materials Science of Polymers for Engineers, Hanser Publishers, Munich (2003), p. 132.*)

shear thinning. Figure 9.21 shows that the viscosity of Polystyrol™ decreases as the strain rate increases, and the figure also shows that the viscosity decreases as the temperature increases. Polystyrol is a trade name for a high-impact polystyrene (HIPS). The molecular-level explanation for shear thinning is that at low shear rates, the LCMs have more time to become entangled and bond strongly. Conversely, at high shear strain rates, the bonding is not as strong, and the additional stress required for an increased strain rate is small. Shear thinning is important in the processing of polymers, and it allows for high shear strain rates during processes such as injection molding. If the exponent (*n*) in Equation 9.2 for non-Newtonian liquids is greater than 1, then as the shear strain rate increases, the slope of a plot of applied shear stress versus strain rate increases relative to the value at lower strain rates, as shown in Figure 9.20. These polymers are called shear-thickening polymers. This behavior is observed in some polymers with ionic cross-link bonds, or ionomers.

The creep curves for the thermoplastic polymers PMMA and PP are shown for several different stresses at 20°C in Figure 9.22. These creep curves are similar to the curve shown in Figure 9.4. Thermoplastic polymers strain harden with plastic strain, and thermal vibrations activate plastic strain in polymers. However, the strain hardening in polymers does not anneal with an increase in temperature; however, thermoplastic polymers do soften as the temperature increases. The magnitude of creep strain in polymers is dependent upon temperature. At temperatures below the glass transition temperature, polymers are rigid and creep very little. However, at temperatures above the glass transition temperature, the creep rate in polymers increases rapidly with an increase in temperature. The mechanism of creep in polymers is similar to the plastic deformation of polymers in a tensile test. In noncrystalline thermoplastic polymers, the plastic strain is due to the sliding of LCMs relative to each other. Plastic strain in polymers with partial crystallinity causes the crystalline areas to be pulled apart, rotated, and elongated; and the LCMs are oriented in the direction of the tensile stress, as shown in Figure 6.8. Crystalline areas creep at a lower rate than do noncrystalline areas in the same polymer. The plastic strain rate of thermoplastic polymers is modeled with Equation 9.6.

Thermoset polymers can also creep via the sliding of LCMs in areas where there are no covalent bonds connecting the LCMs. However, the amount of creep strain is less than for thermoplastic polymers, because the three-dimensional and cross-link covalent bonds do not creep, and they limit the amount of creep.

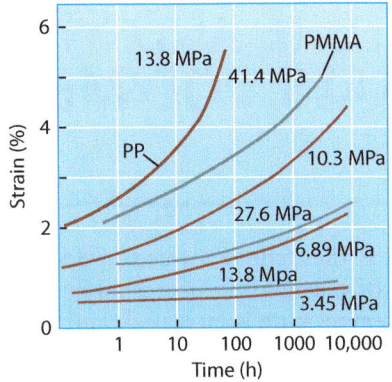

Figure 9.22 Creep strain in percent as a function of the logarithm of time in hours (h) for PMMA (blue) and PP (red) at several different stresses, at 20°C. (*Based on Askeland, D.R., Fulay, P.P., and Wright, W.J., The Science and Engineering of Materials, 6th ed., Cengage Learning, Stamford, CT. (2011), p. 627.*)

In addition to the types of presentations of data discussed above, polymer scientists and engineers use the creep modulus (E_c), which is the applied stress (σ) divided by the total strain (ε), as shown in Equation 9.16.

$$E_c = \frac{\sigma}{\varepsilon}$$ **9.16**

Figure 9.23 shows the creep modulus for some polymers. The creep modulus decreases with time because the total strain increases with time, and the stress is assumed to be constant.

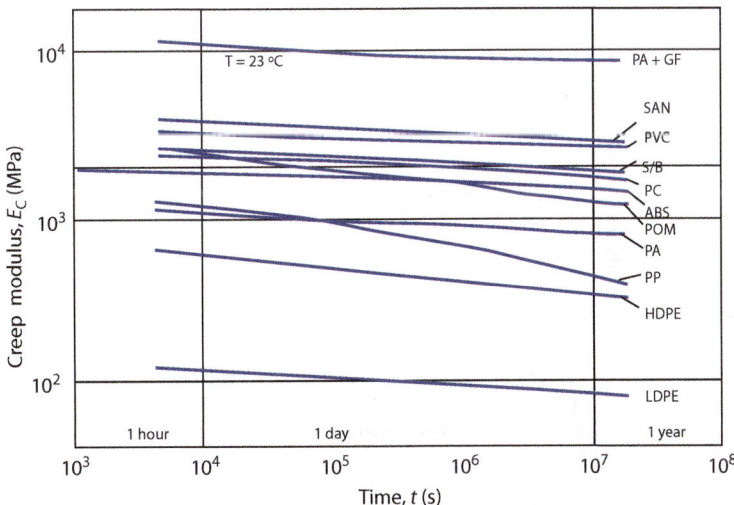

Figure 9.23 The creep modulus for some polymers and PA (nylon) + GF (glass fibers) as a function of time at 23°C. (*Based on Osswald, T.A., and Menges G., Materials Science of Polymers for Engineers, Hanser Publishers, Munich (2003), p. 391.*)

9.5   TEMPERATURE AND STRESS-STRAIN CURVES IN METALS, POLYMERS, AND CERAMICS

We have shown that plastic strain is thermally activated in addition to being activated by stress. As temperature increases, the amount of plastic strain increases in a tensile test. This results in a decrease in the yield stress in metals as the temperature is increased, and also an increase in the strain to failure, as shown schematically in Figure 9.24 for BCC metals, such as iron. BCC metals are not close packed, and HCP metals have only the basal close-packed plane. BCC and HCP metals become brittle at low temperatures because of the limited amount of dislocation glide and cross-slip. In FCC polycrystalline metals, high ductility is maintained down to temperatures approaching absolute zero, because dislocation glide and cross-slip occurs on many possible slip systems at low temperatures in this close-packed structure. In FCC metals, the yield strength does increase as temperature decreases. Polycrystalline ceramics are brittle at room temperature; however, there is some ductility at sufficiently high temperatures, typically above half the melting temperature.

   In polymers there is also a decrease in the yield stress as temperature is increased, and also an increase in the elongation to failure, as shown in Figure 9.25 for PMMA. In Figure 9.25, to convert the elongation in percent to strain, divide by 100. An increase in temperature results in more thermal energy to break the van der Waals bonds between LCMs and allow the LCMs to slide past each other. Amorphous polymers are brittle below the glass transition temperature, because there is insufficient thermal energy to break the van der Waals bonds between different LCMs.

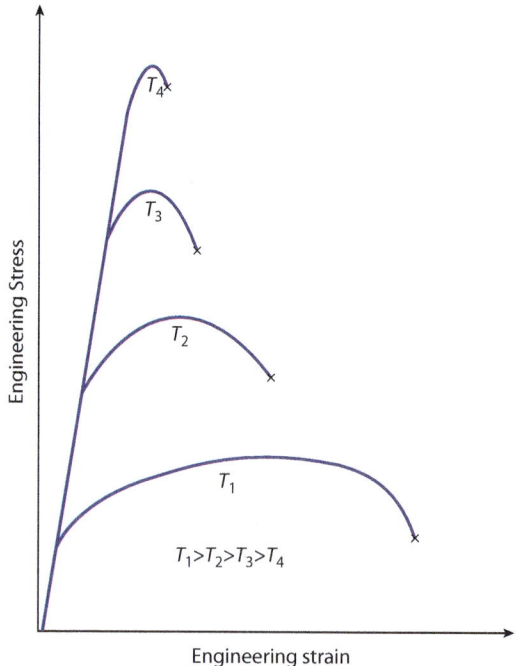

Figure 9.24   A schematic of the engineering stress strain diagrams for a BCC metal such as iron at different temperatures.

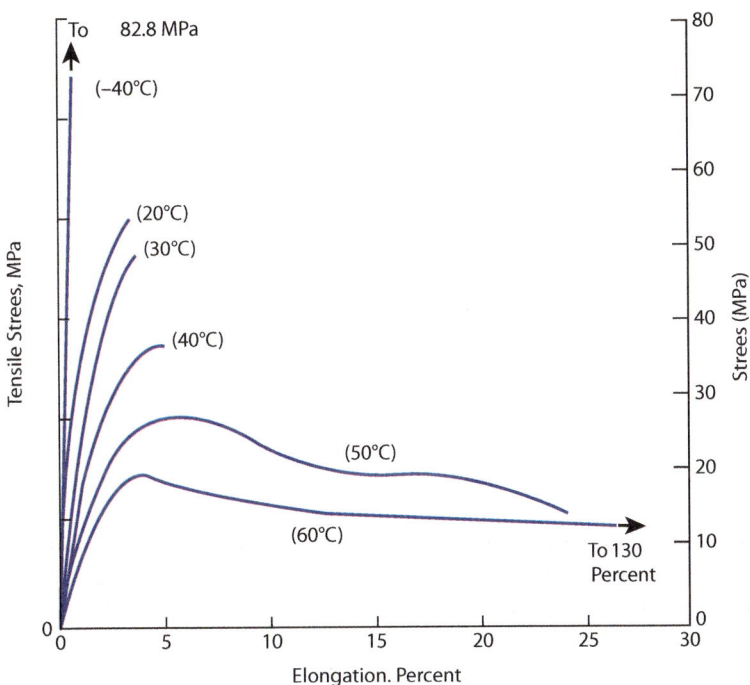

Figure 9.25 Tensile stress versus elongation in percent for PMMA at temperatures of 60°C, 50°C, 40°C, 30°C, 20°C, −40°C. (*Based on Carswell, T.S. and Nason, H.K., Effect of Environmental Conditions on the Mechanical Properties of Organic Plastics, in Symposium on Plastics, ASTM Philadelphia, PA (1944), p. 25.*)

9.6 VISCOELASTIC STRAIN MODELS OF ELASTIC AND PLASTIC STRAIN

The total strain of a material is a sum of the elastic and plastic strain, and elastic and plastic strain each result from different mechanisms. Elastic strain in solids results from the stretching or bending of interatomic bonds. The model of elastic deformation is a spring that recovers its deformation when unloaded, as shown schematically in Figure 9.26a. For an elastic solid, the elastic strain (ε_e) is given by Equation 9.7, and the elastic strain rate ($d\varepsilon_e/dt$) is given by Equation 9.17.

$$\frac{d\varepsilon_e}{dt} = \frac{1}{E}\frac{d\sigma}{dt}$$ **9.17**

There is an elastic strain rate only if the stress changes with time.

There are a variety of mechanisms for plastic strain depending upon the material and temperature, as discussed in this chapter. A mechanical model for all types of plastic strain in a solid is a dashpot, as is shown in Figure 9.26b. A dashpot is a cylinder containing fluid with a plunger. The plunger allows fluid to flow through the clearance between the plunger and the cylinder, or through an orifice in the plunger. In a dashpot there is no recovery of deformation of the plunger if the applied stress is reduced to 0. The plastic strain rate ($d\varepsilon_p/dt$) is given by Equation 9.5 for a viscous liquid.

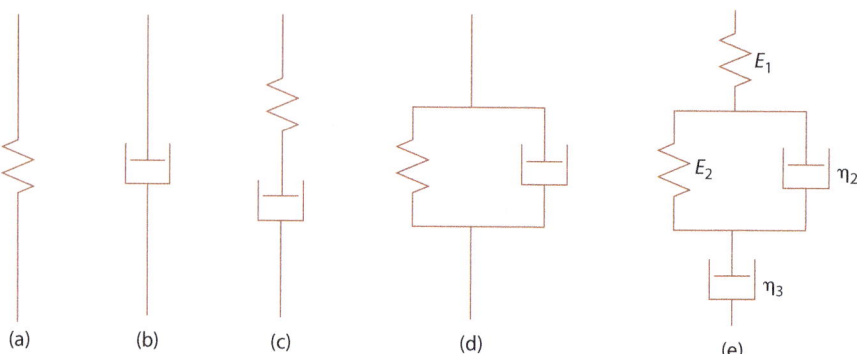

Figure 9.26 Mechanical analogs of strain in materials. (a) Elastic strain in a spring. (b) Plastic strain in a dashpot. (c) A Maxwell model of a solid with a spring and a dashpot in series. (d) A Voigt model of a solid with a spring and dashpot in parallel. (e) A four-element viscoelastic model with a spring and dashpot in series and in parallel.

Viscoelastic models combine elastic and plastic (viscous) strain-rate components to predict the strain rate observed in a material. For example, what combination of elastic and plastic strain components is necessary to obtain the strain-versus-time curve in Figure 9.4? A **Maxwell solid** is a series combination of a spring and dashpot, as shown in Figure 9.26c. In a series arrangement, the spring and dashpot have the same stress, or isostress. If a constant stress is applied to a Maxwell solid at zero time, the spring elastically strains instantly. The elastic strain is given by Equation 9.7 and shown in Figure 9.27. The plastic strain provided by the dashpot is obtained by integrating Equation 9.5 over time to obtain the total strain shown in Figure 9.27.

$$\varepsilon_p = \frac{\sigma^n}{3\eta} t \qquad\qquad \textbf{9.18}$$

The plastic strain in a Maxwell solid accumulates linearly with time when subject to a constant stress, as shown in Figure 9.27. If the stress on the Maxwell solid is reduced to 0 at time $t = t_1$, the elastic strain in the spring is recovered. However, the plastic strain (ε_p) accumulated by the dashpot is not recovered. The Maxwell solid does not have stage I transient creep strain.

In a **Voigt solid**, the spring and dashpot are combined in parallel, as shown in Figure 9.25d. In a parallel arrangement, the spring and the dashpot have the same strain, or isostrain. The shock absorber in an automobile is a parallel arrangement of a spring and a dashpot. If a stress is applied to a Voigt solid at 0 time, and then the stress is held constant, then the strain-versus-time plot appears as in Figure 9.28. There

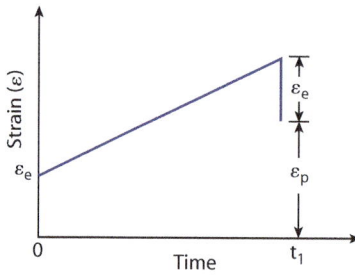

Figure 9.27 Strain (ε) as a function of time (t) for a Maxwell solid subject to an applied stress. The strain at time $t = 0$ is all elastic and equal to ε_e. Plastic strain of magnitude ε_p accumulates with time. At time t_1 the specimen is unloaded and the elastic strain (ε_e) is recovered.

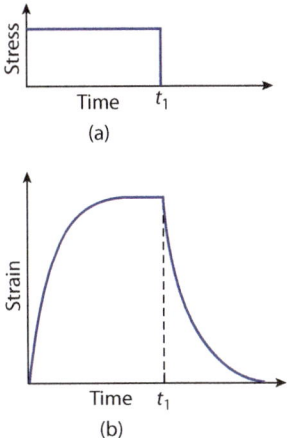

Figure 9.28 (a) The stress applied to a Voigt solid as a function of time. (b) The strain in a Voigt solid as a function of time for the stress profile in part (a).

is no initial elastic deformation, because the dashpot has no elongation at 0 time as shown by Equation 9.18. The spring must elongate along with the dashpot, and the dashpot elongates with time. However, the strain from the dashpot as a function of time is not linear, because as the spring strains, it supports more stress as it strains and the dashpot experiences less stress. Therefore, the rate of strain in the dashpot decreases with time until it becomes 0 when the spring carries all of the stress. If the Voigt solid is unloaded at time t_1, strain is recovered with time, as shown in Figure 9.28, because the spring contracts. With sufficient time all of the strain is recovered, because the spring eventually fully recovers to its original length once the load is removed. Although all the strain is recovered, the strain recovery is time dependent. A Voigt solid is anelastic.

In a **four-element model** a Maxwell and a Voigt solid are connected in series, as shown in Figure 9.26e, and the strain-versus-time plot for this material appears as in Figure 9.29. The plot in Figure 9.29 is

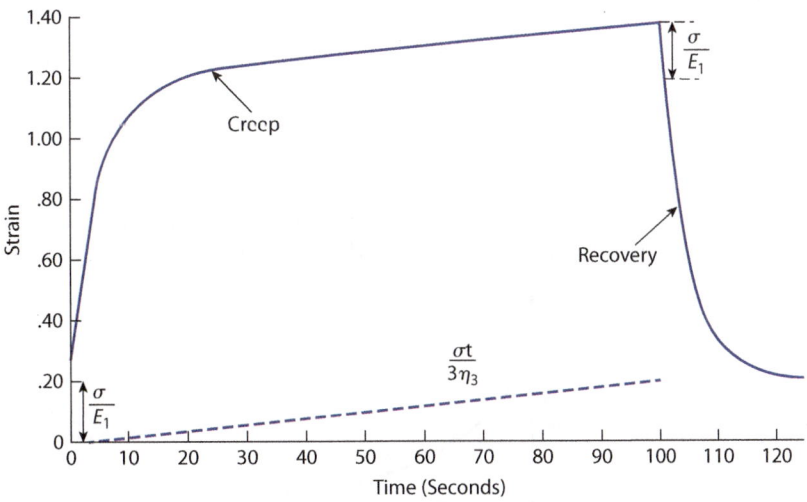

Figure 9.29 The creep response of a four-element model, with the elastic constant E_1 equal to 500 MPa, E_2 equal to 100 MPa, η_2 equal to 500 MPa · s, η_3 equal to 16.7 GPa · s; and an applied stress of $\sigma = 100$ MPa. (*Based on Nielsen, L., Mechanical Properties of Polymers, Litton Educational Publishing, Van Nostrand Reinhold N.Y. (1962), p. 51.*)

more like the creep curve in Figure 9.4 than Figures 9.27 and 9.28b. In Figure 9.29 the stress is applied at time 0, and the stress is removed at 100 s.

For the polymer in Figure 9.29, compare the stage II creep rate determined from the plot of strain versus time, to that determined using Equation 9.5 and the properties given in the figure caption.

Solution

From the plot in Figure 9.29, the steady-state stage II strain in 100 seconds is 0.20, resulting in a strain rate of

$$\frac{d\varepsilon_p}{dt} = 0.20 \times 10^{-2}\,\text{s}^{-1}$$

To calculate the plastic strain rate using Equation 9.5, the applied stress is 100 MPa. The dashpot corresponding to η_3 with a viscosity equal to 16.7 GPa produces the stage II creep rate.

$$\frac{d\varepsilon_p}{dt} = \frac{\sigma}{3\eta} = \frac{100\ \text{MPa}}{3(16.76 \times 10^3\ \text{MPa} \cdot \text{s})} = 2 \times 10^{-3}\,\text{s}^{-1}$$

The creep rate from the plot and from the calculation agree.

9.7   STRESS RELAXATION

Stress relaxation is the reduction of stress in a material subject to a constant strain. In stress relaxation, a constant strain (ε) is applied to a specimen at $t = 0$, and the stress at $t = 0$ is measured to be (σ_0). After a period of time (t), the stress is observed to drop to $\sigma(t)$, as shown in Figure 9.30. The strain (ε) at $t = 0$ is all elastic, because it requires time for the accumulation of any plastic strain according to Equation 9.18. The value of the initial stress (σ_0) is calculated from the elastic modulus and Hooke's law, as shown in Equation 9.19.

$$\sigma_0 = E\varepsilon \tag{9.19}$$

To calculate the stress as a function of time, assume that the material behaves like a Maxwell solid, or like elastic and plastic components in series. When the Maxwell solid is deformed, the initial strain is all

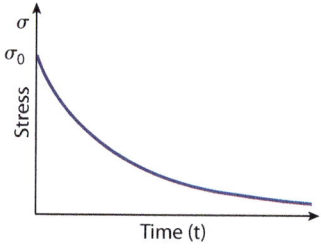

Figure 9.30   The stress (σ) as a function of time (t) in a Maxwell solid subject to a fixed strain. The stress at $t = 0$ is equal to σ_0, and the stress relaxes, or decreases, exponentially with time.

elastic. Then with time, the spring relaxes or elastic strain decreases, because the dashpot provides plastic strain with time. One equation for this analysis is that the total strain (ε) is the sum of the elastic and plastic strain, and the total strain is a constant, as shown in Equation 9.20.

$$\varepsilon = \varepsilon_e + \varepsilon_p = \text{Constant} \qquad \textbf{9.20}$$

Also, Equation 9.21 shows that the sum of the elastic and plastic strain rates is 0, because the total strain is constant at ε.

$$\frac{d\varepsilon}{dt} = 0 = \frac{d\varepsilon_e}{dt} + \frac{d\varepsilon_p}{dt} \qquad \textbf{9.21}$$

Equation 9.20 shows that every increment of plastic strain reduces the elastic strain by an equal but opposite increment, because the sum of the strains is constant. Assuming that the plastic part of the material responds as a Newtonian fluid ($n = 1$), and inserting $n = 1$ into Equation 9.5, then inserting Equations 9.5 and 9.17 into Equation 9.21 results in Equation 9.22.

$$0 = \frac{1}{E}\frac{d\sigma}{dt} + \frac{\sigma}{3\eta} \qquad \textbf{9.22}$$

Rearranging terms in Equation 9.22 results in Equation 9.23.

$$\frac{d\sigma}{\sigma} = -\frac{E}{3\eta}dt \qquad \textbf{9.23}$$

Integrating Equation 9.23 results in

$$\ln \sigma = -\frac{E}{3\eta}t + C \qquad \textbf{9.24}$$

where C is a constant. Using the boundary condition that at $t = 0$, $\sigma = \sigma_0$ results in $C = \ln \sigma_0$, and Equation 9.25.

$$\ln \sigma = -\frac{E}{3\eta}t + \ln \sigma_0 \qquad \textbf{9.25}$$

Rearranging the terms in Equation 9.25 results in Equation 9.26.

$$\ln \frac{\sigma}{\sigma_0} = -\frac{E}{3\eta}t \qquad \textbf{9.26}$$

Taking the exponential of each side of Equation 9.26 results in Equation 9.27.

$$\frac{\sigma}{\sigma_0} = \exp - \left(\frac{E}{3\eta}t\right) \qquad \textbf{9.27}$$

At $t = 0$, the stress is equal to σ_0, and as time increases, σ decreases exponentially, as schematically shown in Figure 9.30. The term $(E/3\eta)$ in Equation 9.27 is the relaxation time (t_R). Making this substitution results in Equation 9.28.

$$\frac{\sigma}{\sigma_0} = \exp\left(-\frac{t}{t_R}\right) \qquad \textbf{9.28}$$

When the time (t) is equal to the relaxation time (t_R), the stress (σ) is equal to $\sigma_0 \exp(-1)$ according to Equation 9.28. It would have been difficult to derive Equation 9.27 without a deformation model such as the Maxwell solid.

Example Problem 9.4

A polymer specimen is strained at time $t = 0$ by 5×10^{-4}, and this requires an initial stress of 3×10^7 Pa. This strain is constant for ten days, and the stress on the polymer has dropped to 1×10^7 Pa. Assume that this polymer can be modeled as a Maxwell solid with the viscous component behaving as a Newtonian fluid.

(a) What is the elastic modulus and viscosity of this material?

(b) What is the stress in the part if the strain is constant for 30 days?

Solution

a) The elastic modulus is related to the initial elastic strain at $t = 0$.

$$E = \frac{\sigma}{\varepsilon} = \frac{3 \times 10^7 \, \text{Pa}}{5 \times 10^{-4}} = 0.6 \times 10^{11} \, \text{Pa}$$

The viscosity comes from the decay of the stress by using Equation 9.27.

$$10 \text{ days} = 240 \text{ hours} = 14{,}400 \text{ minutes} = 8.64 \times 10^5 \text{ seconds}$$

$$\frac{\sigma}{\sigma_0} = \frac{1 \times 10^7 \, \text{Pa}}{3 \times 10^7 \, \text{Pa}} = \frac{1}{3} = \exp - \left(\frac{E}{3\eta}t\right) = \exp - \left(\frac{0.6 \times 10^{11} \, \text{Pa}}{3\eta}(8.64 \times 10^5 \text{ s})\right)$$

Take the natural logarithm of the above equation.

$$-1.1 = -\frac{0.6 \times 10^{11} \, \text{Pa}}{3\eta}(8.64 \times 10^5 \text{ s})$$

Solve for the viscosity.

$$3\eta = \frac{0.6 \times 10^{11} \, \text{Pa}}{1.1}(8.64 \times 10^5 \text{ s}) = 4.7 \times 10^{16} \, \text{Pa} \cdot \text{s}$$

$$\eta = 1.57 \times 10^{16} \, \text{Pa} \cdot \text{s}$$

b) To determine the stress after 30 days, use Equation 9.27 with the known values of the elastic modulus (E) and viscosity (η); and insert the number of seconds in 30 days, which is equal to 25.92×10^5 seconds.

$$\frac{\sigma}{\sigma_0} = \frac{\sigma}{3 \times 10^7 \, \text{Pa}} = \exp - \left(\frac{E}{3\eta}t\right) = \exp - \left(\frac{0.6 \times 10^{11} \, \text{Pa}}{4.7 \times 10^{16} \, \text{Pa} \cdot \text{s}}(25.92 \times 10^5 \text{ s})\right)$$

Solve for the stress.

$$\sigma = (3 \times 10^7 \, \text{Pa}) \exp(-3.3) = 3 \times 10^7 \, \text{Pa}(0.036) = 0.11 \times 10^7 \, \text{Pa}$$

Summary

- Elastic strain at a particular stress does not change with time, because the atoms respond so quickly to changes in stress that the elastic strain seems instantaneous.

- The elastic modulus decreases as the temperature increases, unless there is a phase transformation, and it starts to decrease rapidly as the melting temperature is approached. The reason for this is that at high temperatures, the atoms are on the average farther apart than at low temperatures; therefore, the bonding is weaker.

- Viscosity is a property that relates the plastic shear strain rate to shear stress in liquids and super-cooled liquids, such as glasses and polymers. The plastic shear strain rate of a Newtonian liquid is equal to the applied shear stress divided by the viscosity. In non-Newtonian liquids, the plastic strain rate raised to a power not equal to 1 is equal to the applied shear stress divided by the viscosity. The temperature dependence of the plastic shear strain rate of liquids and noncrystalline super-cooled liquids is given by an Arrhenius-type equation.

- Creep strain is time-dependent plastic strain. In stage I creep, there is a nonlinear transient plastic strain rate that is inversely related to time and increases with increased stress and temperature. Stage II, or steady-state creep, occurs in crystalline solids when the strain-hardening rate is equal to the rate of softening due to annealing. In crystalline solids the stage II creep rate is given by an Arrhenius-type equation multiplied by the stress to an exponential power. Stage III creep is an increased creep rate that results in rupture. The inverse of the time to rupture also follows an Arrhenius relationship times a function of stress. The time to rupture, stress, and temperature are combined into plots such as the Larson-Miller parameter for design purposes.

- At temperatures below half of the melting temperature in crystalline metals, the primary mechanism of creep strain is dislocation glide. At temperatures of approximately half of the melting point or higher, the climb of edge dislocations over obstacles makes a significant contribution to creep strain in metals. At temperatures higher than the equi-cohesive temperature, the grain boundaries are weaker than the crystalline material, and grain boundary sliding contributes to creep strain. Also, at elevated temperatures diffusion of atoms occurs along the grain boundaries, resulting in Coble creep. Because of grain boundary sliding and the general weakness of grain boundaries at elevated temperatures, high-performance gas-turbine blades are made from directionally solidified metals or from single crystals. At temperatures above half the melting temperature in a metal solid subject to tension, there is transport of material by vacancy diffusion, and atoms move in the direction of the tension to elongate the specimen by Herring-Nabarro creep.

- Because dislocation motion is difficult at low temperature in covalently bonded materials, it is necessary to increase the temperature in ceramics to approximately 50% of the absolute melting temperature before dislocation glide occurs. At high temperatures the mechanisms of stage II creep in crystalline ceramics are similar to those in metals, and the equations for predicting creep strain are the same as in metals.

- The various creep mechanisms for metals and ceramics are displayed on deformation mechanism maps.

- Superplasticity is when metals and ceramics demonstrate extremely high plastic strains, of hundreds to thousands of percent. Normally superplasticity occurs at temperatures above 50% of the absolute melting temperature, at strain rates on the order of $10^{-3}\,\text{s}^{-1}$, and in materials that are highly sensitive to the strain rate. Superplasticity is a result of high-temperature grain boundary sliding and the diffusional flow of atoms.

- Creep strain in glass results from the breaking of bonds between different SiO_2 tetrahedra, which allows the tetrahedra of silicon and oxygen to rotate or move and rebond. The equations for predicting creep strain in glass are based upon viscoelastic models that relate strain to viscosity, stress, and temperature.
- The plastic strain rate and creep strain in polymers are modeled with viscoelastic equations. In noncrystalline thermoplastic polymers, the plastic strain is due to the sliding of LCMs relative to each other. Plastic strain in polymers with partial crystallinity causes the crystalline areas to be pulled apart, rotated, and elongated; and the LCMs are oriented in the direction of the tensile stress.
- In a tensile test, there is a decrease in the yield stress in metals and thermoplastic polymers as temperature is increased, and an increase in the strain to failure. Similar behavior is observed in ceramics at temperatures in excess of half the melting temperature.
- Stress relaxation is the reduction in stress in a part or specimen subject to a constant strain. Stress relaxation is modeled with viscoelastic models such as the Maxwell solid.

Supplemental Reading Subjects and Authors

Full references are listed at the end of the book.

General:	*Askeland, Fulay, and Wright*
Creep and plastic strain in metals:	*Caddell; Courtney; Dieter; Hertzberg; Sims, Stoloff, and Hagel; Donachie and Donachie*
Creep and plastic strain in glass and ceramics:	*Carter and Norton*
Creep and plastic strain in polymers:	*Kinney; McCrum, Buckley, and Bucknall; Osswald and Menges; Schultz*

Homework

Concept Questions

1. The elastic modulus in most materials increases or decreases as the temperature increases.

2. In a Newtonian fluid, the plastic shear strain rate is equal to the applied shear stress divided by the _____.

3. In a Newtonian fluid, the applied shear stress divided by the viscosity is equal to the plastic shear strain rate raised to the power of _____.

4. If the natural log of the viscosity is plotted as a function of the inverse of the absolute temperature, the slope of the line is equal to the _____ of activation for plastic strain divided by Boltzmann's constant.

5. Time-dependent plastic strain is _____ strain.

6. In stage I creep, the strain rate is _____ related to time.

7. Stage II creep in metals begins when the strain-hardening rate is equal to the rate of _____.

8. If the log of the creep strain rate is plotted as a function of the log of the applied stress, the slope is equal to the _____ of the applied stress.

9. The Larson-Miller parameter is equal to the absolute temperature times the log of the time to _____ plus a constant.

10. In crystalline materials, at temperatures below one-half the melting temperature, plastic strain is due to dislocation _____.

11. In FCC metals at low temperatures, screw dislocations can circumvent obstacles by _____ _____.

12. The process of an edge dislocation combining with a vacancy to circumvent an obstacle is dislocation _____.

13. At temperatures above the equi-cohesive temperature, the crystalline areas are _____ than the grain boundaries.

14. At temperatures above the equi-cohesive temperature, the creep mechanism of grain boundary _____ is observed.

15. In a turbine blade with grain boundaries parallel to the blade axis, the shear stress on the grain boundaries of a rotating turbine blade is equal to _____.

16. The change in shape due to transport of atoms along grain boundaries is called _____ creep.

17. The change in shape due to transport of atoms through the bulk of the material is called _____-_____creep.

18. In covalently bonded materials the creep mechanism of dislocation glide generally starts at a temperature of _____ percent of the melting temperature in kelvin.

19. Superplastic materials are highly sensitive to the strain _____.

20. If a material is strain-rate sensitive, when the strain rate increases, the applied _____ must increase drastically.

21. The working point for the soda-lime glass in Figure 9.3 is approximately _____ K.

22. The softening point for the soda-lime glass in Figure 9.3 is approximately _____ K.

23. The annealing point for the soda-lime glass in Figure 9.3 is approximately _____ K.

24. In thermoset polymers, the cross-link and three-dimensional covalent bonds (do or do not) creep.

25. As temperature increases in metals and polymers, the yield stress _____.

26. A viscoelastic model of a solid with a series spring and dashpot is a _____ solid.

27. In a Maxwell solid, when a stress is applied at $t = 0$, the total strain is equal to the _____ strain.

28. In a Maxwell solid, when the stress is removed after a period of time, residual strain is equal to the _____ strain.

29. A viscoelastic model of a solid with a parallel spring and dashpot is a _____ solid.

30. In a Voigt solid, if a tensile stress is applied at $t = 0$, the total strain at $t = 0$ is equal to _____.

31. In a Voigt solid with an elastic modulus of 1 GPa, if a tensile stress of 1 MPa is applied, the total strain after a long time is equal to _____.

32. In a Voigt solid with an elastic modulus of 1 GPa, if a stress of 1 MPa is applied for a long time and then removed for a long time, the total strain is equal to _____.

33. In stress relaxation, a constant _____ is applied to the specimen.

34. In stress relaxation, the elastic strain rate is equal to _____ the plastic strain rate.

35. The relaxation time for tensile stress relaxation is equal to the elastic modulus divided by three times the _____.

Engineer in Training–Style Questions

1. In a Newtonian fluid, the viscosity is equal to
    (a)   The applied shear stress divided by the shear strain
    (b)   The applied shear stress times the shear strain rate
    (c)   The shear strain rate divided by the applied shear stress
    (d)   The applied shear stress divided by the shear strain rate

2. The viscosity of a liquid with increasing temperature is
    (a)   The inverse of an Arrhenius relationship
    (b)   An Arrhenius relationship
    (c)   A linear increase
    (d)   A linear decrease

3. The stage II creep rate in metals and ceramics at constant stress and temperature is
    (a)   Transient and decreasing
    (b)   Increasing toward rupture
    (c)   A constant
    (d)   Parabolic

4. At temperatures well below half of the melting temperature, the primary mechanism of creep strain in metals is
    (a)   Coble creep
    (b)   Dislocation glide
    (c)   Dislocation creep
    (d)   Herring-Nabarro vacancy flux

5. At temperatures above 0.80 of the melting temperature and at very low stress, the primary mechanism of creep strain in metals is
    (a)   Coble creep
    (b)   Dislocation glide
    (c)   Dislocation creep
    (d)   Herring-Nabarro vacancy flux

6. At a temperature of 1090°C, the superalloys with the highest strength in a 1000-hour stress rupture test are
    (a)   Small-grained polycrystals
    (b)   Strain-hardened polycrystals
    (c)   Single crystals
    (d)   Directionally solidified polycrystals

7. Which of the following is not a characteristic of superplasticity?
    (a)   Grains elongate in the direction of flow
    (b)   Temperatures are above half of the melting temperature
    (c)   The material is highly strain-rate sensitive
    (d)   A strain rate of approximately 10^{-3} s^{-1}

8. In a shear-thinning material, the applied shear stress divided by the viscosity is equal to the shear strain rate to a power that is equal to
    (a)   1
    (b)   A number less than 1
    (c)   A number more than 1
    (d)   2

9. If a polymer is to be extruded at a high shear strain rate, which of the following material characteristics is most desirable?
  (a) Shear thickening
  (b) Newtonian fluid
  (c) Shear thinning
  (d) Elastic

10. Which of the following materials would you predict to be ductile at 50 K?
  (a) Pure BCC iron
  (b) Pure FCC copper
  (c) Pure HCP titanium
  (d) PE

11. An anelastic material can be modeled with a
  (a) Dashpot
  (b) Maxwell solid
  (c) Voigt solid
  (d) Four-element solid

12. In stress relaxation the plastic strain
  (a) Is constant
  (b) Increases with time
  (c) Decreases with time
  (d) Is equal to the negative of the elastic stain

Problems

Problem 9.1: What is the enthalpy of activation for viscous flow in the soda-lime glass shown in Figure 9.3, in the temperature range from 700 to 800 K?

Problem 9.2: For the zirconium metal shown in Figure 9.5, at a logarithm of the applied stress of 7.0 and a temperature of 1000 K, analytically determine the stage II strain rate, assuming that the constants determined in Example Problem 9.1 are applicable.

Problem 9.3: In Figure 9.5, for the temperature 973 K, the power of the stress term changes between the logarithm of stress of 7 and the logarithm of stress of 7.3. Determine the value of the power n at these two values of stress, using the data points given to either side of these stresses.

Problem 9.4: The activation enthalpy for creep is determined in Example Problem 9.1 between the temperatures of 1083 K and 1053 K. Determine the activation enthalpy between 973 K and 1053 K using the stress of $\log \sigma = 7.10$.

Problem 9.5: The following steady-state stage II creep rates were recorded for a nickel-base superalloy at a constant tensile stress of 200 MPa.

T	891 K	913 K
$\dot{\varepsilon}\,s^{-1}$	1×10^{-7}	2×10^{-7}

It is known that the exponent of the stress in the creep-rate equation is $n = 5$.

(a) What would be the total stage II creep strain for this alloy at 913 K and a tensile stress of 200 MPa in 1 year of service?

(b) Predict the stage II creep rate of this alloy at 891 K and a tensile stress of 300 MPa.

(c) Determine the enthalpy of activation for creep in eV/atom for this alloy.

Problem 9.6: The following steady-state stage II creep rates were recorded for a nickel-base superalloy:

$\dot{\varepsilon}\ s^{-1}$	Temperature	Stress
2.8×10^{-7}	811 K	21 MPa
2.8×10^{-5}	811 K	60 MPa
4.4×10^{-7}	700 K	60 MPa

(a) Evaluate all of the terms in the stage II creep-rate equation. Use the strain rate at 811 K and 60 MPa to evaluate the term A.

(b) Determine the stage II creep rate for a stress of 40 MPa and a temperature of 800 K.

Problem 9.7: Determine the power of the stress term in the creep equation for titania using the following data recorded at a temperature of 1250 K and at stresses of 24 MPa and 83 MPa.

At $\sigma_1 = 24$ MPa and $T = 1250$ K, the strain rate is $0.202 \times 10^{-6}\,s^{-1}$.

At $\sigma_2 = 83$ MPa and $T = 1250$ K, the strain rate is $0.234 \times 10^{-5}\,s^{-1}$.

Problem 9.8: Determine the activation enthalpy for creep in titania at a stress of 41 MPa if at a temperature $T_1 = 1250$ K the creep rate is $10^{-6}\ s^{-1}$ and at a temperature $T_2 = 1111$ K the creep rate is $10^{-7.6}$ $s^{-1} = 0.251 \times 10^{-7} s^{-1}$.

Problem 9.9: For silver at a tensile stress of 300 MPa and a temperature of 200°C, what is the primary creep mechanism?

Problem 9.10: For silver at a tensile stress of 30 MPa and a temperature of 715°C, what is the primary creep mechanism?

Problem 9.11: For silver at a tensile stress of 7 MPa and a temperature of 839°C, there are two equally contributing creep mechanisms. What are they?

Problem 9.12: For germanium at a tensile stress of 410 MPa and a temperature of 332°C, what is the primary creep mechanism? The shear modulus of germanium is 41 GPa.

Problem 9.13: For germanium at a tensile stress of 41 MPa and a temperature of 695°C, what is the primary creep mechanism? The shear modulus of germanium is 41 GPa.

Problem 9.14: A Udimet 700 turbine blade is exposed to a temperature of 1200 K at a tensile stress of 140 MPa? How many hours is the turbine blade expected to survive these conditions?

Problem 9.15: A part for a land-based gas-turbine made of Haynes 25 must carry a tensile stress of 140 MPa (20 ksi). For a design lifetime of 10 years, what is the allowable temperature for this part, based only upon high-temperature creep rupture?

Problem 9.16: A turbine blade made of Udimet 700 must be able to operate at a temperature of 838°C for a time of 10,000 h between inspections. What is the maximum tensile stress for this blade based upon high-temperature creep rupture?

Problem 9.17: For the polymer in Example Problem 9.4, calculate the total strain after 30 days if the material is subject to a creep test with a constant stress of 3×10^7 Pa. Assume that this material behaves as a Maxwell solid, and the viscous part of this material behaves as a Newtonian fluid.

Problem 9.18: Your company has developed a new polymer, and you are asked to determine (a) its elastic modulus and (b) its viscosity. To measure these properties, you choose to conduct a stress-relaxation test at 300 K. At 300 K a specimen of the new polymer is strained at $t = 0$ by an amount of 3×10^{-3}, and this requires an initial stress of 6×10^6 Pa. This strain is constant for 100 hours, and the stress on the polymer has dropped to 3×10^6 Pa. Assume that this polymer can be modeled as a Maxwell solid, with the viscous component behaving as a Newtonian fluid.

Problem 9.19: A soda-lime silica glass part that is 1×10^{-1} m long is deformed by a fixed amount of 1×10^{-4} m at a temperature of 667 K. At 667 K the elastic modulus of the soda-lime glass is 4×10^9 Pa, and the viscosity is 1×10^{14} Pa · s. Using a Maxwell model of this material, which is a spring and dashpot in series, and assuming that the fluid in the dashpot behaves as a Newtonian fluid, determine the stress in the material when:

(a) The displacement is first applied, at $t = 0$

(b) The stress after one day

Problem 9.20: A soda-lime glass tube made out of the material in Figure 9.3 is in service in a chemical-processing plant at 395°C, and it is subject to a tensile stress of 10 MPa. The elastic modulus of the glass at this temperature is equal to 40 GPa. Use a Maxwell solid model to answer the following, and assume that the glass behaves as a Newtonian fluid ($n = 1$).

(a) At the instant the glass tube is placed in service, at $t = 0$, what is the strain of the tube?

(b) After one year, what is the total strain of the tube?

Problem 9.21: (a) Develop an equation for the time at which the plastic (viscous) strain is equal to the elastic strain in a Maxwell solid (series spring and dashpot) subject to a stress (σ) where the viscous component is Newtonian ($n = 1$).

(b) Determine the time (t) when the plastic strain is equal to the elastic strain in quartz subject to a stress of 35 MPa at 1250 K, if the viscosity is equal to 1×10^{14} Pa · s and the elastic modulus is equal to 70×10^9 Pa, assuming that the quartz behaves as a Maxwell solid.

(c) Determine the total strain at the time (t).

Problem 9.22: A cube of polymer that is 0.05 m on each side is to be held in place with a clamp. The clamp is designed to compress the 0.05-m-thick polymer by a constant amount of 0.0001 m to provide the clamping force. The maximum clamping force that is allowable is 1×10^4 N. To keep the polymer block securely clamped, the clamping force cannot fall below 0.5×10^4 N for one year, which is the expected life for this device. Assume that this polymer behaves as a Maxwell solid and that the viscous component is Newtonian.

(a) What is the maximum possible elastic modulus in the polymer that will result in the maximum allowable clamping force?

(b) Assuming that a polymer is selected that has the highest allowable elastic modulus, what is the minimum allowable viscosity?

The damaging effect of corrosion resulting from the ocean environment shows on this once useful boat.

Photo by C. M. Gilmore

The goals of this chapter are to understand

- Which materials are subject to oxidation
- The mechanism of oxidation in metals and ceramics
- The conditions for the formation of a protective oxide to passivate metal surfaces
- The equations expressing the oxidation rate in metals
- The conditions for penetration of solvents into polymers
- The effect of solvents on the mechanical properties of polymers
- The effects of weathering on polymers
- Basic electrochemistry applied to corrosion, electroprocessing, and fuel cells
- The various types of corrosion
- Electromachining, electropolishing, and electroplating
- The operation of batteries and fuel cells

Chapter 10

Oxidation, Degradation, Corrosion, Electroprocessing, Batteries, and Fuel Cells

10.1 INTRODUCTION

Why are fuel cells covered in the same chapter as corrosion? The corrosion of a metal is an electrochemical process, and so is the operation of a fuel cell. Other electrochemical applications discussed in this chapter include batteries, electroplating, electropolishing, and electromachining. Oxidation of metals, the weathering of polymers, and the corrosion of metals all cause degradation of the mechanical properties of materials. Engineers must understand the effects of environment on the materials used in a design. The structure for a high-performance wind electrical generator may be sufficiently strong when it is installed on a platform in the ocean, but will the structure be sufficiently strong five years later?

10.2 THE REACTION OF METALS AND CERAMICS WITH OXYGEN

The structural materials where reaction with oxygen is important include metals as well as covalently bonded materials, such as carbon, boron, silicon, silicon nitride, and silicon carbide. The reaction of metals and ceramics with oxygen limits the high-temperature

use of these materials and constrains design. Metals with high melting temperatures, such as molybdenum and niobium, are structurally strong at high temperatures; however, they readily react with oxygen to form oxides with such poor mechanical properties that they cannot be utilized in high-temperature oxidizing environments. If it were possible, designers of high-temperature gas turbines would increase the operating temperature; however, there are not metals with higher melting temperatures than nickel alloys that are also resistant to reaction with oxygen at elevated temperature. The nickel in gas turbines is alloyed with chromium and aluminum to form protective oxides of these elements.

Materials react with oxygen when the Gibbs free energy of the material oxide is less than that of the material plus oxygen gas. The only metal that has a positive change in the Gibbs free energy when reacting with oxygen is gold. Gold does not oxidize, but all other metals do. Carbon forms a gas of either carbon monoxide or carbon dioxide when it oxidizes. Materials with ionic bonding, such as chlorides and fluorides, do not oxidize, because the Gibbs free energy change for the formation of the oxide is positive. Most oxides are not subject to oxidation, although oxides, such as FeO, can be further oxidized to Fe_2O_3.

Fortunately, materials that react with oxygen do not transform instantly to an oxide. In some metals, the oxide coating forms a protective barrier between the material and oxygen in the environment, as shown in Figure 10.1. For example, aluminum is protected from further oxidation in dry air by a layer of Al_2O_3 that forms on the aluminum surface The Al_2O_3 is protective on the aluminum surface because once the Al_2O_3 layer forms, for more oxide to form, the oxygen and aluminum atoms must come together to react. The layer of Al_2O_3 impedes the reaction of aluminum with oxygen in air. As we will see in the discussion in this section, Al_2O_3 is a barrier to diffusion of oxygen and aluminum ions.

Covalently bonded silicon also forms a protective oxide of silica (SiO_2) when oxidized. Silicon is used to coat other covalently bonded materials that do not form protective oxides, such as carbon and boron. When these materials are coated with silicon, the silicon on the surface oxidizes to form a protective layer of SiO_2 on the surface that prevents further oxidation of the carbon or boron beneath.

The degree of protection an oxide offers to an underlying metal is a function of the rate of diffusion of metal ions or oxygen ions through the oxide and the mechanical integrity of the oxide.

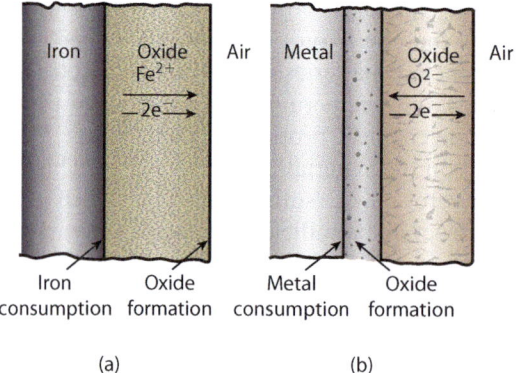

Figure 10.1   (a) The oxidation of a material, such as iron (Fe), where the iron ion (Fe^{2+}) and two electrons ($2e^-$) diffuse through the oxide layer to form oxide at the air-oxide interface. (b) The oxidation of a material where the oxygen ion (O^{2-}) diffuses through the oxide layer to form oxide at the material-oxide interface. Two electrons ($2e^-$) must also diffuse in the opposite direction.

In general, the slower the rate of diffusion of ions, the more protective is the oxide. The relative rates of diffusion of metal ions and oxygen ions determines on which surface the oxide-forming reaction will occur. A schematic of the oxidation process for a metal, such as iron, is shown in Figure 10.1a. In the case of iron, the rate of Fe^{2+} diffusion through Fe_2O_3 is higher than the rate of O^{2-} diffusion through Fe_2O_3, so new oxide forms at the outer surface of the Fe_2O_3. Once a layer of oxide forms on the metal, for further oxidation it is necessary for iron ions (Fe^{2+}) and two electrons ($2e^-$) to diffuse through the oxide surface where they react with oxygen to form new oxide. As the oxide grows in thickness, the rate of oxidation decreases in a protective oxide, because diffusion of the iron ions through the thicker oxide is necessary for further oxide growth. Although iron is shown in Figure 10.1a, there are other metals such as nickel, copper, chromium, and cobalt whose ions are faster diffusers than oxygen ions. In metal oxides, such as tantalum, niobium, hafnium, titanium, and zirconium, it is the oxygen ion that diffuses faster rather than the metal ion, as shown in Figure 10.1b.

For an oxide to be protective of a metal surface, it must be continuous and without porosity. This is most likely to occur if the volume of the oxide formed is approximately equal to the volume of the metal atoms that formed the oxide. If the volume of the oxide is smaller than the volume of the metal consumed, the resulting oxide can have gaps or porosity, and it is not protective. If the oxide has a much larger volume than the metal consumed, compressive stresses in the oxide can result in the oxide breaking off from the metal surface in small chips in a process called spalling.

The ratio of the volume of the oxide formed to the volume of the metal atoms that formed the oxide is the **Pilling-Bedworth ratio** (*PBR*), as shown in Equation 10.1. If the *PBR* is approximately 1, the oxide has a better chance of sufficient mechanical integrity to be protective.

$$PBR = \frac{\text{volume of oxide formed}}{\text{volume of metal consumed}} \qquad \textbf{10.1}$$

Some metals with protective oxides and their *PBR*s include Cu (1.68), Al (1.29), Cr (1.99), Co (1.99), and Pd (1.60).

In general, if the metal has a *PBR* of much less than 1 or more than 2, the oxide is not protective. Some nonprotective metal oxides, followed by their *PBRs*, include Li (0.57), Na (0.57), K (0.45), Mg (0.81), Ta (2.47), Mo (3.40), W (3.40), and V (3.18). Nonprotective oxides are either porous, such as MgO, or they crack, such as Ta_2O_5, or they adhere so poorly to the metal surfaces as to spall off leaving bare metal exposed once again to the environment. The rate of oxidation is not decreased significantly by these oxides with poor mechanical integrity.

Example Problem 10.1

Determine the *PBR* for the oxidation of aluminum to form Al_2O_3 that has a density of 3.97 g/cm³.

Solution

Work this problem on a molar basis. Therefore, 2 moles of Al after oxidation results in 1 mole of Al_2O_3. The mass of 2 moles of Al is equal to twice the molar mass of Al.

$$m_{2Al} = 2 \text{ moles} \left(26.98 \, \frac{g}{\text{mole}}\right) = 53.96 \text{ g}$$

The density of aluminum from Appendix B is equal to 2.71 g/cm³. The volume of 2 moles of aluminum is equal to the mass of 2 moles of Al divided by the density of Al.

$$V_{2Al} = \frac{m_{2Al}}{\rho_{Al}} = \frac{53.96 \text{ g}}{2.71 \text{ g/cm}^3} = 19.91 \text{ cm}^3$$

The mass of 1 mole of Al_2O_3 is equal to

$$m_{Al_2O_3} = 2 \text{ moles of Al} \left(26.98 \frac{\text{g}}{\text{mole}}\right) + 3 \text{ moles of O} \left(16.00 \frac{\text{g}}{\text{mole}}\right)$$

$$m_{Al_2O_3} = 53.96 \text{ g} + 48 \text{ g} = 101.96 \text{ g}$$

The volume of 1 mole of Al_2O_3 is the mass divided by the density.

$$V_{Al_2O_3} = \frac{m_{Al_2O_3}}{\rho_{Al_2O_3}} = \frac{101.96 \text{ g}}{3.97 \text{ g/cm}^3} = 25.68 \text{ cm}^3$$

$$PBR = \frac{V_{Al_2O_3}}{V_{2Al}} = \frac{25.68 \text{ cc}}{19.91 \text{ cc}} = 1.29$$

This *PBR* is more than 1 but not excessively more, and the aluminum oxide layer is protective.

The *PBR* is not the sole determining factor for establishing if an oxide is protective. For example, the *PBR* of iron oxidizing to FeO is 1.7, and based upon the results for other metals, we would expect that FeO is a protective oxide. Rust, or ferric hydroxide $Fe(OH)_3$, forms when iron oxidizes in the presence of water, and this is not protective on iron. Other factors that are important in protective oxide formation on metals are that the oxide must have strong bonding to the metal, resistance to brittle fracture at the operating temperature, low electrical conductivity to impede electron transport, and a thermal expansion coefficient similar to that of the base metal. Note that all of these properties impact the rate of transport through the oxide either through pores, cracks, or spallation, or by diffusion through an intact oxide layer.

As a material oxidizes, it gains oxygen atoms, the weight increases, and the oxide thickness increases. The growth of the oxide is reported either as the oxide thickness or as the directly related weight gain per unit area. Experimentally, it is observed that protective oxides follow either parabolic or logarithmic growth laws where the oxide weight gain per unit area (w) is given by Equation 10.2 for parabolic growth as a function of time (t):

$$w = A_1 t^{1/2} \qquad\qquad \textbf{10.2}$$

or by Equation 10.3 for logarithmic growth.

$$w = A_2 (\ln A_3 t + A_4) \qquad\qquad \textbf{10.3}$$

The A_i in Equations 10.2, 10.3, and 10.4 are temperature-dependent constants. Nonprotective oxides follow a linear growth law, as shown in Equation 10.4.

$$w = A_5 t \qquad\qquad \textbf{10.4}$$

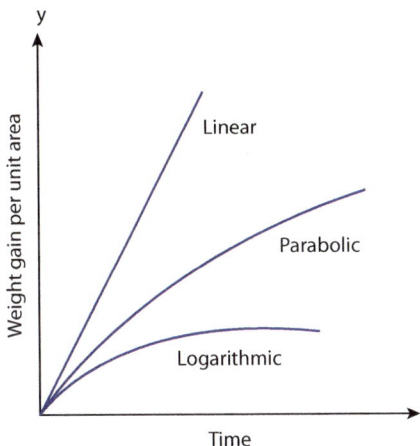

Figure 10.2   The weight gain per unit area as a function of time for materials with linear, parabolic, and logarithmic oxide growth dependence.

Figure 10.2 is a schematic of the weight gain per unit area versus the time for Equations 10.2 through 10.4. As the material oxidizes, it gains oxygen atoms, and the weight increases.

Example Problem 10.2

According to Deal and Grove (Deal, B., and Grove, A., *J. Appl. Phys.* 36 (1965), 3770), the oxidation of Si in dry oxygen at 1 atmosphere of pressure follows the relation shown in Equation 10.5, where h is the oxide thickness.

$$h^2 + Ah = B(t + t_0) \qquad \textbf{10.5}$$

At 1000°C, $A = 0.165 \times 10^{-6}$ m, $B = 3.25 \times 10^{-18}$ m^2/s, and $t_0 = 1224$ s. Calculate the time necessary to form a 100-nm-thick oxide film on Si in dry oxygen at 1000°C.

Solution

$$h = 100 \text{ nm} = 100 \times 10^{-9} \text{m} = 1.00 \times 10^{-7} \text{m}$$

Substituting this thickness into Equation 10.5 results in

$$1.00 \times 10^{-14} \text{m}^2 + (0.165 \times 10^{-6} \text{m})\,(1.00 \times 10^{-7} \text{m}) = 3.25 \times 10^{-18} \text{m}^2/\text{s}\,(t + 1224 \text{ s})$$

$$1.00 \times 10^{-14} \text{m}^2 + 1.65 \times 10^{-14} \text{m}^2 = t(3.25 \times 10^{-18} \text{m}^2/\text{s}) + 0.40 \times 10^{-14} \text{m}^2$$

$$2.25 \times 10^{-14} \text{m}^2 = t(3.25 \times 10^{-18} \text{m}^2/\text{s})$$

$$t = \frac{2.25 \times 10^{-14} \text{m}^2}{3.25 \times 10^{-18} \text{m}^2/\text{s}} = 0.69 \times 10^4 \text{ s} = 1.91 \text{ h}$$

Note: A. Grove was a president and CEO of Intel Corporation. He did original work on the oxidation of Si and the diffusion of impurities into Si.

Metals, such as iron, that do not form protective oxides can be made more resistant to oxidation by alloying. Iron is alloyed with chromium to form a protective chromium oxide on stainless steel. In superalloys, nickel is alloyed with both aluminum and chromium to form protective oxides.

Another approach to protecting against oxidation in many engineering applications is the use of coatings. The paint on an automobile is a coating that protects steel from the formation of rust. A chromium coating is deposited on steel used in automobiles because the chromium forms a protective coating. Titanium-aluminum coatings on titanium increase the service temperature for titanium gas-turbine blades. Coatings of partially stabilized zirconia are applied to superalloys to increase the operating temperature for turbine blades in high-temperature gas turbines. The carbon-carbon fiber-composite tiles on the space shuttle are coated with silicon carbide to protect the carbon from oxidation during reentry into the atmosphere. The silicon in the silicon carbide reacts with oxygen to form a layer of SiO_2 that protects the carbon-carbon tiles.

10.3   DEGRADATION OF POLYMERS

10.3.1   Polymer Solvent Absorption and Permeation

Section 4.8 presented the theory of solubility of solvents in polymers and the permeation of atoms and molecules into polymers. This paragraph is a brief summary of the important conclusions from Section 4.8 that relate to the degradation of polymers. The permeation of molecules or atoms is the penetration of these into polymers, and it depends upon both the solubility of the molecule or atom into the polymer and the diffusivity of the molecule or atom into the polymer. If the molecule or atom is soluble in the polymer and the diffusivity is high, then the molecule or atom permeates through the polymer. However, if the diffusivity and solubility of a molecule or atom is low, the polymer is not permeated. Large solvent molecules tend to have low diffusivity in polymers and do not permeate; smaller molecules and atoms have higher diffusion rates and do permeate. Solvents that have a similar solubility parameter as a polymer are likely to be absorbed into the polymer. If the bonding between LCMs is polar, then polar molecules are likely to be absorbed into the polymer. For example, nylon has permanent dipoles on the LCM, and water has permanent dipoles on its molecule. Nylon absorbs water, resulting in swelling and softening. PE and PTFE are not polar molecules, and they do not absorb water. PTFE is one of the most resistant polymers to chemical attack by polar molecules. If the molecular structure of the solvent and LCM are similar, the solvent is likely to be adsorbed into the polymer. PS has a benzene ring replacing a hydrogen atom on the LCM, and benzene is soluble in PS.

If atoms or molecules are soluble in a polymer and permeate through the polymer, they can degrade its mechanical properties. Figure 10.3 shows the effect of three liquids on the time to rupture in PVC at different stresses. Tubes of PVC were filled with the liquids indicated under pressure. The pressure produces hoop stresses around the circumference of the tube. The term *hoop stress* comes from the steel barrel hoops that hold wooden barrels together and resist the internal pressure in the barrel. Isopropanol decreases the stress for a 1000-hour rupture life to approximately 13.8 MPa, whereas for water, the stress is nearly 34.5 MPa. Figure 10.4 shows the strain to fracture of polyphenylene oxide in different solvents as function of the solubility parameter. If the solubility parameter of the solvent

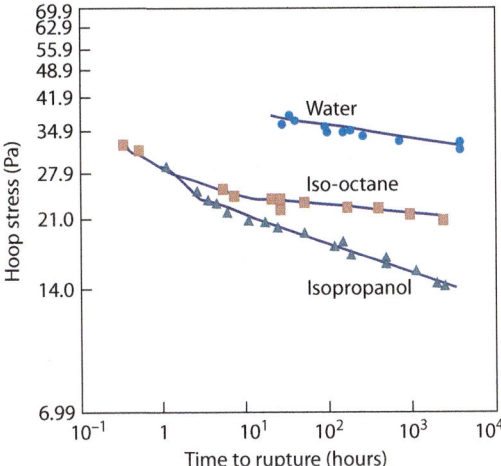

Figure 10.3   The effect of different liquids on the stress rupture life as a function of hoop stress for PVC pipe at 23°C. (*Based on Riddell, M.N., Plast. Eng., 40 (4) (1974), p. 71.*)

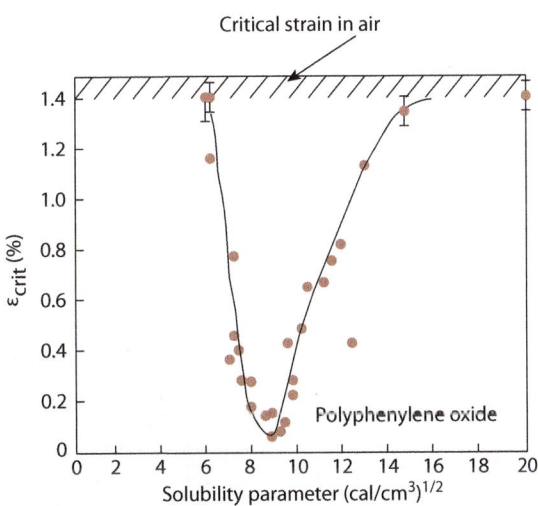

Figure 10.4   The strain to fracture for the polymer polyphenylene oxide in different solvents, as a function of the solvent solubility parameter. (*Based on Osswald, T.A., and Menges G., Materials Science of Polymers for Engineers, Hanser Publishers, Munich (2003), p. 513.*)

is within ± 1 (cal/cm^3)$^{0.5}$, which is approximately equal to ± 2 (MPa)$^{0.5}$, of the solubility parameter of the polymer, mixing of the solvent and polymer is expected, and the strain to fracture of the polymer is significantly reduced. The solubility parameter of polyphenylene oxide is 8.65 (cal/cm^3)$^{0.5}$, or 19.3 (MPa)$^{0.5}$, and the strain to fracture is reduced by an order of magnitude when the solubility parameter of the solvent has this same value.

10.3.2   Degradation of Polymers by Irradiation

When UV light is absorbed in a polymer, it breaks bonds between LCMs and along LCMs, resulting in a degradation of the material's mechanical properties. Photons of ultraviolet (UV) light are bundles of electromagnetic radiation with energy from approximately 1 eV to 1000 eV. Chemical-bonding energies are typically between 2 and 10 eV per atom pair. If the photons of UV light have a higher energy than the bond energy, they can break the chemical bonds in a polymer. If the bonds of the LCMs are broken, the molecular mass of the polymer is reduced and physical properties change. Also, it is observed that if a polymer is irradiated with UV light in the presence of oxygen, a chemical reaction can occur between the polymers and oxygen.

Weathering of a polymer is the combination of irradiation with light, including UV, and the reaction with atoms and molecules in the environment, such as oxygen, nitrogen, and water. Weathering causes cracks to form in polymers, as shown in Figure 10.5, reducing their resistance to fracture. Figure 10.5 shows a section of a U channel made from HDPE that was exposed to an ocean-air environment for approximately 15 years. Figure 10.6 shows the relative impact strength as a function of time for PVC subject to weathering in various geographic locations. After five years the impact strength is reduced to 5% of the original value in Singapore, whereas in Germany the impact strength is 90% of the original value. What can be concluded from these results is that the warm temperatures, high humidity, and high amounts of sunlight in Singapore degrade the fracture resistance of polymers more than in Germany, where there are cooler temperatures, lower humidity, and less sunlight. We will discuss impact strength in Chapter 11 on fatigue and fracture.

Figure 10.5   A photograph of weathering of a U channel made of HDPE that had been exposed to ocean air for approximately 15 years. The dark lines are cracks. The bottom of the U channel runs horizontally across the center of the photograph. (*Photo by C. M. Gilmore*)

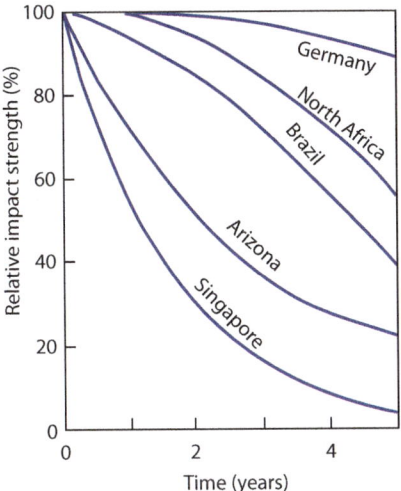

Figure 10.6 The relative impact strength as a function of time for PVC subject to weathering, in various geographic locations. (*Based on Osswald, T.A., and Menges G., Materials Science of Polymers for Engineers, Hanser Publishers, Munich (2003), p. 508.*)

10.4 ELECTROCHEMISTRY FUNDAMENTALS

An **electrochemical reaction** is one that occurs between an electrode of an electron conductor (a metal or semiconductor) and an ion conductor (electrolyte) that involves electron transfer between an electrode and the electrolyte. Electrochemical reactions occur during the corrosion of metals, during electroplating, electropolishing, and electromachining, in batteries, and in fuel cells. An **electrolyte** is any material that conducts ions. There are liquid electrolytes, such as acids and bases; there are solid electrolytes with ionic bonding, such as zirconia; and there are polymer electrolytes. We will discuss ceramic and polymer electrolytes in fuel cells in Section 10.4.12.

The effects of an electrochemical reaction are observed if we use iron screws to attach a copper sheet over a hole in a boat that is used in the ocean. After a few years, the boat sinks, because the iron (Fe) screws dissolve in saltwater, allowing the copper (Cu) sheet to fall off the boat. After we retrieve the copper sheet, inspection reveals that it was unaffected by the saltwater, even though the iron screws have dissolved.

We can study our boat-patch failure in the laboratory with the experiment shown in Figure 10.7, where iron and copper are immersed in a saltwater solution and connected electrically through a voltmeter. You can conduct this experiment in your chemistry or materials science lab. Mix some salt and water inside a nonmetallic container. Partially immerse a piece of copper and a piece of iron in the saltwater, leaving part of each piece above the water line. Connect the exposed ends of the iron and copper to a voltmeter, and you will read a few tenths of a volt (V), with the iron charged positively and the copper charged negatively. If the iron and copper are in electrical contact in the saltwater, the iron electrode dissolves according to the reaction given in Equation 10.6, and the copper electrode does not dissolve.

$$Fe \rightarrow Fe^{2+} + 2e^-$$

10.6

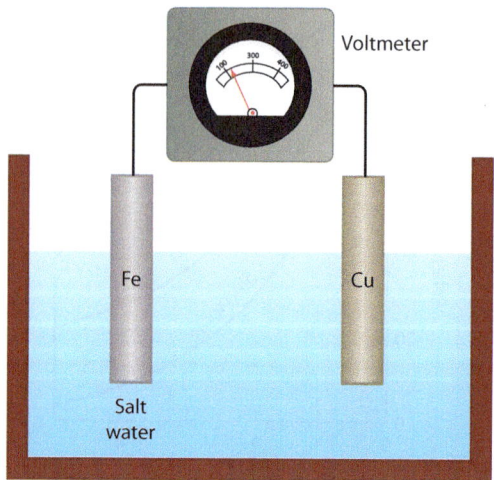

Figure 10.7 A piece of iron and a piece of copper are partially submerged in saltwater and connected to a voltmeter. A voltage is measured, and the iron slowly dissolves in the saltwater.

When an atom loses electrons, as the iron does in Equation 10.6, this is called an **oxidation reaction**. Oxidation occurs at the **anode** of an electrochemical cell. Oxidation does not necessarily involve oxygen.

10.4.1 Standard Electrode Potentials

The experiment to measure the **standard electrode potentials**, or voltages, of materials, such as iron, is shown schematically in Figure 10.8, and the standard electrode potentials (V_0) are presented in Table 10.1. The value of the standard electrode potential of the iron on the left of Figure 10.8 is measured in a

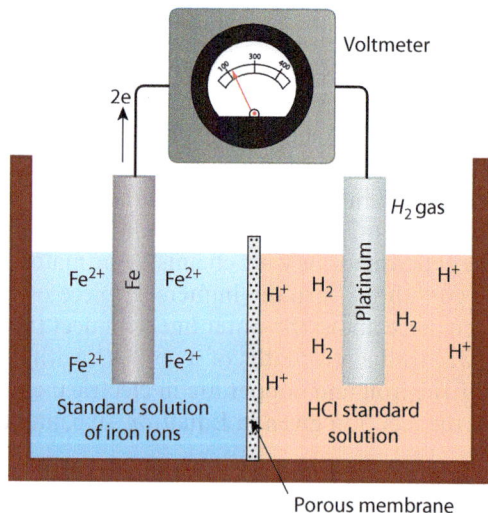

Figure 10.8 A schematic of an electrochemical cell used to measure standard electrode potentials using iron as an example.

Table 10.1	**The Standard-Electrode Potentials (V_0) for Some Reduction Reactions at 25°C**	
	Reaction	Electrode Potential V_0 (Volts)
Anodic ↑	$Li^+ + e^- \rightarrow Li$	−3.05
	$Mg^{2+} + 2e^- \rightarrow Mg$	−2.37
	$Al^{3+} + 3e^- \rightarrow Al$	−1.66
	$Ti^{2+} + 2e^- \rightarrow Ti$	−1.63
	$Mn^{2+} + 2e^- \rightarrow Mn$	−1.63
	$Zn^{2+} + 2e^- \rightarrow Zn$	−0.76
	$Cr^{3+} + 3e^- \rightarrow Cr$	−0.74
	$Fe^{2+} + 2e^- \rightarrow Fe$	−0.44
	$Ni^{2+} + 2e^- \rightarrow Ni$	−0.25
	$Sn^{2+} + 2e^- \rightarrow Sn$	−0.14
	$Pb^{2+} + 2e^- \rightarrow Pb$	−0.13
	$2H^+ + 2e^- \rightarrow H_2$	0.00
	$Cu^{2+} + 2e^- \rightarrow Cu$	+0.34
	$O_2 + 2H_2O + 4e^- \rightarrow 4(OH)^-$	+0.40
	$Ag^+ + e^- \rightarrow Ag$	+0.80
	$Pt^{4+} + 4e^- \rightarrow Pt$	+1.20
	$O_2 + 4H^+ + 4e^- \rightarrow 2H_2O$	+1.23
Cathodic ↓	$Au^{3+} + 3e^- \rightarrow Au$	+1.50

Based on data from Askeland, D.R., Fulay, P.P., and Wright, W.J., The Science and Engineering of Materials, 6th ed., Cengage Learning, Stamford, CT. (2011), p. 858.

1-molar solution, which is 1 mole of its own ions in 1×10^{-3} cubic meters (1000 cm³) of solution. The iron ion concentration results from dissolving a material such as iron sulfate in water. The standard reference electrode on the right is a hydrogen electrode, not a platinum electrode. The **reference electrode** is a 1-molar solution of hydrogen ions (H^+) in HCl in contact with hydrogen gas at 1 atmosphere of pressure. The platinum in the reference electrode does not participate in the reaction. Platinum is inert to the reaction; it serves as a surface where the hydrogen gas bubbles nucleate.

If the metal in the electrode on the left is iron, the iron goes into solution as Fe^{2+}. The two electrons in Equation 10.6 are conducted through the system to the reference electrode, where they combine with H^+ ions in solution to form hydrogen gas at the reference electrode, according to the reaction given in Equation 10.7.

$$2H^+ + 2e^- \rightarrow H_2 \qquad \textbf{10.7}$$

A **reduction reaction** occurs when an ion gains electrons, as shown in Equation 10.7. Reduction reactions occur at the **cathode**.

The solutions in the cell are separated by a semi-permeable membrane that allows migration ions between the two solutions. The current due to electron flow must be balanced by a flow of ions between the two cells through the porous membrane. Otherwise the solutions would become charged and the reaction would soon stop.

Each side of the cell is considered a half cell. By convention, the standard electrode potentials in Table 10.1 are tabulated for the reduction reactions that proceed in the direction indicated by the arrow.

Also, by convention the hydrogen reduction reaction is taken as 0 V, because this is the reference electrode. All voltages are measured relative to a reference electrode. Oxidation reactions are the reverse reactions to those listed in Table 10.1. The oxidation reaction of $Fe \rightarrow 2e^- + Fe^{2+}$ results in a half-cell potential of $+0.44$ V. Note that the half-cell potential for this oxidation reaction is of the opposite sign of the half-cell potential for the reduction reaction because the oxidation is the reverse of the reduction reaction listed in Table 10.1. This voltage is called the **half-cell potential** because the voltage generated on this electrode is the voltage generated by one-half of the complete electrochemical cell. The potential of the full electrochemical cell is the sum of the voltage resulting from the half-cell reaction taking place on the anode (oxidation) and the half-cell reaction taking place on the cathode (reduction). In the case of our example of an electrochemical cell consisting of an iron electrode and a standard hydrogen electrode, the full cell potential is $+0.44$ V, which is the sum of the voltage generated by half-cell reaction on the anode ($+0.44$ V) and the voltage generated by the half-cell reaction on the cathode (0 V). Materials with a more positive standard electrode potential for the reduction reaction are less reactive, or more noble, and materials with a more negative standard electrode potential are more reactive.

The reduction reaction for copper, shown in Equation 10.8, has a more positive standard electrode potential of $+0.34$ V relative to 0 V for the reduction reaction hydrogen.

$$Cu^{2+} + 2e^- \rightarrow Cu \qquad\qquad \textbf{10.8}$$

If copper is substituted for iron in the electrochemical cell shown in Figure 10.8 in an electrolyte of a 1-molar solution of Cu^{2+} ions, the Cu^{2+} ions are reduced and electrodeposited at the copper electrode, and hydrogen is oxidized at the standard electrode according to the reaction $H_2 \rightarrow 2e^- + 2H^+$. We will discuss electrodeposition in more detail in Section 10.4.10.

Figure 10.9 shows a cell with an iron electrode in a molar solution of Fe^{2+} ions and a copper electrode in a molar solution of Cu^{2+} ions. In this cell, as shown in Table 10.1, copper is the more

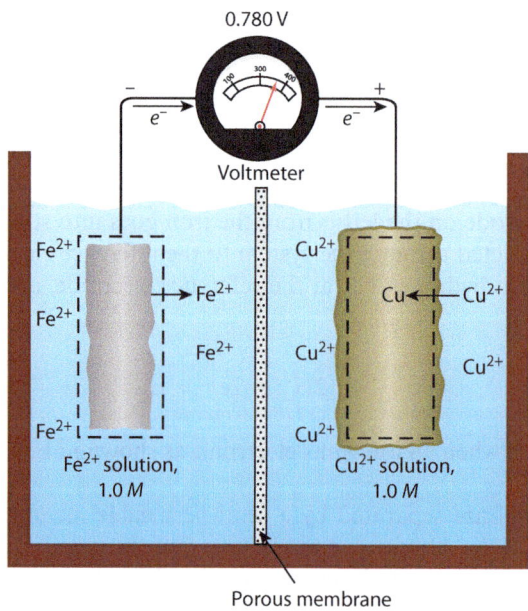

Figure 10.9   A schematic of Fe and Cu in an electrochemical cell in a 1-molar solution of their own ions. The dashed line is the original size of the specimen.

noble electrode, with a more electropositive standard electrode potential of +0.34 V, and iron is more electronegative, with a standard electrode potential of −0.44 V. Therefore, copper is reduced or deposited on the electrode according to the reduction reaction of Equation 10.8, with the generation of +0.34 V, as shown schematically in Figure 10.9; and iron is oxidized or dissolved according to the reaction in Equation 10.6, with the generation of −(−0.44 V) = +0.44 V. The cell develops a voltage of 0.34 V + 0.44 V = 0.78 V. The cell voltage (V) is the sum of the voltage of the reduction reaction and the voltage of the oxidation reaction, as shown in Equation 10.9.

$$V = V_{red} + V_{oxid} \qquad \textbf{10.9}$$

Now we can see why the iron screws dissolved on our boat. The iron is more reactive than the copper, and the iron went into solution when in electrical contact with copper in the electrolyte of salt water.

10.4.2 Standard Electrode Potentials and Gibbs Free Energy

The voltage developed in an electrochemical cell is directly related to the Gibbs free energy change of the reaction. The Gibbs free energy is the energy that is free or available to do work at constant temperature and pressure. The standard-electrode potentials are determined under equilibrium conditions at room temperature (25°C) and 1 atmosphere of pressure. For the oxidation of iron, according to Equation 10.6, two electrons go from the iron electrode to the reference electrode. The work (ΔW) that is required to take these two electrons from the reference electrode in Figure 10.8 back to the iron electrode over the standard electrode potential of V_0 volts is calculated in Equation 10.10.

$$\Delta W = 2eV_0 \qquad \textbf{10.10}$$

The change in the Gibbs free energy per ion for the oxidation reaction at the iron electrode (ΔG) is the negative of ΔW, because it is the change in the Gibbs free energy (ΔG) that causes the creation of the voltage V_0 and the transport of the two electrons to the reference electrode, as shown in Equation 10.11. In Appendix 5A it is shown that the change in Gibbs free energy is equal to the energy available to do work at constant temperature and pressure.

$$\Delta G = -\Delta W = -2eV_0 \qquad \textbf{10.11}$$

We can make this equation more general by replacing the 2 with the charge of an ion (Z). If ΔG is calculated on a per-mole basis, the change in the Gibbs free energy is written as Equation 10.12.

$$\Delta G = -ZFV \qquad \textbf{10.12}$$

The Faraday constant (F) is the charge in coulombs on a mole of positive ions of unit charge, and F is equal to the magnitude of the charge of an electron times Avogadro's number: 96,488 coulombs (C) per mole of ions.

Example Problem 10.3

Calculate the Gibbs free energy change when a mole of iron dissolves into a 1-molar solution of Fe^{2+} ions in a standard electrochemical cell at 25°C.

Solution

The oxidation reaction under consideration is $Fe \rightarrow Fe^{2+} + 2e^-$. From Table 10.1, the potential for this reaction is +0.44 V. This is the negative of the value in Table 10.1, because the oxidation reaction $Fe \rightarrow Fe^{2+} + 2e^-$ is the reverse of the reduction reaction in Table 10.1. The reaction of $2H^+ + 2e^- \rightarrow H_2$ at the reference electrode produces no voltage, because the standard-electrode potential for this reaction is 0. From Equation 10.12, the change in the Gibbs free energy is

$$\Delta G = -ZFV = -2\frac{\text{electrons}}{\text{ion}}\left(9.65 \times 10^4 \frac{C}{\text{mole of electrons}}\right)(0.44 \text{ V})$$

$$\Delta G = -8.49 \times 10^4 \frac{J}{\text{mole of ions}}$$

The reaction of $Fe \rightarrow Fe^{2+} + 2e^-$ occurs spontaneously, because the Gibbs free energy is decreased by this reaction. It follows therefore that for a spontaneous reaction, the cell voltage is positive. If the calculated voltage is negative, the wrong assumption was made for the reactions at the anode and cathode. A negative voltage should not occur because this would result in an increase in Gibbs free energy, as demonstrated by Equation 10.12.

Example Problem 10.4

Copper is the metal electrode in a molar solution of Cu^{2+} ions in a standard cell at 25°C, as shown in Figure 10.8. (a) What reactions occur at the electrodes? (b) What voltage is developed in the cell? (c) What is the Gibbs free energy change per copper ion involved in the reaction?

Solution

a) Since the copper reduction reaction $Cu^{2+} + 2e^- \rightarrow Cu$ is more electropositive than the hydrogen reduction reaction $2H^+ + 2e^- \rightarrow H_2$, the copper does not go into solution. Copper deposits on the copper electrode according to the reaction $Cu^{2+} + 2e^- \rightarrow Cu$, and at the reference electrode the oxidation reaction is $H_2 \rightarrow 2H^+ + 2e^-$. Hydrogen ions are produced from hydrogen gas.

b) The voltage developed is the voltage for the reduction reaction $Cu^{2+} + 2e^- \rightarrow Cu$, or +0.34 V. The hydrogen oxidation reaction contributes 0 voltage to the cell voltage.

c) The Gibbs free energy change is given by Equation 10.12.

$$\Delta G = -ZFV$$

$$\Delta G = -ZFV = -2\frac{\text{electrons}}{\text{ion}}\left(9.65 \times 10^4 \frac{C}{\text{mole of electrons}}\right)(0.34 \text{ V})$$

$$\Delta G = -6.56 \times 10^4 \frac{J}{\text{mole of ions}}$$

This change in the Gibbs free energy is negative; therefore, this reaction is spontaneous.

10.4.3 Electrode Potentials and the Nernst Equation

All of the analysis above is for the ions in a standard 1-molar solution at 25°C. If the solutions are not 1-molar, then the half-cell voltage (V) for the reduction reaction at an electrode is calculated from the standard voltage with 1-molar solutions (V_0) with the **Nernst equation**, shown in Equation 10.13:

$$V = V_0 + \frac{RT}{ZF} \ln C \qquad \textbf{10.13}$$

where R is the universal gas constant (8.31 J/mole · K); T is the absolute temperature of the cell, in kelvin; F is the Faraday constant; and C is the relative concentration of ions at the electrode. Since the standard potentials are measured at 25°C, we assume this temperature for all problems in this chapter. Inserting all of the constants into Equation 10.13 results in Equation 10.14.

$$V = V_0 + \frac{0.0257}{Z} \ln C = V_0 + \frac{0.0592}{Z} \log C \qquad \textbf{10.14}$$

Because V_0 is by convention tabulated for the reduction reaction, Equations 10.13 and 10.14 are applied to the voltage of the reduction reaction. If the material is actually oxidizing, first apply these equations to the reduction reaction, then for the oxidation reaction the sign for the adjusted voltage is the negative of that calculated for the reduction reaction.

Example Problem 10.5

Copper is the metal electrode in a solution with 31.77 g of Cu^{2+} ions in 1×10^{-3} cubic meters of solution, and the other electrode is a 1-molar standard hydrogen electrode at 25°C.
(a) What are the reduction-reaction voltages?
(b) What reactions occur at each of the electrodes?
(c) What voltage is developed in the cell?

Solution

a) We first calculate the reduction reaction voltages, and the electrode with the more positive potential is reduced and the electrode with the more negative potential is oxidized. The copper reduction reaction is $Cu^{2+} + 2e^- \rightarrow Cu$. In a 1-molar solution of Cu^{2+}, there are 63.54 g of Cu^{2+}, which comes from the molar mass of copper. If there are 31.77 g of Cu^{2+} in solution, then the relative concentration of Cu^{2+} at the copper electrode is

$$C_{Cu} = \frac{31.77 \text{ g}}{63.54 \text{ g}} = 0.5$$

The concentration Cu^{2+} at the copper electrode is a 0.5-molar solution, and at the hydrogen electrode the concentration of H^+ is a 1-molar solution. The value of the half-cell voltage at the copper electrode with 1-molar solutions (V_0) is equal to +0.340 V. The value of the half-cell voltage for the reduction reaction at the copper electrode in a 0.5-molar solution is obtained from Equation 10.14. Inserting values into Equation 10.14,

$$V_{Cu} = V_0(Cu) + \frac{0.0592 \text{ V}}{Z} \log C = 0.34 \text{ V} + \frac{0.0592 \text{ V}}{2} \log 0.5$$

$$V_{Cu} = 0.34 \text{ V} + 0.0296 (-0.30) \text{ V} = 0.34 \text{ V} - 0.01 \text{ V} = 0.33 \text{ V}$$

The hydrogen electrode is 1-molar, and its half-cell voltage is 0.0.

b) Because the half-cell potential at the copper electrode is more positive than the half-cell potential at the hydrogen electrode, the reaction at the copper electrode is reduction:

$$Cu^{2+} + 2e^- \rightarrow Cu$$

The reaction at the hydrogen electrode is oxidation:

$$H_2 \rightarrow 2H^+ + 2e^-$$

Therefore, the copper electrode is the cathode and the hydrogen electrode is the anode.

c) The total voltage measured is the sum of the half-cell voltages for the reduction and oxidation reaction.

$$V = V_{red} + V_{oxid} = 0.33 \text{ V} + 0.0 \text{ V} = 0.33 \text{ V}$$

If the concentration of hydrogen ions is not 1-molar, the half-cell potential for the reduction reaction at the hydrogen electrode is calculated with Equation 10.14. If the temperature is not 25°C, then Equation 10.13 is utilized with electrode potentials determined at the specified temperature. If the second electrode is not a hydrogen electrode, and the solution is not standard, then the voltage from the second electrode is determined with Equation 10.14.

Example Problem 10.6

A copper metal electrode is in a solution of 12.7 g of Cu^{2+} ions in 1×10^{-3} m³ of solution, and the other electrode is an iron electrode in a solution of 67 g of Fe^{2+} ions in 1×10^{-3} m³ of solution at 25°C.
(a) What are the reduction reaction voltages?
(b) What reactions occur at the two electrodes?
(c) What voltage is developed in the cell?

Solution

a) In a 1-molar solution of Cu^{2+} ions, there are 63.54 g of Cu^{2+} ions. If there are 12.7 g of Cu^{2+} ions in solution, then the molar concentration at the copper electrode is

$$C_{Cu} = \frac{12.7 \text{ g}}{63.54 \text{ g}} = 0.2$$

The concentration of Cu^{2+} ions at the copper electrode is a 0.2-molar solution. The value of the half-cell voltage at the copper electrode with a 1-molar solution (V_0) is equal to +0.340 V. The half-cell voltage from the reduction reaction at the copper electrode in a 0.2-molar solution is calculated from Equation 10.14.

$$V_{Cu} = V_0(Cu) + \frac{0.0592}{Z} \log C = 0.34 \text{ V} + \frac{0.0592 \text{ V}}{2} \log 0.2$$

$$V_{Cu} = 0.34 \text{ V} + 0.0296(-0.70) \text{ V} = 0.34 \text{ V} - 0.02 \text{ V} = 0.32 \text{ V}$$

In a 1-molar solution of Fe^{2+} ions, there are 55.847 g of iron. If there are 67 g of Fe^{2+} ions in solution, then the molar concentration of Fe^{2+} ions at the iron electrode is

$$C_{Fe} = \frac{67 \text{ g}}{55.85 \text{ g}} = 1.2$$

At the iron electrode, the concentration is a 1.2-molar solution. The value of the half-cell voltage for the reduction reaction of $Fe^{2+} + 2e^- \rightarrow Fe$ at the iron electrode with a 1-molar solution (V_0) is equal to -0.440 V. The value of the half-cell voltage of the reduction reaction for an iron electrode in a 1.2-molar solution is obtained from Equation 10.14.

$$V_{Fe} = V_0(Fe) + \frac{0.0592 \text{ V}}{Z} \log C = -0.44 \text{ V} + \frac{0.0592 \text{ V}}{2} \log 1.2$$

$$V_{Fe} = -0.44 \text{ V} + 0.0296\,(0.08) \text{ V} = -0.44 \text{ V} + 0.002 \text{ V} \approx -0.44 \text{ V}$$

b) The copper reduction reaction $Cu^{2+} + 2e^- \rightarrow Cu$ is more electropositive than the iron reduction reaction of $Fe^{2+} + 2e^- \rightarrow Fe$. Copper deposits on the copper electrode (cathode) according to the reduction reaction $Cu^{2+} + 2e^- \rightarrow Cu$, and at the anode the oxidation reaction is $Fe \rightarrow Fe^{2+} + 2e^-$.

c) The total voltage measured is the sum of the voltage from the two half-cells with reduction and oxidation reactions. The iron is oxidized in this cell with copper, and its half-cell voltage is the negative of the reduction-cell voltage.

$$V = V_{red} + V_{oxid} = V_{Cu} + V_{Fe} = 0.32 \text{ V} + 0.44 \text{ V} = 0.76 \text{ V}$$

10.4.4 Rates of Electrode Change Due to Electrochemical Processes

Figure 10.9 shows iron being removed from the anode and copper being deposited at the cathode of an electrochemical cell. In an electrochemical cell, the mass (m_{ch}) of material removed from the anode or deposited at the cathode is obtained from the cell current (I). In the example of the copper and iron electrochemical cell, shown in Figure 10.9, for every two electrons conducted through the circuit, one iron atom goes into solution at the anode and one copper atom is deposited on the cathode. The current (I) in amperes is the number of coulombs per second through the circuit, and the current times the time (t) is the total number of coulombs. The mass gain or loss at either electrode is calculated from Equations 10.15a and 10.15b, with dimensions inserted into Equation 10.15a.

$$\frac{m_{ch}\,\text{g}}{M \text{ g/mole}}\left(Z\,\frac{\text{mole of electrons}}{\text{mole of ions}}\right) F\frac{\text{C}}{\text{mole of electrons}} = I\frac{\text{C}}{\text{s}} t \text{ s} \qquad \textbf{10.15a}$$

$$\frac{m_{ch}}{M} ZF = It = iAt \qquad \textbf{10.15b}$$

M is the molar mass, Z is the charge on the ions at the electrode of interest, and F is the Faraday constant. The current (I) through the circuit is due to a current density (i) at the electrode, in amperes per square meter of electrode, times the area (A) of the electrode. Equation 10.15 applies to any electrochemical process.

The rate of thinning or of thickening of a piece of material can be predicted from the mass change if the change in dimension is uniform. If the change in dimension is expressed in mills (0.001 inches) per year (mpy) the result is Equation 10.16.

$$mpy = \frac{534 m_{ch}}{\rho A t} \qquad \textbf{10.16}$$

The mass loss or gain Equation 10.16 is in milligrams, ρ is the density of the material removed or added in grams per cubic centimeter, A is the area of the specimen in square inches, and t is the exposure time in hours. Equation 10.17 gives a metric version of Equation 10.16, where mm is millimeters, A is the area of the specimen in square centimeters, and the remaining units are the same as for Equation 10.16.

$$\frac{mm}{y} = \frac{87.6m_{ch}}{\rho At}$$

10.17

Equations 10.16 and 10.17 apply to processes such as uniform corrosion, electroplating, electropolishing, and uniform galvanic corrosion.

10.5   CORROSION

Corrosion is the degradation of a material because of electrochemical reactions with its environment. With corrosion, we consider only the types of material degradation that involve electrolytes, the creation of ions, and electron transfer. The annual cost in the United States resulting from corrosion was estimated in 1998 to be $276,000,000,000 (276 billion dollars) per year. Corrosion is something that must be considered in many engineering designs. There are several different types of corrosion, including uniform corrosion, galvanic corrosion, crevice corrosion, pitting corrosion, intergranular corrosion, selective leaching, erosion corrosion, and stress corrosion. We will consider stress-corrosion cracking in Chapter 11, after discussing the concepts of fracture.

Corrosion occurs primarily in metals because the conduction of electrons is required in the corrosion process. Ceramics have better resistance to corrosion than metals. Ceramics, such as MgO, can dissolve in an electrolyte, such as seawater; however this is not considered to be corrosion because it is not an electrochemical reaction. Although polymers can certainly degrade as a result of exposure to an electrolyte, such as seawater, the term *corrosion* is not typically used to describe the environmental degradation of polymers.

10.5.1   Uniform Corrosion

Uniform corrosion is a form of corrosion that occurs uniformly over a material that is not in contact with any other material. Silverware is uniformly corroded when it is tarnished, or an iron pipe in saltwater is uniformly rusted. Uniform corrosion is due to the material reacting with the environment. The rate of uniform corrosion can normally be predicted by Equations 10.15, 10.16, and 10.17. If tests of the material are conducted in the appropriate environment, engineers can design for this type of corrosion and select materials that provide the desired lifetime.

The rusting of iron that is not in contact with any other metal, such as our dog Rusty in Figure 10.10, is an example of uniform corrosion. Iron rusts in the presence of water and oxygen according to the reaction given in Equation 10.18.

$$2Fe + \frac{3}{2}O_2 + 3H_2O \rightarrow 2Fe(OH)_3$$

10.18

The compound $Fe(OH)_3$ is rust. The rusting of iron is a form of uniform corrosion where the rust does not dissolve; however, rust is structurally useless.

Figure 10.10   A photograph of our dog Rusty who is uniformly corroded. He is made of low-carbon steel that was originally the color of the shiny uncoated steel nails that you see at the hardware store. (*Photo by C. M. Gilmore.*)

Example Problem 10.7

Pure copper uniformly corrodes at the rate of 10 g/m^2 per day in water with 3 weight percent salt. Calculate the current density at the copper surface. Assume that the copper oxidation reaction is $Cu \rightarrow Cu^{2+} + 2e^-$.

Solution

The current density is a result of the two electrons transferred from the copper atom as it goes into solution as an ion. Since current density is requested, Equation 10.15b applies.

$$\frac{m_{ch}}{M} ZF = It = iAt$$

The known values are $Z = 2$, $F = 9.65 \times 10^4$ C/mole, $M_{Cu} = 63.55$ g/mole, $A = 1$ m^2.
The time is the number of seconds in a day.

$$t = 1\,d = \frac{24\,h}{1\,d}\left(\frac{3600\,s}{1\,h}\right) = 8.64 \times 10^4\,s$$

In 8.46×10^4 s, the mass loss is 10 g. Inserting these values into Equation 10.15b,

$$\frac{m_{ch}}{M} = iAt = \frac{10\,g}{63.55\,g/mole\,of\,Cu}\left(2\,\frac{moles\,of\,electrons}{mole\,of\,Cu}\right)9.65 \times 10^4\,\frac{C}{mole\,of\,electrons}$$

$$iAt = 3.04 \times 10^4\,C = i(1\,m^2)(8.64 \times 10^4\,s)$$

$$i = \frac{3.04 \times 10^4\,C}{(1\,m^2)(8.64 \times 10^4\,s)} = 0.35\frac{C}{m^2 \cdot s} = 0.35\frac{A}{m^2}$$

10.5.2   Galvanic Corrosion

Galvanic corrosion is the dissolution of a metal that occurs when it is electrically connected to a more noble metal in an electrolyte. This is the corrosion that dissolved the iron screws that attached the copper plate to our boat. Metallic elements that have a more negative standard-electrode potential in Table 10.1 are more reactive than elements with a more positive standard-electrode potential. The more reactive elements are dissolved or corroded when in contact with a more electropositive or noble element. However, the standard electrode potential values in Table 10.1 are only for pure elemental metals in a molar solution of their own ions under controlled conditions. Most design applications are with metal alloys, and many of the applications are in saltwater or other corrosive environments in uncontrolled conditions. The galvanic series in Table 10.2 gives the relative galvanic corrosion resistance of commercial

Table 10.2	The Galvanic Series of Relative Material Resistance to Corrosion in Seawater
Active. Anodic End	Magnesium and Mg alloys
	Zinc
	Galvanized steel
	5052 aluminum
	3003 aluminum
	1100 aluminum
	Alclad
	Cadmium
	2024 aluminum
	Low-carbon steel
	Cast iron
	50% Pb–50% Sn solder
	316 starless steel (active)
	Lead
	Tin
	Cu-40% Zn brass
	Nickel-based alloys (active)
	Copper
	Cu-30% Ni alloy
	Nickel-based alloys (passive)
	Stainless steels (passive)
	Silver
	Titanium **(passive)**
	Graphite
	Gold
	Platinum
	Noble, Cathodic End

Based on data from ASM Metals Handbook, Vol. 10, 8th ed., ASM International, 1975.

metal alloys in saltwater. Any material at the top of Table 10.2 is galvanically corroded when in contact with a material at the bottom of the table, and the material at the bottom of the table does not corrode when in contact with a material at the top of the table. The metals marked "passive" have a protective oxide, and metals without their protective oxides are marked "active."

If a metal is not in electrical contact with another metal, it can still be corroded in an electrolyte as long as the creation of new metal ions in solution decreases the Gibbs free energy of the system of metal and solution. However, it is not galvanic corrosion if there is not an electrical connection to a second metal electrode.

Our iron screws would last much longer if they held an iron sheet because then the corrosion would have been uniform corrosion rather than the galvanic corrosion. Galvanic corrosion accelerates the disintegration of the iron screws in contact with the copper sheet relative to uniform corrosion because of the decrease in Gibbs free energy that occurs when the anode material goes into solution and the cathode material is deposited.

The accelerated corrosion of iron screws with a small area holding down a large copper plate is an example of the important area effect of galvanic corrosion. The area effect is that if two metals are in contact in an electrolyte, the anode material that is corroded should be large in comparison to the cathode material that is not corroded, as shown by Example Problem 10.8.

Example Problem 10.8

The iron screws holding down our boat patch have a total surface area of 1 cm², and they are holding down a copper plate that has a surface area that is 100 cm². In an experiment with an iron anode and a copper cathode with the same areas as our boat patch, there is a corrosion current (I) of 50 μA in saltwater.

(a) What is the weight removed from the iron screws in one year assuming that the surface area of the screws remains 1 cm²?

(b) What thickness of iron is removed from the screws in one year?

Solution

a) The mass of iron removed is calculated with Equation 10.15.

$$\frac{m_{ch}}{M} ZF = It = iAt$$

The known values are $Z = 2$, $F = 9.65 \times 10^4$ C/mole, $M_{Fe} = 55.85$ g/mole, $A = 1$ cm².
The time is the number of seconds in a year.

$$t = 1\,y = \frac{365\,d}{y}\left(\frac{24\,h}{d}\right)\frac{3600\,s}{1\,h} = 31.54 \times 10^6\,s$$

Solving for the unknown mass change results in

$$m_{ch} = \frac{MIt}{ZF} = \frac{55.85\text{ g/mole of Fe}(50 \times 10^{-6}\,A)(31.54 \times 10^6\,s)}{(2\text{ moles of electrons/mole of Fe})(9.65 \times 10^4\text{ C/mole of electrons})} = 0.46\text{ g}$$

b) The thickness lost in mm is given by Equation 10.17, with the mass change expressed in milligrams (mg), the area in cm², and the time in hours (h).

$$\frac{mm}{y} = \frac{87.6 m_{ch}}{\rho A t}$$

$$m_{ch} = 460\text{ mg}$$

The time in hours for one year is

$$t = 1\,y = 31.54 \times 10^6\,s\,\frac{1h}{3600\,s} = 8.76 \times 10^3\,h$$

The density of iron is 7.87 g/cm³.

$$\frac{mm}{y} = \frac{87.6m_{ch}}{\rho At} = \frac{87.6(460\ mg)}{7.87\ g/cm^3(1\ cm^2)(8.76 \times 10^3\ h)} = 584 \times 10^{-3}\ \frac{mm}{y} = 0.58\ \frac{mm}{y}$$

Screw heads are only several mm thick; therefore, in several years the screw head is dissolved. You can conduct the experiment to determine a corrosion current in a chemistry or materials lab with a piece of iron, such as a nail, a piece of copper wire, an ampere meter that can read micro amperes, and a beaker of saltwater.

Observe that in the calculation of the mm/year in Example Problem 10.8, the area of the iron anode is in the denominator. If the iron and the copper are reversed and the screws are copper and the iron is the plate, and assuming that the current is the same, then the rate at which the iron plate thickness decreases is only 0.58×10^{-2} mm/year. The large surface area anode thickness decreases at a slower rate than a small area anode.

10.5.3   Crevice Corrosion

Crevice corrosion is intense localized corrosion that occurs when a material exposed to a corrosive solution, such as an electrolyte, has areas of restricted exposure, such as in crevices, under bolt or rivet heads, between riveted plates, or under gaskets. Figure 10.11 shows crevice corrosion on the surface of a stainless-steel flange.

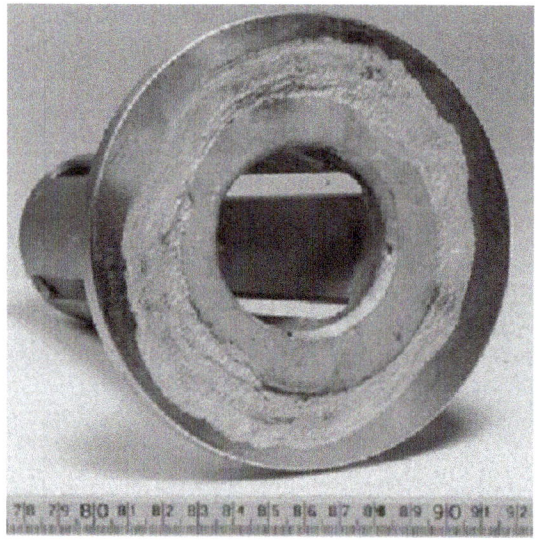

Figure 10.11   Crevice corrosion under a seal on a 316 stainless-steel flange used in a steam condenser system in flowing seawater for two years at less than 40°C. (*ASM Handbooks Online, http://www.asmmaterials.info, ASM International, 2004.*)

Rust formation on iron requires the presence of oxygen and water as shown in Equation 10.18. In a crevice with restricted flow the supply of oxygen is limited and can be depleted by reacting with iron to form rust. However, iron ions continue to go into solution. The depletion of oxygen results in an excess of iron ions (Fe^{2+}) in solution. Chlorine ions (Cl^-) are then attracted into the crevice by the excess iron ions, resulting in an increased concentration of chlorine ions in the crevice. A high concentration of chlorine ions in solution accelerates iron going into solution, resulting in crevice corrosion.

Ways to combat crevice corrosion include the application of welds that close up all crevices rather than using bolts or rivets, the closing of crevices with a filler such as caulking, the creation of drainage for at-risk crevices, the use of nonabsorbing gaskets such as Teflon rather than paper, and the removal of any deposits on the material that restrict the flow of the corrosive solution to the material.

10.5.4   Pitting Corrosion

Pitting corrosion is characterized by localized holes (pits) on the surface of a material, as contrasted with uniform corrosion. Figure 10.12 shows a piece of metal that has suffered pitting corrosion.

Pitting can begin when there are local anomalies in the surface of the material, such as scratches, defects in the material, or local changes in material chemistry. Pitting corrosion is a consequence of changes in local chemistry in the corrosion pit. When metal ions (M^+) from the part go into solution at the anomaly, this attracts corrosive negative ions, such as Cl^-, that are in the corrosive environment, as shown in Figure 10.13. A high concentration of Cl^- ions accelerates the dissolution of metals, such as iron, creating the pit and attracting even more Cl^- ions.

Procedures to combat pitting corrosion include using polished metal surfaces and selecting materials that are known to be more resistant to pitting. For example, stainless steels with high percentages of both chromium and nickel are more resistant to pitting corrosion than stainless steel with lower percentages of these elements.

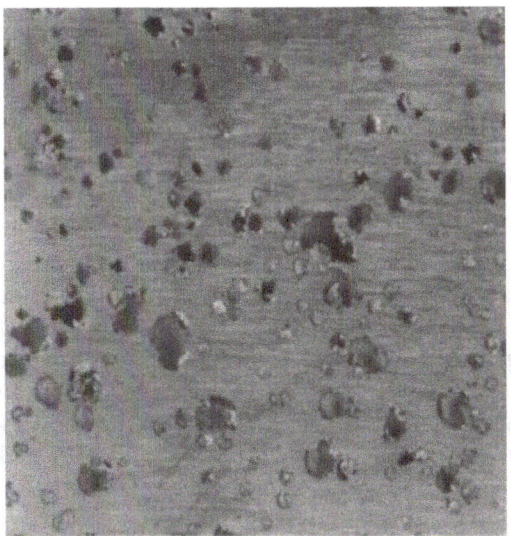

Figure 10.12   Pitting corrosion of a metal. (ASM Handbooks Online, http://www.asmmaterials.info, ASM International, 2004.)

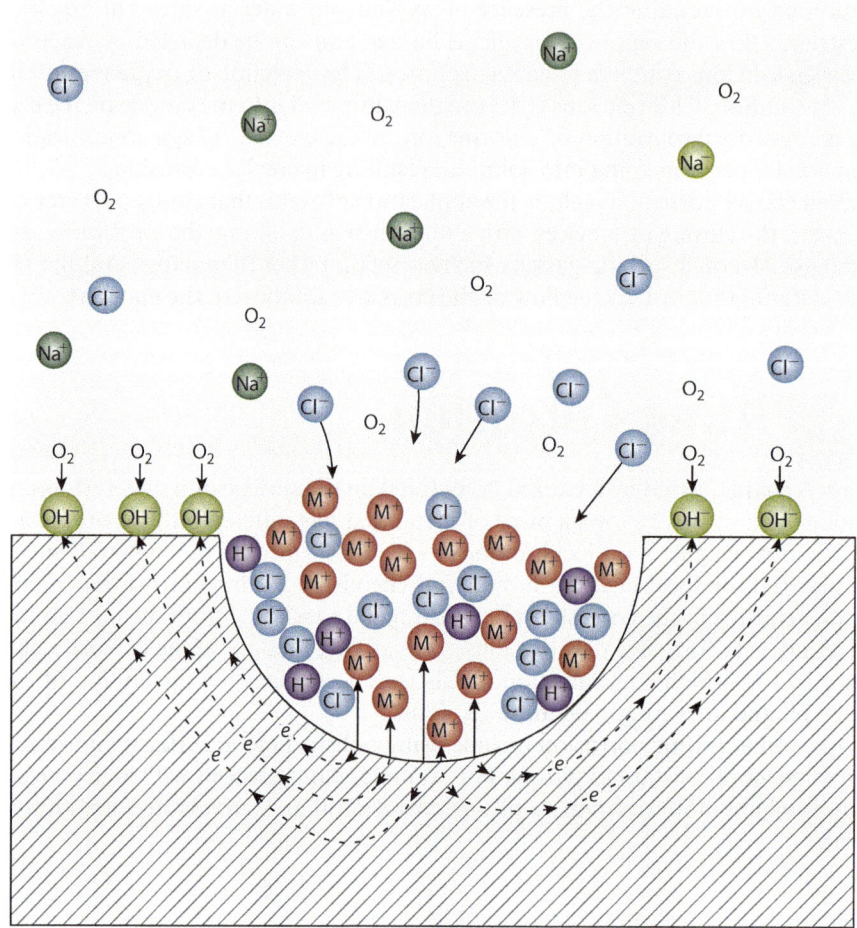

Figure 10.13   A schematic of pitting corrosion of a metal in seawater showing the high concentration of chlorine ions (Cl⁻) in the pit and the resulting accelerated dissolution of metal in the pit. (*Based on Fontana, M. G., Corrosion Engineering, McGraw-Hill, N.Y. (1986), p. 67.*)

10.5.5   Intergranular Corrosion

Intergranular corrosion is present when more corrosion occurs along the grain boundaries than within the grains of a metal. Intergranular corrosion is usually caused by a difference in the chemical composition of the grain boundaries from the grain interiors. Chemical differences can result from improper heat treatment of the metal, or from welding. For example, 18-8 stainless steel is normally a corrosion-resistant stainless steel. However, if it is heat treated in the temperature range from 510°C to 788°C, chromium carbide ($Cr_{23}C_6$) forms in the grain boundaries. The formation of chromium carbide reduces the amount of chromium in solid solution in the vicinity of the grain boundaries, which in turn reduces the corrosion resistance of the material at the grain boundaries. This temperature range should be avoided during any heat treatment of 18-8 stainless steel, and low-carbon stainless steel should be used to avoid carbide formation. For example, to make 18-8 stainless steel resistant to intergranular corrosion, the carbon content has been reduced from 0.20 to 0.08 weight percent. However, these low-carbon steels can readily pick up carbon during processes, such as welding, and by such simple procedures as cleaning the steel in a carbon-based solvent.

10.5.6   Selective Leaching

Selective leaching is the selective removal of one element from a metal alloy. The most common example of selective leaching is the removal of zinc from brass alloys, such as an alloy of 30 weight percent zinc and 70 weight percent copper, that are used in applications such as heat exchangers. If the hot water in a power-plant boiler is acidic, it can result in removal of zinc from the brass, and the brass becomes porous and reduced in strength. One way to avoid this problem is to use brass with less zinc; for example, red brass (15 weight percent zinc and 85 weight percent copper) is immune to selective leaching. Selective leaching also occurs in other alloy systems that are subject to acidic environments; for example, aluminum is leached out of aluminum bronze, and silicon is removed from silicon bronzes.

10.5.7   Erosion Corrosion

Erosion corrosion results when the relative motion between the corrosive environment and a metal results in an accelerated rate of corrosion relative to uniform corrosion. An example where erosion corrosion could occur is on the propeller of a ship, where the propeller has a high velocity when going through corrosive seawater. However, there are many other examples where erosion corrosion can occur, such as at elbows in piping systems, pump impellers, heat-exchanger tubing, turbine blades, and nozzles. Erosion corrosion often occurs because the coating that protects passivated metals, such as the Cr_2O_3 coating on stainless steel, is eroded away by the high-velocity impact of the atoms and molecules in the corrosive environment. This high-velocity impact by water molecules leaves the metal unprotected and subject to accelerated corrosion.

A particular form of erosion corrosion is called cavitation. Cavitation occurs when bubbles form in a corrosive liquid, and then they collapse. The collapse of the bubble sends a shock wave through the liquid, with pressures as high as 400 MPa, that impacts the material.

Ways to prevent erosion corrosion include the selection of metals with more protective oxides, such as titanium alloys; design improvements to prevent direct impact of corrosive environments onto the metal; the application of coatings that are more resistant to erosion corrosion; and cathodic protection, which is discussed next in Section 10.5.8.

10.5.8   Designing to Prevent Corrosion

Several processes and design procedures can inhibit corrosion:

- Avoid designs that use metals widely separated in the galvanic series that are in electrical contact in the presence of an electrolyte.
- Plate a reactive metal with a more noble metal. We can plate iron with a more noble metal such as tin, copper, gold, or silver. Silver-plated iron is used as dinner knives, forks, and spoons. Also, iron is plated with tin in food cans. With this type of protection, a scratch in the noble metal that exposes the more active metal below allows galvanic corrosion to concentrate in the exposed interface.
- Plate the metal to be protected with a less noble metal. If we place an at-risk structural metal in contact with a less noble metal, the less noble metal is dissolved and the more noble structural metal is protected. For example, galvanized steel is coated with zinc. The zinc is a **sacrificial anode**

that dissolves in time. Unlike coating with a noble metal, the zinc need not fully coat the steel to be effective. The steel and zinc only need to be electrically connected. Ships with steel hulls have attached blocks of zinc below the water line that are replaced once they are dissolved. The entire ship hull is not coated with zinc. This technique does not work for food containers, because we do not want our food contaminated with zinc ions.

- **Passivate** the surface of the metal to be protected by forming a protective oxide. Titanium behaves like a noble metal according to the galvanic series in Table 10.2; however, bare titanium metal is highly reactive. When titanium is exposed to air, it rapidly forms a protective oxide of TiO_2, and this passivates the surface. The protective oxide of TiO_2 provides a barrier that prevents further exposure of the titanium metal to the environment. This protection makes titanium and its alloys relatively inert in seawater and other corrosive environments. If the TiO_2 protective oxide is scratched away and exposes bare metallic Ti, the Ti reacts with oxygen in the environment, and a new protective layer of TiO_2 forms rapidly on the exposed Ti. In this way, damaged Ti parts "self-heal" with a new layer of TiO_2. Anodized aluminum has a protective oxide layer of Al_2O_3, and bare metallic Al also forms a protective layer of Al_2O_3. However, Al_2O_3 is not stable in the presence of Cl^- ions, and oxidized aluminum corrodes in a saltwater environment. By contrast, TiO_2 is relatively stable in the presence of Cl^- ions, which is why Ti is one of the best structural metals to resist corrosion in saltwater. In stainless steel, at least 11 weight percent Cr is added to iron to form a stable Cr_2O_3 oxide in order to passivate the stainless-steel surface. The stainless steels that are labeled "passive" in Table 10.2 are passivated with an oxide coating. The active 316 stainless steel does not have an oxide coating; however, it can be passivated by exposing it to oxygen.
- **Cathodically protect** the structure by applying a negative charge to the structure. If we attach iron that is in a corrosive environment to the negative terminal of a battery of sufficient voltage, the Fe^{2+} is attracted to the negative terminal, and it does not go into solution. Steel railroad rails, bridges, and building structures are all cathodically protected. If you look under a steel highway bridge, you will usually find a battery attached to the steel structure. Many of the batteries are now charged by small solar panels that you see when crossing the bridge. Typically the batteries are 12 volts.
- Apply an **inert coating**. If the iron to be protected is isolated from the corrosive environment, then the iron does not corrode. This is the role of paint on automobiles and ships. However, a scratch in the coating will expose the bare metal, although in this case accelerated galvanic corrosion is not a risk as it is in noble metal coatings.

Example Problem 10.9

Galvanized steel is coated with zinc, and steel ships are protected with a sacrificial anode of zinc. The effect of zinc on steel in an electrolyte is simulated in an electrochemical cell similar to Figure 10.9; there is an iron electrode in a molar solution of Fe^{2+} ions and a zinc electrode in a molar solution of Zn^{2+} ions.
(a) What reactions occur at the electrodes?
(b) What voltage is developed in the cell?
(c) What is the Gibbs free energy change per mole of ions involved in the reaction?

Solution

a) Since the iron reduction reaction is more electropositive than the zinc reduction reaction, the iron deposits on the iron electrode according to the reaction $2e + Fe^{2+} \rightarrow Fe$, and zinc is dissolved at the zinc electrode according to the reaction $Zn \rightarrow 2e + Zn^{2+}$.

b) The voltage developed is the sum of the voltage for the reduction reaction of $2e^- + Fe^{2+} \rightarrow Fe$, which is -0.440 V, and the oxidation reaction of $Zn \rightarrow 2e^- + Zn^{2+}$ is $+0.763$ V; the net voltage is $+0.323$ V.

c) The Gibbs free energy change per mole of ions is given by Equation 10.12.

$$\Delta G = -ZFV$$

$$\Delta G = -ZFV = -2\frac{\text{electrons}}{\text{ion}}\left(9.65 \times 10^4 \frac{\text{C}}{\text{mole of electrons}}\right)0.323 \text{ V}$$

$\Delta G = -6.233 \times 10^4$ J/mole of ions. This change in the Gibbs free energy is negative, and this reaction is spontaneous. The iron is protected when in contact with zinc in a corrosive environment.

10.6 ELECTROMACHINING, ELECTROPOLISHING, AND ELECTROPLATING

In Section 10.4.1, we discussed anodic and cathodic reactions and described how the reduction of a metal at a cathode results in the deposition of metal on the cathode. Similarly, the oxidation of a metal at an anode results in the dissolution of metal from the anode. These electrochemical processes can be exploited in metals processing. One such electrochemistry-based metals processing technique is **electromachining**. Electromachining is the removal of material in a specific area of a part by using electrochemical techniques. For example, a hole can be electrochemically drilled in an iron plate by means of an electrochemical cell in which iron is the anode and the cathode is a platinum wire pointing at the surface where the hole is to be drilled. A voltage is then applied to the iron, with the positive terminal attached to the iron and the negative terminal to the platinum. This setup creates nonequilibrium conditions, and it accelerates the metal removal process. The amount of material removed is calculated with Equation 10.15.

Electropolishing is the uniform removal of metal from a part in an electrochemical cell to produce a smooth surface. The metal to be polished is the anode, and the cathode is large enough to obtain uniform removal of the anode metal. For example, to uniformly electropolish a cast metal statue, it is necessary to surround the statue with the cathode. A voltage is applied to the electrochemical cell to increase the rate of metal removal, or to force a reaction that would not occur without an applied voltage. For example, noble metals such as silver and gold are electromachined and electropolished if sufficient voltage is applied to the noble metal anode charging it positive, such that their ions are forced into solution.

Electroplating is the deposition of metal onto an electrode in an electrochemical cell. Example Problem 10.4 described the electroplating of copper on the cathode of a standard cell. However, in a standard cell, when copper is deposited at the cathode, the copper must be replenished in the solution to maintain a constant concentration of Cu^{2+} in the solution. If the anode is copper, then as copper is deposited at the cathode, Cu^{2+} goes into solution at the anode, and a constant concentration of copper is maintained. A voltage must be applied for the deposition of copper to occur at the cathode if the anode is copper. Typical voltages are from 1 to 10 V, and typical current densities are from 0.03 to 1 A/cm². High-current densities can result in porous deposits; however, agitation of the solution, movement of the part, and increasing the solution temperature allows for higher-current densities.

Chrome plating is also often conducted with a chrome anode in a chromic acid solution. There are facilities that specialize in chrome plating of steel parts for classic and modified automobiles, such as the one shown in Figure 10.14.

Figure 10.14   This modified Model T Ford has many chrome plated parts. (© *Images-USA / Alamy*)

Ions that would not normally electrodeposit under equilibrium conditions are forced to deposit by application of a sufficiently large voltage. For example, zinc is electroplated on steel to produce galvanized steel by immersing a steel cathode in a solution of Zn^{2+} ions with a zinc anode. A voltage is applied to the electrodes with the negative terminal attached to the steel cathode and the positive terminal attached to the zinc anode. The negative charge on the cathode attracts the Zn^{2+} ions to the surface of the steel, forcing the electroplating. The zinc anode replenishes the Zn^{2+} deposited on the steel.

In electroplating, the mass of material predicted by Equation 10.15b assumes that the electroplating process is 100% efficient. Actual amounts deposited are less than this, and the reduced amount deposited is attributed to current efficiency. If the current efficiency is 50%, then the amount of material deposited is 50% of the amount predicted by Equation 10.15b. Efficiencies as low as 5% are observed, but 30% is more typical.

Example Problem 10.10

It is desired to electroplate copper to a thickness of 0.01 cm onto a 10 cm² piece of iron for corrosion protection and for decorative appearance. The iron is placed in an electrochemical cell with copper ions in solution. The iron electrode is connected to a direct-current power supply that charges the iron negative and a copper electrode positive. Through experimentation it is found that a current of 1 ampere produces a high-quality copper coating. Determine the time required to produce the desired-thickness coating, assuming that the process is 100% efficient.

Solution

Equation 10.15b applies to this problem. The current is given, and the mass of material needed is calculated from the density and volume of material. All other terms in Equation 10.15b are then known except the time required.

Assume that the copper reduction reaction at the iron electrode is $Cu^{2+} + 2e^- \rightarrow Cu$ and $Z = 2$. The density of copper is $\rho_{Cu} = 8.96$ g/cm³. The volume (V) of copper is the area times the thickness $V = 10$ cm² $(0.01$ cm$) = 0.1$ cm³. The mass of copper deposited is

$$m_{Cu} = V\rho_{Cu} = 0.10 \text{ cm}^3(8.96 \text{ g/cm}^3) = 0.9 \text{ g}$$

Other known values are $F = 9.65 \times 10^4$ C/mole and $M_{Cu} = 63.55$ g/mole, Inserting these values into Equation 10.15b,

$$\frac{m_{Cu}}{M} ZF = \frac{0.9 \text{ g}}{63.55 \text{ g/mole of Cu}} \left(2 \frac{\text{moles of electrons}}{\text{mole of Cu}}\right) 9.65 \times 10^4 \frac{\text{C}}{\text{mole of electrons}}$$

$$\frac{m_{Cu}}{M} ZF = 0.273 \times 10^4 \text{ C} = It = 1 \text{ A}(t)$$

Solving for the time (t),

$$t = 0.273 \times 10^4 \text{ s} = 2730 \text{ s}$$

10.7   BATTERIES

The examples above have shown that electrochemical reactions generate current and voltage; therefore, they are a source of energy, as shown by Equations 10.11 and 10.12. A **battery** is one or more electrochemical cells designed to produce electrical energy. Lead-acid batteries, such as those found in automobiles, are electrochemical cells. Observe in Table 10.1 that a voltage is developed by using lead (Pb) and hydrogen in an acid as electrodes in an electrochemical cell: however, a lead-acid battery does not use lead and hydrogen electrodes. In a lead-acid battery, the anode is lead and the cathode is lead oxide (PbO₂). The electrolyte is sulfuric acid (H_2SO_4). The reaction at the anode is given by Equation 10.19.

$$Pb + H_2SO_4 \rightarrow PbSO_4 + 2H^+ + 2e^- \qquad \textbf{10.19}$$

Equation 10.20 shows the cathode reaction.

$$PbO_2 + H_2SO_4 + 2H^+ + 2e^- \rightarrow PbSO_4 + 2H_2O \qquad \textbf{10.20}$$

The result of the electrochemical reactions is the production of $PbSO_4$ and H_2O, and each electrochemical cell produces approximately 2.10 V. A 12-V battery is obtained by connecting six cells in series. The reactions in Equations 10.19 and 10.20 are reversed when the battery is recharged, depositing Pb on the anode and PbO₂ on the cathode.

10.8 FUEL CELLS

A **fuel cell**, shown in Figure 10.15, is an electrochemical cell that consumes a fuel and directly produces electrical energy. A hydrogen-oxygen fuel cell oxidizes hydrogen to produce water, which is the reverse of the electrolysis of water that produces hydrogen and oxygen. It has been proposed that fuel cells can replace internal-combustion engines and gas turbines as energy sources in automobiles, trucks, and power stations. One such successful application is the power for forklift vehicles, because the hydrogen refueling station is in the warehouse where the forklift is operating, and refueling takes only minutes. Fuel cells are currently used to power satellites; they provide regular power and emergency power for hospitals, telecommunication centers, homes, laboratories, ships, and commercial buildings. Portable fuel cell power supplies are used for military applications, cell phones, computers, and digital cameras. The five main types of fuel cells, shown in Table 10.3, are described in the subsequent paragraphs of this section.

In a fuel cell, shown in Figure 10.15, the main components are the electrodes and the electrolyte. The electrodes must be able to perform the following tasks: conduct electrons, provide a surface for the electrode reactions, allow diffusion of appropriate ions, and be stable in the environment. The electrodes in the acid, alkaline, and proton-exchange membrane (PEM) fuel cells were initially porous platinum.

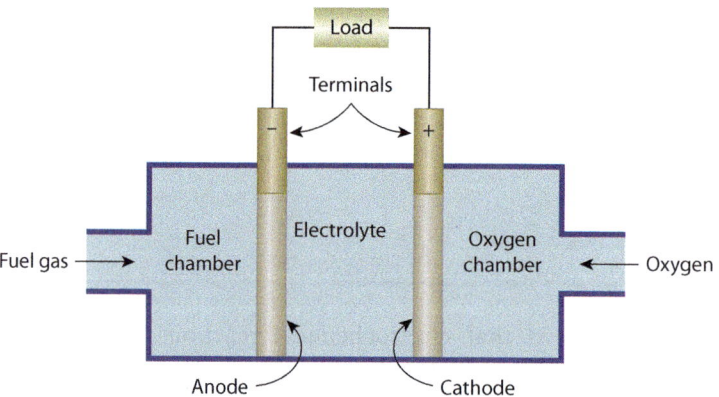

Figure 10.15   A schematic of a fuel cell consisting of a fuel supply, an anode and cathode in an electrolyte, and an oxygen supply.

Table 10.3	Fuel-Cell Types, the Mobile Ion, Operating Temperature in °C (T), and Applications		
Fuel Cell Type	Mobile Ion	T	Applications
Acid	H^+	220	Small power plants of 200 kW
Alkaline	OH^-	50–200	Space vehicles, including *Apollo, Gemini,* and the space shuttle
Molten carbonate	CO_3^{-2}	650	Medium to large power plants of 1 MW
PEM[1]	H^+	50–100	Vehicles and small power plants
Solid oxide	O^{-2}	600–1000	Small to large power plants

1   Proton-exchange membrane

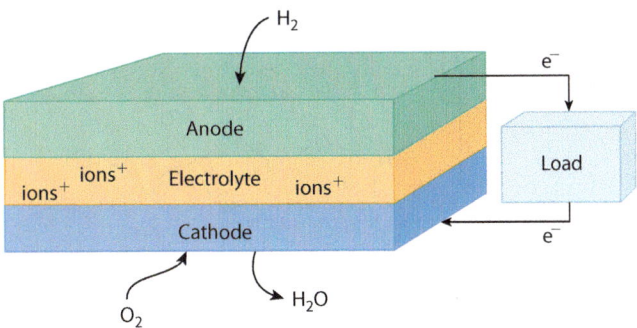

Figure 10.16 A block diagram of an acid electrolyte hydrogen fuel cell. (*Based on http://en.wikipedia.org/wiki/File:Fuel_Cell_Block_Diagram.svg*)

However, because of the expense of platinum, electrodes made of carbon particles coated with small particles of platinum have been developed. The platinum-coated carbon particles are supported on a carbon paper or cloth. The electrolyte must conduct the mobile ions of the fuel cell and provide a high resistance to electrons so that the electrons pass through the load from the anode to the cathode.

An acid electrolyte fuel cell with hydrogen fuel is shown in Figure 10.16. In this fuel cell, hydrogen ions and electrons are produced at the anode, according to Reaction 10.21:

$$2H_2 \rightarrow 4H^+ + 4e^- \qquad\qquad \textbf{10.21}$$

The hydrogen ions diffuse through the acid electrolyte; however, the acid electrolyte has a high resistance for electron transfer. Phosphoric acid is the primary acid used in acid electrolyte fuel cells, and the hydrogen ions are mobile in a liquid acid. The electrons pass through the load of motors, lights, and so on to the cathode. Oxygen introduced at the cathode reacts with the hydrogen ions in Reaction 10.22, producing water, a standard-electrode potential from Table 10.1 of 1.23 V, and an energy of 1.23 eV for each hydrogen ion consumed.

$$O_2 + 4H^+ + 4e^- \rightarrow 2H_2O \qquad\qquad \textbf{10.22}$$

In an alkaline electrolyte hydrogen fuel cell, hydroxide ions (OH^-) are created at the cathode by the reaction of oxygen with water and with electrons that are created at the anode and pass through the load, as shown in Reaction 10.23.

$$O_2 + 4e^- + 2H_2O \rightarrow 4OH^- \qquad\qquad \textbf{10.23}$$

The hydroxide ions diffuse through an alkaline electrolyte, such as an aqueous solution of KOH, to the anode. At the anode, the OH^- ions react, according to Reaction 10.24, with hydrogen to produce water and the four electrons needed for Reaction 10.23.

$$2H_2 + 4OH^- \rightarrow 4H_2O + 4e^- \qquad\qquad \textbf{10.24}$$

A PEM fuel cell uses a polymer membrane as the electrolyte. The cathode and anode are usually thin composite sheets of the small-particle, platinum-coated carbon discussed above. The anode and cathode

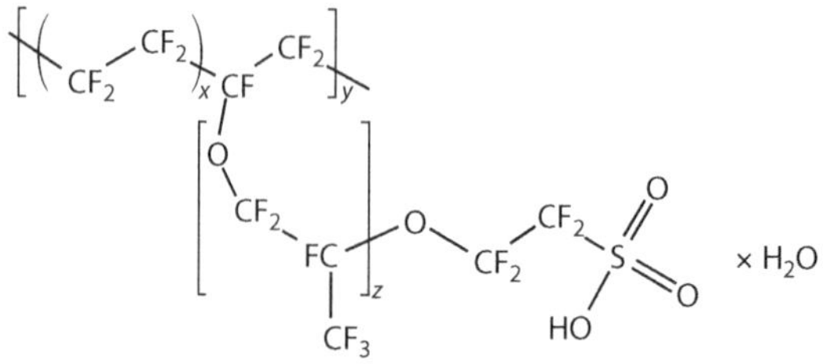

Figure 10.17 A possible molecular structure of an LCM ending in an HSO_3 group for use as an electrolyte in a PEM fuel cell. The x, y, and z give the number of times this unit is repeated. (*Based on http://upload.wikimedia.org/wikipedia/commons/thumb/3/3d/Nafion2.svg/200px-Nafion2.svg.png*)

reactions are the same as in the acid electrolyte fuel cell. The electrolyte is a polymer sheet that conducts hydrogen ions and has a high resistance for electron transfer. One of the polymer electrolytes is a modified polytetrafluoroethylene (PTFE) shown in Figure 10.17. The PTFE is modified by inserting oxygen atoms in the long-chain molecule and attaching side branches that end in a HSO_3 group. Normally PTFE is hydrophobic and it repels water. However, the HSO_3 group is hydrophyllic and it attracts water molecules, which creates a small region that functions similar to sulfuric acid in the PEM. The hydrogen ions diffuse from one hydrophyllic area to another through the membrane. The membrane is well hydrated to conduct hydrogen ions readily. The membranes are as thin as 50×10^{-6} m, resulting in a very thin cell with a low density that is appropriate for vehicles. The low temperature of operation of 50–100°C is suited to intermittent power sources.

In a solid-oxide fuel cell (SOFC) the electrolyte is an oxide, and several oxides are being considered, including ZrO_2, CeO_2, $LaGaO_3$ and $BaZrO_3$. In an SOFC, negative oxygen ions are created at the cathode, according to Reaction 10.25, by an oxygen molecule (O_2) combining with four electrons that are created at the anode and pass through the load.

$$O_2 + 4e^- \rightarrow 2O^{2-}$$ **10.25**

The oxygen ions diffuse through the ceramic-oxide electrolyte to react with hydrogen fuel at the anode, according to Reaction 10.26, producing water and the four electrons needed for Reaction 10.25.

$$2H_2 + 2O^{2-} \rightarrow 2H_2O + 4e^-$$ **10.26**

Solid oxide fuel cells can also use carbon monoxide (CO) as a fuel. Oxygen is introduced at the cathode and oxygen ions are produced according to Reaction 10.25. At the anode the CO reacts with the oxygen ions to produce carbon dioxide according to Reaction 10.27.

$$2CO + 2O^{2-} \rightarrow 2CO_2 + 4e^-$$ **10.27**

At high temperatures of 600–1000°C, the oxides used in SOFCs have a high conductivity for oxygen by oxygen-vacancy diffusion. In addition, the oxides are doped, or alloyed, with other oxides or elements that create additional oxygen vacancies. The electrodes in an SOFC are usually a porous mixture of a ceramic, such as ZrO_2, and nickel. The nickel acts as a catalytic surface for chemical reactions, and the nickel provides electron conductivity. Since an SOFC is operated at high temperatures of 600–1000°C, it is not suitable for intermittent operation in a vehicle. The proper applications are for continuous operation as a power source. Advantages of the high-temperature operation are that high-temperature exit gases and cooling fluids can be used as a source of heat for buildings, to drive turbines, and to assist in the separation of hydrogen from fuel sources such as natural gas.

The discussion above focused on hydrogen and carbon monoxide as the fuel for fuel cells. However, fuel cells can operate with other fuels, including methane, methanol, and natural gas. Most large fuel cell applications use natural gas. One fuel cell fuel that is considered to be renewable is anerobic digester gas (ADG). ADG is a gas composed primarily of methane and carbon dioxide that is emitted from wastewater treatment plants, from crop and animal waste, from breweries, and from agricultural processing facilities that use anerobic digestion to break down biodegradable materials.

Example Problem 10.11

What is the maximum amount of work in joules that can result from the reaction of 1 mole of hydrogen atoms in a hydrogen-oxygen fuel cell that operates with the chemical reaction in Equation 10.22?

Solution

The Gibbs free energy change is given by Equation 10.12, and this is the amount of work that is possible at constant temperature and pressure. The voltage generated is $+1.23$ V according to Table 10.1, and there is one electron generated with a charge of $-e$ per hydrogen atom.

$$\Delta W = \Delta G = -ZFV_0 = -\left(-1\frac{\text{electron}}{\text{ion}}\right)\left(9.65 \times 10^4 \frac{\text{C}}{\text{mol}}\right)(1.23 \text{ V})$$

$$\Delta W = 11.87 \times 10^4 \text{ J/mole of hydrogen atoms}$$

This work is produced, and the only product of the reaction is water. This is why there is interest in fuel cell power. However, the power necessary for initially producing the hydrogen and oxygen can create environmental pollution. For a fuel cell to be truly pollution free, the source of power for hydrogen and oxygen production must also be pollution free.

Summary

- All metals except for gold react with oxygen, and oxygen reacts with ceramics such as silicon, carbides, and nitrides. The reaction of polymers with oxygen is part of weathering. The reaction of structural materials with oxygen in high-temperature applications can significantly limit the lifetime of parts. Materials react with oxygen when the Gibbs free energy of the material oxide is less than that of the material plus oxygen gas.

- For an oxide coating to grow, electrons and either material ions or oxygen ions must diffuse through the oxide.

- For an oxide to be protective on a surface, it must be continuous and without porosity. For metals this is most likely to occur if the volume of the oxide formed is approximately equal to the volume of the metal atoms that formed the oxide, or if the Pilling-Bedworth ratio is equal to approximately 1.

- Experimentally, it is found that protective oxides follow either parabolic or logarithmic growth laws with respect to time, and nonprotective oxides follow a linear-growth law.

- When a protective oxide forms on a metal and then the metal-plus-oxide coating resists corrosion, it is called passivated. Passivation is why titanium resists corrosion in saltwater and other corrosive environments.

- Metals, such as steel, that do not form protective oxides are modified by alloying to enable them to form protective oxides. In stainless steel, iron is alloyed with chromium to form protective chromium oxide.

- Coatings made of metals that are passivated by formation of a protective oxide or metal oxides are utilized to protect base metals that are not resistant to oxidation in high-temperature applications.

- The mechanical properties of a polymer are degraded by the absorption of atoms or molecules that are soluble in the polymer.

- When UV light is absorbed in a polymer, it breaks bonds between LCMs and along LCMs, resulting in a degradation of the mechanical properties of the polymer.

- Weathering of a polymer is the combination of irradiation with light, including UV, and the reaction with atoms and molecules in the environment, such as oxygen, nitrogen, and water.

- Corrosion is the deterioration of a material because of electrochemical reactions with its environment. Corrosion is primarily a concern with metals. Ceramics are more resistant to corrosion than metals. Although polymers are degraded by the environment, the term *corrosion* is not applied to polymers.

- When an atom loses electrons, this is an oxidation reaction. Oxidation occurs at the anode of an electrochemical cell, and the anode is dissolved in the electrolyte. When an ion gains electrons, this is called a reduction reaction. The reduction reaction occurs at the cathode, and ions from the electrolyte are deposited at the cathode of an electrochemical cell.

- The value of the standard-electrode potential of a material is measured in a 1-molar solution of its own ions, relative to a reference electrode of a 1-molar solution of hydrogen ions. The voltage in an electrochemical cell is the sum of the voltage of the reduction reaction and of the oxidation reaction. If the solutions are not 1-molar, then the half-cell voltage for the reduction reaction is calculated with the Nernst equation.

- The voltage developed in an electrochemical cell is directly related to the Gibbs free energy change of the reaction.

- In an electrochemical cell, the mass of material removed or deposited at the electrodes is a function of the cell current.

- The types of corrosion include uniform, galvanic, crevice, pitting, intergranular, selective leaching, and erosion.

- Galvanic corrosion is the dissolution of a metal that occurs when it is electrically connected to a more noble metal in an electrolyte. The galvanic series gives the relative galvanic corrosion resistance of commercial metal alloys in saltwater. The effects of galvanic corrosion are minimized in a design by selecting materials close together in the galvanic series, by having a large area anode and a small area cathode, by coating reactive materials with a more noble metal, by attaching a sacrificial anode to a structural metal, by selecting metals that are passivated in the corrosive environment, by applying cathodic protection, and by applying an inert coating.

- It is possible to remove metal from an anode and to deposit metal on a cathode in electrochemical cells. In electromachining, metal is removed in a specific area of an anode by applying a voltage between the anode and a cathode localized to the area of metal removal. In electropolishing, material is removed uniformly over the anode with a cathode of appropriate size to produce a smooth surface. Electroplating is the deposition of a metal onto a cathode. A voltage can be applied to the cell to force or accelerate reactions.

- Batteries are electrochemical cells designed to produce electrical energy.

- Electrochemical reactions generate current and voltage; therefore, they are a source of energy. A fuel cell consumes a fuel and produces electrical energy. Possible fuels include hydrogen, methane, methanol, carbon monoxide, natural gas, and anerobic digester gas. The electrolytes for fuel cells include acid, alkaline, molten carbonate, proton-exchange membrane, and solid oxide. The only products of the fuel cell reaction are energy and either water or carbon dioxide.

Supplemental Reading Subjects and Authors

Full references are listed at the end of the book.

General:	*Askeland, Fulay, and Wright*
Photographs of corrosion in metals:	ASM Handbook
Corrosion and oxidation:	*Fontana; McCafferty; Uhlig and Revie*
Elecrochemistry:	*Bockris and Reddy; Crow*
Electroplating:	*Lowenheim*
Fuel cells:	*Chang; Larminie and Dicks*
Environmental effects on polymers:	*McCrum, Buckley, and Bucknall; Osswald and Menges*

Homework

Concept Questions

1. The only metal that does not react with oxygen under ambient conditions is _____.

2. For an oxide to be protective, the ratio of the volume of the oxide formed to the volume of the metal atoms that formed the oxide should be approximately _____.

3. If an oxide is not protective, the weight gain as a function of time is _____.

4. Iron is passivated by the addition of _____in stainless steel.

5. The _____ of a polymer is the combination of irradiation with light, including UV, and the reaction with atoms and molecules in the environment such as oxygen, nitrogen, and water.

6. The _____electrode is a 1-molar solution of hydrogen ions (H^+) in HCl in contact with hydrogen gas at 1 atmosphere of pressure.

7. When an ion gains electrons, this is called a(n) _____ reaction.

8. A(n) _____ is the charge in coulombs on a mole of positive ions of unit charge.

9. _____ is the degradation of a material because of electrochemical reactions with its environment.

10. _____ corrosion is the dissolution of a metal that occurs when it is electrically connected to a more noble metal in an electrolyte.

11. In cathodic protection of a metal structure, the structure is connected to the _____ terminal of a battery.

12. When the surface of the reactive metal titanium is passivated, _____ forms on the surface of the metal.

13. When iron is galvanized, it is coated with _____.

14. _____ corrosion is intense, localized corrosion that occurs when a material has areas of restricted exposure to an electrolyte.

15. _____ corrosion is usually caused by a difference in the chemical composition of the grain boundaries from the grain interiors.

16. Selective _____ is the selective removal of one element from a metal alloy.

17. _____ is the removal of material in a specific area of a part by using electrochemical techniques.

18. _____ is uniform removal of metal from a part to produce a smooth surface by using electrochemical techniques.

19. _____ is when metal is deposited onto an electrode in an electrochemical cell.

20. A _____ _____ is an electrochemical cell that consumes a fuel and directly produces electrical energy.

Engineer in Training–Style Questions

1. Which of the following materials would not react with oxygen?
   (a) Titanium
   (b) Stainless steel
   (c) Silicon carbide
   (d) Alumina

2. Which of the following is not a known growth rate for oxides on metals?
   (a) Exponential
   (b) Logarithmic
   (c) Parabolic
   (d) Linear

3. Which of the following is not known to result in a high probability of a solvent penetrating a polymer?
   (a) Similar solubility parameters for the solvent and polymer
   (b) A similar molecular mass for the solvent and polymer molecules
   (c) Similar bond types between molecules of the solvent and polymer
   (d) A similar molecular structure on the solvent and polymer molecules

4. Which of the following metals should not be in electrical contact with a critical structure made of low-carbon steel that is in contact with seawater?
    (a)   Cadmium
    (b)   2024 aluminum
    (c)   Copper
    (d)   Magnesium

5. Which of the following would not protect low-carbon steel from corrosion in contact with an electrolyte?
    (a)   Coat the steel completely with copper
    (b)   Attach the steel to the positive terminal of a 12-V battery
    (c)   Attach a piece of zinc to the steel in the area exposed to electrolyte
    (d)   Completely paint the steel

Problems

Problem 10.1: According to Deal and Grove (B. Deal and A. Grove, *J. Appl. Phys.*, 36 (1965), 3770), the thickness (h) of the oxide on Si in dry oxygen at 1 atmosphere of pressure follows the relation

$$h^2 + Ah = B(t + t_0)$$

and that at 1000°C, $A = 0.165 \times 10^{-6}$ m, $B = 3.25 \times 10^{-18}$ m²/s, and $t_0 = 1224$ s.

(a) Calculate the thickness of oxide film that will form on Si in dry oxygen at 1000°C in 1 hour.
(b) Compare your result with Example Problem 10.2 to see if the growth rate is decreasing with time, as would be expected for parabolic growth.

Problem 10.2: Determine the *PBR* for the oxidation of nickel to form NiO that has a density of 7.45 g/cm³.

Problem 10.3: Magnesium is the metal in a molar solution of Mg^{2+} ions with a reference electrode. (a) What reactions take place at the electrodes? (b) What voltage is developed in the standard cell? (c) Calculate the Gibbs free energy change when 1 mole of magnesium dissolves into a standard solution through the reaction $Mg \rightarrow 2e^- + Mg^{2+}$ with a standard-reference electrode.

Problem 10.4: Silver is the metal electrode in a molar solution of Ag^+ ions in a standard cell with a reference electrode. (a) What reactions occur at the two electrodes? (b) What voltage is developed in the cell? (c) What is the Gibbs free energy change per silver ion involved in the reaction?

Problem 10.5: In an electrochemical cell, there is a tin electrode in a molar solution of Sn^{2+} ions and an iron electrode in a molar solution of Fe^{2+} ions. (a) What reactions occur at the two electrodes? (b) What voltage is developed in the cell? (c) What is the Gibbs free energy change per mole of ions involved in the reaction?

Problem 10.6: A copper metal electrode is in a solution of 635.4 g of Cu^{2+} ions in 1×10^{-3} m³ of solution, and the other electrode is an iron electrode in a solution of 0.5585 g of Fe^{2+} ions in 1×10^{-3} m³ of solution at 25°C. (a) What reactions occur at the two electrodes? (b) What voltage is developed in the cell?

Problem 10.7: A copper 50 weight percent nickel alloy uniformly corrodes at the rate of 0.53 g/m² per day in water with 3 weight percent salt. Calculate the current density at the alloy surface. Assume that the copper oxidation reaction is $Cu \rightarrow Cu^{2+} + 2e^-$ and for nickel it is $Ni \rightarrow Ni^{2+} + 2e^-$. Note the much slower corrosion rate for this 50%Cu−50%Ni alloy in comparison to the pure copper in Example Problem 10.7. These corrosion rates are experimental. This shows the excellent corrosion resistance of copper-nickel alloys of this composition.

Problem 10.8: An iron bolt is galvanized with a coating of zinc 0.03 cm thick to protect the iron in a saltwater environment. The bolt has an area of 2 cm² and is used to secure an iron plate with an area of 100 cm² that is exposed to saltwater. In an experiment in seawater with a zinc-coated iron anode of 2 cm² and an iron plate cathode with an area of 100 cm², a corrosion current of 2 milliamperes (mA) is measured. Predict the expected lifetime in years or days of the zinc coating in this application assuming that the zinc coating is uniformly corroded and that the current efficiency is 100%.

Problem 10.9: Zinc is to be electroplated onto both sides of an iron sheet that is 20 cm² as a galvanized sacrificial anode. It is desired to electroplate the zinc to a thickness of 0.025 mm. It is found that a current of 20 A produces a zinc coating of sufficient quality for galvanized iron. Determine the time required to produce the desired coating, assuming 100% efficiency.

Problem 10.10: It is decided to electropolish a complex cast copper sculpture, because the surface is too complex for mechanical polishing, and the as-cast surface does not have the required appearance. Through experimentation with a piece of copper, it is found that a current density of 1 A/cm² produces the required appearance. It is estimated that the entire surface of the sculpture is 100 cm², and it is decided to electropolish the sculpture at 100 A of current. No more than 0.0001 m of copper should be removed from the sculpture, so that the design shape is not altered. Determine the maximum time for electropolishing if the process is 100% efficient.

Problem 10.11: An electrodeposition process of plating zinc onto steel for corrosion protection is expected to save energy over dipping steel into molten zinc (hot dipping), because it is necessary to keep the hot dip zinc molten continuously; whereas with electrodeposition it is only necessary to run a current through the electrolyte during the electrodeposition process. The electrodeposition should take no more than 4 minutes to be competitive with hot dipping. With the electrodeposition equipment and agitation of the part and the solution, it is found that a current density of 1 A/cm² gives a coating of zinc that is satisfactory. What thickness and mass of zinc can be electroplated per square centimeter of steel in 4 minutes if the process is 100% efficient?

Problem 10.12: In the chrome plating of steel automotive parts, it is found that it is possible to have a high-quality coating with a current density of 0.4 A/cm² by heating the chromic acid solution to 65°C and using a chrome anode. Assume that chrome goes into solution as Cr^{3+}. (a) Predict the thickness of chrome that is deposited in 1 hour, assuming that the process is 100% efficient. (b) Predict the thickness that is deposited in 1 hour if the process is 10% efficient.

The disintegration of the space shuttle *Columbia* is the most dramatic recent disaster resulting from material fracture. On January 16, 2003, approximately 82 seconds after liftoff, a suitcase-size piece of foam thermal insulation broke off the rocket external fuel tank, striking the leading edge of *Columbia's* left wing. The leading edge of the wing is covered with reinforced carbon-carbon (RCC) tiles that we discuss in Chapters 1 and 12. The impact fractured the RCC tile, creating a hole possibly as large as 25 cm in the wing. During reentry, the leading edge of the wing reaches temperatures of 1650°C. Without the protection of the RCC tile, the interior metal structure of the wing was exposed to these temperatures. During reentry, Columbia disintegrated, with the loss of seven astronauts. In this chapter we determine the conditions that can lead to fracture, and we learn how to design to avoid fracture.

NASA

The goals of this chapter are to understand

- The theory of brittle fracture
- The theory of fracture with localized surface plastic strain
- The relationship between plastic strain and fracture strength
- The criteria for a valid critical stress intensity factor
- The effect of temperature and environment on fracture strength
- The use of fracture mechanics in design
- The nondestructive tests available to detect cracks
- The application of fracture mechanics to the failure of materials in a tensile test
- The relationship between cyclic stress and the number of cycles to fatigue failure
- The probabilistic nature of fatigue failure
- The relationship between localized plastic strain and fatigue-crack initiation
- The relationship between fracture mechanics and crack-propagation cycles to failure
- The application of the Weibull statistical distribution to fracture and fatigue data

Chapter 11

Fracture and Fatigue

11.1 INTRODUCTION

Fracture of a material occurs when a material separates into at least two pieces, or when the material separates to the point that it has insignificant load-carrying capacity. Figure 11.1 gives an example of the fracture of materials, where the Liberty tanker ship *Schenectady* is shown nearly separated in half. The ship split while sitting at the dock in calm water that was at a temperature of about 4°C, with an air temperature of −3°C. Reports state that the fracture was sudden and created an explosive sound that was heard for miles. The Liberty ships were World War II cargo and tanker ships produced with continuous-welded steel construction. Because the monolithic steel welds wrapped completely around the ship, many of the vessels broke completely into two pieces. Previously, ships were constructed with individual steel plates riveted to a central frame; a crack could extend only to the end of its steel plate. Of 2710 Liberty ships built, 1500 experienced significant fracture, and 12 broke in half.

The sinking of the Liberty ships resulted in a focused effort to understand the fracture of materials and the development of a theory of fracture that is applied to design. The catastrophic failure of the space shuttle *Columbia* due to fracture of the RCC tiles on the wing shows that even though brittle fracture is now understood, fractures still occur, and the possibility of fracture should be considered in the design of critical structures.

Figure 11.1   The Liberty ship *Schenectady* fractured nearly in two on January 16, 1943, while sitting in calm water at the dock. (U.S. GPO)

In this chapter, the fracture of materials with a crack subject to a continuously increasing stress is treated first. In a fracture test the maximum stress that the cracked specimen can withstand before fracture is determined. The second half of this chapter treats **fatigue failure**, where cracks develop and grow from multiple-stress applications.

11.2   FRACTURE

11.2.1   Theory of Brittle Fracture

A material is brittle when it fractures without plastic strain. A. A. Griffith initiated the theory of the fracture of brittle materials. He observed that if glass fibers are drawn from a melt without damaging their surfaces in any way, and then immediately tested in uniaxial tension, these fibers have an incredibly high fracture stress (σ_f) of approximately 7 GPa. The glass fibers fail in brittle fracture without any plastic deformation. However, other forms of glass break at lower stresses. The S-glass fibers in fiberglass have a fracture stress of 5 GPa. If we test ordinary window glass, the tensile fracture stress drops to only 69 MPa. How can the same material have such different fracture stresses? Griffith concluded that the different fracture stresses are due to different-size defects, and for glass the defects are cracks. Freshly drawn glass filaments have nearly perfect surfaces; any defects or cracks are of atomic dimensions. Glass fibers that are used for fiberglass have small surface cracks due to abrasion and reaction with the environment, with resulting cracks of nanometer dimensions, as is demonstrated in Example Problem 11.1. Normal window glass has much larger cracks and defects along the edges that are fractions of a millimeter in size.

Figure 11.2a shows a material with an edge crack of length *a*, and Figure 11.2b shows an internal crack of length 2*a*. The reason that the internal crack is given the length 2*a* is that it has two crack tips; the length *a* is associated with each crack tip. The edge crack of length *a* results if the material in Figure 11.2b is cut in half, resulting in two edge cracks, each of length *a*.

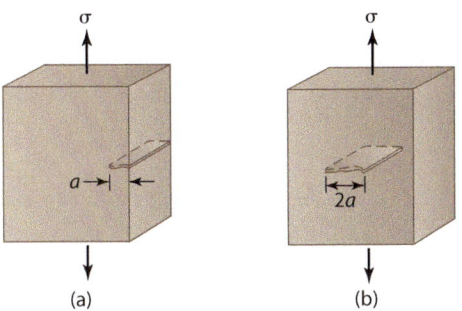

Figure 11.2 A schematic representation of (a) an edge or surface crack of depth a in a large plate, and (b) an interior crack of total length $2a$ in a large plate. (*Based on Askeland, D. R., Fulay, P. P., and Wright, W. J.,* The Science and Engineering of Materials, *6th ed., Cengage Learning, Stamford, CT (2011), p. 248.*)

If a stress (σ) is applied to a material of volume (V) with no crack, the strain energy (U_0) in this material is given by Equation 11.1.

$$U_0 = \frac{\sigma^2 V}{2E} \qquad\qquad \textbf{11.1}$$

If an internal crack of length $2a$ is introduced into the material, the interatomic bonds along the crack surface are broken, and this relaxes the stress along the edges of the crack. No stress is maintained across the crack, because of the broken atomic bonds. Figure 11.3 demonstrates how a cut in a sheet of rubber opens up when the sheet is stretched, relaxing the stress above and below the cut. Above and below the cut, the rubber is not stretched. If you push on the rubber above and below the cut with your finger the rubber easily deforms, it is floppy; whereas the rubber away from the cut is stretched and stiff.

A theoretical analysis of an infinite plate with a central crack of length $2a$ shows that the amount of strain energy relaxed (U_r) due to the presence of the crack is calculated with Equation 11.2:

$$U_r = -\frac{\sigma^2}{2E}(2\pi a^2 t) \qquad\qquad \textbf{11.2}$$

where t is the thickness of the specimen. The negative sign indicates that this is a strain-energy decrease in the material. In Equation 11.2, it is as if a volume of material with a radius of $2^{1/2}a$ and thickness t is fully relaxed to 0 stress. When the crack is present, not only is the stress relaxed, but also two new surfaces of area $2(2at)$ are present that have energy γ_s per unit area. There is an increase in energy (U_s) due to the presence of the two surfaces, calculated with Equation 11.3.

$$U_s = 2(2at\gamma_s) = 4at\gamma_s \qquad\qquad \textbf{11.3}$$

Equation 11.4 gives the total energy (U) for the stressed specimen of volume V with a crack of length $2a$ inside the specimen extending through the thickness t.

$$U = U_0 - \frac{\pi \sigma^2 a^2 t}{E} + 4at\gamma_s \qquad\qquad \textbf{11.4}$$

For a fixed amount of stress and for very small crack lengths, the negative stress-relaxation term is smaller than the positive surface-energy term, because a^2 is much smaller than a. For very small crack sizes an increase in crack length is stable, because the surface-energy term $4at\gamma_s$ increases more rapidly than the relaxation-energy term $(-\pi\sigma^2 a^2 t/E)$. However, when the crack size is larger, at some critical crack length (a) when the crack grows by length da, the differential of the stress-relaxation term $(-2\pi\sigma^2 tada/E)$

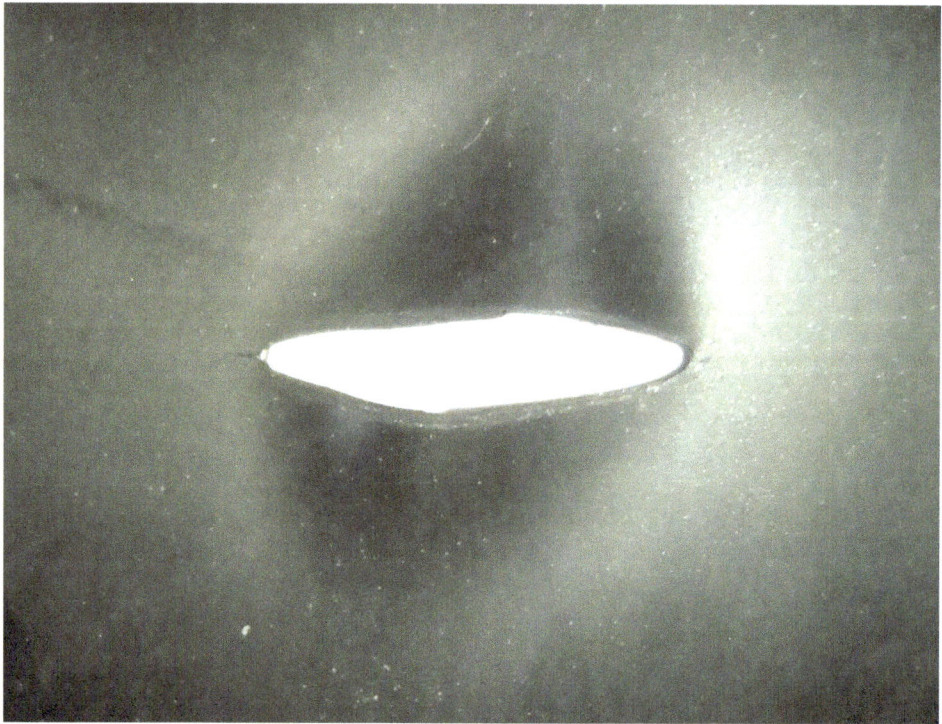

Figure 11.3   A cut in a rubber sheet with an applied tensile stress, showing the relaxation of stress in the rubber around the cut. (*Photo by C. M. Gilmore*)

becomes greater in magnitude than the differential of the surface-energy term ($4t\gamma_s da$). At this point the crack is unstable, because an increase in crack length by da reduces the internal energy. When the combination of the stress (σ) and the crack length (a) reach a critical value, the increment of energy necessary to grow the crack by an increment (da) is equal to 0, and the specimen fractures. Analytically, Equation 11.5 gives this condition as

$$\frac{dU}{da} = 0 = -\frac{2\pi\sigma_f^2 at}{E} + 4t\gamma_s \qquad\qquad \textbf{11.5}$$

Collecting all of the specimen variables on the left side of Equation 11.5 and all of the material constants on the right side results in Equation 11.6.

$$\sigma_f(\pi a)^{1/2} = (2E\gamma_s)^{1/2} \qquad\qquad \textbf{11.6}$$

Also, the second derivative of the total energy (U) with respect to a (d^2U/da^2) is negative, indicating that total strain energy is a maximum. This condition corresponds to the maximum strain energy the specimen with a crack of length a can absorb. If the stress on the specimen is increased, the specimen is unstable, and rapid crack propagation occurs. This is the condition for fracture at stress σ_f for a crack length of a. Solving Equation 11.6 for the fracture stress (σ_f) produces Equation 11.7.

$$\sigma_f = \left(\frac{2E\gamma_s}{\pi a}\right)^{1/2} \qquad\qquad \textbf{11.7}$$

Equation 11.7 shows that as the crack length (a) increases, the fracture stress (σ_f) decreases, because $E\gamma_s$ is a constant. The fracture stress is not a material constant; it depends upon the length of the defects or cracks in the material. The larger the cracks, the lower the fracture stress, which explains why glass can have different fracture strengths depending upon the type of glass.

If the material has no cracks, then the crack length (a) approaches the lattice parameter. Using MgO as an example with a lattice parameter of 0.4212 nm, a surface energy of 1.5 J/m², and an elastic modulus of 300 GPa, Equation 11.7 results in a theoretical strength of 26 GPa. However, the tensile strength of pressed and sintered MgO that has many small voids that act as cracks is approximately 105 MPa. The strength of MgO with many small cracks is more than two orders of magnitude less than the strength of a material with no crack. Whiskers of the ceramic alumina have a tensile strength of approximately 21 GPa, as shown in Table 7.3. This is close to the theoretical result for MgO. The calculation of theoretical strength is performed for MgO because of its cubic structure with a single lattice parameter. Alumina is hexagonal with two lattice parameters, and this introduces complexity into the calculation.

In Equation 11.6, all of the terms on the right-hand side of the equation are material constants. The product of these constants is also a constant. The **critical-stress intensity factor** (K_{Ic}) given in Equation 11.8 is also a material constant, where Y is a geometrical factor.

$$K_{Ic} = Y\sigma_f(\pi a)^{1/2} = Y(2E\gamma_s)^{1/2} \qquad \textbf{11.8}$$

Because cracks can be in a different orientation relative to the specimen geometry and to the applied stress, a geometrical factor Y is introduced into Equation 11.8. For the internal crack in Figure 11.2b, the value of Y is 1.0, and for the edge crack shown in Figure 11.2a, the value of Y is 1.1.

This analysis applies to all brittle materials including ceramics, concrete, brittle polymers, and brittle metals such as gray cast iron. K_{Ic} is also called the fracture toughness. However, fracture toughness can be confused with toughness. Toughness is the area under the stress-strain curve to fracture, resulting from a tensile test of a smooth specimen. To avoid this confusion, this textbook uses the term critical-stress intensity factor for K_{Ic}.

Example Problem 11.1

Silica glass fibers in fiberglass have a fracture stress of 0.5×10^4 MPa, and ordinary window glass has a fracture stress of 69 MPa. Fracture in both of these materials is brittle. If K_{Ic} of silica glass is 0.79 MPa · m$^{1/2}$, what length defect is responsible for initiating fracture in each of these forms of silica glass? Assume that the crack is a surface crack.

Solution

The relationship between fracture stress and crack or defect length is given in Equation 11.8 as

$$K_{Ic} = Y\sigma_f(\pi a)^{1/2}$$

Since surface cracks are assumed, $Y = 1.1$ and $K_{Ic} = 0.79$ MPa · m$^{1/2}$.
For the glass fibers, $\sigma_f = 5000$ MPa

$$K_{Ic} = 0.79 \text{ MPa} \cdot \text{m}^{1/2} = (1.1)(5000 \text{ MPa})(\pi a)^{1/2}$$

Solving for the crack length (a),

$$(\pi a)^{1/2} = \frac{0.79 \text{ MPa} \cdot \text{m}^{1/2}}{5500 \text{ MPa}} = 1.4 \times 10^{-4} \text{ m}^{1/2}$$

$$(\pi a) = 2.0 \times 10^{-8} \text{ m}$$

$$a = 0.64 \times 10^{-8} \text{ m} = 6.4 \text{ nm}$$

The crack in the silica glass fibers is of nanometer dimensions.
For the bulk glass, $\sigma_f = 69$ MPa

$$K_{Ic} = 0.79 \text{ MPa} \cdot \text{m}^{1/2} = (1.1)(69 \text{ MPa})(\pi a)^{1/2}$$

Solving for the crack length (a),

$$(\pi a)^{1/2} = \frac{0.79 \text{ MPa} \cdot \text{m}^{1/2}}{76 \text{ MPa}} = 1.0 \times 10^{-2} \text{ m}^{1/2}$$

$$(\pi a) = 1.0 \times 10^{-4} \text{ m}$$

$$a = 0.32 \times 10^{-4} \text{ m}$$

The crack in bulk glass is nearly four orders of magnitude larger than for the silica glass fibers.

Tensile stress applied to a crack, as shown in Figures 11.2 and 11.4a, opens the crack. This kind of fracture is called mode I fracture, and the critical-stress intensity related to mode I fracture is K_{Ic}. Shear stress can also be applied to a crack, as shown in Figures 11.4b and 11.4c for mode II and III fracture, with the respective values of K_{IIc} and K_{IIIc}. The tensile crack opening K_{Ic} is the most frequently utilized critical-stress intensity factor.

11.2.2　Fracture of Brittle Materials: Ceramics, Metals, and Polymers

Ceramic materials and glass are brittle. Table 11.1 presents values of the K_{Ic} for some brittle ceramic materials at room temperature.

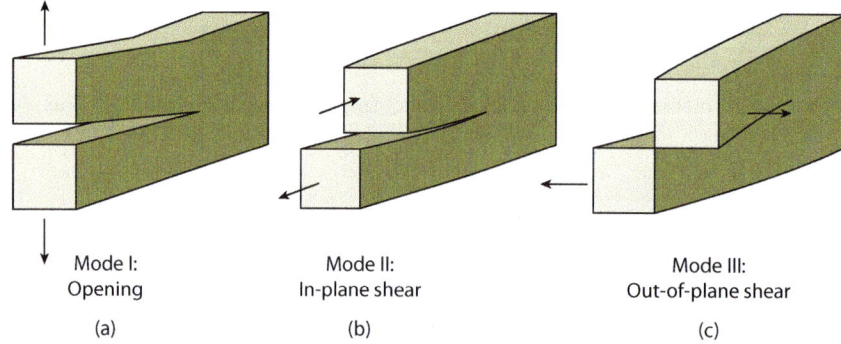

Mode I:
Opening
(a)

Mode II:
In-plane shear
(b)

Mode III:
Out-of-plane shear
(c)

Figure 11.4　The three primary modes of fracture. (a) In mode I, the crack surface is perpendicular to the applied tensile force. (b) In mode II, shear stress propagates the crack parallel to the direction of the applied shear force. (c) In mode III, the crack propagates the crack perpendicular to the direction the applied shear force. *(Based on http://upload.wikimedia.org/wikipedia/commons/e/e7/Fracture_modes_v2.svg)*

Table 11.1	K_{Ic} for Some Brittle Ceramic Materials at Room Temperature

	$K_{Ic}(\mathrm{MPa \cdot m^{1/2}})$
Silicate glasses	0.7–0.9
Glass ceramics	~2.5
Single crystal NaCl	~0.3
Single crystal Si	~0.6
Single crystal MgO	~1
Single crystal SiC	1.5
Sintered, hot-pressed SiC	4–6
Single crystal Al_2O_3	
(0001)	4.5
$(10\bar{1}0)$	3.1
$(10\bar{1}2)$	2.4
$(11\bar{2}0)$	2.4
Polycrystalline Al_2O_3	3.5–4
Al_2O_3-Al composites	6–11
Reaction-bonded Si_3N_4	2.5–3.5
Sintered, hot-pressed, and gas-pressure sintered Si_3N_4	6–11
Cubic stabilized ZrO_2	~2.8
MgO-partially stabilized zirconia (PSZ)	9–12
Tetragonal zirconia polycrystals (Y-TZP, Ce-TZP)	6–12
Al_2O_3-ZrO_2 composites	6.5–13
Single crystal WC	~2

Reprinted from Chaing, Y-M., Birnie D. B., and Kingery, W. D., *Physical Ceramics—Principles for Ceramic Science and Engineering,* John Wiley & Sons, N.Y. (1997), p. 484.

Partially stabilized zirconia (PSZ) and tetragonal zirconia polycrystals (TZP) have a high K_{Ic} value relative to that of stabilized cubic zirconia and many of the other ceramic materials. This value of K_{Ic}, along with a high melting temperature of 2715°C, is why PSZ and TZP are considered for applications such as cylinder liners for high-temperature gasoline and diesel engines.

The K_{Ic} value is relatively high in PSZ because of a stress-induced martensitic transformation in the partially stabilized zirconia. Pure zirconia has three possible crystal structures, as shown in the zirconia (ZrO_2)-calcia (CaO) phase diagram in Figure 5.17. ZrO_2 is cubic at temperatures from 2370°C to its melting temperature. From 2370°C down to 1170°C, pure zirconia is tetragonal, and at temperatures below 1170°C, pure zirconia transforms to a monoclinic phase through a martensitic phase transformation. When compounds, such as calcia, ceria, magnesia, or yttria, are added to zirconia, the temperature for the transformation to the tetragonal and monoclinic phases is decreased. If approximately 10 molar percent MgO is added to zirconia, and the mixture is given a specialized heat treatment, a partially stabilized microstructure results at room temperature that has a cubic-zirconia matrix and precipitate particles of tetragonal zirconia. The material in the tetragonal structure at the tip of a crack is induced by stress to transform martensitically into the monoclinic structure,

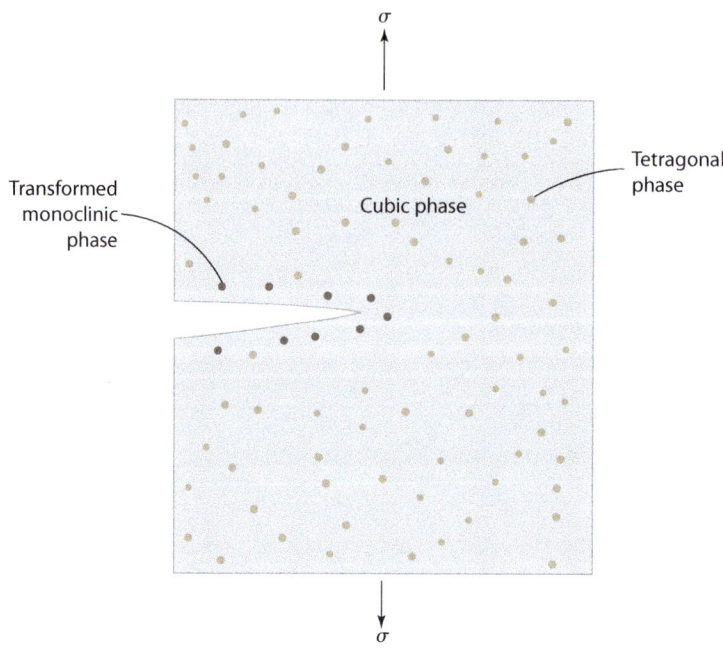

Figure 11.5 A schematic of a heat-treated, partially stabilized zirconia specimen with a crack subject to a tensile stress (σ). The matrix is cubic zirconia, with the tetragonal-phase particles in the bulk material shown as lightly shaded circles. The stress at the crack tip induces particles of the tetragonal phase to transform martensitically into the monoclinic phase, shown as dark circles, causing compressive stresses at the crack tip and absorbing energy. (*Based on Courtney, T. H.,* Mechanical Behavior of Materials, *2nd ed.,* McGraw-Hill, N.Y. (2000), p. 469.)

as shown in the schematic of Figure 11.5. The strain energy increases the energy of the tetragonal phase making it less stable, and the applied stress assists the movement of atoms into the monoclinic phase that has a lower energy at lower temperatures. Also, the movement of atoms by distance (Δx) during the transformation while subject to an applied force (F) absorbs energy $F \Delta x$. The absorbed energy ($F \Delta x$) adds a new positive term to the energy for a material with a crack in Equation 11.4, and it results in a larger value for K_{Ic}. Additionally, compressive stresses are created at the crack tip where the monoclinic phase is present, because of a 4.7% volume expansion during the martensitic transformation from the tetragonal to the monoclinic phase. The compressive state of stress at the crack tip suppresses crack propagation. Cracks propagate as a result of tensile or shear stresses, not compressive stresses.

In TZP, sufficient compounds, such as MgO, are added to the zirconia to stabilize the tetragonal phase at the operating temperature. However, a martensitic transformation of the tetragonal phase to the monoclinic phase is induced by stress at the crack tip in the same manner as described for PSZ. Compressive stress is also developed at the crack tip as a result of the 4.7% expansion where the monoclinic phase forms from the tetragonal phase.

Some metals fracture in a brittle manner, such as BCC metals at low temperatures, very high-strength tool steels, and white and gray cast iron. However, many metals with a crack will fracture with some localized plastic strain along the fracture surface, as discussed in Section 11.2.3. Some polymers are brittle, such as PMMA at low temperature, epoxy, and rubber. However, many polymers also fracture with some localized plastic strain along the fracture surface.

11.2.3 Fracture with Localized Surface Plastic Strain

The analysis for brittle fracture is extended to materials that fracture with plastic strain limited to the fracture surface, while the bulk of the specimen remains elastic. The following analysis applies to all materials that meet this condition. If the plastic deformation is only along the fracture surface and not in the bulk of the specimen, there is a plastic work per unit area (γ_p) associated with the fracture surface. We can then replace the surface energy (γ_s) in Equations 11.7 and 11.8 with ($\gamma_s + \gamma_p$), as shown in Equations 11.9 and 11.10.

$$\sigma_f = \left(\frac{2E(\gamma_s + \gamma_p)}{\pi a} \right)^{1/2} \qquad\qquad \textbf{11.9}$$

$$K_{Ic} = Y\sigma_f(\pi a)^{1/2} = Y2E(\gamma_s + \gamma_p)^{1/2} \qquad\qquad \textbf{11.10}$$

The surface-energy term also appears in Equations 11.9 and 11.10 because there is still a surface associated with the crack. However, usually the plastic work per unit area is much greater than the surface energy per unit area ($\gamma_s << \gamma_p$).

The compact tension specimen, shown in Figure 11.6, is often used to determine the value of K_{Ic} for metals. Figure 11.6 shows the loading of a compact tension specimen. A notch is machined into the specimen, and then the specimen is cycled to create a sharp fatigue crack of known length (a). Then the specimen is loaded to fracture. Specimens for determination of K_{Ic} have also been designed that have

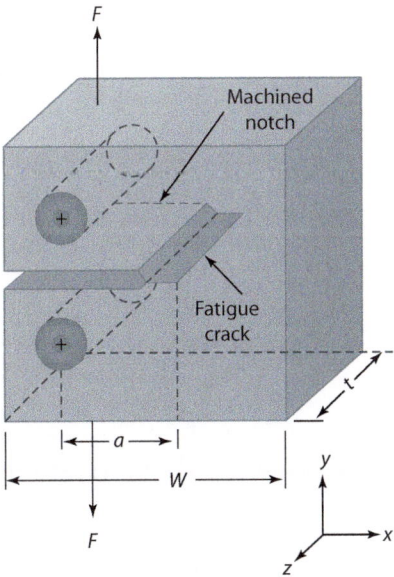

Figure 11.6 The design of the compact tension specimen to determine K_{Ic}. (*Based on Smith, W. F., and Hashemi, J., Foundations of Materials Science and Engineering, 5th ed., McGraw-Hill, N.Y. (2010), p. 290.*)

a center crack, and another design uses an edge crack and three-point bending. With the specimen in Figure 11.6, the K_{Ic} is determined with Equation 11.11, where F_c is the load at fracture.

$$K_{Ic} = \frac{F_c}{Wt}(a)^{1/2}\left[(29.6 - 185.5\left(\frac{a}{W}\right) + 655.7\left(\frac{a}{W}\right)^2 - 1017\left(\frac{a}{W}\right)^3 + 638.9\left(\frac{a}{W}\right)^4\right] \qquad \textbf{11.11}$$

11.2.4   Fracture of Metals

The K_{Ic} for metals, such as the high-strength steels, titanium alloys, and aluminum alloys shown in Table 11.2, is much greater than for the brittle materials shown in Table 11.1. The main reason for this difference is the magnitude of the surface plastic work (γ_p) in metals in comparison to the magnitude of the surface energy (γ_s) that is the only term present for brittle ceramics. Brittle ceramics do not have any surface plastic work at room temperature. In Table 11.2, the aluminum alloys have the lowest values of K_{Ic}. One reason for this is the low elastic modulus of 70 GPa for aluminum in comparison to 116 GPa

Table 11.2   The Room-Temperature K_{Ic}, Yield Stress σ_y, and Critical Crack Length a_c at the Yield Stress for Some Metals

Material	K_{Ic} MPa$\sqrt{m}$	σ_y MPa	a_c mm
2014-T651	24.2	455	3.6
2024-T3	~44	345	~21
2024-T851	26.4	455	4.3
7075-T651	24.2	495	3.0
7178-T651	23.1	570	2.1
7178-T7651	33	490	5.8
Ti-6A1-4V	115.4	910	20.5
Ti-6A1-4V	55	1035	3.6
4340	98.9	860	16.8
4340	60.4	1515	2
4335 + V	72.5	1340	3.7
17-7PH	76.9	1435	3.6
15-7Mo	49.5	1415	1.5
H-11	38.5	1790	<0.6
H-11	27.5	2070	0.23
350 Maraging	55	1550	1.6
350 Maraging	38.5	2240	<0.4
52100	~14.3	2070	~0.06

Reprinted from Hertzberg, R. W., *Deformation and Fracture Mechanics of Engineering Materials*, 3rd ed., John Wiley & Sons, N.Y. (1989), p. 304.

and 211 GPa for titanium and steel, respectively. Equation 11.10 shows that the critical-stress intensity factor is linearly related to the elastic modulus.

Within each type of material, there is a wide range of values of K_{Ic}. For example, 4340 steel with a yield strength of 1550 MPa has a K_{Ic} of 60.4 MPa $\cdot$ m$^{1/2}$. For 4340 steel with a lower yield strength of 860 MPa, the value of K_{Ic} is 98.9 MPa $\cdot$ m$^{1/2}$. The higher value of K_{Ic} results in a higher fracture stress for the same length of crack a, according to Equation 11.10. From Equation 11.9, the fracture stress is increased if the plastic work per unit area (γ_p) is increased. Conversely, if the yield strength of a material is increased, the ductility decreases, and the fracture strength decreases for a given crack length a, as demonstrated in Example Problem 11.2.

Example Problem 11.2

A bolt with a 1-mm-deep thread will attach a jet engine to the air frame. Assume that the thread is a 1-mm-deep surface crack. Compare the fracture stress for the bolt if it is to be made from the two different 4340 steels shown in Table 11.2.

Solution

The relationship between crack length and fracture stress is given by Equation 11.10.

$$K_{Ic} = Y\sigma_f(\pi a)^{1/2}$$

For the 4340 steel with the yield strength of 1515 MPa, the K_{Ic} is

$$K_{Ic} = 60.40 \text{ MPa} \cdot \text{m}^{1/2} = Y\sigma_f(\pi 0.001 \text{ m})^{1/2} = 1.1\sigma_f(0.056 \text{ m}^{1/2}) = 0.062 \text{ m}^{1/2}\sigma_f$$

Solving for the fracture stress,

$$\sigma_f = \frac{60.4 \text{ MPa} \cdot \text{m}^{1/2}}{0.062 \text{ m}^{1/2}} = 974 \text{ MPa}$$

Note that the fracture stress for this steel with a thread is only 63% of the yield stress. This material would fracture before yielding.

For the steel with a yield strength of 860 MPa, the value of the K_{Ic} is

$$K_{Ic} = 98.9 \text{ MPa} \cdot \text{m}^{1/2} = Y\sigma_f(\pi 0.001 \text{ m})^{1/2} = 1.1\sigma_f(0.056 \text{ m}^{1/2}) = 0.062 \text{ m}^{1/2}\sigma_f$$

Solving for the fracture stress,

$$\sigma_f = \frac{98.9 \text{ MPa} \cdot \text{m}^{1/2}}{0.062 \text{ m}^{1/2}} = 1595 \text{ MPa}$$

The steel with the lower yield strength has a higher fracture strength than the steel with the higher yield strength. The low yield strength steel yields before it fractures in this application. The higher-yield-strength steel fractures before the stress reaches the yield strength. The lower-yield-strength steel is a better selection for this design if the yield strength is sufficiently high, because it is best to have yield before fracture.

Figure 11.7a shows microvoids resulting from plastic deformation on the fracture surface of a metal. The coalescence of microvoids on the fracture surface indicates a ductile material with a high energy absorption. High-carbon steels at low temperature have **transgranular** cleavage of the grains

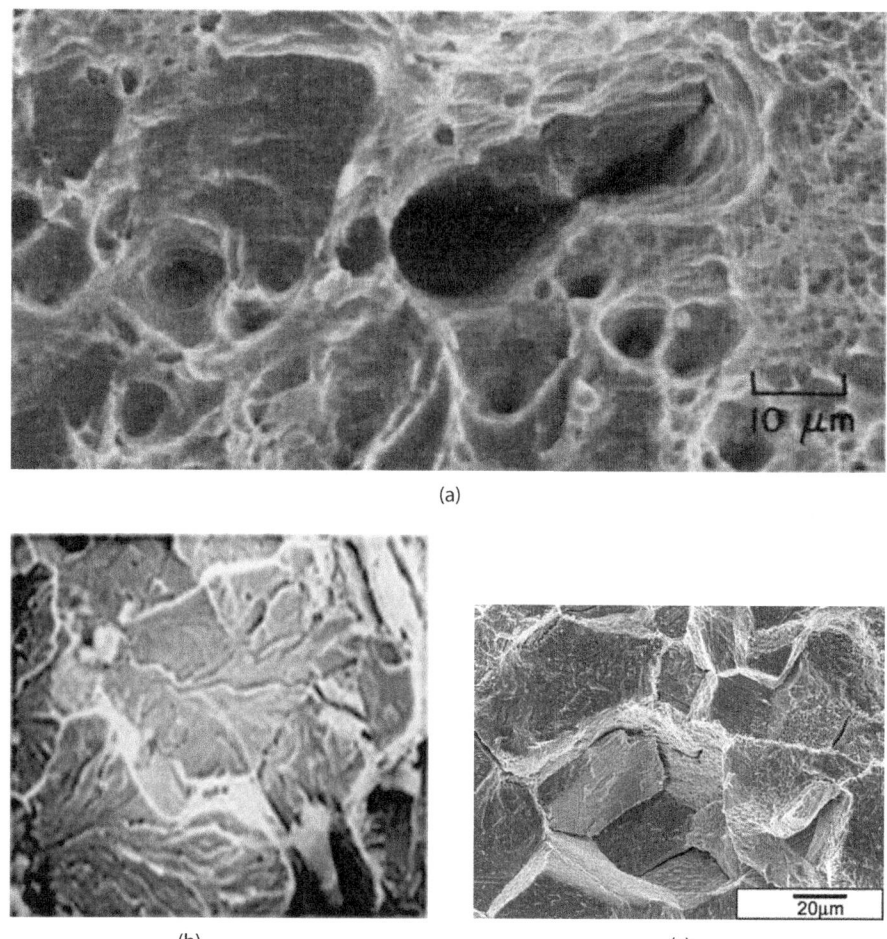

(a)

(b)   (c)

Figure 11.7   Scanning electron microscope images of fracture surfaces. (a) Microvoids on the fracture surface of the ductile alloy Cu-27 weight percent Ni-9 weight percent Fe. (b) Transgranular cleavage in a brittle carbon steel. (c) Intergranular fracture in nickel-aluminum bronze. (*ASM Handbooks Online, http://www.asmmaterials.info, ASM International, 2004*).

on the fracture surface, as shown in Figure 11.7b. In transgranular cleavage, the crack fractures across the crystalline grains on the cleavage planes similar to how salt crystals or diamonds are cleaved. Transgranular cleavage does not absorb as much energy as microvoid coalescence does. A metal with transgranular cleavage has a lower K_{Ic} than does the same composition metal that fractures with microvoid coalescence. **Intergranular** fracture, shown in Figure 11.7c, is a mode of fracture in metals that absorbs even less energy. In intergranular fracture, the fracture surface follows the grain boundaries. Intergranular fracture occurs in metals that have brittle grain boundaries. The flat surfaces in Figure 11.7c are grain boundary surfaces. In precipitation-hardened alloys, brittle grain boundaries can result from improper heat treatment, as demonstrated in Figure 7.18b. Grain boundaries can also become brittle if a metal, such as steel, is contaminated with metals such as tin or lead. The tin and lead concentrate in the grain boundaries of steel and induce brittleness.

11.2.5   Fracture of Polymers

Polymers have much lower K_{Ic} values than do metals, as shown in Table 11.3 for polymers and Table 11.2 for metals. One reason for this difference is that the elastic modulus (E) of polymers is much less than for metals, and Equation 11.8 shows that K_{Ic} is linearly related to the elastic modulus. Also, the surface energy of polymers is nearly two orders of magnitude less than for metals, as shown in Table 3.2.

In polymers, the weak van der Waals bond between different LCMs in comparison to the covalent bonds along the LCMs results in different mechanisms of fracture for polymers than is present in metals and ceramics. When a polymer is subject to sufficient tensile stress, a **craze** forms perpendicular to the tensile axis when the weak van der Waals bonds that hold different LCMs together are broken, as shown in Figure 11.8a, but the covalent bonds along the long chain molecules are not broken. The term *craze* is also applied to small cracks observed in ceramics; however, in ceramics there is no atomic bond between the two surfaces of the crack. Polymer LCMs, or fibrils, still hold the material together that cross the craze in a polymer. The polymer is not significantly weakened by the formation of a craze, because it is only the weak dipole bonds that are broken. Figure 11.8b shows how a craze becomes a microvoid when some of the LCMs break. Microvoids then link to form a crack, as shown in Figure 11.8c. In ductile polymers microvoids are observed on the fracture surface.

In brittle polymers, such as PMMA at room temperature, the initial part of the fracture surface is glassy smooth. This smooth section is called the mirror area. In later stages of growth, the crack changes from one crazed area to another, producing a rough surface.

Table 11.3 The Room-Temperature K_{Ic} for Some Polymers	
Polymer	$K_{Ic}(\text{J} \cdot \text{cm}^{-3} \cdot \text{m}^{1/2})$
epoxy	0.6
polyester thermoset	0.6
polystyrenes	0.7–1.1
high-impact polystyrenes	1–2
poly(methyl methacrylate)s	0.7–1.6
poly(ether sulfone)	1.2
acrylonitrile-butadiene-styrene	2.0
polycarbonate	2.2
poly(vinyl chloride)s	2–4
polyamide (nylon 6,6)	2.5–3
polyethylenes	1–6
polypropylenes	3–4.5
polyoxymethylene	4
poly(ethylene terephthalate)	5

Note: $\text{J} \cdot \text{cm}^{-3} \cdot \text{m}^{1/2} = \text{MPa} \cdot \text{m}^{1/2}$
Reprinted from Brostow, W., Kubat, J., and Kubat, M., *Mechanical Properties, Physical Properties of Polymers Handbook*, Mark, J. E., ed., AIP Press, Woodbury, N.Y. (1996), p . 331.

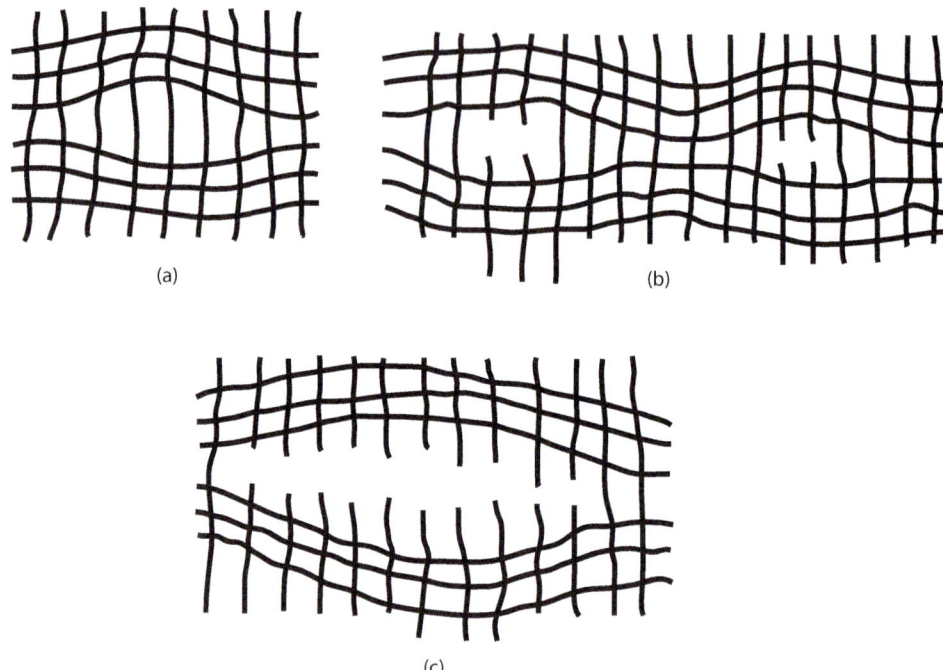

Figure 11.8 (a) A craze in a polymer will form perpendicular to a tensile stress. For this illustration, the tensile stress is vertical on the page. (b) Microvoids form when the tensile stress breaks some LCMs. (c) A crack forms when several microvoids link.

11.2.6 The Effect of Specimen Thickness on the Critical Stress Intensity Factor

If a fracture test reveals plastic deformation on more than just the fracture surface, then the analysis discussed above is not valid, because the plastic deformation is not a surface work per unit area. This scenario happens when the material is very ductile or if the fracture specimen is very thin. Figure 11.9 shows (a) thin and (b) thick specimens fractured in tension. For the thick specimen, the fracture surface is perpendicular to the tensile load, as it is for a brittle material. In the thick specimen the plastic deformation is limited to the fracture surface. If the thickness of the specimen is reduced, the percentage of fracture surface normal to the tensile axis is reduced, and the relative amount of plastic deformation is increased. In very thin specimens the fracture surfaces are at 45° to the tensile axis, and they meet in the center at a chisel edge, as shown in Figure 11.9a. There is plastic deformation throughout the material in the vicinity of the fracture surface in the specimen in Figure 11.9a.

The increased plastic deformation in the bulk of the thinner specimen results in an increased amount of work required to fracture the specimen per unit volume of material in comparison to the thick specimen. The increased work in the thinner specimen makes it appear that the value of the plastic work term (γ_p) in Equation 11.10 is increased, and the critical-stress intensity factor is increased. Figure 11.10 is a schematic of the value of the stress intensity factor for fracture as a function of thickness. The most conservative and accepted value of K_{Ic} is obtained from a fracture test where the fracture surface is normal to the tensile loading direction. All other higher values of measured stress intensity

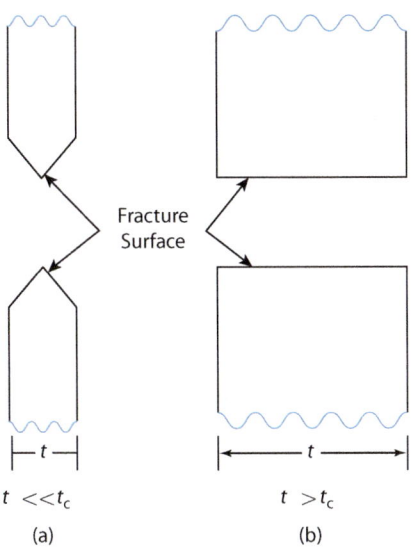

Figure 11.9 (a) A schematic of the 45° chisel edge fracture surface of a specimen with thickness much less than a critical thickness t_c. (b) A schematic of the flat fracture surface of a specimen with thickness greater than a critical thickness t_c.

factors are labeled K_c and called stress intensity factor, but not critical-stress intensity factor and not K_{Ic}. K_c values are not valid values of K_{Ic}. The condition for a valid K_{Ic} is that the thickness (t) shown in Figure 11.6 meets the condition described in Equation 11.12.

$$t > 2.5\left(\frac{K_{Ic}}{\sigma_y}\right)^2$$

11.12

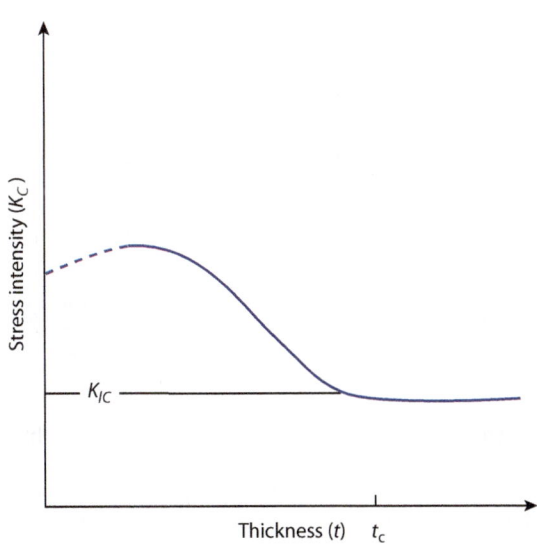

Figure 11.10 A schematic of the stress intensity required for fracture, measured as a function of thickness. Valid values of K_{Ic} are when the thickness is greater than a critical thickness (t_c).

When the fracture surface is normal to the tensile axis, this is called plane-strain fracture, because there is no significant plastic strain in the thickness direction (z). All of the strain is in the x-y plane identified in Figure 11.6. The term *plane stress* is applied to when the fracture surface is at 45° to the tensile axis and forms a chisel edge. There is no applied stress in the z direction; all of the stress is in the x-y plane.

11.2.7   The Effect of Temperature on K_{Ic}

The value of K_{Ic} increases with an increase in temperature, because the total amount of plastic work at fracture increases with temperature. As shown in Equation 11.10, as the surface plastic work increases, the value of K_{Ic} increases. The amount of plastic work increases with increasing temperature, because at elevated temperatures there is more thermally activated dislocation motion in crystalline materials and more chain sliding in polymers.

The temperature dependence of the energy to fracture a specimen is often studied with a Charpy or Izod impact test, schematically shown in Figure 11.11a. The specimen orientations for the Charpy and Izod tests are shown below the pendulum in Figure 11.11a. Details of the Charpy and Izod specimen shape are shown in Figure 11.11b. For each of these tests, a pendulum with a known initial potential energy at an initial height (h_0) strikes a specimen with a machined notch. When the pendulum strikes the specimen, the bending of the specimen results in a tensile stress on the notch, and the pendulum has sufficient energy to fracture the specimen. The fracture of the specimen absorbs some energy from the pendulum. The pendulum swings to a final measured height (h_f). The energy absorbed by the specimen from the pendulum is determined from the difference in heights, and the energy absorbed is displayed on the test machine.

Because the Charpy and Izod specimens are not as thick as required in Equation 11.12 of a valid K_{Ic} test specimen, the Charpy and Izod fracture has a greater percentage of material with plastic

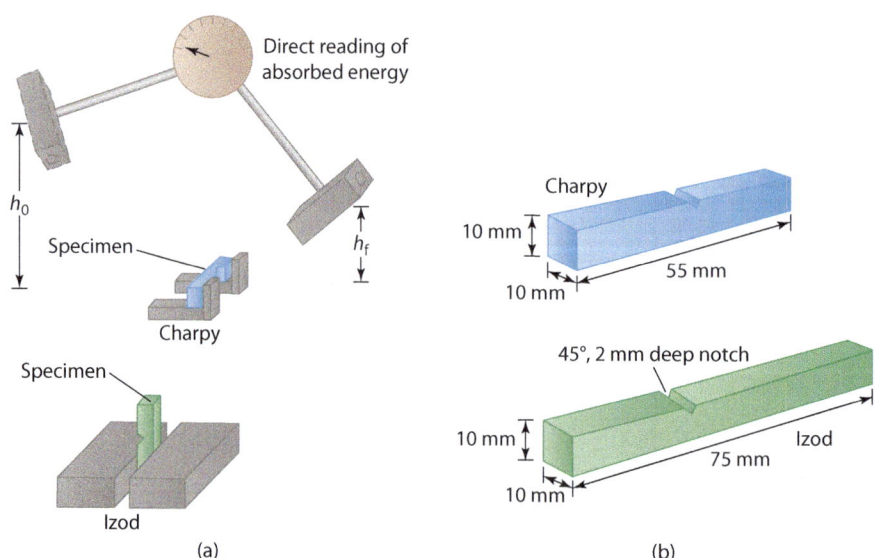

Figure 11.11   (a) The Charpy and Izod impact testers. In the top-left figure, the impact hammer attached to a pendulum starts at height h_0 and finishes at height h_f after fracturing the specimen. The bottom of (a) shows the specimen-mounting configurations for the Charpy and Izod tests. (b) The design of the specimen for the Charpy (top) and Izod (bottom) impact fracture test. (*Based on Askeland, D. R., Fulay, P. P., and Wright, W. J., The Science and Engineering of Materials, 6th ed., Cengage Learning, Stamford, CT (2011), p. 227.*)

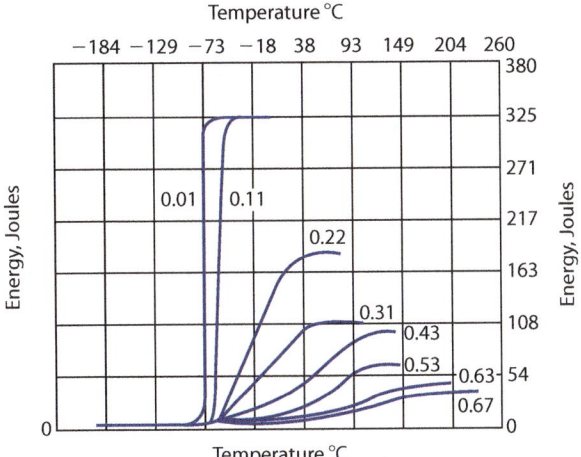

Figure 11.12   The impact energy for plain-carbon steels of varying carbon content, as a function of temperature. (*Based on Rineholt, J. A., and Harris, W. J.,* Trans. American Society for Metals, *43 (1951), p. 1175.*)

deformation at a given temperature than would a valid K_{Ic} fracture specimen. Therefore, the Charpy and Izod fracture results cannot be used to obtain K_{Ic} and the design stresses that result from application of Equation 11.10. However, the Charpy and Izod results provide an indication when fracture could be a problem in an application if the energy absorption is low. The Charpy impact energies of various steels are shown in Figure 11.12 as a function of temperature. Note that structural steels, such as 1022 and 1031, lose nearly 50% of their energy-absorbing capacity at 273 K (0°C) relative to higher temperatures, such as 373 K (100°C). Zero degrees celsius is a common temperature for applications of structural steel. The low energy absorption of higher-percentage carbon steels at temperatures close to 0°C is one reason why the Liberty ships, shown in Figure 11.1, had brittle fracture in cold water and cold air.

One definition of the **ductile-to-brittle transition temperature** (T_{DBTT}) is the temperature where the energy absorbing capacity of the material is reduced by 50% from fully ductile fracture. To determine the T_{DBTT} measure the upper-shelf energy and the lower-shelf energy; these measurements are where the energy absorbed becomes horizontal at high and low temperatures, respectively. To determine the T_{DBTT}, we determine where the energy absorption is 50% of the difference between the energy absorbed at the upper shelf and the energy absorbed at the lower shelf. For example, the energy absorbed at the uppershelf for the 1022 steel is 180 J, and the energy absorbed at the lower shelf is approximately 10 J. The difference is 170 J and 50% of the difference is 85 J. If 85 J is subtracted from the energy absorbed at the upper shelf of 180 J, the result is 95 J. The T_{DBTT} is the temperature where the 1022 steel absorbs 95 J or approximately 250 K (−23°C).

Another way to measure the T_{DBTT} is to determine the temperature where 50% of the fracture is ductile and 50% is brittle. Ductile fracture produces a dull gray surface because the microvoids, shown in Figure 11.7, trap the light that strikes the surface. Ductile fracture absorbs more energy than does transgranular cleavage or intergranular fracture. Transgranular cleavage, shown in Figure 11.7 and intergranular fracture, shown in Figure 11.7 produce bright reflective surfaces, because the light reflects from the flat crystalline cleavage planes or from the grain boundaries. The T_{DBTT} is the temperature when 50% of the fracture surface is dull resulting from plastic strain and 50% of the fracture surface is bright resulting from transgranular cleavage or intergranular fracture. The percentages of ductile fracture and of transgranular cleavage are determined by visual inspection or by using microscopic inspection techniques.

The T_{DBTT} determined by different techniques may not agree exactly; however, each measurement should give an indication when brittle fracture and low fracture stresses could become a problem. Figure 11.12 shows that fracture tests must be conducted down to the minimum temperature that would be experienced in a design application because very large decreases in energy absorption are possible with very small temperature changes.

The reason that steel experiences a ductile-to-brittle transition is shown schematically in Figure 11.13a, where the yield stress and the fracture stress of steel is shown as a function of temperature. As the temperature increases, plastic strain is activated at a lower stress, which results in a decrease in the yield stress. Equation 11.10 shows that more surface plastic strain increases the fracture stress. At high temperatures, the yield stress is lower than the fracture stress. The steel specimen deforms plastically before it fractures, and the fracture is ductile. At lower temperatures, the fracture stress is lower than the yield stress, and the specimen fractures before it yields, meaning that this material is brittle. A transformation from ductile to brittle fracture occurs when the yield stress and fracture stress cross when plotted as a function of temperature.

Figure 11.13b shows the yield strength and fracture strength for high-purity FCC metals such as copper, nickel, aluminum, or austenitic stainless steel as a function of temperature. The fracture strength

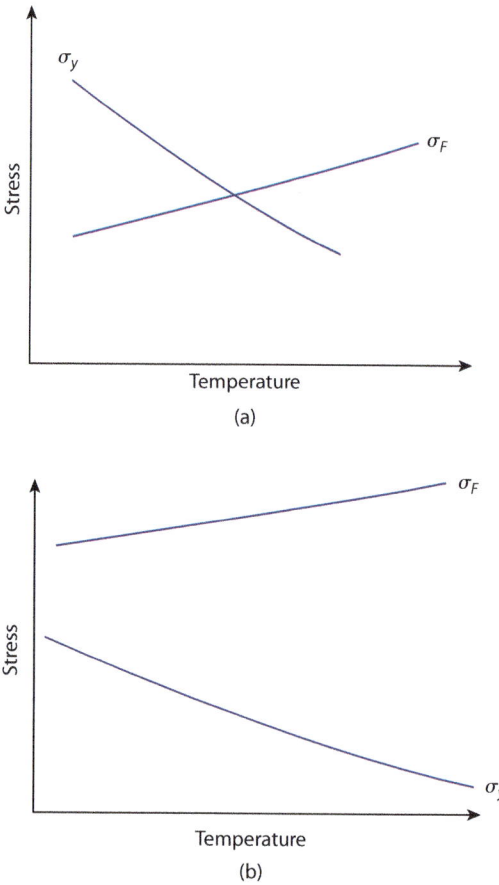

Figure 11.13 (a) The yield and fracture strength as a function of temperature for a BCC metal such as carbon steel. (b) The yield and fracture strength as a function of temperature for a high purity FCC metal such as copper, nickel, aluminum, or austenitic stainless steel.

is greater than the yield strength at all temperatures. Such a material does not experience a ductile-to-brittle transition. Highly alloyed and highly strengthened FCC metals can have a T_{DBTT} because the yield strength is increased, and it can intersect the fracture stress.

11.2.8 Environmental Effects on the Fracture of Metals, Polymers, and Ceramics

If the material is in a hostile environment, the value of K_{Ic} is reduced. For a metal, corrosion can result in the production of brittle reaction products at the tip of a crack, and the corroded metal fractures with a lower K_{Ic}. Also, corrosion can result in the metal ions going into solution, causing the crack to grow in length in a manner similar to pit formation shown in Figure 10.13. A combination of stress and corrosion reducing the K_{Ic} of a metal is called **stress corrosion cracking**. The symbol for the critical-stress intensity under conditions of stress corrosion cracking is K_{ISCC}. Figure 11.14 shows the value of the stress intensity factor for a titanium alloy in saltwater as a function of time. The corrosive effects require time, and the value of K_{ISCC} is taken as the value where there is no more time dependence to K_{ISCC}, which is approximately 38 MPa · m$^{1/2}$ for this titanium alloy. The value of K_{Ic} in air for this titanium alloy is typically 100 MPa · m$^{1/2}$. Environments that reduce the critical-stress intensity for metals include water, saltwater, acids, hydrogen, neutron irradiation, and liquid metals such as mercury.

In polymers, the K_{Ic} is reduced by oxygen, water, acids, solvents, and ultraviolet radiation. Corrosive environments break the bonding between LCMs at the crack tip, resulting in crack growth and a loss of strength. Figure 10.6 shows the percent reduction in impact strength of PVC as a function of time at various locations. Reductions in impact strength by more than 90% are possible in PVC after 4 years in a warm and humid environment such as Singapore.

Ceramics have their K_{Ic} lowered by reactions at the crack tip in environments such as water, saltwater, acids, and bases. These environments can produce new compounds at the crack tip. Also, ions in ceramics can go into solution in solvents, such as water, because ceramics such as MgO have a significant amount of ionic bonding. Because environments in addition to corrosion, such as UV radiation, can reduce the value of K_{Ic}, the general term *environmental-assisted cracking* is utilized for the critical-stress intensity (K_{IEAC}) in environments that reduce the fracture stress. If any material is to be applied in an environment that can reduce the value of K_{Ic} over time, and a fracture-avoiding design is required, then the application designer should utilize a valid K_{IEAC} for the application environment when making any design calculations.

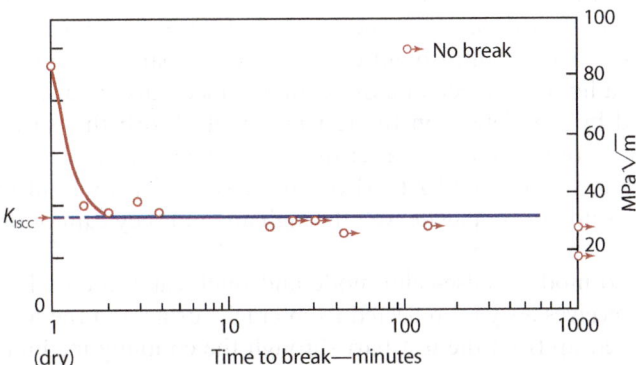

Figure 11.14 The time to failure at various values of stress-intensity factor (K_I) for the titanium alloy Ti-8Al-1Mo-1V in a solution of 3.5% NaCl in water. (*Based on Brown, B. F.*, Naval Engineers Journal, *78(3) (1966), p. 457.*)

11.2.9   Fracture Mechanics and Design

Values of K_{Ic}, such as those in Tables 11.1 to 11.3, are applied to design by using nondestructive testing to measure the crack length (*a*) in a part or structure, and then using Equation 11.10 to calculate the fracture stress from the known value of K_{Ic}. Some of the possible nondestructive tests (NDTs) are summarized below and include visual inspection, dye penetrant, X-ray and gamma-ray penetration, ultrasound, eddy currents, and magnetic particles. If no crack is found with nondestructive inspection, we must assume that one or more cracks are present that are just below the minimum detectable crack length possible with the inspection technique available. We must assume the cracks are present because we cannot prove they are not present.

To obtain the maximum fracture resistance for a structure, designers should select the material with the highest value of K_{Ic} that meets the strength requirements. A higher yield strength than is required generally means lower ductility and less fracture resistance. Also, the final product should have the minimum possible crack lengths. This requirement means careful machining that does not introduce surface cracks; the selection of materials low in defects that act as internal cracks, such as voids or inclusions; and careful nondestructive inspection. The next section describes the primary NDTs capable of detecting cracks.

11.2.10   Nondestructive Tests

Dye penetrant (DP) detects surface cracks and other defects by their absorption of dye. DP is a rapid, simple, inexpensive, and sensitive NDT for observing surface cracks. In DPNDT, the surface is first cleaned, and then a liquid dye that penetrates into cracks is coated onto the surface of the part. After the DP is given ample time to penetrate any cracks, the DP on the surface is removed. Any DP that has penetrated cracks remains. A developer is then applied that draws the penetrant to the surface of the part and enhances the visibility of the DP. Various colored dyes are available that provide contrast with colored parts for visual inspection, and fluorescent DP is viewed in ultraviolet light. Any type of material can be tested with DP; however only surface cracks are observed.

Ultrasonic NDT (UNDT) uses the absorption and reflection of ultrasonic waves to detect cracks, porosity, and changes in density. UNDT is one of the most widely utilized tests for the detection of cracks. Ultrasonic waves are sound waves of a frequency above 20,000 hertz (Hz). Most ultrasonic waves for NDT are generated with a piezoelectric transducer (PZT). A PZT is a crystal that converts a voltage into an expansion or contraction of the PZT crystal. PZT crystals are covered in Chapter 16. A high-frequency voltage applied to the PZT crystal generates high-frequency deformation waves in the crystal. The deformation is transmitted into ultrasound waves in the test part through a coupling medium. The coupling medium can be a liquid or a gel, and dry coupling mediums are also available. The ultrasonic wave intensity is reduced by any defects in the test part that absorb the waves and produce heat or that scatter the waves. Any pores or cracks scatter the ultrasonic waves, and some of the scattered wave intensity is reflected back to the original PZT. Also, a decrease in density results in a reduction in wave intensity. Ultrasonic waves do not propagate well through air, and they cannot propagate at all through a vacuum.

UNDT operates in two modes: pulse-echo mode and pitch-catch mode. In pulse-echo mode, one PZT sends a signal and receives a signal reflected from crack surfaces in the test part. A PZT can also convert deformation picked up from the test part through the coupling medium into a voltage that is electronically converted into a signal that reveals the presence of the crack. In the pitch-catch mode, a signal is sent by one PZT and detected by another PZT. In this mode, the second PZT detects a decrease in signal intensity when defects are in between the pitching and catching PZTs. UNDT signals are compared with signals from well-characterized cracks to obtain crack lengths. UNDT can detect internal cracks, but cracks that are parallel to the wave propagation direction are difficult to observe. Surface waves can be generated to observe surface cracks. UNDT can be used on any material. For example, medical ultrasonography is the imaging of human tissue, fetuses, organs, and arteries with UNDT.

Eddy current (EC) NDT uses changes in the eddy currents induced in a material to detect cracks, changes in chemistry, and porosity. An example of an EC detector is the metal detectors that you might see people using at the beach. In ECNDT, an alternating magnetic field from a coil induces ECs in a conductive part. The EC in the conductive part changes the magnetic field, which is detected with a second receiver coil. A crack in a metal part reduces the EC in the part, and this is detected in the receiver coil. ECNDT can detect both surface and internal cracks, changes in chemical composition, and surface quality. ECNDT can only test conductors such as metals and graphite. ECNDT is the best technique for detecting cracks in uniform continuous conducting materials, such as metal pipe, extrusions, or railroad rails. ECNDT does not contact the part, and there is no alteration of the part. ECNDT can detect cracks in conductors that are buried inside nonconductors, such as steel reinforcing rods in concrete and graphite fibers in epoxy resin.

Radiographic NDT (RNDT) detects differences in material density with both X-rays and γ-rays. Anyone who has had a broken bone or a decayed tooth X-rayed has observed how this technique detects material defects. X-rays and γ-rays are both electromagnetic radiation, as is visible light, and they are discussed in Chapter 18. X-rays are photons that have energies from approximately 100 eV to 10^5 eV, and γ-rays have energies from 10^5 eV to 10^8 eV. X-rays are generated by directing high-energy electrons at a target, which excites the target electrons out of their normal orbitals. X-rays are generated when other electrons fill the vacant orbitals. γ-rays are emitted by radioactive elements. One of the most common radioactive sources is Co^{60}, which emits γ-rays of 1.17 and 1.33 MeV as a result of radioactive decay. Several other radioactive sources include Cs^{137} (0.66 MeV) and Ir^{192} (0.31, 0.47, 0.60 MeV and some additional energies).

The energy of an X-ray is absorbed in a material by exciting electrons out of their normal orbitals, and low-energy γ-rays are absorbed by the same mechanism; however, high-energy γ-rays are also absorbed by the nucleus. Materials with a high concentration of electrons and therefore a high density strongly absorb X-rays. The energy required for RNDT depends upon the application. RNDT of a thin polymer part requires a low-energy X-ray, such as 10,000 eV. However, if it is desired to observe possible radioactive material in an unopened train box car, γ-rays with 10^6 eV might be required. The higher the energy of the radiation, the deeper the penetration is into the material. Photographic film was the original detector of X-rays and γ-rays; however, now electronic detector systems are available that give instant images. Cracks and changes in density are observed visually on the film or on electronic screens. A crack, a void, or a material of low density does not absorb X-rays as strongly as a dense material does. A crack appears dark on film because the crack allows a greater intensity of X-rays to reach the film. A tooth cavity appears dark, and a metal filling is white because it absorbs more of the X-rays. RNDT can be used on any material of any shape. It can detect both surface and internal cracks; however, it is hard to observe very tightly closed cracks and cracks whose surfaces are perpendicular to the X-ray beam. **Computed tomography** (CT) is a computed three-dimensional radiographic image created by analyzing images of a specimen from different angles.

Magnetic particle inspection (MPI) detects changes in the magnetic field in a ferromagnetic material due to the presence of cracks and other defects. MPI is an economical technique for detecting surface cracks in ferromagnetic materials, such as BCC steel. Ferromagnetic materials are covered in Chapter 17. In MPI, a ferromagnetic part is magnetized in one direction. However, any defect disturbs the direction of magnetization. A crack or pore is detected by covering the surface of the part with fine dry ferromagnetic particles that become oriented in the direction of magnetization by the magnetic field of the part. However, at the crack or pore, the magnetic particles are disturbed, and this is observed visually. Magnetic particles are available in red, gray, and black, and fluorescent magnetic particles are also available. MPI is simple to use, it is inexpensive, parts of nearly any shape can be tested, and it is very sensitive to surface cracks. However, MPI can be used only on ferromagnetic materials, such as BCC steel, and it can detect only surface cracks and defects.

In a **proof test**, the part with a known value of K_{Ic} is tested to a specified stress (σ), and if the part does not fail, this guarantees that there is not a crack of length greater than a determined by Equation 11.10. We must still assume that there are one or more cracks that are just an increment shorter than (a), because the proof test cannot prove that such a crack length does not exist. Unfortunately, if the part fails the proof test, the part is destroyed. A proof test is conducted on critical parts where failure during service is not acceptable, such as in the rocket motor casings used to propel the shuttle into space. Proof testing is also valuable when the part is to be subject to repeated stress that produces fatigue-crack propagation. The assumed crack length resulting from the proof test is used to calculate a safe fatigue life or an inspection period, as shown in Section 11.3.5.

Example Problem 11.3

Nondestructive testing is conducted on a critical steel part, and no crack is found. The minimum detectable crack length for internal cracks for the NDT technique is 0.2 mm. The part is made from the 4340 steel in Table 11.2 with a yield strength of 1515 MPa. Based upon the NDT, what is the maximum allowable stress on this part, assuming that a safety factor of 2 is applied to both the fracture stress and yield stress for this design?

Solution

Since the NDT minimum detectable crack length is 0.2 mm, we must assume at least one crack exists in the steel that is just an increment smaller than 0.2 mm or essentially equal to 0.2 mm. For an internal crack, $2a$ is equal to 0.2 mm, or a is equal to 0.1 mm. Table 11.2 shows the K_{Ic} for this material is 60.4 MPa · m$^{1/2}$. The relationship among K_{Ic}, crack length, and fracture strength is given by Equation 11.10. For an internal crack, Y is equal to 1.

$$K_{Ic} = Y\sigma_f(\pi a)^{1/2} = \sigma_f([\pi][0.0001 \text{ m}])^{1/2} = 0.018 \text{ m}^{1/2}\sigma_f = 60.40 \text{ MPa} \cdot \text{m}^{1/2}$$

Solving for the fracture stress,

$$\sigma_f = \frac{60.4 \text{ MPa} \cdot \text{m}^{1/2}}{0.018 \text{ m}^{1/2}} = 3407 \text{ MPa}$$

This fracture stress is above the yield stress and above the ultimate tensile strength, which means that the steel should not fracture until the ultimate tensile strength has been exceeded. Applying the safety factor of 2 to the fracture stress results in a maximum applied stress of 1704 MPa. Applying the safety factor of 2 to the yield stress results in

$$\sigma_{des} = \frac{\sigma_{ys}}{2} = \frac{1515}{2} \text{ MPa} = 758 \text{ MPa}$$

The design stress is 758 MPa determined by the safety factor applied to the lowest stress, which is the yield stress in this problem.

Example Problem 11.4

A part made of the 4340 steel in Table 11.2 with a yield strength of 1515 MPa is to be proof tested. The engineer in charge decides that since the steel has a sharp upper yield stress of 1515 MPa, the part can be proof tested to 90% of the upper yield stress without producing any permanent deformation to the part. The part passes the proof test. What is the maximum-length internal crack possible in the part?

Solution

The relationship among K_{Ic}, crack length, and fracture strength is given by Equation 11.10, and for an internal crack, $Y = 1$. The part is stressed to 90% of 1515 MPa, which is 1364 MPa. Since this is the maximum applied stress, it is assumed that if the stress is increased by the smallest of increments, the part would fracture. Even though the part did not fail, it is assumed that 1364 MPa is equal to the fracture stress, because it has not been proven that the part would not fail if the stress was increased by the smallest of increments above 1364 MPa.

$$K_{Ic} = Y\sigma_f(\pi a)^{1/2} = 1364 \text{ MPa}(\pi a)^{1/2} = 60.40 \text{ MPa} \cdot \text{m}^{1/2}$$

Solving for the maximum internal crack length that could be present in the part,

$$(\pi a)^{1/2} = \frac{60.4 \text{ MPa} \cdot \text{m}^{1/2}}{1364 \text{ MPa}} = 0.044 \text{ m}^{1/2}$$

$$\pi a = 0.002 \text{ m}$$

$$a = 0.0006 \text{ m}$$

$$2a = 0.0012 \text{ m, or } 1.2 \text{ mm}$$

This is the maximum crack length possible in the material, because any larger crack would have caused failure at a stress lower than 1364 MPa.

11.2.11 Fracture Mechanics and the Fracture of Smooth Tensile Test Specimens

The most common form of mechanical-property test is the tensile test with a smooth specimen. Fracture-mechanics analysis applies to the fracture of smooth tensile specimens when there is not ductile failure of the specimen resulting from necking and extensive plastic strain. Modes I, II, and III of fracture of smooth tensile specimens are shown in Figure 11.15. These modes of fracture are not the same as those shown in Figure 11.4, where a major crack is present in the specimen.

Mode I fracture, shown in Figures 11.15a and 11.15b, is brittle fracture; there is no plastic deformation prior to fracture, and fracture occurs at the mode I "x" in the stress-strain diagram in Figure 11.16. The fracture initiates at a defect in the material when the combination of stress (σ) and defect length (a) satisfies Equation 11.10. This is the mode of fracture that occurs in brittle materials, such as glass, ceramics, BCC metals at low temperature, and gray and white cast iron. The fracture starts at a crack, as shown in Figure 11.15a. In polycrystalline materials, the crack propagation can be transgranular cleavage, as shown in Figure 11.15a, or intergranular crack propagation, as shown in Figure 11.15b. Figure 11.7b shows the appearance of the fracture surface for transgranular cleavage, and Figure 11.7c shows the fracture surface appearance for intergranular fracture. The appearance of these fracture surfaces applies

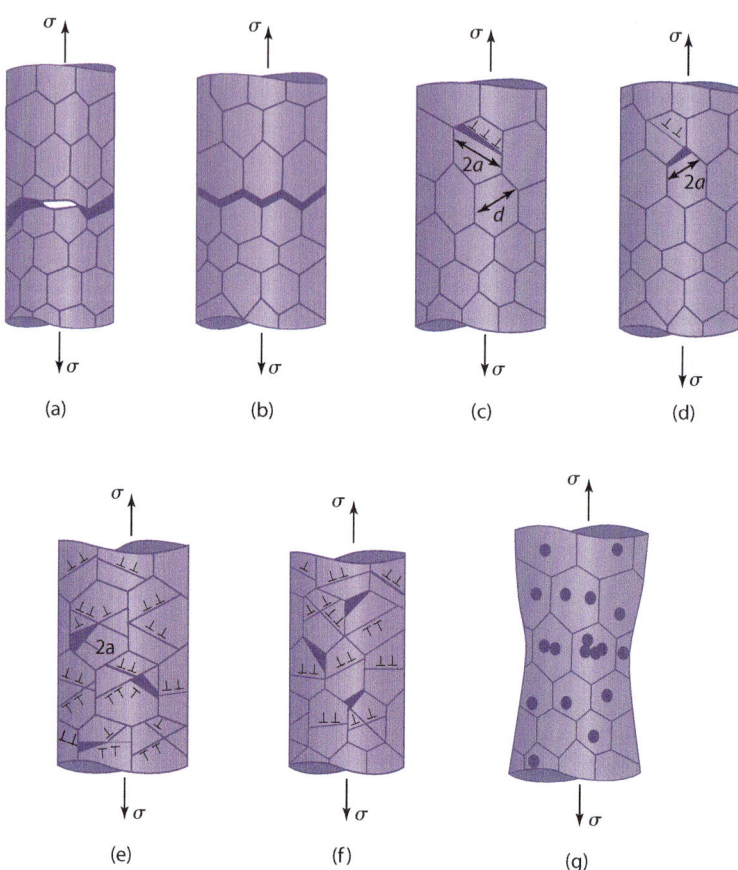

Figure 11.15 Mode I brittle fracture is due to the presence of a critical-length defect shown in (a) as the white area resulting in transgranular cleavage crack propagation (purple area) and in (b) intergranular crack propagation (the dark boundary area) results in fracture. In mode II fracture, isolated plastic strain creates in (c) a transgranular cleavage crack (dark area with dislocation symbols ⊥) that grows to a critical size to cause transgranular fracture. In (d) an intergranular crack (dark boundary area) is formed by plastic strain and grows to a critical size to cause intergranular fracture. In mode III fracture, extensive plastic strain (indicated by dislocation symbols ⊥⊥⊥) results in the initiation of many cracks; as shown in (e) transgranular cleavage cracks form, grow, and link until a critical crack length is reached, and in (f) intergranular cracks form, grow, and link until a critical crack length is reached. (g) Extensive plastic strain results in the creation of microvoids that coalesce to cause ductile fracture. (*Based on Courtney, T. H., Mechanical Behavior of Materials, 2nd ed., McGraw-Hill, N.Y. (2000), p. 481.*)

to modes I, II, and III. If the material is amorphous, such as glass or amorphous polymers, then the crack propagates through the amorphous material in a direction perpendicular to the tensile stress. In mode I fracture, there is no reduction in cross-sectional area or necking of the tensile specimen.

   Mode II fracture, shown in Figures 11.15c and 11.15d, occurs in metals when there is no defect of sufficient length present to cause fracture prior to localized plastic deformation. The localized plastic strain results in the formation of a crack that is of sufficient length (a) to cause fracture at the stress (σ) in accordance with the fracture mechanics criterion in Equation 11.10. In polycrystalline materials, the crack can be transgranular as in Figure 11.15c, or intergranular as in Figure 11.15d. In mode II failure, there is no significant reduction in area of the specimen after fracture. The stress-strain diagram appears

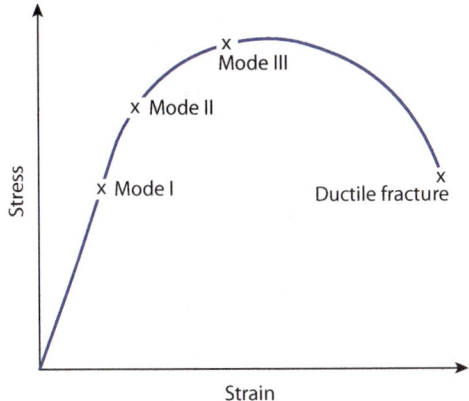

Figure 11.16 A schematic of a stress-strain diagram. The "x" marks where a specimen fails in mode I, II, III, or ductile fracture.

as shown in Figure 11.16, with a small amount of plastic strain before fracture at the "x" for mode II. Mode II and III fracture do not apply to polycrystalline ceramics.

In mode III fracture of metals, shown in Figures 11.15e and 11.15f, there is no crack of sufficient length to cause fracture in the material prior to the initiation of extensive plastic strain throughout the specimen. As a result of extensive plastic strain in polycrystalline metals, transgranular microcracks can form as in Figure 11.15e or intergranular microcracks can form as in Figure 11.15f. In mode III fracture, many small cracks can link to form a larger critical-length crack (a) at the stress (σ) that satisfies the fracture mechanics criterion Equation 11.10. The stress-strain diagram appears as in Figure 11.16 with fracture at the "x". In mode III the load-carrying capacity does not decrease before fracture. In mode III fracture, the area of the specimen reduces by 1 to 10% prior to fracture; however, no necking occurs.

Ductile fracture of metals, shown in Figure 11.15g, occurs when there is extensive plastic deformation throughout the specimen. In ductile fracture a neck forms that reduces the cross-sectional area of the specimen. The stress-strain diagram of these materials shows an ultimate tensile strength, as demonstrated in Figure 11.16, followed by a decrease in the load-carrying capacity. Failure occurs by plastic strain rather than fracture defined by Equation 11.10. In very ductile materials, the reduction in area approaches 100%, and the failure is due to the extreme reduction in area. In metals, such as medium-strength steels, titanium, and aluminum alloys, there is a smaller reduction in area of approximately 10 to 30%. In steel, voids form around inclusions in the steel. Inclusions are particles of impurities, such as iron sulfide. The interface between the inclusions and the steel is weaker than the steel, and the interface fractures. A small void then opens at the inclusion-steel interface. Voids link together to form a crack that eventually leads to failure in the neck region.

A round tensile specimen of ductile steel will usually fail such that the fracture surfaces have a cup-and-cone shape, as shown in Figure 11.17. The voids form at the center of the specimen, and at the outer edge of the specimen, the specimen pulls apart by shear. The presence of the surface inclined at 45° to the tensile axis (a shear lip) is evidence of shear strain. A greater amount of 45° surface on a fractured tensile test specimen indicates more plastic strain to failure. The fracture surface in some metal alloys is in one plane at 45° to the tensile axis, which is called a slant shear fracture. Some high-strength aluminum-alloy tensile specimens with a rectangular cross section fail by slant shear fracture.

Smooth polymer tensile specimens also demonstrate mode I, II, and III failure; however, the transgranular and intergranular differentiation does not apply to polymers. Brittle polymers, such as PMMA at low temperature and epoxy, fracture without significant plastic strain in mode I. The cracks can be surface imperfections due to machining or processing or they can be internal defects. Mode II fracture occurs in polymers when a small amount of plastic strain produces microcracks that grow to a

Figure 11.17 The cup and cone fracture of a ductile aluminum alloy. The cup fracture surface is on the right and the cone fracture surface is on the left. © *sciencephotos / Alamy*

critical length, as demonstrated in Figure 11.8. In mode III fracture of polymers, extensive plastic strain results in the linking of microvoids and cracks to produce a critical-length crack. In the stress-strain diagram for a thermoplastic polymer in Figure 6.23, mode I fracture occurs before the sharp yield point, mode II fracture occurs immediately after the sharp yield point, mode III fracture occurs after significant plastic strain, and ductile fracture occurs after extensive plastic strain.

Ductile failure occurs in polymers such as PE tested at room temperature, and PMMA tested at elevated temperatures. PE is one of the most ductile polymers; when tested at room temperature, the reduction in area approaches 100% due to plastic flow resulting from LCM sliding. For thermoplastic polymers, lower temperatures and higher strengths favor craze, microvoid, and crack formation leading to mode III, II, and I failure as temperature is reduced. Higher temperatures and lower strength favor the sliding of LCMs and ductile failure.

 ## 11.3 FATIGUE FAILURE

(*Mike Peters, Mother Goose and Grimm, Image 50815*) MOTHER GOOSE&GRIMM ©1994 Grimmy, Inc. Distributed by King Features Inc. Mike Peters © King Features Inc.

11.3.1   Introduction

Fatigue failure occurs when a part fails as a result of the application of many stress cycles. An example of a fatigue failure is the shaft shown in Figure 11.18, which has a keyway that transferred torque from the shaft to a wheel. The rotating shaft survived many cycles, but eventually a crack initiated at the base of the keyway at the location marked "Origin." The site of crack initiation is the origin of the clamshell markings or striations on the fracture surface. These markings are produced at the tip of the propagating fatigue crack as the applied stress goes through a cycle, and they are evidence of fatigue-crack propagation. An investigator of failures, such as airline or train crashes, looks for striations on the surface of fractured parts to determine if the failure occurred as a result of a single high stress or if it was the result of many stress applications that caused fatigue failure. The crack grows in length until fracture occurs, according to the fracture-mechanics criterion given by Equation 11.10.

Figure 11.18   The fracture surface of a rotating shaft that failed in fatigue. The crack initiation site, labeled as "Origin," is located in the corner of a keyway. The fatigue crack then propagated from the origin, producing striations or clam shell markings on the fracture surface until final rupture occurred. (*Reprinted from Wulpi, D.J., Understanding How Components Fail, 1985, ASM International.*)

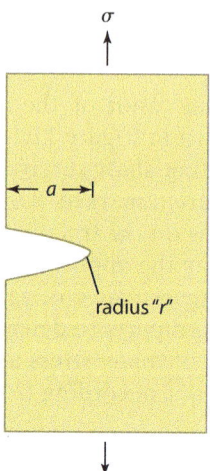

Figure 11.19   A crack with a depth a and radius of curvature r resulting in a stress concentration. *(Based on Askeland, D. R., Fulay, P. P., and Wright, W. J., The Science and Engineering of Materials, 6th ed., Cengage Learning, Stamford, CT (2011), p. 252.)*

Cracks often initiate at a discontinuity, such as a keyway, because a discontinuity increases the stress locally, producing a stress concentration. Equation 11.13 shows that the **stress-concentration factor** (K_s) is the ratio of the maximum stress at the discontinuity (σ_{max}) to the average stress (σ_{av}) in the specimen without the discontinuity.

$$K_s = \frac{\sigma_{max}}{\sigma_{av}}$$ **11.13**

The stress-concentration factor for a crack is related to the length of the crack (a) and the radius of curvature of the crack tip (r), as shown in Figure 11.19 and Equation 11.14.

$$K_s \approx 2\sqrt{\frac{a}{r}}$$ **11.14**

11.3.2   Fatigue Tests of Smooth Metal and Polymer Specimens

In fatigue testing, multiple stress cycles that simulate the service conditions are applied until failure occurs in a number of cycles (N_f). Some of the possible cyclic stress applications are shown in Figure 11.20. In Figure 11.20a, the stress goes from tension to an equal amount of compression; this is a fully reversed stress. The mean stress (σ_m) is given by Equation 11.15, and it is equal to zero for a fully reversed stress.

$$\sigma_m = \frac{\sigma_{max} + \sigma_{min}}{2}$$ **11.15**

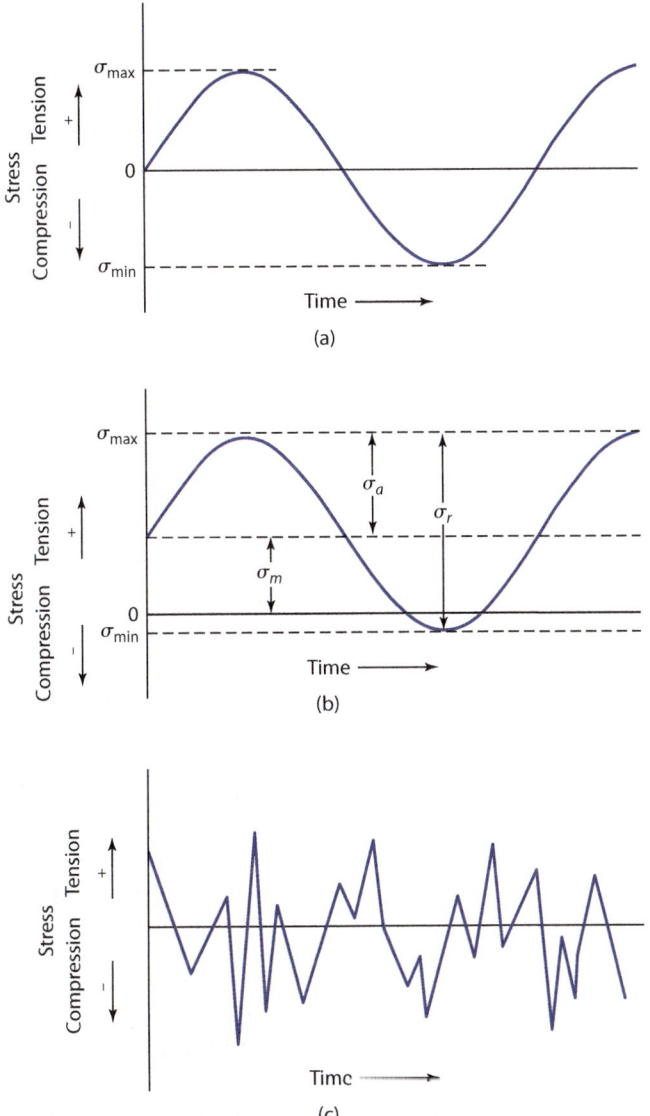

Figure 11.20 The variation of stress with time. (a) A fully reversed stress cycle where the tensile and compressive stress are equal in magnitude but of opposite sign. (b) A stress cycle where the tensile stress is greater than the magnitude of the compressive stress. (c) Random stresses applied to a specimen. (*Based on Callister, W. D., Materials science and Engineering, An Introduction, 6th ed., John Wiley & Sons, N.Y. (2003), p. 212.*)

The stress range (σ_r) is given by Equation 11.16,

$$\sigma_r = \sigma_{max} - \sigma_{min}$$ **11.16**

and the stress amplitude (σ_a) is given by Equation 11.17.

$$\sigma_a = \frac{\sigma_{max} - \sigma_{min}}{2} = \frac{\sigma_r}{2}$$ **11.17**

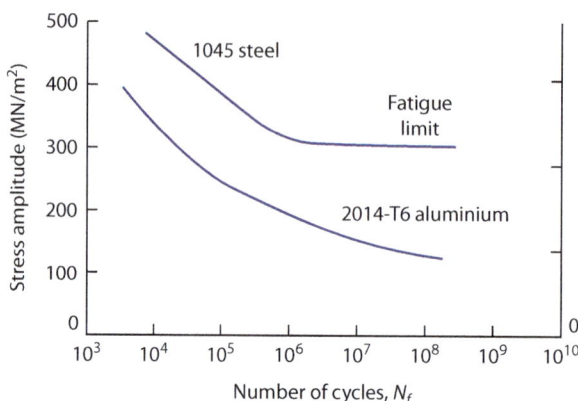

Figure 11.21   The number of cycles to failure as a function of stress amplitude for 1045 steel and 2014-T6 aluminum. The 1045 steel shows a fatigue endurance limit, and the aluminum alloy does not. (*Based on Courtney, T. H., Mechanical Behavior of Materials, 2nd ed., McGraw-Hill, N.Y. (2000), p. 577.*)

Figure 11.20b shows a sinusoidal cycle where the tension and compression stresses are not equal, and Figure 11.20c shows the application of random stresses. In cyclic stressing of a part, its life depends upon the mean stress and the stress range.

Fatigue tests are run with various types of cyclic load applications, including tension-compression, torsion, and bending. In Figure 11.21 the average number of cycles to failure (N_f) is plotted as a function of the stress amplitude (σ_a) for two metals: 1045 steel and 2014-T6 aluminum. This data is called an S-N plot for the stress and number of cycles to failure. Unless otherwise noted, the stress cycle is a fully reversed stress cycle.

As the stress amplitude increases, the average number of cycles to failure decreases, and the slope of the S-N plot is high at high stress amplitude. Fatigue failure with a high stress amplitude and low number of cycles to failure is called low-cycle fatigue. At high stress amplitudes, there is both elastic and plastic deformation during each cycle. For example, the 1045 steel in Figure 11.21 fails in 10^4 cycles at a stress amplitude of 475 MPa. The yield stress for 1045 steel is in the range of 450 to 650 MPa. At a stress amplitude of 475 MPa, there could be a significant amount of plastic strain in each cycle, depending upon the actual yield stress. If the stress amplitude is reduced, the average number of cycles to failure increases.

For the 1045 steel, there is a significant change in slope of the S-N plot at stress amplitudes less than 350 MPa. Lower stresses result in a large number of cycles to failure called high-cycle fatigue. In high-cycle fatigue, most the strain is elastic; however, there must be a small amount of plastic strain, because a truly elastic material should never fail in fatigue. In the 1045 steel, when the stress is reduced to below 300 MPa, the stress amplitude versus the number-of-cycles curve becomes horizontal and approaches infinity; this is called the **fatigue endurance limit**. The presence of a fatigue endurance limit means that for lower stresses, such as 250 MPa, the number of cycles to failure is infinite for practical purposes. Steels with a sharp yield stress have an endurance limit. Materials with a gradual transition from elastic to plastic deformation are more likely to behave like the 2014-T6 aluminum, with a gradual increase in fatigue life as the stress amplitude is reduced, but the S-N plot is not horizontal. Table 11.4 gives the stress for the number of cycles to failure (fatigue life) indicated for each type of material for fully reversed stress cycles. Although the stress is listed as the endurance limit, some of these materials, such as the aluminum alloys, behave in a manner similar to the 2014-T6 aluminum alloy, and the S-N plot is not horizontal. The results in Table 11.4 are the allowable

Table 11.4 The Tensile Strength (σ_u), Yield Strength (σ_y), and Stress Amplitude (σ_f) for the Indicated Life for Some Metal Alloys

Material	Condition	σ_u MPa	σ_y MPa	σ_f MPa
Steel Alloys[a] (Endurance limit based on 10^7 cycles)				
1015	Cold drawn—0%	455	275	240
1015	Cold drawn—60%	710	605	350
1040	Cold drawn—0%	670	405	345
1040	Cold drawn—50%	965	855	410
4340	Annealed	745	475	340
4340	Q & T (204°C)	1950	1640	480
4340	Q & T (427°C)	1530	1380	470
4340	Q & T (538°C)	1260	1170	670
HY140	Q & T (538°C)	1030	980	480
D6AC	Q & T (260°C)	2000	1720	690
9Ni–4Co–0.25C	Q & T (315°C)	1930	1760	620
300M	—	2000	1670	800
Aluminum Alloys[b] (Endurance limit based on 5×10^8 cycles)				
1100-0		90	34	34
2014-T6		483	414	124
2024-T3		483	345	138
6061-T6		310	276	97
7075-T6		572	503	159
Titanium Alloys[c] (Endurance limit based on 10^7 cycles)				
Ti–6Al–4V		1035	885	515
Ti–6Al–2Sn–4Zr–2Mo		895	825	485
Ti–5Al–2Sn–2Zr–4Mo–4Cr		1185	1130	675
Copper Alloys[c] (Endurance limit based on 10^5 cycles)				
70Cu–30Zn Brass	Hard	524	435	145
90Cu–10Zn	Hard	420	370	160
Magnesium Alloys[c] (Endurance limit based on 10^8 cycles)				
HK31A-T6	—	215	110	62–83
AZ91A	—	235	160	69–96

[a]*Structural Alloys Handbook*, Mechanical Properties Data Center, Traverse City, MI, 1977.

[b]*Aluminum Standards and Data 1976*, The Aluminum Association, New York, 1976. (See source for restrictions on use of data in design.)

[c]*Materials Engineering* 94(6) (Dec. 1981), Penton/IPC Publication, Cleveland, OH.

Reprinted from Hertzberg, R. W., *Deformation and Fracture Mechanics of Engineering Materials*, 3rd ed., John Wiley & Sons, New York (1989), p. 477.

stresses for the number of cycles to failure indicated. If the stress is reduced from the indicated value, it should not be assumed that the fatigue life is infinite. Note that the stress amplitude for the indicated life is typically from 35 to 60% of the yield stress. If a part is to be subject to multiple-stress applications, the applied stress should be well below the yield stress. This is one reason why in many designs the applied stress is not allowed to exceed 50% of the yield stress. However, if there are to be many such cycles, 50% of the yield stress may not be sufficiently conservative.

Figure 11.22 shows that the shape of the *S-N* plot for polymers is similar in shape to that of metals. However, in the polymer PS there is a very low-cycle-fatigue region of the *S-N* plot that has a low slope. The PS tests were run at stresses close to the ultimate tensile strength of 36 to 52 MPa. The applied stress in these fatigue tests is sufficient to form crazes in the polymer. The crazes act as microvoid-initiation sites that grow to form microcracks and cracks. Craze formation limits the stress that the polymer can withstand for a given number of cycles to failure. Crazes form in polymers but not in metals. Craze formation in polymers is one difference between the fatigue behavior of metals and polymers.

The application of a given stress cycle to a specimen does not have a unique number of cycles to failure. Rather, the number of cycles to failure has a distribution, as shown in Figure 11.23, where applied stress and the number of cycles to failure is plotted for different probabilities of failure (*P*). The average number of cycles to failure corresponds to the number of cycles for a 50% failure rate at the designated stress. For example, the 7075-T6 alloy in Figure 11.23 has a 50% probability of failure within 10^6 cycles if the stress amplitude is 250 MPa. Additionally, there is a 10% probability that the 7075-T6 alloy fails within 10^5 cycles at a stress amplitude of 250 MPa. A design for fatigue life must take into consideration the probability of failure. If the average life is utilized as the lifetime of the part, that means that 50% of the parts will fail within the design life. For applications such as aircraft, this failure rate is unacceptable. In the design of critical parts, the probability of failure

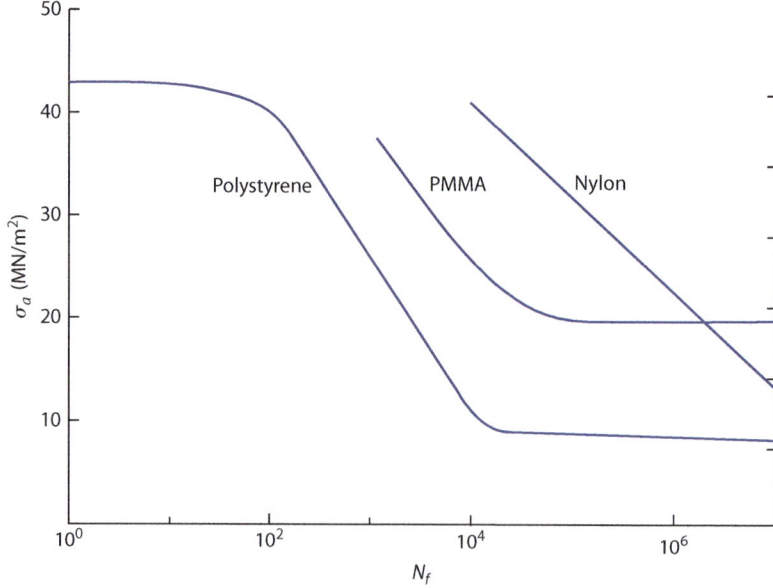

Figure 11.22   The cycles to failure as a function of stress amplitude for PS, PMMA, and nylon. (*Data for nylon and PMMA from Riddell, M. N., Koo, G. P., and O'Toole, J. L., Polymer Engr. Sc., 6(1966), 363. Data for polystyrene from Beardmore, P., and Rabinowitz, S., Treat. Matls. Sc. and Tech., 6 (1957), 267. Figure reprinted from Courtney, T. H., Mechanical Behavior of Materials, 2nd ed., McGraw-Hill, N.Y. (2000), p. 607.*)

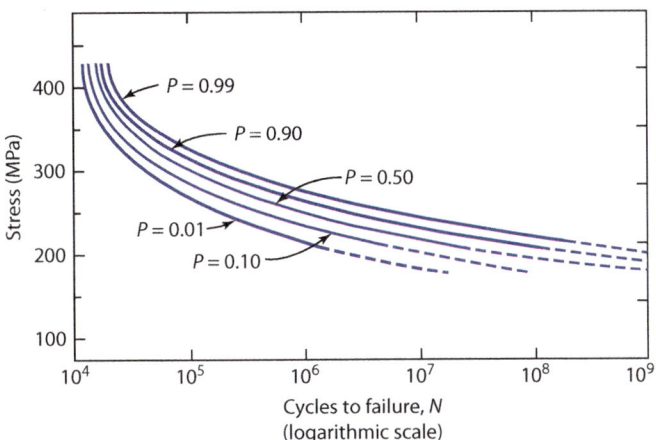

Figure 11.23 Probability of failure (P) as a function of stress amplitude and cycles to failure for the aluminum alloy 7075-T6. *(Based on Sinclair, G. M., and Dolan T. J., Trans. ASME, 75 (1953), 867.)*

during the design lifetime should be 1% or lower. However, in design it is often not practical to have a 0% probability of failure, because the part may have to be very large and heavy to have applied stresses that are very low. The statistics of failure are covered in Section 11.5.

Example Problem 11.5

For the aluminum alloy 7075-T6 in Figure 11.23, determine the stress amplitude that has a 90% probability of surviving 10^6 cycles.

Solution

A 90% probability of survival is a 10% probability of failure. In Figure 11.23, a vertical line drawn at 10^6 cycles intersects the line representing a 10% probability of failure, at an alternating stress of approximately 240 MPa.

11.3.3 Fatigue-Crack Initiation in Metals and Polymers

Fatigue failure can occur in materials that are stressed at a fraction of the yield stress, as shown in Table 11.4. For example, the 7075-T6 aluminum yield stress is 505 MPa; however, there is a 50% probability of failure in 10^8 cycles for this alloy with an alternating stress of only 200 MPa. In PMMA the yield strength is in the range of 54 to 73 MPa; however, there is a 50% probability that fatigue failure can occur in only 10^4 cycles at a stress of 20 MPa. Why do materials fail in fatigue at stresses below the yield stress? With elastic strain, atoms return to their original atomic positions when unloaded. A truly elastic material will not fail from fatigue. The presence of fatigue failure in metals and polymers indicates that some

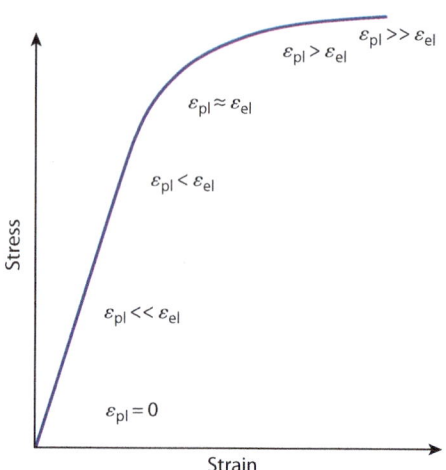

Figure 11.24   The relationship of elastic-strain and plastic-strain magnitudes to the stress-strain curve.

nonreversible, or plastic, process occurs during the cyclic stressing. There must be nonelastic deformation in the 7075-T6 alloy, even at 200 MPa, and in the PMMA at 20 MPa. Since plastic strain is not observed macroscopically at these low stresses in a conventional tensile stress-strain test, the plastic strain must be very small or microplastic. However, with repeated cycles the microplastic strain accumulates and eventually causes fatigue failure. Even a high-strength aluminum alloy, such as 7075-T6, has some weak areas, and plastic deformation begins and concentrates in these areas. The observation that fatigue failure occurs at stresses as low as 35% of the yield stress in 7075-T6 aluminum demonstrates that the material is not 100% elastic at these stresses.

The relative magnitudes of the elastic and plastic strain in a stress-strain test of a material with an apparent elastic region and a yield stress are schematically shown in Figure 11.24. At very low stresses, such as 0 to 20% of the yield stress, some structural materials are truly elastic ($\varepsilon_{pl} = 0$), and these materials will not fail from fatigue at these very low stresses unless a crack or some other major defect is already present in the material. At higher stresses, such as 20 to 50% of the yield stress in a tensile test, there is plastic strain, but the plastic strain is much less than the magnitude of the elastic strain ($\varepsilon_{pl} << \varepsilon_{el}$), and the plastic strain is not observable in a tensile test. This small plastic strain is referred to as microplastic strain. Structural materials can fail in fatigue at these low stresses because of accumulation of microplastic strain. At still higher stresses in a tensile test, such as 50 to 90% of the yield stress, the magnitude of the plastic strain is larger; however, it is still less than the magnitude of the elastic strain ($\varepsilon_{pl} < \varepsilon_{el}$). At the yield stress the amount of plastic strain is comparable in magnitude to the elastic strain ($\varepsilon_{pl} \cong \varepsilon_{el}$). At stresses larger than the yield stress, the plastic strain is larger than the elastic strain ($\varepsilon_{pl} > \varepsilon_{el}$). At stresses comparable to the ultimate tensile strength in a ductile metal, the plastic strain is much larger than the elastic strain ($\varepsilon_{pl} >> \varepsilon_{el}$). Figure 11.24 does not imply that the plastic strain in a tensile test is the same as the plastic strain in cyclic fatigue, because it is known that the development of strain hardening is different for a tensile test than it is for cyclic stresses. Figure 11.24 explains why materials fatigue at stresses significantly below the yield stress, where the material appears to be elastic in a tensile test.

During cyclic deformation, plastic strain accumulates. In tests of polished, smooth specimens, it is assumed that there are no cracks in the specimen at the beginning of the test. However, if plastic deformation concentrates in a local weak area of the material, fatigue cracks are initiated. Fatigue cracks in ductile crystalline metals are initiated by the mechanism shown in Figure 11.25 for a specimen subject to alternating shear stress in the directions of the arrows. During cyclic deformation, if very

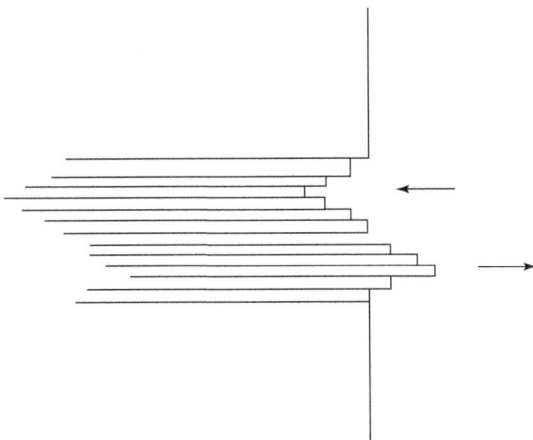

Figure 11.25 The formation of intrusions and extrusions on a metal surface, due to plastic deformation by dislocation motion on slip planes resulting from an alternating shear stress in the directions of the arrows. (*Based on Wood, W. A., The Study of Metal Structures and Their Properties, Pergamon Press, Elmsford, N.Y. (1971), p. 266.*)

small amounts of microplastic deformation are not reversed in each cycle of loading, then localized plastic deformation causes intrusions and extrusions on the specimen surface. The intrusions and extrusions develop with the number of cycles due to the accumulation of plastic strain, and intrusions can eventually become a crack.

In polymers, there is a large range of local bond strengths between LCMs, depending upon the proximity and orientation of neighboring LCMs. There are some particularly weak areas where the LCMs are distant from each other. The weak areas open up to form a craze, as shown in Figure 11.8. Cycling transforms the craze to a microvoid, where the crossing LCMs have been broken. Microvoids grow and link to form microcracks that grow with cycles into cracks.

With smooth, polished specimens of ductile metals and polymers, the fatigue life to failure (N_f) is the sum of the number of cycles to initiate a crack (N_i) and the cycles of to propagate a crack to failure (N_p). Fatigue-crack initiation in ductile metals and polymers is the result of localized plastic deformation. In ceramic materials, plastic deformation does not occur at low temperatures; thus there should not be any crack initiation by plastic-deformation processes. In smooth, polished polycrystalline ceramics, it is assumed that there are microcracks and voids present that can grow to form cracks and propagate to failure. The number of cycles to failure in smooth, polished polycrystalline ceramics is the number of cycles in crack propagation.

11.3.4   Improving Fatigue Life

It is possible through proper design and processing to increase the number of cycles required to initiate a crack (N_i). Surface defects such as scratches or tool marks act as crack-initiation sites where fatigue cracks form and these defects decrease N_i. A smooth or polished surface is less likely to initiate cracks resulting in an increase in N_i. Likewise, the material should be internally uniform, since inclusions, voids, and other internal defects are fatigue-crack initiation sites that decrease N_i. Materials with a high yield strength relative to the applied stress are less likely to fail from fatigue because there is a small amount of plastic strain in these materials. Case hardening is a process that increases the yield strength of the surface of a metal by diffusing or implanting atoms such as carbon or nitrogen into the surface. Case hardening produces a strong elastic outer shell that resists crack initiation, and a ductile core is maintained to resist fracture. Case hardening also produces compressive stresses at the surface, because additional atoms are diffused into the surface. Cracks do not propagate if the stresses at the crack tip are compressive. For this reason, the surface of a part is often produced with compressive stresses. Compressive surface stresses are also produced with processes that plastically deform the surface, such as **shot peening** and by certain thermal treatments. In shot peening a metal part, small hard spheres are shot at the surface of the metal part at high velocity. The high-velocity spheres plastically deform the surface of the metal part and produce compressive stresses at the metal surface. Tempering glass, covered in Section 7.3.3, is an example of a thermal treatment that produces a surface compressive stress.

The strengthening mechanisms of metals should produce a microscopically uniform metal, without any weak or strong regions. For example, in a precipitation-hardened metal, it is better to have small precipitate particles with a uniformly small spacing than larger particles with large spacing and low-strength metal in between. However, the precipitate particles must be sufficiently large that they are stable when subject to cyclic stresses. If the precipitate particles are too small, they can be cut by dislocations, and they can become smaller than the minimum stable particle size and dissolve. Cutting of precipitate particles allows weak regions to form in a metal, and the weak regions become sites of fatigue-crack initiation in an otherwise strong matrix. Figure 7.13 shows precipitate particles that have been sheared by dislocations. Also, the interface between the large particles, such as the Fe_3C plates in pearlite, and the surrounding α-phase matrix are subject to cracking during the fatigue cycling of plain-carbon steels.

Similarly, the interface region in composite materials is an area where fatigue cracks initiate, because of the large discontinuity in the elastic modulus that exists at the interface between a high-modulus-reinforcing phase and a low-modulus matrix. Also, there is often weak bonding between the matrix and the reinforcing phase in composite materials.

11.3.5   Fatigue-Crack Propagation

In crack-propagation studies, the crack length (a) is measured experimentally as a function of the number of cycles (N), as shown in Figure 11.26.

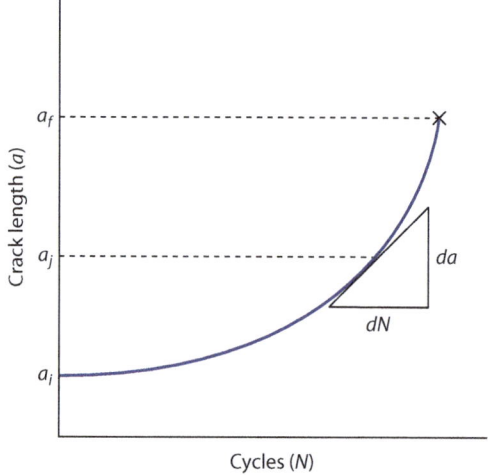

Figure 11.26   A schematic of crack length (a) as a function of the number of applied cycles (N) from an initial crack length a_i to a crack length at failure of a_f. The crack-propagation rate when the crack is of length a_j is the slope (da/dN) of the plot of a versus N.

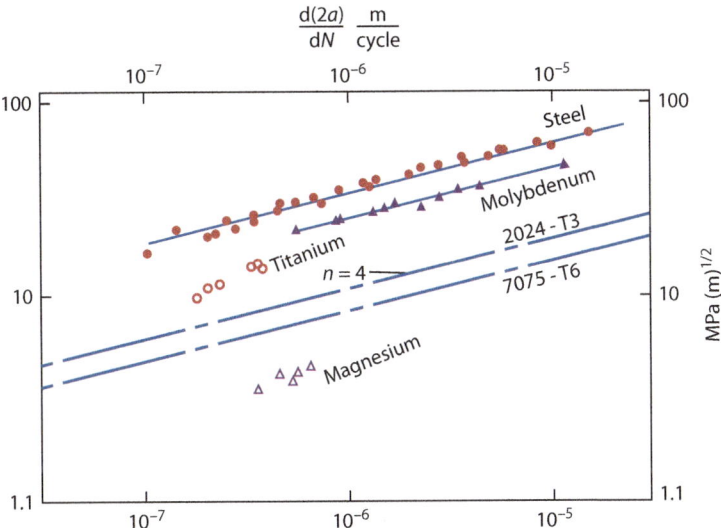

Figure 11.27 The fatigue-crack propagation rate ($d2a/dN$) as a function of the stress-intensity range ΔK for some metals. A crack length of $2a$ is used for internal cracks. (*Based on Paris, P. C.*, Fatigue-An Interdisciplinary Approach, *10th Sagamore Conference, Syracuse University Press, Syracuse, N.Y. (1964), p. 107.*)

Analysis of propagating cracks has shown that if the log of the crack propagation rate (da/dN) is plotted versus the log of the stress intensity range (ΔK) that a straight line results for an extensive range of crack propagation rates, as shown in Figure 11.27 for several metals, in Figure 11.28 for several polymers, and in Figure 11.29 for the ceramic partially stabilized zirconia. The stress intensity range (ΔK) is given by Equation 11.18:

$$\Delta K = Y\sigma_r(\pi a)^{1/2} = Y\Delta\sigma(\pi a)^{1/2} \qquad \textbf{11.18}$$

and $\Delta\sigma$ is the stress range (σ_r). In Figure 11.29, the units of the vertical axis are crack-propagation rate in meters per second; this equals the rate in meters per cycle times the cyclic frequency in cycles per second. Also, in Figure 11.29 the bottom horizontal axis is K_{max}. It is possible to convert from K_{max} to ΔK with the stress ratio ($\sigma_{min}/\sigma_{max}$) given for each test.

If the plot of log(da/dN) versus log(ΔK) is linear, with a slope equal to n, then Equation 11.19 follows from the equation of a straight line:

$$\log \frac{da}{dN} = n \log \Delta K + \log A$$

$$\textbf{11.19}$$

where A is a constant. Removing the logarithms from Equation 11.19 and inserting ΔK from Equation 11.18 results in Equation 11.20, which is called the Paris equation, where Y is assumed to be 1.

$$\frac{da}{dN} = A \Delta K^n = A \Delta\sigma^n(\pi a)^{n/2} \qquad \textbf{11.20}$$

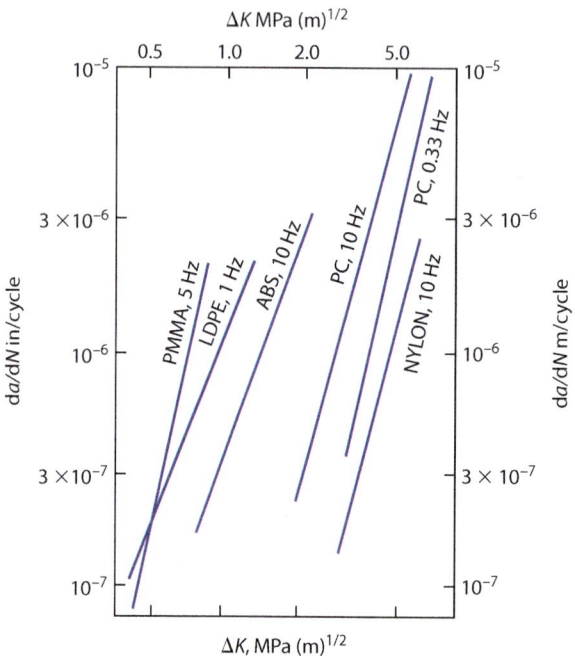

Figure 11.28   The fatigue-crack propagation rate (*da/dN*) as a function of the stress-intensity range ΔK for some polymers. (*Based on Hertzberg, R. W., Nordberg, H., and Manson, J. A., J. Matls. Sc., 5 (1970), p. 521.*)

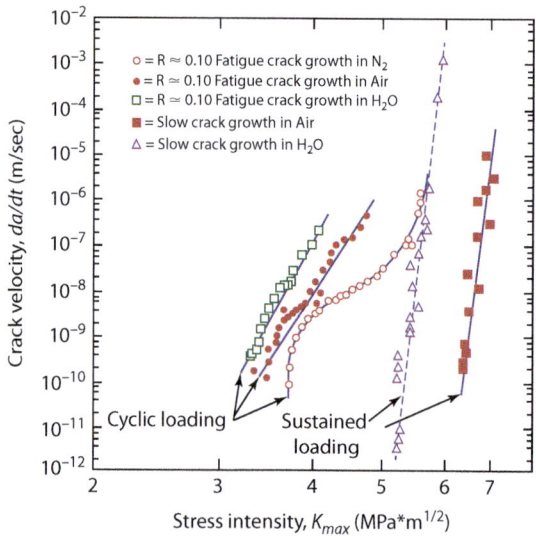

Figure 11.29   Crack velocity (*da/dt*) as a function of maximum stress intensity on the bottom axis for zirconia plus 9 weight percent MgO subject to cyclic and sustained stress. (*Based on Ritchie, R. O., Materials Science and Engineering, A103 (1988), p. 27.*)

Example Problem 11.6

For the 7075-T6 alloy in Figure 11.27, determine the power n and the pre-exponential constant A in the crack-propagation rate Equation 11.20.

Solution

There are two unknowns; therefore, it is necessary to solve two equations. Select two points with a significant separation, such as 10^{-4} mm/cycle (10^{-7} m/cycle) and 10^{-2} mm/cycle (10^{-5} m/cycle). The values of ΔK for these two crack-propagation rates are $10^{0.7}$ MPa $\cdot$ m$^{1/2}$ and $10^{1.25}$ MPa $\cdot$ m$^{1/2}$, or 5.0 MPa $\cdot$ m$^{1/2}$ and 17.8 MPa $\cdot$ m$^{1/2}$, respectively. Set up two equations with these values:

1. 10^{-7} m/cycle $= A(5.0 \text{ Mpa} \cdot \text{m}^{1/2})^n$
2. 10^{-5} m/cycle $= A(17.8 \text{ Mpa} \cdot \text{m}^{1/2})^n$

Dividing equation 1 by equation 2 eliminates A and allows for the solution of n.

$$\frac{10^{-7} \text{ m/cycle}}{10^{-5} \text{ m/cycle}} = 10^{-2} = \frac{(5.0 \text{ MPa} \cdot \text{m}^{1/2})^n}{(17.8 \text{ MPa} \cdot \text{m}^{1/2})^n} = (0.28)^n$$

Taking the log of this equation results in

$$-2 = n \log (0.28)$$

Solving for n,

$$n = \frac{-2}{\log 0.28} = 3.63$$

The pre-exponential constant A is solved for by substituting n into either equation 1 or 2. Choosing equation 1,

$$10^{-7} \text{ m/cycle} = A(5.0 \text{ MPa} \cdot \text{m}^{1/2})^{3.63}$$

$$10^{-7} \text{ m/cycle} = A(3.45 \times 10^2 \text{ MPa}^{1.815} \cdot \text{m}^{1.815})$$

The units of A are going to be strange because of the power 1.815. Instead of carrying through the power for the units for A, just say it is in units of MPa $\cdot$ m. Solving for A,

$$A = 0.29 \times 10^{-9} \text{ MPa} \cdot \text{m units}$$

The surface of a part that has failed from fatigue has a sequence of striations that are shown at high magnification in Figure 11.30. Each striation corresponds to the position of the crack tip in a particular cycle.

The striations on the fatigue fracture surface are created in ductile metals and polymers by the proposed mechanism shown in Figure 11.31. During the tensile part of the cycle, shown in Figure 11.31b, plastic deformation occurs at the crack tip extending the crack by x, and eventually blunting the crack tip and arresting its growth as shown in Figure 11.31c. Then, during the compressive part of the cycle, the crack tip is resharpened as shown in Figure 11.31d. In Figure 11.31e the crack has grown by x relative to Figure 11.31a. This sequence of events creates the striations on the fracture surface, and it results in the incremental crack growth on the fatigue fracture surface.

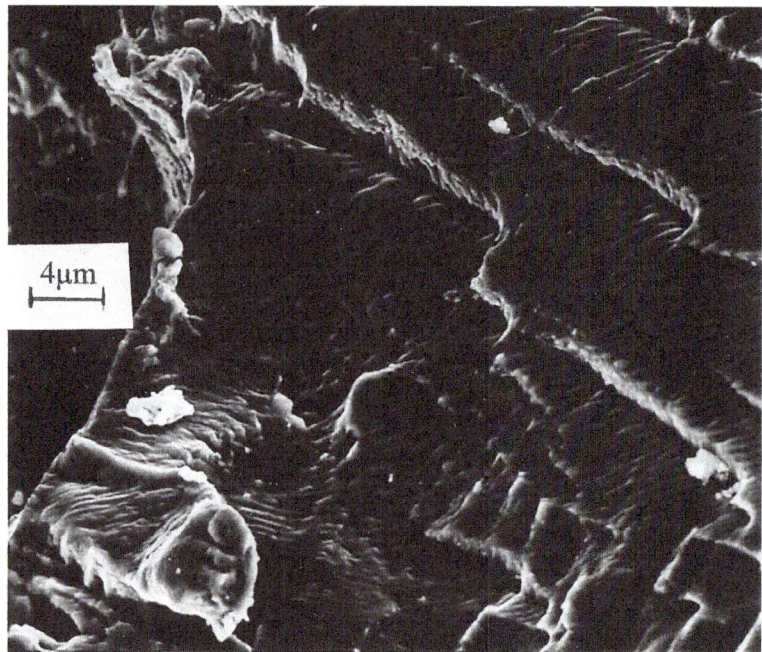

Figure 11.30 Fatigue striations showing increments of fatigue-crack growth on a commercial purity titanium specimen during low cycle fatigue. (*Photo by C. M. Gilmore*)

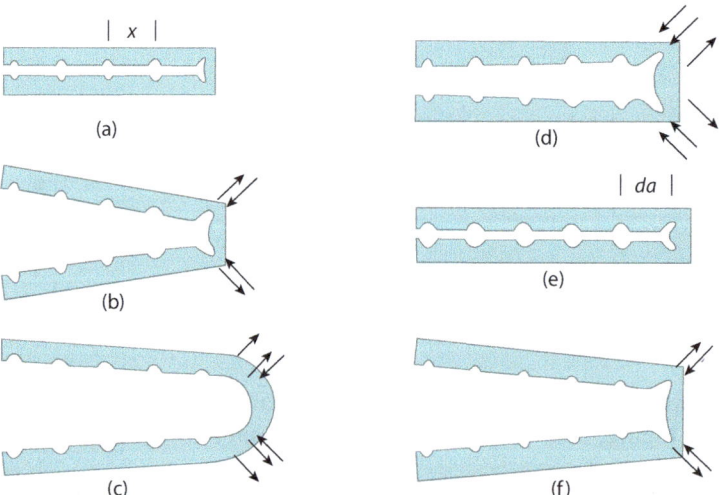

Figure 11.31 The mechanism of fatigue striation formation in ductile materials. (a) A fatigue crack with striation spacing x. (b) A tensile applied stress opens the crack. (c) Plastic deformation extends and then blunts the crack. (d) Stress reversal closes and resharpens the crack. (e) Plastic deformation has closed and resharpened the crack and the crack is now grown by length da from the original length in (a). (f) Applied tensile stress opens the crack again. (*Based on Laird, C. M., "The Influence of Metallurgical Structure on the Mechanisms of Fatigue Crack Propagation," in* Fatigue Crack Propagation, *ASTM STP 415, Philadelphia, PA (1966), p. 136,.)*

The striations on the fracture surface are like a history book of the stresses applied to the part and assist investigators in determining how a part failed. The crack-propagation rate for failed specimens, such as those in Figure 11.30, is calculated for one cycle of loading ($dN = 1$) with the measurement of the spacing between striations (da). Once the crack-propagation rate is determined, the only term in Equation 11.20 that is unknown is the stress range ($\Delta\sigma$). With the known crack-propagation rate, the value of ΔK can be determined from a figure such as 11.27. The size of the crack (a) is measured on the fracture surface. If ΔK and a are known, the value of $\Delta\sigma$ is deduced from Equation 11.18. The need to measure the crack size (a) and the spacing between striations (da) to determine the stress range is why after any serious structural failure, the investigators try to retrieve the failed parts or resulting fragments. The history of the stresses applied to the part and leading to the failure can be deduced from the striations on the fracture surface.

Example Problem 11.7

Assume that fatigue striations are found on the fracture surface of a part from an airplane that had crashed that are spaced at 1.0 μm per cycle. The part is made from 7075-T6 aluminum, and it is determined that the crack-propagation rate data shown in Figure 11.27 was taken under conditions that are representative of the load cycle and environment expected for this part. At the location on the part where the fatigue striations are observed, the macroscopic length of the internal crack is $2a = 0.02$ m. What was the stress range applied to the alloy at the instant the fatigue striations were created on the fracture surface?

Solution

The spacing between striations is 1.0 μm per cycle, or 1×10^{-6} m/cycle, on the scale used in Figure 11.27. The spacing from one striation to another corresponds to one cycle, or $dN = 1$. Locating 1×10^{-6} m/cycle on the crack-propagation axis, this corresponds to a ΔK value of 9 MPa $\cdot$ m$^{1/2}$ for the alloy 7075-T6. We are also informed that the observed crack length at the location where these striations are measured is $2a = 0.02$ m, or $a = 0.01$ m. Equation 11.18 relates ΔK, $\Delta\sigma$, and a, and it is assumed that Y is equal to 1.

$$\Delta K = \Delta\sigma(\pi a)^{1/2} = 9 \text{ MPa} \cdot \text{m}^{1/2} = \Delta\sigma(\pi 0.01 \text{ m})^{1/2}$$

Solving for the stress range $\Delta\sigma$,

$$\Delta\sigma = \frac{9 \text{ MPa} \cdot \text{m}^{1/2}}{(\pi 0.01 \text{ m})^{1/2}} = \frac{9 \text{ MPa} \cdot \text{m}^{1/2}}{0.177 \text{ m}^{1/2}} = 51 \text{ MPa}$$

This is a small stress range, since the yield stress of 7075-T6 aluminum is approximately 505 MPa. The low stress range indicates that the airplane likely failed in high cycle fatigue and not due to a high stress overload.

11.3.6 Fatigue-Crack-Propagation Life

Once the plot of log(da/dN) versus log ΔK is known for a specific type of loading and environment, the number of cycles can be calculated for the crack to grow from an initial crack length (a_i) to a final crack length (a_f). Collecting all of the terms that are a function of a in Equation 11.20, and putting them on the right side of the equation, results in Equation 11.21.

$$dN = \left(\frac{1}{A\Delta\sigma^n\pi^{n/2}}\right)a^{-n/2}\, da \qquad\qquad \textbf{11.21}$$

We then integrate dN in Equation 11.21 from 0 cycles to the number of crack-propagation cycles to failure (N_f), and we integrate the right-hand side of Equation 11.21 from the initial crack length (a_i) to the final crack length (a_f) as a function of the crack length (a), which results in Equation 11.22.

$$N_f = \int_0^{N_f} dN = \int_{a_i}^{a_f} \left(\frac{1}{A \Delta \sigma^n \pi^{n/2}} \right) a^{-n/2}\, da = \frac{1}{\left(-\dfrac{n}{2} + 1 \right)} \left(\frac{1}{A \Delta \sigma^n \pi^{n/2}} \right) \left(a_f^{-n/2 + 1} - a_i^{-n/2 + 1} \right) \qquad \textbf{11.22}$$

We assume that the stress range is a constant. The initial crack length (a_i) in Equation 11.22 is the observed crack length, or if nondestructive inspection is performed on the part and no crack is found, then the initial crack length assumed is the minimum detectable crack length for the nondestructive inspection equipment. If the life to fracture is being calculated, then the final crack length is when the maximum stress intensity factor (K_{max}) is equal to K_{Ic}, as shown in Equation 11.23.

$$K_{max} = Y\sigma_{max} (\pi a_f)^{1/2} = K_{Ic} \qquad \textbf{11.23}$$

Solving Equation 11.23 for the crack length at fracture yields Equation 11.24.

$$a_f = \frac{1}{\pi} \left(\frac{K_{Ic}}{Y\sigma_{max}} \right)^2 \qquad \textbf{11.24}$$

To design parts or structures to have the longest fatigue crack-propagation life, the most important factor is to have the minimum possible initial crack length, as demonstrated in Example Problem 11.8. This can be accomplished by selecting material without internal cracks such as inclusions, by good surface preparation, such as polishing and good machining practice, and by good design to eliminate surface discontinuities whenever possible. Also, finished parts should be nondestructively inspected by equipment with the minimum practical detectible crack length. Material should also be selected with the maximum practical K_{Ic}.

Example Problem 11.8

The landing gear of an aircraft is made from steel with a K_{Ic} of 56 MPa $\cdot$ m$^{1/2}$ determined in air at room temperature. Nondestructive inspection has found no surface or internal cracks. The minimum detectable crack length for the nondestructive inspection equipment is $2a = 0.24$ mm. Laboratory tests have shown that the crack growth rate for a stress cycle that approximates the landing and takeoff cycle follows the relationship $da/dN = A(\Delta K)^6$, where $A = 3 \times 10^{-13}$ in units of MPa and meters. The stress cycle is approximated by a stress range of $\Delta\sigma = 160$ MPa with a mean stress of 80 MPa.
(a) With this stress cycle, what is the crack length necessary to cause fracture of the steel, assuming an internal crack in room-temperature air?
(b) What is the number of stress cycles that would produce failure of the part?

Solution

a) With this stress cycle, the stress goes from 0 to 160 MPa. The part should break when it is at the maximum stress of 160 MPa. The value of K_{Ic} is known to be 56 MPa $\cdot$ m$^{1/2}$. From Equation 11.23, we have

$$K_{max} = Y\sigma_{max}(\pi a_f)^{1/2} = K_{Ic} = 56 \text{ MPa} \cdot \text{m}^{1/2} = 160 \text{ MPa} (\pi a_f)^{1/2}$$

Solving for the crack length at fracture,

$$a_f = \frac{1}{\pi}\left(\frac{56\ \text{MPa} \cdot \text{m}^{1/2}}{160\ \text{MPa}}\right)^2 = \frac{1}{\pi}(0.35\ \text{m}^{1/2})^2 = \frac{0.12\ \text{m}}{\pi} = 3.9 \times 10^{-2}\ \text{m}$$

For an internal crack,

$$2a_f = 7.8 \times 10^{-2}\ \text{m, or } 7.8\ \text{cm}$$

b) The stress range is 160 MPa and $n = 6$. Since A is given in units of MPa, the stress must also be in units of MPa. Equation 11.22 results in

$$N_f = \frac{1}{\left(-\dfrac{n}{2}+1\right)}\left(\frac{1}{A\Delta\sigma^n\pi^{n/2}}\right)\left(a_f^{-n/2+1} - a_i^{-n/2+1}\right)$$

$$N_f = \frac{1}{\left(-\dfrac{6}{2}+1\right)}\left(\frac{1}{3 \times 10^{-13}(160\ \text{MPa})^6(\pi)^3}\right)\left((3.9 \times 10^{-2}\ \text{m})^{-6/2+1} - (1.2 \times 10^{-4}\ \text{m})^{-6/2+1}\right)$$

In Equation 11.22 the units of MPa and meters are utilized consistently; however, the units are complex and are dropped from further computations.

$$N_f = -\left(\frac{1}{3.12 \times 10^2}\right)(0.67 \times 10^3 - 0.7 \times 10^8)$$

$$N_f = -3.2 \times 10^{-3}(-0.7 \times 10^8) = 2.2 \times 10^5\ \text{cycles}$$

Note that the most significant contribution to the lifetime of the part comes from the initial crack length determined by nondestructive inspection. The term that came from the final crack length that is related to K_{Ic} made little contribution to the number of cycles to failure. The inspection is the most important part of the calculated fatigue life. For this reason, if it is found that nondestructive test results have been falsified, it is critical to conduct nondestructive tests of the parts immediately.

Many factors can affect the crack-propagation rate, the pre-exponential term (A), and the power (n) that enter into Equation 11.20. Although the stress range ($\Delta\sigma$) enters directly into Equation 11.20, the stress ratio ($\sigma_{max}/\sigma_{min}$) is also an important variable. The data must be for the correct stress ratio. The environment is also an important influence on the crack-propagation rate. Higher temperatures result in the oxidation of metals at the crack tip, which increases the crack-propagation rate. The presence of corrosive environments, hydrogen, liquids, and irradiation makes the material at the crack tip more brittle, leading to increased crack-propagation rates. If crack-propagation rate data is to be used to predict a fatigue lifetime, the test data must replicate the application as closely as possible; even the frequency of the test should be similar to the real load application frequency.

11.4 SUSTAINED-LOAD CRACK PROPAGATION

Figure 11.29 shows the crack-propagation rate for a constant stress (sustained loading) in partially stabilized zirconia. In sustained-load cracking, the stress is maintained constant on a specimen with a crack, and it is observed that the crack grows even though the crack length is below the critical length

given by Equations 11.8 or 11.10. As the crack slowly grows in length with time (t) at a rate (da/dt), this increases the stress-intensity factor (K_I), as shown in Figure 11.29. The crack continues to grow until it reaches the critical length calculated from K_{Ic} if the part is in air or K_{IEAC} if the part is in a corrosive environment. The mechanisms that produce the slow crack growth are the same as those discussed we discussed in Section 11.2.7 for environmentally assisted cracking. Note in Figure 11.29 that for partially stabilized zirconia for a given crack-propagation rate, the value of K_{max} is lower in water than in air. Both zirconia and magnesia have significant amounts of ionic bonding, and ions at the crack tip dissolve into the water, causing crack growth. The material at the crack tip in glasses and ceramics is observed to react with corrosive environments in a manner similar to that for metals.

In metals, the growth of cracks with time is often due to corrosive and environmental reactions at the crack tip. For example, corrosive ions, such as Cl^-, that are in a solution react with metal at the crack tip. The K_{IEAC} of the metal corrosion product is much lower than the K_{Ic} of the metal. The corrosion product, such as a metal chloride, at the crack tip fractures, extending the crack, and new metal is exposed to the corrosive ions. Also, it is possible that a metal chloride dissolves, resulting in metal ions going in solution. The process repeats when new metal is exposed to the solution, and the crack slowly grows until it reaches a critical length.

Environments for polymers, such as oxygen, water, saltwater, solvents, acids, bases, and ultraviolet irradiation, result in the breaking of LCMs in the crack-tip area, producing slow crack growth.

11.5   FAILURE STATISTICS

In a tensile test of a ductile metal, the ultimate tensile strengths of different specimens are not exactly the same, but they are similar because the ultimate tensile strength of a ductile metal depends upon the processes of plastic flow occurring throughout the metal at many locations. Assuming that the different metal specimens are processed with the same procedures and have the same composition, the processes of plastic flow throughout the metal are similar from specimen to specimen. However, not all specimens of a brittle ceramic fail at a similar tensile stress. An example is the small-diameter polycrystalline ceramic fibers of alumina that reinforce composite materials. Their tensile fracture strength can vary by a factor of 4. In a brittle ceramic material, the fracture stress depends upon the length of the largest defect in the specimen. If the length of defects varies significantly from specimen to specimen in a brittle material, then the fracture strength also varies significantly from specimen to specimen.

Weibull proposed that for specimens of the same size subject to an applied stress (σ), the probability of survival is given by $P_s(\sigma)$ in Equation 11.25:

$$P_s(\sigma) = \exp -\left(\frac{\sigma}{\sigma_0}\right)^m \qquad \textbf{11.25}$$

where σ_0 is the characteristic strength of the material, and m is the shape parameter. In tensile tests to determine the probability of survival, a group of N similar specimens are tested to fracture. The tensile fracture strengths are given a rank (n) from 1 for the lowest strength to N for the highest strength. If the nth-ranked specimen broke at a stress of σ, the probability of failing at this stress [$P_f(\sigma)$] is then given by Equation 11.26.

$$P_f(\sigma) = 1 - P_s(\sigma) = \frac{n}{N+1} \qquad \textbf{11.26}$$

Equation 11.25 shows that as the applied stress increases, the probability of survival decreases exponentially. The meaning of the characteristic strength is obtained by setting the applied stress (σ)

equal to the characteristic strength (σ_0) in Equation 11.25. Then the probability of survival is given by Equation 11.27.

$$P_s(\sigma_0) = \exp(-1) = \frac{1}{e} = 0.368 \qquad \text{11.27}$$

When the applied stress is equal to the characteristic strength, the probability of survival is equal to the inverse of e, and the natural-logarithm (ln) of the inverse of $P_s(\sigma_0)$ is equal to 1, as shown in Equation 11.28.

$$\ln \frac{1}{P_s(\sigma_0)} = \ln e = 1 \qquad \text{11.28}$$

Taking the natural log of Equation 11.28 results in Equation 11.29.

$$\ln\left(\ln \frac{1}{P_s(\sigma_0)}\right) = \ln 1 = 0 \qquad \text{11.29}$$

The characteristic strength (σ_0) is the stress where the natural log (log) of the inverse of the probability of survival is equal to 0, as shown by Equation 11.29. The shape parameter (m) in Equation 11.25 is determined by taking the natural log of the inverse of the probability of survival, as shown in Equation 11.30.

$$\ln \frac{1}{P_s(\sigma)} = \left(\frac{\sigma}{\sigma_0}\right)^m \qquad \text{11.30}$$

Taking the natural log of Equation 11.30 results in Equation 11.31.

$$\ln\left(\ln \frac{1}{P_s(\sigma)}\right) = m \ln \sigma - m \ln \sigma_0 \qquad \text{11.31}$$

If the left side of Equation 11.31 is plotted as a function of the natural log of the stress (σ), the slope is equal to the shape parameter (m), as shown in Figure 11.32. The values of the shape parameter (m) vary from low values, such as 3 for the tensile fracture strength of concrete blocks and alumina fibers, to

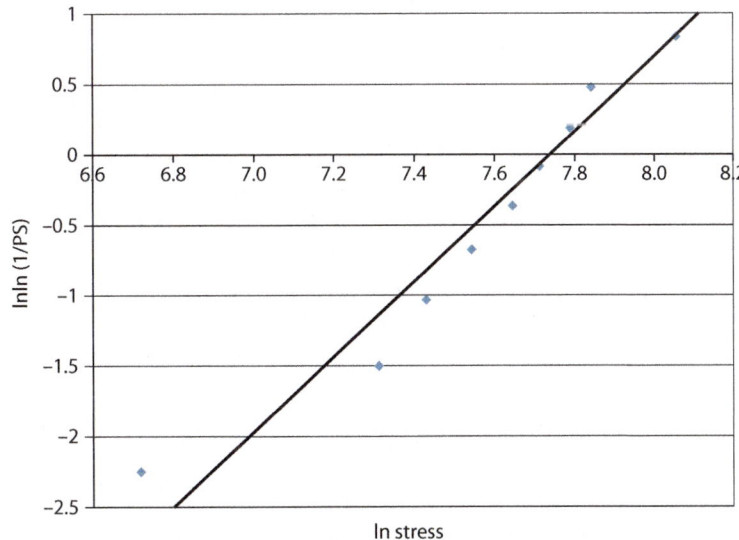

Figure 11.32 A plot of the natural-log log of the inverse of the probability of survival, as a function of the natural log of the failure stress in MPa for nine alumina fiber specimens.

100 for the tensile fracture strength of steel. A small value of the shape parameter (m) indicates a large amount of scatter in the strength test results, and a large value of the shape parameter (m) indicates very similar strength test results for all the specimens.

Example Problem 11.9

Nine alumina filaments of the same length and the same diameter are tested to failure in tension. The ranked tensile strengths are 828, 1496, 1685, 1886, 2089, 2237, 2410, 2543, and 3151 MPa. (a) Plot these data points to see if there is a linear relation between the natural log of the strength and the natural-log log of the inverse of the probability of survival. (b) Determine the characteristic strength and the Weibull shape parameter for these alumina specimens. (c) What is the stress for a 90% probability of survival?

Solution

a) The probability of survival for the ranked failure strengths is given by Equation 11.26, where n is the rank number and $N = 9$. Table 11.5 presents the calculations in Equation 11.31 necessary to form Figure 11.32. From Table 11.5, the $\ln[\ln(1/P_s)]$ is plotted in Figure 11.32 as a function of the natural-log of the stress $[\ln(\sigma)]$.

b) The characteristic strength is the stress where the plot of the data is equal to 0 on the y axis.

$$\ln \sigma_0 = 7.73$$

$$\sigma_0 = 2276 \text{ MPa}$$

From the data plot, the slope (m) is equal to

$$m = \frac{1.0 + 2.5}{8.1 - 6.8} = \frac{3.5}{1.3} = 2.7$$

c) The stress for a 90% probability of survival for these alumina fibers is 828 MPa, from the first line in the table.

Table 11.5 The Specimen Rank (n), the Probability of Survival (P_s), the Inverse of the Probability of Survival ($1/P_s$), $\ln(1/P_s)$, $\ln[\ln(1/P_s)]$, the Stress (σ), and the Natural-log of the Stress $[\ln(\sigma)]$

Rank = n	$P_s = 1 - \dfrac{n}{N+1}$	$1/P_s$	$\ln(1/P_s)$	$\ln[\ln(1/P_s)]$	σ MPa	$\ln(\sigma)$
1	0.9	1.111111	0.105361	−2.25037	828	6.719013
2	0.8	1.25	0.223144	−1.49994	1496	7.31055
3	0.7	1.428571	0.356675	−1.03093	1685	7.429521
4	0.6	1.666667	0.510826	−0.67173	1886	7.542213
5	0.5	2	0.693147	−0.36651	2089	7.644441
6	0.4	2.5	0.916291	−0.08742	2237	7.712891
7	0.3	3.333333	1.203973	0.185627	2410	7.787382
8	0.2	5	1.609438	0.475885	2543	7.8411
9	0.1	10	2.302585	0.834032	3151	8.055475

The Weibull distribution is also used to analyze fatigue failures. In fatigue there are two possible variables: the applied cyclic stress and the number of cycles to failure. Since the Weibull distribution has only one variable, we must reduce the fatigue problem to a single variable. In fatigue testing, it is normal to hold the applied alternating stress at a fixed value of (σ_a) and to determine the variable number of cycles to failure (N) for a group of specimens. By ranking the cycles to failure at the applied alternating stress (σ_a), the probability of surviving N cycles, or $P_s(N)$, is determined where N is the variable. The Weibull distribution for fatigue failure with the applied alternating stress (σ_a) is then given by Equation 11.32:

$$P_s(N) = \exp -\left(\frac{N}{N_a(\sigma_a)}\right)^{m(\sigma_a)} \qquad \textbf{11.32}$$

where $N_a(\sigma_a)$ is the characteristic life for an applied alternating stress of σ_a. When N is equal to the characteristic life, the probability of survival is $1/e$. The shape parameter $m(\sigma_a)$ is a function of the applied alternating stress (σ_a).

Summary

- Fracture occurs when a material separates into at least two pieces, or when the material separates to the point that it has insignificant load-carrying capacity.

- Fracture of a material with an existing crack occurs when the combination of the applied stress and the crack length reaches a critical value where the increase in energy necessary to grow the crack by an increment is equal to 0. The energy to propagate the crack comes from the relaxation of stress around the crack. The critical-stress intensity factor (K_{Ic}) when fracture occurs is equal to the applied stress times the square root of the crack length times a geometrical factor.

- Fracture mechanics analysis is applied to brittle materials and to materials with plastic strain localized to the fracture surface. Fracture mechanics analysis is applied to ceramics, metals, and polymers that meet this condition.

- Partially stabilized zirconia (PSZ) and tetragonal zirconia polycrystal (TZP) ceramics have a relatively high value for K_{Ic} because of a stress-induced martensitic transformation from the tetragonal phase to the monoclinic phase of zirconia that absorbs energy during crack propagation and produces compressive stresses at the crack tip. PSZ and TZP ceramics are being considered for high-temperature applications requiring some fracture resistance.

- Structural metal alloys with a high K_{Ic} have localized plastic strain during fracture, a relatively high elastic modulus, and relatively high surface energy. Ceramics have low values for K_{Ic} in comparison because they do not have localized plastic-strain energy during fracture. Polymers have a low K_{Ic} because they have a low elastic modulus and low surface energy. Polymers do have localized plastic strain during fracture.

- The most conservative and accepted value of K_{Ic} is obtained from a fracture test in thick specimens where the fracture surface is normal to the tensile loading direction.

- The value of K_{Ic} increases with temperature, because the amount of plastic work at fracture increases with temperature. The temperature dependence of fracture is often studied with the Charpy or Izod impact tests. These tests do not give values of K_{Ic} that can be used for design. The ductile-to-brittle transition temperature (T_{DBTT}) is where 50% of the fracture is ductile and 50% is brittle. The ductile-to-brittle transition is explained by the increase in yield stress and the decrease in fracture stress as temperature decreases. When the fracture stress is less than the yield stress, the material fracture is brittle.

- Fracture mechanics analysis is applied to design by measuring an existing crack length with nondestructive testing, or by using the minimum-detectable crack length. The fracture stress is then calculated from a known value of K_{Ic}. Some of the possible nondestructive tests (NDTs) include visual inspection, dye penetrant, X-ray and gamma-ray penetration, ultrasound, eddy currents, and magnetic particles.

- Fracture mechanics analysis applies to the fracture of smooth tensile specimens when there is not ductile failure resulting from necking and extensive plastic strain.

- The stress-concentration factor (K_s) is the ratio of the maximum stress at a discontinuity to the average stress in the specimen without the discontinuity.

- A fatigue failure is when a specimen fails as a result of the application of multiple stress cycles. In fatigue testing, multiple stress cycles that simulate the expected service conditions are applied until failure occurs at the number of cycles (N_f). Fatigue failure with a high stress amplitude and low number of cycles to failure is called low-cycle fatigue, and fatigue failure with a low stress amplitude and high number of cycles to failure is called high-cycle fatigue. When the plot of the stress amplitude versus the number of cycles to failure becomes horizontal and approaches infinity, this stress amplitude is the fatigue endurance limit.

- The number of cycles to failure in fatigue has a statistical distribution.

- At applied stresses below the yield stress, microplastic strain is present in metals and polymers. Fatigue-crack initiation in ductile metals and polymers is the result of the accumulation of localized plastic deformation. In polycrystalline ceramics, we can assume that microcracks, voids, or cracks are present that can propagate.

- With smooth polished specimens of ductile metals and polymers, the fatigue life to failure is the sum of the cycles to initiation of a crack plus the cycles of crack propagation to failure. For a polycrystalline ceramic, the number of cycles to failure is the number of cycles in crack propagation, because we must assume that small cracks already exist in the ceramic.

- In most materials, the log of the crack-propagation rate versus the log of the stress-intensity range forms a straight line for an extensive range of crack-propagation rates.

- In the analysis of a fatigue failure, the crack-propagation rate is calculated from the striation spacing on a fatigue fracture surface. The applied stress range is calculated from data for crack-propagation rates as a function of stress-intensity range and the crack length.

- The number of cycles to failure in crack propagation is calculated by integrating the equation for the crack propagation rate from an initial crack length to a final crack length.

- The most significant contribution to the design lifetime of a part results from having a small initial crack length or a small minimum detectable crack size resulting from nondestructive testing.

- The probability of a specimen surviving an applied stress is calculated with the Weibull distribution. The parameters for the Weibull distribution are obtained from plots of the probability of survival versus the fracture stress. The Weibull distribution is applied to fatigue failure by having the number of cycles to failure as the variable at a fixed stress amplitude.

Supplemental Reading Subjects and Authors

Full references are listed at the end of the book.

General:	*Askeland, Fulay, and Wright*
Fracture of materials:	*Courtney; Dieter; Hertzberg; Myers and Chawla*
Fracture of metals:	*Reed-Hill and Abbaschian; Tetelman and McEvily*
Fracture of ceramics:	*Carter and Norton; Chaing, Birnie, and Kingery*
Fracture of polymers:	*Hearle; Kinney; McCrum, Buckley, and Bucknall; Osswald and Menges; Rudin; Schultz; Strong; van Krevelen; Winding and Hiatt*
Fatigue of metals:	*Courtney; Hertzberg; Wood; Weibull*
Nondestructive testing:	*Shull*
Statistics of failure:	*Gumbel*

Homework

Concept Questions

1. A material is _____ if it fractures without plastic strain.

2. From a fracture mechanics viewpoint, the reason freshly drawn glass fibers are stronger than window glass is that window glass has _____ that decrease its strength.

3. When a brittle material reaches the fracture stress, the increase in energy necessary to extend a crack in the material is equal to _____.

4. The energy to extend a crack in a stressed material comes from _____ of stresses along the crack surface.

5. In the shear mode II of fracture, the crack propagates _____ to the forces that produce the shear stress.

6. In partially stabilized zirconia, particles of the tetragonal phase transform to the _____ phase at the crack tip when subject to a sufficient stress.

7. For a given steel, there is a crack of length a in the steel, and the yield strength of the steel is increased by heat treatment. The fracture strength of this steel is (increased or decreased).

8. The appearance of _____ coalescence on the fracture surface indicates a ductile metal with high energy absorption.

9. In _____ cleavage, the material fractures across the crystalline grains on the cleavage planes.

10. In _____ fracture, the fracture surface follows the grain boundaries.

11. In a craze, the _____ bonds that hold different LCMs together are broken.

12. An accepted value of K_{Ic} is obtained from a fracture test where the fracture surface is _____ to the tensile loading direction.

13. One measure of the ductile-to-brittle transition temperature is the temperature where _____ percent of the fracture is brittle.

14. In a(n) _____ test, the part is subject to a specified stress, and if the part does not fail, the maximum possible crack length is then calculated.

15. In mode I fracture of a smooth tensile specimen, there is _____ percent plastic strain prior to fracture.

16. _____ failure is when the specimen fails as a result of the application of many stress cycles.

17. Dye penetrant is an effective method to detect _____ cracks.

18. In ultrasonic nondestructive testing with the pulse-echo technique, a crack is detected by the _____ultrasonic waves.

19. Magnetic particle inspection can detect only _____ cracks.

20. Eddy-current NDT is well suited for detecting cracks in conducting materials that are _____.

21. X-ray and γ-ray NDT detect differences in _____ in materials.

22. _____ on a fracture surface are an indication of fatigue-crack propagation.

23. The stress-_____ factor is the ratio of the maximum stress at a discontinuity to the average stress in the specimen without the discontinuity.

24. Fatigue failure with a high stress amplitude is _____ -cycle fatigue.

25. If the stress amplitude in a fatigue test is below the endurance limit, the number of cycles to failure approaches _____.

26. The number of cycles to failure in polycrystalline ceramics is the number of cycles in crack _____.

27. In high-cycle fatigue of a smooth polymer specimen, the initiation of a crack is most likely to begin with _____ formation.

28. In a design using the aluminum alloy 7075-T6, would it be safe to assume that a part could withstand an infinite number of stress applications that are only 50% of the yield stress? (yes or no)

29. The log of the fatigue-crack propagation rate for most materials is linearly related to the log of the stress _____ range to a power.

30. The characteristic strength of a group of tensile specimens corresponds to a probability of survival equal to _____.

Engineer in Training–Style Questions

1. The K_{Ic} factor for a brittle material is equal to the product of constants for a specimen of a certain geometry. Which of the following is not one of these constants?
   (a) The elastic modulus
   (b) The fracture stress
   (c) The surface energy
   (d) A geometrical factor

2. The ductile-to-brittle transition temperature for the 1043 steel in Figure 11.12 is approximately
    (a) 225 K
    (b) 275 K
    (c) 325 K
    (d) 375 K

3. Which of the following NDI tests cannot be conducted on a ceramic material?
    (a) Eddy current
    (b) Dye penetrant
    (c) X-ray
    (d) Ultrasonic pulse echo

4. Which of the following NDI tests cannot detect an internal crack?
    (a) Eddy current
    (b) Dye penetrant
    (c) X-ray
    (d) Ultrasonic pulse echo

5. What mode of fracture is present for a smooth tensile specimen when a small amount of plastic strain creates a crack that is of sufficient length to cause fracture?
    (a) Mode I
    (b) Mode II
    (c) Mode III
    (d) Ductile

6. For most materials, the fatigue endurance limit is what percentage of the yield stress?
    (a) 60 to 80
    (b) 35 to 60
    (c) 25 to 35
    (d) 15 to 25

7. A crack is 1 mm long and has a 1-nm radius of curvature at the crack tip. What is the magnitude of the stress concentration factor for this crack?
    (a) 2
    (b) 20
    (c) 200
    (d) 2000

8. For the 7075-T6 alloy in Figure 11.23, the probability of surviving 10^6 cycles with a stress amplitude of 210 MPa is equal to
    (a) 99%
    (b) 90%
    (c) 50%
    (d) 10%

9. Which of the following procedures would not be expected to increase the fatigue-crack initiation life of a plain low-carbon steel?
    (a) Case hardening
    (b) Shot peening
    (c) Decreasing the grain size
    (d) Annealing

10. The best way to increase the calculated fatigue-crack propagation cycles to failure of a part is
     (a) To select a material with a high K_{Ic}
     (b) To ensure that the part has a very small initial crack length with nondestructive testing equipment
     (c) To polish the surface of the part
     (d) To case harden the part

Problems

Problem 11.1: The tensile strength of 99.8% dense sintered alumina is 205 MPa, in comparison to a tensile strength of 125 MPa for 85% dense sintered alumina. The value of the K_{Ic} for the alumina is 4 MPa · m$^{1/2}$. Compare the defect lengths in these two alumina materials, assuming that the fracture is caused by existing internal cracks.

Problem 11.2: A brittle PMMA specimen that has a K_{Ic} of 0.7 MPa · m$^{1/2}$ fractures at a tensile stress of 50 MPa. Observation of the fracture surface indicates that the fracture started at the specimen edge. Predict the length of the crack that initiated the fracture.

Problem 11.3: You observe a 1-cm-long crack in the middle of the glass windshield of your car. Compare the expected fracture strength of your defective glass windshield to that of a new one that has a tensile fracture strength of 69 MPa. Assume that the K_{Ic} of glass is 0.84 MPa · m$^{1/2}$.

Problem 11.4: A concrete tensile test specimen fractures at a tensile stress of 4 MPa. Estimate the length of internal cracks in this concrete specimen if this type of concrete has a K_{Ic} of 0.3 MPa × m$^{1/2}$.

Problem 11.5: A part is to be made from the 4340 steel in Table 11.2 that has a yield strength of 1515 MPa. This design requires that the calculated fracture stress must be at least twice as large as the yield stress, to ensure that fracture does not occur before yield. What is the minimum detectable crack length capability required of the NDT equipment to ensure this condition?

Problem 11.6: A critical aircraft part made from 7075-T651 aluminum will be proof tested to 90% of the yield strength of 496 MPa, to ensure that no large internal defects are present in the part. If the part passes the proof test, what can we assume is the largest internal crack in the part?

Problem 11.7: For a BCC metal, show items (a) to (c) on one figure. For simplicity, assume that all of the plots are linear. These plots should show general trends and do not need to be for a specific alloy.

  (a) Show how the yield stress changes as a function of homologous temperature (T/T_m). Label this as σ_{y1}.

  (b) Show how the fracture stress changes as a function of T/T_m. Label this as σ_f. Assume that the yield stress is increased by $\Delta\sigma$ for all temperatures to a value that you label as σ_{y2}.

  (c) Show in your figure how the temperature of the transition from brittle fracture to ductile fracture changes when the yield strength of the BCC metal is increased by $\Delta\sigma$.

Problem 11.8: For the aluminum alloy 7075-T6, shown in Figure 11.23, a stress amplitude of 250 MPa is applied for 10^6 cycles to a group of 100 specimens. How many of these specimens should fail in this test?

Problem 11.9: For the PMMA in Figure 11.28, determine the power n and the pre-exponential constant A for the crack-propagation rate in Equation 11.20. It may be helpful to set up a ΔK scale that is 10^x to obtain the values of ΔK.

Problem 11.10: Assume that the fatigue striations just before final fracture on the part in Example Problem 11.7 had increased to 1×10^{-5} meters per cycle when the internal crack length had grown to $2a = 0.03$ m. Assume that the data in Figure 11.27 is for a minimum stress of 0. (a) Determine the stress intensity range that created these striations, and compare the value of K_{max} to the value of K_{Ic}. (b) Predict the stress range applied to this part just before failure. (c) Did the part fail by fatigue fracture due to propagation of a crack to a critical size, or by an overload exceeding the ultimate tensile strength?

Problem 11.11: A steel part must withstand 4000 cycles of a stress of 0 to 500 MPa. The steel has a K_{Ic} of 200 MPa · m$^{1/2}$. Laboratory tests have shown that the crack growth rate for this stress cycle follows the relationship $da/dN = A(\Delta K)^6$, where A is equal to 2.5×10^{-14} in units of MPa and meters.

(a) With this stress cycle, what is the critical crack length for failure of the steel, assuming an internal crack with $Y = 1$?

(b) What is the minimum detectable crack length required of the nondestructive inspection equipment to guarantee the 4000-cycle fatigue life?

Problem 11.12: A large critical part is made of steel for which $K_{Ic} = 60$ MPa · m$^{1/2}$ determined in air at room temperature. Nondestructive inspection has found no cracks in the part. The minimum detectable crack length for the nondestructive inspection equipment is $2a = 0.20 \times 10^{-3}$ m. The part is to be subject to a stress range of $\Delta\sigma = 300$ MPa about a mean stress of 150 MPa. Laboratory tests in air at room temperature show that the crack growth rate in this material follows the relationship $da/dN = 1 \times 10^{-13} \Delta K^6$ in units of MPa and meters.

(a) Predict the crack length at fracture, assuming an internal crack with $Y = 1$.

(b) Calculate the lifetime of this part based upon fatigue-crack propagation in air at room temperature.

Problem 11.13: Nine bulk silicon carbide specimens of the same size were tested and the following strengths were obtained: 22, 26, 33, 50, 29, 56, 58, 41, and 43 MPa. (a) Plot these data to evaluate the characteristic strength and the Weibull modulus. (b) What is the stress for a 90% probability of survival for this material?

Problem 11.14: A series of ceramic specimens are tested in tension. We know that this type of material has a Weibull shape parameter of 5, but the strength depends upon the processing treatment. In a particular batch of specimens, we observe that 50% of the specimens fractured at a stress of 100 MPa or less. What is the characteristic strength of this batch of material?

A photograph of a Formula 1 race car. In 1981 the McLaren MP4 was the first Formula 1 race car designed with a carbon-fiber-reinforced epoxy chassis. Now all Formula 1 cars use this technology, as do some high-performance sports cars. It is predicted that conventional automobiles will eventually use a carbon-fiber-reinforced chassis.

The goals of this chapter are to understand

- Reinforcing and matrix materials that are available
- The reason why small-diameter filaments have such a high strength
- Isostrain and isostress models of composite materials
- The models of thermal and electrical conductivity
- The layering of woven fabric to obtain quasi-isotropic properties in the fabric plane
- The construction of panels to resist bending
- The mechanical properties of short-fiber composites
- The strength and fracture of continuous- and short-fiber composites
- The origin and analysis of thermal stresses
- Processing procedures unique to continuous-fiber-reinforced composites
- The structure and properties of construction composites: wood, concrete, and asphalt

Chapter

12

Composite Materials

12.1 INTRODUCTION

In a **composite material** at least two different materials are combined into a material where each material remains distinct. Most composite materials have a **matrix** phase that is continuous and surrounds the second phase of particles, plates, or fibers. Usually the second phase provides unique properties to the composite material. For example, silver particles provide conductivity to a resistor made of silver and epoxy, and carbon fibers provide strength and stiffness to an epoxy-carbon fiber fishing rod. Some composite materials are human made, such as fiberglass, but wood is a natural composite made of cellulose crystals and lignin resin. Other composite materials include steel-reinforced concrete, carbide cutting tools, oxide-dispersion-strengthened superalloys, and the reinforced carbon-carbon tiles on the space shuttle. Oxide-dispersion-strengthened superalloys and carbide cutting tools are examples of particle composite materials. We discussed the strength of these materials in Section 7.2.4. Some of the structural composite materials in an Airbus 380 aircraft are shown in Figure 12.1.

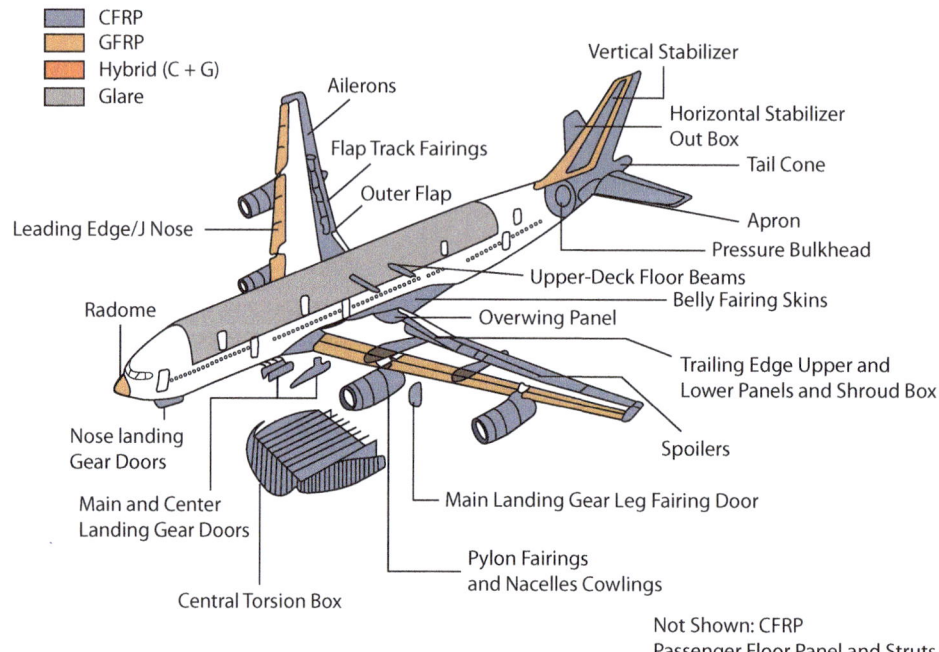

Figure 12.1 The composite material parts in an Airbus 380 including carbon fiber reinforced plastic (CFRP), glass fiber reinforced plastic (GFRP), combined carbon and glass reinforced plastic (Hybrid C+G), and Glare. GLARE is Glass fibers Laminated with Aluminum sheets to Reinforce Epoxy. (*FAA, New Large Aircraft Composite Fire Fighting, http://www.airporttech.tc.faa.gov/safety/bagot2.asp*)

12.2   REINFORCEMENT MATERIALS

Composite materials can be made of small-diameter filaments of a very high strength, which is one reason why designers are interested in composites. For example, some fine glass and PE filaments are nearly 100 times as strong as bulk glass and bulk PE. A composite material can incorporate these strong filaments as a reinforcing material. The properties of some high-strength reinforcing whiskers, fibers, and metal wires are listed in Table 12.1. One of the least-expensive fibers is glass, which is extensively used in fiberglass composites. E-glass was originally developed for electronic applications. S2-glass is stronger than E-glass and more expensive. Glass fibers are of high strength, but they have a significantly higher density than do other filaments such as carbon, Kevlar, and OUHMWPE.

The low density, high tensile strength, and high modulus of filaments, such as carbon, Kevlar, and OUHMWPE, give these materials a very high specific strength and modulus. The specific strength and specific modulus of some reinforcing fibers is plotted in Figure 12.2, along with ranges for bulk metals and polymers. The PE in Figure 12.2 is OUHMWPE. Materials in the top right corner of Figure 12.2 have a high specific strength and modulus, and materials in the lower left corner have a low specific strength and modulus.

One of the most widely utilized reinforcing fibers in advanced composites is carbon. Carbon fibers are used as the primary reinforcement fiber in the Boeing 787, in the chassis of Formula 1 race cars, and in the chassis of high-performance sport cars. Carbon fibers have a combination of very high strength and relatively low density, but they are brittle. Carbon fibers are a mixture of amorphous carbon and small-grained polycrystalline graphite, with the graphite plates oriented primarily along the fiber axis.

Table 12.1 The Typical Properties of Reinforcing Materials, Including the Diameter (*D*), Density (*ρ*), Axial Young's Modulus (*E*), Poisson's Ratio (*ν*), Tensile Strength (TS), Failure Strain (FS), and Axial Coefficient of Thermal Expansion (ACTE)

Material	D 10^{-6} m	ρ g/cm^3	E GPa	ν	TS GPa	FS %	ACTE 10^{-6} K^{-1}
Whiskers							
Graphite		2.2	700		20		
Al$_2$O$_3$		4.0	430		21		
Si$_3$N$_4$		3.2	360		14		
SiC		3.2	480	0.3	21		4.0
Fibers							
E-glass	3–20	2.4	72.4	0.22	3.45	2.6	4.9
S2-glass	10–20	2.4	86.9		4.59		
Boron	142	2.5	400	0.20	2.80	0.7	5.0
HS[1] carbon	8	1.75	230	0.20	3.4	1.1	−0.4
HM[2] carbon	8	1.95	380	0.20	2.4	0.6	−0.7
UHM[3] carbon	8	2.0	520	0.20	1.0	0.2	−1.1
SiC	15	2.6	190	0.20	2.0	1.0	6.5
Saffil^{TM}[4]	3	3.4	300	0.26	2.0	0.7	7.0
FP^{TM} (Al$_2$O$_3$)	20	3.9	380	0.26	2.0	0.4	8.5
Kevlar 49	12	1.44	130	0.35	3.6	2.8	−2
OUHMWPE	38	0.97	117		2.6	3.5	
Cellulose		1.0	80	0.3	2.0	3.0	
Metal wires							
HS steel		7.9	210	0.29	2.4		
Tungsten		19.4	410		4.0		

1 HS—High strength
2 HM—High modulus
3 UHM—Ultra high modulus
4 Saffil^{TM}—Al$_2$O$_3$

OUHMWPE and Kevlar are two highly oriented polymer fibers that are utilized in composite materials. OUHMWPE is stronger than high-strength steel wire, has a low density, and has some ductility to failure. However, OUHMWPE does not bond well to epoxy because of its low surface energy. Kevlar has a density lower than that of glass, it has some ductility, and it bonds reasonably well to epoxy. Kevlar is used in composite materials to provide fracture resistance in addition to strength. Both OUHMWPE and Kevlar fibers are used in bullet-proof vests. The fibers in bullet proof vests are woven into fabric, but they are not in a composite material.

Note in Table 12.1 the incredible tensile strength of whiskers, such as graphite. Graphite whiskers are more than eight times as strong as high-strength steel wires. Whiskers are very expensive, and they are applied only where high performance justifies the high price. Unfortunately, the length of a whisker is only about 0.01 m, and whiskers are not at present made in continuous-fiber form.

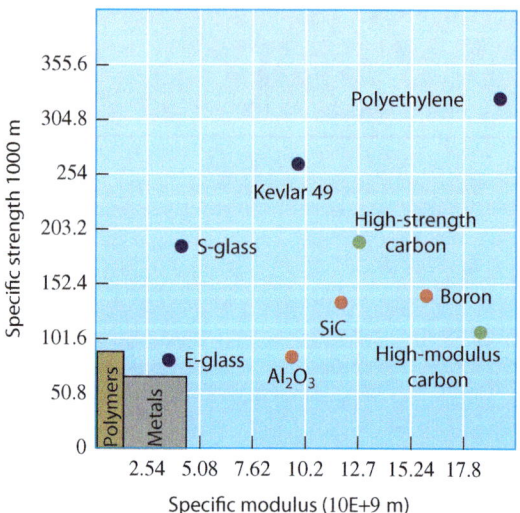

Figure 12.2 The specific strength and specific modulus of some reinforcing fibers. Bulk polymers and metals are shown in the lower left corner. *(Based on Askeland, D.R., Fulay, P.P., and Wright, W.J., The Science and Engineering of Materials, 6th ed., Cengage Learning, Stamford, CT. (2011), p. 668.)*

Carbon nanotubes, discussed in Section 7.5, are one of the most recent high-strength, high-modulus materials developed. Carbon nanotubes in a bundle, shown in Figure 1.11, and in continuous fiber are just beginning to appear in applications. Present carbon nanotube composite applications include parts for wind turbines and high-performance sporting equipment, such as bicycles, skis, arrows, surfboards, and hockey sticks.

12.3   MATRIX MATERIALS

The properties of the matrix material are the bulk properties that are presented in Chapter 8. In structural composite materials, the matrix primarily holds the strong reinforcing phase into the shape of the part. Epoxy and polyester thermoset polymers, presented in Table 8.18, are extensively used in composite fabrication, because the liquid resin and liquid hardener can be mixed to form a liquid at room temperature that is impregnated into the fiber phase. The mixed epoxy or polyester hardens in a few hours and forms a rigid solid. Carbon-fiber-reinforced epoxy is being selected for many high-performance applications in aircraft, automobiles, and recreational equipment. The selection of carbon-reinforced epoxy for the fuselage in the Boeing 787 is an example of the acceptance of this material.

Thermoplastic polymers and metals must be melted for the liquid matrix material to impregnate the fibers. Thermoplastic polymers are reinforced with continuous fibers, chopped fibers, and particles such as silica and wood particles. Polyetheretherketone (PEEK), in Table 8.17, is a thermoplastic polymer whose 334°C melting temperature is one of the highest, and it has applications as a matrix material for elevated-temperature applications. The melting temperature of the matrix limits the applications of composites with a thermoplastic polymer matrix.

Aluminum, cobalt, magnesium, nickel, and titanium metal are used as the matrix for metal-matrix composites. Cutting tools such as drills, saws, and tool bits are made from a ductile metal matrix, such as cobalt, and hard particles such as tungsten carbide or diamond. Composites with a matrix of nickel and nickel-based alloys are dispersion strengthened with particles of thoria (ThO_2) and yttria (Y_2O_3). These alloys are utilized in high-temperature gas turbines, because of the high-temperature strength provided by the ceramic dispersions. The properties of Inconel 6000, which is strengthened with yttria, are presented in Table 8.10. Particles in a metal matrix can also be used to decrease density. Ford Motor Company has developed an aluminum-matrix drive shaft with particles of boron carbide that allows for a higher-speed rotation in high-performance automobiles and race cars. Similar alloys have been used in high-performance bicycles.

The F-22 fighter aircraft used a titanium matrix reinforced with continuous silicon carbide fibers for a nozzle actuator in the turbines, and the F-16 fighter aircraft used a similar material for a landing-gear component. Aluminum reinforced with boron fibers was used in tubular struts for the space shuttle, and an aluminum high-gain antenna boom reinforced with graphite fiber is used in the *Hubble* space telescope. In most of these applications, the composite was selected to minimize deflection. However, metal matrix composites with continuous fiber reinforcement have not found extensive application, because of the cost of fabrication relative to that of discontinuous-fiber composites. Discontinuous-fiber-reinforced composites use short fibers, and some of the standard metallurgical processes are used with these composite materials. The mechanical properties of discontinuous-fiber-reinforced composites are not equal to those of continuous-fiber-reinforced composites; however, they can be of higher yield strength than the metal matrix, and they can have isotropic properties.

The primary reason that ceramic matrices are reinforced with a second phase, such as fibers, filaments, and rods, is that ceramic matrix composites have increased resistance to fracture in comparison to the ceramic alone. The increased resistance to fracture is because the fibers, filaments, or rods pulling out from the ceramic matrix absorb energy due to friction between the fiber and the matrix. Composites made of a carbon matrix and carbon fibers are used as ablation heat-shield material in outer space reentry vehicles, in rocket motors, as components in gas turbines, and as aircraft brake material. One of the outstanding properties of carbon-carbon composites is that they retain their strength in temperatures in excess of 2000°C. Concrete is a ceramic matrix that can be reinforced with steel bars to form a composite of reinforced concrete, which has increased resistance to brittle fracture relative to concrete.

Composite materials are produced to have properties that are not available in monolithic materials. An example is carbon-fiber-epoxy used to make fishing rods. Today we cannot produce a metal fishing rod as strong and light as a carbon-fiber rod. The reason for this is the incredible strength and low density of carbon fibers in an epoxy matrix. A fishing rod is designed so that the carbon fibers, which are much stronger than the epoxy, support most of the stress and minimize the stress on the epoxy. The analysis below shows how this is achieved and how composite materials should be designed.

12.4 ANALYSIS OF UNIAXIAL CONTINUOUS-REINFORCED COMPOSITES WITH LOADS APPLIED PARALLEL TO THE REINFORCEMENT

A composite material with a strong reinforcing phase and a weak matrix is strong if the reinforcing phase carries most of the stress. In this section we find that the phase with the highest elastic modulus carries the highest stress. The carbon-fiber fishing rod provides an example of how a uniaxial composite material is structured. The rod consists of parallel carbon fibers along the rod axis that are embedded in a matrix of epoxy, which holds the carbon fibers together. The epoxy is of very low strength relative to that of the carbon fibers, as shown in Table 12.1 and Table 8.18. A schematic of such a material is shown in Figure 12.3.

If a force (F) is applied axially along the direction of the reinforcing fibers, the force is carried by both the fibers (F_f) and the matrix (F_m), as shown in Equation 12.1.

$$F = F_f + F_m$$

12.1

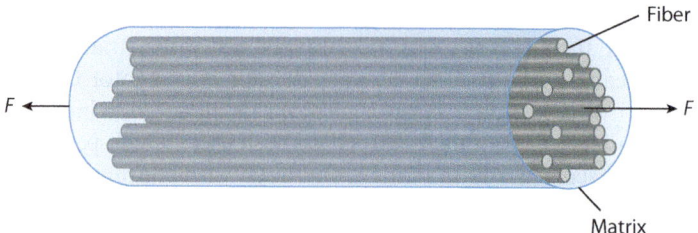

Figure 12.3 A force (F) is applied parallel to the axis of the uniaxial continuous reinforcing fibers.

The force in each component of the composite is equal to the stress in that component times the area of that component. Rewriting Equation 12.1 in terms of stress and area results in Equation 12.2.

$$\sigma A = \sigma_f A_f + \sigma_m A_m \qquad \textbf{12.2}$$

The average stress in the composite is σ, A is the total cross-sectional area of the composite, σ_f is the stress in the fibers, A_f is the total cross-sectional area of all the fibers, σ_m is the stress in the matrix material, and A_m is the total cross-sectional area of the matrix material. The fiber stress (σ_f) and matrix stress (σ_m) represent the actual stress in these materials; however, the composite stress (σ) is the average stress from the material components; it does not represent any actual stress in the composite. Assume that the fibers and matrix are both continuous and equal to the length of the composite (L), as shown in Figure 12.3. Multiplying each side of Equation 12.2 by the length (L) results in Equation 12.3.

$$\sigma A L = \sigma_f A_f L + \sigma_m A_m L \qquad \textbf{12.3}$$

The term AL is the total volume (V) of the composite, $A_f L$ is the total volume (V_f) of the fibers, and $A_m L$ is the total volume (V_m) of the matrix. Equation 12.3 is rewritten in terms of volume in Equation 12.4.

$$\sigma V = \sigma_f V_f + \sigma_m V_m \qquad \textbf{12.4}$$

Dividing each side of Equation 12.4 by the total volume (V) results in Equation 12.5

$$\sigma = \sigma_f (V_f / V) + \sigma_m (V_m / V) = \sigma_f v_f + \sigma_m v_m \qquad \textbf{12.5}$$

where v_f is the volume fraction of fibers and v_m is the volume fraction of matrix. Equation 12.5 is called the **rule of mixtures**, which shows that the physical property of a mixture of materials is a function of the volume fraction of the component materials. To make a strong composite material, the strong fibers must carry most of the load, and the volume fraction of fibers should be high.

To determine the values of the stresses in the fibers, the matrix, and the composite, we make the **isostrain** assumption that all three have equal strains (ε). The isostrain assumption applies when the force is parallel to the axis of the reinforcement. This analysis also applies if the reinforcing phase is in the form of plates that are parallel to the loading axis. The assumption of isostrain is reasonable if there is good bonding between the fibers and the matrix and there is no slippage of the fibers in the matrix. If each term in Equation 12.5 is divided by the strain ($\varepsilon = \varepsilon_f = \varepsilon_m$), the result is that the composite elastic modulus in the fiber direction (E_c) is calculated by the rule of mixtures, as shown in Equation 12.6.

$$\frac{\sigma}{\varepsilon} = E_c = \frac{\sigma_f}{\varepsilon_f} v_f + \frac{\sigma_m}{\varepsilon_m} v_m = E_f v_f + E_m v_m \qquad \textbf{12.6}$$

We can determine E_c if we know the volume fractions and elastic moduli of the fibers and the matrix. If the fibers have a high modulus in comparison to that of the matrix, then a composite with a high modulus of elasticity is produced if the volume fraction of fibers is high. However, there is a limit to the volume fraction of fibers of approximately 75%. Higher-volume fractions of fibers result in decreased strength, because there is insufficient matrix material to bond to all of the fibers, and voids form between the fibers where there is insufficient material. The voids can then act as cracks.

Equation 12.7 gives the strain in the composite.

$$\varepsilon = \frac{\sigma}{E_c}$$ **12.7**

With the isostrain assumption, the stress in the fibers and the matrix is determined from Equations 12.8 and 12.9.

$$\sigma_f = \varepsilon E_f$$ **12.8**

$$\sigma_m = \varepsilon E_m$$ **12.9**

Example Problem 12.1

A uniaxial composite material is made into a circular rod with a 1.27-cm diameter from 70 volume percent continuous carbon fibers and 30 volume percent epoxy. The rod is subject to an axial force of 100,000 N. The carbon fibers have an elastic modulus of 500 GPa and an ultimate tensile strength of 3 GPa. The epoxy has an elastic modulus of 5 GPa and an ultimate tensile strength of 0.05 GPa.
(a) Determine the elastic modulus in the axial direction of this composite material.
(b) Determine the axial stress and strain in the composite material.
(c) Determine the axial stress in the carbon fibers and in the epoxy matrix.
(d) Can all of the components of this composite withstand these stresses?

Solution

a) The elastic modulus of the composite material is given by the rule of mixtures in Equation 12.6.

$$E_c = E_f v_f + E_m v_m = 500 \text{ GPa } (0.70) + 5 \text{ GPa } (0.30) = 350 \text{ GPa} + 1.5 \text{ GPa} = 352 \text{ GPa}$$

b) The average composite stress is determined from the load and the composite cross sectional area.

$$\sigma = \frac{F}{A} = \frac{100{,}000 \text{ N}}{\pi [6.35 \times 10^{-3} \text{ m}]^2} = \frac{1 \times 10^5 \text{ N}}{1.27 \times 10^{-4} \text{ m}^2} - 0.79 \times 10^9 \frac{\text{N}}{\text{m}^2}$$

From the composite elastic modulus and the composite stress, we can determine the strain in the composite by using Equation 12.7:

$$\varepsilon = \frac{\sigma}{E_c} = \frac{790 \times 10^6 \text{ N/m}^2}{352 \times 10^9 \text{ N/m}^2} = 2.24 \times 10^{-3}$$

c) We assume the isostrain model for axial loading to calculate the stress in the fibers and in the matrix with Equations 12.8 and 12.9.

$$\sigma_f = \varepsilon E_f = (2.24 \times 10^{-3})(500 \times 10^9 \text{ Pa}) = 1120 \times 10^6 \text{ Pa} = 1.12 \text{ GPa}$$

$$\sigma_m = \varepsilon E_m = (2.24 \times 10^{-3})(5 \times 10^9 \text{ Pa}) = 11.2 \times 10^6 \text{ Pa} = 0.0112 \text{ GPa}$$

d) The fiber stress is less than the minimum tensile strength for the carbon fibers of 3 GPa. The fibers should be sufficiently strong. The minimum tensile strength for epoxy is 0.05 GPa. The matrix should also be sufficiently strong. The fibers have 100 times the stress as the matrix. The stress ratio in the isostrain model is equal to the ratio of the elastic moduli.

12.5   ANALYSIS OF UNIAXIAL CONTINUOUS-REINFORCEMENT COMPOSITES WITH LOADS APPLIED PERPENDICULAR TO THE REINFORCEMENT

Assume that a composite material is made with layers of two different materials (*a*) and (*b*) that are subject to a force (*F*) that is perpendicular to the layers of reinforcement, as shown in Figure 12.4. In this analysis the composite material is fabricated from slabs of material that extend entirely across the material. This material configuration simplifies the mathematics. However, at the end we will discuss how this model may apply to a matrix and fibers.

Each layer (*a* and *b*) must support the same force; each layer has the same cross-sectional area, and the stress must be the same in each layer. This is the **isostress model**, and it is shown mathematically by Equation 12.10:

$$\sigma_c = \sigma_a = \sigma_b = \frac{F}{A_c} \qquad \textbf{12.10}$$

where σ_c is the composite stress, σ_a is the stress in the *a*-type material, and σ_b is the stress in the *b*-type material. With the force applied as in Figure 12.4, the strains add, and they add in proportion to the volume fraction of each phase or the rule of mixtures, as shown in Equation 12.11:

$$\varepsilon_c = v_a \varepsilon_a + v_b \varepsilon_b \qquad \textbf{12.11}$$

where ε_c is the strain in the composite, v_a is the volume fraction of the *a* phase, ε_a is the strain in the *a* phase, and ε_b is the strain in the *b* phase. The elastic modulus of the composite (E_c) is then given as Equation 12.12.

$$E_c = \frac{\sigma_c}{\varepsilon_c} = \frac{\sigma_c}{v_a \varepsilon_a + v_b \varepsilon_b} = \frac{\sigma_c}{v_a \dfrac{\sigma_a}{E_a} + v_b \dfrac{\sigma_b}{E_b}} \qquad \textbf{12.12}$$

The stress cancels out of the right side of Equation 12.12, because $\sigma_a = \sigma_b = \sigma_c$. If Equation 12.12 is then multiplied in the numerator and denominator by $E_a E_b$, the result is shown in Equation 12.13.

$$E_c = \frac{\sigma_c}{\varepsilon_c} = \frac{E_a E_b}{v_a E_b + v_b E_a} \qquad \textbf{12.13}$$

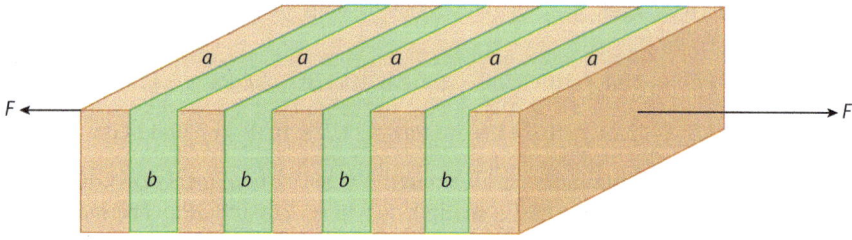

Figure 12.4   A composite material made of layers of two different materials (*a*) and (*b*) that are subject to a force (*F*) that is perpendicular to the layers of reinforcement.

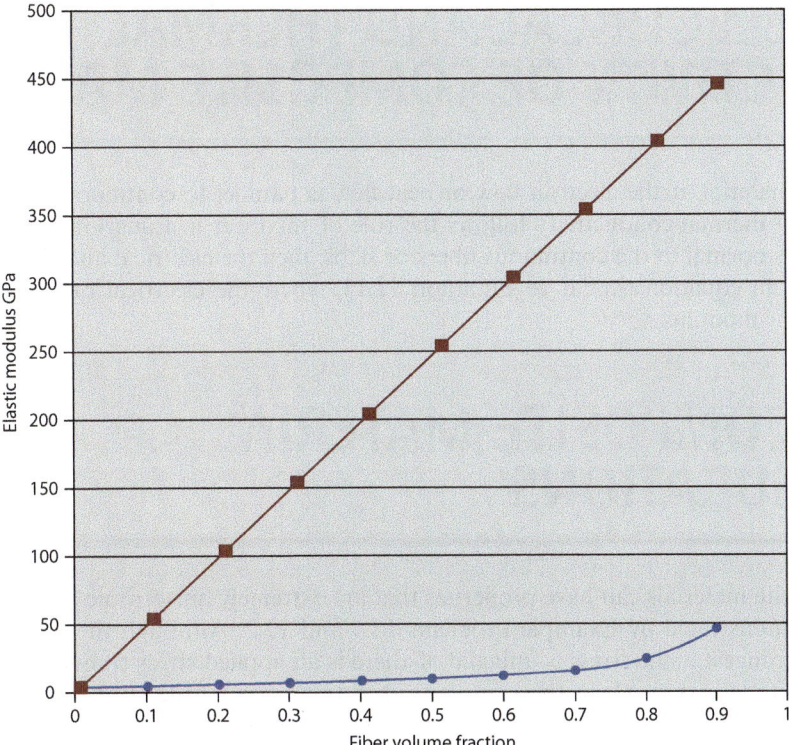

Figure 12.5 The composite elastic modulus parallel (maroon) and perpendicular (blue) to the fibers plotted as a function of fiber volume fraction for the fibers and matrix in Example Problem 12.1.

Although the equation for the isostress elastic modulus is developed for slabs that are perpendicular to the applied force (F), this model gives an approximation to the elastic modulus in a direction perpendicular to the fibers (transverse) in a uniaxial composite material. For a discussion of more advanced models of the transverse elastic modulus in a uniaxial composite material, see books such as that by Hull and Clyne.

For a uniaxial composite material with a reinforcing phase that has a higher elastic modulus than the matrix does, the isostrain elastic modulus is always larger than the isostress elastic modulus, as shown in Figure 12.5. In Figure 12.5 the composite elastic modulus parallel and perpendicular to the fibers is calculated as a function of fiber-volume fraction for the fibers and matrix in Example Problem 12.1.

Example Problem 12.2

For the composite material in Example Problem 12.1, calculate the elastic modulus in the direction perpendicular to the fiber axis, assuming that the composite follows the isostress model.

Solution

The isostress modulus is given by Equation 12.13. Taking a to be the fibers and b to be the matrix results in

$$E_c = \frac{E_a E_b}{v_a E_b + v_b E_a} = \frac{(500 \text{ GPa})(5 \text{ GPa})}{(0.7)(5 \text{ GPa}) + (0.3)(500 \text{ GPa})} = \frac{2500 \text{ GPa}}{3.5 + 150} = 16.3 \text{ GPa}$$

In comparison, the axial elastic modulus is 352 GPa, or nearly 22 times as large. This composite is much stiffer in the axial direction than in the transverse direction.

12.6 ELECTRICAL AND THERMAL CONDUCTIVITY OF COMPOSITE MATERIALS

In a composite material, if the electron flow or heat flow is parallel to continuous fibers or slabs, then the electrical and thermal conductivity follows the rule of mixtures in Equation 12.5. If the electrical or thermal flow is normal to the continuous fibers or slabs, then the electrical and thermal conductivity is modeled with an equation similar to Equation 12.13, where the electrical or thermal conductivity replaces the elastic modulus.

12.7 COMPOSITE-MATERIAL CONFIGURATIONS

Uniaxial composite materials can have properties that are extremely anisotropic or unequal in different directions, as demonstrated by Example Problems 12.1 and 12.2. Although the uniaxial arrangement of fibers is the strongest if the stress is uniaxial, if there is an applied stress transverse to the fibers, the uniaxial composite is very weak in the transverse direction. The tensile strength of a uniaxial fiber-reinforced composite material in the direction perpendicular to the fibers is in general less than that of the matrix material, and it can be significantly less if the interface bonding between the matrix and the reinforcing phase is weak. For this reason, if there are multiaxial loads, the fiber orientations must also be multiaxial.

To achieve multiaxial fiber orientations, the fibers are woven into a fabric with fibers oriented at 90° relative to each other, as shown in Figure 12.6. The individual fibers, as many as 10,000, are bundled together into a **roving**, and the rovings are woven into a fabric. In Figure 12.6 an individual roving in the pictured E-glass cloth is approximately 1 mm across. A larger roving results in a stiffer final product, and a smaller roving gives a better surface finish and greater flexibility to the fabric.

The layers of fabric are placed at different orientations, as shown in Figure 12.7, to produce a quasi-isotropic composite material within the plane of the fabric. In the final step the layers of fabric are impregnated with matrix material to produce the composite material. The composite is very weak perpendicular to the layers of fabric, because there are no fibers in this direction. It is possible to weave three-dimensional fabric, but it is expensive and difficult.

Mat is produced by laying fibers in random directions, as shown in Figure 12.8. A composite material made from mat and a matrix material is not as strong as a composite material made from the same volume fraction of woven fabric.

Multiple layers of composite material, shown in Figure 12.7, or composite mat resist bending if they are attached with adhesive to each side of a lightweight hollow-core honeycomb structure, as shown in Figure 12.9. Composite panels are made with the configuration of materials shown in Figure 12.9.

The techniques for analyzing the stress and strains in the fibers and matrix of a multilayered composite are covered in most books on composite materials, such as that of Hull and Clyne. Also, computer programs have been developed to analyze layered composite materials. An interesting result that comes from the analysis of composite materials with fibers that are not parallel to the loading axis is demonstrated in Figure 12.10. The macro models in Figure 12.10 are made of clear polyurethane elastomer with metal rods embedded as reinforcement. A tensile stress is applied to each specimen where the hook and spring attach at the top of each specimen. In Figures 12.10a and 12.10d, the reinforcement

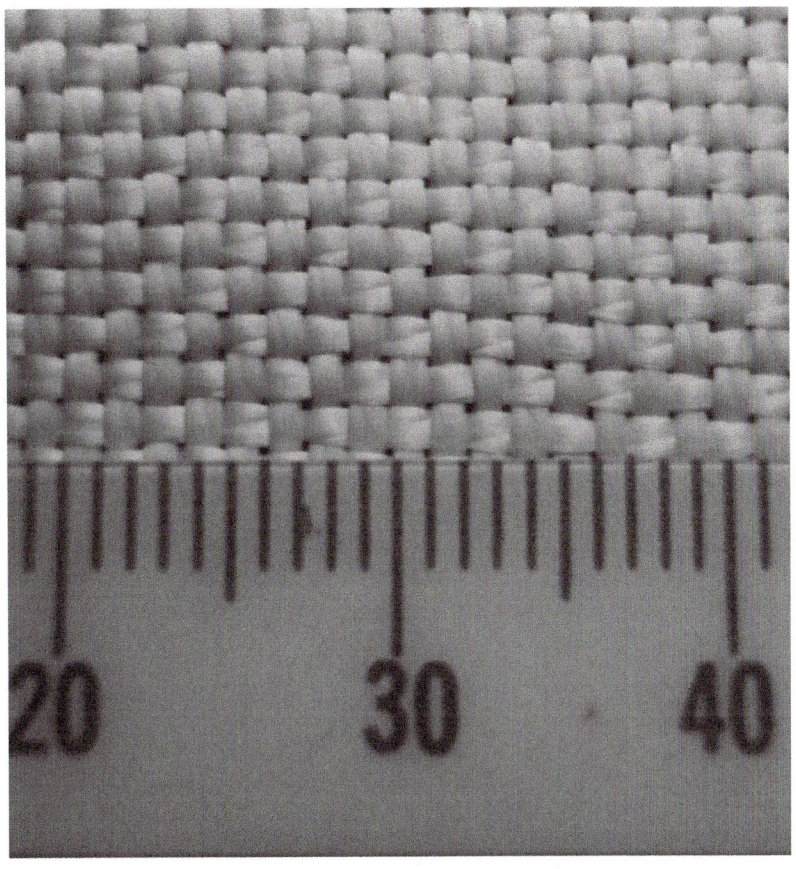

Figure 12.6 A photograph of E-glass cloth. The scale is in millimeters. (*Photo by C.M. Gilmore*)

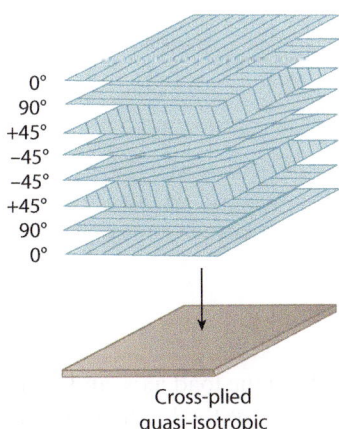

0°
90°
+45°
−45°
−45°
+45°
90°
0°

Cross-plied
quasi-isotropic

Figure 12.7 An example of possible orientations of fabric weaves to obtain a quasi-isotropic composite in the plane of the fabric. (*Based on Askeland, D.R., Fulay, P.P., and Wright, W.J., The Science and Engineering of Materials, 6th ed., Cengage Learning, Stamford, CT. (2011), p. 667.*)

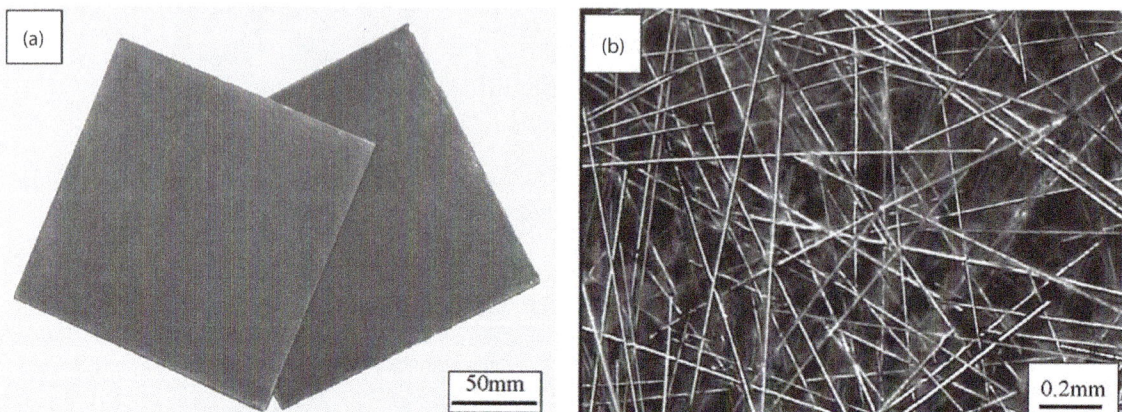

Figure 12.8 (a) A photograph of composite material specimens made of chopped glass coated fibers of the amorphous metal $Fe_{69}Co_{10}Si_8B_{13}$ in a polyamine matrix. These materials have potential radar absorbing applications. (b) An optical micrograph showing the orientation of the chopped fibers. (*Reprinted from Materials Science and Engineering: B, Volume 175, Issue, Zhihao Zhang,Chengduo Wang,Yanhong Zhang,Jianxin Xie, "Microwave absorbing properties of composites filled with glass-coated Fe_{69}, Co_{10}, Si_8, B_{13} amorphous microwire," Pages 233–237, Copyright 2010, with permission from Elsevier.*)

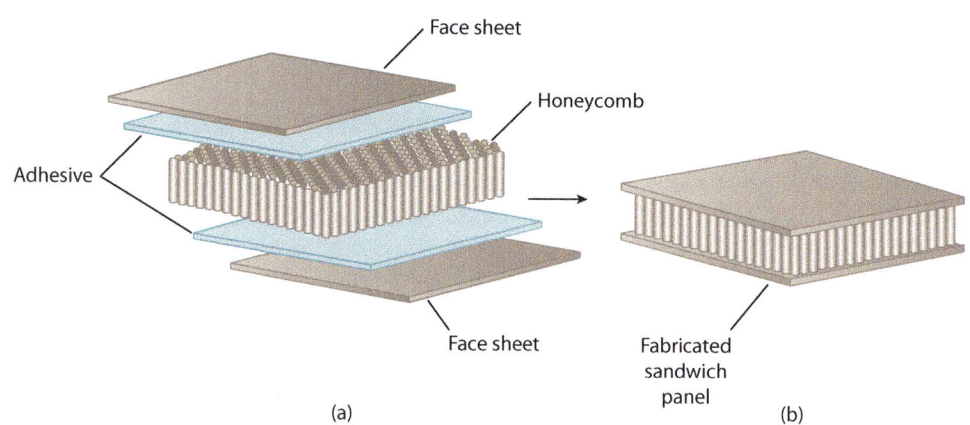

Figure 12.9 (a) The components of a honeycomb-sandwich composite panel consisting of two face sheets of composite material, adhesive, and the honeycomb material. (b) The assembled honeycomb sandwich panel. (*Based on Askeland, D.R., Fulay, P.P., and Wright, W.J., The Science and Engineering of Materials, 6th ed., Cengage Learning, Stamford, CT. (2011), p. 688.*)

is perpendicular and parallel respectively to the load axis, and no net shear of the specimen is produced. In Figure 12.10c, the reinforcement is at 30° to the loading axis, and the composite material is sheared. However, Figure 12.10b shows that when the reinforcement is at 60° from the load axis, there is no shear of the specimen. Composite materials that have fibers oriented at 0, ±60°, and ±90° to the load axis are **balanced composites**. A balanced composite is one that produces no shear strains when stressed in tension. If the layers of fibers are at angles of 0°, 60°, 120°, 120°, 60°, 0°, the composite is balanced and

Figure 12.10 Composite models made of clear polyurethane elastomer with metal reinforcing rods pulled in tension. In relation to the load direction, the reinforcement is (a) perpendicular, (b) at 60°. (c) at 30°. (d) parallel. *(Hull, D. and Clyne, T.W., An Introduction to Composite Materials 2nd ed. Cambridge University Press, Cambridge (1996) pg. 91. Reprinted with the permission of Cambridge University Press.)*

symmetric. A composite is **symmetric** if the layers have a mirror image on the other side of a mirror plane. The composite layering of 0°, 60°, 120°, 120°, 60°, 0° is symmetric about a mirror plane between the 120° layers because the sequence 120°, 60°, 0° is a mirror image of 0°, 60°, 120°. Analysis shows that a layering of fibers at ±45° is also balanced.

12.8 ANALYSIS OF SHORT-FIBER COMPOSITES

In the analysis and discussion presented in Sections 12.4 and 12.5, the fibers are continuous. However, many composites are made with short or chopped fibers because of the ease of fabrication. Composite materials made of continuous-fiber woven fabric are often fabricated by hand, because it is difficult to automate the process of layering the fabric and impregnating it with polymer matrix material. Chopped fibers are mixed with liquid matrix material and sprayed or cast in liquid form, or deformed into shape with normal processing procedures for thermoplastic polymers, resulting in lower fabrication costs. Additionally, whiskers, which have the highest strength of any material form, are produced only as short fibers. A randomly oriented chopped-fiber composite is schematically shown in Figure 12.11.

The discussion that follows applies to a weak low-elastic-modulus matrix reinforced with strong high-modulus reinforcement. Short-fiber-reinforced composites are not as strong as continuous-fiber composites, because at the end of each fiber, the tensile stress in the fiber drops to 0. The equations for the stress and strain developed for continuous fibers cannot be applied to short fibers, because we must assume that the strain is constant in the reinforcement in the isostrain model and that the stress is constant in the isostress model of composite materials. These assumptions are not valid for short-fiber composite materials.

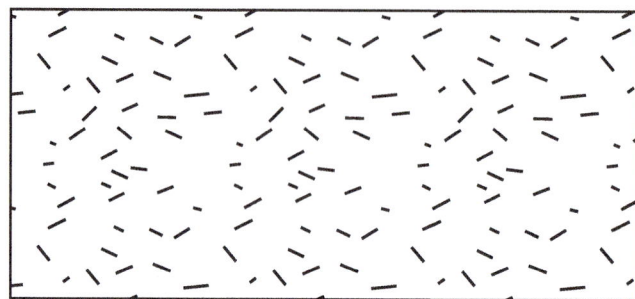

Figure 12.11   A schematic of a randomly oriented short-fiber composite material.

The goal of the analysis for a short-fiber composite material is to calculate the stresses in the fiber, the average stress carried by the composite material, the elastic modulus of the composite, and the composite strain. All of these are determined as a function of the volume fraction of fiber, the length of the fiber, and the diameter of the fiber. A schematic of short fibers in a matrix is shown in a cylinder of composite material in Figure 12.12a with an axial applied tensile stress (σ_c). We assume that the fibers are all aligned parallel to the axis of the specimen and to the loading direction. A small increment of fiber from x to $x + dx$ is shown in Figure 12.12b with matrix-fiber shear stresses (τ_{mf}) and fiber tensile stress (σ_f). How τ_{mf} and σ_f result from the application of σ_c is explained with the help of Figure 12.12c.

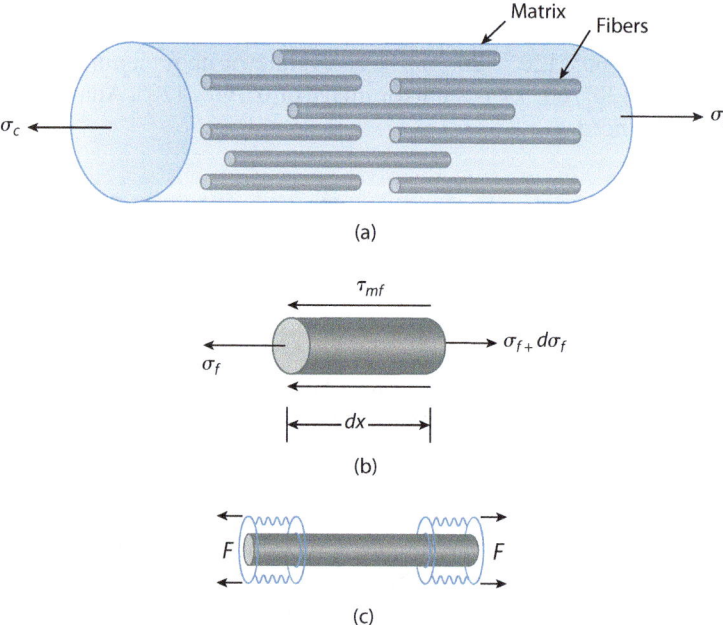

Figure 12.12   (a) A schematic showing short fibers in a matrix with an applied stress σ_c. (b) A section of short fiber of length dx. At the location x along the fiber axis, the fiber stress is σ_f, and at the location $x + dx$, the fiber stress is $\sigma_f + d\sigma_f$ due to the interface shear stress τ_{mf}. (c) A composite material specimen with clear polymer ends and a rod running through the two polymer ends. The polymer ends are subject to a tensile force F. The rod is continuous from one end of the specimen to the other.

The composite material specimen in Figure 12.12c has clear polymer ends that are threaded so that when the specimen is pulled in tension, the force (F) is initially applied to only the polymer material through the threads. The rod of strong reinforcing material extends through each polymer end piece; however, the tensile force is not applied directly to the rod. For equilibrium the central section of the rod not surrounded by polymer must have a tensile force F. No tensile force is applied directly to the rod; the tensile force in the rod results from shear stresses between the polymer ends and the rod. The tensile force in the rod is 0 at both ends of the rod and increases to F, where the polymer terminates.

The shear stress on the surface of the fiber due to the tensile stress in the matrix at the position x is τ_{mf}, as shown in Figure 12.12b. The 0 position for x is at the center of the fiber. At the position x, the tensile stress in the fiber is σ_f and the tensile stress at the position $x + dx$ is equal to $\sigma_f + d\sigma_f$. Since the element of fiber shown in Figure 12.12b is in equilibrium, the forces on the element sum to 0, as shown in Equation 12.14a.

$$\frac{\pi d_f^2}{4} d\sigma_f - \tau_{mf}(\pi d_f)\, dx = 0$$

12.14a

Rearranging terms results in Equation 12.14b

$$d\sigma_f = \frac{4\tau_{mf}\, dx}{d_f}$$

12.14b

It is necessary to establish the conditions for the integration of Equation 12.14b to calculate the tensile stress in the fiber at any point. In a short-fiber composite, the circumferential surface at each end of the fiber has the maximum shear stress (τ_{mf}), and at the center of the fiber τ_{mf} is 0. As with the specimen in Figure 12.12c, it is assumed that each end of the fiber has zero tensile stress (σ_f). It is possible to include a small σ_f at the two ends of the fiber if there is strong fiber-polymer bonding. At the center of the fiber σ_f is a maximum.

As the short-fiber composite stress is increased, τ_{mf} at the ends of the fibers eventually reaches an instability point. The instability can be due to the matrix-fiber interface fracture or due to the matrix yielding. It is assumed that this is the maximum shear stress (τ_{mf}^*) that is developed at the matrix-fiber interface. The instability starts at the ends of the fiber and moves toward the center of the fiber as the tensile stress on the composite is further increased. The maximum stress that can be developed in a fiber corresponds to when the matrix-fiber shear stress all along the fiber is equal to τ_{mf}^*. If the fiber is sufficiently long, it is possible for the center of the fiber ($x = 0$) to reach the same stress ($\sigma_f^c = \varepsilon_c E_f$) that is present in a continuous-fiber composite at a composite strain of ε_c, as shown in Figure 12.13a.

The length of fiber where only the center of the fiber length reaches continuous-fiber stress is the critical fiber length (l_c). At the critical fiber length, it is assumed that the tensile stress at the fiber end is 0, and that the fiber stress linearly increases to σ_f^c at the center of the fiber length. For a critical-length fiber, Equation 12.14b is integrated from the center of the fiber length ($x = 0$) where the tensile stress is σ_f^c to the end of the fiber at $x = l_c/2$, where the tensile stress is 0. The result is Equation 12.15.

$$\sigma_f^c = \frac{4\tau_{mf}^* l_c}{2d_f} = \frac{2\tau_{mf}^* l_c}{d_f}$$

12.15

The critical fiber length divided by the fiber diameter is the **critical-aspect ratio** (l_c/d_f). Solving for the critical-aspect ratio from Equation 12.15 results in Equation 12.16.

$$\frac{l_c}{d_f} = \frac{\sigma_f^c}{2\tau_{mf}^*}$$

12.16

With Equation 12.16, we can calculate the average stress carried by the short fibers. If the fiber is longer than the critical length, there is a central section of the fiber that has a stress equal to σ_f^c, as shown

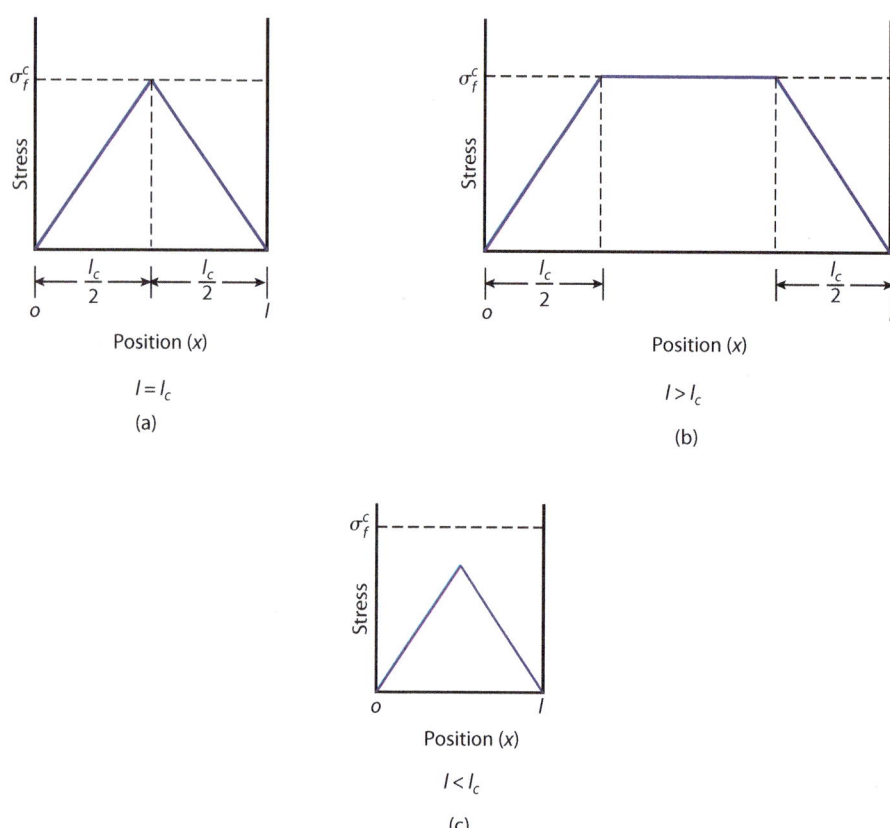

Figure 12.13   Plots of the fiber stress as a function of position along the fiber for short fibers that are (a) equal to the critical length, (b) greater than the critical length, and (c) less than the critical length.

in Figure 12.13b, and over the distance of $l_c/2$ from both ends of the fiber, the fiber stress drops from σ_f^c to 0. The average stress in the fiber is calculated with the help of Figure 12.13b. A fraction of the fiber equal to $(l - l_c)/l$ has the stress σ_f^c, and a fraction of the fiber equal to l_c/l has an average fiber stress equal to $\sigma_f^c/2$. Summing the fractions of length times the average stress in that length gives the average fiber stress $(\overline{\sigma}_f)$, as shown in Equation 12.17.

$$\overline{\sigma}_f = \left(\frac{l - l_c}{l}\right)\sigma_f^c + \frac{l_c}{2l}\,\sigma_f^c = \sigma_f^c\left(1 - \frac{l_c}{2l}\right)$$ **12.17**

If the fibers are shorter than the critical length, then the center of the fiber only reaches the stress of $\sigma_f^c(l/l_c)$, as shown in Figure 12.13c, and the average stress in the fibers $(\overline{\sigma}_f)$ is equal to half of the maximum stress, as shown in Equation 12.18.

$$\overline{\sigma}_f = \frac{l}{2l_c}\sigma_f^c$$ **12.18**

The composite stress for any length of short fibers is given by Equation 12.19.

$$\sigma_c = v_f\overline{\sigma}_f + v_m\sigma_m$$ **12.19**

The average fiber stress ($\overline{\sigma}_f$) is given by Equation 12.17 if $l > l_c$, and it is given by Equation 12.18 if $l < l_c$, and both Equations 12.17 and 12.18 work if $l = l_c$. The matrix stress (σ_m) in Equation 12.19 is equal to $\varepsilon_c E_m$. We assume that the strain in the matrix at locations away from the fibers is equal to the composite strain. This is similar to the isostrain assumption; however, the strain in the fibers is not equal to the matrix strain except in the center of those fibers with $l > l_c$.

The composite elastic modulus (E_c) is determined by writing Equation 12.19 in terms of elastic moduli and strain. Equation 12.20 is for fibers that are longer than the critical length.

$$\varepsilon_c E_c = v_f \left(1 - \frac{l_c}{2l}\right) \varepsilon_c E_f + v_m \varepsilon_c E_m \qquad \textbf{12.20}$$

Canceling the composite strain from each term in Equation 12.20 gives the composite elastic modulus, as shown in Equation 12.21.

$$E_c = v_f \left(1 - \frac{l_c}{2l}\right) E_f + v_m E_m \qquad \textbf{12.21}$$

In a similar manner, if the fibers are less than the critical length, the elastic modulus is derived with Equation 12.22.

$$E_c = v_f \frac{l}{2l_c} E_f + v_m E_m \qquad \textbf{12.22}$$

The analysis above is an approximation that gives good results, but a more rigorous analysis is the shear-lag model presented in Hull and Clyne. Figure 12.14 shows shear-lag model analysis of the axial fiber stress for a chopped-fiber composite with 30 volume percent glass fibers oriented parallel to the load axis in a polyester matrix subject to an axial strain of 0.001. In Figure 12.14, the length of the fiber is $2L$, the center of the fiber is taken as 0, and the fiber aspect ratio (s) is L/r, where r is the fiber radius. In

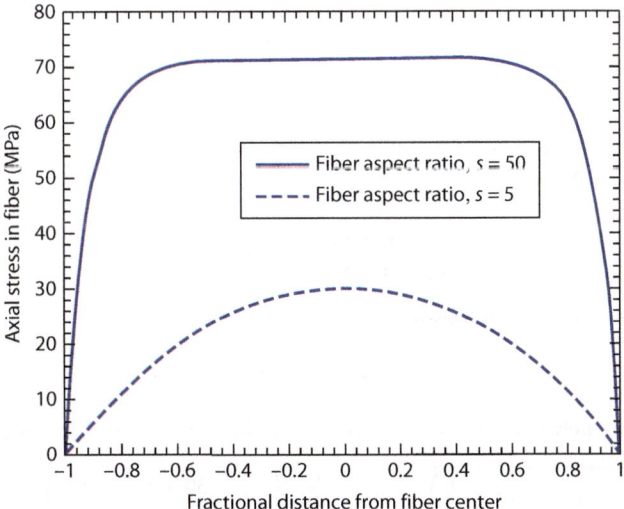

Figure 12.14 A shear-lag model calculation of axial-fiber tensile stress, as a function of fractional distance from the fiber center at 0 to the ends at +1 and –1. The composite materials have 30% glass fibers in a polyester matrix subject to an axial tensile strain of 0.001 for two fiber aspect ratios (s) of 50 and 5. (*Based on Hull, D. and Clyne, T.W., An Introduction to Composite Materials 2nd ed. Cambridge University Press, Cambridge (1996), p. 110.*)

this calculation, the axial stress at the ends of the fiber is assumed to be 0. In Figure 12.14, the fiber with an aspect ratio of 50 carries a stress in the center plateau region of the fiber equal to that carried by the fibers in a continuous-fiber composite with the same volume fraction of fiber material. The short fiber with an aspect ratio of 5 does not have a stress in the fibers equal to that in a continuous-fiber composite at any location. The load-carrying capacity of a composite with the aspect ratio of 5 is much less than the load-carrying capacity of the fiber with an aspect ratio of 50. The lower stress in the low-aspect-ratio fiber means that the matrix is subject to a higher stress than for a high-aspect-ratio fiber composite at the same stress, as shown in Equation 12.1. For the same volume fraction of fiber as a continuous-fiber composite, a chopped-fiber composite cannot carry as much load as a continuous-fiber composite can, because the stress is reduced at the end of the fibers. Fibers with a larger aspect ratio (s) have more of the fiber that carries a stress equal to that of a continuous fiber composite.

Example Problem 12.3

A composite-material specimen is made from 50 volume percent epoxy resin and 50 volume percent chopped E-glass fibers that are oriented parallel to the specimen axis. The E-glass fibers have a length of 300×10^{-6} m, a diameter of 10×10^{-6} m, and an elastic modulus of 76×10^9 Pa. The epoxy matrix has an elastic modulus of 4×10^9 Pa. Measurements have shown that the interface of the epoxy and E-glass has a shear strength of 50×10^6 Pa. In a tensile test, this composite material is linear elastic up to a strain of 4×10^{-3}, where a sharp decrease in elastic modulus is observed. Answer the following for a composite strain of 4×10^{-3}:
(a) What is the critical length?
(b) What is the average stress in an E-glass fiber?
(c) What is the composite-material stress?
(d) What is the composite-material elastic modulus?

Solution

a) The critical length is obtained from Equation 12.16.

$$\frac{l_c}{d_f} = \frac{\sigma_f^c}{2\tau_{mf}^*}$$

$$l_c = \frac{d_f \sigma_f^c}{2\tau_{mf}^*} = \frac{d_f(E_f)(\varepsilon_c)}{2\tau_{mf}^*} = \frac{10 \times 10^{-6}\ \text{m}(76 \times 10^9\ \text{N/m}^2)(4 \times 10^{-3})}{2(50 \times 10^6\ \text{N/m}^2)}$$

$$l_c = \frac{3040\ \text{N/m}}{100 \times 10^6\ \text{N/m}^2} = 30.4 \times 10^{-6}\ \text{m}$$

The fiber stress if this were a continuous-fiber composite is

$$\sigma_f^c = \varepsilon_c E_f = (76 \times 10^9\ \text{N/m}^2)(4 \times 10^{-3}) = 304 \times 10^6\ \text{N/m}^2$$

b) The fibers are approximately ten times the critical length. The average fiber stress is given by Equation 12.17.

$$\overline{\sigma}_f = \sigma_f^c\left(1 - \frac{l_c}{2l}\right) = 304 \times 10^6 \frac{\text{N}}{\text{m}^2}\left(1 - \frac{30.4 \times 10^{-6}\ \text{m}}{2(300 \times 10^{-6}\ \text{m})}\right)$$

$$\overline{\sigma}_f = 304 \times 10^6 \frac{\text{N}}{\text{m}^2}(1 - 0.05) = 289 \times 10^6 \frac{\text{N}}{\text{m}^2}$$

For a fiber that is ten times the critical length, the average fiber stress is only reduced by approximately 5% from the continuous fiber stress.

c) The composite stress is calculated from Equation 12.19. The matrix stress is the matrix elastic modulus times the composite strain.

$$\sigma_c = v_f \bar{\sigma}_f + v_m \sigma_m = v_f \bar{\sigma}_f + v_m E_m \varepsilon_c$$

$$\sigma_c = 0.5(289 \times 10^6 \text{ Pa}) + 0.5(4 \times 10^9 \text{ Pa})(4 \times 10^{-3}) = 145 \times 10^6 \text{ Pa} + 8.0 \times 10^6 \text{ Pa}$$

$$\sigma_c = 153 \times 10^6 \text{ Pa}$$

d) The composite elastic modulus is calculated from

$$E_c = \frac{\sigma_c}{\varepsilon_c} = \frac{153 \times 10^6 \text{ Pa}}{4 \times 10^{-3}} = 38.3 \times 10^9 \text{ Pa}$$

12.9 COMPOSITE FAILURE

12.9.1 Axial Tensile Failure of Uniaxial Continuous Fiber Composites

Even though the fibers are much stronger than the matrix, it is not necessarily true that the matrix fails first in a tension test of a composite material. In axial loading parallel to the fibers of a uniaxial continuous-composite material, the material with the lowest strain to fracture will fail first. This is because both the matrix and the fibers are at the same strain, but they are not at the same stress. Fracture stresses and strains for fibers are presented in Table 12.1. For matrix materials the fracture stress and strains are those of the bulk material, and those properties are presented in numerous tables in Chapter 8. The axial tensile strength of some uniaxial continuous-composite materials is presented in Table 12.2.

Brittle fibers and a brittle matrix   In the discussion of the failure of composite materials, we can assume that the fracture of the fibers occurs in a brittle manner; this is true for most composite material fibers, such as glass, carbon, boron, alumina, and silicon-carbide. Ductile fibers, such as OUHMWPE, Kevlar, and metallic wires, are an exception. The fracture of composite materials with a brittle matrix, such as thermoset polymers, ceramics, and carbon, are considered first and shown schematically in Figure 12.15a and 12.15b. In Figure 12.15, the composite axial stress (σ), the fiber stress (σ_f), and the matrix stress (σ_m) are plotted as a function of strain (ε). If the brittle matrix cracks at a lower strain than the fibers do, the maximum composite strain without an abrupt change in elastic modulus is the strain at the ultimate tensile strength of the matrix (ε_{mu}). The maximum stress for the composite without any change in elastic modulus is σ_1 in Figure 12.15a. This stress is determined by using ε_{mu} to calculate the fiber and matrix stress and inserting these stresses into the rule of mixtures in Equation 12.5, assuming that both the fibers and the matrix are elastic. Once the matrix is cracked across the entire cross section, the matrix stress is 0, and the fibers carry all of the stress. Then the composite stress is the

Table 12.2	Typical Mechanical Properties for Four Unidirectional Composite Materials with an Epoxy Matrix and the Indicated Fiber Reinforcement, Modulus Values are in GPa, and Strength Values are in MPa.			
Direction/Property	E-Glass	Kevlar	Carbon	S-Glass
Axial				
Modulus	44.8	75.8	145	56
Tensile strength	1124	1241	1517	1980
Compression strength	896	276	2068	626
Transverse				
Modulus	11.0	5.52	10.3	11.4
Tensile strength	31	14	48	31
Compression strength	138	55	138	138
Shear (in the fiber plane)				
Modulus	4.14	2.41	5.52	4.48
Strength	71.7	34.5	82.7	71.7

Based on data from the Assessment of Research Needs for Wind Turbine Rotor Materials, National Academy Press, 37 (1991), http://www.nap.edu/openbook.php?isbn=0309044790

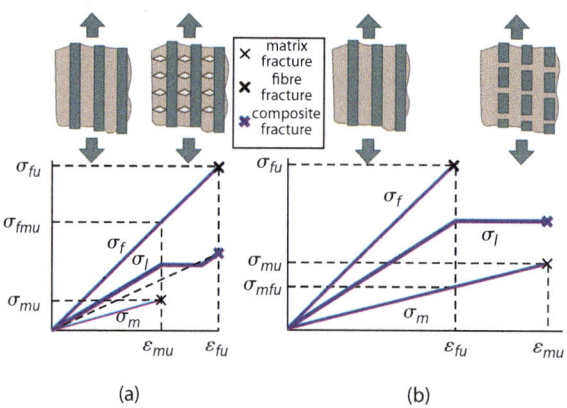

(a)   (b)

Figure 12.15 Schematics of the stress-strain plot to fracture for a composite with a brittle matrix and fibers. The composite stress is the thick line, the fiber stress is the medium line, and the matrix is the thin line. (a) The brittle matrix fractures before the reinforcement, as shown in the material schematic at the top. The dashed line with a slope is the product of $v_f\sigma_f$. (b) A brittle matrix where the fibers fracture before the matrix, as shown in the material schematic at the top. (*Based on Hull, D. and Clyne, T.W., An Introduction to Composite Materials 2nd ed. Cambridge University Press, Cambridge (1996), p. 160.*)

product of the volume fraction of fibers times the stress in the fibers ($v_f\sigma_f$); this product is the dashed line in Figure 12.15a. The composite stress increases to the fracture strain of the fibers (ε_{fu}).

The stress-strain plot for a composite with fiber fracture at a lower strain than the brittle matrix is shown in Figure 12.15b. The maximum composite stress without an abrupt change in elastic modulus is calculated with Equation 12.5 using the strain-for-fiber fracture (ε_{fu}). After the continuous fibers start to fracture, the composite continues to carry a stress, because the fractured long fibers behave like the short fibers we discussed in Section 12.8. With increased strain, the composite stress does not increase because the fibers are fracturing and the force carried by the fibers is decreasing. However, the matrix stress increases as more fibers are fractured, until the matrix fracture strain is reached.

Brittle fibers and a ductile matrix

If the matrix material has a yield strain, the shape of the stress-strain curve for the composite depends upon the magnitude of the yield strain relative to the fracture stress of the reinforcing phase. Matrix materials with a yield strain include metals and thermoplastic polymers. If the fibers fracture at a lower strain than the yield strain of the matrix, the stress-strain curve for the composite and the analysis for the maximum composite stress without an abrupt change in elastic modulus is the same as in Figure 12.15b, where the fibers fractured at a lower strain than the matrix did. The composite-material instability occurs at the fracture strain of the fibers (ε_{fu}).

If the matrix material yields at a lower strain than the fiber fracture strain, the composite is linear elastic up to the yield strain of the matrix (ε_{my}), with the elastic modulus given by the rule of mixtures in Equation 12.6. After yield of the matrix, the stress of the composite continues to increase with strain, but at a reduced effective modulus, as shown in Figure 12.16. The magnitude of the reduced effective modulus depends upon the amount of strain-hardening in the matrix material. If there is no strain hardening in the matrix, then the composite modulus (E_c) after yield of the matrix is equal to $E_f v_f$, because the matrix is then making no contribution to the composite elastic modulus. The composite stress is given by the rule of mixtures in Equation 12.5, with the matrix contribution given by a constant value of $\sigma_{my} v_m$, where σ_{my} is the matrix yield stress, and the fiber contribution is given by $\varepsilon E_f v_f$. Once the fiber fracture strain (ε_{fu}) is reached, the continuous fibers begin to fracture and then behave as finite-length fibers. The strain continues to increase, but there is no increase in the composite stress as the fibers continue to fracture. As the fibers fracture, they carry less stress; thus the stress on the matrix must increase since the composite stress remains constant. The composite fractures at a strain less than the matrix fracture strain, because the cracks in the fibers are stress concentrations that increase the local strain. The dashed line for σ_1 indicates that the extent of the strain at the stress σ_1 is uncertain.

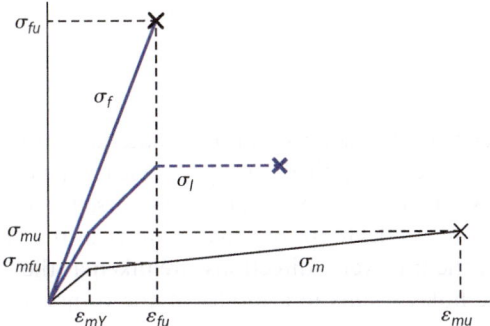

Figure 12.16 The stress-strain plot for a composite material with a ductile matrix that yields at a strain (ε_{my}) that is less than the fracture strain of the fibers (ε_{fu}). The composite stress is the thick line, the fiber stress is the medium line, and the matrix is the thin line. (*Based on Hull, D. and Clyne, T.W., An Introduction to Composite Materials 2nd ed. Cambridge University Press, Cambridge (1996), p. 172.*)

Example Problem 12.3

A uniaxial composite material is made from 70 volume percent continuous carbon fibers and 30 volume percent epoxy. The carbon fibers have an elastic modulus of 500 GPa, and the epoxy has an elastic modulus of 5 GPa. The carbon fibers have a tensile strength of 3 GPa, and the epoxy matrix material has an ultimate tensile strength of 0.05 GPa. Determine the following:
(a) The maximum strain without instability for the composite. Assume that both the fiber and matrix are brittle and linear elastic to failure.
(b) The composite stress at instability.

Solution

a) We must first determine the strain at fracture for both the fibers and the matrix. The strain that causes the fibers to fail is

$$\varepsilon_{fu} = \frac{\sigma_{fu}}{E_f} = \frac{3 \text{ GPa}}{500 \text{ GPa}} = 0.006$$

The strain that causes the matrix to fracture is equal to

$$\varepsilon_{mu} = \frac{\sigma_{mu}}{E_m} = \frac{0.05 \text{ GPa}}{5 \text{ GPa}} = 0.01$$

The high-strength fibers fracture at a lower strain than the matrix does; the fibers fracture first at a strain of 0.006.

b) The stress in the epoxy when the fibers fracture is calculated from the elastic modulus of the epoxy of 5 GPa and the strain at fracture of the fibers of 0.006.

$$\sigma_m = E_m\varepsilon_m = 5 \text{ GPa}(0.006) = 0.03 \text{ GPa} = 30 \text{ MPa}$$

The stress in the fibers upon fracture is 3 GPa. The stress at the instability in the composite is now determined with the isostrain model in Equation 12.5.

$$\sigma_1 = \sigma_f v_f + \sigma_m v_m = 3000 \text{ MPa}(0.7) + 30 \text{ MPa}(0.3) = 2100 \text{ MPa} + 9 \text{ MPa} = 2109 \text{ MPa}$$

12.9.2   Transverse Tensile Failure of Uniaxial Continuous Fiber Composites

A transverse tensile stress occurs when a uniaxial composite material is stressed normal to the fiber axis in tension. When a transverse tensile stress is applied, the transverse tensile failure stress (σ_{2u}) is less than the stress to fail the matrix alone. As shown in Table 12.2, the transverse tensile strength is much less than the axial tensile strength. The direction parallel to the fiber axis is taken as the "1" direction, and the directions normal to the axial direction, or the transverse directions, are taken as the "2" and "3" directions.

Assuming that the fibers are holes results in a model of the transverse tensile strength for uniaxial composite materials that produces predictions in reasonable agreement with the experimental results. Equation 12.23 shows the result of a geometric analysis for circular fibers, or holes, in a square array.

$$\sigma_{2u} = \sigma_{mu}\left[1 - 2\left(\frac{v_f}{\pi}\right)^{0.5}\right] \qquad \textbf{12.23}$$

For a composite material with 50% uniaxial continuous carbon fibers in an epoxy matrix that has an ultimate tensile strength of 50 MPa, determine the transverse tensile strength.

Solution

The transverse tensile strength is given by Equation 12.23.

$$\sigma_{2u} = \sigma_{mu}\left[1 - 2\left(\frac{v_f}{\pi}\right)^{0.5}\right] = 50 \text{ MPa} \left[1 - 2\left(\frac{0.5}{\pi}\right)^{0.5}\right] = 50 \text{ MPa} [1 - 0.80] = 10 \text{ MPa}$$

The tensile strength of the 50% fiber composite in the transverse direction is only 20% of the ultimate tensile strength of the matrix. This is why fibers in composite materials are placed in multiple directions if there is to be any significant stress in the transverse direction. Only in cases of pure tensile stress or pure bending should uniaxial composite materials be applied.

12.9.3   Compression Failure of Uniaxial Continuous Fiber Composites

Uniaxial composite materials are strong in compression when the stress is applied parallel to the fiber axis, as shown in Table 12.2. There are several modes of failure in compressive loading parallel to the fiber axis. We can observe that the fibers buckle in unison to form a kink band, as shown in Figure 12.17. The higher is the volume fraction of fibers, the more constraint or support that is provided by neighboring fibers to prevent buckling, and the stronger is the composite.

Figure 12.17   A photograph of a wood specimen subject to a vertical compressive force showing local collapse of the cellulose fibers that are parallel to the vertical direction. (*Photo by C. M. Gilmore*)

Another mode of failure present with compression stresses parallel to the fibers of a uniaxial composite is crack formation at the fiber-matrix interface. There are tensile strains due to Poisson's ratio that are in a direction perpendicular, or transverse, to the fiber-matrix interface that pull the fibers and matrix materials apart, forming cracks. The bond between the fibers and the matrix is relatively weak, because the bond results from weak van der Waals bonds. As shown in Table 12.2, uniaxial composite materials stressed in transverse tension are very weak. The transverse strains produced by a compressive stress parallel to the fibers due to Poisson's ratio results in cracks similar to those produced by a transverse tensile stress. Figure 12.18 shows a wood specimen that has been compressed parallel to the cellulose fibers, producing both kink bands in the cellulose fibers and cracking parallel to the cellulose fiber axis. It is difficult to model the compressive strength of composite materials, because of the complex failure modes.

As shown in Table 12.2, the transverse compressive strength of a uniaxial composite material is much less than for compression parallel to the fiber axis. If a uniaxial composite material with a ductile matrix is compressed normal to the fiber axis (transverse compression) and the matrix yields, and the fibers can shear past each other. This is observed in the compression of wood specimens. The fibers do provide compressive strength to the matrix similar to dispersion strengthening. If there is good bonding of the fiber-matrix interface, the strength of a ductile matrix composite in transverse compression is comparable to the compressive strength of the dispersion-reinforced matrix.

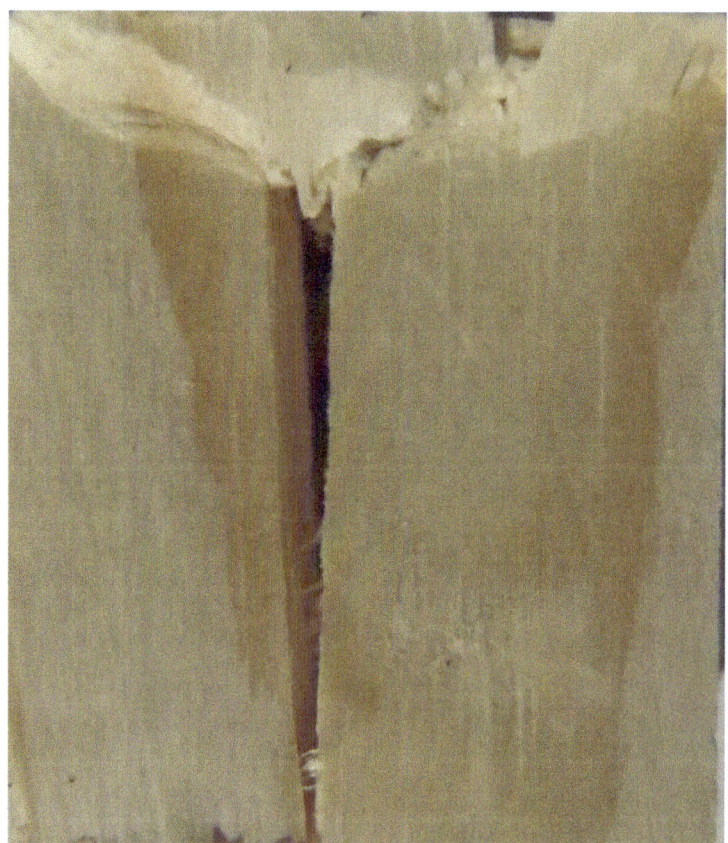

Figure 12.18   A wood specimen compressed with a vertical force showing local collapse of cellulose fibers at the top of the specimen on both sides of the "V"- shaped material section. Also, a vertical crack formed up the center of the specimen resulting from transverse strain due to Poisson's ratio. (*Photo by C. M. Gilmore*)

12.9.4 Shear Failure of Uniaxial Continuous Fiber Composites

The shear strength of a uniaxial composite material depends upon the direction of the applied shear stress relative to the fiber axis. Uniaxial composite materials are weak in shear if the shear stress is applied in the direction of the fibers called in-plane-shear, because this shears the weak matrix-fiber interface. Cracks develop in areas with relatively high percentages of matrix material. The shear strengths presented in Table 12.2 are significantly less than the axial tensile strength, and the in-plane-shear strength is approximately twice the magnitude of the transverse tensile strength.

12.9.5 Failure in Short-Fiber Composites

All of the modes of failure observed in long-fiber composites discussed above can also be observed in short-fiber composites. However, for short-fiber composites another mode of failure is debonding of the fiber and matrix around the fiber circumference, due to interface shear. The shear stress at this interface reaches its maximum at the ends of the fibers. For short-fiber composites with fibers oriented parallel to the stress axis, cracks form when the ends of the fibers shear, or pull out, relative to the matrix material.

12.9.6 The Critical-Stress Intensity Factor for Composite Materials

K_{Ic} at room temperature for several composite materials is presented in Table 12.3. Observe that the value of the critical-stress intensity factor of a composite material such as high-tensile-strength (HTS) carbon-reinforced epoxy is 42 MPa · m$^{1/2}$. Carbon-fiber-reinforced epoxy is made from epoxy that alone has a critical-stress intensity factor of only 0.3 to 0.5 MPa · m$^{1/2}$, and the carbon fibers are brittle. A rule of mixtures could never produce this composite critical-stress intensity factor of 42 MPa · m$^{1/2}$. It is proposed that the primary energy-absorbing mechanism in the fracture of composite materials with a brittle matrix and brittle fibers is the pullout of fibers from the matrix. The pullout of fibers is shown in a composite with a glass matrix and SiC fibers in Figure 12.19. As the fibers pull out of the matrix during fracture, there is a frictional force between the matrix and the fibers, which increases the energy required

| Table 12.3 | The Critical Stress Intensity Factor (K_{Ic}) for Some Unidirectional Composite Materials with an Epoxy Matrix. The Volume Fraction of Fibers (v_f) is Indicated in Parentheses after the Fiber Type. |

Fiber (v_f)	K_{Ic} (MPa · m$^{1/2}$)
HM-Carbon (0.6)	24
HTS-Carbon (0.7)	42
Boron (0.7)	47

Data from Beaumont, P.W.R., Schultz, J., and Friedrich, K., Failure Analysis of Composite Materials, CRC Press, Boca Raton, FL., (1990), p. 99.

Figure 12.19 A scanning electron micrograph of pullout of SiC fibers at the fracture surface of a glass-matrix composite. (*ASM Handbook Volume 9, Metallography and Microstructures, 2004, ASM International, in ASM Handbooks Online, http://www.asmmaterials.info, ASM International, 2004.*)

for fracture and the critical-stress intensity factor. This is similar to the reinforcing of concrete with steel rebars that have ribs along the surface to increase the pullout energy. Fibers can toughen brittle materials to increase their resistance to fracture.

12.10 THERMAL STRESSES IN COMPOSITE MATERIALS

The difference in thermal expansion coefficients of fibers in Table 12.1 and matrix materials in Table 4.2 results in thermal stresses in composite materials that are subject to a temperature change. **Thermal stresses** are one form of internal stress. Internal stresses are stresses that are not caused by an external force. For example, the axial thermal expansion coefficient for high-strength carbon fiber from Table 12.1 is -0.4×10^{-6} K^{-1}, whereas for an epoxy matrix the thermal expansion coefficient from Table 4.2 is 55×10^{-6} K^{-1}. The large difference on thermal expansion coefficient can lead to significant thermal stresses in the composite material. The thermal expansion coefficients in Table 12.1 are given in units of K^{-1} and in Table 4.2 in units of °C^{-1}. Numerically the units K^{-1} are equal to °C^{-1}, because in Equation 4.12 for the thermal expansion coefficient the temperature appears as an increment of temperature (dT). An increment of 1 K is equal to an increment of 1°C.

Temperature change can occur during service and during composite fabrication. Composite materials with a metal or thermoplastic matrix are fabricated by melting the matrix material, and then infiltrating the liquid matrix into the fiber reinforcement. The matrix and the reinforcement are stress free at the melting temperature of the matrix, but as the matrix cools, thermal stresses develop because of the difference in thermal expansion coefficients of the fiber and matrix materials. Composite materials with a thermoset matrix are often cured at an elevated temperature, such as 200°C, where the thermoset matrix hardens. The matrix and fibers are stress free at the cure temperature. Then, the composite material is cooled to room temperature, and the thermal stresses develop during the cooling.

The magnitude of the elastic stress and strain developed as a result of a temperature change ΔT is evaluated from the coefficient of thermal expansion and the elastic moduli of the materials. Figures 12.20a and 12.20b show a slab of reinforcing material (or fiber) and a slab of matrix material of the same length at the initial temperature of T_1. The length is set equal to one so that changes in length are numerically equal to the magnitude of the strain. Most composite materials are made of many layers of reinforcement or many fibers; however, this slab model provides a useful representation. In Figure 12.20c the reinforcing material and the matrix are heated and allowed to expand independent of each other as though they are separated. The separated matrix and fiber strain by amounts $\alpha_m \Delta T$ and $\alpha_f \Delta T$, respectively. However, in the composite material the matrix and fiber are bonded together, and the final length of the matrix and fiber are the same, as shown in Figure 12.20d. Forcing the lengths to be the same after the temperature change ΔT produces an elastic strain in the matrix (ε_m) and an elastic strain in the fiber (ε_f). Figure 12.20d shows that the thermal strain of the composite material (ε_c) is given by $\alpha_c \Delta T$, where α_c is the thermal expansion coefficient of the composite material.

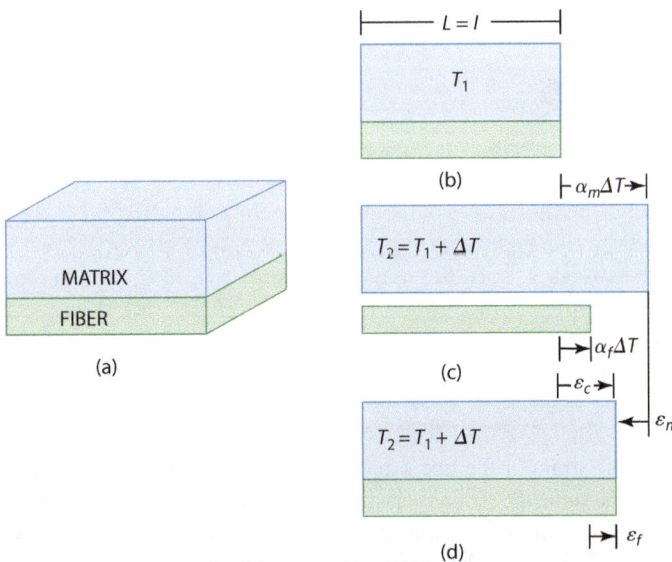

Figure 12.20 (a) A slab model of thermal expansion of a composite material with continuous fiber reinforcement (green) and matrix (blue) in three dimensions. (b) The length of the matrix and reinforcement are equal to unity for the composite material at ambient temperature T_1. (c) Separate matrix and reinforcement expand to different lengths at a higher temperature T_2. (d) In a composite material, the fiber and matrix materials must expand to the same length at temperature T_2, resulting in the thermal strains indicated by the arrows.

Equations 12.24 and 12.25 for the thermal expansion coefficient of the composite material are deduced from Figure 12.20.

$$\varepsilon_c = \alpha_c \Delta T = \alpha_m \Delta T + \varepsilon_m \qquad \textbf{12.24}$$

$$\varepsilon_c = \alpha_c \Delta T = \alpha_f \Delta T + \varepsilon_f \qquad \textbf{12.25}$$

If Equation 12.24 is subtracted from Equation 12.25, the result is an equilibrium equation for the strain, as shown in Equation 12.26.

$$(\alpha_f - \alpha_m)\Delta T + (\varepsilon_f - \varepsilon_m) = 0 \qquad \textbf{12.26}$$

Since there is no applied stress on the composite material, the sum of the thermal stresses in the fiber and in the matrix is 0, as shown in Equation 12.27. The stresses add in proportion to the volume of material to satisfy force equilibrium.

$$\sigma_c = v_f \sigma_f + v_m \sigma_m = 0 \qquad \textbf{12.27}$$

Assuming that all of the strains associated with the thermal stresses are elastic, Equation 12.27 is written as Equation 12.28.

$$v_f E_f \varepsilon_f + v_m E_m \varepsilon_m = 0 \qquad \textbf{12.28}$$

Through algebraic manipulation of Equations 21.24 through 21.28, the composite thermal expansion coefficient parallel to the reinforcement axis is determined, as shown in Equation 12.29.

$$\alpha_c = \frac{\alpha_m v_m E_m + \alpha_f v_f E_f}{v_m E_m + v_f E_f} \qquad \textbf{12.29}$$

Example Problem 12.5

A uniaxial long-fiber composite shaft is made of 66.6% SiC fibers and 33.3% epoxy resin. The composite is cured at 175°C (448 K) and then cooled to 25°C (298 K). The elastic modulus of the SiC fibers is 400 GPa, and the thermal expansion coefficient is 4×10^{-6} °C^{-1}. The elastic modulus of the epoxy is 4 GPa, and the thermal expansion coefficient is 60×10^{-6} °C^{-1}.
(a) Calculate the internal stresses, both the magnitude and sign, in the matrix and in the fibers after the cured composite is cooled from 175°C (448 K) to 25°C (298 K). Assume that each material is elastic at all temperatures after the cure and that there is good interface bonding.
(b) Calculate the composite strain in cooling from 175°C (448 K) to 25°C (298 K).

Solution

There are two unknowns: the strain in the fibers and the strain in the matrix. If we know the strain, we can calculate the stress, since we assume the strain is elastic. Once we know the strain in the fiber and in the matrix, the composite strain is calculated from Equation 12.24 or 21.25. Since there are two unknowns, we must solve two equations. One equation comes from the observation that there is no applied stress on the composite.

$$\sigma_c = v_f \sigma_f + v_m \sigma_m = 0$$

$$0 = 0.66\sigma_f + 0.33\sigma_m$$

$$2\sigma_f = -\sigma_m$$

Since all of the strains are assumed elastic, we can relate these internal stresses to the fiber and matrix strains through Hooke's law.

$$\varepsilon_f = \frac{\sigma_f}{E_f} \text{ and } \varepsilon_m = \frac{\sigma_m}{E_m} = \frac{-2\sigma_f}{E_m}$$

Equation 12.26 relates the elastic strain to the thermal strain.

$$(\alpha_f - \alpha_m)\Delta T + (\varepsilon_f - \varepsilon_m) = 0$$

Inserting the relations for fiber and matrix strain above, only the fiber stress is unknown.

$$(4 \times 10^{-6} \text{ K}^{-1} - 60 \times 10^{-6} \text{ K}^{-1})(298 \text{ K} - 448 \text{ K}) + \sigma_f\left(\frac{1}{E_f} - \frac{-2}{E_m}\right) = 0$$

$$(-56 \times 10^{-6} \text{ K}^{-1})(-150 \text{ K}) + \sigma_f\left(\frac{1}{400 \times 10^9 \text{ Pa}} + \frac{2}{4 \times 10^9 \text{ Pa}}\right) = 0$$

$$8400 \times 10^{-6} + \sigma_f[0.0025 \times 10^{-9} \text{ Pa}^{-1} + 2(0.25 \times 10^{-9} \text{ Pa}^{-1})] = 0$$

Solving for the fiber stress,

$$\sigma_f = \frac{-8.4 \times 10^{-3}}{0.503 \times 10^{-9} \text{ Pa}^{-1}} = -16.7 \times 10^6 \text{ Pa}$$

and the matrix stress,

$$\sigma_m = 33.4 \times 10^6 \text{ Pa}$$

The composite strain is now solved for from Equations 12.24 or 12.25.

$$\varepsilon_c = \alpha_m \Delta T + \varepsilon_m$$

$$\varepsilon_c = (60 \times 10^{-6} \text{ K}^{-1})(-150 \text{ K}) + \frac{33.4 \text{ MPa}}{4000 \text{ MPa}} = -9 \times 10^{-3} + 8.35 \times 10^{-3} = -0.65 \times 10^{-3}$$

The fibers have an internal compressive stress, and the epoxy matrix has an internal tensile stress. In Table 8.18 the tensile strength of epoxy is listed as 30–60 MPa. The tensile stress in this composite is in the range of the tensile strength of epoxy. The compressive stress on the fibers is relatively small.

12.11 PROCESSING OF CONTINUOUS-FIBER COMPOSITES

There are many different ways that continuous-fiber composites are produced, and the mode of production depends upon the application. The processing is much different for a uniaxial composite fishing rod than it is for the skin of an aircraft or for a kayak. This section covers some of the most common fabrication techniques for continuous-fiber composites. As shown in Table 12.1, one common factor is that many of the filaments used in composite materials are typically 10 μm in diameter. This allows the filaments to be flexible, but many filaments must be joined together to form the composite material.

The processing of short-fiber composites with a polymer matrix is the same as the processing of polymers with particle fillers. Short-fiber composites with a metal matrix are processed with the same procedures as for dispersion-strengthened metals. We will cover these subjects in Chapter 13.

12.11.1   Processing of Uniaxial Continuous-Fiber Composites

Products such as fishing rods and golf-club shafts are made of uniaxial composite materials. One method of fabricating these products is by **pultrusion**, shown in Figure 12.21. In pultrusion, the fibers are pulled from a roll into a bath of liquid polymer, such as epoxy, and the fibers and liquid are formed into the desired shape by a die. If the matrix is a thermoset polymer, such as epoxy, the shaped composite is cured at elevated temperature in an oven to accelerate the hardening process. Pultrusion is most suited to producing continuous composite materials of constant shape, such as an I-beam or rod.

Composite products such as pressure vessels, liquid storage tanks, and rocket motor casings are made by filament winding, shown in Figure 12.22. In filament winding, the reinforcing material is in the form of multifilament tape. The tape is wrapped around a rotating **mandrel**, which is a form of the desired shape. The location of the wrap on the mandrel is determined by the position of the tape source. The wrapped mandrel can be dipped in liquid matrix material to make the composite, or the tape can be prepeg.

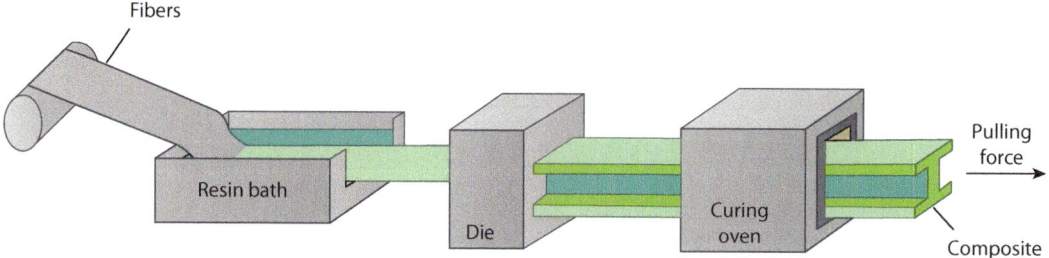

Figure 12.21   The steps in producing a composite material by pultrusion. (*Based on Askeland, D.R., Fulay, P.P., and Wright, W.J., The Science and Engineering of Materials, 6th ed., Cengage Learning, Stamford, CT. (2011), p. 676.*)

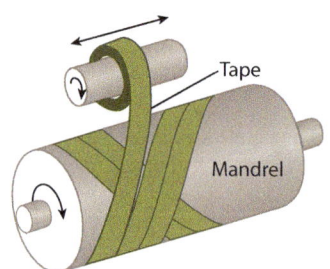

Figure 12.22   In filament winding, reinforcing tape is unwound from a roll and wound onto a mandrel shaped like the desired product. The mandrel rotates while the tape source moves to place the tape in the desired location. The mandrel is removed from the composite after the composite is hardened. (*Based on Askeland, D.R., Fulay, P.P., and Wright, W.J., The Science and Engineering of Materials, 6th ed., Cengage Learning, Stamford, CT. (2011), p. 676.*)

In **prepeg-tape**, liquid epoxy or polyester is mixed with hardener that is then impregnated into the tape fibers. The prepeg-tape is immediately refrigerated to suppress hardening. When the prepreg-tape is to be wrapped, it is removed from refrigeration and immediately wrapped, and it begins to harden. In filament winding the tape can be uniaxial fibers or it can be a narrow woven tape.

12.11.2 Processing of Woven Continuous-Fiber Composites

The starting material for production of a woven-fiber reinforced composite is a fabric, such as that shown in Figure 12.6. Because the filaments are very small in diameter, the woven cloth is flexible and can be formed into a shape by the techniques shown in Figure 12.23. In many products the desired layers and orientations of woven cloth are placed in molds by hand. Then, epoxy or polyester is impregnated into the fabric, and rollers are used to form the shape and spread the matrix material, as shown in Figure 12.23a. Figure 12.23b shows how the shape is formed with air pressure, and Figure 12.23c shows how the shape is formed with a solid die.

Vacuum bagging, shown in Figure 12.24, is used for high-performance composite materials. In vacuum bagging, the composite material in the mold with the liquid thermoset polymer matrix is placed in an airtight

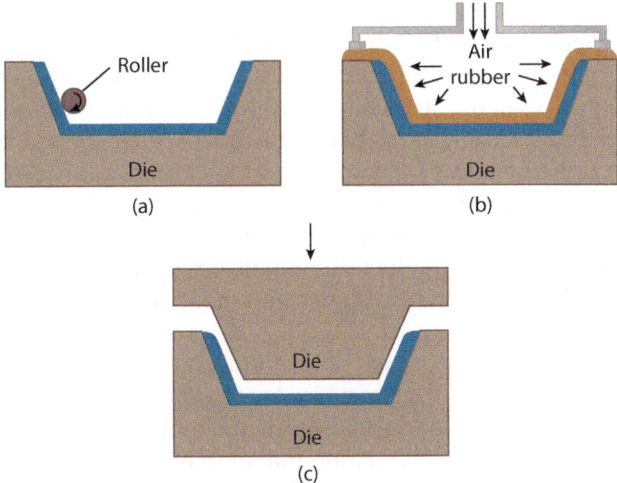

Figure 12.23 Techniques for shaping woven-cloth composite materials. The part is the dark blue shape. (a) Hand layup, shaping, and impregnation of liquid thermoset matrix material into a mold or die with hand-held devices such as a roller. (b) Forming the shape with air pressure. (c) Forming the shape with a solid die. (*Based on Askeland, D.R., Fulay, P.P., and Wright, W.J., The Science and Engineering of Materials, 6th ed., Cengage Learning, Stamford, CT. (2011), p. 676.*)

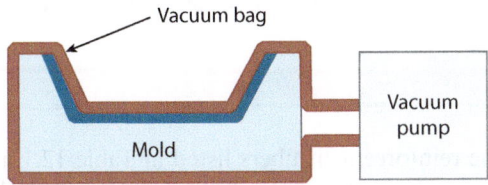

Figure 12.24 Vacuum-bag processing of composite materials. The vacuum bag is brown, and the part is the dark blue shape.

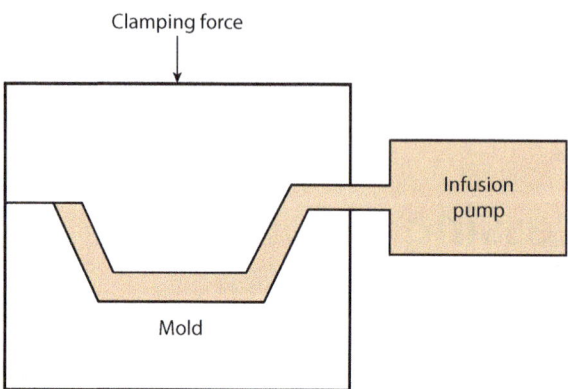

Figure 12.25. A schematic of the resin transfer molding process. The infusion pump injects polymer material (brown) into the mold cavity. The mold cavity can contain reinforcement material or small reinforcing material particles that are mixed with the injected polymer material.

plastic bag. Then a vacuum pump line is inserted and sealed inside the bag. The vacuum inside the bag results in air pressure on the outside of the bag forcing the bag onto the composite material and into the shape of the mold. The pressure on the composite also squeezes any extra liquid matrix material out of the composite. The vacuum on the composite removes air bubbles from the liquid matrix, and it removes water vapor from the fabric surface. After the thermoset polymer matrix hardens, the part is removed from the vacuum bag.

Figure 12.25 is a schematic of **resin transfer molding** (RTM). In RTM the fabric is placed in the mold and the mold is closed under pressure. The liquid matrix material is then injected into the mold with the infusion pump. The mold can also be under vacuum to assist the injection of liquid matrix material into the woven cloth. Large panels for auto bodies and aircraft are made in this way. Shapes formed from thermoset matrix composite materials are usually heated in an oven or an autoclave to accelerate the hardening process and to form a stronger product. Randomly oriented mat can be used in the processes shown in Figures 12.23 through 12.25; however, composites made from mat are not as strong as those made from woven fabric.

In many products, the layers of woven cloth in the composite have different properties and functions. If the composite is to have human contact, such as a kayak or a surfboard, a very fine weave of polyester or nylon cloth at the surface provides a good surface finish and a ductile surface layer. Brittle filaments at the surface, such as glass or carbon, can fracture and stick out, resulting in a rough surface that produces skin abrasions. Fabric with coarse roving should be used for interior layers to provide strength and stiffness. Coarse roving at the surface results in a rough surface. Interior layers of carbon provide overall strength, and a honeycomb structure provides bending strength.

12.12 RECYCLING OF FIBER-REINFORCED COMPOSITES

Composites made with most of the reinforcement fibers listed in Table 12.1 are not recyclable, because the fiber is not easily separated from the matrix even if the matrix material is recyclable. Thermoset matrix materials cannot be recycled except as filler for products such as asphalt. Thermoplastic matrix materials can be recycled if the thermoplastic can be separated from the reinforcing material by melting.

A composite that could be recycled is PE matrix with OUHMWPE fibers, because this composite could be chopped and melted to form pellets of 100% PE. Composites with cellulose fibers can be recycled, because the cellulose fibers are biodegradable. For this reason, consideration is being given to the use of composites with fibers of cellulose from wood, flax, and pineapple leaves. The polymer matrix can be recycled once the fiber is biodegraded. Cellulose fibers have good strength in comparison to E-glass fiber, as shown in Table 12.1.

12.13   COMPOSITES FOR CONSTRUCTION

12.13.1   Wood

Wood is a uniaxial composite material that is frequently utilized in compression and bending. The strength of wood varies with direction because of its structure. Trees grow with annual rings, as shown in Figure 12.26. Each ring has a section of early wood that forms during the spring, and late wood that forms during the summer, shown in Figure 12.26b. Figures 12.26c and 12.26d show that wood is made of elongated cellulose cells that are bonded together with **lignin**. Lignin is a phenolic polymer resin that is three-dimensionally cross-linked. The cellulose cells in early wood are larger than the cells in late wood. The cellulose cells are crystalline and covalently bonded. They have an ultimate tensile strength of approximately 2 GPa, an elastic modulus of 80 GPa, and a failure strain of approximately 3%, as shown in Table 12.1. The rays in Figure 12.26a are radial cells that transport nutrients and water in the radial direction in the tree.

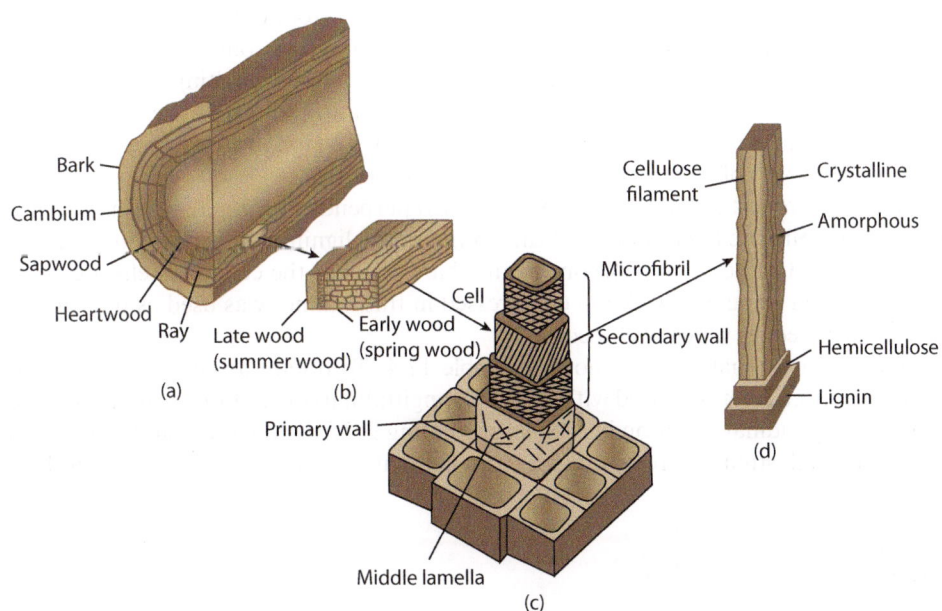

Figure 12.26   The structure of wood. (a) A schematic of a cut section from a tree. (b) Layers of early wood and late wood. (c) A wood cell. (d) Crystalline cellulose and amorphous lignin. (*Based on Askeland, D.R., Fulay, P.P., and Wright, W.J., The Science and Engineering of Materials, 6th ed., Cengage Learning, Stamford, CT. (2011), p. 699.*)

Table 12.4	Mechanical Properties of Selected Woods, Elastic Modulus in GPa (10^6 psi), and Strength in MPa (10^3 psi)				
Wood	E^1	TSL[2]	TSR[3]	CSL[4]	CSR[5]
Pine	8.3(1.2)	73(10.6)	2.1(0.31)	33(4.8)	3.0(0.44)
Fir	14(2.0)	78(11.3)	2.7(0.39)	38(5.5)	4.2(0.61)
Maple	10(1.5)	108(15.7)	7.6(1.1)	54(7.8)	10(1.5)
Oak	13(1.8)	78(11.3)	6.5(0.94)	43(6.2)	5.6(0.81)

Strength data based on F. F. Wangaard, "Wood: Its Structure and Properties," J. Educ. Models for Mat. Sci. and Engr., Vol. 3, No. 3, 1979.
1 Longitudinal tensile elastic modulus.
2 Tensile strength in the longitudinal direction (TSL).
3 Tensile strength in the radial direction (TSR).
4 Compressive strength in the longitudinal direction (CSL).
5 Compressive strength in the radial direction (CSR).

Wood boards are normally cut so that the cellulose fibers are oriented parallel to the long direction of the board. This is the longitudinal (L) direction. The radial direction (R) is from the center of the tree to the outer diameter, and the transverse direction is perpendicular to the radial and longitudinal directions. The elastic modulus and strength properties of selected woods in different directions are presented in Table 12.4.

The tensile strength in the longitudinal direction (TSL) is at least ten times as large as it is in the radial direction (TSR). This is due to the high strength of the cellulose cells that carry the stress in axial tension, in comparison to the weak lignin-cellulose interface that fractures with transverse tension.

The yield of wood due to longitudinal compressive forces (CSL) is initially due to localized collapse of cellulose fibers similar to that observed in Figure 12.17, resulting from the combination of compression of the long slender columns of cellulose and shear stresses at 45° to the compression axis. At later stages of failure, cracks appear parallel to the fiber axis due to tensile strains resulting from Poisson's ratio, as shown in Figure 12.18. The strains resulting from Poisson's ratio pull apart the weak cellulose-lignin interface, even though there is no tensile stress in this direction. The longitudinal compressive strength of wood is approximately 50% of the tensile longitudinal strength.

When wood is compressed to failure with loads applied perpendicular to the fiber axis or in the radial direction (CSR), the cellulose fibers slide past each other in the lignin matrix, resulting in a low CSR. If the wood is compressed sufficiently, it becomes quite hard because the cellulose cells are collapsed and the density of the wood is increased. Wood compressed in this manner was used as the edge material in snow skis before metal edges were developed.

Wood is an anisotropic material, as shown in Table 12.4. To obtain a sheet of wood that is quasi-isotropic within the sheet, layers of wood with different longitudinal orientation are glued together to form plywood with the longitudinal direction in each layer rotating 90° from the previous layer. Microlaminate structural beams of high strength are now made from laminates of very thin layers of wood.

12.13.2 Concrete

Concrete is a composite with a matrix of hydrated **Portland cement**; the reinforcement is an aggregate (stones) mixture sized from fine sand to coarse rocks. Portland cement is made with a starting mixture of 60 to 65% lime (CaO), 20 to 25% silica (SiO_2), 7 to 12% alumina (Al_2O_3), and 7 to

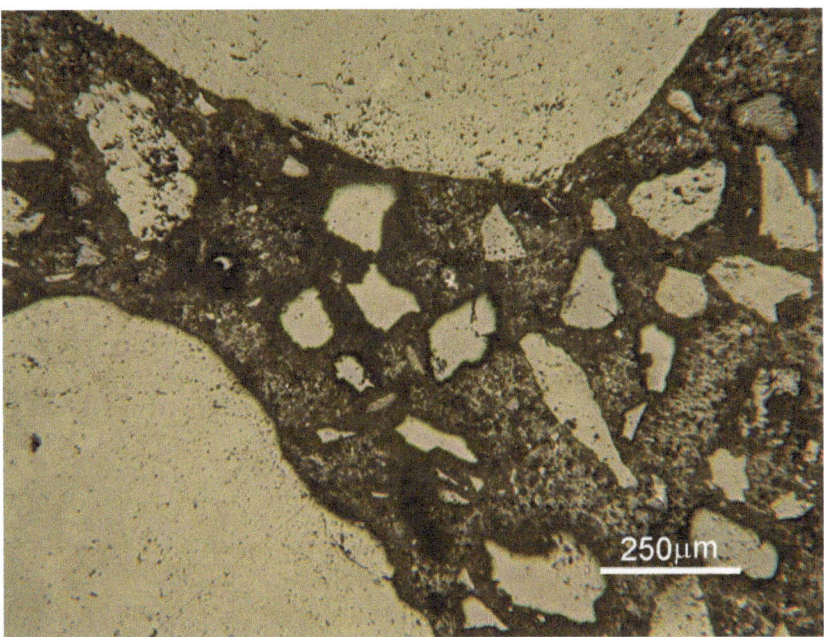

Figure 12.27   A Micrograph of concrete with light aggregate and dark Portland cement. (*Dr. Kevin M. Knowles*)

12% iron oxide (Fe_2O_3). These components are ground into powder, blended, and then heated to a temperature of approximately 1500°C. The components react with each other to form four primary compounds: tricalcium silicate ($3CaO\text{-}SiO_2$), dicalcium silicate ($2CaO\text{-}SiO_2$), tricalcium aluminate ($3CaO\text{-}Al_2O_3$), and tricalcium aluminoferrite ($3CaO\text{-}Al_2O_3\text{-}Fe_2O_3$). This material is then ground and mixed with gypsum ($CaSO_4\text{-}2H_2O$) to control the curing of the cement. All of these compounds have primarily ionic bonding, with some component of covalent bonding. Portland cement is a ceramic material.

The Portland cement is mixed with water and aggregate to form a concrete slurry. The aggregate is a mixture of hard particles sized from grains of sand to rocks. The rocks provide compressive strength to the concrete, and the sand fills the spaces between the rocks. The Portland cement matrix binds the aggregate together. The cement must fully coat all of the aggregate; approximately 15 volume percent cement is typical. The appearance of a cross-section of concrete is shown in Figure 12.27. The large light particles are the coarse aggregate, which is surrounded by smaller light aggregate particles bonded together with the dark Portland cement.

As the concrete slurry cures, reactions occur between the cement and the water to form hydrates. An example of a hydration reaction is

$$2(3CaO \cdot SiO_2) + 6H_2O \rightarrow 3CaO \cdot 2SiO_2 \cdot 3H_2O + 3Ca(OH)_2$$

Other hydration reactions occur with the other components of the cement. The hydration reactions result in the cement bonding to the aggregate to form solid concrete. The strength of concrete as a function of time and of the water-to-cement ratio is shown in Figure 12.28. The greater the time for a given amount of water-to-cement ratio, the greater is the amount of the hydration reaction, and the higher is the strength. The hydration reactions continue after 28 days; however, a significant amount of the reaction occurs within this time. The strength shown in Figure 12.28 is in MPa. Amounts of water above 0.3 parts water to cement by weight in the initial composition result in lower strength, because the excess

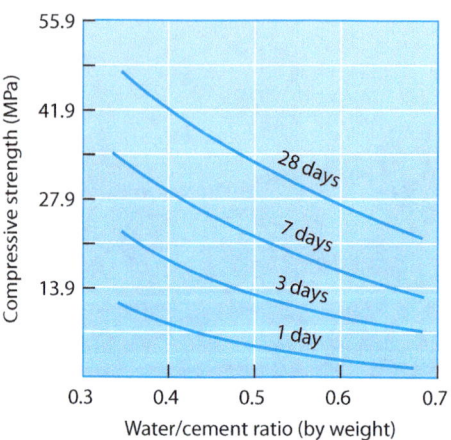

Figure 12.28   The compressive strength of concrete as a function of the water-to-cement ratio for different cure times. *(Based on Askeland, D.R., Fulay, P.P., and Wright, W.J., The Science and Engineering of Materials, 6th ed., Cengage Learning, Stamford, CT. (2011), p. 706.)*

water evaporates, leaving porosity in the concrete. Portland cement is hydraulic, meaning that it can cure under water.

Concrete is a brittle-elastic material. Concrete fails in tension by the propagation of a crack perpendicular to the tensile load axis. The weak interface between the aggregate and the cement fractures, and this fractured interface acts as a crack. Also, pores in the concrete act as cracks. There is a range to the mechanical properties of concrete because of the many different possible variables in the preparation of concrete. The compressive elastic modulus is in the range from 17 GPa to 40 GPa. The most important property is the compressive strength, which has a range of 14 MPa to 40 MPa. The tensile strength of concrete, which has a range from 1.4 MPa to 4 MPa, is one-tenth of the compressive strength; this is true for many ceramic materials.

In compression, concrete fails by shear at an angle to the compressive load axis, due to the propagation of shear cracks along the cement aggregate interfaces. Also, cracks form parallel to the applied compressive force. This cracking is due to tensile strains resulting from the Poisson's ratio that are perpendicular to the compressive force.

Concrete for construction is normally reinforced with ductile steel rebar and steel mesh to increase the energy absorption to fracture. Figure 1.5 shows steel reinforcing bars that are used in concrete construction.

12.13.3   Asphalt

Asphalt, used for paving roads, is a composite with a matrix of bitumen and particles of aggregate. **Bitumen** is a relatively soft thermoplastic polymer obtained from petroleum. The aggregate is usually similar to the aggregate in concrete. There should be a range of sizes of aggregate to obtain a high aggregate packing factor. To achieve the maximum hardness of the asphalt surface, the volume fraction of aggregate should be as high as possible and still have complete bonding between the bitumen and the aggregate.

Recycled materials can be used in asphalt for roads. The Federal Highway Administration has found that asphalt with up to 15% recycled glass used as aggregate provides good wear in highway surface asphalt. Asphalt with up to 25% glass could potentially be used in base layers. For over 30 years the state of California has used asphalt with 18 to 22% shredded used tires as part of the aggregate to produce rubberized asphalt for roads. The ASTM Standard D6114 is for testing rubberized asphalt.

Summary

- Composite materials combine two or more different materials into a resulting product where each original material retains distinct.

- Many composite materials are reinforced with very high-strength filaments of materials that are very small in diameter, such as glass, Kevlar, OUHMWPE, and carbon.

- Most composite materials consist of a matrix material that is continuous and that surrounds the reinforcing phase of particles or fibers.

- Epoxy and polyester thermoset polymers are extensively used in composite fabrication, because the liquid resin and liquid hardener can be mixed to form a liquid at room temperature that is then impregnated into the fibers. The mixture hardens in a few hours and forms a rigid solid. Thermoplastic polymers and metals are melted for the liquid matrix material to impregnate the fibers.

- Ceramic matrix composites with fiber reinforcement have increased resistance to fracture, because the fiber pullout from the ceramic matrix provides an energy-absorbing mechanism during fracture in otherwise brittle ceramic materials.

- In a uniaxial composite material where the reinforcing phase is parallel to the loading axis, the isostrain model is assumed to apply. In the isostrain model, the composite elastic modulus is calculated by the rule of mixtures, and the composite strain is calculated from the applied stress divided by the composite elastic modulus. The stress in the reinforcing phase and the matrix is calculated from the product of the composite strain and the elastic modulus of each phase.

- In a uniaxial composite material where the reinforcing phase is perpendicular to the loading axis, the isostress model is assumed to apply. In the isostress model, the strains sum according to the rule of mixtures. For a uniaxial composite material with a reinforcing phase with a higher elastic modulus than that of the matrix, the isostrain elastic modulus is always larger than the isostress elastic modulus.

- In a composite material, if the electron flow or heat flow is parallel to the reinforcement, then the electrical and thermal conductivity follow the rule of mixtures. If the electrical or thermal flow is normal to the reinforcement, then the electrical and thermal conductivity are modeled with the same form of equation as the elastic modulus for the isostress model.

- To obtain quasi-isotropic properties in a plane, fibers are woven into a fabric with fibers at 0° and 90° orientations. The layers of fabric are oriented at angles of ±45° or ±60° to produce a quasi-isotropic composite material. The composite is very weak perpendicular to the layers of fabric. Composite mat is produced by laying fibers in random directions. Composite materials made of mat are not as strong as those made of woven fabric. Multiple layers of composite material or composite mat resist bending if they are attached with adhesive to each side of a lightweight hollow-core honeycomb structure.

- Composite materials that have fibers oriented at 0° and ±60°, or 0° and ±45° to the load axis, are balanced. A balanced composite is one that produces no shear strains when stressed in tension. If the layers of fibers are at angles that are symmetric about the center of the layering, such as 0°, 60°, 120°, 120°, 60°, and 0°, the composite is symmetric.

- Many composite-material products are made with short or chopped fibers, because of their ease and lower cost of fabrication relative to those of continuous-fiber composites. Chopped fibers are mixed with a liquid polymer and sprayed or cast as a liquid, or deformed with normal polymer processing procedures. For the same volume fraction of reinforcement as a continuous fiber, a short-fiber composite cannot carry as much load as a continuous-fiber composite, because the stress in the fibers is reduced at the end of the fibers. If the fiber aspect ratio is approximately 10, the reduction in the load-carrying capacity is approximately 5%.

- At the critical length, the axial stress at the center of a short fiber is equal to the value in a continuous-fiber composite and decreases to 0 at the end of the fiber.

- A model that provides reasonable agreement with experimental results for the transverse strength of uniaxial composite materials assumes that the fibers are holes.

- Two modes of failure in a uniaxial composite material compressed parallel to the fiber axis are (1) that fibers buckle in unison to form a kink band, and (2) cracks form at the fiber matrix interface due to strains resulting from Poisson's ratio.

- The transverse compressive strength of a uniaxial composite material is much less than for compression parallel to the fiber axis. The fibers provide transverse compressive strength to the matrix similar to dispersion strengthening.

- Uniaxial composite materials are weak in shear if the shear stress is applied in the direction of the fibers, because the stress shears the weak matrix-fiber interface. The in-plane shear strength is approximately twice the magnitude of the transverse tensile strength.

- In short-fiber composites, all of the modes of failure observed in long-fiber composites are observed. In addition, there is debonding of the fiber and matrix at the fiber ends, due to fiber pullout from the matrix material.

- The difference in thermal-expansion coefficients of reinforcement and matrix materials results in internal stresses in composite materials that are subject to temperature change.

- Processing procedures for uniaxial continuous-fiber composite materials include pultrusion and filament winding.

- Prepeg is liquid epoxy or polyester resin mixed with hardener that is then impregnated into uniaxial or woven reinforcing material. The composite material is immediately refrigerated to suppress hardening. Prepreg is removed from refrigeration and immediately processed, because it hardens at room temperature.

- Processing procedures for shaping woven continuous-fiber reinforced composites include hand layup into a mold, air-pressure forming in a mold, resin transfer molding, and vacuum bagging.

- Wood is a uniaxial composite material that is frequently utilized in construction for its compression and in bending properties. The strength of wood comes from crystalline cellulose cells that are embedded in the polymer lignin. The mechanical properties of wood are highly anisotropic, and differently oriented layers of wood in plywood result in a quasi-isotropic material in the plane of the board.

- Concrete is a composite with a matrix of hydrated Portland cement. The reinforcement is an aggregate with particle sizes from fine sand to coarse rocks. When Portland cement and aggregate are mixed with water, hydration reactions bond the Portland cement to the aggregate forming solid concrete. Concrete is a brittle-elastic material, and the tensile strength is one-tenth of the compressive strength. Concrete for construction is reinforced with ductile steel rebar and steel mesh to increase the energy absorption at fracture.

- Asphalt is a composite with a matrix of bitumen and particles of aggregate. Recycled glass and shredded rubber tires are incorporated into asphalt for use in roads.

Supplemental Reading Subjects and Authors

Full references are listed at the end of the book.

General:	*Askeland, Fulay, and Wright; Courtney*
Composite materials:	*Chawla; Hull and Clyne; Schwartz*
Metal matrix materials:	*Reed-Hill and Abbaschian*
Ceramic matrix materials:	*Carter and Norton; Chaing, Birnie, and Kingery*
Polymer matrix materials:	*Hearle; Kinney; McCrum, Buckley, and Bucknall; Osswald and Menges; Rudin; Schultz; van Krevelen; and Winding and Hiatt*

Homework

Concept Questions

1. If a force is applied parallel to a reinforcing phase of a composite material, we can assume that the _____ is equal in the matrix and the reinforcing phase.

2. From a fracture mechanics viewpoint, small-diameter brittle filaments have a higher strength than bulk material does, because the smaller-diameter material has a smaller _____.

3. Carbon fibers are made of amorphous carbon and polycrystalline _____ whose plates are oriented along the fiber axls.

4. The _____ material in a composite is continuous and surrounds the reinforcing phase of particles or fibers.

5. In epoxy and polyester thermoset polymers, liquid resin and liquid _____ are mixed to form a liquid that transforms to a solid.

6. The primary reason for adding fiber reinforcement to a ceramic matrix is to increase the resistance to _____.

7. If a force is applied parallel to a reinforcing phase of a composite material, the elastic modulus is calculated with the rule of _____.

8. If a force is applied perpendicular to a reinforcing phase in the shape of slabs in a composite material, we can assume that the _____ is equal in the alternating slabs of material.

9. Composite _____ is produced by laying fibers in random directions in a plane.

10. A _____ composite is one that produces no shear strains when stressed in tension.

11. When a tensile stress is applied to a composite material with short fibers, the tensile stress in the matrix is transferred into the fibers by _____ stress between the matrix and the fiber.

12. In a short-fiber composite subject to a tensile stress parallel to the fibers, at the center of the fiber the matrix-fiber interface shear stress is equal to _____.

13. At the critical fiber length in a short fiber composite, the fiber stress at the center of the fiber length is equal to the fiber stress in a _____ fiber composite.

14. The critical fiber length divided by the fiber _____ is equal to the critical aspect ratio.

15. In axial loading parallel to the fibers of a uniaxial continuous composite material, it is the material with the lowest _____ to fracture that fails first.

16. A satisfactory model of the transverse strength of uniaxial composite materials is to assume that the fibers are _____in the composite material.

17. When fiber-reinforced composite materials are compressed along the fiber axis, it is observed that the fibers can buckle in unison to form a _____ band.

18. When a compressive stress is applied parallel to the fibers of a composite material, cracks form at the fiber-matrix interface due to _____ effect strain.

19. The in-plane shear strength of a uniaxial composite material is approximately _____ the magnitude of the transverse tensile strength.

20. _____ stresses are stresses in a material that are not caused by an external applied force.

21. If there is no external applied force, the sum of the internal stresses in a composite is equal to _____.

22. The composite material processing technique where fibers are pulled from a roll into a bath of liquid polymer, formed into the desired shape by a die, and cured in an oven is called _____.

23. In filament _____, the reinforcing material is wrapped around a rotating mandrel of the desired shape.

24. In _____ _____ _____, the reinforcing fabric is placed in the mold and liquid matrix material is injected into the mold under pressure.

25. The longitudinal strength of wood is due to the high strength of crystalline _____ cells.

26. Concrete is a composite with a matrix of hydrated _____ cement, and the reinforcement is aggregate with sizes ranging from fine sand to coarse rocks.

27. When Portland cement and aggregate are mixed with water, _____ reactions bond the Portland cement to the aggregate, thereby forming solid concrete.

28. _____ is a composite with a matrix of bitumen and particles of aggregate.

Engineer in Training–Style Questions

1. The reinforcing material with the highest strength available for composites is
   (a) Whiskers
   (b) Graphite fibers
   (c) SiC fibers
   (d) OUHMWPE

2. Which of the following fibers has the highest specific tensile strength?
   (a) Kevlar
   (b) OUHMWPE
   (c) SiC
   (d) Boron

3. Which of the following fiber degree layering sequences produces a balanced symmetric composite material?
    (a) 0/60/120/0/60/120
    (b) 0/90/90/0
    (c) 0/60/120/120/60/0
    (d) 0/45/135/0/45/135

4. The lowest-strength direction and loading for a uniaxial-fiber reinforced composite materials is
    (a) In the fiber plane shear
    (b) Axial compression
    (c) Transverse tension
    (d) Transverse compression

5. In a composite material with a brittle matrix and brittle fibers, if the strain to fracture of the fibers is greater than the strain to fracture of the matrix, which of the following is not true?
    (a) The slope of the composite stress-strain curve has a discontinuity at the fracture strain of the matrix.
    (b) The stress in the composite increases after the matrix is fractured and is given by $\varepsilon E_f v_f$.
    (c) The composite fractures at the fracture strain of the fibers.
    (d) The composite fractures at the fracture strain of the matrix.

6. Which of the following processes would be the most likely process for making low-cost continuous-fiber composite fenders for a truck?
    (a) Resin transfer molding
    (b) Vacuum bagging
    (c) Pultrusion
    (d) Filament winding

7. Which of the following processes would be the most likely process for making continuous-fiber composite material for a Formula 1 race car chassis?
    (a) Resin transfer molding
    (b) Vacuum bagging
    (c) Pultrusion
    (d) Filament winding

8. Which of the following processes would be the most likely process for making composite materials for the shafts of golf clubs?
    (a) Resin transfer molding
    (b) Vacuum bagging
    (c) Pultrusion
    (d) Filament winding

9. Which of the following processes would be the most likely process for making continuous-fiber composite material for a high-pressure natural gas tank?
    (a) Resin transfer molding
    (b) Vacuum bagging
    (c) Pultrusion
    (d) Filament winding

10. Wood has the highest strength for which of the following types of loading?
    (a) Tension in the longitudinal direction
    (b) Tension in the transverse direction
    (c) Compression in the longitudinal direction
    (d) Compression in the transverse direction

Problems

Problem 12.1: In Example Problem 12.1, a uniaxial composite material is made into a circular rod with a 1.27-cm diameter from 70 volume percent continuous carbon fibers and 30 volume percent epoxy. The rod is subject to an axial force of 100,000 N. The composite material in Example Problem 12.1 is to be replaced with a less expensive composite made of 70 volume percent continuous E-glass fibers and 30 volume percent epoxy. The elastic moduli are 5 GPa for the epoxy resin and 72.4 GPa for the E-glass.

    (a) Compare the elastic modulus, composite strain, fiber and matrix stresses, and density of this composite with the carbon epoxy composite in Example Problem 12.1. Use the density of UHM carbon, and assume the density of the epoxy is 1.2 g/cm^3. (b) Can both the E-glass fiber and matrix withstand the applied force?

Problem 12.2: For the E-glass composite material in Problem 12.1, predict the elastic modulus in the direction perpendicular to the fiber axis.

Problem 12.3: Design a composite material made of epoxy and uniaxial E-glass fibers that has an axial elastic modulus of 50 GPa. The epoxy resin has an elastic modulus of 5 GPa, and the E-glass fibers have an elastic modulus of 72.4 GPa. (a) What is the volume fraction of fibers and matrix in this composite? (b) What is the transverse elastic modulus of this material?

Problem 12.4: For the E-glass and epoxy composite material in Problem 12.1, calculate the thermal conductivity (k_T) in the parallel and transverse directions to the fiber axis. The thermal conductivity of E-glass is 13 $Wm^{-1}K^{-1}$ and for epoxy it is 0.1 $Wm^{-1}K^{-1}$.

Problem 12.5: Design a composite material made of epoxy and uniaxial E-glass fibers that has an axial thermal conductivity of 8 $Wm^{-1}K^{-1}$. The thermal conductivity of E-glass is 13 $Wm^{-1}K^{-1}$, and for epoxy it is 0.1 $Wm^{-1}K^{-1}$.

    (a) What is the volume fraction of fibers and matrix in this composite?

    (b) What is the transverse thermal conductivity of this material?

Problem 12.6: Concrete has a tensile strength of 3×10^6 Pa and an elastic modulus of 30×10^9 Pa. Because the concrete fractures in a brittle manner, it is to be reinforced with 10 volume percent uniaxial 1020 steel rebar that has a yield strength of 205×10^6 Pa, and the steel rods have an elastic modulus of 211×10^9 Pa.

    (a) What is the maximum strain that this composite can withstand without the concrete cracking or the rebar yielding?

    (b) What is the maximum tensile stress applied parallel to the rebar axis that can be withstood by this reinforced concrete before the concrete cracks or the rebar yields? Assume that there is good bonding between the concrete and the rebar.

Problem 12.7: Estimate the transverse tensile strength of the concrete in Problem 12.6.

Problem 12.8: A uniaxial composite material is made from 75 volume percent continuous Kevlar fibers and 25 volume percent epoxy. The Kevlar fibers have an elastic modulus of 131×10^9 Pa and a tensile strength of 4.0×10^9 Pa. The epoxy resin has an elastic modulus of 5×10^9 Pa and a tensile strength of 50×10^6 Pa. The composite must withstand an axial stress of 0.5×10^9 Pa and a transverse stress of 4×10^6 Pa. The axial and transverse stresses are not applied simultaneously. Assume that both the kevlar fibers and the epoxy matrix are elastic up to their failure.

(a) Determine the elastic modulus in the axial direction of this composite material.

(b) Determine the elastic modulus in the transverse direction of this composite material.

(c) Determine the axial strain in the composite due to the axial stress alone.

(d) Determine the stresses in the fibers and in the matrix for the axial stress.

(e) Can each of the components of this composite withstand the axial stress, and if not, what is the fracture stress?

(f) What is the transverse fracture stress? Can the composite withstand the transverse stress applied alone?

Problem 12.9: A uniaxial composite material is made from 75 volume percent continuous Kevlar fibers and 25 volume percent epoxy. The Kevlar fibers have an elastic modulus of 131×10^9 Pa and a tensile strength of 4.0×10^9 Pa. The epoxy resin has an elastic modulus of 5×10^9 Pa and a tensile strength of 50×10^6 Pa.

(a) In a uniaxial test to failure, what component will fail first, and at what strain?

(b) What is the composite stress at fracture of the first component?

Assume that both the Kevlar fibers and the epoxy matrix are elastic up to their failures.

Problem 12.10: Commercial-purity titanium reinforced with 50% uniaxial continuous silicon carbide fibers has been proposed for use as a fan-blade material in gas turbines, because the high elastic modulus will reduce blade deflection. The yield strength of the commercial-purity titanium is 414 MPa, and the elastic modulus is 170 GPa. Assume that the titanium is perfectly plastic after yield; thus the yield and tensile strength are equal. The strain at fracture of the titanium is 25%. The tensile strength of the silicon carbide fibers is 3.9 GPa, and the elastic modulus is 400 GPa.

(a) Calculate the strain where the first change in elastic modulus occurs in this composite material for a uniaxial tensile stress parallel to the fibers.

(b) Calculate the composite stress where the change in elastic modulus occurs.

(c) Calculate the elastic modulus at low strain, and if the titanium yields before the fibers fail, calculate the elastic modulus after the titanium yields.

(d) Calculate the fracture strength of the composite.

Problem 12.11: A composite material is made from a 7075-T6 aluminum matrix with 50% uniaxial alumina fibers. The yield strength of the aluminum is 500 MPa, and the elastic modulus is 70 GPa. The tensile strength of the alumina fibers is 1.4 GPa, and the elastic modulus is 380 GPa.

(a) Calculate the strain where the first change in elastic modulus occurs for this composite material in an axial tensile test.

(b) Calculate the composite stress where the change in elastic modulus occurs.

Problem 12.12: A composite material is made from a 5052 aluminum matrix with 60% uniaxial continuous boron fibers. The yield strength of the aluminum is 200 MPa, and the elastic modulus is 70 GPa. Assume that the matrix material is perfectly plastic after yield (elastic-perfectly plastic); therefore the ultimate tensile strength for the aluminum is also 200 MPa, and the strain at fracture of the matrix material is 15%. The tensile strength of the boron fibers is 2.8 GPa, and the elastic modulus is 400 GPa. Assume that the boron fibers are brittle.

(a) Calculate the strain where the first change in elastic modulus occurs in this composite material for a uniaxial tensile stress parallel to the fibers.

(b) Calculate the composite stress where the initial change in elastic modulus occurs.

(c) If the matrix yields before the fibers fracture, calculate the elastic modulus before and after yield of the matrix.

(d) Calculate the fracture strength of the composite.

Problem 12.13: A composite material has a matrix of epoxy with an elastic modulus of 5 GPa and a reinforcement of E-glass fibers with an elastic modulus of 72.4 GPa. Use a computer program or spread sheet to calculate the elastic modulus in both the axial and transverse directions as a function of volume fraction of fibers. Use increments of 0.1 volume fraction, and terminate the calculations at a fiber volume fraction of 0.8.

Problem 12.14: A composite material is made from 45 volume percent aramid (Kevlar) chopped fiber and 55 volume percent polycarbonate matrix. The aramid fibers have a length of 6×10^{-4} m, a diameter of 12×10^{-6} m, an elastic modulus of 130×10^9 Pa, and a tensile strength of 3.6×10^9 Pa. The polycarbonate matrix has an elastic modulus of 2.4×10^9 Pa. Measurements have shown that the aramid-polycarbonate interface has a shear strength of 32×10^6 Pa. The tensile strength of the polycarbonate is 80 MPa. For safety, it is decided that the maximum stress in the matrix of this composite should be 60 MPa. Assume that the aramid fibers are oriented along the tensile axis.

(a) What is the strain in the composite when the matrix is at a stress of 60 MPa? Use this as the design strain.

(b) What would be the fiber stress if this was a continuous-fiber composite, and is this fiber stress less than the tensile strength of the fibers?

(c) What is the critical fiber length?

(d) What is the average stress in the chopped aramid fiber?

(e) What is the composite-material stress?

Problem 12.15: The critical-stress intensity factor K_{Ic} for a glass-fiber reinforced epoxy matrix composite is 50 MPa(m)$^{1/2}$, and the tensile strength is 500 MPa. In a design application, it is expected that the maximum stress is 400 MPa. What size crack could produce fracture of this composite? Assume that the geometrical factor Y is equal to 1.0.

Problem 12.16: A composite material is made of 50% uniaxial glass fibers and 50% epoxy resin. The epoxy matrix is cured at 150°C and then cooled to 25°C. The elastic modulus of the glass fibers is 76 GPa, and the thermal expansion coefficient is 5×10^{-6} °C^{-1}. The elastic modulus of the epoxy resin is 4 GPa, and the thermal expansion coefficient is 60×10^{-6} °C^{-1}.

(a) Calculate the thermal stresses, both the magnitude and sign, in the matrix and in the fibers after the cured composite is cooled to room temperature. Assume that each material is elastic.

(b) Calculate the composite strain in cooling from 150°C to 25°C.

Problem 12.17: Carbon-fiber reinforced epoxy is being used as the outer surface material in high-performance aircraft where the surface temperatures are elevated because of the friction with air. The maximum-use temperature for epoxy resin is approximately 260°C. A composite sheet is produced from 75 volume percent carbon fiber and 25 volume percent epoxy resin. Assume that the epoxy and carbon are stress free at room temperature, and the composite is heated from room temperature (22°C) to 260°C during service. The elastic modulus of the carbon fiber is 380 GPa, the ultimate tensile strength is 2.4GPa, and the axial thermal expansion coefficient is $-0.7 \times 10^{-6} \, K^{-1}$. The elastic modulus of the epoxy resin is 4 GPa, the ultimate tensile strength is 0.07 GPa, and the thermal expansion coefficient is $60 \times 10^{-6} \, K^{-1}$.

(a) Calculate the thermal stresses, both the magnitude and sign, in the epoxy matrix and in the carbon fibers at 260°C. Assume that each material is elastic at all temperatures, and that there is good interface bonding.

(b) Calculate the composite strain in heating from 22°C to 260°C.

(c) Is there any component of this design that might cause a problem? Comment on the results in comparison to the data provided. If data is not available on these components, justify your evaluation of this design based upon your knowledge of these materials.

Steel is hot-rolled when it is desired to have large reductions in thickness without an increase in hardness.

Digital Vision/Getty Images

The goals of this chapter are to understand

- The processes for deforming ductile metals into their final shapes
- Superplastic forming
- The techniques for casting metals
- The techniques for making single crystals of metals and ceramics
- Powder techniques for hard metals and ceramics
- The techniques of welding, brazing, and soldering for metals and ceramics
- The processes for forming brittle ceramics
- The processes for forming and joining plastics

Chapter 13

Materials Processing

13.1 INTRODUCTION

In this chapter we discuss how to form materials into desired shapes by deformation, casting, powder, particle, and joining techniques. We also study how single crystals are made. We do not cover how the material is originally produced, such as how steel is originally created in a blast furnace. We will assume that the materials are available for purchase, and it is the role of the engineer to form the material into the desired shape of a part. Also, we do not consider the processing operations that cut the material, such as drilling and milling. Plans for processing the material into shape should be included in the design process, because it must be possible to manufacture the part at a cost that allows for a profit. Also, if the selected material is processed inappropriately, the part may not meet design specifications, because the material lacks the expected properties. For example, a high-strength aluminum alloy is purchased in the T6 condition. If the alloy is heated to form it into shape, the alloy will age, and if it overages, its strength will decrease. The engineer must be aware of the effects of processing on the properties of a material.

13.2   PROCESSING OF METALS

Ductile metals are deformed into shape without melting the metal. Metals that are superplastic experience extremely large strains, such as 1000%. Superplastic forming (SPF) occurs at temperatures typically over half of the absolute melting point of the metal. Metals, such as magnesium and lead, that melt at very low temperatures are cast into shape. Also, some very hard metals that cannot be deformation processed, such as superalloys, are cast into shape. Other very hard metals that are not easily deformed or cast into shape are produced by powder metallurgy techniques.

13.2.1   Deformation Processing of Metals

The processes for deforming a solid ductile metal into a final shape include rolling, forging, extrusion, drawing, and stamping, as shown in Figure 13.1. **Rolling** deforms the material in compression between two rolls to reduce the thickness, as shown in Figure 13.1a. Bar, sheet, and other open shapes such as I-beams, U-channels, and railroad rails are rolled with appropriately shaped rolls. A shape that closes back on itself cannot be rolled. For example, the letter "C" is a shape that closes back on itself.

Forged metals, as shown in Figure 13.1b, are pressed or pounded into a three-dimensional shape. In an open-die forge, the part is deformed but no particular shape is imprinted on the forge surfaces. A die provides shape to a material as a result of force. Open-die forging can be a preliminary step to other forms of shaping. In a closed-die forge, the shape of the part is imprinted into the dies. Forging produces parts with a high fracture strength, because the forces are primarily compressive and do not form cracks in the metal. Parts that require high strength, such as tools and automobile piston rods, are forged.

In **extrusion**, the material is forced through an opening in a die into the desired shape. Figure 13.1c shows direct extrusion on the top and indirect extrusion on the bottom. Direct extrusion is like squeezing toothpaste out of a tube. Seamless pipe, tubes, bars, and shapes such as "U" and "I" shapes are extruded. A shape that closes back on itself can be extruded.

In **drawing**, the metal is pulled through a die to reduce its cross section, as shown in Figure 13.1d. Wires, rods, and tubes are drawn. Drawing can be conducted only on metals that have a high degree of work hardening. The stress in the drawn metal must be below its yield stress for dimensional stability of the drawn metal. However, the stress in the supply metal must be above its yield stress for it to flow through the die. This is possible if there is a high amount of strain-hardening in the metal. The drawing force applied to the drawn metal is transmitted to the supply metal to produce the stress on the supply metal.

In **stamping**, sheet metal is pressed into a die with a punch, as shown in Figure 13.1e. The steel fenders and door panels of automobiles are stamped. A consideration in stamping is spring-back. When the force on the die is released, spring-back elastic recovery of the strain makes the shape of the part different from the shape of the die. Spring-back must be considered in the shape of the die.

13.2.2   Cold-Working of Metals

Ductile metals, such as pure copper, aluminum, and low-carbon steel, are **cold-worked** by deformation processing at temperatures below the recrystallization temperature or below 50% of

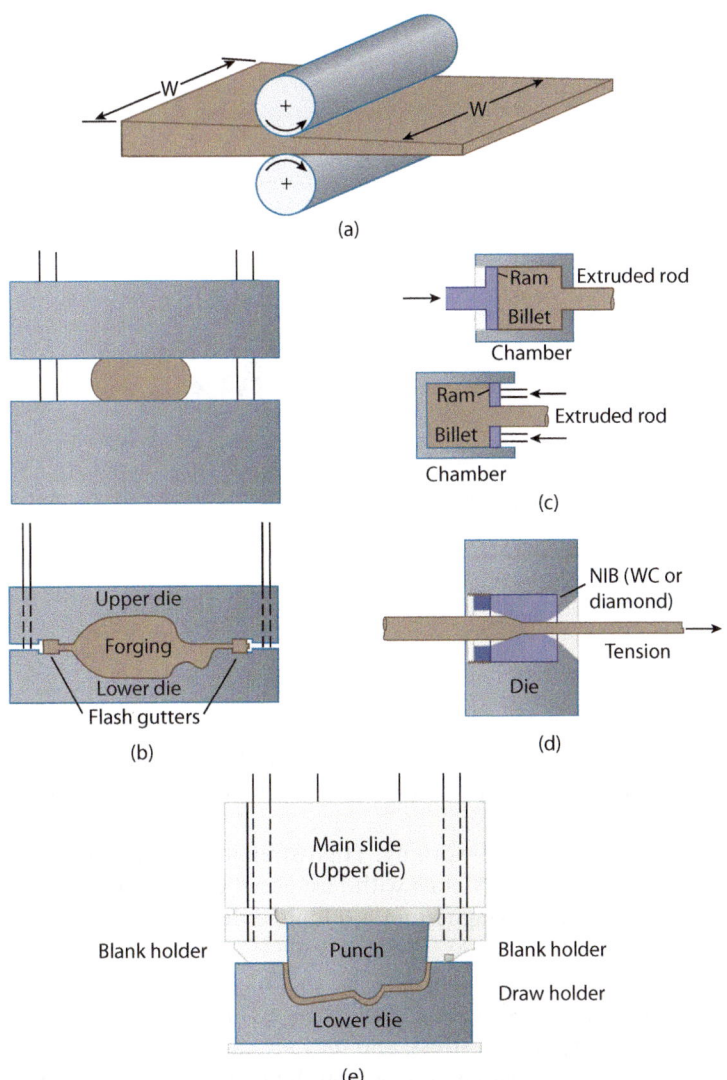

Figure 13.1 Schematics of metal-deformation processing. The metal being processed is the brown material. (a) Rolling. (b) The upper part of 13.1b is an open-die forge, and the lower figure is a closed-die forge. (c) The upper figure is direct extrusion, and the lower figure is indirect extrusion. (d) Drawing. (e) Stamping. (*Based on Meyers, M. A., and Chawla, K. K., Mechanical Behaviors of Materials, 2nd Edition. Cambridge University Press, 2009, Fig. 6.1.*)

the melting temperature. Straight copper tubing, called hard copper, is cold-worked. Cold-work results in strain hardening of the metal, and the ductility is decreased. Cold-work is used when a small dimension change as well as an increase in strength are necessary. With cold-working, excellent dimensional tolerances and good surface finishes are possible. The steel sheet in automobile bodies is cold-rolled, and the yield strength is increased by nearly a factor of 10. A schematic of the deformation of the material and the grains resulting from cold-rolling is shown in

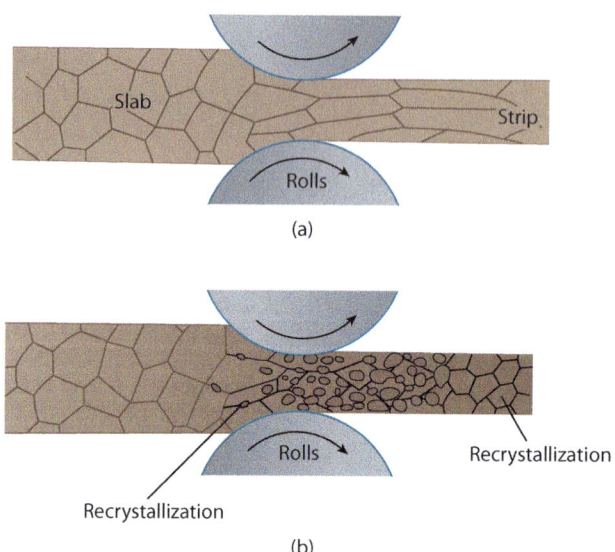

Figure 13.2   (a) The appearance of grains as a result of cold-rolling a metal into strip. (b) The appearance of grains as a result of hot-rolling a metal with recrystallization during and after rolling. (*Adapted from Courtney, T. H., Mechanical Behavior of Materials, 2nd ed., McGraw-Hill, New York (2000), p. 342.*)

Figure 13.2a. In cold-work, the grains are deformed in proportion to the overall dimensional change of the metal. At elevated working temperatures that are below the recrystallization temperature, recovery can occur in deformed grains.

Some metals are not suitable for cold-working. High-strength metals, such as titanium alloys and HSLA steel, are too hard to cold-work. Metals with hexagonal crystal structures, such as magnesium, have a limited number of slip systems. Magnesium has limited ductility at low temperatures, and only a small degree of cold-work is possible.

Cold-working also produces residual stresses in the material. For example, rolling produces compressive stresses in the surface of a sheet and tensile stresses in the center of the thickness of the sheet. The **residual stresses** are a form of internal stress, and they are called "residual" because they are present after the cold-working process. The residual stresses can be of advantage. For example, fatigue cracks are less likely to form when compressive stresses are present. The design engineer can take advantage of the residual stresses if the part is designed so that the outer surface has compressive stresses as a result of the processing.

Cold-work results in texturing and preferred orientations. During cold-work of metals, grains become deformed, as shown in Figure 13.2a, resulting in **texture**. In metals, plastic strain occurs by slip primarily on the most closely packed planes in the most closely packed directions. The tendency is for the most closely packed direction to be rotated into the direction of the deformation producing texture. Preferred orientation is not eliminated by recrystallization.

Texture and preferred orientation result in anisotropic properties in metals, including the elastic modulus, yield strength, ultimate tensile strength, toughness, and fracture toughness. For example, in α-iron the elastic modulus in the <111> direction is 276 GPa, in the <100> direction it is 129 GPa, and in polycrystalline α-iron with random grain orientations it is 209 GPa. In rolled iron, one preferred orientation that is observed is with the <100> direction parallel to the rolling axis. This orientation results in a material with a low elastic modulus parallel to the rolling direction. Also, according to Kaufman, Shilling, and Nelson, the value of the K_{Ic} for a 0.045-m-thick wrought plate of 7075-T651 aluminum is 29.7 MPa(m)$^{1/2}$ for a crack with its surface perpendicular to the rolling direction and propagating in the transverse direction of the plate, and K_{Ic} is 17.8 MPa(m)$^{1/2}$ for a crack with its surface perpendicular to the thickness of the plate and propagating in the transverse direction of the plate.

Cold-work decreases the electrical conductivity of metals, because cold-work increases the dislocation density, and dislocations decrease the conductivity by scattering electrons. However, electrical wire for power transmission is cold-drawn to increase its tensile strength, because the decrease in conductivity is acceptable.

If large deformations are necessary to produce a metal product by cold-work, it is necessary to anneal the metal if the metal becomes too hard during cold-work. Annealing softens the metal and recovers ductility.

13.2.3   Hot-Working of Metals

Metals are **hot-worked** at temperatures above the recrystallization temperature, as shown in Figure 13.2b, or above 50% of the absolute melting temperature. At elevated temperature, the forces necessary to produce deformation are decreased, and there is an increase in the plastic-strain rate. If a metal is too strong to be cold-worked, or if extensive deformation is required, the metal is hot-worked. Figure 13.2b shows a metal that begins to recrystallize during deformation and continues to recrystallize after deformation. Metals that undergo recrystallization during deformation include FCC metals such as nickel, copper, brass, and γ-iron.

Some metals exhibit high recovery rates during hot-working, and as a result they do not recrystallize. Metals that undergo high recovery rates during hot-working have grains that are elongated in the working direction, as shown in Figure 13.2a. These metals include FCC metals such as aluminum, BCC metals such as α-iron, and HCP metals such as zirconium. Recovery also occurs in the hot metal after it is deformed. The recovery after deformation is the same as the recovery that would occur in a furnace.

During hot-working, the metal does not work-harden, because it recovers or recrystallizes and softens at the same time it is work-hardened by plastic deformation. There is no strength increase or ductility decrease as a result of hot-working. However, the strength of the metal can be altered by changes in grain size. A thick ingot can be continuously reduced to a thin sheet by hot-rolling. Even if hot-worked metals are recrystallized, they are not isotropic. The working process elongates second-phase particles and inclusions in the working direction. Hot-working in compression can close voids and diffusion-bond the surfaces, effectively eliminating voids and improving the fracture strength.

Hot-working metals that are subject to oxidation results in a surface oxide scale that must be removed. This is often accomplished with pickling, which is an acid treatment. It is more difficult to maintain close dimensional tolerances in hot-working if surface scale must be removed, and consideration must be given to thermal contraction upon cooling. Metals that are highly reactive with oxygen, such as beryllium and tungsten, are hot-worked in inert-gas atmospheres.

13.2.4   Superplastic Forming of Metals and Ceramics

Superplastic forming (SPF) is the superplastic deformation of metals and ceramics into a desired shape. SPF of titanium alloys has been adopted for the formation of aircraft skins and frames, because of the difficulty of forming titanium by other deformation processes. With SPF, complex shapes can be formed

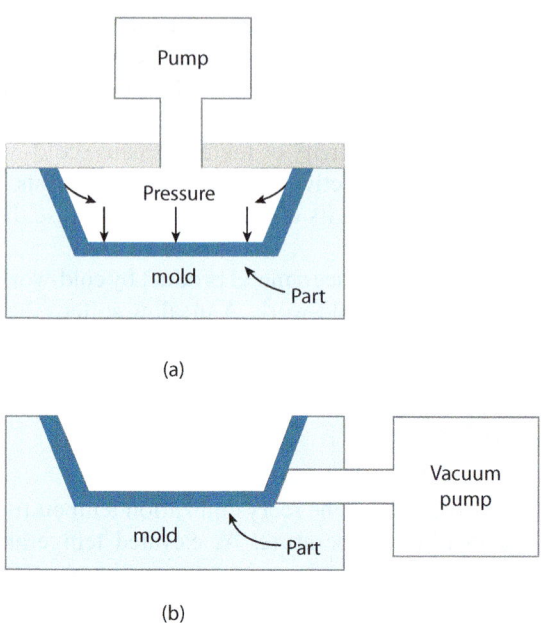

Figure 13.3 (a) Superplastic forming (SPF) with an assist from gas pressure provided by a high-pressure gas pump. The part is the dark blue shape, which starts as a flat plate that is deformed into a U shape in this cross section. (b) SPF combined with vacuum forming. Seals are required to maintain the pressure or vacuum.

that are near final dimensions with strains that are large. We discussed the theory of superplasticity and the necessary conditions in Section 9.4.4. Some aluminum, iron, and titanium alloys and ceramics are superplastic at temperatures above 50% of the absolute melting temperature, and there is a list of some of these in Table 9.1. The superplastic strains can be as large as 4900%. Superplasticity is a result of high-temperature deformation similar to grain boundary sliding and the diffusional flow of atoms along grain boundaries that produces high-temperature creep. The strain rates in SPF are typically on the order of $10^{-3}s^{-1}$. This is one of the disadvantages of SPF. The times for the part to creep into shape can be several hours. The time for reaching the final shape can be accelerated by combining SPF with vacuum forming or with high-pressure gas, as shown in Figure 13.3. Both vacuum and an inert gas reduce the adsorption of oxygen in titanium. Also, the high temperatures required for SPF of high-melting-point metals and ceramics are a disadvantage. However, because of the high temperature, SPF is often combined with diffusion bonding, which we will discuss in Section 13.2.8.

13.2.5 Casting of Metals

In **casting**, the material is formed into shape as a liquid in a mold, and then the liquid is cooled to form a solid. A mold provides shape to a liquid. Some metals that are cast have a low melting temperature, such as lead, zinc, aluminum, and magnesium. Other metals, such as cast iron and some turbine-blade materials, are cast because they are too hard for deformation processing. Also, products with complex shapes that cannot be deformation processed are cast. A schematic of various casting techniques is shown in Figure 13.4. The casting techniques for metals include sand, permanent mold, die, and investment. In **sand casting**, shown in Figures 13.4a and 13.4b, the mold is made from sand mixed with a binder that is packed around a pattern of the desired shape. The mold is then split, as

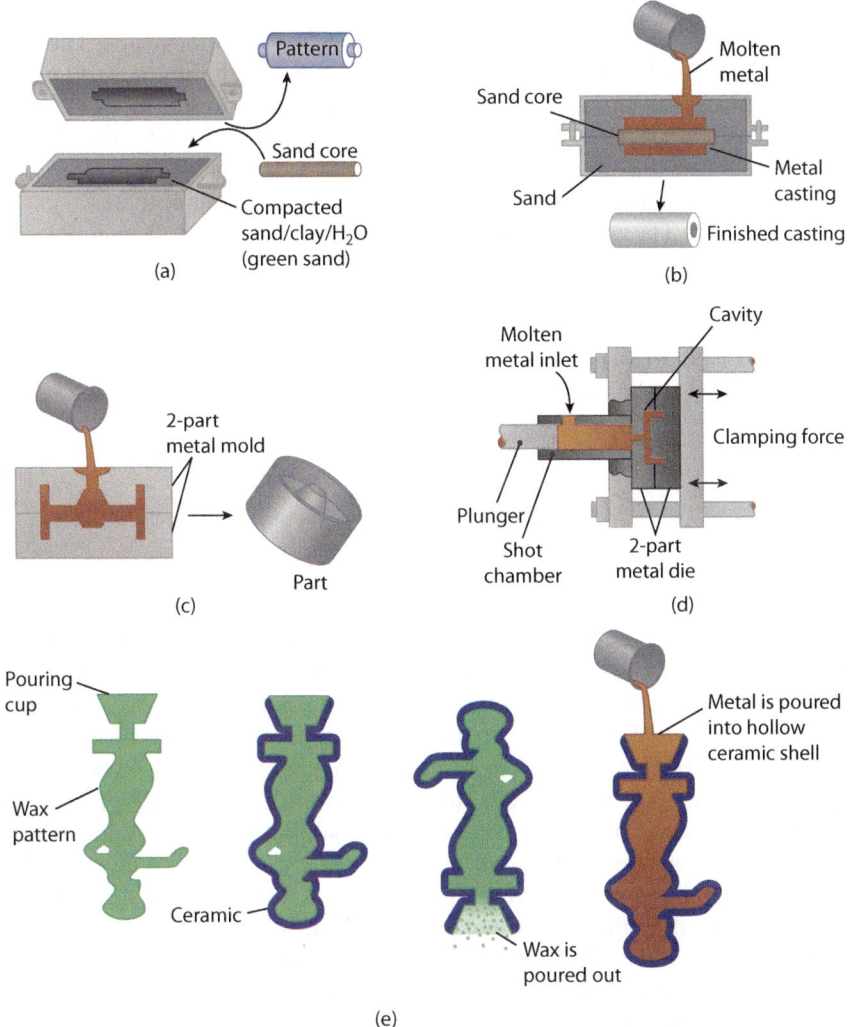

Figure 13.4 The different casting procedures for metals. (a) A sand mold and pattern to form the specimen shape. (b) Pouring liquid metal into a sand mold with pattern removed. (c) A permanent mold and the resulting part. (d) Die casting with metal dies. (e) Investment casting starts with a wax or polymer pattern that is encased in ceramic except at the top. Heat removes the wax or polymer and sinters the ceramic. Liquid metal is then poured into the ceramic mold. (*Based on Askeland, D.R., Fulay, P.P., and Wright, W.J., The Science and Engineering of Materials, 6th ed., Cengage Learning, Stamford, CT. (2011), p. 352.*)

shown in Figure 13.4a. The pattern is removed to create the space into which the liquid metal is poured, as shown in Figure 13.4b. In sand casting, it must be possible to remove the pattern without disturbing the mold, which limits the shapes that can be formed. The sand mold is usually destroyed when the cast part is removed.

Permanent metal molds like those shown in Figure 13.4c are reusable; otherwise casting in permanent metal molds is similar to sand casting. The mold must be made of a material that melts at a temperature significantly higher than that of the metal for the part. Pistons for internal-combustion engines are cast in permanent molds. If a mold is permanent, the part must be easy to remove from it. This limits the shapes that are made with permanent molds. Die casting of metals is shown in Figure 13.4d. In die casting, liquid metal is injected under pressure into dies that have a space with the desired shape. The dies are reused to produce many parts. The reusable dies and part shapes have the same limitations as reusable molds do.

Complex shapes that cannot be easily removed from a mold are **investment** cast. Investment casting, shown in Figure 13.3e, includes the lost-wax technique. In investment casting, the shape of the desired part is first made from a low-melting-point material, such as wax, polymer, or metal. The low-melting-temperature material in the desired shape is then coated by a slurry of ceramic material in water. The slurry is dried to form a green mold. The green mold with the low-melting-temperature material is heated to a high temperature, which sinters the ceramic mold and melts the low-melting-temperature material. The liquid flows out of the mold, leaving a space that is the desired shape of the part, along with pathways for the flow of metal. Liquid metal for the part is poured into the mold and allowed to solidify into the desired shape. The ceramic mold must be destroyed to remove the part. Very complex shapes can be investment cast as long as a path can be created for liquid metal to fill the space. Some gas-turbine blades are investment cast.

13.2.6   Powder Techniques for Metals and Ceramics

Metals that are too hard to be plastically deformed, and metals and ceramics that have a very high melting temperature, are processed into shape by powder techniques. Powder techniques are necessary when metals are mixed with ceramics, such as for carbide cutting tools made from cobalt and tungsten carbide. Powders are collections of very small particles. Some of the alloys for high-temperature gas turbines are produced with powder techniques, particularly those alloys with a dispersion of the ceramics thoria (ThO_2) or yittria (Y_2O_3).

In powder processing, the powder of the metals and ceramics are initially pressed into the desired shape forming a **green compact**. Then, the green compact is subject to a high temperature so the powder particles bond to each other, as shown in Figure 13.5, by sintering. Sintering is discussed in Section 7.3.2. The particle surfaces are replaced with a grain or phase boundary that is of lower energy. Also, diffusion allows atoms to move and reduce the sizes of voids, as schematically shown in Figure 13.5. However, it is extremely difficult to eliminate all of the voids in a powder-processed part, as demonstrated by the sintered ceramic in Figure 13.6. Voids in powder-processed parts act as cracks. To maximize the fracture toughness of the final part, the size of the voids must be reduced to the minimum size possible. For reactive metals, the sintering process is conducted in an inert-gas atmosphere. Some ceramics such as carbides also react with oxygen and are processed in inert gas. Oxides are processed in air. In **hot isostatic pressing** (HIP), the metal or ceramic is heated under pressure, and in **hot-pressing**, the metal or ceramic is heated and uniaxially

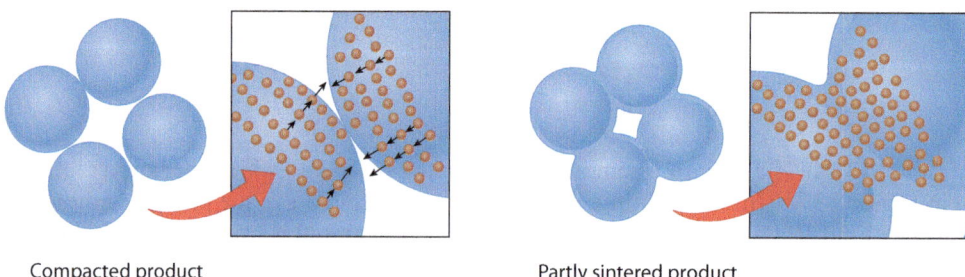

Compacted product   Partly sintered product

Figure 13.5   A schematic of sintering. On the left, particles of metal or ceramic are compacted. On the right, the metal or ceramic is heated, and the resulting diffusion of atoms (orange circles) eliminates the surface between the particles, bonding the particles together. (*Based on Askeland, D.R., Fulay, P.P., and Wright, W.J., The Science and Engineering of Materials, 6th ed., Cengage Learning, Stamford, CT. (2011), p. 183.*)

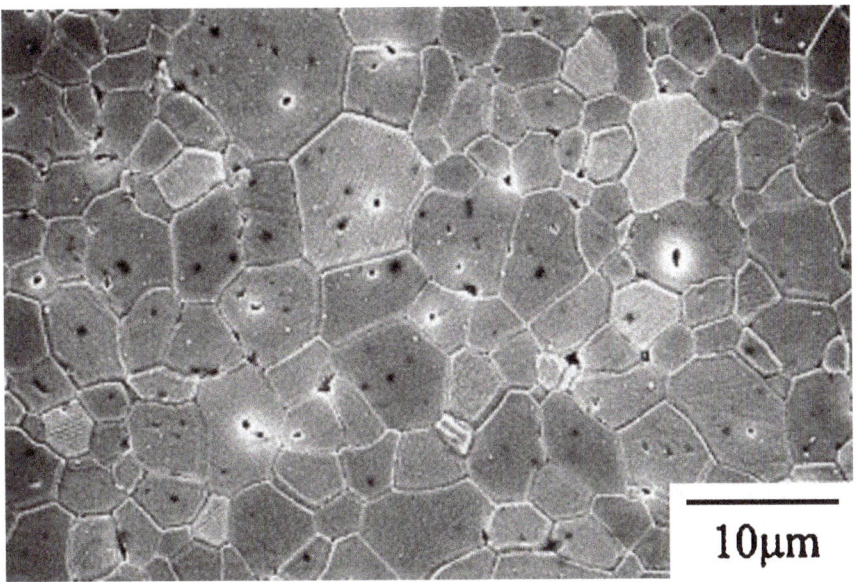

Figure 13.6 A photomicrograph of pressed and sintered alumina, showing the powder particles and voids. *(Reprinted from Acta Materialia, Volume 50, Issue 19, O.-S. Kwon,S.-H. Hong,J.-H. Lee,U.-J. Chung,D.-Y. Kim,N.M. Hwang, "Microstructural evolution during sintering of TiO2/SiO2-doped alumina: mechanism of anisotropic abnormal grain growth," Pages 4865–4872, Copyright 2002, with permission from Elsevier.)*

compressed. In **liquid-phase sintering**, a small amount of material in a mixture of materials it melted so that it wets the solid phases. The liquid phase assists in bonding particles and in increasing density.

13.2.7   Single Crystal Growth of Metals and Ceramics

Single crystals are utilized in applications, including the silicon in integrated electrical circuits discussed in Chapter 16, ruby and *pn*-junction laser materials discussed in Chapter 18, high-performance photovoltaic solar cells discussed in Chapter 18, and the high-performance gas-turbine blades discussed in Section 9.4.1. In the **Bridgeman-Stockbarger** technique of single crystal growth, shown schematically in Figure 13.7, a crucible with a sharp tip at the bottom containing liquid metal is lowered out of the

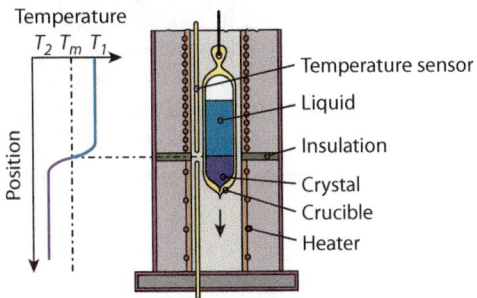

Figure 13.7   A schematic of the Bridgeman-Stockbarger technique of growing single crystals. *(Based on https:// commons.wikimedia.org/wiki/File:Bridgman-Stockbarger-Verfahren.svg)*

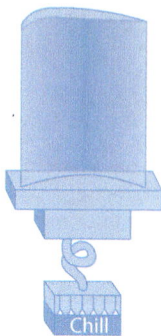

Figure 13.8   A schematic of the growth of a single crystal turbine blade. (*Based on Askeland, D.R., Fulay, P.P., and Wright, W.J., The Science and Engineering of Materials, 6th ed., Cengage Learning, Stamford, CT. (2011), p. 357.*)

high-temperature zone of a furnace, to a temperature below the liquid-solid equilibrium temperature. The crucible is made from a very stable high-temperature material, such as alumina. The metal in the sharp tip solidifies first as the crucible is lowered into the low-temperature part of the furnace. It is desired that only one crystal initially nucleate in the sharp tip of the crucible, and that only this single crystal grows as the crucible is lowered into the lower-temperature region of the furnace. The Bridgeman-Stockbarger technique is primarily utilized for metals and other low-melting-temperature materials, because the crystal material is in a crucible that must not melt or react with the growing crystal.

Single crystal turbine blades and vanes are formed with a modification of the Bridgeman-Stockbarger technique. An investment casting is made in the shape of the turbine blade, along with the loop segment below the blade, as shown in Figure 13.8. The investment casting containing liquid metal is lowered through the furnace, as shown in Figure 13.7. Metal in the area marked "Chill" in Figure 13.8 is cooled below the melting temperature, and many crystals solidify. However, the loop segment is designed so that only one crystal should grow through the loop as the investment casting is lowered through the furnace. The result is a single crystal turbine blade, as shown in Figure 9.13. Turbine vanes are also made as single crystals.

Another form of single crystal growth is the **Czochralski** technique (CT), shown in Figure 13.9. In the CT, a small single crystal, or seed, is lowered into molten material that is the same as or very similar

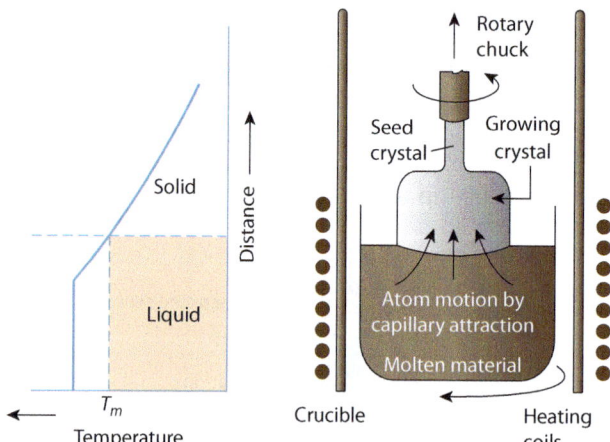

Figure 13.9   A schematic of the Czochralski technique (CT) for growing single crystals, showing the liquid material, the growing solid single crystal, and the seed crystal. On the left of the figure is the temperature as a function of position through the furnace. (*Adapted from Askeland, D. R., Fulay, P. P., and Wright, W. J., The Science and Engineering of Materials, 6th ed., Cengage Learning, Stamford, CT (2011), p. 745.*)

Figure 13.10   Single crystals of silicon grown by the Czochralski technique (CT). (*Peter Sobolev / Shutterstock.com*)

to the seed. The seed is then drawn out of the liquid into a section of the furnace that is just below the melting temperature (T_m) of the material. Furnace temperatures are shown as a function of position on the left. Capillary attraction draws liquid onto the seed crystal. The growth of the new material is in crystallographic alignment with the atoms in the seed crystal. The alignment of new crystal material with existing crystalline material is called **epitaxial** growth. This technique is used to grow single crystals of metals, silicon, and ruby. One advantage of this technique is that the growing crystal is not in contact with a crucible. Single crystals of silicon are shown in Figure 13.10. Single crystals up to 30 cm in diameter are produced by this technique. A crystal, such as shown in Figure 13.10, is sliced into wafers and then processed to produce integrated circuits for electronic systems and solar cells. Complex shapes, such as turbine blades and vanes, are not produced with the Czochralski technique.

13.2.8   Joining of Metals: Welding, Brazing, and Soldering

Sometimes it is not possible or profitable to produce the desired product in one solid piece. In these cases parts are joined together to produce the final desired product. The processes of joining materials we will cover in this section include welding, brazing, and soldering. Processes such as riveting and bolting are not covered in this book. In **fusion welding**, shown schematically in Figure 13.11, the metal parts are melted in the vicinity of the joint, and molten filler metal can be added to the weld. **The fusion zone** where molten metal solidifies to join the two metal pieces is indicated in Figure 13.11b. Filler metal in the form of a welding rod is used to provide a uniform surface. The filler metal is similar to the base metal. There is a **heat-affected zone** (HAZ) where the metal is not melted but the heating is sufficient to alter the metal properties. If the metal was cold-worked, the strength is reduced in the HAZ. If there is precipitation hardening, the precipitates in the HAZ are changed. For this reason, it is not recommended to weld highly cold-worked or precipitation hardened alloys. Aluminum alloys that are precipitation hardened, such as the 2000 and 7000 series, are not recommended for welding. Solid-solution-strengthened alloys are suitable for welding, because the solid solution atoms remain in the HAZ. The heating of the weld material can be with an oxyacetylene torch, an electron beam, or a laser.

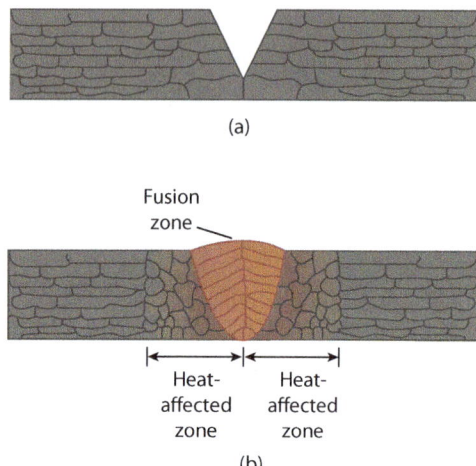

(a)

(b)

Figure 13.11   (a) Two parts to be joined by fusion welding. (b) Two parts welded together, showing the fusion zone and the heat-affected zone. (*Based on Askeland, D.R., Fulay, P.P., and Wright, W.J., The Science and Engineering of Materials, 6th ed., Cengage Learning, Stamford, CT. (2011), p. 361.*)

For all of these three techniques, the filler metal is a separate rod. In metallic electric arc welding, the metallic electrode is consumed and provides the filler metal. Usually, the welded metals are of the same type, although dissimilar metals can be welded. Aluminum and titanium alloys must be welded in an inert-gas atmosphere, because these metals rapidly react with oxygen when heated to the melting temperature.

Another form of fusion welding is **friction welding**, where one part is spun at a high frequency and the other part is static or spun in the opposite direction. The two parts are then forced together, and friction from the two parts rubbing against each other heats and melts the metals in the contact area, forming a weld. In friction welding there is no filler metal. An example of friction welding is to spin a steel rod and force it onto a steel disk rotating in the opposite direction, thereby welding the end of the rod to the center of the disk.

In **resistance welding**, or spot welding, shown schematically in Figure 13.12a, a current is passed through electrodes and through a small area, or spot, of two sheets of metal to be joined. The electrodes are pressed against the sheet metal during the welding process, as shown in Figure 13.12b. The heating occurs primarily at the interface between the two sheets of metal, because the interface has a higher resistance. This results in the two sheets of metal being fused at the spot. The result of the spot weld is a spot-fusion zone between the sheets, as shown in Figure 13.12c. In spot welding, there is no filler metal. The only preparation is to assure that the surfaces to be welded are clean. Steel sheet metal is the primary metal that is spot welded. Spot welding is used extensively in the fabrication of the steel chassis of automobiles. Metals that require an inert-gas environment because of oxidation, such as aluminum and titanium, are not spot welded.

Brazing and soldering are joining process where the two different parts made of metals or ceramics are joined without melting the parts. Brazing is the higher-temperature process, and soldering is a lower-temperature process. A melted filler material bonds the two parts together in both brazing and soldering. For example, two pieces of steel can be brazed with a brass filler metal. The brass-filler rod is melted, but the two pieces of steel do not melt. In brazing, there is diffusional mixing at the interface of the filling material and the parts that are brazed.

Solder is the term used to describe the process of joining two parts with low-melting-temperature solders where there is no significant diffusional mixing of the solder with the parts. Solders are made from mixtures of atoms that form a low-melting-temperature eutectic. In soldering, there may be some diffusion at the solder-part interface, but it is not significant. Lead-tin solders are used in electronic systems to join copper conductors to copper contacts. The lead-tin phase diagram presented in Figure 5.11

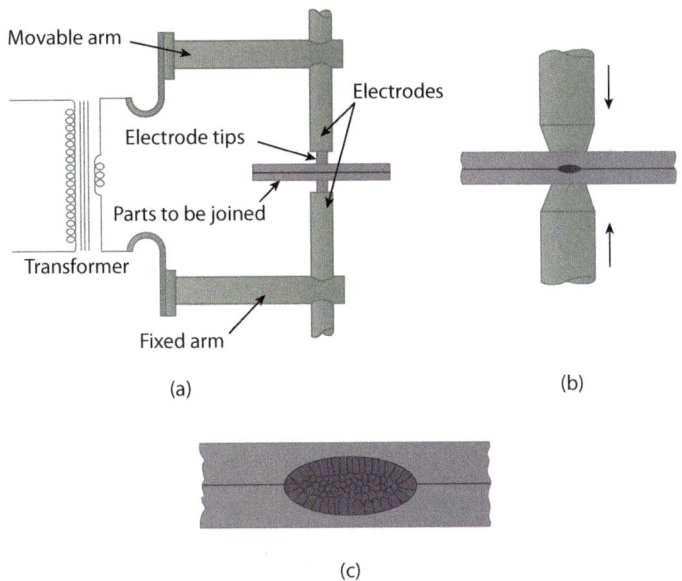

Figure 13.12 (a) A schematic of a resistance welding, or spot welding, machine. (b) A detail of the copper electrodes that carry the current and of the two sheets of metal that are welded under pressure, and the location of the spot weld (dark area). (c) A schematic of a spot weld in two sheets of metal showing the grain structure. *(Based on Timings, R.L., Engineering Materials 2nd ed. Addison Wesley Longman Ltd. Essex, England (1998), p. 272.)*

shows that the eutectic composition of lead and tin melts at 183°C. The Ag-Cu phase diagram in Figure 5.8 shows the melting temperature of the eutectic Ag-Cu composition is 780°C. The use of the higher melting Ag-Cu alloys is called both brazing and hard soldering. Lead cannot be used to solder domestic water systems because of lead poisoning; therefore, solders of Ag + Cu + Bi and other elements have been developed that have melting temperatures similar to those of lead-tin eutectics. In the soldering and brazing of metals, a **flux** is utilized that removes the oxide layer on the metal surfaces and allows the filler metal to wet the solid surfaces of the two metals to be joined.

Diffusion bonding of two materials is shown schematically in Figure 13.13. Diffusion bonding is often combined with superplastic forming of titanium aircraft structures. In diffusion bonding, the two metal parts are polished to a surface roughness of better than 0.4 μm, because it is necessary to

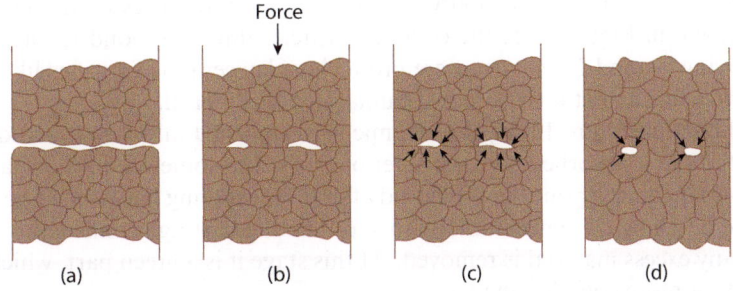

Figure 13.13 The steps in diffusion bonding. (a) Polished surfaces are brought together. (b) A force and high temperature are applied. (c) The parts bond along the interface by diffusion and void sizes are reduced. (d) Over time, the void sizes continue to reduce, and grain growth occurs. *(Based on Askeland, D.R., Fulay, P.P., and Wright, W.J., The Science and Engineering of Materials, 6th ed., Cengage Learning, Stamford, CT. (2011), p. 187.)*

minimize voids and have maximum contact along the interface. The parts are then pressed together with uniaxial force or hydrostatic pressure and heated to temperatures from 50 to 80% of the absolute melting temperature. The force or pressure deforms the surface to produce better contact, but it does not exceed the yield strength of the metal. Hydrostatic pressure can be much higher than uniaxial pressure, because hydrostatic pressure does not change the shape of the parts. The combination of pressure and high temperature results in diffusion that bonds the two materials together, and along with creep closes voids along the interface. Diffusion bonding of metals is normally conducted in a vacuum or in an inert-gas atmosphere to minimize oxidation. Diffusion bonding is also used to weld metals to ceramics, and ceramics to ceramics, as discussed in Section 13.3.3.

13.3   PROCESSING CERAMICS

Because ceramics have very high melting temperatures, casting liquid ceramics into the shape of a part is impractical for most engineering projects. Additionally, because a polycrystalline ceramic is brittle at temperatures below 50% of its melting temperature, deformation processing of ceramics is not practical. Because of these limitations, most ceramics are processed as powders. Ceramic parts are made by the same powder procedures of pressing and heating as for the metals we discussed in Section 13.2.6. A micrograph of a powder-processed aluminum oxide is shown in Figure 13.6. Some ceramics are suitable for SPF, as shown in Table 9.1. Because of the high melting temperature of ceramics, fusion welding is difficult; however, ceramics are joined by brazing and soldering.

Glass and glass-ceramics have lower melting temperatures than do ceramics and are processed as liquids. Glass and glass-ceramics are fusion welded, brazed, and soldered.

13.3.1   Slip Processing of Ceramics

A **slip** is a mixture of small particles, water, and other additives that give the slip the desired flow properties. The processing procedures applied to a ceramic slip include slip casting, tape casting, extrusion, and injection molding. The use of a slip originated with the formation of clay into shapes. Clay is primarily a mixture of ceramic particles of aluminosilicates. Clay is **hydroplastic**, meaning that when it is combined with water, it is easily plastically deformed. When ceramic particles are in a slip, the mixture is easily formed into a shape. The shaped ceramic slip is then dried, and it is called a green part. The green ceramic part is then sintered at temperatures from 900°C to 1400°C. The sintering evaporates the water and other additives, and, as shown in Figure 13.5, the ceramic particles diffusion bond to each other to provide strength. Humans living around 27,000 BCE were producing clay ceramic figurines like the one shown in Figure 1.1, and clay ceramics are the first known human-made material.

In slip casting, shown in Figure 13.14, the ceramic slip is poured into a mold made from plaster of Paris. The water in the slip is absorbed by the plaster of Paris, and some water evaporates. A thin-walled ceramic object is made by pouring out the slip liquid after a thin coating has formed on the mold surface; a solid piece is made by filling the entire mold with slip. The slip casting is removed from the mold after partial drying, and any excess material is removed. At this stage it is a green part, which is then sintered. Split molds are used for more complex shapes.

Tape casting is used to produce thin sheets of ceramic material primarily for use in electronic systems as capacitors and electronic packaging. In tape casting, shown in Figure 13.15, the slip is spread on a continuous tape of the width of the desired ceramic. Porous tape, such as paper, is often used to absorb water out of the slip. The thickness of the ceramic slip is controlled by a precision blade or gate. The tape is dried, cut into green tiles, and then sintered.

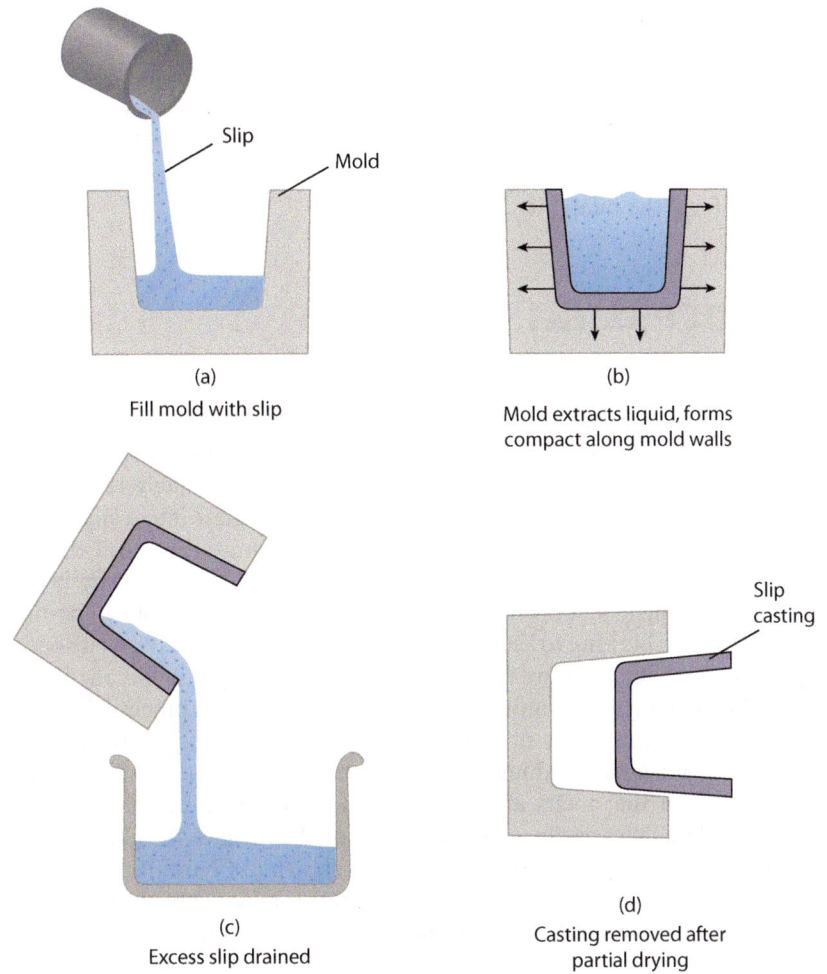

Slip

Mold

(a)
Fill mold with slip

(b)
Mold extracts liquid, forms
compact along mold walls

(c)
Excess slip drained

Slip
casting

(d)
Casting removed after
partial drying

Figure 13.14 A schematic of slip casting. (*Based on Modern Ceramic Engineering, by D. W. Richerson, Marcel Dekker. (1992), p. 462.*)

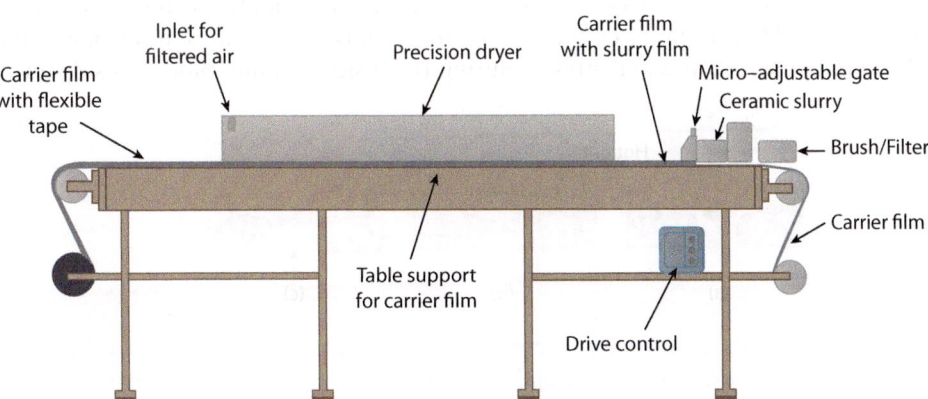

Carrier film
with flexible
tape

Inlet for
filtered air

Precision dryer

Carrier film
with slurry film

Micro–adjustable gate
Ceramic slurry

Brush/Filter

Carrier film

Table support
for carrier film

Drive control

Figure 13.15 A schematic of tape casting. (*Based on Schwartz, M., Handbook of Structural Ceramics, McGraw-Hill Inc. N.Y. (1992), p. 5.17.*)

The ceramic slip can be processed in most of the ways that a ductile metal or polymer is processed. Ceramic slip is extruded into bar and other shapes in a facility similar to that shown in Figure 13.1c for metals. After extrusion the ceramic slip is dried and sintered. In another process similar to the die casting of metals, shown in Figure 13.4d, or the injection molding of polymers, shown in Figure 13.20, the ceramic slip is injected under pressure into a mold or a die of the desired shape. The slip is dried in the die to form a green part. The green part is removed from the die, excess material is removed, and the green part is sintered to form the final part.

13.3.2 Processing Glass and Glass-Ceramics

Glass and glass-ceramics flow at temperatures lower than crystalline ceramic materials do, and glass is formed by plastic flow at temperatures equal the working temperature or higher. The viscosity where glass is worked is equal to approximately 10^3 Pa · s or 10^4 poise, as we discussed in Section 9.4.5. For example, the soda-lime silica glass in Figure 9.3 has a working temperature of approximately 1250 K or 977°C. Glass at the working temperature is formed by many techniques that force the glass into a shape. You have probably seen a glass blower turn a glob of glass on the end of a blowpipe into a beautiful vase. The glass glob is heated in a furnace to the working temperature, and then the glob is expanded by blowing through the hollow blowpipe, like blowing up a balloon. The vase is shaped by rotating the blowpipe and pressing on the hot glass with steel blades while blowing to keep the vase expanded. Mechanical devices have been developed that blow glass at the working temperature into a mold to form a simple shape; these devices replace the glass blower. Glass at the working temperature is pressed into shape by the process shown in Figure 13.16.

Glass plate is made by rolling glass at the working temperature between hot rolls, as shown in Figure 13.17a. In the float process of making plate glass, shown in Figure 13.17b, molten glass is spread over the surface of liquid tin that melts at 232°C. At the hot end of the furnace the temperature is sufficiently high that the glass flows easily. At the cool end of the furnace the glass is a solid.

Glass fibers are made by drawing fibers from a device, as shown in Figure 13.18. Molten glass is drawn through a small orifice, such as Pt dies, and the fiber diameter is reduced by pulling on the fiber while at the working temperature. The fine solid fibers are then wrapped on a roll.

13.3.3 Joining Ceramics

Ceramics are joined to other ceramics and metals via techniques similar to those that join metals to metals: with fusion welding using laser beams, electron beams, electric arcs, ultrasound, and friction. However, the development of thermal stresses during the fusion welding process results in fracture of

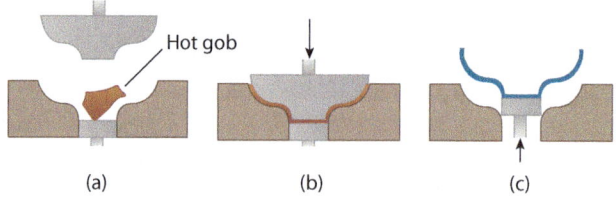

Figure 13.16 The steps in pressing glass. (a) A hot gob of glass is put into the mold. (b) The hot glass gob is pressed into the mold and given a shape. (c) The molded glass cools to form a solid and is then ejected. (*Based on Askeland, D.R., Fulay, P.P., and Wright, W.J., The Science and Engineering of Materials, 6th ed., Cengage Learning, Stamford, CT. (2011), p. 587.*)

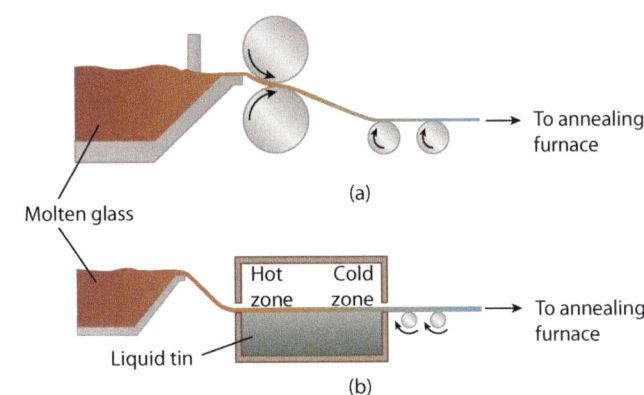

Figure 13.17 Techniques for making plate glass. (a) Forming plate glass by rolling. (b) Forming plate glass by spreading liquid glass over liquid tin. (*Based on Askeland, D.R., Fulay, P.P., and Wright, W.J., The Science and Engineering of Materials, 6th ed., Cengage Learning, Stamford, CT. (2011), p. 586.*)

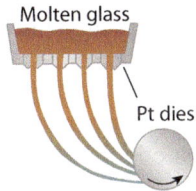

Figure 13.18 The process for producing glass fibers. (*Based on Askeland, D.R., Fulay, P.P., and Wright, W.J., The Science and Engineering of Materials, 6th ed., Cengage Learning, Stamford, CT. (2011), p. 587.*)

the ceramic in many instances. For this reason, brazing is more frequently used for joining ceramics than is fusion welding. Metal brazes for ceramics have been developed based upon alloys with major constituents of Sn, Cu, Ag, Au, and Ti. For a metal to wet a ceramic surface during brazing, the surface energy of the ceramic must be higher than that of the metal. As shown in Table 3.2, some metals, such as tin, have lower surface energies than some ceramics, and brazes have been developed based upon Sn. However, many metals have higher surface energies than ceramics and do not wet and bond to the ceramic surface. Various procedures have been developed to overcome this problem. One approach is to use metals that react with the ceramic surface. For example, in titanium alloy brazes the titanium reacts with Al_2O_3 and replaces aluminum atoms in the alumina ceramic, providing a good bond at the interface. Another approach uses a metal with a low surface energy to coat the ceramic, and then metals with higher surface energies are used as filler metals.

Glass solders with low melting temperatures have been developed to join ceramics to ceramics, and ceramics to metals. Most low-temperature glass solders are based upon mixtures of B_2O_3-PbO-ZnO or B_2O_3-PbO-SiO_2. Glass solders with higher melting temperatures are based upon Al_2O_3 mixed with other ceramics, such as CaO, SiO_2, MgO, and MnO.

Ceramics are bonded to other ceramics and to metal by diffusion bonding, as we discussed in Section 13.2.8. To promote bonding, a foil of precious metals, aluminum, or a transition metal is often placed between the parts to be diffusion bonded.

13.4   PROCESSING OF POLYMERS

13.4.1   Deformation Processing of Thermoplastic Polymers

Thermoplastic polymers have relatively low melting temperatures if the polymer is semicrystalline or glass transition temperatures if the polymer is amorphous, and they are easily deformed at temperatures approaching the melting temperature or above the glass transition temperature. The processing procedures we will discuss in this section heat the thermoplastic to a temperature where it flows easily, and then the polymer is deformed into shape by the application of force. The starting material of granular or pelletized plastic is fused into a solid piece during processing under heat and force, and the voids between the particles are eliminated. The primary processes for producing thermoplastic plastic parts with force at elevated temperature include extrusion, injection molding, blow molding, thermoforming, and calendaring.

Extrusion, shown in Figure 13.19, is used to make plastic pipes, bars, U-channels, and most any continuous shape. Polymer extrusion is similar to metal extrusion, which we discussed in Section 13.2.1. In polymer extrusion, the plastic and any additives are loaded into a hopper, heated, and forced into a die by a screw mechanism. The die forms the desired shape. The extruded shape is cooled, and the puller assists in pulling the plastic through the die. It is also possible to have dies that produce thin sheets and that coat wire with plastic insulation.

In **injection molding**, a heated thermoplastic is forced into a permanent mold cavity , as shown in Figure 13.20. The plastic can be injected by a screw mechanism as shown in Figure 13.20, or if the plastic is injected by a piston it is called ram injection molding. Complex shapes are made with injection molding; however, the part must be removable from the mold. Split molds allow the removal of parts with complex shapes.

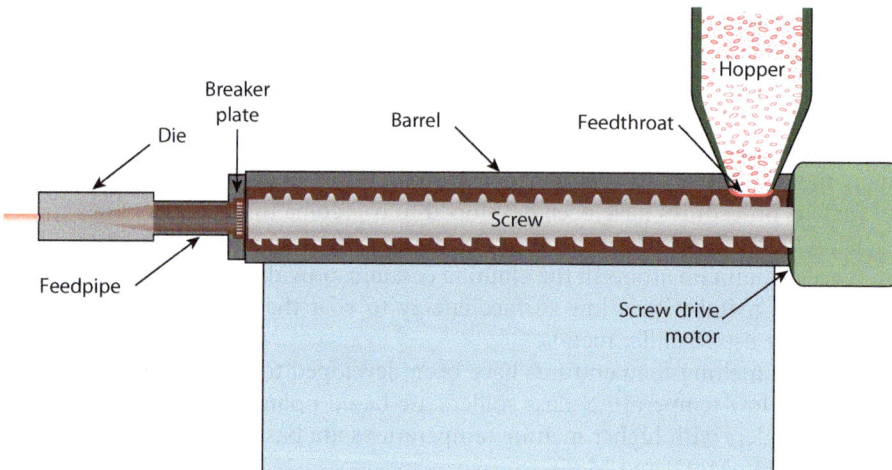

Figure 13.19   A plastic (the red shaded area) extrusion facility. (*Based on http://en.wikipedia.org/wiki/Plastics_ extrusion*)

Injection Molding

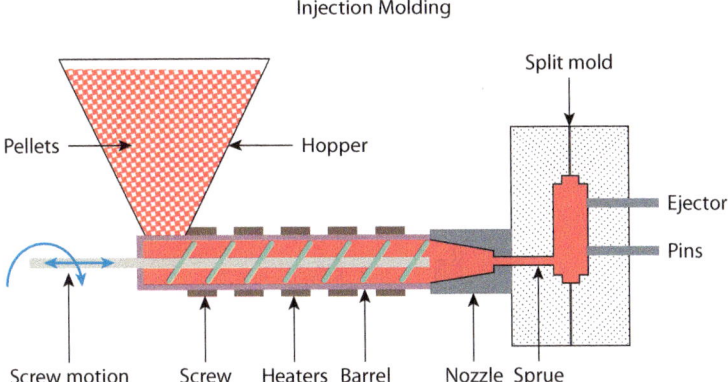

Figure 13.20 Injection molding of plastic parts. Heaters are indicated around the injection chamber. The split mold opens after injection of the plastic for part ejection. (*http://www.substech.com/dokuwiki/doku.php?id=injection_molding_of_polymers*)

Shapes such as plastic bottles are made by **blow molding**. A blow-molded bottle is made in a split mold. The two basic forms of blow molding are dependent upon how the preform of the bottle, which is called a **parison**, is formed. In injection blow molding, shown in Figure 13.21, a parison with a closed end is formed by injection molding in a separate previous process. The heated parison is lowered into a split mold, and air is blown into the hot parison, expanding the parison into the shape of the bottle. The plastic is cooled by the mold, and the shape is ejected by opening the mold.

In extrusion blow molding, an extruder forms a hot tube that is the parison. The parison leaving the extruder is lowered so that it extends through the split mold. The open end of the parison is pinched by closure of the mold. Air is blown into the parison, expanding the plastic into the shape of the bottle. Extrusion blow molding can be a continuous operation.

Thermoforming is the process of using heat and force to deform a sheet of plastic into a shape, as shown in Figure 13.22. The applied force in Figure 13.22 is from a vacuum, but it is also possible to use air pressure or a plug of the desired contour. When a plug is used to form the part, the plastic sheet and the plug are heated. Some objects that are thermoformed include egg cartons, product packaging, bathtubs, automobile instrument panels, and truck fenders. Both thick sheets of plastic and film are thermoformed. **Calendaring** is the rolling of hot plastic into a sheet, in a process similar to the hot rolling of metals that is shown in Figure 13.1.

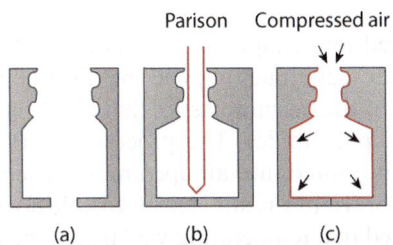

Parison   Compressed air

(a)   (b)   (c)

Figure 13.21 A schematic of injection blow-molding. (a) The open mold. (b) The mold is closed, and the heated parison is lowered into the mold. (c) Air is blown into the hot parison, expanding it into the shape of the bottle. (*Based on Askeland, D.R., Fulay, P.P., and Wright, W.J., The Science and Engineering of Materials, 6th ed., Cengage Learning, Stamford, CT. (2011), p. 641.*)

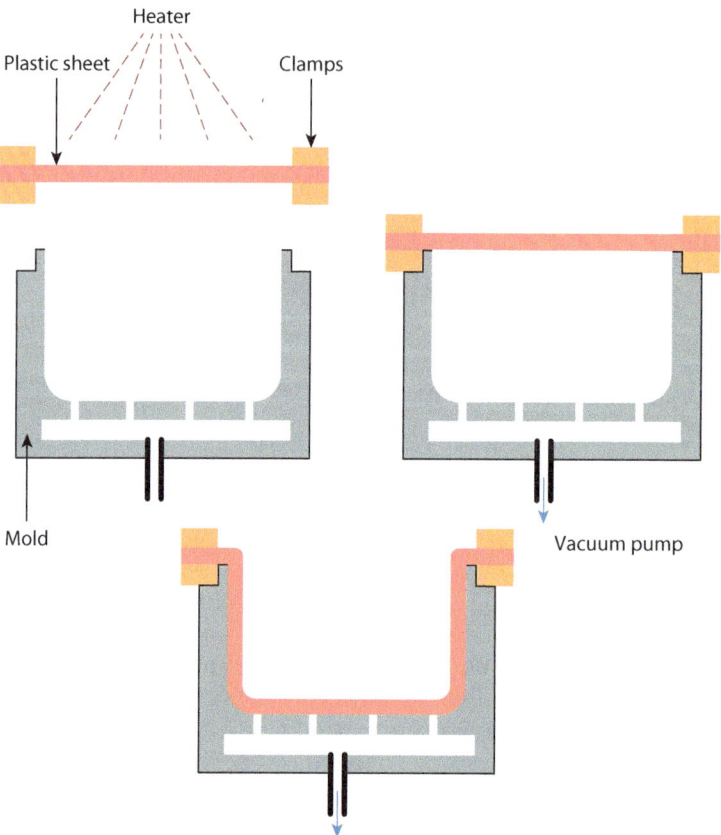

Figure 13.22 Vacuum-assisted thermoforming. The plastic sheet is pink. The sequence of steps goes from the top left where the plastic is heated. In the top right the plastic is lowered onto the mold. In the bottom center figure the heated plastic sheet is pulled into the mold and into shape by the vacuum. (*Based on http://www.substech.com/ dokuwiki/lib/exe/detail.php?id=thermoforming&cache=cache&media=vacuum_thermoforming.png*)

13.4.2   Low-Temperature Forming Techniques for Thermoplastic Polymers

Thermoplastic polymers are also formed into shape by the processes of **cold-forming** and **solid-phase forming**, which are similar to the powder techniques applied metals and ceramics. Many of the names applied to the process are similar. Some of the thermoplastics that are formed at low temperature include ABS, polycarbonate, PTFE, PVC, and polyamides. The processes start with granular thermoplastic material. In these processes the plastic is poured into an open mold. The mold is similar in appearance to the mold for compression molding, shown in Figure 13.23a. In cold-forming, the mold is not heated. In solid-phase forming the mold is heated to a temperature well below the polymer melting temperature or the glass transition temperature. After pouring the plastic into the mold, the mold is closed and the thermoplastic polymer is pressed. Pressures on the material in the mold are sufficient to press the granular particles into intimate contact, eliminate most voids, and form the polymer into a green compact. As with metals and ceramics, the green compact part is low in strength. The green compact is then sintered to increase its strength. During sintering the polymer is heated but not melted. LCMs

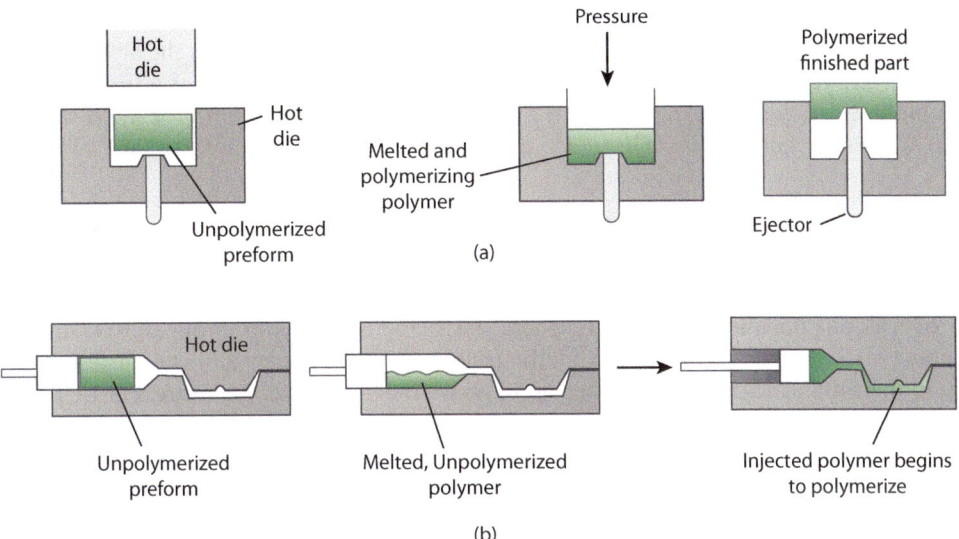

Figure 13.23   (a) Compression molding and (b) transfer molding of thermoset plastics. (*Based on Askeland, D.R., Fulay, P.P., and Wright, W.J., The Science and Engineering of Materials, 6th ed., Cengage Learning, Stamford, CT. (2011), p. 642.*)

diffuse across particle boundaries to bond the particles together. The sintering can be done in the mold or in a separate process. Thermal and other internal stresses in parts are relieved by annealing the polymers at a temperature lower than the sintering temperature.

13.4.3 Deformation Processing of Thermoset Polymers

The main deformation processes for producing parts from thermosetting polymers are compression molding and transfer molding, shown in Figures 13.23a and b. In both of these processes, the resin and the hardener are premixed to form a **molding compound**. The molding compound is a semisolid that has a dough-like consistency, and it flows easily when it is cold. The molding compound is stored at low temperature to suppress curing until it is to be formed. The molding compound is forced into shape by pressure, and heat is applied in the die. The hot die must remain closed until the thermoset polymer is sufficiently cured for the part to be removed. Subsequent curing can also occur at room temperature or elevated temperature. In **compression molding**, shown in Figure 13.23a, the molding compound is placed in an open hot die; then it is compressed and heated. Some of the objects that are compression molded include safety helmets, rubber mounts, and automobile body panels.

In **transfer molding**, shown in Figure 13.23b, the molding compound is injected into a hot die, where it cures. Transfer molding is used to make more complex parts and parts that require a higher tolerance than with compression molding; otherwise similar parts are made by both processes. A variation of transfer molding is **reaction injection molding** (RIM). In RIM, the resin and the hardener are injected into a mixer and then immediately injected into the hot die under pressure.

Automobile tires are made with several of the methods discussed in this section. The tread, sidewall, and liner all have different rubber compositions. These rubber compositions are first prepared with sulfur and extruded into the general shape that is needed to fabricate the tire. At this stage the rubber is not cured, and it has the consistency of gum. Steel belts and sidewall fabric are sandwiched between layers of rubber in a calendar. The components of the tire, including the rubber, steel belts, and sidewall fabric, are assembled in a tire-building machine (TBM). The TBM is a drum that allows the layering of the tire components. The product of the TBM is a green tire that has the general shape of a tire. The green tire is placed in a mold, the tire is pressurized from the inside to force the rubber into the mold shape and to produce the tread design, and the tire is heated to approximately 177°C to vulcanize the rubber.

13.4.4   Casting Polymers

The casting of a liquid polymer is a process where the shape is formed without significant pressure. Thermoset polymers are liquid just after mixing the resin with the hardener and can be cast into shape. **Hot-melt** liquid thermoplastic polymers are heated to above the melting temperature. Plastic particles or powder suspended in a solvent or dissolved in a solvent form a liquid that is cast. The solvent is either evaporated or reacts with the polymer to form a solid.

The most common form of casting is **mold casting**, where the liquid is poured into an open mold in a process similar to permanent mold casting of metals, shown in Figure 13.4c. Because there is little pressure in casting, the mold can be made of relatively low-strength material such as epoxy, silicone, or plaster of Paris. Metals are also used for more permanent molds. It is common to apply a vacuum to a mold to remove voids in the plastic and to ensure that the liquid fills the mold. After the part solidifies, it is removed from the mold.

A process similar to mold casting is **embedding**, where a part, such as an electrical device, is lowered into a mold filled with liquid plastic, and the plastic completely surrounds the part. After the plastic solidifies, the part embedded in plastic is removed from the mold. This technique is used to protect sensitive electrical devices from the environment, and roller-blade wheels are produced by embedding the bearings in urethane contained in a roller-shaped mold.

In **dip casting**, a part to be coated with plastic is dipped into liquid plastic and then removed. The plastic hardens after removal. The plastic coating on the handles of electrical pliers is dip coated.

In **slush casting**, liquid plastic is poured into a mold cavity. The mold cavity has the outer shape of the desired product, such as the outer shape of a glove. The mold is closed and rotated so that the liquid plastic coats the entire inside of the mold. When the proper amount of plastic has solidified on the mold surface, the remaining excess liquid plastic is poured out of the mold. After solidification, a complex shape can be removed from the mold if it is flexible. Plastic items that are open inside, such as boots and gloves, are made by slush casting.

Sheets of plastic are **cell cast** in a mold of two sheets of polished glass that are sealed on the edges with gaskets, and liquid plastic is poured into the mold. Plastic sheet is continuously cast by pouring liquid plastic between two rotating belts of stainless steel that are sealed on the edges.

13.4.5   Joining Plastics

Two ways of joining plastics are by the use of adhesives and fusion bonding. With **adhesive joining**, the plastic part is not melted. Adhesives cover the surface of the polymers to be joined and adhere with van der Waals forces. This requires the surface energy of the adhesive to be less than the surface energy of the polymer, and some polymers, such as PE, have a very low surface energy. Both thermoplastic and thermosetting polymers are joined with adhesives. The adhesive materials include epoxies, polyurethanes, acrylics, cyanoacrylates, anaerobics, silicones, phenolics, and urea formaldehydes. Some of these materials are discussed in more detail in Section 8.3. Epoxies are two-part adhesives, consisting of a resin and a hardener. Polyurethanes combine a polyol and an isocyanate, and the polyol can be water. If the polyol is water, the polyurethane cures on contact with water vapor in the air. Acrylics cure by polymerization of two components; one component is applied to one surface and the other component to the other surface to be joined. The acrylic cures when the two surfaces are brought together. Cyanoacrylates (superglues) are single-component systems that cure on contact with water vapor in the air. Anaerobic adhesives are single-component adhesives that polymerize in the absence of oxygen. The oxygen is eliminated when the two surfaces are pressed together with the adhesive between them. Oxygen is then excluded from the surface, and the adhesive cures. Single-component silicone systems cure on contact with water vapor in the air. Two-component silicones cure upon mixing of the components. Phenolics and urea formaldehydes are available in single- or two-part systems. These adhesives are used for wood and plywood.

Thermoplastic polymers such as PE, nylon, and polypropylene are melted and used as hot-melt adhesives and applied to the surfaces to be joined. Only the adhesive is melted, not the part. A hot-melt glue gun uses this form of joining. In industrial applications, hot air is used to melt a plastic adhesive rod.

In fusion bonding, thermoplastic polymer parts are heated to softening along the surfaces to be joined, and then pressed together. The surfaces are bonded by the mixing of LCMs at the interface of the parts to be joined. The surfaces are heated by hot air, ultrasound, radiofrequency, or friction. Hot-air heating uses an industrial device similar to a hair drier. In ultrasonic welding, high-frequency sound waves are directed at the surfaces of thermoplastic materials to be welded. The sound waves activate thermal vibrations that heat the material to the softening temperature, and the surfaces are pressed together to bond. In radiofrequency (rf) welding, the rf waves cause dipoles on the polymer molecules to oscillate at the radio frequency, and this heats the material. The heated materials are pressed together when they are softened by the heat. In friction welding, the heat necessary to melt the materials is provided by the relative motion of one part relative to another. For example, a polymer rod is spun inside a hole in a polymer plate, and the friction of the rod against the hole melts the surfaces of the rod and the hole. Once the spinning is stopped, the rod and the plate cool and are bonded. All of the bonding procedures based upon heat to soften the polymer work only on thermoplastic materials. Thermosetting materials are bonded with adhesives or by mechanical means.

Summary

- The processes for deforming a solid ductile metal into a final shape include rolling, forging, extrusion, drawing, and stamping. Metals are hot-worked at temperatures above the recrystallization temperature or above 50% of the melting temperature, and metals are cold-worked below these temperatures.

- Some metals and ceramics are superplastic formed at temperatures above 50% of the melting temperature if they have a very fine grain size and a very high strain-hardening rate.

- In casting, a metal is formed into shape in the liquid state in a mold, and then the liquid is cooled to form a solid. Some metals that are cast have a low melting temperature, such as lead, zinc, aluminum, and magnesium. Other metals, such as cast iron and some turbine-blade materials, are cast because they are too hard for deformation processing. The casting techniques for metals include sand mold, permanent mold, die, and investment.

- Metals that are too hard to be deformed plastically and that have a very high melting temperature are processed into shape by powder techniques. Powder techniques are also used for ceramics and for ceramic-metal mixtures. Powder techniques consist of pressing powder into the desired shape to form a green compact, and heating the green compact to sinter the particles into a solid part. In liquid-phase sintering, a component of the mixture of particles is melted to assist in bonding particles together.

- Single crystals of metals and ceramics are produced with the Bridgeman-Stockbarger technique and with the Czochralski technique. Complex shapes, such as turbine blades and vanes, are produced with the Bridgeman-Stockbarger technique. The Czochralski technique is used to produce cylindrical single crystals such as silicon for electronics and ruby for lasers.

- The processes of joining metals and ceramics include fusion welding, brazing, soldering, and diffusion bonding. In fusion welding, the material at the joint is melted, and some welding processes require a filler metal. Metal parts are frequently fusion welded; however, the high temperatures necessary to melt ceramics result in thermal stresses that often fracture ceramics. In brazing, the parts to be joined are not melted. A brazing filler material is melted and diffuses into the parts to be joined. Brazes are available for both metals and ceramics. In soldering the filler material is melted, and it wets the surface of the parts to be joined, but there is insignificant diffusion between the solder and the parts. Metals are soldered with Pb-Sn and Ag-Cu solders, and glass solders have been developed for ceramics. In diffusion bonding parts to be joined are polished, pressed together, and heated to 50 to 80% of the absolute melting temperature. Reactive metals are heated in a vacuum or inert gas. The parts bond by the diffusion of atoms between the parts.

- The primary processes for shaping thermoplastic polymer parts with force at elevated temperatures include extrusion, injection molding, blow molding, thermoforming, and calendaring.

- Granular thermoplastic polymers are formed into green compacts at room temperature in cold-forming or at slightly elevated temperatures in solid-phase forming, and pressure is applied in both processes to form a green compact. The green compacts are sintered to increase their strength, and they are annealed if internal stresses are present.

- The main deformation processes for producing parts from thermosetting polymers are compression molding and transfer molding. In both of these procedures, premixed resin and hardener form a molding compound. In compression molding the molding compound is placed in an open hot die and forced into shape by pressure resulting from closing the die. In transfer molding, the molding compound is injected into a hot die, and the molding compound cures in the hot die.

- The casting of a liquid polymer forms a shape without significant pressure. Thermoset polymers are cast just after mixing the resin with the hardener. Thermoplastic polymers are cast at temperatures above the melting temperature; these are hot-melt plastics. Plastic particles or powder are also suspended in a solvent or dissolved in a solvent to form a fluid that is cast. The solvent is either evaporated or reacts with the polymer to form a solid. The casting methods include mold, embed, dip, slush, cell, and continuous.

- Two ways of joining plastics are by the use of adhesives and fusion bonding. With adhesive joining the plastic part is not melted. Adhesives cover the surface of the polymers to be joined and adhere with van der Waals forces. The adhesive materials include epoxies, polyurethanes, acrylics, cyanoacrylates, anaerobics, silicones, phenolics, and urea formaldehydes. In fusion bonding, thermoplastic polymer parts are heated to softening along the surfaces to be joined, and then pressed together. The surfaces are fusion bonded by the mixing of LCMs at the interface of the parts to be joined. In fusion bonding the surfaces are heated by hot air, ultrasound, radiofrequency, or friction.

Supplemental Reading Subjects and Authors

Full references are listed at the end of the book.

General:	*Askeland, Fulay, and Wright*
Processing metals:	*Courtney; Dieter; Kaufman, Shilling, and Nelson*
Processing ceramics:	*Kingery, Bowen, and Uhlmann; Schwartz*
Processing polymers:	*Osswald and Menges*

Homework

Concept Questions

1. In hot-work, a metal is deformed at temperatures that are either above the _____ temperature, or above 50% of the absolute melting temperature.

2. In powder fabrication of metals and ceramics, when the pressed powder part is sintered, the powder particles bond to each other, and voids are reduced by the _____ of atoms.

3. Metals that do not recrystallize during hot-working have a high _____ rate.

4. In _____, the material is forced through an opening in a die into the desired shape.

5. For metals and ceramics to be superplastic, they must have a high _____ hardening rate.

6. The method of joining metals or ceramics where the parts are not melted but there is diffusion of the filler material into the parts to be joined is called _____.

7. In the Czochralski technique of single crystal growth, the crystal growth is initiated by epitaxial attachment of atoms to a(n) _____ crystal.

8. The crystallographic alignment of new and existing crystalline material during crystal growth is called _____ growth.

9. The combination of water with small ceramic particles is a(n) _____.

10. After a powder-processed part is pressed, and before it is sintered, it is called a(n) _____ compact.

11. Metal solders have low melting temperatures relative to those of the two components in the solder, because metal solders have _____ phase diagrams.

12. The primary method of joining thermosetting polymers is with _____.

13. _____ is the process of forming a sheet or film of thermoplastic polymer into a shape with heat and force.

14. The process of expanding a parison with air pressure to form a bottle is called _____ molding.

15. The process of injecting a thermoset polymer into a hot die is called _____ molding.

16. In cold-forming of plastics, the green compact is _____ to produce the final product.

17. The process where a plastic sheet is formed by rolling is called _____.

18. The process of casting a sheet of plastic in a mold formed by two sheets of glass is called _____ casting.

19. The type of adhesive that cures by the exclusion of oxygen is called _____.

Engineer in Training–Style Questions

1. Which of the following deformation processes is best suited to producing a continuous strip with a "C" shape from metal that has a low strain-hardening rate?
   (a) Rolling
   (b) Extrusion
   (c) Drawing
   (d) Forging

2. For a metal to be drawn into shape, it must have a high
   (a) Tensile strength
   (b) Yield strength
   (c) Strain-hardening rate
   (d) Ductility

3. Which of the following deformation processes is best suited to producing automobile engine piston rods?
   (a) Rolling
   (b) Extrusion
   (c) Drawing
   (d) Forging

4. Which of the following is not true about cold-work?
   (a) It increases strength.
   (b) It produces a good surface finish.
   (c) It increases ductility.
   (d) It provides good dimensional tolerance.

5. Which of the following metals is least likely to be suitable for wire drawing?
   (a) Magnesium
   (b) Aluminum
   (c) Copper
   (d) Iron

6. Metals that have a high recovery rate during hot-working
   (a) Readily recrystallize
   (b) Do not recrystallize
   (c) Do not have texture
   (d) Do not have preferred orientation

7. Which of the following metals is most likely to recrystallize during hot-working?
   (a) Aluminum
   (b) α-Iron
   (c) Cartridge brass
   (d) Zirconium

8. Which of the following is not an advantage of superplastic forming?
   (a) The part is near its final dimensions.
   (b) A high strain is possible.
   (c) Parts can be diffusion bonded.
   (d) High strain rates are possible.

9. Which casting technique is appropriate for a very complex metal shape that cannot be removed from a mold?
   - (a) Sand mold
   - (b) Permanent mold
   - (c) Die
   - (d) Investment

10. Which forming technique is best suited to producing cobalt-tungsten carbide cutting tools?
    - (a) Hot isostatic pressing of powder
    - (b) Investment casting
    - (c) Forging
    - (d) Superplastic forming

11. The technique for producing single crystals of turbine blades is by a modification of
    - (a) Investment casting
    - (b) Directional solidification
    - (c) The Bridgeman-Stockbarger technique
    - (d) The Czochralski technique

12. Which of the following aluminum alloys is not recommended for welding?
    - (a) 3003
    - (b) 5052
    - (c) 6061
    - (d) 7075

13. Which of the following joining techniques is difficult to apply to ceramic materials?
    - (a) Fusion welding
    - (b) Brazing
    - (c) Soldering
    - (d) Diffusion bonding

14. Which of the following procedures does not apply to the deformation processing of thermoplastic polymers?
    - (a) Ram injection molding
    - (b) Blow molding
    - (c) Thermoforming
    - (d) Compression molding

15. A process for producing a product like a rubber glove is
    - (a) Injection molding
    - (b) Extrusion
    - (c) Slush casting
    - (d) Calendaring

16. Which of the following polymers cannot be joined by fusion welding?
    - (a) PE
    - (b) PVC
    - (c) Epoxy
    - (d) PTFE

The Boeing 787 aircraft is the first large commercial airplane whose fuselage and wings are made of carbon-fiber-reinforced epoxy. This chapter develops a methodology to justify the decision to use carbon-fiber-reinforced epoxy for this design, based upon mechanical properties and cost.

© Robert Schlesinger/dpa/Corbis

The goals of this chapter are to understand

- The contributions to the cost of a part
- The relative cost of various materials
- The relative cost of manufacturing processes
- The method for calculating the performance per unit cost
- The method for calculating the relative merit of materials for weight-critical designs
- Design requirements and materials for hip implants
- Sustainable supplies of materials
- Materials that can be recycled or are biodegradable
- Sources of mechanical-property data and cost

Chapter 14

Material Selection

14.1 MATERIALS, MECHANICAL PROPERTIES, AND DESIGN

In previous chapters, we have studied the mechanical and physical properties of materials, and we have learned about the materials that are available for engineering designs. In Chapter 8, we discussed ceramics, metals, and polymers, and in Chapter 12 we covered composites. This chapter discusses databases that contain a more extensive coverage of materials. It is estimated that there may be more than 100,000 materials available for a design. The engineer selecting a material for a design should be aware of all of the potential materials. This chapter discusses how an engineer should select one or more suitable materials from all of these potential candidates. The material selection should be part of the design, and as numerical as possible. In previous chapters, we have calculated numerical values for yield strength, tensile strength, ductility, fracture toughness, fatigue life, creep rupture life, and so on. This data is part of the numerical analysis of materials selection, along with cost.

During the design, the engineer is responsible for knowing what physical properties of the material are important.

- Is the amount of strain important? If so, what elastic modulus for the material gives the appropriate strain, as discussed in Section 6.3?

- If high stresses are present, does the material have sufficient yield strength? Is the critical-stress intensity (K_{Ic}) of the material sufficient to prevent fracture at maximum stress, as discussed in Section 11.2.9?

- If the stress varies with time, is the fatigue life sufficient to avoid failure within the specified design lifetime, as discussed in Section 11.3?

- If the temperature of the application is high, is the creep resistance of the material sufficient at the design temperature, as discussed in Section 9.4?

- If the application is in a corrosive environment, is the material resistant to that corrosive environment, as discussed in Section 10.5?

The engineer also must know how the performance of the material is affected by the design. Sharp corners, threads, and keyways act as stress concentrators that locally increase the stress, which can lead to crack formation and to fracture if the critical-stress intensity factor is not sufficient. If the stress varies with time, stress concentrations can result in fatigue-crack initiation, fatigue-crack propagation, and material failure as discussed in Section 11.3. If there is a corrosive environment present, the contact of dissimilar metals in the design may result in galvanic corrosion.

The engineer selecting the material must also know how the material is to be manufactured into a part, as discussed in Chapter 13, and if the manufacturing process changes the mechanical properties of the material in the part. For example, if the design requires a high-strength precipitation-hardened aluminum alloy, these alloys are not suitable for welding, because in the heat-affected zone of the weld, the mechanical properties of the alloy are altered and the yield strength is decreased. Other processes such as rolling, drawing, or forging increase the yield strength of a ductile metal, such as low-carbon steel, and they decrease the ductility and the critical-stress intensity factor.

The engineering properties related to material selection are covered in the previous sections of this book. What we have not discussed previously is cost, which is a necessary consideration when selecting materials. Here, we begin an in-depth look into how to include cost with performance in design and materials selection.

14.2   COST ANALYSIS

Many factors contribute to the initial cost of a product to a buyer, as shown in Table 14.1. These factors include the cost related to producing this particular product, fixed costs that are outside of the control of the engineer selecting the material, and the profit for the manufacturer.

Table 14.1   Costs Contributing to the Purchase Price of a Product

Costs of Production	Fixed Costs	Manufacturer's Profit
Design	Research and development	
Materials	Administration	
Manufacture	Sales and marketing	
Inspection	Factory overhead	
Liability		

The selection of materials can affect the cost of the product design, manufacture, inspection, and liability. For example, selection of a high-strength, low-alloy steel for a steering control arm in an automobile to replace a low-strength, low-carbon steel results in a lower-weight part; and this allows the dimensions to be reduced. However, the part must be redesigned, resulting in a redesign cost. It is more expensive to manufacture the part made from high-strength steel, because higher forces or higher temperatures are necessary to deform the material into the proper shape. High-strength steel may have a lower K_{Ic} than does low-strength steel. If the stresses are higher in a high-strength steel with a lower K_{Ic}, the critical-size crack that could cause failure in the high-strength steel is smaller than for the low-strength steel. Therefore, inspection procedures for the high-strength steel must detect smaller defects than is necessary for the low-strength steel; a change in the inspection procedures and possibly some new inspection equipment may be required to accomplish this.

The engineer must also determine if the cost of liability should be considered. For noncritical parts, such as the push buttons on the car radio, it may not be necessary to consider liability. In contrast, failure of a steering control arm would result in the loss of steering control and may cause a fatal accident. If it is found that the failure was due to poor design, material selection, manufacturing, or inspection, the manufacturer could be subject to lawsuits. When designing a critical part, such as the automobile steering control arm, the potential cost of failure must be measured not only in terms of lawsuits, but also in terms of repair costs and lives lost.

For example, the author of this textbook was a consultant in a case involving the failure of automobile steering control arms. It was found that the control arms of the manufacturer's cheapest models had the lowest failure incidence, and the most expensive models had the highest failure incidence. Investigation revealed that the same steering control arms were used for all of the models, to save cost. The more expensive models had higher weights and more stress on the control arms. This was a case where buyers did not receive a higher-quality product by paying more money. Lawsuits did result from the failure of these control arms, and lives were lost. To avoid this scenario, the engineers had options when designing the control arms. For example, they could have redesigned the control arms used in the heavier automobiles, to increase the amount of stress these parts could handle without failure. Or, if the company wanted to use the same model control arm for every vehicle (regardless of weight), the engineers could have designed a control arm that could tolerate use in the heaviest car model. This would make the part safe for all the cars in the product line.

In this chapter, we calculate only the material cost. All of the other costs listed in Table 14.1 are beyond the scope of this textbook. After the design engineer considers all of the costs in Table 14.1, and then adds manufacturer's profit to that value, the final price of the product must still be attractive to a customer, or else the product will not sell.

14.2.1   Material Cost and Performance

There are different levels of material cost for material producers, distributors, and retailers. Engineers normally purchase their materials from a distributor if buying large amounts of material, or from a retailer if purchasing small quantities. It is quite difficult to find listed prices of materials, because the price of materials can change rapidly in periods of inflation or deflation. Also, some distributors do not want the competition to know their prices. If the engineer knows the shape of the desired material and how much is desired for an order, then they can ask for a quote of material price from distributors. Some retailers do list their product prices on the internet. One retailer that lists metal prices is McMaster-Carr. Table 14.2 presents recent costs per cubic meter of some engineering metals from this site, based on a rod that is 0.0254 m in diameter and 1.83 m in length.

The metals listed in Table 14.2 are for standard-purity materials, but some applications require higher-purity materials. For example, the steel used to make submarines is very low in carbon and other

Table 14.2	Cost per Cubic Meter of Some Engineering Metals

Metals	Cost 10^4 $/m^3$
Aluminum Alloys	
2024	6.57
4032	9.44
6061	3.76
7075	8.20
Copper Alloys	
CP[1] Copper (C11000)	21.62
Naval brass	19.32
Nickel Alloys	
CP nickel	111.69
Monel 400	82.25
Hastelloy X	80.17
Inconel 625	82.25
Steel	
1045	4.38
1117	3.98
4340	10.78
8620	4.66
Stainless Steel	
410 (ferritic)	8.41
304 (austenitic)	10.37
15-5 PH[2]	21.50
17-4 PH	13.20
Titanium Alloys	
CP Ti (R50500)	50.32
Ti-6Al-4V (R56400)	65.80
Tungsten	
CP[3]	221.72

1   CP = commercial purity
2   PH = precipitation hardened
3   Price based on a length of 0.305 m

interstitials. This results in steel that has a high K_{Ic} at low temperatures, which allows submarine hulls to resist fracture in deep cold water, even when under the polar ice cap. Similarly, the titanium alloys used for the rotors in high-temperature gas turbines require a low interstitial content. These low-interstitial alloys are more expensive than the standard-purity metals listed in Table 14.2.

Table 14.2 shows why many metal products are made of carbon steel. Carbon steel is one of the cheapest metals per m³, and it has good mechanical properties. This is why automobile frames and building frames are primarily made from carbon steel. However, decreasing the weight of an automobile increases its fuel

Table 14.3	The Cost in U.S. Dollars per Kilogram and per Cubic Meter for Some Engineering Plastics	
Material	$10^4$$/m^3	$/kg
ABS	3.20	26.70
Acetal	9.65	67.96
Nylon	9.41	82.54
PEEK	36.59	281.46
LDPE	1.35	14.67
HDPE	1.16	12.21
UHMWPE	0.94	10.10
PP	1.46	16.04
PUR	7.17	59.75
PVC	1.60	11.43
Phenolic	13.11	93.64
Teflon	4.66	21.47

efficiency. One way that automobiles can be made lighter is by replacing carbon steel with a high-strength steel alloy, such as 4340; however, some of the alloy elements for steel include aluminum, chromium, molybdenum, and nickel, which increase the cost of steel. Automobiles can also be made significantly lighter by replacing steel of density 55.8×10^3 kg/m^3 with aluminum of density 27×10^3 kg/m^3. However, Table 14.2 shows that aluminum is more expensive than steel. Steel is in general stiffer, stronger, and has a higher critical-stress intensity factor than does aluminum. So how does the design engineer decide what material to choose when deciding among carbon steel, high-strength steel, or a high-strength aluminum alloy? Procedures for helping engineers make this choice are presented below.

Table 14.3 presents the cost of some engineering plastics. These are recent prices for rod with a circular or square cross section with a diameter or width, respectively, of 2.54×10^{-2} m per unit length of 0.305 m listed at *www.totalplastics.com*. The cost per kilogram is the cost per cubic meter divided by the density.

The prices in Table 14.3 are for small quantities of material cut to length. Table 14.3 shows that there is more than an order of magnitude of difference in the cost of polymer materials from UHMWPE to PEEK.

Several Web sites list composite-material prices. The Web site *www.totalplastics.com* has the price of fiberglass-epoxy composite products, and the Web site *www.dragonplate.com* has prices for carbon-epoxy products. Table 14.4 presents prices for glass-epoxy rod with a 0.0254 m diameter and 1.22 m length, and carbon-fiber rod 0.00635 m in diameter and 1.22 m in length.

Table 14.4	The Cost of Some Composite Materials
Material	Cost $10^4$$/m^3
Figerglass-epoxy (G-10)	28.26
Carbon fiber-epoxy	38.84
Concrete	0.013

G-10 fiberglass is a standard of the National Electrical Manufacturers Association. The cheapest material per unit volume in Tables 14.2 to 14.4 is concrete. This is why many large projects such as bridges, dams, foundations, and building walls are made with steel-reinforced concrete. If the objective is to fill a space with material at low cost, concrete is the best choice.

14.2.2   Manufacturing Cost

Material cost is only one of the costs to consider when choosing a material. It must also be possible to manufacture the product at a cost that allows it to be sold with the desired profit. The manufacturing cost is required to transform the initial purchased material into the final desired product. In manufacturing, the initial material is processed into the desired final product. For example, many product shapes made from fiberglass are fabricated by hand, which makes the product very expensive to produce.

The lowest manufacturing cost may result if the design allows for the use of a material that is in an available standard shape. For example, ceramics, metals, and polymers are available in shapes such as a circular or square rod, and in sheets. If these standard shapes can be used as the starting material, and if a minimum of additional processing is required, this may provide an economical design. However, many parts are too complex in shape to be made from standard material shapes, and the initial material must be formed into the desired shape through flow processing. The material can be formed into a new shape as a solid, liquid, or powder. We covered the processes to form the material shape in Chapter 13.

Flow processes change the shape of an initial material into the shape of the final product, by flow of a solid or of a liquid. Processes that change shape through the flow of solid metals include forging, rolling, drawing, and extrusion. Metals can also be formed into shape in the liquid state by casting; however, casting is in general more expensive than forming the metal shape in the solid state. Sand casting and investment casting produce one part per casting, and they are labor intensive and expensive. Also, the energy required to melt the metal is significant when melting high-temperature metals. However, the high-temperature gas-turbine alloys are so strong that they cannot be flow processed as solids; instead, they are investment cast. Metals are also cast in permanent molds and die cast. Permanent molds and dies are expensive to produce, and they have a limited lifetime; however, many parts are produced at reasonable prices by these processes, such as automobile engine pistons. Powder metallurgy techniques are used when the metal is so hard that it is not easily deformed as a solid, and when the metal is not easily melted. Some examples include metal-ceramic mixtures (cermets) and metals that are dispersion strengthened with ceramics, such as nickel thoria. Powder metallurgy techniques are usually very expensive, for to produce a material without voids requires pressing the material at a high temperature for a long time, and also in an inert atmosphere if the material is reactive with oxygen.

The initial material for deformation-processing thermoplastics is usually solid pellets that are formed into shape by extrusion, injection molding, thermoforming, calendaring, or blow molding. Since the softening temperature of polymers is low and the strength is low, the energy required to process the polymer is modest. Also, the equipment to deformation process polymers does not have to be as robust as that for metals, and as a result this equipment is less expensive. All of these factors make it less expensive to deformation process polymers than metals. Polymers are also cast, and the low melting or glass transition temperature makes this a relatively inexpensive process. For these reasons, the forming of polymers is of modest expense relative to the forming of metals.

The initial material for thermoset polymers is either liquid that is cast, or molding compound that is compression molded or transfer molded. All of these processes are of relatively low expense in comparison to metal processing.

Ceramic solids are brittle at low temperatures, and they have very high melting temperatures. Therefore, ceramics are not usually flow formed into shape as solids or liquids, except in superplastic forming.

Glasses are an exception, in that they are melted and formed by the processes of blowing, drawing, pressing, and hot-rolling. Crystalline ceramic parts are usually produced by powder techniques. Powder ceramics are pressed and sintered into solid specimens. To produce ceramics with a low void content, such as 0.1%, is expensive. Ceramics are also formed into shape by hydroplastic forming techniques with a slip by extrusion and slip casting. Slip forming is less expensive than powder techniques.

If the part cannot be completely formed into the desired shape by flow processes, then it can be machined into shape. Machining processes include milling, turning on a lathe, sawing, and drilling. These processes are not discussed in this book, because in general these processes do not affect the mechanical properties of the material; however, the machining process can result in crack formation at keyways and sharp edges that reduce the fracture strength of materials. Machining is usually an expensive process. It is normally cheaper to form the final shape by flow processes; however, in many cases, such as threads, keyways, and bolt holes, machining is a necessary process. Most metals and polymers can be machined, except for those that are very soft, such as lead and LDPE, and some very hard metals. Very soft materials are not easily machined, because they plastically deform rather than have the material removed by a cutting tool. Also, very hard materials, such as superalloys, are not easily machined, because they damage the cutting tools. If machining is to be a significant portion of the processing, then the engineer should consider a material that is easily machined.

The final process in manufacturing the product is **fabrication**. Fabrication is the process of joining together individual material parts into the final product. An example of fabrication is the production of an automobile body. The body is made primarily out of sheet steel that is joined by mechanical, metallurgical, or chemical methods. The mechanical methods are bolting, screwing, riveting, and clipping. The metallurgical methods of joining parts are welding, brazing, and soldering. The chemical method of joining parts is with adhesives. We covered the processes of joining metals and polymers by welding, brazing, and soldering in Chapter 13. Fabrication processes such as riveting, bolting, screwing, and the use of adhesives are not usually expensive, and they do not alter the properties of the material. Welding and brazing are more expensive, and the process can alter the material in the heat-affected zone. Polymer parts are fabricated into an assembly with adhesives, bolting, screwing, clipping, and welding. Because ceramics are brittle, there are a limited number of mechanical fabrication techniques. Ceramics can be bolted or screwed as long as the ceramic is in compression and threads are not in the ceramic. Ceramics melt at very high temperatures, and they are brittle; therefore, ceramics are not usually fusion welded. Ceramics are brazed and soldered, as we discussed in Chapter 13. Also, adhesive joining is not possible in high-temperature ceramic applications, because most adhesives are polymer based and cannot withstand high temperatures.

14.2.3   Other Cost

Another cost for the engineer to consider is the cost of ownership. The cost of ownership includes the purchase price, operation, maintenance, repairs, inspection, and insurance. Although the cost of ownership does not affect the cost of producing the product, it does affect product sales, profits, and the value of the product. The **value** of the product is the extent to which the product meets the performance criteria. If the customer feels that the product is a good value for the specified price, then the customer will purchase the product. There are many examples of how value affects consumer purchases. Certain automobiles are perceived by customers to have excellent repair records, low maintenance costs, low operational costs, and good resale prices, and these automobiles are perceived to be of good value. This perception allows these manufacturers to charge a premium for their automobiles, increasing the price to the owner, which makes these companies more profitable than other automobile companies.

The operational cost of automobiles is very important to sales when the price of fuel increases. When fuel is expensive, automobile companies cannot sell large pickup trucks and sport utility vehicles at normal prices; they must be heavily discounted to sell. Automobiles that have excellent performance records in crash tests have lower insurance rates that decrease the cost of ownership and increase the safety of the owner. These two features increase the value of the automobile to safety-conscious potential purchasers. The cost to maintain an automobile affects its value; lower maintenance cost results in an automobile with greater value.

The Boeing 787 Dreamliner is another example of good value. The aircraft has been lightened through the use of graphite-reinforced composites and improvements in engine design. It is expected that these improvements will significantly improve the aircraft's fuel efficiency, and with the currently high price of jet fuel, this aircraft is selling well. An example where inspection is an important cost of ownership is in commercial aircraft. Aircraft are designed to have regular inspection periods. If the inspection period is extended through use of superior materials or designs, this is a cost savings. Also, if the aircraft is designed such that the inspection is less expensive, then this is a savings to the cost of ownership and an increase in value.

In this chapter, we do not analyze the cost of ownership; however, it is the engineer's responsibility to provide a product that maximizes product value while allowing for the desired company profit. The engineer can achieve this through intelligent design and selection of materials.

14.3 MATERIAL SELECTION PROCEDURES

In considering materials for a design, the engineer should initially include all possible materials. Before selecting a material, the engineer must have the design requirements. Let us assume that the part is a rod in tension in an automobile. The design requirements are that the material that must have an elastic modulus (E) of at least 100×10^9 Pa, a tensile yield strength (σ_y) of at least 1.0×10^9 Pa, a critical-stress intensity factor (K_{Ic}) of at least 40 MPa $\cdot$ m$^{1/2}$, and a density (ρ) of not more than 8×10^3 kg/m^3. In addition, we desire the most cost-effective material that meets the design requirements. We construct Table 14.5 to see what classes of materials from the universe of all materials might satisfy the design requirements.

The data for completing Table 14.5 comes from previous chapters. The decision of "Accept" means that at least some materials in this class can meet the design specifications, based upon data presented in previous chapters; this class of material is accepted for further consideration. The decision of "Reject" means that no

Table 14.5   The Suitability of Classes of Materials for the Design Requirements

Material	Design Requirements					
	E	σ_y	K_{Ic}	ρ	Cost	Decision
Ceramics	H	F	F	A	High	Reject
Polymers	F	F	F	H	A	Reject
Aluminum alloys	F	F	F	H	High	Reject
Titanium alloys	A	A	A	H	High	Accept
Steel alloys	A	A	A	A	A	Accept
Composite materials	A	A	A	H	High	Accept

H—Exceeds the requirement

A—Meets requirement, acceptable

F—Does not meet requirement

materials of this class meet all of the design requirements. The elastic moduli for some metals, ceramics, and polymers are presented in Table 6.1. The yield strength of some metals is presented in Table 6.3, and the yield strength of some polymers is presented in Table 6.5. The tensile strength of ceramics is presented in Table 6.6. The K_{Ic} of ceramics, metals, and polymers are presented in Tables 11.1, 11.2, and 11.3, respectively. The values for the elastic modulus and tensile strength of some uniaxial composite materials are presented in Table 12.2, and values for the critical-stress intensity factor are presented in Table 12.3. From Appendix A, the density of nickel and copper are both above 8×10^3 kg/m³, and alloys of these metals are very expensive. These two factors eliminate these metals from consideration. Ceramics are of a high elastic modulus; for example, fully dense aluminum oxide has an elastic modulus of 415×10^9 Pa, but the tensile strength is only 0.2×10^9 Pa for 99.8% dense aluminum oxide, and the K_{Ic} for aluminum oxide is only 3.5 MPa · m$^{1/2}$. Ceramics must be rejected from this application because of their low tensile strength and their low K_{Ic}. Polymers are rejected because they have low values of elastic modulus, yield strength, and K_{Ic}. Aluminum alloys have an elastic modulus of approximately 70×10^9 Pa, the yield strengths are a maximum of 70×10^9 Pa, and the value of K_{Ic} for high-strength alloys is typically 25 MPa · m$^{1/2}$. Aluminum alloys are rejected because none of the mechanical properties are sufficiently high. Titanium alloys have an elastic modulus of approximately 115×10^9 Pa, the yield strengths are a maximum of 120×10^9 Pa, and the values, of K_{Ic} for high-strength alloys is typically 50 MPa · m$^{1/2}$, and the density is approximately 4.5×10^3 kg/m³. Titanium alloys may satisfy the design requirements. The elastic modulus of iron is 211×10^9 Pa, some steel alloys have yield strengths of over 1×10^9 Pa and a K_{Ic} of over 50 MPa · m$^{1/2}$, and the density of iron is 7.86×10^3 kg/m³. Steel alloys may satisfy the design requirements. The composite material of carbon-fiber-reinforced epoxy may also meet the design requirements. Example Problem 12.3 shows that in a carbon-epoxy composite, the brittle fibers fracture before the brittle matrix does; therefore, the tensile strength and the yield strength are the same, as shown in Figure 12.15b. The data in Tables 12.2 and 12.3 for carbon-epoxy composites are for a unidirectional composite material, and the properties in the transverse direction may not meet the design requirements, as is demonstrated by the low transverse elastic modulus and strength of this composite material. However, the design is for a bar in tension, and unidirectional composite materials are very efficient for this type of application. We accept carbon-epoxy composite materials for further consideration.

Now we consider the cost of the materials that may meet the design requirements. In selecting a material for a design, consider as many materials as possible. However, for this analysis, only materials in Tables 14.2 through 14.4, and materials for which there is sufficient mechanical-property data in previous chapters, are considered. Table 14.6 presents the mechanical-property data and costs for these materials.

In Table 14.6, the values of σ_y and K_{Ic} for metals are from Table 11.2, and the values of E and ρ for the metal alloys are from *www.matweb.com*. For large quantities of these materials, the price would be

Table 14.6 The Elastic Modulus (E), Yield Strength (σ_y), Critical-Stress Intensity Factor (K_{Ic}), Density (ρ), and Cost per Unit Volume (C_V) for Candidate Materials

Material	E 10^9 Pa	σ_y 10^6 Pa	K_{Ic} MPa · m$^{1/2}$	ρ 10^3 kg/m³	C_V 10^4 \$/m³
Ti-6Al-4V	113.8	910	115.4	4.43	65.80
Ti-6Al-4V	113.8	1035	55	4.43	65.80
4340	205	860	98.9	7.85	10.78
4340	205	1515	60.4	7.85	10.78
17-7PH	204	1435	76.9	7.83	16.55
15-7Mo	204	1415	49.5	7.83	21.50
Carbon-epoxy	138	1447	40	1.67	38.84

This mechanical property and cost data should not be used for actual design or material selection.

lower, but this should give a good idea of the relative price. The cost per cubic meter is based upon data from McMaster-Carr for a rod that is 1 inch in diameter, and the cost of uniaxial carbon-epoxy rod was obtained from the Web site of Dragonplate. These estimates would be confirmed by quotes if these candidate materials make it to the next phase of consideration. The yield strength of the composite material is taken as equal to the tensile strength, since the carbon-fiber-reinforced epoxy does not yield.

14.3.1 Material Selection for Performance per Unit of Cost

An approach to cost analysis for designs that are not weight critical is presented in the Fulmer Materials Optimizer, by Waterman. Table 14.7 gives formulas for the performance per unit of cost for the elastic modulus (E), yield strength (σ_y), and critical-stress intensity factor (K_{Ic}) for different loading conditions. For example, a rod in tension has a design requirement that the applied force not result in yield or fracture. The performance in yield strength per unit cost is given by σ_y/C_V. As shown by the formula, the performance per unit cost is increased by reducing C_V, or by increasing the performance in the form of a higher σ_y, E, and K_{Ic}. If ultimate tensile strength is more important than the yield strength, then ultimate tensile strength replaces yield strength in Table 14.7. If the cost per unit weight (C_W) is given, then the cost per unit volume (C_V) is C_W times the density (ρ). Table 14.7 presents some loading conditions that are not considered in this book in anticipation that engineers will encounter these loading conditions in other courses. For loading conditions that are not considered in Table 14.7, see Waterman or Charles et al.

We now insert mechanical property data and cost per unit volume from Table 14.6 into the formulas in Table 14.7 to produce Table 14.8 for a rod in tension with the assumption that weight is not critical except that the density must be less than $8 \times 10^3 \, \text{kg/m}^3$.

Since the units for K_{Ic} are different than for E and for σ_y in Table 14.8, relative values of each term are obtained by dividing each number in a column in Table 14.8 by the largest number in that column to produce dimensionless relative performance per unit cost for each material and property as presented in Table 14.9.

Table 14.7	Formulas for Performance per Unit Cost for Different Loading Conditions		
Loading Conditions	**Performance per Unit Cost Formulas**		
	Stiffness	**Ductile Yield**	**Brittle Fracture**
Rod in tension	E/C_V	σ_y/C_V	K_{Ic}/C_V
Short column in compression	E/C_V	σ_y/C_V	K_{Ic}/C_V
Thin-walled pipe or pressure vessel under internal pressure	E/C_V	σ_y/C_V	K_{Ic}/C_V
Rod in bending	$E^{1/2}/C_V$	$\sigma_y^{2/3}/C_V$	$K_{Ic}^{2/3}/C_V$
Plate in bending	$E^{1/3}/C_V$	$\sigma_y^{1/2}/C_V$	$K_{Ic}^{1/2}/C_V$
Slender column in buckling	$E^{1/2}/C_V$		
Block in shear	G/C_V	τ_y/C_V	K_{IIc}/C_V

Table 14.8	The Elastic Modulus (E) in Units of 10^9 Pa, Yield Strength (σ_y) in Units of 10^6 Pa, and the Critical-Stress Intensity Factor (K_{Ic}) in Units of MPa(m)$^{1/2}$ Each Divided by cost per Unit Volume (C_V) in Units of $10^4\$/m^3$, Using Data from Table 14.6		
Material	E/C_V	σ_y/C_V	K_{Ic}/C_V
Ti-6Al-4V	1.73	13.8	1.75
Ti-6Al-4V	1.73	15.7	0.84
4340	19.0	79.8	9.17
4340	19.0	141	5.60
17-7PH	12.3	86.7	4.65
15-7Mo	9.49	65.8	2.30
Carbon-epoxy	3.55	37.3	1.03

If we assume that each property is of equal importance to the design, then the values of relative performance per unit cost of a given material are summed and divided by the number of performance parameters. If one parameter is more important than others, then weighting factors can be applied and the denominator adjusted to give a perfect score of 1.0. The relative performance per unit cost (RPC) of each material for three equally weighted performance parameters is calculated from Equation 14.1 and shown in Table 14.10.

$$RPC = \frac{1}{3}\left(\frac{RE}{C_V} + \frac{R\sigma_y}{C_V} + \frac{RK_{Ic}}{C_V}\right)$$

14.1

In this analysis, the 4340 steel with the highest yield strength gives the highest performance per unit of cost. One reason for this is that the yield strength is three times the design requirement. If the engineer decides that avoiding fracture is more important than avoiding yield, the 4340 steel with the lower yield strength and higher value of K_{Ic} could be used instead. Both of these steels would cost the same amount for the material, assuming that the heat-treatment costs are similar. The steel with the higher yield strength should provide higher resistance to fatigue-crack initiation but lower resistance to fatigue-crack propagation. The engineer must decide which of these physical properties is most important. If this design is for a plate in bending instead of a rod in tension, then the cost analysis is based upon $E^{1/3}/C_V$, $\sigma_y^{1/2}/C_V$, and $K_{Ic}^{1/2}/C_V$, as shown in Table 14.7. For further discussion of this cost analysis, see Waterman.

Table 14.9	The Relative Elastic Modulus (RE), Yield Strength ($R\sigma_y$), and Critical-Stress Intensity Factor (RK_{Ic}) Each Divided by Cost per Unit Volume (C_V), Using Data from Table 14.8		
Material	RE/C_V	$R\sigma_y/C_V$	RK_{Ic}/C_V
Ti-6Al-4V	0.09	0.10	0.19
Ti-6Al-4V	0.09	0.11	0.09
4340	1.0	0.57	1.0
4340	1.0	1.0	0.61
17-7PH	0.65	0.62	0.51
15-7Mo	0.50	0.47	0.25
Carbon-epoxy	0.19	0.27	0.11

Table 14.10	The Relative Performance per Unit of Cost of Each Material in Table 14.9

Material	RPC
Ti-6Al-4V	0.13
Ti-6Al-4V	0.10
4340	0.86
4340	0.87
17-7PH	0.59
15-7Mo	0.41
Carbon-epoxy	0.19

14.3.2 Material Selection Procedures for Weight-Critical Design

When weight is critical, such as in aircraft, high-performance sporting equipment, and high-efficiency automobiles, the material with the highest specific modulus or specific strength is the most efficient material. The specific elastic modulus is E/ρ, and in a similar way the specific strength and specific critical-stress intensity factor are found by dividing each of these terms by the density. Dividing each term in Table 14.6 by the density results in the specific mechanical properties presented in Table 14.11, and C_V/ρ is the cost per unit weight. Table 14.12 presents the relative values from Table 14.11.

If it is assumed that each of the design requirements is of equal importance and that cost is of equal importance to performance, then the relative merit (RM) of each candidate material is given by Equation 14.2.

$$RM = \frac{1}{4}\left(\frac{RE}{\rho} + \frac{R\sigma_y}{\rho} + \frac{RK_{Ic}}{\rho} + 1 - \frac{RC_V}{\rho}\right) \qquad \textbf{14.2}$$

The relative-merit contribution from cost is given by $1 - RC_V/\rho$, because the lowest relative-cost material is the most desired. The results of applying Equation 14.2 to the candidate materials are shown in Table 14.13.

Table 14.11	Values of the Specific Elastic Modulus, Yield Strength, Critical-Stress Intensity Factor, and Cost per Kilogram for the Candidate Materials in Table 14.6

Material	E/ρ	σ_y/ρ	K_{Ic}/ρ	C_V/ρ
Ti-6Al-4V	25.7	205	26.0	14.85
Ti-6Al-4V	25.7	234	12.4	14.85
4340	26.1	110	12.6	1.37
4340	26.1	193	7.70	1.37
17-7PH	26.1	183	9.82	2.11
15-7Mo	26.1	181	6.32	2.75
Carbon-epoxy	82.6	866	24.0	23.26

Table 14.12	Relative Values of the Specific Elastic Modulus, Specific Yield Strength, and Specific Critical-Stress Intensity Factor, and Cost per Cubic Meter Divided by the Density from Table 14.11 for Candidate Materials			
Material	RE/ρ	$R\sigma_y/\rho$	RK_{Ic}/ρ	RC_v/ρ
Ti-6Al-4V	0.31	0.24	1.0	0.64
Ti-6Al-4V	0.31	0.27	0.48	0.64
4340	0.32	0.13	0.48	0.06
4340	0.32	0.22	0.30	0.06
17-7PH	0.32	0.21	0.38	0.09
15-7Mo	0.32	0.21	0.24	0.12
Carbon-epoxy	1.0	1.0	0.92	1.0

Table 14.13 shows that on a relative basis, the carbon-fiber-reinforced epoxy material has the highest merit for a weight-critical design. The high merit is why carbon-reinforced epoxy is appearing in many high-performance applications where weight is critical. Race cars and high-performance automobiles already use carbon-reinforced-epoxy for applications such as the chassis. Carbon-fiber-reinforced epoxy is used in other high-performance applications, such as aircraft parts, fishing rods, tennis rackets, and racing boats of all types. As we discussed in Chapters 1 and 12 and also saw at the beginning of this chapter, Boeing is using carbon-fiber-reinforced epoxy for the fuselage of the 787 Dreamliner. Based upon mechanical properties and material cost, the carbon-fiber-reinforced epoxy clearly has the highest relative merit in Table 14.13. However, producing an airplane with this material has been a challenge. Using a relatively new material such as carbon-fiber-reinforced epoxy for a design has meant a significant development cost. The material properties and manufacturing processes of carbon-fiber-reinforced epoxy are not as well known as for standard materials, such as 7075-T6 aluminum. For example, the joining process for carbon-fiber-reinforced epoxy is not as well established as for aluminum. Carbon-fiber-reinforced epoxy cannot be welded, but it can be joined with adhesives; however, adhesives are not as strong as carbon-fiber-reinforced epoxy. Carbon-fiber-reinforced epoxy can be mechanically joined, but one must be careful not to crush the material or allow shear along the fiber-epoxy interface. Development and manufacturing cost could significantly increase the cost of using carbon-fiber-reinforced epoxy in a new design, and these costs are not included in Table 14.13.

In the past, subsonic commercial aircraft fuselages were primarily made from high-strength aluminum alloys that are not considered in this analysis, but they are considered in Problem 14.1. For supersonic

Table 14.13	The Relative Merit (*RM*) of the Candidate Materials for the Design Requirements
Material	RM
Ti-6Al-4V	0.48
Ti-6Al-4V	0.35
4340	0.47
4340	0.44
17-7PH	0.45
15-7Mo	0.41
Carbon-epoxy	0.73

aircraft at speeds in excess of 2.5 times the speed of sound, titanium alloys are used for the fuselage because of their high-temperature strength. For further discussions of the selection and use of materials in aircraft structures, see Charles et al.

14.3.3   Material Selection for an Anchor Stem of a Total Artificial Hip Replacement

In a total hip replacement, the bones of the hip joint are replaced by the mechanical system shown schematically in Figure 14.1. The femoral stem is fixed into the femur with polymer bone cement. Attached to the stem is a head that fits into a socket (acetabular component) that is attached to the pelvis by screws or bone cement or a combination of both. The head is often made of the same material as the stem, although the head is in some cases made of other materials, such as polycrystalline alumina. The socket is usually made of metal. The liner between the socket and head, shown in Figure 14.1a, allows for ease of motion and lubrication. The liner is made of materials such as UHMWPE, polyoxymethylene homopolymer, polycrystalline alumina, and turboplastic graphite.

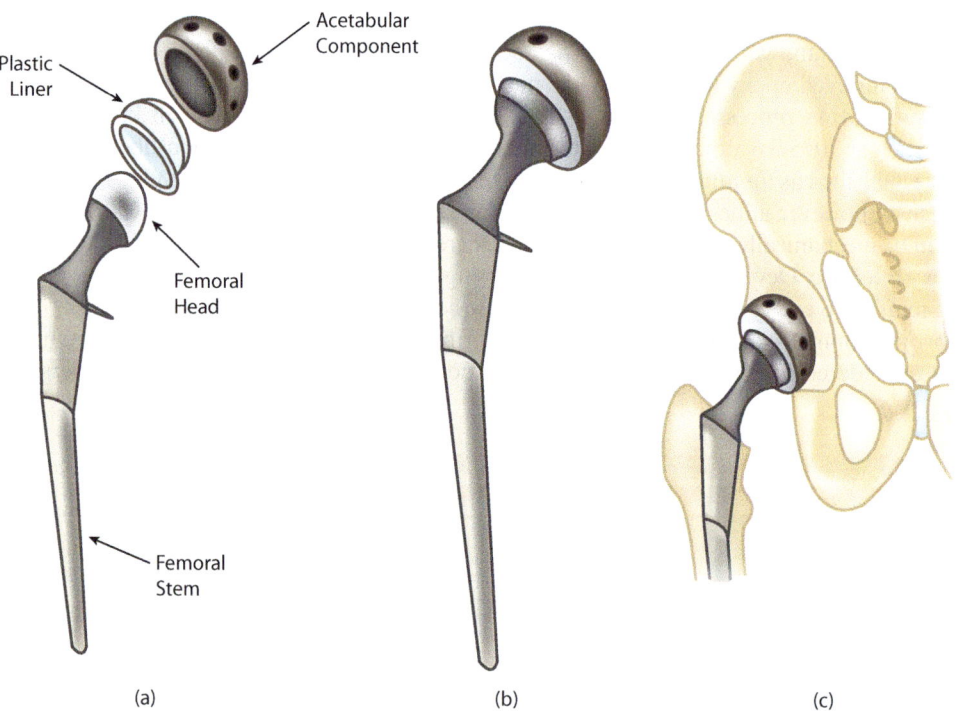

Figure 14.1   (a) The components of an artificial total hip replacement. (b) The assembled components of the hip replacement. (c) The artificial hip replacement joining the pelvis to the femur. (*Based on http://orthoinfo.aaos.org/topic. cfm?topic=a00377*)

The stem material has the following requirements:

- Biocompatible and nontoxic
- Resistant to degradation in the body environment
- Minimum yield strength of 500 MPa to prevent permanent deformation
- Minimum fatigue strength of 400 MPa for 10^7 cycles in the body environment
- Ultimate tensile strength of 650 MPa
- Ductility of at least 8% to prevent brittle fracture
- High specific strength to reduce the weight of the component

The required stem yield strength of 500 MPa rules out all polymers. The requirement of 8% ductility eliminates all ceramics from consideration. The remaining materials are metals and composites. Various stainless steels have been used as stem materials, and 316L is now the most widely used stainless steel for the hip stem. However, 316 L is subject to crevice corrosion in body fluids. Cobalt-based alloys, such as Co-27Cr-5Mo-0.3C and, more recently, Co-20Cr-35Ni-9.5Mo-0.01C, were developed as stem materials for joint replacement, because these alloys are more resistant to crevice corrosion than 316 L stainless steel is, and they also have excellent biocompatibility. The Co-27Cr-5Mo-0.3C alloy is normally used in the cast condition, and Co-20Cr-35Ni-9.5Mo-0.01C is normally used in the wrought condition.

Clinical tests have shown that titanium alloys are the most biocompatible and corrosion resistant of the metal alloys available for body implants. The most widely used titanium alloy for implants is Ti-6Al-4V extra-low-interstitial (ELI) grade. The surface treatment is important for high-strength Ti-6Al-4V alloys because of notch sensitivity. It has been found that implantation of the Ti-6Al-4V surface with nitrogen ions reduces wear-accelerated corrosion. The mechanical properties, densities, and costs of 316L, Ti-6Al-4V, and the Co-Cr-Mo alloy are shown in Table 14.14. It is not to be expected that the Ti-6Al-4V ELI grade is cheaper than the standard grade presented in Table 14.2; however, these prices were the quoted prices. Under normal circumstances, expect the ELI grade to be more expensive.

For a hip stem, the primary loading is bending where the head is attached to the stem, and in this location the stem is in the shape of a rod. We assume that this hip transplant is for a robust person in good physical condition; therefore, we can assume that weight of the stem is not critical. If this was a frail person weight could be critical. Table 14.15 gives the performance per unit cost for each material in Table 14.14. The 480 MPa yield strength of the annealed 316 L stainless steel listed is below the 500-MPa requirement. With a little less annealing, or with a little cold-work, this alloy would have a

| Table 14.14 | Mechanical Properties Including Yield Strength (σ_y) in MPa, Ultimate Tensile Strength (σ_u), Elongation at Fracture (ε_f), Elastic Modulus (E), Critical-Stress Intensity Factor (K_{Ic}), Density (ρ), and Cost (C_v) of Some Alloys for the Stem in Total Hip Replacement |

Alloy	σ_y MPa	σ_u MPa	ε_f%	E GPa	K_{Ic} MPa $\cdot$ m$^{\frac{1}{2}}$	ρ 10^3 kg/m^3	C_v 10^4\$/m^3
316L	480	860	37	196	260	8.0	15.4
Co-Cr-Mo	862	1089	8	235	126	8.3	33
Ti-6-4-ELI	790	860	15	114	100	4.43	40

The mechanical-property data is from various sources. The cost data is from various Internet sites for a 1-inch-diameter rod. This mechanical property and cost data should not be used for design or material selection.

Table 14.15	The Elastic Modulus (E) and Yield Strength (σ_y) in Units of 10^9 Pa, and Critical-Stress Intensity Factor (K_{Ic}) in Units of MPa · m$^{1/2}$ Each Divided by Cost per Unit Volume (C_V) in Units of $10^3$$/m^3, using Data from Table 14.14		
Material	$E^{1/2}/C_V$	$\sigma_y^{2/3}/C_V$	$K_{Ic}^{2/3}/C_V$
316 L	2.85	3.99	2.65
Co-Cr-Mo	1.47	2.75	0.76
Ti-6Al-4V ELI	0.95	2.14	0.54

higher yield strength and sufficient ductility and resistance to fracture; therefore this material is considered in the analysis. The critical-stress intensity factor in Table 14.14 is determined in air, but it should be determined in a solution simulating the body environment for this application. The relative performance per unit of cost for each material is presented in Table 14.16.

The relative performance for cost (*RPC*) of each material for three performance parameters is calculated with Equation 14.3 and shown in Table 14.17.

$$RPC = (RE^{1/2}/C_V + R\sigma_y^{2/3}/C_V + RK_{Ic}^{2/3}/C_V)/3 \qquad \textbf{14.3}$$

The 316 L stainless steel clearly gives the highest performance per unit cost of these three materials for these parameters. However, the engineer must also consider other factors that are not included in this analysis. For example, crevice corrosion is not considered. If the implant is to go into a person where it is expected that the implant will not be in use for an extended time, crevice corrosion may not be a problem. However, if the implant is expected to be in place for many years, it may be preferable to choose a material not subject to crevice corrosion.

In this analysis, a high elastic modulus makes a positive contribution to performance, because it results in less bending of the stem under load. However, failures of the bonding of the stem to the bone have been attributed to the large difference in elastic modulus of the stem relative to bone. Bone has an elastic modulus of 17.2 GPa, which is less than one-tenth the value for the potential stem metals. If the bonding of the bone to the stem is not perfect, the strains in the bone and stem are different, and shear strain develops at the bone-stem interface. Stems with a lower elastic modulus may fail less often if this is the cause of failure. Factors such as stem-bone elastic-modulus mismatch must be be considered if it is confirmed that a stem material with a high elastic modulus leads to failure, and the contribution of elastic modulus to performance would have to be reanalyzed.

Polymer-matrix composite materials with reinforcement by filaments, such as Kevlar, are being studied for use as implant materials. With composite materials, it should be possible to develop a stem with mechanical properties more closely matched to those of bone. If the elastic modulus of the bone and the stem are the same, then the strains at the bone-stem interface are matched when subject to the same stress. A composite stem that has a core made of the Co-Cr-Mo alloy surrounded with the polymer PEEK has

Table 14.16	Relative Elastic Modulus, Yield Strength, and Critical Stress-Intensity Terms Using Data from Table 14.15		
Material	$RE^{1/2}/C_V$	$R\sigma_y^{2/3}/C_V$	$RK_{Ic}^{2/3}/C_V$
316 L	1.0	1.0	1.0
Co-Cr-Mo	0.51	0.69	0.29
Ti-6Al-4V ELI	0.33	0.54	0.20

| Table 14.17 | The Relative Performance per Unit of Cost of Each Material in Table 14.16 |

Material	RPC
316 L	1.0
Co-Cr-Mo	0.50
Ti-6Al-4V ELI	0.36

been developed. One of the purposes of this stem is to reduce the elastic modulus of the stem so that it is more compatible with the bone material in the femur.

The selection of materials for implant in the body is based upon criteria that we discuss in this book, such as stiffness, yield strength, fracture, fatigue, and corrosion. Materials that are available for implantation are similar to materials discussed in this book. The reaction of these materials with the body environment, and the body's response to these materials must also be considered. These subjects are treated in a course in biomaterials.

14.4 MATERIALS SELECTION FOR SUSTAINABILITY

Sustainability is the capacity to endure. The sustainable use of materials is the use of materials such that the supply of the material endures. Plants are a sustainable source of material, because the plant can be grown again. The use of ore from the earth to make a material is not a sustainable source of material, because once the ore is processed, it cannot be replaced. The **recycling** of materials improves the sustainability of material sources. Recycling is the processing of an existing product into a material that can be reused to make a new product. In some cases, such as aluminum, the recycled material is the same as the original material. However, in the case of plastics, the recycled material may be different than the original material, as discussed in Section 14.4.3. A completely sustainable system of material use is an idealization that probably will never be achieved; however, there are forces to increase the degree of sustainable use of materials. One of the most important factors is cost. If the recycling of materials is cost effective, companies will do it voluntarily. In the year 2011 in North America, where there are no requirements for the use of recycled material, approximately 66×10^9 kg of iron were recycled, as were 45×10^9 kg of paper, 3.6×10^9 kg of aluminum, 2.7×10^9 kg of glass, and 1.8×10^9 kg of plastics.

Recycling of materials has other advantages. For example, according to the Aluminum Association, recycling of aluminum results in a 95% savings in energy in comparison to producing aluminum from bauxite ore, and recycling reduces air pollution. If materials are not recycled, the natural sources of materials will deplete more rapidly, and the quality of the sources will be reduced. For example, high-grade ore is mined first, and lower-grade ore is left for the future. Lower-grade ore is more expensive to mine and convert into a material. In addition, the amount of material in landfills is reduced by recycling material.

If the costs of materials are reduced by recycling, there will be a greater market for the materials. In addition, some customers find value in a product that is made from recycled material or that is produced in a manner that reduces pollution compared with how similar products are made. Also, some governments are now requiring that materials be recycled, in an effort to have a more sustainable material supply. The European Union has enacted a directive that by 2015 automobiles must be made of 95% recyclable materials, of which 85% can be recovered through mechanical recycling or reuse and 10% recovered through energy recovery or thermal recycling.

14.4.1   Sustainable Use of Iron

Iron is the most widely recycled material by weight, for several reasons. Ferrous (BCC) iron is ferromagnetic, and large magnets can readily separate it from other materials. The steel in automobiles is nearly 100% recycled. The basic oxygen furnace (BOF), discussed in Section 1.1, that smelts iron ore into iron uses approximately 30% scrap iron. The iron from a BOF is relatively pure and can be used for any purpose that requires iron. The electric arc furnace (EAF) uses 100% scrap iron to produce recycled iron. The iron from an EAF contains sufficient impurities from the recycled source that the iron is not suitable for cold-working into a thin sheet for use in appliances and automobile panels. EAF steel is used for steel-reinforcing bars, I-beams, thick plate, and other forms of steel that do not require extensive cold-work. Iron can be recycled repeatedly, but the lowest-impurity iron comes from the smelting of iron ore in a BOF. To produce recycled steel in an EAF requires approximately 60% of the energy of steel made in a BOF, and there is a similar reduction in air pollution.

14.4.2   Sustainable Use of Nonferrous Metals

Approximately 33% of aluminum, 40% of copper, 35% of lead, 30% of zinc, and 60% of austenitic stainless steel produced in the United States is from recycled material. It is estimated that 75% of all aluminum produced since 1888 is still in use. From 80 to 90% of all austenitic stainless steel is recycled, and metals can be recycled repeatedly. However, impurities are usually present in recycled metal, and as a result high-purity metals are usually made from ore. Much of the supply of metals is easily identified for recycling. Small fishing boats, canoes, and aircraft skin are made of aluminum. Wire is made of copper or aluminum. Kitchen sinks are made of stainless steel. Also, factories produce waste of known composition as a result of machining metals.

Metals that are part of municipal waste must be separated. Municipal waste is first shredded. Ferromagnetic metals are separated first with permanent magnets. Nonmagnetic metals are detected and separated from the remaining mixed waste with eddy currents. We discussed eddy-current systems for nondestructive testing in Section 11.2.9. The nonmagnetic mixed waste is spread on a conveyor belt. A strong alternating magnetic field induces eddy currents and magnetic forces in the nonmagnetic metals that pull them from the conveyor belt into storage bins. Nonmetals are unaffected by the alternating magnetic field and continue on the conveyor belt. The nonmagnetic metals in municipal waste are separated from each other and from other waste by density, using liquid flotation and high-pressure air. The separated metals are pressed into blocks and melted to form new metal. The molten metal is chemically analyzed, and alloy additions are made to produce the desired metal alloy.

14.4.3   Sustainable Use of Plastics

The sustainable use of plastics has a number of obstacles. Most plastics are made from petroleum, which is a not a renewable resource. It is estimated that only 24% of plastic bottles are recycled. It is very difficult to separate different polymers from each other in waste, such as HDPE from PVC and epoxy. If scrap thermoplastic polymers are mixed and melted, most polymers that are solid at room temperature do not mix with other polymers, even when melted, because of the very small entropy of mixing, as we discussed in Section 4.8.2. Melting different thermoplastic polymers followed by solidification results in a solid that has volumes of different materials separated by weak phase boundaries.

Because most polymers do not mix in the liquid state, it is necessary to sort different thermoplastic polymers to produce a product that has uniform properties through the volume. Polymer products are hand-sorted according to the triangular, three-arrow recycle symbol with a number in the center, and the number indicates the polymer type. The numbers and polymer types are 01-polyethylene terephthalate (PET), 02-HDPE, 03-PVC, 04-LDPE, 05-PP, 06-PS, 07-other (PC, PTFE, and so on) 09-ABS. Plastics are sorted, cleaned, cut into flakes, and melted to form plastic pellets. The recycled plastic is often not used in the same manner as the original polymer. PET is the material in clear beverage containers, and much the recycled PET is used to produce polyester fiber. Many polymers including HDPE, PVC, PP, PS, and ABS are mixed with 30 to 70% wood flour to produce pellets of plastic wood. The plastic wood pellets are produced into lumber shapes primarily by extrusion, as we discussed in Section 13.4.1.

Polymers can be recycled with **pyrolysis**. In pyrolysis, polymers are heated in an inert atmosphere to temperatures such as 500°C, where the LCMs are broken into shorter molecules. Pyrolysis produces gas and liquid that can be used as fuel and a solid residue. In some cases, the molecules resulting from pyrolysis can be used to make a new polymer.

An alternative to recycling a polymer is to produce a biodegradable polymer that is disposed of in a landfill, and to replace the polymer with a new polymer from a sustainable source. **Biodegradation** is the breakdown of a polymer into simple molecules by bacteria. Most polymers, such as PE and PP, are not biodegradable, and they are primarily produced from petroleum, of which we have a limited supply. PE can be made from the fermentation of biomass to produce ethanol, which is converted into ethylene and then into PE. Several biodegradable thermoplastic polymers have recently been developed from sustainable supplies. Polyhydroxybutyrate (PHB), also known as biopol, is in the polyester family, and it is produced by plants and microorganisms. A company is producing PHB with the bacteria in municipal waste water. Among other applications, PHB is made into internal medical sutures, because PHB biodegrades inside the body, and PHB is nontoxic and biocompatible. PHB has a tensile strength of 40 MPa, which is comparable to that of PP. Polylactic acid (PLA) is a recently developed biodegradable polymer produced by the bacterial fermentation of biomass from corn, cane sugar, and sugar beets. PLA is a semicrystalline polymer, with a melting temperature of 175°C, a density of 1.3 g/cm^3, a tensile strength of 10 to 60 MPa, and an elastic modulus from 0.35 to 16 GPa. The range of properties in PLA is a result of different molecular masses and crystallinities, and PLA can be drawn into a fiber. PLA can be thermally depolymerized through pyrolysis to form highly purified lactic acid, which is a raw material for making new PLA. PLA has been used to make plastic drinking cups, fabric for woven shirts, microwave trays, medical sutures, drug delivery devices, agricultural mulch cover film, trash bags, food packaging, and disposable tableware. Harvestform™ is a PUR-type polymer produced from soy oil, a sustainable source. Harvestform has preservatives to stop it from biodegrading, but without the preservatives it would be biodegradable. As we will discuss in Section 14.4.5, Harvestform is used as a matrix in composite materials.

Thermoset polymers and are not easily recycled, because they cannot be melted. The primary method to recycle thermoset polymers is to grind the polymer and any reinforcement materials into particles that are used as filler in plastics, concrete, and asphalt. Rubber in tires is an extensively utilized thermoset polymer that presents problems for sustainable use. Ground-up tire rubber, called crumb, is added to asphalt, concrete, and to surface material in basketball and tennis courts. Tires and other thermoset polymers are shredded and pyrolyzed to produce gas and a liquid that can be used as fuel.

14.4.4   Sustainable Use of Ceramics and Glass

The original source of ceramics is from ores such as silica, zirconia, titania, and so on. Glass is the primary ceramic material that is recycled and it comes from silica with the addition of other ceramics.

Glass can be continuously recycled if it is adequately separated. The primary source of recycled glass is containers, and approximately 80% of glass containers are made into new containers. Other uses of recycled glass include fiberglass insulation, glass countertops, decorative bricks and tiles, filler in asphalt and concrete, abrasives, and water-filtration media.

Concrete for recycling is crushed and used in a manner similar to how gravel and rocks are used. Crushed concrete is also used as aggregate for new concrete. Large pieces of concrete are used as riprap to protect shorelines against wave action from water.

There is very little recycling of other ceramics except when manufacturers of ceramics recycle their own waste products and return them to the process cycle.

14.4.5   Sustainable Use of Composite Materials

The primary reuse of composite materials with a thermoset polymer matrix is to grind the composite and use it as filler in other plastics, asphalt, or concrete. Composite materials with a thermoset polymer matrix can be pyrolyzed, as discussed above. Polymer fibers are pyrolyzed, but glass and carbon fibers are part of the solid residue.

Composite materials with a thermoplastic matrix and reinforcing fibers are also difficult to recycle, because it is difficult to separate the thermoplastic matrix from the reinforcing fibers. One way to make a composite recyclable by melting is to use highly oriented thermoplastic-polymer reinforcing fibers in a matrix of the same polymer. A composite material made of PP fibers in a PP matrix is being investigated as a composite for applications in the automotive industry. The PP fibers have a density of 0.9 g/cm³, a tensile strength of 650 MPa, and an elastic modulus of 18 GPa.

An alternative to recycling is to use composite materials made from a sustainable source, such as plants or microorganisms, that are biodegradable and can be disposed of in a landfill. According to the Henry Ford Museum, Ford Motor Company produced an automobile in 1941 that had composite body panels made of a matrix of phenolic (phenol formaldehyde) reinforced with soybean fiber. Figure 14.2 is a photograph of Henry Ford attacking the automobile with an axe to demonstrate its durability. It is reported that the axe had a rubber cover over the blade, but it is still a severe test. Henry Ford wanted to combine the fruits of industry with agriculture. Also, there was a shortage of iron at the time, because iron was needed to manufacture military equipment for World War II. The entry of the United States into World War II in 1941 suspended all automobile production, including the experimental program on the soybean car, and Henry Ford died two years later.

In 1994 Mercedes Benz introduced plastic interior door panels reinforced with jute in its E-Class automobiles. Jute is a natural biodegradable fiber, with a density of 1.3 g/cm³, a tensile strength of 400 to 700 MPa, and an elastic modulus of 26.5 GPa. Mercedes Benz A-Class automobiles now use flax-reinforced PP underbody structural components. Flax is a natural biodegradable fiber, with a density of 1.2 g/cm³, a tensile strength of 2000 MPa, and an elastic modulus of 85 GPa. John Deere and Co. has replaced steel panels on its hay balers with soy-based PUR Harvestform reinforced with glass fibers. A composite material with a matrix and reinforcement made of biodegradable polymers from sustainable sources is a natural progression from the existing composite materials.

Reinforced carbon-carbon (RCC) composites are a potential sustainable-use material. The matrix of RCC composites is made from carbonized furan. The thermosetting polymer furan is obtained from biomass; a major source is oat hulls. The graphite fibers in RCC composites are made by carbonizing polymer filaments. If a sustainable source of polymers filaments is developed, RCC composites would be a sustainable-use material system. RCC composites presently have applications as heat-shield materials for the space reentry vehicles and rockets, brake and clutch liners, rocket motor parts, jet engine components, and biomedical implants.

Figure 14.2   As a demonstration of product durability, Henry Ford used an axe to strike his experimental automobile made with phenolic resin reinforced with soybean fiber. (*Ford Motor Company. Engineering Photographic Department, Image P.188.28273. From the collections of The Henry Ford.*)

14.5   MATERIAL DATABASES

14.5.1   Graphical Material Databases

When beginning the materials selection process for a design, a good set of databases to consult are the graphical presentations of Ashby. Ashby has developed plots of mechanical properties as a function of relative cost per cubic meter for classes of materials, as shown in Figures 14.3 and 14.4 for elastic modulus and tensile strength. Relative cost in these figures is defined as the ratio of the cost of the material per unit weight, divided by the cost of plain-carbon steel rod per unit weight. The use of relative cost is an attempt to take price inflation out of the data; however, the relative cost of different materials can change with time. The dimensionless relative cost is multiplied by the density in 10^6 grams per cubic meter (Mg/m^3) to give relative cost per cubic meter. This data can be used as preliminary cost and mechanical property data. Individual materials, such as 4340 steel, are not presented; only classes of materials are presented. In Figures 14.3 through 14.7, GFRP stands for glass-fiber-reinforced plastic, KFRP stands for Kevlar-fiber reinforced plastic, and CFRP stands for carbon-fiber-reinforced plastic.

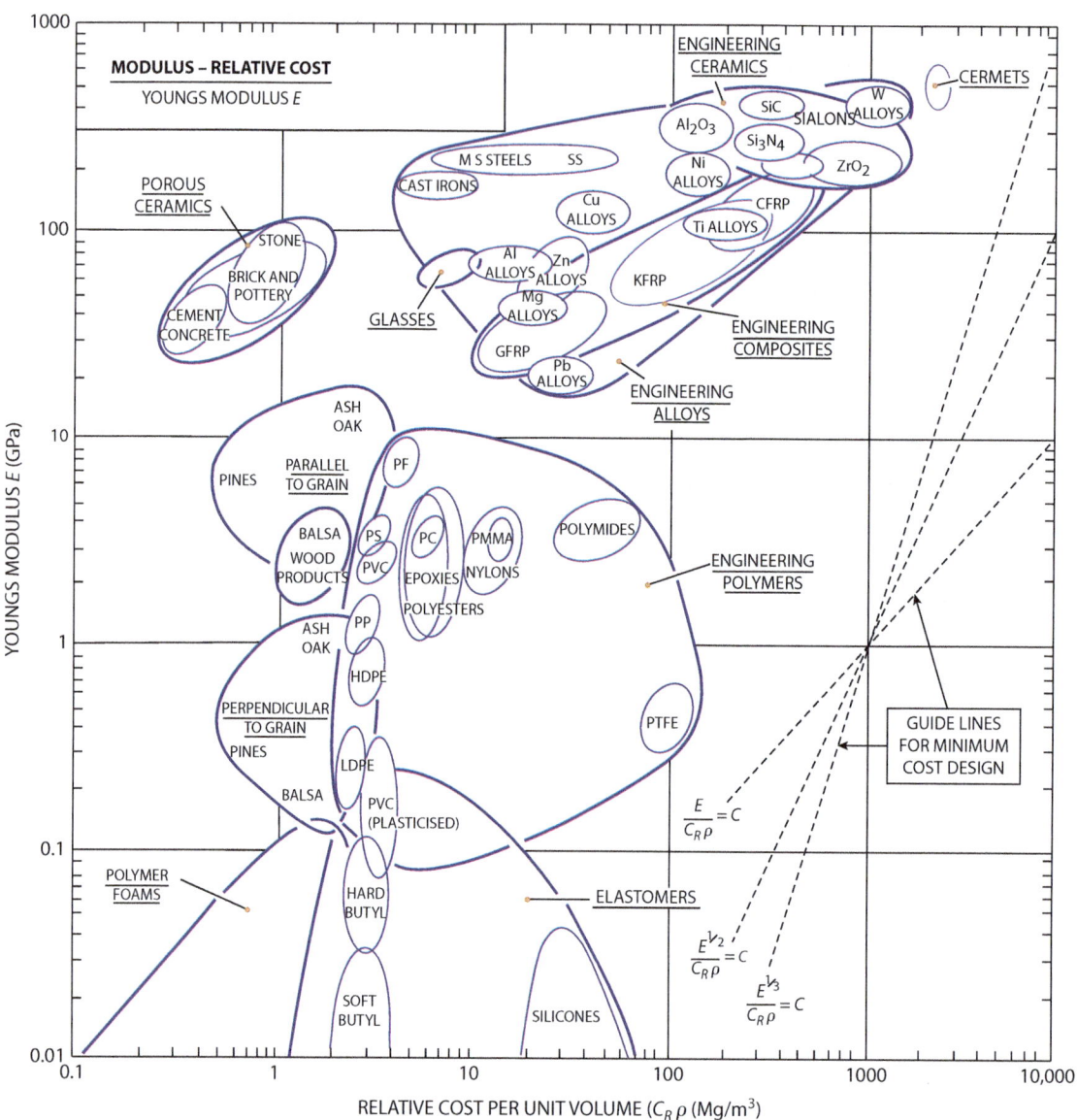

Figure 14.3 A log-log plot of Young's modulus (E) in GPa, as a function of cost per unit weight relative to that of plain-carbon steel (C_R) multiplied by the density (ρ) in Mg/m³, for a variety of material classes. (*Based on Ashby, M.F., Materials Selection in Mechanical Design, Pergamon Press, Terrytown, N.Y. (1992), p. 50.*)

The dashed lines on the lower right of Figures 14.3 and 14.4 have a slope equal to the power of the mechanical properties given in Table 14.7, because both the x and y scales are logarithmic. To find the materials with the highest performance per unit cost per unit volume, slide the dashed line with the mechanical property to the appropriate power, from right to left, at constant slope. For example, to find the highest elastic modulus per unit of cost for a rod in tension or compression, slide the dashed line corresponding to $E/C_R\rho$ to the left, keeping the slope constant. Materials along this dotted line at any location have a slope of 1. If the dashed line $E/C_R\rho$ is moved to the far left in Figure 14.3 with a slope of 1, it is found that concrete, brick, stone, and so on have the highest elastic modulus per unit of cost per unit volume.

Ashby has also developed graphical presentations of elastic modulus, yield strength, and critical-stress intensity factor as a function of density for classes of materials, as shown in Figures 14.5 through 14.7. This data is used to provide preliminary input for weight-critical designs. As an example, in Figure 14.5 the elastic modulus is plotted in GPa (10^9 Pa) as a function of density in Mg/m^3 ($10^6 g/m^3$) on logarithmic scales. In the example in Section 14.3, the requirements are for an elastic modulus (E) of at least 100×10^9 Pa and a density (ρ) of not more than $8 \times 10^3 kg/m^3$. Drawing a horizontal line at 100 GPa and a vertical line at the density (ρ) of 8 Mg/m^3, the possible candidate materials based upon this property are above the 100 GPa line and to the left of the 8Mg/m^3 line. The possible materials based only upon elastic modulus include uniaxial carbon-fiber-reinforced polymers (CFRP uni-ply), glasses and pottery,

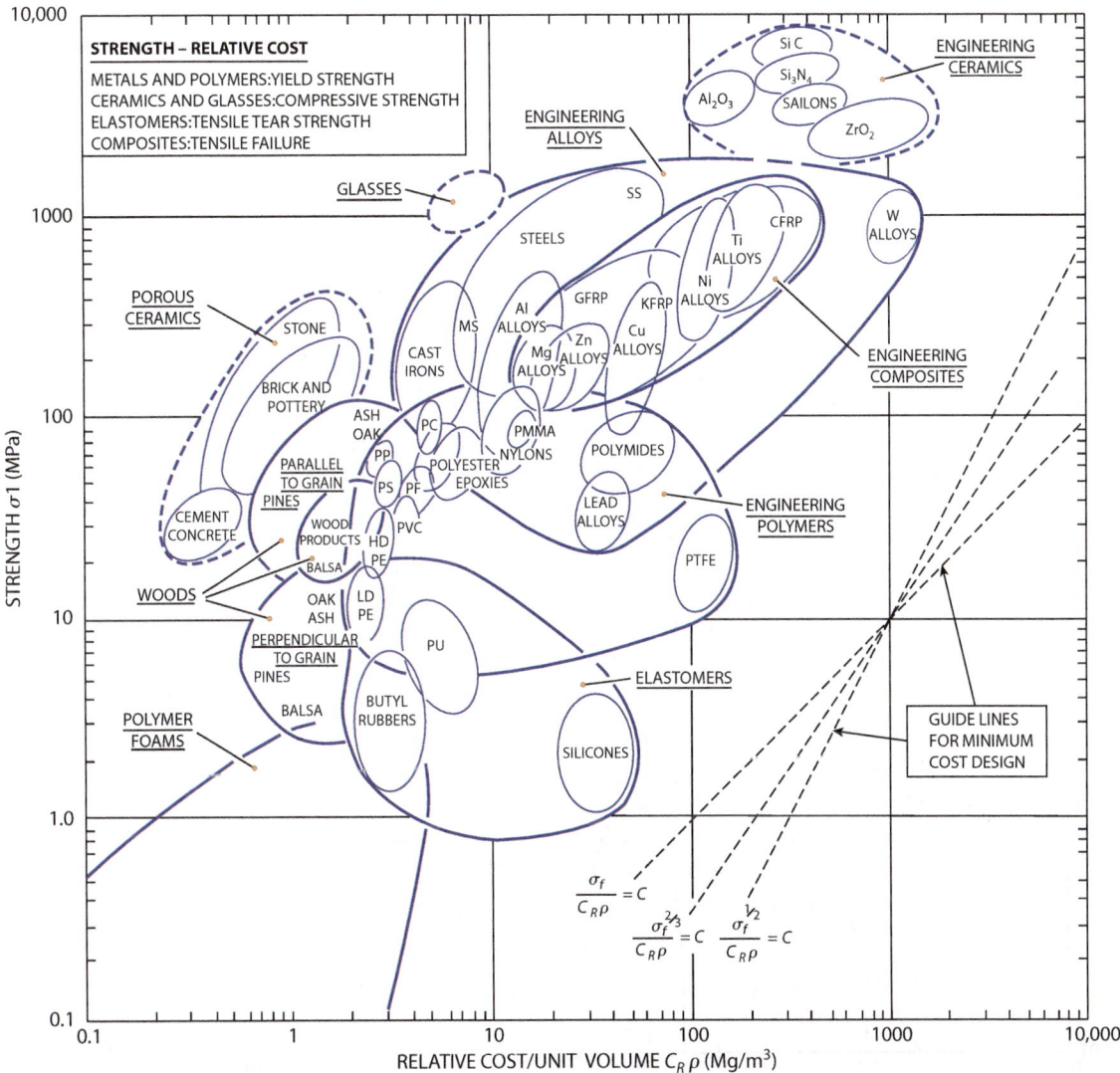

Figure 14.4 A log-log plot of strength in MPa (yield strength for metals and polymers, compressive strength for ceramics, tear strength for elastomers, and tensile strength for composites), as a function of cost per unit weight relative to that of plain-carbon steel (C_R) multiplied by the density (ρ) in Mg/m^3 for a variety of material classes. (*Based on Ashby, M.F., Materials Selection in Mechanical Design, Pergamon Press, Terrytown, N.Y. (1992), p. 51.*)

most ceramics, Ti alloys, steels, and engineering ceramics. Glasses, pottery, and engineering ceramics are eliminated by the requirement of a critical-stress intensity factor of at least 40 MPa · m$^{1/2}$. These graphs allow the engineer to focus quickly on the materials that may meet the elastic modulus and density design requirements; however, final mechanical properties and cost analysis should be based upon actual material properties and quotes of the possible materials. The dashed guidelines in Figures 14.5 to 14.7 are for minimum-weight designs. To find the materials with the highest specific mechanical property, slide the dashed line with the appropriate slope from right to left, at a constant slope. Note that the logarithmic

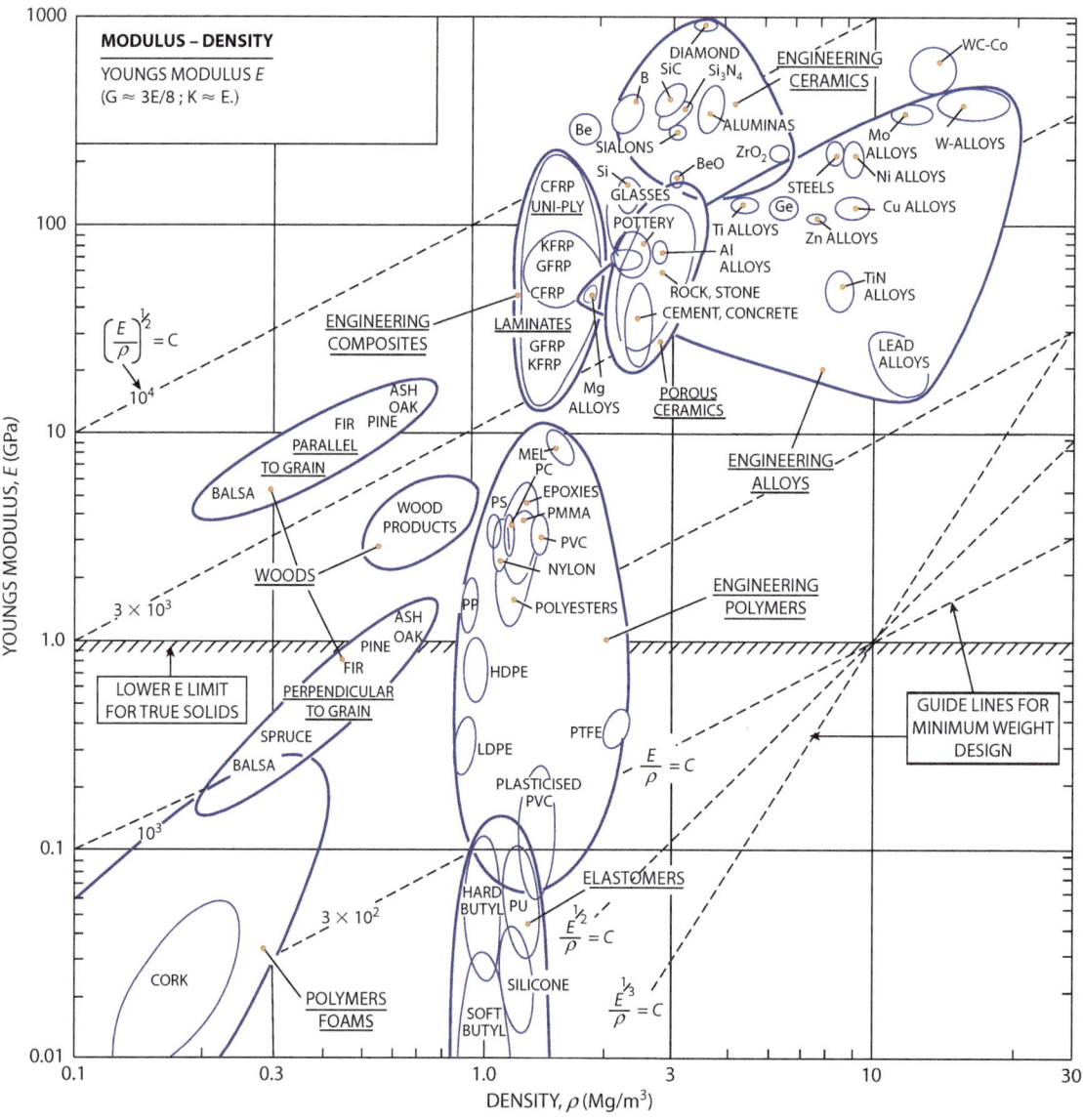

Figure 14.5 A log-log plot of Young's modulus (E) in GPa, as a function of density in Mg/m^3 for a variety of material classes. (*Based on Ashby, M.F., Materials Selection in Mechanical Design, Pergamon Press, Terrytown, N.Y. (1992), p. 28.*)

scales in Figures 14.5 through 14.7 are different for *x* than for *y*. For this reason, the slope of 1 is not at 45°. To get the correct slope, you must use the numbers along the scale. For example, to find the highest specific elastic modulus for a short column in compression, slide the dashed line corresponding to E/ρ to the left, keeping the slope constant. Materials along this dashed line have a constant value of E/ρ or specific elastic modulus, and the material to the extreme left along the dashed line has the highest specific elastic modulus. Diamond has the highest specific elastic modulus of any material.

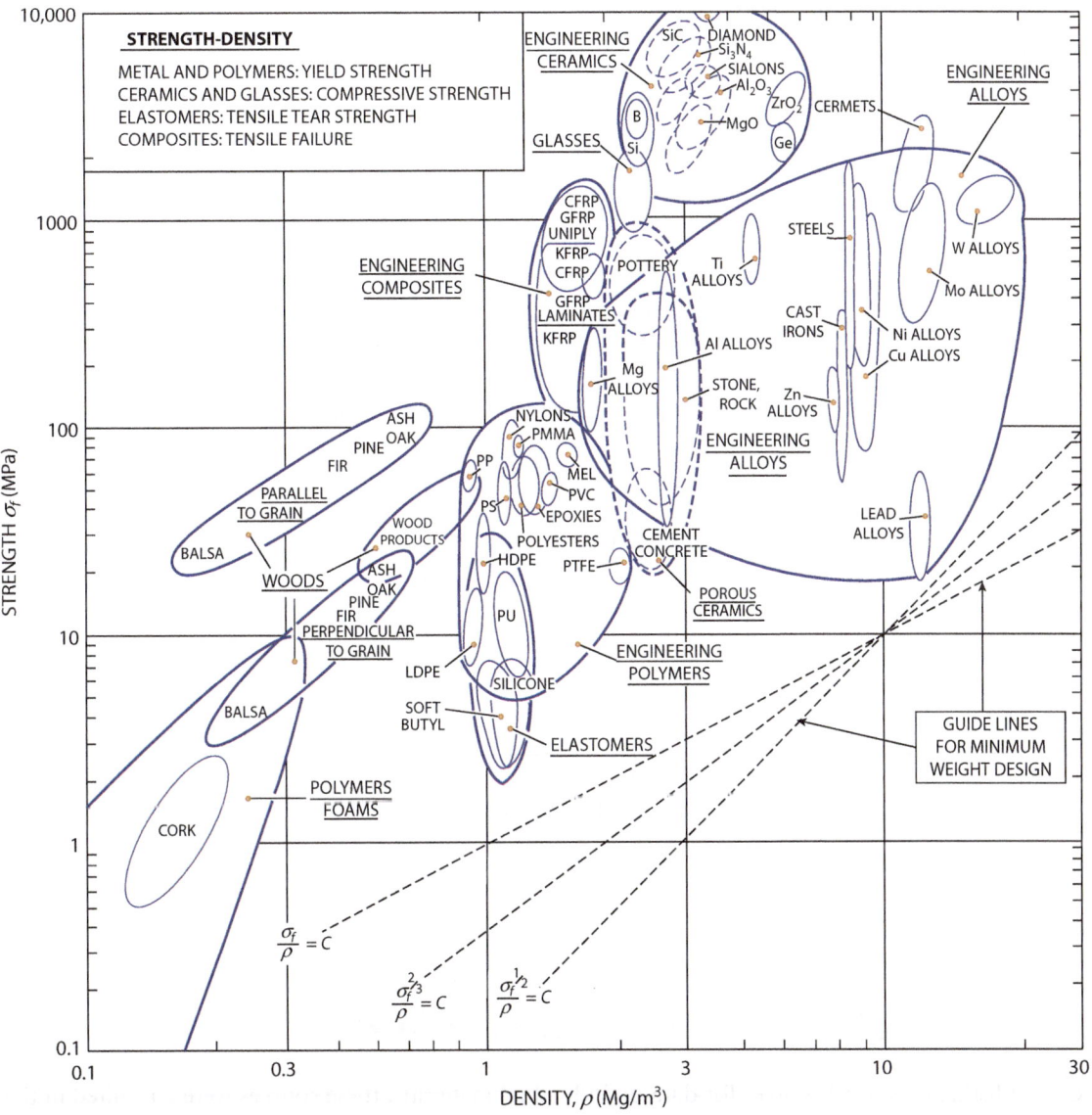

Figure 14.6 A log-log plot of strength (yield strength for metals and polymers, compressive strength for ceramics, tear strength for elastomers, and tensile strength for composites) in MPa, as a function of density in Mg/m³ for a variety of material classes. (*Based on Ashby, M.F., Materials Selection in Mechanical Design, Pergamon Press, Terrytown, N.Y. (1992), p. 30.*)

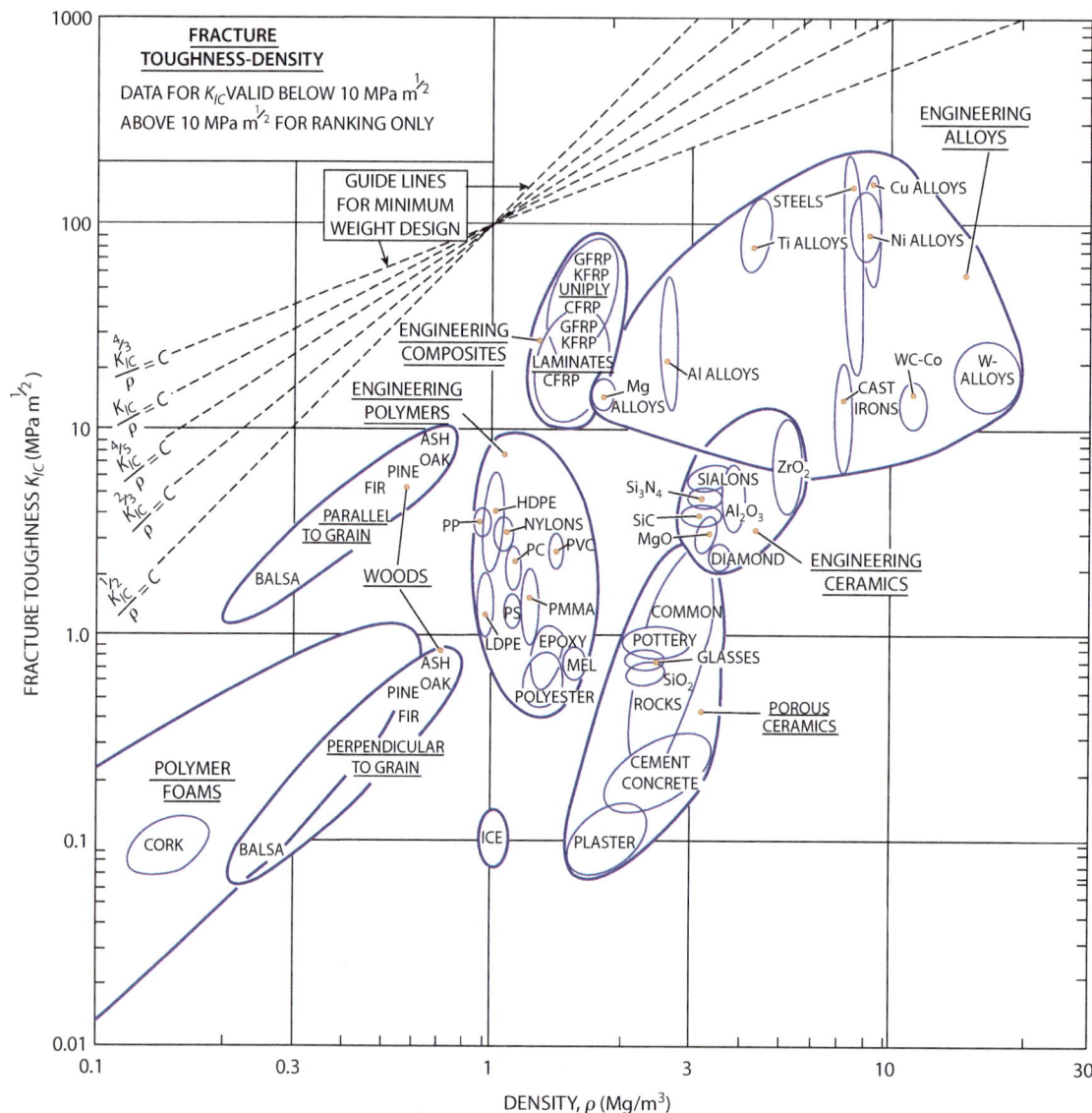

Figure 14.7   A log-log plot of fracture toughness (K_{Ic}) in MPa · m$^{1/2}$ as a function of density in Mg/m^3 for a variety of material classes. (*Based on Ashby, M.F., Materials Selection in Mechanical Design, Pergamon Press, Terrytown, N.Y. (1992), p. 32.*)

14.5.2 Printed Material Databases

Below is a listing of several sources for data on individual materials; these sources were consulted in the preparation of this book.

- **Metals Handbook**

  ASM International

  Materials Park, Ohio 44073, and the Web site

http://www.asm-intl.org

The *Metals Handbook* is available in many university libraries, and the online version is available through many university libraries.

■ **Handbook of Structural Ceramics**

Schwartz, M.,
McGraw-Hill, Inc., New York (1992)

■ **Smithell's Metals Reference Book**

eds. Gale, W. F., and Totemeier, T. C., Elsevier Butterworth-Heinemann, Oxford (2004). Other editions may be available in your library.

■ **Physical Properties of Polymers Handbook**

Mark, J. E., ed.,
AIP Press, Woodbury, NY (1996)

14.5.3 Internet Material Databases

■ www.matweb.com

The more than 69,000 materials in this free database include ceramics, composites, fibers, metals, natural materials, semiconductors, thermoset and thermoplastic polymers, and wood. Data includes density, hardness, tensile strength, elongation, reduction in area, elastic modulus, Poisson's ratio, bulk modulus, tensile modulus, shear modulus, electrical resistivity, linear coefficient of thermal expansion, heat capacity, thermal conductivity, melting point, chemistry, and water absorption of polymers.

■ www.granta.co.uk

This database charges a fee for use. It has four levels of use: Educational levels 1, 2, and 3; and industrial level. The database is the graphical approach of Ashby and covers most materials.

■ www.aluminum.org

This is the Web site of the Aluminum Association. It has basic information about aluminum, alloy chemical compositions, industry standards, design information, recycling information, and many more subjects.

■ www.copper.org

This Web site provides information about copper and copper alloys. The mechanical-property data includes tensile strength, yield strength, percent elongation at fracture, hardness, shear yield strength, fatigue strength, and impact strength. The Web site has information about copper and copper alloys, including microstructure, processing, physical properties, corrosion, and applications.

■ www.steel.org

This is the Web site of the American Iron and Steel Institute. It has basic information about steel, alloy chemical compositions, industry standards, design, recycling, and many more subjects.

■ www.engineershandbook.com

This Web site has a variety of information available about design, including tables with mechanical-property data for steels, cast irons, aluminum alloys, magnesium alloys, titanium alloys, and copper alloys. The mechanical-property data includes yield strength, tensile strength, percent elongation to fracture, reduction in area, and hardness. The Web site provides information about the characteristics of polymers, their coefficients of thermal expansion, and applications of polymers; however, mechanical-property data is not provided. The Web site also provides information about other reference tables, manufacturing methods, rapid prototyping, engineering software, and mechanical components.

■ www.mcmaster.com

This site has prices for a variety of mechanical items, including but not limited to fasteners, tubing, fittings, power transmission items, abrasives and polishing supplies, pipe, sawing tools, cutting tools, and raw materials.

■ http://stores.ebay.com/online-metal-supply/

This Web site sells small quantities of metals and polymers, including aluminum, brass, bronze, copper, cobalt, magnesium, plastics, phenolics, steel, stainless steel, and titanium. The materials are either listed with purchase prices, or they are being auctioned and then sold to the highest bidder. Configurations of these materials include rods, bars, and extruded shapes.

■ www.ptonline.com

This Web site is operated by Plastics Technology and covers polymers, elastomers, phenolics, and epoxies. The Web site has a materials database that includes density, yield strength, ultimate tensile strength, impact energy, elongation at fracture, coefficient of thermal expansion, water absorption, and mold shrinkage. The Web site has a database of machineries used to process polymers. It also has an extensive list of polymer prices; however, these prices are for large quantities of bulk material, such as truckloads or 10,000 kg. These are not the prices that would be paid for finished material. The Web site allows limited free access to data and full use of the site for an annual fee.

■ www.professionalplastics.com

This Web site has an extensive listing of polymers for sale with prices listed for polymers, such as ABS, acetal, acrylic, PVC, nylon, PEEK, phenolics, G-10/F-4 glass-epoxy composites, and Macor® machinable ceramic. Some of the forms are rod, sheet, tubing, and films. The Web site also provides some physical properties for polymers, including tensile strength, compressive strength, flexural strength, modulus of elasticity, shear strength, impact energy, hardness, density, water absorption, dielectric strength, dissipation factor, dielectric constant, resistance, and flame resistance.

■ www.totalplastics.com

This Web site has plastic product forms of sheet, rod, tube, film, and tape. Plastics are listed for various functions or properties, including abrasion resistance, bullet penetration, chemical resistance, corrosion resistance, flame resistance, high temperature, high strength, medical grade, FDA approved materials, military specifications, impact resistance, weathering, clarity, and wear resistance. Actual physical properties are not listed. A recent check of this Web site indicates that prices are now only by quote.

■ www.materialdatacenter.com

This Web site has information about plastic and bioplastics. By entering a material name, such as HDPE, different manufacturers of HDPE are listed, along with their products and the company description of each product's physical properties. Properties include density, flow temperature, yield strength, flexural strength, flexural modulus, elongation to fracture, and impact strength. There is an applications database where the polymer used for various applications are listed. Also included are a biomaterials data center and literature about biomaterials. This is a new Web site, and some of the databases are under construction.

Summary

- Material selection should be part of the design process. It is the engineer's responsibility to know what physical properties of the material are important for the design. The engineer selecting the material must know how the material is to be manufactured into a part and if the manufacturing process changes the mechanical properties of the material.

- The initial cost of a product to a buyer includes the cost related to producing this particular product, fixed costs that are outside of the control of the engineer selecting the material, and the profit for the manufacturer.

- Material cost can be obtained by a quote from distributors, and some retailers list their costs for materials on the Internet. In designs where weight is not critical, the material cost per unit volume is used in cost analysis, and in weight-critical designs, material cost per unit weight is used in the analysis.

- The engineer should consider the cost of ownership of a product, which includes the purchase, operation, maintenance, repairs, inspection, and insurance costs. The cost of ownership affects the value of the product, and value can affect the selling price of the product.

- The performance per unit cost of a mechanical property for a particular type of loading is the physical property to a power appropriate to that type of loading, divided by the cost per unit volume. The relative performance for a mechanical property per unit cost (RPC) is obtained by dividing the performance per unit cost for each material by the largest value in that mechanical property category to produce a dimensionless relative RPC for each material. The total relative RPC for a material is obtained by summing the RPCs for all of the mechanical properties and dividing by the number of RPCs.

- For designs that are weight critical, relative specific mechanical properties are summed along with 1 minus the relative cost per unit weight, to determine the relative merit of a material.

- The sustainable use of materials is the use of materials such that the supply of the material endures. Plants are a sustainable source, and ore is not a sustainable source of material.

- The recycling of materials improves the sustainability of material sources. The primary materials that are recycled are metals, plastics, paper, glass, and concrete. Metals and glass can be continuously recycled. For continuous recycling, different metals and different colors and types of glass must be separated. Recycled polymers are usually used for products such as plastic wood. Polymers must be separated for recycling, because different liquid polymers generally do not mix. Recycled concrete is used in applications similar to those of gravel and rocks. Recycling of materials reduces material cost, energy use, air pollution, and the size of landfills.

- A sustainable cycle for polymers is to produce a biodegradable polymer that is disposed of in a landfill, and to replace the polymer with a new polymer from a sustainable source. A few biodegradable polymers from sustainable sources have been developed.

- It is difficult to recycle composites made of different matrix and reinforcement materials. Thermoplastic composites made with the same material for the matrix and reinforcing fibers can be recycled. PP is one such candidate composite material for recycling. Composites made with a matrix of biodegradable polymer from a sustainable source reinforced with fibers of the same material, or reinforced with natural fibers, are sustainable composite materials.

- Graphical plots of mechanical properties versus density, and mechanical properties versus cost, are available for classes of materials for initial screening of suitable materials for a design. Mechanical properties of specific materials are available in handbooks and on Internet sites. Cost data for specific materials is available on a few Internet sites; however, the best cost data comes from quotes provided by material distributors.

Supplemental Reading Subjects and Authors

Full references are listed at the end of the book.

General:	*Askeland, Fulay, and Wright*
Materials selection:	*Ashby; Charles, Crane, and Furness; Lewis; Waterman; Waterman and Ashby*
Hip stem implants:	*Akhavan, Mathiesen, and Schulte; Glassman, Crownshield, Schenck, and Herberts*

Homework

Concept Questions

1. The _____ of the product is the extent to which the product meets the performance criteria.

2. _____ is the process of joining together individual material parts into the final product.

3. In designs, where weight is not critical, the material cost is in units of cost per unit _____.

4. In weight-critical designs, the stiffness requirement for the relative merit uses the relative _____ elastic modulus.

5. For human body implants, one of the problems with 316 L stainless steel is _____ corrosion.

6. _____ is the capacity to endure.

7. _____ is the processing of an existing product into a material that can be used to make a new product.

8. The electric arc furnace uses _____ percent scrap steel.

9. Nonmagnetic metals in municipal waste are separated with _____ _____.

10. Most polymers that are solid at room temperature do not mix with other polymers even when melted, because of the very high _____ of mixing.

11. _____ is the breakdown of a polymer into simple molecules by bacteria.

12. In _____, polymers are heated in the absence of oxygen to temperatures where the LCMs are broken into shorter molecules.

Engineer in Training–Style Questions

1. Which of the following is not considered a cost of production?
   (a) Research and development
   (b) Design
   (c) Materials
   (d) Liability

2. If a product requires a high elastic modulus, which class of materials can be eliminated from consideration?
   (a) Metals
   (b) Polymers
   (c) Ceramics
   (d) Composites

3. The performance per unit cost for stiffness of a plate in bending that is not weight critical is given by
   - (a) E/C_V
   - (b) $E^{2/3}/C_V$
   - (c) $E^{1/2}/C_V$
   - (d) $E^{1/3}/C_V$

4. For an anchor stem of a hip implant, the requirement of 8% ductility rules out what class of materials?
   - (a) Metals
   - (b) Polymers
   - (c) Ceramics
   - (d) Composites

5. Polymer products with a triangular three-arrow recycle symbol with the number "02" in the center are made of
   - (a) Polyethylene terephthalate (PET)
   - (b) HDPE
   - (c) PVC
   - (d) PP

6. Which of the following composite materials could be recycled by melting the shredded composite and form recycled pellets with uniform properties?
   - (a) Epoxy matrix and carbon fibers
   - (b) PP matrix and glass fibers
   - (c) PP matrix and PP fibers
   - (d) PP matrix and PE fibers

7. Which of the following is not a sustainable source of materials?
   - (a) Ore
   - (b) Plants
   - (c) Municipal waste
   - (d) Microorganisms

8. What class of material is the least expensive per unit volume that for certain has an elastic modulus of over 100 GPa?
   - (a) GFRP
   - (b) Concrete
   - (c) Cast iron
   - (d) Steel

9. What class of metals provides candidates with the lowest price per unit volume and a yield strength of over 1000 MPa?
   - (a) Aluminum alloys
   - (b) Steels
   - (c) Titanium alloys
   - (d) Cast iron

10. For a rod in tension, what material class provides the highest specific critical-stress intensity factor?
    - (a) Steels
    - (b) Titanium alloys
    - (c) Aluminum alloys
    - (d) Uni-ply composites (GFRP, CFRP, and KFRP)

11. For a plate in bending, what material provides the highest specific elastic modulus?
    (a) Uni-ply CFRP
    (b) Balsa wood parallel to grain
    (c) Beryllium
    (d) Diamond

Problems

Problem 14.1: Conduct an analysis to select a material for the skin of an airplane wing. The primary loading of an airplane wing is in bending. During flight, there is low pressure on the top of the wing and high pressure on the wing bottom, bending the wing up. When the plane is stationary on the ground, the weight of the wing bends the wing down. Important parameters for a wing are a high specific elastic modulus, specific tensile yield strength, and specific resistance to fracture. The operating temperature for a wing depends upon the speed of the plane and the altitude. Assume that our plane is subsonic and that the temperature is below 100°C.

(a) Explain what classes of materials can be eliminated from the analysis, based upon the important parameters.

(b) Explain why some materials within acceptable classes can be eliminated based upon important parameters. Why can steel and magnesium alloys be eliminated?

(c) Conduct a numerical analysis of the relative merit of possible materials using the aluminum alloy 7075-T651, the two conditions of Ti-Al-6V, and carbon-epoxy. These materials are selected because aircraft wings are made from each of these materials.

The following table presents the elastic modulus (E), yield strength (σ_y), critical-stress intensity factor (K_{Ic}), density (ρ), and cost (C_V) of candidate materials.

Material	E (Gpa)	σ_y (MPa)	K_{Ic} (MPa · m$^{\frac{1}{2}}$)	ρ (10^3 kg/m^3)	C_V (10^4\$/m^3)
Ti-6Al-4V	113.8	910	115.4	4.43	161
Ti-6Al-4V	113.8	1035	55	4.43	161
7075-T651	71.7	495	24.2	2.81	12.6
CFRP	138	1447	40	1.67	48.84

These prices are listed prices for metal plate measuring $\frac{1}{8}$ in. by 12 in. by 12 in. The carbon-epoxy is layered laminate from Dragonplate for a sheet of dimensions 0.3175 cm thick by 0.305 m square.

(d) Based upon your results, discuss why Boeing has chosen to use carbon-epoxy as the skin material in the wings of the Boeing 787.

(e) Why don't more subsonic commercial aircraft producers use this material?

Problem 14.2: For the stem of the total hip replacement we discussed in Section 14.3.3, assume that the hip replacement is to go into a lightweight, frail person, and therefore the weight of the stem is critical. Use the data in Table 14.14 to determine the material with the highest relative merit for this application.

An optical microscope image with polarized light of polymer spherulites. Polymer spherulites are semi-crystalline regions that grow radially, as discussed in Section 3.4.3.

US NSF

The goals of this chapter are to understand

- Crystal structure determination by diffraction of X-rays, electrons, and neutrons
- Strain and stress determination in crystals with diffraction
- Chemical analysis by X-ray fluorescence spectroscopy
- X-ray absorption spectroscopy and applications
- Optical, electron, and probe microscopy

Chapter 15

Experimental Methods

15.1 INTRODUCTION

This chapter presents some of the experimental techniques available to determine chemical composition, crystal structure, surface and internal defects, phases, and precipitates. The experimental methods to determine mechanical properties are discussed in association with those subjects, because the mechanical tests are integral to the mechanical properties.

15.2 CRYSTAL-STRUCTURE ANALYSIS BY DIFFRACTION

The crystal structure of a material is determined with diffraction techniques, including the Bravais lattice type, the lattice parameters of the unit cell, and the atom positions. X-ray diffraction is the most widely utilized technique to determine the crystal structure of materials. High-energy electron diffraction in a transmission electron microscope is used to identify the structure of phases, such as precipitates in thin films, and the structure of small particles. Low-energy electron diffraction is used to determine the structure of surfaces on materials. The structure of a surface can be different than the structure of the bulk material. Neutron diffraction is used to determine the structure of materials with light elements, such as hydrogen, compounds with elements that are adjacent in the periodic table such as boron-carbide or silicon-aluminum, and magnetic materials. The Internet site for the International Union of Crystallography (*http://www.iucr.org/education /resources*) offers online course packages and simulators for crystal-structure analysis.

15.2.1   X-Ray Diffraction

X-ray diffraction is used to determine the crystal structure of metals, ceramics, polymers, and even DNA, because the wavelengths of X-rays are comparable to the interatomic distances. **X-rays** are electromagnetic radiation with wavelengths typically from 0.1 to 1 nm. One advantage of X-ray diffraction is that it is normally performed in air, because air does not significantly attenuate a high-energy X-ray beam.

Another advantage is that X-rays are generated by X-ray tubes that are relatively simple devices. In an X-ray tube, electrons are emitted from a hot filament cathode by thermionic emission and accelerated to an anode by a high voltage, such as 50,000 V. **Thermionic** emission is heating the material to such a high temperature that electrons are emitted by the material. Thermionic emission is covered in Section 16.3.1. There are two types of X-rays that come from an X-ray tube: **continuous white radiation** and **characteristic radiation**, as shown in Figure 15.1. The continuous white radiation is energy emitted as the high-energy electrons lose energy in inelastic collisions with electrons in the anode material.

Characteristic X-rays are emitted by atoms in the anode. Electrons of sufficient energy can excite atomic electrons in the anode from the $1s$, $2s$, $2p$, . . . energy levels. When an electron is excited out of

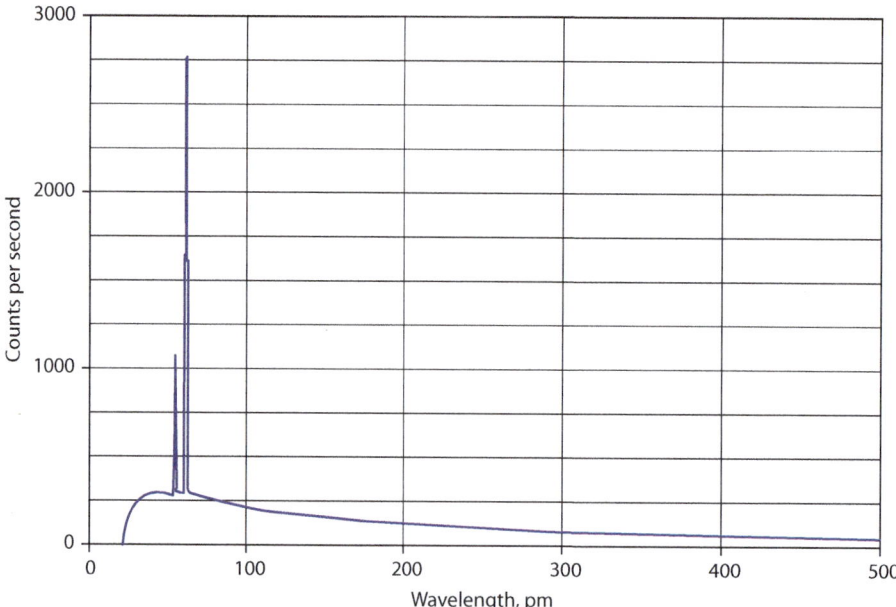

Figure 15.1   The intensity in counts per second from a rhodium X-ray tube with an operating voltage of 60 kilovolts (kV), as a function of wavelength in picometers (pm) (10^{-12} m). The sharp peaks are characteristic X-rays. The smooth intensity starting at approximately 25 pm and continuing to 500 pm is the continuous or white radiation. (*Based on http://upload.wikimedia.org/wikipedia/commons/thumb/5/5c/TubeSpectrum.jpg/800px-TubeSpectrum.jpg*)

a normal electron orbital, the atom is in an excited state. The excited states of the atom are given the designation **K, L, M, and N** when the principal quantum number (n) of the electron missing from the atom is, respectively, 1, 2, 3, and 4. The K excited state corresponds to the energy of the atom with an electron of $n = 1$ being removed from an atom. A K excited state is immediately followed by an electron from one of the higher-energy levels of the atom filling the unoccupied $1s$ electron orbital, and the emission of a photon with an energy equal to the difference in energy of the atom before and after the electron transition. If the wavelength of the photon emitted is between 0.1 and 1 nm, it is an X-ray. The relationship between photon energy (E_p) and wavelength (λ) is shown in Equation 15.1

$$E_p = \frac{hc}{\lambda}$$

15.1

where h is Planck's constant (6.63×10^{-34} J · s) and c is the speed of light (3×10^8 m/s).

The **Laporte selection rule** resulting from quantum mechanics governs which electrons can make the transitions. The angular-momentum quantum number (l) must change by +1 or −1. A $2s$ electron with $l = 0$ cannot make the transition to a vacant $1s$ orbital, because there is no change in l. A $2p$ electron with $l = 1$ can make the transition to the vacant $1s$ orbital with the emission of a K_α X-ray. The K_α X-ray has the energy of an atom with a missing $1s$ electron minus the energy of an atom with a missing $2p$ electron. The emitted photon is named after the excited state that resulted in the photon (K), and the subscript α indicates that the electron that filled the excited state came from the $2p$ level. The $2p$ electrons can have l quantum numbers of both 0 and 1. K_α X-rays are emitted by electrons coming from the $2p$ level with $l = 1$. There is coupling between the electron angular momentum and the spin with the result that the spin quantum number of + or − 1/2 either adds to or subtracts from the angular momentum quantum number of $l = 1$ producing net electron quantum numbers of 3/2 and 1/2. The different net quantum numbers result in a splitting of the K_α line into $K_{\alpha1}$ and $K_{\alpha2}$ with the net quantum number 3/2 producing the higher energy $K_{\alpha1}$ line. Similar arguments hold for the $K_{\beta1}$ and $L_{\alpha1}$ lines. However, in many cases it is not possible to resolve the energy difference between $K_{\alpha1}$ and $K_{\alpha2}$, and then the combined $K_{\alpha1}$ and $K_{\alpha2}$ X-rays are called K_α. When a $3p$ electron from the atom fills the empty $1s$ orbital, the designation of the emitted photon is K_β. In most cases, it is not possible to resolve the different energies of electrons coming from the $3p$ level with different l values, and in these cases of the X-rays resulting from $3p$ to $1s$ electron transitions are called K_β. An $L_{\alpha1}$ photon results when a $3p$ electron with $l = 1$ fills an empty $2s$ orbital. The wavelength in angstroms (10^{-10} m) of the $K_{\alpha1}$, $K_{\alpha2}$, $K_{\beta1}$, and $L_{\alpha1}$ emission lines from the elements are presented in Table 15.1. Absorption edges are discussed in Section 15.3.2.

Table 15.1	The Wavelengths in Angstroms (10^{-10} m) of Characteristic Emission Lines and Absorption Edges of Some Elements					
Element	Z	K_α (weighted average)*	$K_{\alpha2}$ very strong	$K_{\alpha1}$ very strong	$K_{\beta1}$ weak	K edge	$L_{\alpha1}$ very strong
Na	11		11.909	11.909	11.617		
Mg	12		9.8889	9.8889	9.558	9.5117	
Al	13		8.33916	8.33669	7.981	7.9511	
Si	14		7.12773	7.12528	6.7681	6.7446	
P	15		6.1549	6.1549	5.8038	5.7866	
S	16		5.37471	5.37196	5.03169	5.0182	
Cl	17		4.73050	4.72760	4.4031	4.3969	
Ar	18		4.19456	4.19162	. . . .	3.8707	

(Continued)

Table 15.1   *(Continued)*

Element	Z	K_α (weighted average)*	$K_{\alpha 2}$ strong	$K_{\alpha 1}$ very strong	$K_{\beta 1}$ weak	K edge	$L_{\alpha 1}$ very strong
K	19		3.74462	3.74122	3.4538	3.43645	
Ca	20		3.36159	3.35825	3.0896	3.07016	
Sc	21		3.03452	3.03114	2.7795	2.7573	
Ti	22		2.75207	2.74841	2.51381	2.49730	
V	23		2.50729	2.50348	2.28434	2.26902	
Cr	24	2.29092	2.29351	2.28962	2.08480	2.07012	
Mn	25		2.10568	2.10175	1.91015	1.89636	
Fe	26	1.93728	1.93991	1.93597	1.75653	1.74334	
Co	27	1.79021	1.79278	1.78892	1.62075	1.60811	
Ni	28		1.66169	1.65784	1.50010	1.48802	
Cu	29	1.54178	1.54433	1.54051	1.39217	1.38043	13.357
Zn	30		1.43894	1.43511	1.29522	1.28329	12.282
Ga	31		1.34394	1.34003	1.20784	1.19567	11.313
Ge	32		1.25797	1.25401	1.12889	1.11652	10.456
As	33		1.17981	1.17581	1.05726	1.04497	9.671
Se	34		1.10875	1.10471	0.99212	0.97977	8.990
Br	35		1.04376	1.03969	0.93273	0.91994	8.375
Kr	36		0.9841	0.9801	0.87845	0.86546	
Rb	37		0.92963	0.92551	0.82863	0.81549	7.3181
Sr	38		0.87938	0.875214	0.78288	0.76969	6.8625
Y	39		0.83300	0.82879	0.74068	0.72762	6.4485
Zr	40		0.79010	0.78588	0.701695	0.68877	6.0702
Nb	41		0.75040	0.74615	0.66572	0.65291	5.7240
Mo	42	0.71069	0.713543	0.70926	0.632253	0.61977	5.40625
Tc	43		0.676	0.673	0.602		
Ru	44		0.64736	0.64304	0.57246	0.56047	4.84552
Rh	45		0.617610	0.613245	0.54559	0.53378	4.59727
Pd	46		0.589801	0.585415	0.52052	0.50915	4.36760
Ag	47		0.563775	0.559363	0.49701	0.48582	4.15412
Cd	48		0.53941	0.53498	0.475078	0.46409	3.95628
In	49		0.51652	0.51209	0.454514	0.44387	3.77191
Sn	50		0.49502	0.49056	0.435216	0.42468	3.59987
Sb	51		0.47479	0.470322	0.417060	0.40663	3.43915
Te	52		0.455751	0.451263	0.399972	0.38972	3.28909
I	53		0.437805	0.433293	0.383884	0.37379	3.14849
Xe	54		0.42043	0.41596	0.36846	0.35849	. . . .

(Continued)

Table 15.1 (Continued)

Element	Z	K_α (weighted average)*	$K_{\alpha2}$ strong	$K_{\alpha1}$ very strong	$K_{\beta1}$ weak	K edge	$L_{\alpha1}$ very strong
Cs	55		0.404812	0.400268	0.354347	0.34473	2.8920
Ba	56		0.389646	0.385089	0.340789	0.33137	2.7752
La	57		0.375279	0.370709	0.327959	0.31842	2.6651
Ce	58		0.361665	0.357075	0.315792	0.30647	2.5612
Pr	59		0.348728	0.344122	0.304238	0.29516	2.4627
Nd	60		0.356487	0.331822	0.293274	0.28451	2.3701
Pm	61		0.3249	0.3207	0.28209	. . . .	2.2827
Sm	62		0.31365	0.30895	0.27305	0.26462	2.1994
Eu	63		0.30326	0.29850	0.26360	0.25551	2.1206
Gd	64		0.29320	0.28840	0.25445	0.24680	2.0460
Tb	65		0.28343	0.27876	0.24601	0.23840	1.9755
Dy	66		0.27430	0.26957	0.23758	0.23046	1.90875
Ho	67		0.26552	0.26083	. . . .	0.22290	1.8447
Er	68		0.25716	0.25248	0.22260	0.21565	1.78428
Tm	69		0.24911	0.24436	0.21530	0.2089	1.7263
Yb	70		0.24147	0.23676	0.20876	0.20223	1.6719
Lu	71		0.23405	0.22928	0.20212	0.19583	1.61943
Hf	72		0.22699	0.22218	0.19554	0.18981	1.56955
Ta	73		0.220290	0.215484	0.190076	0.18393	1.52187
W	74		0.213813	0.208992	0.184363	0.17837	1.47635
Re	75		0.207598	0.202778	0.178870	0.17311	1.43286
Os	76		0.201626	0.196783	0.173607	0.16780	1.39113
Ir	77		0.195889	0.191033	0.168533	0.16286	1.35130
Pt	78		0.190372	0.185504	0.163664	0.15816	1.31298
Au	79		0.185064	0.180185	0.158971	0.15344	1.27639
Hg	80		. . . .	. . . .	. . . .	0.14923	1.24114
Tl	81		0.175028	0.170131	0.150133	0.14470	1.20735
Pb	82		0.170285	0.165364	0.145980	0.14077	1.17504
Bi	83		0.165704	0.160777	0.141941	0.13706	1.14385
Th	90		0.137820	0.132806	0.117389	0.11293	0.95598
U	92		0.130962	0.125940	0.111386	0.1068	0.91053

Based on data from Cullity, B. D., *Elements of X-Ray Diffraction*, Addison-Wesley Pub. Co., Inc., Reading, MA (1956), p. 464.

* In averaging, $K_{\alpha1}$ is given twice the weight of $K_{\alpha2}$.

Diffraction intensities for all types of radiation, including X-rays, occur when Bragg's law, as shown in Equation 15.2, is obeyed. Diffraction is shown schematically in Figure 15.2, where crystal planes with Miller indices (*hkl*) have a spacing of d_{hkl}.

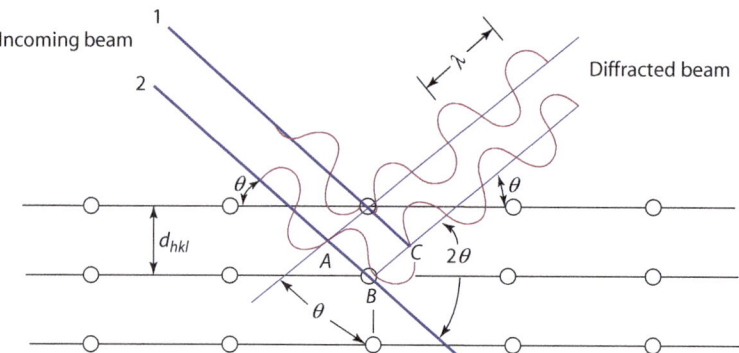

Figure 15.2 A schematic of the configuration for the diffraction of monochromatic X-rays of wavelength λ, at an incident and diffracted angle of θ from a crystal with interplanar spacing of d_{hkl}.

Monochromatic X-rays have a single wavelength (λ). One source of monochromatic X-rays is the characteristic X-rays emitted by an X-ray tube, shown in Figure 15.1. A parallel beam of X-rays is produced by selecting only those X-rays that pass through a series of apertures, or through a **collimator**. A collimator is a device that limits the X-rays to a beam where the X-rays all travel parallel to each other. Two pinholes in lead sheets separated by a distance form a collimator for a circular X-ray beam, and two slits or apertures collimate a rectangular beam. The angle (θ) is selected by rotating the crystal to the angle θ in front of the fixed X-ray beam, and the detector is rotated to the angle θ relative to the crystal; or, equivalently, the detector is rotated to an angle of 2θ relative to the incident X-ray beam. X-rays are scattered by the electrons around atoms in different (hkl) planes. The scattered X-rays are in phase with each other, and they constructively add if Equation 15.2, Bragg's law of diffraction, is satisfied:

$$n\lambda = 2d_{hkl} \sin \theta, \qquad\qquad\qquad \textbf{15.2}$$

where n is an integer that gives the order of the reflection. When Equation 15.2 is satisfied, the length $AB + BC$ in Figure 15.2 is equal to $n\lambda$, and AB and BC are each equal to $d_{hkl} \sin \theta$. Then the X-rays scattered from successive planes in the crystal are in phase with each other and sum to form a high intensity. A value of n equal to 1 is a first-order reflection, and it corresponds to the condition when the length $AB + BC$ corresponds to 1 wavelength. A value of n equal to 2 is a second-order reflection that corresponds to the condition when the length $AB + BC$ corresponds to 2 wavelengths. If the scattering does not satisfy Equation 15.2, the scattered X-rays from different planes interfere with each other, and a low scattered intensity results.

A powder X-ray **diffractometer** is shown in Figure 15.3. In a powder X-ray diffractometer, the specimen is in the form of a powder or a polycrystal. A diffractometer is a mechanical device that allows the angle of the sample relative to an incident X-ray beam (θ) and of the detector angle (2θ) to be set at a known value, or to be scanned continuously through known values, and for the diffraction intensity to be recorded electronically as a function of the value of θ. In a powder or a polycrystal specimen, there are many orientations of crystals or grains in the specimen, and it is probable that some of the grains are oriented at the proper value of θ when the (hkl) planes satisfy Bragg's law in Equation 15.2. The X-ray tube is the source of the X-rays.

Figure 15.4 is a plot of diffracted intensity as a function of the detector angle (2θ) for tungsten metal. To obtain this data in a powder diffractometer, the angle of the specimen (θ) is scanned with a goniometer from a low angle to a high angle approaching 90° relative to the incident X-ray beam, and the detector simultaneously scans at an angle of 2θ relative to the incident X-ray beam. A **goniometer** is a device that allows the angle of the specimen relative to the incident beam be changed and measured. When the angle θ is such that Equation 15.2 is satisfied, a diffraction intensity peak is observed.

In a **Debye-Scherrer camera**, shown schematically in Figure 15.5 on page 666, powder of the material in a fine glass tube or a polycrystalline rod is utilized as the specimen. A collimated X-ray beam is

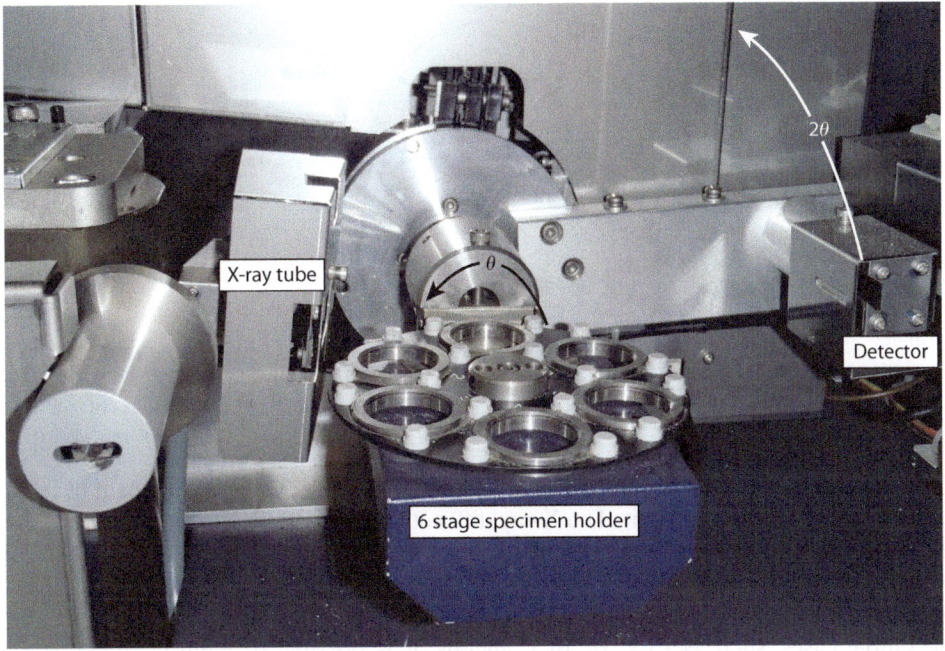

Figure 15.3 A Photograph of an X-ray diffractometer. showing the X-ray tube, specimen holders, and the detector. (Photo by C. M. Gilmore)

incident upon the specimen. The specimen is rotated continuously in front of the X-ray beam to ensure that some of the grains are oriented at the proper value of θ when the (hkl) planes satisfy Bragg's law in Equation 15.2. Then, the X-ray beam is diffracted at an angle of 2θ relative to the incident X-ray beam, as shown in Figure 15.5. The diffracted X-rays are recorded on film that forms a narrow cylinder around the inside surface of the camera. The angles of the diffracted beams are determined geometrically by the

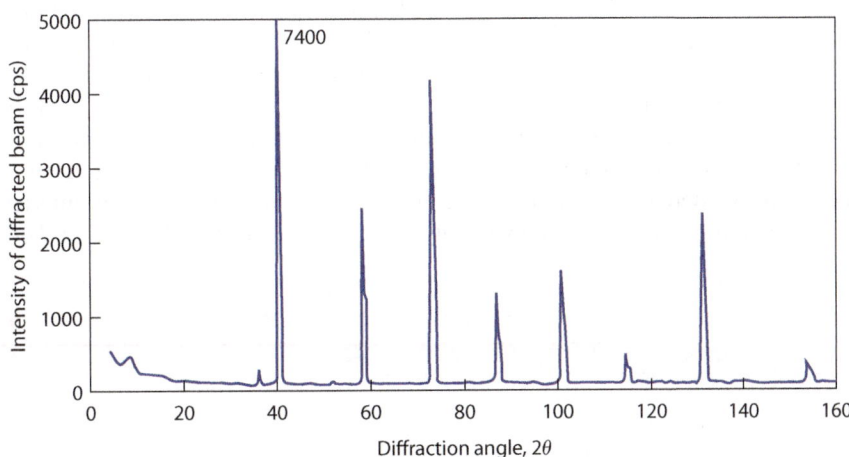

Figure 15.4 The diffraction intensities in counts per second, as a function of the detector angle (2θ) from a tungsten specimen, using Cu K_α radiation. (*Based on Guy, A.G. Introduction to Materials Science, McGraw-Hill, N.Y. (1972), p. 47.*)

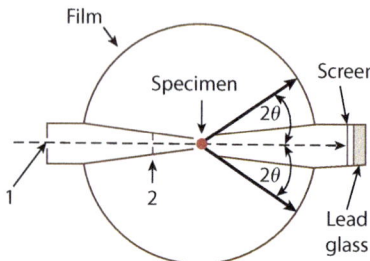

Figure 15.5   A schematic of a Debye-Scherrer powder camera. Monochromatic X-rays enter the camera at 1 and are collimated at 2. The specimen is a powder in a thin glass tube, or it is a polycrystalline rod. The X-rays are diffracted at angles 2θ from the incident beam when Bragg's law is satisfied. The fluorescent screen and leaded glass are for viewing the shadow of the specimen in the X-ray beam, to ensure that the specimen is centered in the X-ray beam. (*Adapted from Barrett, C., and Massalski, T. B., Structure of Metals, 3rd ed., Pergamon, New York (1980), p. 119.*)

distance from the 0-angle position, which corresponds to the position where the incident beam strikes the film after passing through the specimen. The X-ray diffraction forms circles around the X-ray beam at an angle of 2θ, because powder crystallites can diffract out of the plane shown in Figure 15.5 as long as the crystallite is oriented at an angle of θ relative to the incident X-ray beam; however, in the Debye-Scherrer camera, only a narrow strip of film records diffracted X-rays.

In some simple cases, such as FCC or BCC metals, it is possible to identify the crystal structure from the peak intensities that are present from a powder X-ray diffractometer or from a Debye-Scherrer camera. The X-ray diffraction peaks of a cubic crystal are identified by calculating the value of the interplanar spacing (d_{hkl}) for each peak intensity from the measured values of θ and from the known value of λ. Then, the values of the Miller indices (hkl) for each peak intensity are deduced from the relationship between d_{hkl} and $h, k,$ and l for cubic crystals shown in Equation 15.3:

$$d_{hkl} = \frac{a}{(h^2 + k^2 + l^2)^{1/2}} \qquad \textbf{15.3}$$

where a is the lattice parameter of the cubic crystal. From the peak intensities present, we can determine if the cubic crystal is simple, BCC, or FCC from the rules shown in Table 15.2.

For example, in BCC crystals the (200) peak is present because the sum $2 + 0 + 0$ is equal to the even number 2; however, the (100) diffraction peak is absent because this sum is equal to the odd number 1. X-ray diffraction peaks are referred to by the set of planes (hkl) that is responsible for the diffraction. For FCC crystals, unmixed means the values for $h, k,$ and l are all even or all odd. The (131) peak is present in FCC crystals because all of the numbers are odd (unmixed); however, the (123) peak is absent because there is a mixture of even and odd numbers. The peaks that are absent from the diffraction patterns of BCC and FCC crystals is due to the body-centered and face-centered atoms. These atoms scatter X-rays that are out of phase with the X-rays scattered from atoms at the cube corners, and this results in interference of the diffracted X-rays. A simple cubic crystal that only has atoms at the cube corners diffracts all (hkl) intensities.

Table 15.2	Rules Governing the Presence or Absence of Diffraction Peaks in Cubic Crystals	
Cubic Bravais Lattice	**Peaks Present**	**Peaks Absent**
Simple	All	None
BCC	$(h + k + l)$ = even number	$(h + k + l)$ = odd number
FCC	hkl unmixed	hkl mixed

Example Problem 15.1

For a polycrystalline copper specimen with a lattice parameter of 0.363 nm, determine the diffraction angle for the first seven peaks observed in a diffractometer with copper K_α X-rays of wavelength 0.1542 nm.

Solution

The possible reflections from FCC copper where the values of h, k, and l are unmixed are listed below:

(hkl)	$(h^2+k^2+l^2)$	d_{hkl} nm	$\sin\theta$	θ degrees
(111)	3	0.2096	0.3679	21.59
(200)	4	0.1815	0.4248	25.14
(220)	8	0.1283	0.6007	36.92
(311)	11	0.1094	0.7044	44.78
(222)	12	0.1048	0.7358	47.37
(400)	16	0.0908	0.8496	58.17
(331)	19	0.0833	0.9258	67.79

The value of d_{hkl} comes from Equation 15.3. For the (111) peak, d_{111} is given by

$$d_{111} = \frac{a}{(h^2 + k^2 + l^2)^{1/2}} = \frac{0.363 \text{ nm}}{(1^2 + 1^2 + 1^2)^{1/2}} = \frac{0.363 \text{ nm}}{(3)^{1/2}} = \frac{0.363 \text{ nm}}{1.732} = 0.210 \text{ nm}$$

The value of $\sin\theta$ then comes from Equation 15.2 with $n = 1$.

$$\sin\theta = \frac{n\lambda}{2d_{hkl}} = \frac{0.1542 \text{ nm}}{2(0.210 \text{ nm})} = 0.368$$

$$\theta = 21.6 \text{ degrees}$$

The same procedure is repeated for each of the peaks.

Note in Example Problem 15.1 that for the FCC crystal, the sum $(h^2 + k^2 + l^2)$ is in the sequence 3, 4, 8, 11, 12, 16, 19, ... This sequence is characteristic of the FCC structure, and it is used to determine if the FCC structure is present from the X-ray diffraction pattern. For peak intensities present from a BCC structure, this sequence is 2, 4, 6, 8, 10, 12, 14, 16, 18, ... For peak intensities present from a simple cubic crystal, the sequence is 1, 2, 3, 4, 5, 6, 8, 9, 10, ... The BCC and the simple cubic sequence appear similar, if the BCC sequence is divided by 2. However, the simple cubic sequence does not have a 7, where as the BCC sequence has a 14.

Example Problem 15.2

A specimen is 90 atom percent iron and 10 atom percent Ni. With sufficient FCC nickel added to BCC iron, it is possible to produce FCC alloys. The results of an X-ray diffractometer scan of this alloy with cobalt K_α X-ray radiation of wavelength 0.179 nm gives peak intensities at the following angles: 26.1°, 38.6°, 49.8°, 61.9°, and 80.5°.
(a) Is this iron BCC or FCC?
(b) What is the lattice parameter?

Solution

a)   From Equations 15.2 and 15.3, solve for $\sin\theta$ in terms of the values of h, k, and l.

$$\sin\theta = \frac{n\lambda}{2d_{hkl}} = \frac{n\lambda(h^2 + k^2 + l^2)^{1/2}}{2a}$$

Taking the square of the above equation and taking $n = 1$, the result is

$$\sin^2\theta = \frac{\lambda^2(h^2 + k^2 + l^2)}{4a^2}$$

All of the terms on the right side of the above equation are constant for the different peaks, except h, k, and l. Therefore, the sum of the squares of the possible h, k, and l values should be in the same sequence as the square of the values of the sine of the diffraction angles. The squares of the $\sin\theta$ are 0.194, 0.389, 0.583, 0.778, and 0.972. Dividing each of these by 0.194 produces the sequence 1, 2, 3, 4, 5. It is known that iron does not form in the simple cubic system. Simple cubic is eliminated as a possibility. The sequence of $(h^2 + k^2 + l^2)$ must be 2, 4, 6, 8, and 10. The alloy has the BCC structure.

b) To solve for the lattice parameter a, use one of the peaks, such as the (110) peak, and the equation

$$\sin^2\theta = \frac{\lambda^2(h^2 + k^2 + l^2)}{4a^2}$$

$$0.194 = \frac{(0.179\ \text{nm})^2(1^2 + 1^2 + 0^2)}{4a^2} = \frac{0.032\ \text{nm}^2\,(2)}{4a^2}$$

Solving for a^2 and then a,

$$a^2 = \frac{0.032\ \text{nm}^2}{2(0.194)} = 0.0826\ \text{nm}^2$$

$$a = 0.287\ \text{nm}$$

For a more precise lattice parameter, determine the lattice parameter from each of the peak intensities, and then plot the value of a determined for each peak intensity as a function of the value of $\sin^2\theta$ used to calculate a. The most accurate value of the lattice parameter (a) results from an extrapolation of $\sin^2\theta$ to 1.0.

The **Laue X-ray technique**, shown in Figure 15.6, allows the study of the crystal symmetry, the determination of the crystal class, and the orientation of single crystals. The sample is normally a single crystal for this technique, which is used in forward transmission and in back-reflection. The stationary crystal is irradiated by all of the X-rays coming from the X-ray tube. These X-rays consist of continuous or white radiation and the characteristic peaks, shown in Figure 15.1.

In the Laue technique, the crystal planes with interplanar spacing (d_{hkl}) that are at an angle θ relative to the incident beam diffract X-rays of a wavelength (λ) at an angle 2θ relative to the incident X-ray beam, in accordance with Bragg's law in Equation 15.2. The value of θ is fixed by the stationary crystal, but the wavelength (λ) is variable. The X-rays that are diffracted can be the continuous white radiation and the characteristic X-rays. The Laue technique is not used to determine lattice parameters, because the wavelength of the diffracted X-rays is not known.

Figure 15.7 is a back-reflection Laue photograph of a $FeSi_2$ crystal with the X-ray beam normal to the fourfold-symmetric axis. A fourfold-symmetric axis means that when the X-ray diffraction pattern is rotated by $360/4 = 90$ degrees, the diffraction pattern looks exactly the same. The crystal class is determined by rotating the crystal to observe the other symmetry axes. For example, a hexagonal crystal has a

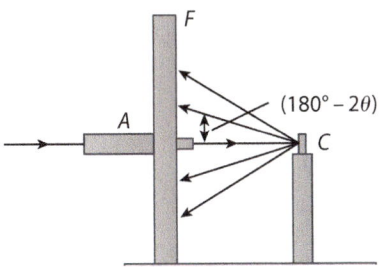

Figure 15.6 A schematic of the back-reflection Laue X-ray technique. The incident X-ray beam, indicated by the arrow pointing to the right, passes through a collimator (A) and through a hole punched into the film (F), and is incident upon the crystal (C). Wavelengths satisfying the Bragg diffraction law are diffracted back, as indicated by the left-pointing arrows, to the film (F). (*Adapted from Cullity, B. D., Elements of X-ray Diffraction, Addison-Wesley Pub. Co., Inc. (1956), p. 141.*)

twofold-symmetric axis at 90° from the sixfold-symmetric axis. By finding the angles between major symmetry axes, the crystal class is established. A cubic crystal has a fourfold-symmetric axis 90° from another fourfold-symmetric axis. The Laue technique is also used to orient a single crystal of unknown orientation into a desired orientation. Crystals to be oriented are mounted on a goniometer that allows adjustment of the

Figure 15.7 A Laue back-reflection photograph from a β FeSi$_2$ crystal showing a four fold symmetry axis. (Reprinted from Thin Solid Films, Volume 515, Issue 22, Y. Hara, M. Tobita, S. Ohuchi, K. Nakaoka, "Growth of plate-type β-FeSi2 single crystals by optimization of composition ratio of source materials," Pages No. 8259–8262, Copyright 2007, with permission from Elsevier.)

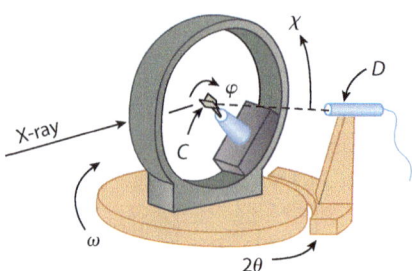

Figure 15.8   A schematic of a single crystal diffractometer showing the X-ray beam incident on a crystal (C). The crystal can be rotated about its axis in the specimen holder by the angle φ, about a vertical axis by the angle ω, and about a horizontal axis through the center of the goniometer by the angle χ. The detector (D) can rotate to the angle 2θ relative to the incident beam. (*Adapted from Massa, W., Crystal Structure Determination, 2nd ed., Springer, Berlin (2004), p. 75.*)

crystal about three axes. If the sample is a single crystal, the Laue technique produces X-ray diffraction spots as shown in Figure 15.7. A polycrystalline specimen produces rings of diffraction, because crystallites can be oriented at the diffraction angle θ for all angles degrees around the X-ray beam.

If a new material is produced or discovered, and powder diffractometer peaks do not match a known crystal structure, then it is necessary to use a single crystal X-ray diffractometer, shown schematically in Figure 15.8, to determine the crystal structure. In a single crystal X-ray diffractometer, the crystal can be rotated about three perpendicular axes by φ, ω, and χ to orient different (hkl) planes in the X-ray beam to satisfy Bragg's law, and the detector is set to the appropriate angle of 2θ.

With single crystal diffraction, it is possible to determine the crystal class, such as cubic, because the diffraction pattern from a single crystal has the symmetry of the crystal, plus the diffraction conditions add in an inversion center. Figure 15.7 shows a diffraction pattern with an inversion center. With an inversion center for each diffraction spot, there is another spot just like it on a line drawn through the center of the photograph to the other side of the photograph. Figure 15.7 could be the diffraction pattern for a crystal with twofold-symmetry with the added inversion center. The crystal class is determined by rotating the crystal to the symmetry axes that identify the crystal class. The interplanar spacings (d_{hkl}) are determined from Bragg's law in Equation 15.2 by using characteristic radiation of known wavelength (λ) that diffracts X-rays at an angle 2θ. The lattice parameters of the crystal are determined with equations similar to Equation 15.3. Equation 15.3 is for cubic crystals only.

To determine the atom positions in an unknown crystal, it is necessary to measure the intensity $I(hkl)$ from as many diffraction peaks as possible with the detector of the single crystal X-ray diffractometer. By mathematical analysis of the diffraction peak intensities the electron density of the crystal is determined, with the added inversion center. Recall that X-rays are scattered by electrons around the atoms. Because of the added inversion center, the analysis does not give unique atom positions, but crystallographers have developed techniques to determine the atom positions in the unit cell. Determining the electron density also gives the atom type at the position in the crystal unit cell.

X-ray diffraction is also used to study defects in a crystal. With monochromatic X-rays and a large perfect single crystal with the atoms all at their perfect crystal positions, the diffraction peak should have the same energy width as the line in the incident radiation, and this is very narrow. However, a crystal is not perfect, and the defects in the crystal increase the energy width of the diffraction peak. Structural defects result in diffraction-peak broadening that is independent of temperature, and lattice vibrations result in peak broadening that is dependent upon temperature. If the crystal is small, this also produces diffraction-peak broadening. Using the diffraction peak intensities and the broadening of the X-ray diffraction lines analytical techniques have been developed that allow evaluation of lattice vibration amplitudes and structural defects in crystals, such as the static displacement of atoms from the perfect-crystal atom positions.

15.2.2   Electron Diffraction

Electrons are diffracted in a manner similar to X-rays, because electrons have both particle and wave characteristics. Electrons are also scattered by the electrons around atoms, but a major difference between X-rays and electrons is that X-rays penetrate more deeply into a material than do electrons of the same energy. Because electron beams are attenuated even in air, electron-diffraction experiments are conducted in a vacuum. A vacuum also keeps the surfaces of specimens and thin-film specimens from contamination by the environment.

The wavelength (λ) of any particle with wave characteristics, such as electrons and neutrons, with momentum (p) is given by the **DeBroglie relationship** that equates the wave momentum (h/λ) and particle momentum (mv), as shown in Equation 15.4.

$$p = \frac{h}{\lambda} = mv \qquad\qquad \textbf{15.4}$$

In Equation 15.4, h is Planck's constant (6.63×10^{-34} J · s), m is the particle mass, and v is the particle velocity. We assume that there is no potential energy for the particle, and the internal energy (E) is equal to the kinetic energy (KE) that is a function of the wavelength, as shown in Equation 15.5.

$$KE = E = \frac{p^2}{2m} = \frac{h^2}{2m\lambda^2} \qquad\qquad \textbf{15.5}$$

The wavelength of the particle from Equations 15.4 and 15.5 is then used to determine the Bragg diffraction conditions in Equation 15.2.

There are two main applications of electron diffraction: high-energy electron diffraction in a **transmission electron microscope** (TEM) and **low-energy electron diffraction** (LEED). A TEM, as we discuss in Section 15.4.3, produces a high-energy beam of electrons that penetrates through thin films of material. The average kinetic energy of the electrons in a TEM is a result of the accelerating voltage. For a TEM with an accelerating voltage of 100 kilovolts (kV), the average electron kinetic energy when the electron first encounters a specimen is 100 kilo electron volts (keV). For a 100-keV electron beam, the average wavelength determined with Equation 15.5 using an electron mass of 9.11×10^{-31} kg is 3.88×10^{-12} m. This wavelength is much shorter than the typical atomic radius and typical interplanar spacing (d_{hkl}). According to Equation 15.2, the diffraction angles (θ) are very small. This is appropriate, because a TEM can accommodate only a small diffraction angle. Electron diffraction in a TEM is used primarily to identify the presence of phases of a known crystal structure in thin films, and to determine the orientation of a particular volume of crystalline material in a sample. Electron diffraction in a TEM can be used to determine an unknown complex crystal structure of small nano dimensioned particles.

Example Problem 15.3

What is the Bragg diffraction angle of the (111) peak from a thin gold film that is FCC with a lattice parameter of 0.407 nm, in a TEM operated at 100,000 V accelerating potential?

Solution

Since the accelerating voltage is 100,000 V, assume that the electrons arriving at the thin film have a kinetic energy of 100,000 eV. It is necessary to use the SI system of units in Equation 15.5; the eV is not part of a unified system of units. Converting 100,000 eV to joules results in

$$100{,}000 \ \text{eV} \left(\frac{1.602 \times 10^{-19} \ \text{J}}{1 \ \text{eV}} \right) = 1.602 \times 10^{-14} \ \text{J}$$

Use Equation 15.5 to determine the square of the wavelength.

$$\lambda^2 = \frac{h^2}{2Em_e} = \frac{(6.63 \times 10^{-34}\,\text{J}\cdot\text{s})^2}{2(1.602 \times 10^{-14}\,\text{J})9.11 \times 10^{-31}\,\text{kg}} = \frac{43.96 \times 10^{-68}\,\text{J}^2\cdot\text{s}^2}{29.19 \times 10^{-45}\,\text{J}\cdot\text{kg}} = 1.506 \times 10^{-23}\,\text{m}^2$$

Solving for the wavelength,

$$\lambda = 3.88 \times 10^{-12}\,\text{m}$$

To use Bragg's law, Equation 15.2, first determine the interplanar spacing for the (111) planes with Equation 15.3.

$$d_{hkl} = \frac{a}{(h^2 + k^2 + l^2)^{1/2}} = \frac{0.407\,\text{nm}}{(1^2 + 1^2 + 1^2)^{1/2}} = \frac{0.407\,\text{nm}}{(3)^{1/2}} = \frac{0.407\,\text{nm}}{1.732} = 0.235\,\text{nm}$$

Now everything in Bragg's law is known, except the diffraction angle, and we assume first-order diffraction.

$$\sin\theta = \frac{n\lambda}{2d_{hkl}} = \frac{3.88 \times 10^{-12}\,\text{m}}{2(0.235 \times 10^{-9}\,\text{m})} = 8.26 \times 10^{-3}$$

$$\theta = 0.47\,\text{degrees}$$

LEED is used to study the structure of surfaces. In LEED, the electron kinetic energies range from 10 to 60 0 eV, and a kinetic energy of 100 eV corresponds to a wavelength of 0.124 nm. This wavelength is typical of interplanar spacings (d_{hkl}). Because the energy of the electrons is low, the penetration distance into materials is very short. For example, the average penetration depth of a 100-eV electron beam is approximately 1 nm. Thus LEED analyzes only the surface of the material. In a LEED system, shown in Figure 15.9, the electrons are thermionically emitted by a hot filament (cathode). The electrons are accelerated to the desired energy by the grid. The electron beam is focused to a fraction of a millimeter by a lens system, indicated by A1, A2, A3, and A4 in Figure 15.9. Electrons from the primary beam are

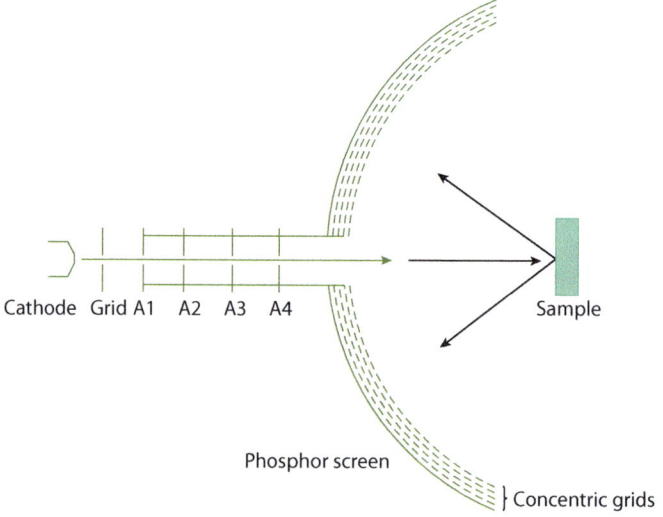

Figure 15.9 A schematic of a low-energy electron diffraction (LEED) system. (*Based on www.en.wikipedia.org/wiki/Low_energy_electron_diffraction*)

back-reflected from the specimen surface, and those back-reflected electrons that satisfy Bragg's law are diffracted at angles 2θ relative to the incident electron beam. Since the accelerating voltage of a LEED system is variable, it is possible to vary the wavelength of the electrons to obtain diffraction conditions. Non-diffracted electrons are eliminated by voltage applied to the concentric grids, shown in Figure 15.9. LEED is usually combined with other surface-analysis techniques, such as electron spectroscopy, to analyze the both the structure and the chemistry of the sample surface. Surface structures can differ from the bulk structure.

15.2.3 Neutron Diffraction

Neutrons are particles, but they also have a wavelength given by Equations 15.4 and 15.5 using the neutron mass of 1.675×10^{-27} kg. A nuclear reactor is a source of neutrons.

Neutron diffraction is applied in situations where X-ray diffraction is not suitable for determining the crystal structure. In X-ray diffraction, the intensity of X-ray scattering from the electrons is proportional to the square of the atomic number. Hydrogen at atomic number 1 has relatively low scattering intensity of 1 in comparison to atomic-number-74 tungsten with a relative scattering intensity of 5476. It is nearly impossible to observe hydrogen atoms in a crystal with tungsten atoms using X-ray diffraction. However, tungsten scatters neutrons approximately 1.6 times more strongly than hydrogen scatters neutrons, allowing hydrogen to be detected in a crystal with tungsten by neutron diffraction. With X-ray diffraction, it is difficult to distinguish atoms that are adjacent in the periodic table, because they have similar numbers of electrons. For example, nickel and cobalt atoms have 28 and 27 electrons, respectively, and the squares of these numbers give the relative scattering intensity of 748 and 729, respectively. However, for neutrons the relative scattering intensities for nickel and cobalt are 13.4 and 1.0, respectively. Neutron diffraction would make it much easier to distinguish nickel and cobalt atom positions in a crystal containing these two elements.

The penetration depth of neutrons into materials is very large, because neutrons are scattered by the nucleus of an atom, and the nucleus is very small in comparison to the atomic size. Therefore, incoming neutrons see mostly open space in passing through a material. Neutrons are diffracted from deep in the material, and they give the structure of the bulk of the material. Neutron-diffraction experiments are conducted in air, because of the small absorption of neutrons in air.

15.2.4 Strain and Stress Determination in Crystals by Diffraction

X-ray diffraction is the most commonly utilized diffraction technique for elastic strain measurement, because of the accuracy of the wavelength of the characteristic X-ray lines and the relative portability of X-ray sources. However, both electrons and neutrons are used for elastic-strain measurement. The elastic strain can be due to an applied stress, and residual strains can be measured. Diffraction is a direct method of measuring residual elastic strain. A simple example demonstrates the diffraction method of strain measurement. Assume that a sample is oriented so that the (100) planes satisfy Bragg's law, the equilibrium lattice parameter (a) of a cubic crystal is known, and there is no applied stress. In diffraction measurements of strain, the gauge length is the interplanar spacing. If there is elastic strain, the unstrained interplanar spacing (d^0_{100}) between the (100) planes is changed by the amount Δd_{100}, and the strain in the [100] direction is given by Equation 15.6:

$$\varepsilon_{[100]} = \frac{\Delta d_{100}}{d^0_{100}} = \frac{\sigma}{E} \qquad\qquad \textbf{15.6}$$

where σ is the residual stress in the [100] direction, and E is the elastic modulus of the sample. Δd_{100} is the value d_{100} measured with either applied or residual stress minus the value d^0_{100} for an unstrained material. This analysis is valid for any set of planes with spacing d_{hkl}. Elastic strain due to an applied stress is also measured by diffraction techniques, but it is most convenient to measure the strain perpendicular to the applied stress. Plastic strain does not change the lattice parameter, and it is not measured by diffraction techniques.

Example Problem 15.4

A steel part has an equilibrium lattice parameter of 0.2866 nm. With X-ray diffraction, it is observed that the (200) peak using chromium K_α X-rays of wavelength 0.2291 nm is at an angle 2θ of 106.46°.
(a) What is the residual strain in this part?
(b) If the elastic modulus of this part is equal to 211 GPa, what is the residual stress in this part?

Solution

a) Use Bragg's law to calculate the observed interplanar spacing d_{200}.

$$d_{200} = \frac{n\lambda}{2 \sin \theta} = \frac{0.2291 \text{ nm}}{2 \sin 53.23} = \frac{0.2291 \text{ nm}}{2(0.8010)} = \frac{0.2291 \text{ nm}}{1.602} = 0.1430 \text{ nm}$$

The unstressed interplanar spacing d^0_{200} is given by Equation 15.3.

$$d^0_{200} = \frac{a}{(2^2 + 0^2 + 0^2)^{1/2}} = \frac{a}{2} = \frac{0.2866 \text{ nm}}{2} = 0.1433 \text{ nm}$$

The change in interplanar spacing Δd_{200} is given by the observed spacing with residual strain minus the unstrained interplanar spacing.

$$\Delta d_{200} = d_{200} - d^0_{200} = 0.1430 \text{ nm} - 0.1433 \text{ nm} = -0.0003 \text{ nm}$$

The strain in the [100] direction is given by Equation 15.6 adjusted for these planes.

$$\varepsilon_{[100]} = \frac{\Delta d_{200}}{d^0_{200}} = \frac{-0.0003 \text{ nm}}{0.2886 \text{ nm}} = -0.0010$$

b) The residual stress in the [100] direction is given by Equation 6.13.

$$\sigma = E\varepsilon_{[100]} = 211 \times 10^9 \text{ Pa}(-1.0 \times 10^{-3}) = -211 \times 10^6 \text{ Pa}$$

The residual stress is compressive in this direction.

15.3 CHEMICAL ANALYSIS BY X-RAY SPECTROSCOPY

15.3.1 X-Ray Fluorescence Spectroscopy

In X-ray **fluorescence** spectroscopy for chemical analysis, the sample is bombarded with either primary X-rays, as shown schematically in Figure 15.10, or with electrons. Primary X-rays are generated in an X-ray tube. The term *fluorescence* applies to the emission of a photon if the time between creation and elimination of the excited state is less than 10^{-8} seconds. For X-rays, this time is typically 10^{-18} seconds. Figure 15.11 shows the X-ray spectrum of intensity versus diffraction angle resulting from an analysis of a stainless-steel specimen with a spectrometer similar to that shown schematically in Figure 15.10. In wavelength dispersion, an analyzing crystal of known lattice parameter is used to separate the various wavelengths of secondary-characteristic radiation emitted by the specimen by X-ray diffraction. The secondary characteristic radiation of wavelength (λ) is diffracted from the analyzing crystal that has a known value of interplanar spacing (d_{hkl}) at an angle 2θ relative to the incident X-ray beam that satisfies Bragg's law. In Figure 15.11 the detector was continuously scanned from a high value of 2θ to a low value, and the diffracted intensity was recorded as a function of 2θ. The X-ray spectrum in Figure 15.11 shows that the characteristic X-ray peaks for Fe, Ni, Cr, Co, Mn, Cu, and W were recorded. The characteristic X-rays have a unique energy that is characteristic of the atom type, because the X-ray energy is dependent upon the energy levels of the atom. Therefore, if an X-ray peak intensity with the energy of the K_α radiation from Fe is observed coming from a sample of unknown chemical composition, this indicates that there is Fe in the specimen. Also the intensity of the emitted characteristic radiation is proportional to the concentration of Fe atoms in the sample, with correction factors. This is the basis for X-ray fluorescence spectroscopy. The wavelengths of the X-ray emission lines from the elements are given in Table 15.1. The sample is usually a solid; however, it can be a liquid or vapor.

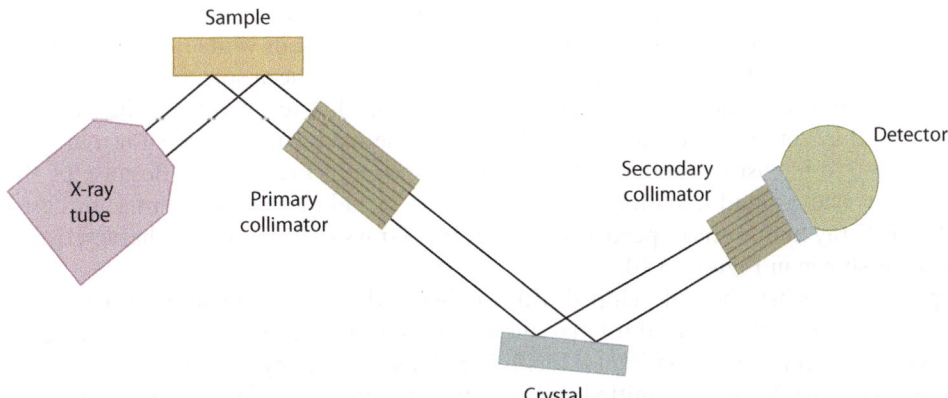

Figure 15.10 A schematic of an X-ray fluorescence spectrometer using diffraction by a crystal (wavelength dispersion) to separate the wavelengths. The X-ray tube creates primary radiation that is incident on the sample. Secondary X-rays are emitted from the sample, and they go through the primary collimator. The analyzing crystal diffracts the secondary X-rays at the Bragg angle to the detector. (*Based on http://upload.wikimedia.org/wikipedia/commons/4/45/DmwdxrfFlatXtalMonochrom.jpg*)

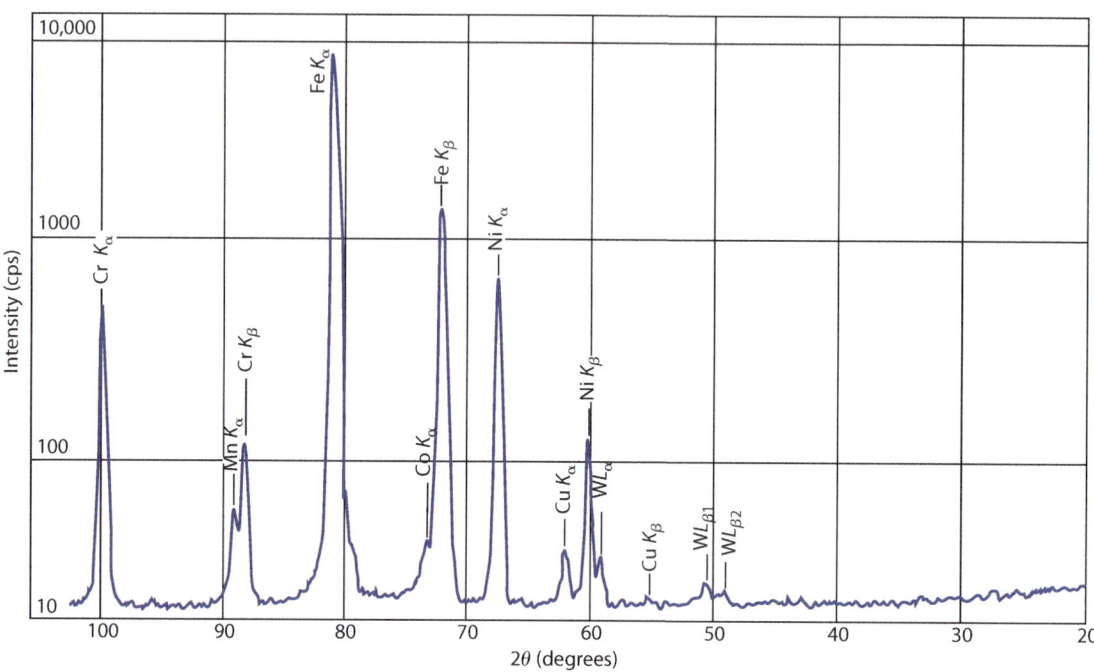

Figure 15.11   An X-ray fluorescence spectrum from a stainless-steel specimen, using a mica-analyzing crystal. *(Based on Cullity, B. D.,* Elements of X-Ray Diffraction, *Addison-Wesley Pub. Co., Inc. (1956), p. 405.)*

Chemical analysis is also conducted by analyzing X-rays generated by electron beams penetrating samples in electron microscopes, and this is discussed in Sections 15.4.2 and 15.4.3. However, if a specimen is bombarded with high-energy electrons, the continuous white radiation shown in Figure 15.1 can mask the characteristic X-rays of elements with low concentrations in the specimen. The spectrum from the X-ray tube in Figure 15.1 is the same spectrum that would be observed in an electron microscope from a pure rhodium sample, using an electron beam of 60 keV. The continuous white radiation in Figure 15.1 is nearly 25% of the K_β peak intensity (the shorter wavelength peak). The absorption of X-rays in a sample does not generate the continuous white background radiation that resulted from the slowing down of electrons. To see this, compare the background intensity generated with X-rays, shown in Figure 15.11, with the background intensity generated with electrons, shown in Figure 15.1. Because of the absence of the continuous background radiation, chemical analysis including small concentrations are normally conducted with X-ray fluorescence spectrometers that use X-rays to generate the characteristic X-rays in the specimen, as shown in Figure 15.10.

X-ray spectrometers have been developed with solid-state detectors capable of measuring the energy of the X-rays from the specimen, in addition to the X-ray intensity. This is referred to as **energy-dispersive X-ray** (EDX) spectroscopy, shown in Figure 15.12. In EDX spectroscopy, primary X-rays are incident on a sample, and secondary X-rays are emitted and they directly enter the EDX detector system. The voltage signal from the detector is in direct proportion to the energy of the X-rays from the sample, and the current is proportional to the intensity (count rate). This signal is amplified and sent to a multichannel analyzer. The multichannel analyzer records the signal intensity in various channels, or bins, in proportion to the X-ray energy, and outputs this signal to a computer. Figure 15.13 shows the EDX spectrum of a sample that was made by pressing a mixture of powders from many different elements. The second-order peaks are electronic signals that have 50% of the actual photon energy.

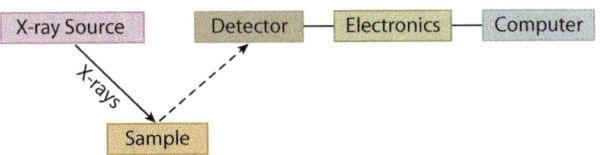

Figure 15.12 A schematic of the components of an energy-dispersive X-ray (EDX) spectroscopy system. (*Based on http://upload.wikimedia.org/wikipedia/commons/2/28/Dmedxrfschematic.jpg*)

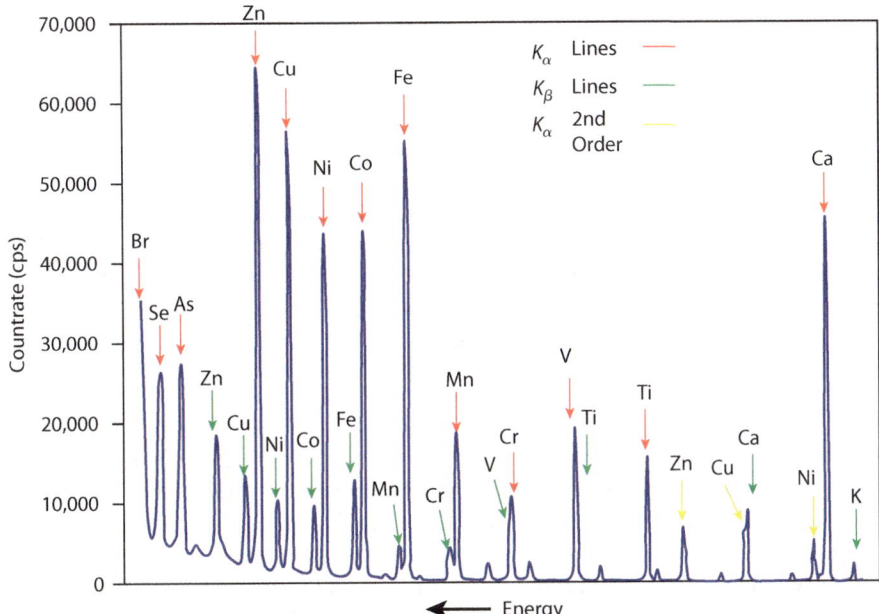

Figure 15.13 An X-ray fluorescence spectrum using an energy-dispersive X-ray (EDX) spectrometer. The specimen is composed of many different elements powders pressed into a sample. (*Based on http://upload.wikimedia. org/wikipedia/commons/thumb/2/28/XRFScan.jpg/800px-XRFScan.jpg*)

As the atomic number decreases, the wavelength of the K_α peak increases and the X-ray energy decreases. Low-energy X-rays are adsorbed extensively in air, and the X-rays from low atomic number elements, typically atomic numbers 13 (Al) and less, are analyzed in a vacuum spectrometer.

15.3.2   X-Ray Absorption

X-ray absorption spectroscopy is utilized for chemical analysis, because an absorption edge is characteristic of the atoms just as the emitted photon is characteristic of the atom. An absorption edge corresponds to the lowest energy that can excite an electron from an atomic electron orbital. The wavelength of the K absorption edge of most elements from atomic number 11 to 92 is presented in Table 15.1. X-ray absorption is covered in more detail in Section 18.4.2.

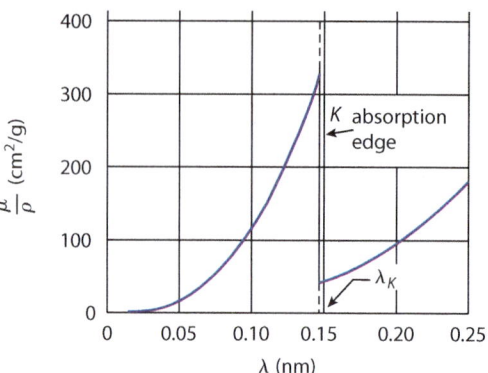

Figure 15.14　The mass absorption coefficient for nickel, plotted as a function of wavelength. (*Based on Cullity, B. D.* Elements of X-Ray Diffraction, *Addison-Wesley Pub. Co., Inc. (1956), p. 11.*)

The K absorption edge for nickel is shown in Figure 15.14 where, the mass absorption coefficient (μ/ρ) is plotted as a function of wavelength (λ). The linear absorption coefficient is μ, and the density is ρ. The mass absorption coefficient is independent of the density of the specimen; it is valid for a solid, liquid, or vapor. If the incident X-ray beam has an intensity I_0, then the intensity of the X-ray beam after penetrating a distance x into the specimen has an intensity I_x, as given by Equation 15.7.

$$I_x = I_0 \exp -\mu x \qquad \textbf{15.7}$$

It is desirable to have a uniform intensity of X-rays of various energies for X-ray absorption analysis. The intensity of radiation transmitted through the specimen is analyzed with either wavelength dispersion or with EDX. Since the transmitted beam is analyzed, the specimen must be thin enough to transmit X-rays. The presence of an absorption edge for an element is evidence that the element is present in the specimen. For example, if an absorption edge, such as that shown in Figure 15.14, is present at a wavelength of 0.1488 nm, then it is certain that Ni is present in the specimen. The magnitude of the edge can be related to the amount of Ni in the specimen with standards and by analytical procedures.

The most uniform source of continuous radiation in the X-ray region is from particle accelerators. As particles lose kinetic energy, they emit radiation just as electrons going through a material emit continuous radiation as they lose energy. The continuous radiation resulting from the energy loss of particles is used for X-ray absorption analysis. Since particle accelerators are not common laboratory devices, this technique is not used for routine chemical analysis. Also, the fine structure associated with the X-ray absorption edge is studied, because it reveals details about the chemical bonding of the atom absorbing the X-ray and its interactions with neighbors.

An application of the absorption edge is to filter out the K_β X-rays and much of the high-energy continuous X-rays coming from an X-ray tube, while allowing the K_α X-rays to penetrate through the filter. This technique produces a more monochromatic X-ray beam for X-ray diffraction experiments, as shown schematically in Figure 15.15. The filter for this technique is made from an element that has an absorption edge with an energy lower (longer wavelength) than that of the K_β X-ray beam coming from the X-ray tube, as shown in Figure 15.15a. The higher-energy, shorter-wavelength K_β X-rays are then absorbed by the filter, as shown in Figure 15.15b. However, the filter absorption edge must be higher in energy (shorter wavelength) than the K_α X-rays in the beam are, to allow the K_α X-rays in the beam to pass through the filter.

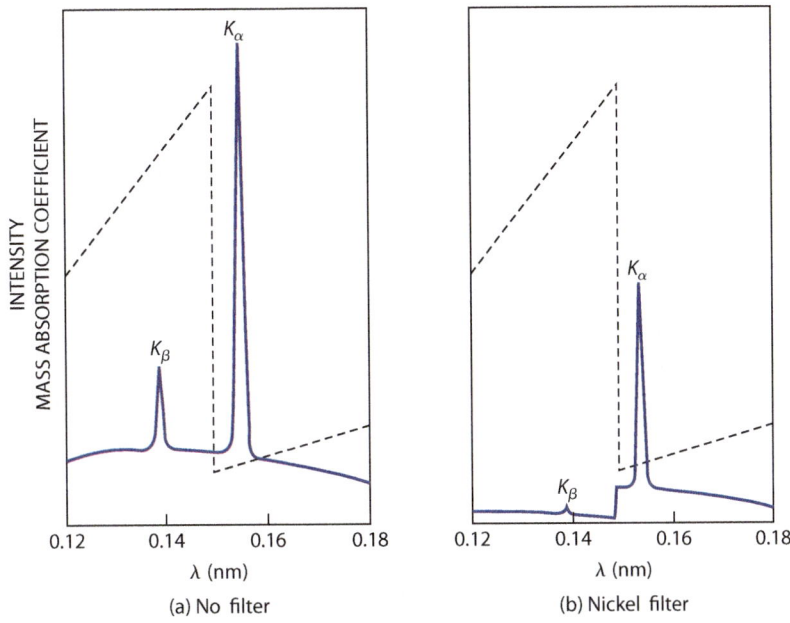

Figure 15.15 (a) A schematic of the intensity of the X-ray spectrum, including the K_α and K_β X-ray line intensity of a copper X-ray tube (solid blue line) without a filter, and the mass absorption coefficient of nickel (dashed line) as a function of wavelength. (b) A schematic of the intensity of the same copper X-ray tube after passing through a nickel filter (solid line) and the mass absorption coefficient of nickel (dashed line) plotted as a function of wavelength. *(Based on Cullity, B. D., Elements of X-Ray Diffraction, Addison-Wesley Pub. Co., Inc. (1956), p. 16.)*

Example Problem 15.5

In X-ray diffraction experiments, the intensity of the K_β characteristic radiation and the intensity of higher-energy white radiation is reduced relative to the intensity of the K_α characteristic radiation. The reduction is accomplished by placing a thin-film filter of an element in front of the exit port on the X-ray tube. If the X-ray tube is copper, determine the element for a filter that will selectively absorb the copper K_β line and transmit the K_α line.

Solution

From Table 15.1 the wavelength of the copper K_α line is 0.1541 nm and the K_β line is 0.1392 nm. If the filter is to transmit the copper K_α line, the absorption edge of the filter must have a wavelength that is shorter (higher energy) than the K_α wavelength of 0.1541 nm. Energies less than the absorption edge (longer wavelength) are selectively transmitted. Nickel is a possibility, with an absorption edge at 0.1488 nm. It is also necessary that the filter absorb the copper K_β at a wavelength of 0.1392 nm, and nickel does this because the K_β wavelength is shorter (higher energy) than the wavelength of the nickel absorption edge. The copper K_β radiation excites $1s$ electrons from the nickel atom and is absorbed. A schematic of how the nickel filter works is shown in Figure 15.15. The wavelength of the K absorption edge of the filter must be between the wavelength of the X-ray tube K_α and K_β lines; often the filter element is one atomic number less than that the X-ray tube element.

15.4 MICROSCOPY

Microscopy is the use of a microscope to study matter. A **microscope** is an instrument that increases the visual resolution relative to that of the human eye. **Resolution** is the minimum distance between two points that can be observed. The human eye has a resolution of about 150 μm. Microscopy techniques are by far the most commonly utilized instruments to characterize materials. Most microscopy techniques reveal the surface topography of specimens from the micro scale (10^{-6} m) to the atomic scale (10^{-10} m); however, a transmission electron microscope (TEM) allows observation of the internal structure of the specimen. The term **microstructure** refers to the character of a material that is observed with dimensions down to approximately 10^{-6} m; **nanostructure** is 10^{-9} m in size, and microscopes that have atomic resolution (10^{-10} m) have been developed. The STM image of the surface of nickel at the beginning of Chapter 2 is of individual atoms. With the proper microscope, dislocations, grain boundaries, twin boundaries, stacking faults, grains, precipitates, second phases, and atoms are viewable.

15.4.1 Optical Microscopy

Optical microscopes view the sample with visible light. A simple optical microscope has a single lens that works like a magnifying glass, whereas a compound optical microscope has two or more lenses. Most compound lens microscopes use one set of lenses to collect the light from the specimen and another set of lenses to focus the light for the eye to observe or for recoding the image. Most optical microscopy of materials is conducted on compound microscopes similar to the one shown in Figure 15.16. Optical microscopes can magnify materials up to approximately 1000 times their original sizes. **Magnification** is the ratio of the resolution possible with the human eye, divided by the resolution with a microscope. The maximum resolution with an optical microscope is approximately 200 nm. Transparent materials are observed in an optical microscope in the transmission mode, where the light passes through the material into the lens system of the microscope. However, the **depth of focus** in an optical microscope is quite small—for example, 0.86 μm at 500 times magnification; therefore, only a thin section of the material is in focus. The depth of focus is the thickness of the specimen that appears in focus. Because of the small depth of focus, it is best to view only thin materials in transmission with an optical microscope.

For opaque materials, it is necessary to view light that is reflected from the material surface. Material surfaces can be observed in their natural state, such as a machined surface; however, due to the small depth of focus for an optical microscope, if the surface is not flat, much of the surface can be out of focus. To observe grains and grain boundaries of a metal, the metal is polished with fine abrasives such as alumina or diamond to obtain a flat surface. The polished surface is then etched with various chemicals, such as acids, to enhance the contrast of features on a surface. The images of grain boundaries in the stainless steel shown in Figure 3.14b and the brass shown in Figure 7.6 are optical micrographs of polished and etched surfaces. Defects in the material, such as grain boundaries or dislocations, etch more rapidly than defect-free material does, because the defects have a higher energy and are less stable. Grain boundaries appear dark because they are recessed, and light does not reflect from them into the microscope as well as it does from the polished surface. Note that different grains of the brass in Figure 7.6b have different appearances, because the grains with different crystallographic orientations etch differently, and they scatter the light with different intensities into the microscope. Because of different material reactions to etching, optical microscopy is utilized to observe the grains in materials, grain boundaries, twin boundaries, phase boundaries, large precipitates, and second phases. We can also observe where dislocations intersect a

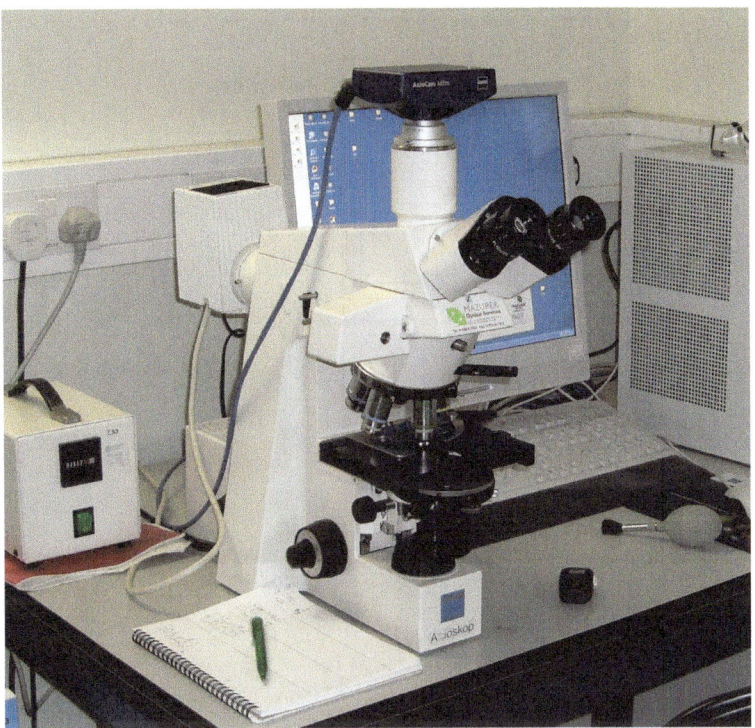

Figure 15.16   An optical microscope with a top-mounted digital camera for recording images. (Courtesy of Richard Wheeler)

surface, because an acid etch preferentially removes material around the dislocation, leaving an etch pit at the site of the dislocation.

The introductory photograph in this chapter was taken with an optical microscope using polarized light. If polarized light is either transmitted or reflected from the surface of an optically active material, the polarization of the transmitted or reflected beam is rotated relative to that of the incident beam. In the technique called cross polarization, an analyzer is placed at 90° from the incident polarized light. Without an optically active specimen, no light passes through the analyzer. Placing an optically active specimen between the polarized light source and the analyzer rotates the polarization of the light and results in an image, such as the spherulites in the introductory photograph.

15.4.2   Scanning Electron Microscopy

In a **scanning electron microscope** (SEM), shown schematically in Figure 15.17, an electron beam with energies from a few hundred eV to 50 keV is scanned across the surface of a sample resulting in an image of the sample surface. Figure 12.19 is an image of a composite fracture surface taken with an SEM. Because electrons are absorbed and scattered by air, the beam column and specimen chamber of an electron microscope is in a vacuum. In electron microscopes, the electron beam is created by heating a filament, such as tungsten or lanthanum hexaboride, with a high electrical current, resulting in thermionic emission of electrons. In Figure 15.17, the electron gun is the electron emitter. The emitted electrons form a beam that is accelerated through the column, toward the sample, and then focused with electromagnetic

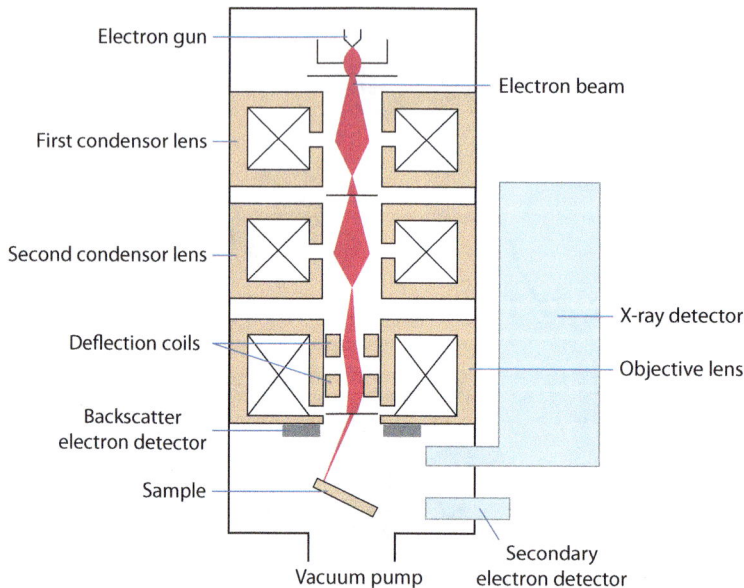

Figure 15.17   A schematic of a scanning electron microscope (SEM) showing the electron beam in red, lenses to focus the beam, deflection coils to scan the electron beam across the sample, and secondary electron and backscatter electron detectors for surface imaging of the sample. An X-ray detector for chemical analysis is shown. *(Based on http://upload.wikimedia.org/wikipedia/commons/thumb/0/0d/Schema_MEB_%28en%29.svg/2000px-Schema_MEB_%28en%29.svg.png)*

lenses to a diameter of 1 to 5 nm. The focused electron beam is scanned across the specimen surface in a line with an electromagnetic deflection coil, and then the electron beam is stepped to scan line after line across the sample surface. This is the same way that an image is produced on the screen of a cathode ray tube in a television or an oscilloscope.

When the electron beam strikes the atoms in the surface of the sample, a number of processes occur. Sufficiently high-energy electrons in the primary electron beam can excite core electrons from the atoms in the sample, producing secondary electrons that are emitted from the surface. Vacant core electron orbitals are subsequently filled by higher-energy electrons, with the emission of characteristic X-rays. The secondary electrons emitted from the surface are typically less than 50 eV, and as a result they primarily come from within a few nanometers of the surface.

Most images from an SEM are produced from the secondary electrons. The intensity of secondary electrons detected results in an intensity on the SEM image screen that is representative of the material in the spot that the primary beam irradiated at that instant. Since the primary electron beam continuously scans across the sample, stepping an increment after each scan, an image of the scanned surface is produced. The intensity of secondary electrons forming the image is primarily dependent upon the height of the spot producing the intensity. If it is a high spot, the intensity of emitted secondary electrons is high, and this appears as a bright spot. If it is a low spot, such as the bottom of an etch pit, the intensity of secondary electrons reaching the detector is low, and this appears as a dark spot. The SEM image is displayed on a computer screen, and the image can be recorded on film or stored in computer memory. The sample image can be magnified up to 20,000 times in a typical research-grade SEM with a resolution of 20 nm, and magnifications of 100,000 times with resolutions down to 1 nm are achieved in a high-resolution SEM. Because the depth of focus of an SEM is hundreds of times that of an optical microscope, the images appear three-dimensional as in the composite-material fracture surface shown in Figure 12.19.

Some of the electrons in the primary electron beam are elastically backscattered from the electrons on the atoms in the sample, and the backscattered electron intensity is dependent upon the number of electrons on the scattering atom. An image of the surface from the backscattered electrons is indicative of the atomic number of the elements on the surface. Backscattered electrons are also diffracted as they pass through and out of the specimen, and the image of these diffracted electrons is used to determine the crystallographic orientation of the surface. The backscattered electrons are detected with a different detector design than are low-energy secondary electrons. In one design shown in Figure 15.17, a detector shaped like a doughnut with a hole in the center is placed above the sample, and the hole in the center of the detector allows the primary electron beam to pass to the sample.

Because the sample surface in an SEM is continuously bombarded with electrons, an insulating surface becomes negatively charged and emits electrons, resulting in a glowing sample image. If insulating surfaces are coated with a thin layer of a conductive material, such as gold, platinum, or graphite, then the charge is conducted to ground. A conductive coating allows for the study of ceramics, polymers, and biological materials in an SEM. You have probably seen images of insects, such as ants, taken with an SEM.

SEMs are often equipped with EDX spectroscopy systems for chemical analysis, as indicated by the X-ray detector in Figure 15.17. EDX systems on an SEM allow for the analysis of the chemistry of small areas observed in the microscope, although the spatial resolution in the chemical-analysis mode is not as good as in the microscopic mode.

An SEM can also be equipped with a detector that can determine the spin-polarization direction of the electrons excited from the sample, which lets us study the direction of magnetization of regions of a sample.

15.4.3 Transmission Electron Microscopy

In a transmission electron microscope (TEM), shown schematically in Figure 15.18, an electron beam with energies from 50 keV to over 1000 keV penetrates through a thin-film sample and produces an image of the interior of the sample. The beam column and specimen chamber of a TEM is also in a vacuum. The electron beam in a TEM is formed by thermionic emission of electrons from a filament. The

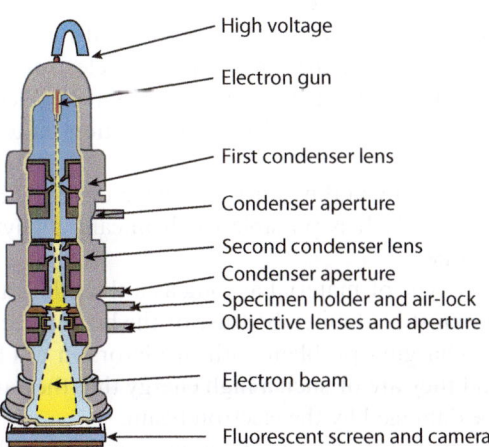

Figure 15.18 A schematic of a transmission electron microscope (TEM), showing the electron gun, lenses, apertures, specimen holder, observation screen, and camera. (*Based on http://upload.wikimedia.org/wikipedia/ commons/2/24/Electron_Microscope.png*)

beam is focused and confined by apertures in a manner similar to how an SEM functions. However, in a conventional TEM the electron beam does not scan. The TEMs in most laboratories have accelerating voltages from 50 to 200 kV; only a few labs have a TEM with accelerating voltages of 1000 kV or higher. In a TEM, the primary electron beam penetrates through the specimen, creating the bright field image on a fluorescent screen. The fluorescent screen can be tilted out of the beam to expose photographic plates, film, or electronic detectors for a permanent record of the image. The specimen thickness that can be penetrated depends upon the beam energy and the desired resolution. For example, a 100-keV electron beam in a TEM can penetrate a nickel specimen of up to 200 nm thick. Higher-energy electron beams can penetrate thicker specimens, but a high resolution requires thin specimens.

The image produced in a TEM is primarily due to electron diffraction and scattering differences in the specimen. If there is a volume of material, such as a grain, that diffracts electrons out of the primary beam, then this volume appears darker than does a volume that transmits the primary beam. Also, if there is a defect in the material, such as a dislocation, and the defect scatters electrons out of the primary electron beam, then the defect appears dark in comparison to the surrounding material, as shown in the TEM of dislocations in iron in Figure 7.1. In a TEM it is possible to observe dislocations, grain boundaries, phase boundaries, twin boundaries, precipitates, and second phases because the defects scatter the electron beam and interfere with diffraction.

It is also possible to observe an image produced by electron beams diffracted from particles such as precipitates. In these images, an electron beam diffracted by the precipitate particle is isolated with a device that blocks all other electron beams. In these images, the precipitate particle is bright and the matrix material is dark. These images are called dark field images, and Figure 7.13 is an example.

A TEM can magnify specimens up to 50 million times their original size, making atomic resolution possible. The resolution in a TEM is approximately 0.17 nm at a beam energy of 100 keV, and 0.07 nm at a beam energy of 1000 keV. A resolution of 0.050 nm has been achieved in a TEM. Depth of focus is not an issue in a TEM, since the specimens must be thin to be observed in transmission, and the specimen thickness is much less than the depth of focus. The entire specimen appears in focus when the image is properly focused.

The crystal structure of a small volume of material in the sample can be determined by electron diffraction, as we saw with Example Problem 15.3. A TEM is used to determine crystal structures of nano sized particles. Also electron diffraction in a TEM is used to identify small volumes of material that have an already determined crystal structure, or to determine the orientation of a small volume of material in a thin-film sample. It is possible to produce an electron diffraction pattern of a small volume of the sample by masking off the remainder of the sample from the electron beam and then blocking the forward transmitted beam. In this way the diffracted electrons from the selected area are observed in a selected-area diffraction pattern. For example, the electron diffraction pattern of a precipitate particle can be obtained in a TEM.

Chemical composition is determined by adding an EDX detector to a TEM in a manner similar to that shown for the SEM in Figure 15.17. It is possible to chemically analyze individual atoms in the highest-resolution electron microscopes.

A TEM can be used to study any type of material as long as a thin film can be made of the material. Materials of a high atomic number must be thinner, because of the higher electron-absorption coefficient in these materials. There is not a charging problem with insulators in a TEM, because the electrons penetrate through the material, and they are of such a high energy that they are not affected by a surface charge. Biological materials can be damaged by the electron beam.

Although a TEM views the full thickness of a specimen, we can observe the surface of a material by making a replica of the surface. The replica is made from a thin layer of polymer that is coated onto the specimen and then peeled away. The replica is an inverse of the surface. A depression in the surface is a peak on the replica. The replica is then coated with a noncorrosive conductor, such as gold, to increase the contrast of the replica. The replica of the surface when viewed in the TEM provides a high-magnification image of the surface morphology.

An electron microscope with an electron beam of sufficient energy for transmission, such as 200 kV, can be equipped with scanning coils similar to those of a SEM, to produce a scanning transmission electron microscope (STEM). STEMs operating in the transmission mode are able to achieve atomic resolution. One advantage of a STEM is that there is a minimum of radiation damage to the specimen, because the electron beam scans across the specimen rather than being continuously fixed on the specimen as a TEM does. This is an advantage for viewing biological material.

STEMs are also able to operate as an SEM by observing the secondary electron emission from a specimen. Recently, atomic resolution with secondary electron emission has been achieved with STEMs operating at 200 kV.

15.4.4   Scanning Tunneling Microscopy

With a **scanning tunneling microscope** (STM), it is possible to observe features on a surface of both metals and semiconductors that are of atomic dimensions, as shown in the image of the nickel surface at the beginning of Chapter 2. A schematic of an STM is shown in Figure 15.19.

An STM has a metallic probe, such as a tungsten wire, that is atomically sharp. The probe is scanned with **piezoelectric transducers** (PZTs) across the surface of the sample in a manner similar to how an SEM works. PZTs are crystalline materials that are deformed by the application of a voltage, and we will

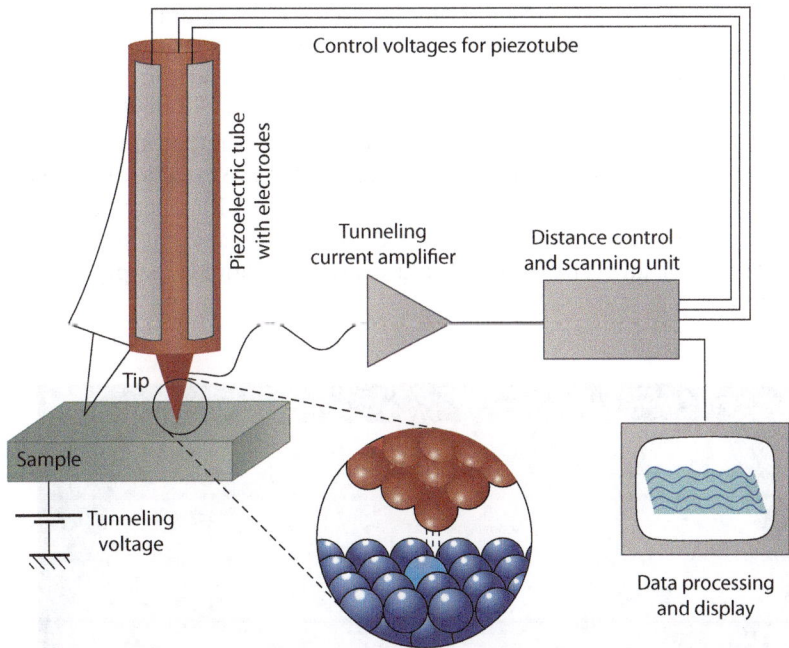

Figure 15.19   A schematic of a scanning tunneling microscope (STM), with an inset of the atomically sharp probe tip, the surface of the sample, the gap between the tip and the sample, and the tunneling current between the tip and the sample indicated in the amplified image. The tube can flex as shown by the outline to the left. (*Based on http://en.wikipedia.org/wiki/File:ScanningTunnelingMicroscope_schematic.png*)

learn more about them in Chapter 16. The probe tip is positioned approximately 1 nm from the sample surface, and this distance is called the gap. A few volts are applied between the tip of the probe and the sample surface. If the probe is biased negatively relative to the sample, then electrons in the probe tunnel across the gap and into the sample. Tunneling is a quantum mechanical phenomena, resulting from a finite probability that the electrons can cross the gap into unfilled electron quantum states of the sample. The tunneling current is exponentially dependent upon the negative of the size of the gap between the probe tip and the sample surface. The size of the gap is maintained at a fixed value by a feedback system that keeps a constant tunneling current between the tip and the sample. Changes in height on the surface created by steps and valleys are observed by the feedback system, which maintains a fixed value of the gap. At a step up on the surface the tip moves up to maintain a constant tunneling current, and this tip displacement is recorded and the height of the step measured. The image that is observed on a computer screen is the tip displacement necessary to maintain a constant tunneling current as a function of the x and y position on the surface. Because the STM depends upon a current flow into or from the sample, this kind of microscope cannot be used to observe electrically insulating materials, such as ceramics, polymers, and biological materials; only conductors and semiconductors are observable with an STM.

With an STM designed for atomic resolution, it is possible to observe images of atomic dimension on a surface, as shown in the STM image of the surface of nickel at the beginning of Chapter 2. Atomic-dimension features are observed on a flat surface because the probability of tunneling into unoccupied electron quantum states changes with position around the atom. Examples of atomic-resolution studies with an STM are atoms and molecules adsorbed onto a surface, nuclei of a few atoms or molecules formed on a surface during growth of crystals using vapor deposition, and the diffusion of atoms across a surface.

The type of atom on a surface can be identified with an STM. The tunneling current is proportional to the probability of tunneling from a negatively charged tip into unoccupied surface quantum states of the positively charged sample. The applied voltage between the probe and the specimen surface can be varied to explore the energy of the unoccupied surface quantum states of the sample. The tunneling current as a function of voltage can be used to identify the type of atoms on a surface.

The tunneling current in an STM has been used to manipulate atoms on a surface. Figure 15.20 shows xenon atoms on the (110) surface of nickel that were manipulated to form the letters "IBM." This was the first time that individual atoms were manipulated to produce a desired product.

An STM can be operated in air, liquids, or a vacuum. STMs that are capable of atomic resolution are operated in a vacuum and require significant vibration isolation. In STMs operated in air, condensation of water onto the specimen surface often interferes with the image, and surface oxidation occurs on reactive specimen surfaces. For an STM operated in a conducting liquid, the probe is coated with an insulator, except for the tip.

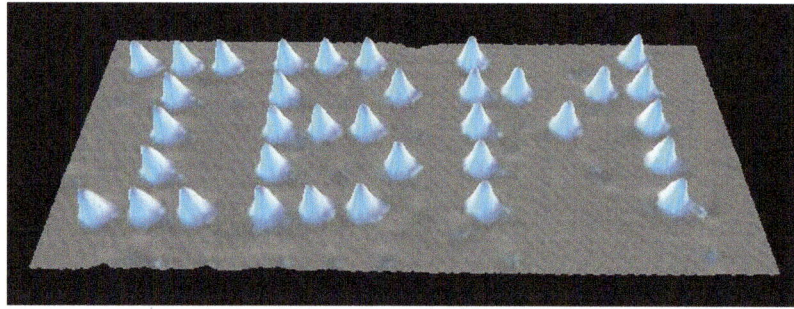

Figure 15.20   Individual xenon atoms manipulated on a (110) surface of nickel with an STM to produce the letters "IBM." The image was originally created by Donald Eigler of IBM. (IBM)

15.4.5   Atomic Force Microscopy

The operation of an **atomic force microscope** (AFM), shown schematically in Figure 15.21, is based upon the interaction force between the atoms at the tip of a sharp probe made of silicon, silicon nitride, or quartz; and the atoms in the sample. Since all atoms have interaction forces with a probe, it is theoretically possible to study any material with an AFM. In the microscope mode of operation, an AFM is used to study surface topography, such as etched grain boundary grooves, steps and clusters of atoms on a surface, and the arrangement of surface atoms. Atomic resolution has been obtained with AFMs operated in vacuum and in liquids. Figure 1.8 is an AFM image of individual carbon atoms in graphene. Figure 15.22 is an AFM image of clusters of atoms of the high temperature superconducting material $YBa_2Cu_3O_{7-\delta}$.

Three modes of AFM operation can produce images of surface contours: contact, dynamic, and dynamic-contact. The contact mode is similar to the operation of a record player and to how you would slide your finger over the bark of a tree to determine its roughness. On a vinyl record, changes in height in the grooves are sensed with a needle probe and converted into sound as the vinyl record is rotated under the fixed needle. When you slide your finger over the bark of a tree, your finger moves up and down with the contour of the tree bark, and from the up-and-down motion of your finger you sense the roughness of the bark surface.

In the contact mode of AFM, the probe touches the surface of the sample with a constant force in the range from 0.1 to 0.5 nN. The interaction force includes both repulsive and attractive forces calculated with equations similar to Equation 2.11. The sample is scanned in x and y coordinates with PZTs relative to the probe (tip) on the end of a cantilever. The position of the probe is determined with a laser beam reflected from a polished surface on the probe. A feedback system maintains a constant position of the probe by having the PZT move the sample in the z direction. The sample is moved down at a peak and up at a valley by the PZT. The image produced is of the x and y coordinates on the sample and the z displacement of the sample, as shown in Figure 15.22, or indicated by color, as

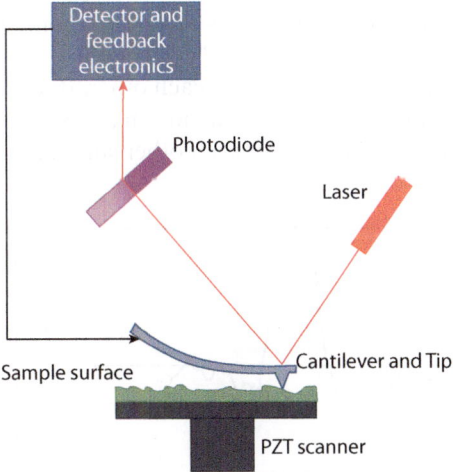

Figure 15.21   A schematic of an AFM. The sample is attached to a piezoelectric transducer (PZT) that scans the sample under a sharp tip attached to a cantilever. The position of the tip is measured by a laser light beam reflected from the end of the cantilever into a photodiode light detector. (*Based on http://en.wikipedia.org/wiki/File:Atomic_force_microscope_block_diagram.svg*)

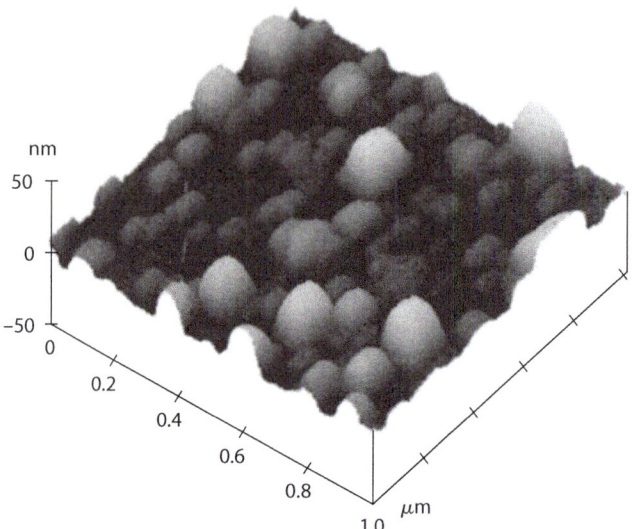

Figure 15.22   An AFM image of clusters of $YBa_2Cu_3O_{7-\delta}$ deposited on an insulating substrate of $SrTiO_3$ by pulsed laser deposition. (With kind permission from Springer Science+Business Media: Applied Physics A, Materials Science and Processing, "Critical dimensions for YBCO islands grown by pulser laser depositions," Vol. 75, (2002), p. 566, Gilmore, C.M., and Kim, J., © Springer-Verlag 2001)

shown in Figure 1.8. It is also possible to record the height along a line scan in the surface (section analysis), as shown in Figure 15.23.

Atomic resolution is not normally obtained with the contact mode of an AFM. In the contact mode, the probe is pressed against the surface, and too many atoms from the probe and the sample interact at once to achieve atomic resolution. Even though contact mode AFM images show the proper atomic periodicity, the image does not show any atomic-size defects. A comparison has been made to what would be observed by sliding over each other the bottoms of two large egg containers that each have spaces for 2 dozen individual eggs. The periodicity of the individual egg containers is observed by the periodicity of the up and down motion as the egg containers slide over each other. But a defect in one of the individual egg containers is not observed, because the up and down motion (z) as a function of x and y is an average from all of the individual egg containers sliding over each other simultaneously.

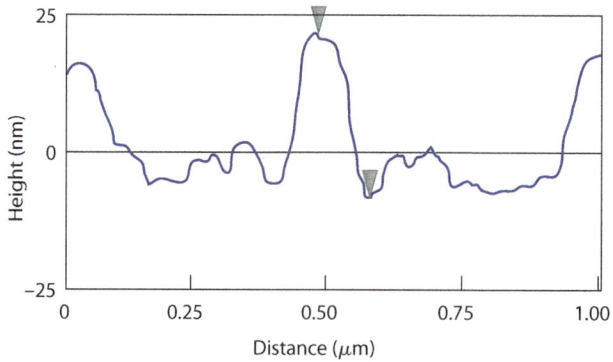

Figure 15.23   An AFM section analysis of the clusters in Figure 15.22. (*Based on Gilmore, C. M., and Kim, J., Applied Physics A, Materials Science and Processing, 75, (2002), p. 566.*)

Atomic resolution with an AFM is observed with the dynamic mode of operation in an ultra-high vacuum. In the dynamic mode of operation, the tip does not touch the sample. The cantilever is vibrated a few nanometers above the sample. The dynamic mode uses either amplitude modulation or frequency modulation. In amplitude modulation, the cantilever is vibrated at a constant frequency as the tip is scanned across sample without touching the surface. The amplitude of vibration is monitored, and the amplitude changes depending upon the force of interaction with the surface. The force of interaction is given by equations similar to 2.11. In frequency modulation, the cantilever is vibrated at constant amplitude. The tip is scanned across sample and the resonant frequency of the cantilever changes depending upon the force of interaction with the surface. The frequency is adjusted to the resonant value with the feedback system. In amplitude and frequency modulation, the tip-to-sample distance and the resulting interaction force is maintained at a constant value with the feedback system. The movement of the sample in the z direction at the position x, y necessary to maintain a constant tip-to-sample distance and the resulting interaction force results in an image of the topography of the surface.

In the dynamic mode, only the long-range, attractive forces between the probe tip and the sample are measured. The short-range, repulsive forces in Equation 2.11 are not sampled, because the probe tip and the sample do not come into contact where the short-range repulsive forces are observed. The magnitude of the interaction force is measured in piconewtons. The dynamic mode of operation is also used on biological materials and other soft samples that would be damaged by contact with a sharp probe tip.

In the dynamic-contact mode, also called the tapping mode, the cantilevered probe is vibrated at its resonant frequency at an amplitude that is about ten times as large as in the dynamic mode. At the higher amplitude of vibration, the tip can intermittently touch the surface of the sample. By tapping the surface the short-range, repulsive forces are sampled. The long-range attractive interaction forces between the probe tip and the sample determine the resonant frequency. The image of the surface topography is produced in a manner similar to that of the dynamic mode. The dynamic-contact mode is analogous to determining the texture of the bark of a tree with your finger by repeatedly touching it in a series of locations, rather than sliding your finger over it with continuous contact.

In the AFM force mode of operation, the force between the tip and the sample is measured as a function of separation. This force is measured as the tip is brought from a large distance into contact with the sample, and the reverse process is also measured. In this way force displacement curves similar to that shown in Figure 2.32 are developed. It is also possible to measure the force between a dynamic cantilevered probe and a single atom or a cluster of atoms on the sample surface by measuring the change in the resonant frequency of the dynamic cantilever. Not only is it possible to measure the overall bond force, the direction of the bond forces can be measured. The chemistry of individual atoms on the surface is identified by their characteristic bonding with the atoms at the tip of the cantilever. Friction force microscopy (FFM) has been developed to measure the lateral forces due to friction resulting from the relative motion between the probe and the sample.

It is possible to manipulate atoms on a surface with an AFM. The atoms are moved by vibrating the dynamic cantilever in the vicinity of the atom to be moved. The ability to move atoms provides the ability to process materials at the atomic level.

An AFM can be operated in air, liquids, or vacuum. In systems operated in air, impurities such as water vapor can condense on the surface and affect the interaction forces between the tip and the sample. The contact mode and dynamic-contact modes of operation can be conducted in air, but the results can be affected by condensed water. It is very difficult to operate in the dynamic mode in air at room temperature, because water vapor condenses between the vibrating probe and the specimen.

Atomic resolution is possible in the dynamic and dynamic-contact modes of AFM operation in vacuum. Atomic resolution has also been achieved in the dynamic-contact mode of operation in liquids.

Summary

- Diffraction techniques are utilized to determine the crystal structure of materials, including the Bravais lattice type, the lattice parameters, and the atom positions. The radiation types that are used for diffraction include X-rays, electrons, and neutrons.

- X-ray diffraction is used to determine the crystal structure of metals, ceramics, polymers, and even DNA, because the wavelengths of X-rays are comparable to interatomic distances. Diffraction intensities for all types of radiation occur when Bragg's law is obeyed. The X-ray powder diffractometer and the Debye-Scherrer camera are used to determine if materials of known crystal structures are present in a sample. The Laue X-ray technique is used to determine the crystal class and for crystal orientation. A single crystal diffractometer is used to determine unknown crystal structures.

- The wavelength of any particle with wave characteristics, such as electrons and neutrons, is given by the DeBroglie relationship, which equates the wave and particle momentum.

- Electron diffraction in a transmission electron microscope (TEM) is used to identify the crystal structure of very small volumes of material and to determine the orientation of a particular volume of crystalline material in a sample. Low-energy electron diffraction (LEED) is usually combined with other surface-analysis techniques, such as electron spectroscopy, to analyze both the structure and the chemistry of the sample surface. Neutron diffraction is used to determine the crystal structure of samples with elements of low atomic number, such as hydrogen, with elements of adjacent atomic number, such as nickel and cobalt, and magnetic elements.

- Diffraction can be used to measure elastic strain in crystalline samples.

- In X-ray fluorescence spectroscopy, the sample analyzed for chemical composition is bombarded with either primary X-rays or with electrons. If the irradiation is primary X-rays, they are absorbed by the atoms in the sample, and the atoms emit secondary X-rays that are characteristic of the elements in the sample. The intensity of the emitted characteristic X-rays is indicative of the concentration of the element emitting the X-rays. The sample is usually a solid; however, it can be a liquid or vapor. Chemical analysis is also conducted by analyzing X-rays generated by electron beams penetrating samples in electron microscopes. However, if a specimen is bombarded with high-energy electrons, the continuous white radiation emitted can mask the characteristic X-rays of trace elements in the specimen. The characteristic X-rays emitted from the sample are separated by diffraction from a crystal (wavelength dispersion) or by electronic detectors and analyzers (energy dispersion).

- X-ray absorption edges are used for chemical analysis of thin films and to study the bonding of atoms. Filters for X-ray tubes to produce a more monochromatic X-ray beam use absorption edges.

- A microscope is an instrument that increases the visual resolution relative to that of the human eye. Microscopes are based upon visible light, electrons, and surface probes.

- The maximum magnification in an optical microscope is typically 1000 times, with a maximum resolution of approximately 200 nm and a depth of focus of less than a μm.

- In a scanning electron microscope (SEM), an electron beam with energies from a few hundred eV to 50 keV scans across the surface of a sample and produces an image of the sample surface. In a typical research-grade SEM, the sample image is magnified up to 20,000 times, with a resolution of 20 nm. Magnifications of 100,000 times, with resolutions down to 1 nm, are achieved in a high-resolution SEM with a depth of focus of hundreds of times that of an optical microscope. Energy-dispersive X-ray (EDX) spectroscopy is added to an SEM for chemical analysis.

- In a transmission electron microscope (TEM), an electron beam with energies from 50 keV to over 1000 keV penetrates through a thin-film sample and produces an image through the sample. The resolution in a TEM is approximately 0.17 nm at a beam energy of 100 keV, and 0.07 nm at an energy of 1000 keV. A resolution of 0.05 nm has been achieved in a TEM. All of the sample is in focus with a TEM, when the image is properly focused. EDX analysis is added to a TEM to identify atom types.

- With a scanning tunneling microscope (STM), it is possible to observe features on a surface of both metals and semiconductors that are of atomic dimensions. The image from an STM is created by the tunneling current between an atomically sharp probe and a conducting surface as a function of the scanned x, y position on the surface. The type of atom on the surface is identified by characteristic tunneling currents between the tip and the sample atoms. Atoms on surfaces are manipulated with the tunneling current. STMs are operated in air, liquids, or a vacuum.

- An atomic force microscope (AFM) uses the interaction force between the atoms at the tip of a sharp probe at the end of a cantilever and the atoms in the sample to control the height (z) of the sample relative to the probe. The image from an AFM is of the height (z) of the sample as a function of the scanned x, y position on the surface. AFMs have three modes of operation: The contact mode has a constant force between the probe and the sample, the dynamic mode vibrates the cantilever above the sample surface and maintains a constant force of interaction between the probe tip and the sample, and the dynamic-contact (tapping) mode vibrates the cantilever above the sample and the probe tip periodically touches the sample. AFMs can be operated in air, liquids, and in vacuum. Atomic resolution is obtained in vacuum with the dynamic and dynamic-contact modes of operation, and in liquids with the dynamic-contact mode. The dynamic mode is useful for biological and other soft samples. AFMs are used to measure the interaction force between the probe and the sample as a function of separation distance, to identify the types of atoms on a surface from characteristic force-separation plots, to measure frictional forces on a surface, and to manipulate atoms on a surface.

Supplemental Reading Subjects and Authors

Full references are listed at the end of the book.

General:	*Askeland, Fulay, and Wright*
X-ray diffraction and spectroscopy:	*Barrett and Massalski; Cullity; Massa*
General microscopy:	*Rochow and Tucker*
Electron microscopy:	*Goldstein and Yakowitz; Hall; Reimer*
Probe microscopy:	*Birdi; Bonnell; Chen*

Homework

Concept Questions

1. _____ are electromagnetic radiation with wavelengths typically from 0.1 nm to 1 nm.

2. Diffraction intensities occur when _____ law of diffraction is obeyed.

3. X-rays are scattered by _____ on the atoms in the sample.

4. In the _____ technique the specimen is in a fixed position relative the X-ray beam and diffraction of continuous radiation occurs when Bragg's law is satisfied.

5. A(n) _____ is a device that allows the angle of a sample relative to a reference to be changed and measured.

6. If the sequence of the values of $\sin^2 \theta$ for the diffraction peaks of a cubic crystal are 3, 4, 8, 11, 12, 16, 19, then the crystal is _____-_____ cubic.

7. The most accurate value of the lattice parameter results from an extrapolation of $\sin^2 \theta$ to _____.

8. If a crystal has a fourfold axis of symmetry, the crystal can be rotated by _____ degrees about this axis and the diffraction pattern will look exactly the same.

9. If a crystal as a sixfold axis of symmetry that is perpendicular to a twofold axis of symmetry, this is a(n) _____ crystal.

10. Defects in crystals increase the energy _____ of the X-ray diffraction lines relative to a perfect crystal.

11. The wavelength of any particle with wave characteristics, such as electrons and neutrons, is given by the _____ relationship that equates the wave and particle momentum.

12. Neutrons are scattered by the _____ of the atoms in a sample.

13. In diffraction measurements of strain, the reference gauge length is the unstressed _____ spacing.

14. If a sample is bombarded with high-energy electrons, there are two forms of X-rays emitted: continuous-wavelength X-rays and _____ X-rays.

15. _____ emission is when a filament is heated to a high temperature and emits electrons.

16. If an X-ray fluorescence spectrometer uses diffraction by a crystal to analyze the spectrum from a sample, this is called _____ dispersion.

17. Fluorescence is the decay of an excited state is less than _____ seconds with the emission of a photon.

18. For an electron transition to be allowed on an atom with a missing core electron, the angular momentum quantum number (l) must change by + or – _____.

19. The L absorption edge corresponds to the excitation of an electron out of an atomic orbital with the principal quantum number of _____.

20. _____ is the minimum distance between two points that can be observed.

21. _____ is the ratio of the resolution possible with the human eye divided by the resolution with a microscope.

22. The backscattered image from an SEM is primarily dependent upon the atomic _____ of the elements in the sample.

23. The surface of a sample for observation in an SEM must be a(n) _____.

24. It is possible to observe the surface of a material in a TEM by making a(n) _____ of the surface.

25. The tunneling current in an STM is _____ dependent upon the negative of the gap between the tip and the specimen.

Engineer in Training–Style Questions

1. The wavelength of an X-ray is between
   (a) 10 and 100 nm
   (b) 1 and 10 nm
   (c) 0.1 to 1 nm
   (d) 0.01 and 0.1 nm

2. Which of the following may not be possible to determine with the powder method of X-ray diffraction?
   (a) The presence of a material in a specimen with a known crystal structure
   (b) The unit cell lattice parameters of a known crystal structure
   (c) If a cubic crystal is simple, body centered, or face centered
   (d) A previously unknown complex crystal structure

3. For a BCC crystal, which of the following X-ray (hkl) diffraction peaks is not present?
   (a) (200)
   (b) (110)
   (c) (130)
   (d) (111)

4. For an FCC crystal, which of the following X-ray (hkl) diffraction peaks is not present?
   (a) (100)
   (b) (111)
   (c) (131)
   (d) (200)

5. To orient a single crystal in an X-ray beam, it would be best to use
   (a) A Debye-Scherrer camera
   (b) The Laue technique
   (c) A powder diffractometer
   (d) A single crystal diffractometer

6. To determine an unknown crystal structure, it would be best to use
   (a) The Debye-Scherrer camera
   (b) The Laue technique
   (c) A powder diffractometer
   (d) A single crystal diffractometer

7. What diffraction technique would you use to determine the atomic positions of hydrogen in uranium hydride?
   (a) X-ray diffraction
   (b) High-energy electron diffraction
   (c) Low-energy electron diffraction
   (d) Neutron diffraction

8. Which form of radiation is not suitable for determining the elastic strain in a crystalline material using diffraction?
   (a) X-rays
   (b) Visible light
   (c) Electrons
   (d) Neutrons

9. What strain(s) can be measured by diffraction techniques?
   (a) Elastic and plastic
   (b) Plastic
   (c) Elastic
   (d) Viscous

10. A K_β X-ray results from which of the following electron transitions in an excited atom?
    (a) $2s \rightarrow 1s$
    (b) $2p \rightarrow 1s$
    (c) $3p \rightarrow 1s$
    (d) $3p \rightarrow 2s$

11. The best filter to remove the K_β X-rays from a Mo X-ray tube that have a wavelength of 0.06322 nm but that transmits the K_α X-rays of wavelength 0.0711 nm is
    (a) Nb with a K edge of 0.06529 nm
    (b) Mo with a K edge of 0.06198 nm
    (c) Y with a K edge of 0.07276 nm
    (d) Ru with a K edge of 0.05605 nm

12. An SEM is not suitable for viewing which of the following materials?
    (a) Metals
    (b) Ceramics
    (c) Biological materials coated with a conductor
    (d) Polymers coated with a conductor

13. In an SEM, the crystallographic orientation of the surface is determined by the diffraction of
    (a) Secondary electrons
    (b) Emitted X-rays
    (c) Backscattered electrons
    (d) The primary electron beam

14. Which of the following microscopes is not capable of resolving individual atoms?
    (a) An SEM operating at 30 kV
    (b) A TEM operating at 200 kV
    (c) An STM
    (d) An AFM

15. Which of the following cannot be used to observe an electrically insulating surface?
    (a) An optical microscope
    (b) A STM
    (c) A TEM
    (d) An AFM

16. Which of the following is not suitable for studying biological material?
    (a) An optical microscope
    (b) A high-voltage TEM
    (c) A STM
    (d) An AFM

17. Frictional forces between a probe and a sample are measured with a variant of which microscope?
    (a) An optical microscope
    (b) A SEM
    (c) A STM
    (d) An AFM

18. Which of the following modes of observation of the image of a sample in an AFM is not expected to produce atomic resolution?
    (a) Contact
    (b) Amplitude-modulated dynamic
    (c) Frequency-modulated dynamic
    (d) Dynamic-contact or tapping

Problems

Problem 15.1 Determine the values of (hkl), the interplanar spacing (d_{hkl}), and the angles of diffraction (θ) for the first eight peaks for BCC iron with a lattice parameter of 0.287 nm, when using molybdenum K_α radiation with a wavelength of 0.0711 nm.

Problem 15.2 A copper-nickel alloy is used in a heat exchanger for a power plant that uses ocean water for cooling. One way to check the consistency of the alloy's composition across batches of material is to determine the lattice parameter of the alloy. We know that copper and nickel are both FCC crystals and that copper and nickel form a limited solid solution, and also that the lattice parameter of the alloy depends upon the composition. To determine the lattice parameter of the alloy, the angles of the diffraction peaks from a polycrystalline piece of the alloy are measured in an X-ray diffractometer with a cobalt X-ray tube. Cobalt is used because the cobalt K_α radiation is of a lower energy than is the absorption edge in either copper or nickel, and no characteristic X-rays will be generated in the copper and nickel. Also, the copper and nickel should be relatively transparent to the cobalt K_α radiation.

With the cobalt K_α radiation, the lowest-angle intense peak, is measured at 25.74 degrees. The lattice parameter of copper is 0.361 nm and for nickel it is 0.352 nm, and the wavelength of cobalt K_α radiation is 0.1790 nm.

(a) What are the Miller indices of this peak, and what is the lattice parameter of this copper-nickel mixture?

(b) Assuming that the lattice parameter varies linearly with composition in this alloy, what is the percent of nickel and copper?

Problem 15.3 A specimen is made of 80 atom % iron and 20 atom % Ni. When sufficient FCC nickel is added to BCC iron, it is possible to produce FCC iron-nickel alloys, which are austenitic. An X-ray diffractometer scan of this specimen with cobalt K_α radiation that has a wavelength of 0.1790 nm results in peak intensities at the following angles: 26.1°, 30.56°, 45.99°, 57.49°, and 61.74°.

(a) Is this iron-nickel alloy BCC or FCC?

(b) Determine the lattice parameter using the 26.1° peak.

Problem 15.4 For the X-ray diffraction pattern in Figure 15.4 produced with copper K_α of wavelength 0.1541 nm, assume that this is some unknown material, but it has been determined that the material is cubic by Laue diffraction.

(a) Prove that the material is BCC.

(b) Determine h, k, and l for each of the lines, but you can ignore the small peak at 35°. This is a K_β peak.

(c) Determine the lattice parameter of this material using the (321) peak.

Problem 15.5 For a more precise lattice parameter, determine the lattice parameter (a) from each of the peak intensities for tungsten in Figure 15.4. Then, plot the value of a determined for each peak intensity as a function of the value of $\sin^2\theta$ used to calculate a. The most accurate value of a results from extrapolation of $\sin^2\theta$ to 1.0.

(a) Present your data table and plot.

(b) What lattice parameter results?

Problem 15.6 A single crystal wafer of silicon is to be used as a substrate for the deposition of thin-metal films that could be affected by the substrate's crystallographic orientation. However, the crystallographic orientation of this particular wafer is not known. The wafer is placed in an X-ray diffractometer with a copper X-ray tube with a K_α X-ray line at 0.1542 nm, and an intense peak is measured at an angle of 29.46°. Silicon is FCC with a lattice parameter of 0.543 nm. What planes are parallel to the surface of this silicon crystal?

Problem 15.7 The TEM diffraction pattern at a 200-kV accelerating potential from a piece of metal taken from a meteorite gives well-defined concentric rings, indicating that the material is polycrystalline and has a very small grain size. Measurement of the rings shows that the angles in degrees are as follows: 0.39°, 0.55°, 0.67°, 0.77°, 0.86°, 0.95°, 1.02°, 1.09°, and 1.16°. Assume that this material is cubic.

(a) Is the crystal structure simple cubic, BCC, or FCC?

(b) Determine the lattice parameter using the 0.39° peak.

(c) What is most likely the major element in this sample?

Problem 15.8 It is desired to conduct a diffraction experiment with neutrons of a wavelength of 0.1 nm. What is the kinetic energy of these neutrons in both joules and electron-volts?

Problem 15.9 A uniaxial carbon-fiber-reinforced composite material is subject to an axial tensile stress. It is desired to determine the actual stress in the graphite fibers using X-ray diffraction, because the calculation of stress equal to force divided by the area of the composite is not valid for the graphite fibers. Graphite fibers have a hexagonal structure, with the basal plane oriented parallel to the fiber axis and the c axis perpendicular to the fiber axis. In the graphite unit cell, there are two sets of parallel graphite planes, with a c lattice parameter of 0.6708 nm. The spacing between individual graphite planes is 0.3354 nm. With X-ray diffraction it is observed that the (0002) peak from the graphite fibers using chromium K_α X-rays of wavelength 0.2291 nm is at an angle 2θ of 39.98°. The in plane elastic modulus of the graphite fibers is 400 GPa and Poisson's ratio is 0.20.

(a) What is the strain in the graphite fibers perpendicular to the fiber axis?

(b) What is the strain in the graphite fibers parallel to the fiber axis?

(c) What is the tensile axial stress in the graphite fibers?

Problem 15.10 In the X-ray chemical analysis of the stainless-steel specimen in Figure 15.11, the Cr K_α line with a wavelength of 0.2291 nm occurs at 99°. What is the interplanar spacing of the crystal used in this wavelength dispersion experiment?

Problem 15.11 Select the best filter for a cobalt X-ray tube that will reduce the intensity of the K_β line and transmit the cobalt K_α line.

(a) What are the wavelengths of the cobalt K_α and K_β lines?

(b) What is your selected filter material?

(c) What is the wavelength of the K absorption edge of your selected filter material?

APPENDIX A

The Electronic Configuration for Some of the Elements

Atomic Number	Element	$1s$	$2s$	$2p$	$3s$	$3p$	$3d$	$4s$	$4p$	$4d$	$4f$	$5s$	$5p$	$5d$	$6s$	$6p$
1	Hydrogen	1														
2	Helium	2														
3	Lithium	2	1													
4	Beryllium	2	2													
5	Boron	2	2	1												
6	Carbon	2	2	2												
7	Nitrogen	2	2	3												
8	Oxygen	2	2	4												
9	Fluorine	2	2	5												
10	Neon	2	2	6												
11	Sodium	2	2	6	1											
12	Magnesium	2	2	6	2											
13	Aluminum	2	2	6	2	1										
14	Silicon	2	2	6	2	2										
15	Phosphorus	2	2	6	2	3										
16	Sulfur	2	2	6	2	4										
17	Chlorine	2	2	6	2	5										
18	Argon	2	2	6	2	6										
19	Potassium	2	2	6	2	6		1								
20	Calcium	2	2	6	2	6		2								
21	Scandium	2	2	6	2	6	1	2								
22	Titanium	2	2	6	2	6	2	2								
23	Vanadium	2	2	6	2	6	3	2								
24	Chromium	2	2	6	2	6	5	1								
25	Manganese	2	2	6	2	6	5	2								
26	Iron	2	2	6	2	6	6	2								
27	Cobalt	2	2	6	2	6	7	2								
28	Nickel	2	2	6	2	6	8	2								
29	Copper	2	2	6	2	6	10	1								
30	Zinc	2	2	6	2	6	10	2								
31	Gallium	2	2	6	2	6	10	2	1							
32	Germanium	2	2	6	2	6	10	2	2							
33	Arsenic	2	2	6	2	6	10	2	3							
34	Selenium	2	2	6	2	6	10	2	4							
35	Bromine	2	2	6	2	6	10	2	5							

(Continued)

Atomic Number	Element	1s	2s	2p	3s	3p	3d	4s	4p	4d	4f	5s	5p	5d	6s	6p
36	Krypton	2	2	6	2	6	10	2	6							
37	Rubidium	2	2	6	2	6	10	2	6		1					
38	Strontium	2	2	6	2	6	10	2	6		2					
39	Yttrium	2	2	6	2	6	10	2	6	1	2					
40	Zirconium	2	2	6	2	6	10	2	6	2	2					
41	Niobium	2	2	6	2	6	10	2	6	4	1					
42	Molybdenum	2	2	6	2	6	10	2	6	5	1					
43	Technetium	2	2	6	2	6	10	2	6	6	1					
44	Ruthenium	2	2	6	2	6	10	2	6	7	1					
45	Rhodium	2	2	6	2	6	10	2	6	8	1					
46	Palladium	2	2	6	2	6	10	2	6	10						
47	Silver	2	2	6	2	6	10	2	6	10		1				
48	Cadmium	2	2	6	2	6	10	2	6	10		2				
49	Indium	2	2	6	2	6	10	2	6	10		2	1			
50	Tin	2	2	6	2	6	10	2	6	10		2	2			
51	Antimony	2	2	6	2	6	10	2	6	10		2	3			
52	Tellurium	2	2	6	2	6	10	2	6	10		2	4			
53	Iodine	2	2	6	2	6	10	2	6	10		2	5			
54	Xenon	2	2	6	2	6	10	2	6	10		2	6			
55	Cesium	2	2	6	2	6	10	2	6	10		2	6		1	
56	Barium	2	2	6	2	6	10	2	6	10		2	6		2	
57	Lanthanum	2	2	6	2	6	10	2	6	10	1	2	6		2	
⋮	⋮	⋮	⋮	⋮	⋮	⋮	⋮	⋮	⋮	⋮	⋮	⋮	⋮		⋮	
71	Lutetium	2	2	6	2	6	10	2	6	10	14	2	6	1	2	
72	Hafnium	2	2	6	2	6	10	2	6	10	14	2	6	2	2	
73	Tantalum	2	2	6	2	6	10	2	6	10	14	2	6	3	2	
74	Tungsten	2	2	6	2	6	10	2	6	10	14	2	6	4	2	
75	Rhenium	2	2	6	2	6	10	2	6	10	14	2	6	5		
76	Osmium	2	2	6	2	6	10	2	6	10	14	2	6	6		
77	Iridium	2	2	6	2	6	10	2	6	10	14	2	6	9		
78	Platinum	2	2	6	2	6	10	2	6	10	14	2	6	9	1	
79	Gold	2	2	6	2	6	10	2	6	10	14	2	6	10	1	
80	Mercury	2	2	6	2	6	10	2	6	10	14	2	6	10	2	
81	Thallium	2	2	6	2	6	10	2	6	10	14	2	6	10	2	1
82	Lead	2	2	6	2	6	10	2	6	10	14	2	6	10	2	2
83	Bismuth	2	2	6	2	6	10	2	6	10	14	2	6	10	2	3
84	Polonium	2	2	6	2	6	10	2	6	10	14	2	6	10	2	4
85	Astatine	2	2	6	2	6	10	2	6	10	14	2	6	10	2	5
86	Radon	2	2	6	2	6	10	2	6	10	14	2	6	10	2	6

Based on data from Askeland and Phule 4th ed. pp. 982–983.

Element	Symbol	Atomic Number	Crystal Structure	Lattice Parameters (Å)	Atomic Mass g/mol	Density (g/cm³)	Melting Temperature (°C)
Aluminum	Al	13	FCC	4.04958	26.981	2.699	660.4
Antimony	Sb	51	hex	$a = 4.307$ $c = 11.273$	121.75	6.697	630.7
Arsenic	As	33	hex	$a = 3.760$ $c = 10.548$	74.9216	5.778	614 subl.
Barium	Ba	56	BCC	5.025	137.3	3.5	729
Beryllium	Be	4	hex	$a = 2.2858$ $c = 3.5842$	9.01	1.848	1290
Bismuth	Bi	83	hex	$a = 4.546$ $c = 11.86$	208.98	9.808	271.4
Boron	B	5	rhomb	$a = 10.12$ $\alpha = 65.5°$	10.81	2.3	2300
Cadmium	Cd	48	HCP	$a = 2.9793$ $c = 5.6181$	112.4	8.642	321.1
Calcium	Ca	20	FCC	5.588	40.08	1.55	839
Cerium	Ce	58	HCP	$a = 3.681$ $c = 11.857$	140.12	6.6803	798
Cesium	Cs	55	BCC	6.13	132.91	1.892	28.6
Chromium	Cr	24	BCC	2.8844	51.996	7.19	1875
Cobalt	Co	27	HCP	$a = 2.5071$ $c = 4.0686$	58.93	8.832	1495
Copper	Cu	29	FCC	3.6151	63.54	8.93	1084.9
Gadolinium	Gd	64	HCP	$a = 3.6336$ $c = 5.7810$	157.25	7.901	1313
Gallium	Ga	31	ortho	$a = 4.5258$ $b = 4.5186$ $c = 7.6570$	69.72	5.904	29.8
Germanium	Ge	32	FCC	5.6575	72.59	5.324	937.4
Gold	Au	79	FCC	4.0786	196.97	19.302	1064.4
Hafnium	Hf	72	HCP	$a = 3.1883$ $c = 5.0422$	178.49	13.31	2227
Indium	In	49	tetra	$a = 3.2517$ $c = 4.9459$	114.82	7.286	156.6
Iridium	Ir	77	FCC	3.84	192.9	22.65	2447
Iron	Fe	26	BCC FCC	2.866 3.589	55.847 (>912°C)	7.87	1538
Lanthanum	La	57	HCP	$a = 3.774$ $c = 12.17$	138.91	6.146	918

(Continued)

Element	Symbol	Atomic Number	Crystal Structure	Lattice Parameters (Å)	Atomic Mass g/mol	Density (g/cm³)	Melting Temperature (°C)
Lead	Pb	82	FCC	4.9489	207.19	11.36	327.4
Lithium	Li	3	BCC	3.5089	6.94	0.534	180.7
Magnesium	Mg	12	HCP	$a = 3.2087$ $c = 5.209$	24.312	1.738	650
Manganese	Mn	25	cubic	8.931	54.938	7.47	1244
Mercury	Hg	80	rhomb	$a = 2.006$ @ $-46°$ C $\alpha = 70°\ 31.7'$	200.59	13.546	-38.9
Molybdenum	Mo	42	BCC	3.1468	95.94	10.22	2610
Nickel	Ni	28	FCC	3.5167	58.71	8.902	1453
Niobium	Nb	41	BCC	3.294	92.91	8.57	2468
Osmium	Os	76	HCP	$a = 2.7341$ $c = 4.3197$	190.2	22.57	2700
Palladium	Pd	46	FCC	3.8902	106.4	12.02	1552
Platinum	Pt	78	FCC	3.9231	195.09	21.45	1769
Potassium	K	19	BCC	5.344	39.09	0.855	63.2
Rhenium	Re	75	HCP	$a = 2.760$ $c = 4.458$	186.21	21.04	3180
Rhodium	Rh	45	FCC	3.796	102.99	12.41	1963
Rubidium	Rb	37	BCC	5.7	85.467	1.532	38.9
Ruthenium	Ru	44	HCP	$a = 2.6987$ $c = 4.2728$	101.07	12.37	2310
Selenium	Se	34	hex	$a = 4.3640$ $c = 4.9594$	78.96	4.809	217
Silicon	Si	14	FCC	5.4307	28.08	2.33	1410
Silver	Ag	47	FCC	4.0862	107.868	10.49	961.9
Sodium	Na	11	BCC	4.2906	22.99	0.967	97.8
Strontium	Sr	38	FCC BCC	6.0849 4.84 (>557°C)	87.62	2.6	768
Tantalum	Ta	73	BCC	3.3026	180.95	16.6	2996
Technetium	Tc	43	HCP	$a = 2.735$ $c = 4.388$	98.9062	11.5	2200
Tellurium	Te	52	hex	$a = 4.4565$ $c = 5.9268$	127.6	6.24	449.5
Thorium	Th	90	FCC	5.086	232	11.72	1775
Tin	Sn	50	FCC	6.4912	118.69	5.765	231.9
Titanium	Ti	22	HCP BCC	$a = 2.9503$ $c = 4.6831$ 3.32 (>882°C)	47.9	4.507	1668
Tungsten	W	74	BCC	3.1652	183.85	19.254	3410

(*Continued*)

(*Continued*)

Element	Symbol	Atomic Number	Crystal Structure	Lattice Parameters (Å)	Atomic Mass g/mol	Density (g/cm³)	Melting Temperature (°C)
Uranium	U	92	ortho	$a = 2.854$ $b = 5.869$ $c = 4.955$	238.03	19.05	1133
Vanadium	V	23	BCC	3.0278	50.941	6.1	1900
Yttrium	Y	39	HCP	$a = 3.648$ $c = 5.732$	88.91	4.469	1522
Zinc	Zn	30	HCP	$a = 2.6648$ $c = 4.9470$	65.38	7.133	420
Zirconium	Zr	40	HCP	$a = 3.2312$ $c = 5.1477$	91.22	6.505	1852
			BCC	3.6090 (>862°C)			

Note that 1 Å = 10^{-8} cm = 0.1 nanometer (nm)
Based on data from Askeland, D.R., Fulay, P.P., and Wright, W.J., The Science and Engineering of Materials, 6th ed., Cengage Learning, Stamford, CT. (2011) pp. 888–889.

Element	Atomic Radius (Å)	Valence	Ionic Radius (Å)
Aluminum	1.432	+3	0.51
Antimony	1.45	+5	0.62
Arsenic	1.15	+5	2.22
Barium	2.176	+2	1.34
Beryllium	1.143	+2	0.35
Bismuth	1.60	+5	0.74
Boron	0.46	+3	0.23
Bromine	1.19	−1	1.96
Cadmium	1.49	+2	0.97
Calcium	1.976	+2	0.99
Carbon	0.77	+4	0.16
Cerium	1.84	+3	1.034
Cesium	2.65	+1	1.67
Chlorine	0.905	−1	1.81
Chromium	1.249	+3	0.63
Cobalt	1.253	+2	0.72
Copper	1.278	+1	0.96
Fluorine	0.6	−1	1.33
Gallium	1.218	+3	0.62
Germanium	1.225	+4	0.53
Gold	1.442	+1	1.37
Hafnium	1.55	+4	0.78
Hydrogen	0.46	+1	1.54
Indium	1.570	+3	0.81
Iodine	1.35	−1	2.20
Iron	1.241 (BCC)	+2	0.74
	1.269 (FCC)	+3	0.64
Lanthanum	1.887	+3	1.016
Lead	1.75	+4	0.84
Lithium	1.519	+1	0.68
Magnesium	1.604	+2	0.66
Manganese	1.12	+2	0.80
		+3	0.66
Mercury	1.55	+2	1.10
Molybdenum	1.363	+4	0.70·

(Continued)

(Continued)

Element	Atomic Radius (Å)	Valence	Ionic Radius (Å)
Nickel	1.243	+2	0.69
Niobium	1.426	+4	0.74
Nitrogen	0.71	+5	0.15
Oxygen	0.60	−2	1.32
Palladium	1.375	+4	0.65
Phosphorus	1.10	+5	0.35
Platinum	1.387	+2	0.80
Potassium	2.314	+1	1.33
Rubidium	2.468	+1	0.70
Selenium	1.15	−2	1.91
Silicon	1.176	+4	0.42
Silver	1.445	+1	1.26
Sodium	1.858	+1	0.97
Strontium	2.151	+2	1.12
Sulfur	1.06	−2	1.84
Tantalum	1.43	+5	0.68
Tellurium	1.40	−2	2.11
Thorium	1.798	+4	1.02
Tin	1.405	+4	0.71
Titanium	1.475	+4	0.68
Tungsten	1.371	+4	0.70
Uranium	1.38	+4	0.97
Vanadium	1.311	+3	0.74
Yttrium	1.824	+3	0.89
Zinc	1.332	+2	0.74
Zirconium	1.616	+4	0.79

Note that 1 Å = 10^{-8} cm = 0.1 nanometer (nm)
Based on data from Askeland, D.R., Fulay, P.P., and Wright, W.J., The Science and Engineering of Materials, 6th ed., Cengage Learning, Stamford, CT. (2011) pp. 891–892.

REFERENCES

Ackerman, F. J., and Koskinas, G. J., *J. Chem. Engr. Data*, 17(1) (1972), 51.

Akhavan, S., Mathiesen, M. M., and Schulte, L., "Clinical and histological results related to a low modulus composite total hip replacement stem", *J. Bone Joint Surg.* 88, 1308 (2006).

Allen, S. M., and Thomas, E. L., *The Structure of Materials*, John Wiley & Sons, New York (1999).

Ashby, M. F., *Materials Selection in Mechanical Design*, Pergamon Press, Tarrytown, NY (1992).

Askeland, D. R., Fulay, P. P., and Wright, W. J., *The Science and Engineering of Materials*, 6th ed., Cengage Learning, Stamford, CT (2011).

Askeland, D. R., and Phule, P. P., *The Science and Engineering of Materials*, 4th ed. Thomson-Brooks/Cole, Pacific Grove, CA (2003).

American Society for Materials Handbook online, ASM International, Materials Park, OH (1992).

American Society for Materials Handbook, Vol. 9, *Metallography and Microstructures*, ASM International, Materials Park, OH (1985).

Bansal, N. P., and Doremus, R. H., *Handbook of Glass Properties*, Academic Press, Orlando, FL (1986).

Bardeen, J., Cooper, L. N., Schrieffer, J. R., "Microscopic Theory of Superconductivity", *Physical Review*, 106(1) Apr. 1957, p. 162.

Barrett, C., and Massalski, T. B., *Structure of Metals*, 3rd revised ed., Pergamon, Tarrytown, NY (1980).

Beadle, W. E., Tsai, J. C. C., and Plummer, R. D., *Quick Reference Manual for Silicon Integrated Circuit Technology*, John Wiley & Sons, New York (1985).

Benjamin, C-K. T., Wang, C., Allen, R., & Bao, Z., "An electrically and mechanically self-healing composite with pressure- and flexion-sensitive properties for electric skin applications", *Nature Nanotechnology*, 7 (2012) 825.

Bernardes, N., *Phys. Rev.* 112 (1958), 1534.

Birdi, K. S., *Scanning Probe Microscopes—Applications in Science and Technology*, CRC Press Boca Raton, FL (2003).

Bockris, J. O'. M., and Reddy, A. K. N., *Modern Electrochemistry*, Vol. 2B, *Electronics in Chemistry, Engineering, Biology, and Environmental Science*, 2nd ed., Kluwer Academic/Plenum Publ., New York (2000).

Boening, H. V., *Polyolefins: Structure and Properties*, Elsevier Press, Lausanne, Switzerland (1966).

Bonnell, D. A., ed. *Scanning Probe Microscopy and Spectroscopy: Theory, Techniques, and Applications*, 2nd ed., Wiley–VCH, New York (2001).

Boyer, H. E. (editor), *Atlas of Stress-Strain Curves*, ASM International, Metals Park, OH (1987).

Brenner, S. S., *J. Appl. Phys.* 27, 1484 (1956) and 28, 1023 (1957).

Brostow, W., Kubat, J., and Kubat, M., *Mechanical Properties, Physical Properties of Polymers Handbook*, Mark, J. E. ed., AIP Press, Woodbury, NY (1996).

Caddell, R. M., *Deformation and Fracture of Solids*, Prentice-Hall, England Cliffs, NJ (1980).

Cahn, R. W., "Recovery and Recrystallization", *Physical Metallurgy*, Cahn, R.W., ed. North-Holland, Amsterdam, 1137 (1970).

Callen, H. B., *Thermodynamics*, John Wiley & Sons, New York (1960).

Carter, C. B., and Norton, M. G., *Ceramic Materials: Science and Engineering*, Springer, New York (2007).

Chaing, Y-M., Birnie D. B., and Kingery, W. D., *Physical Ceramics–Principles for Ceramic Science and Engineering*, John Wiley & Sons, New York (1997).

Chang, S. S. L., *Energy Conversion*, Prentice Hall, Englewood Cliffs, NJ (1964).

Charles, J. A., Crane, F. A. A., and Furness, J. A. G., *Selection and Use of Engineering Materials*, Butterworth Heinemann, Oxford, England (1997).

Chawla, K. K., *Composite Materials: Science and Engineering*, 2nd ed., Springer, New York (1998).

Chen, C. J., *Introduction to Scanning Tunneling Microscopy*, Oxford University Press, New York (1993).

Chikazumi, S., *Physics of Ferromagnetism*, Oxford University Press, Oxford, England (2009).

Chu, P. C. W., "High Temperature Superconductors", *Scientific American*, Sept. (1995).

Courtney, T. H., *Mechanical Behavior of Materials*, McGraw-Hill, New York (1990).

Courtney, T. H., *Mechanical Behavior of Materials*, 2nd ed., McGraw-Hill, New York (1999).

Crangle, J., *Solid State Magnetism*, Van Nostrand, New York (1991).

Crow, D. R., *Principles and Applications of Electrochemistry*, 4th ed., Blackie Academic & Professional, New York (1994).

Cullity, B. D., *Elements of X-Ray Diffraction*, Addison-Wesley Pub. Co. Inc., (1956).

Dieter, G. E., *Mechanical Metallurgy*, 3rd ed., McGraw-Hill Book Co., New York (1986).

Donachie, M. J., and Donachie, S. J., *Superalloys: A Technical Guide*, American Society for Metals International, Metals Park, OH (2002).

Dresselhaus, M. S., Dresselhaus, G., and Avouris, P., eds., *Carbon Nanotubes: Synthesis, Structure, Properties, and Applications*, Springer-Verlag, Berlin (2000).

Eisenstadt, M. M., *Introduction to Mechanical Properties of Materials*, The Macmillan Co., New York (1971).

Evans, A. G., and Langdon, T., *Prog. Mat. Sc.*, 21, 171 (1976).

Fischer, J. C., Johnston, W. G., Thompson, R., and Vreeland, Jr., T., *Dislocations and Mechanical Properties of Crystals*, John Wiley & Sons, New York (1957).

Fontana, M. G., Corrosion Engineering, McGraw-Hill, New York (1986).

Fossheim, K., and Subdϕ, A., *Superconductivity Physics and Applications*, John Wiley & Sons, Chichester, England (2004).

Gersten, J. I., and Smith, F. W., *The Physics and Chemistry of Materials*, John Wiley & Sons, New York (2001).

Glassman, A. H., Crownshield, "R. D., Schenck, R. P., and Herberts, P.", A low stiffness composite biologically fixed prosthesis, *Clin. Orthop.* 393 (2001) 128.

Goldstein, J. I., and Yakowitz, H., eds. *Practical Scanning Electron Microscopy*, Plenum, New York (1975).

Gordon, R. G., and Kim, Y. S., *J. Chem. Phys.* 56 (1972), 3122.

Griffith, A. A., *Phil. Trans. Roy. Soc. London*, A 221 (1920), 163.

Republished in *Trans. ASM*, 61 (1968) 871.

Gumbel, E. J., *Statistics of Extremes*, Columbia Univ. Press, New York (1958). Republished by Dover Publ. Inc., New York (2004).

Gutowski, T. G., ed. *Advanced Composite Manufacturing*, John Wiley & Sons, New York (1997).

Guy, A. G. *Introduction to Materials Science*, McGraw-Hill, New York (1972).

Hadziioannou, G., and van Hutten, P. F., eds., *Semiconducting Polymers*, Wiley–VCH, Weinheim, FRG (2000).

Hall, C. E. *Introduction to Electron Microscopy*, 2nd ed., McGraw-Hill, Highstown, NJ (1966).

Harrison, W. A., *Electronic Structure and the Properties of Solids: The Physics of the Chemical Bond*, W. H. Freeman and Co., San Francisco, CA (1980).

Hayden, H. W., Moffatt, W. G., and Wulff, J., *The Structure and Properties of Materials*, Vol. III, *Mechanical Behavior*, John Wiley & Sons, New York (1965).

Haynes, W. M., ed., *Handbook of Chemistry and Physics*, 91st ed., CRC Press, Boca Raton, FL (2010–2011).

Haynes, W. M., ed., *Handbook of Chemistry and Physics*, 92nd ed., CRC Press, Boca Raton, FL (2011–2012).

Hearle, J. W. S., *Polymers and Their Properties*, Vol. 1, *Fundamentals of Structure and Mechanics*, Ellis Horwood Ltd. Chichester, England (1982).

Heeger, A. J., Scariciftci, N. S., and Namdas, E. B., *Semiconducting and Metallic Polymers*, Oxford University Press, Oxford, England (2010).

Hertzberg, R. W., *Deformation and Fracture Mechanics of Engineering Materials*, John Wiley & Sons, New York (1976).

Hertzberg, R. W., *Deformation and Fracture Mechanics of Engineering Materials*, 3rd ed., John Wiley & Sons, New York (1989).

Hudson, J. B., *Surface Science: An Introduction*, Butterworth-Heinemann, Boston, MA (1992).

Hull, D., and Clyne, T. W., *An Introduction to Composite Materials*, 2nd ed., Cambridge University Press, Cambridge, England (1996).

Hultgren, R., Desai, P. D., Hawkins, D. T., Gleiser, M., and Kelly, K. K., *Selected Values of the Thermodynamic Properties of Binary Alloys*, ASM, Metals Park, OH (1973).

Kasap S. O., *Optoelectronics and Photonics: Principles and Practices*, Prentice Hall, Saddle River, NJ (2001).

Kasap, S. O., *Principles of Electronic Materials and Devices*, 2nd ed., Irwin McGraw-Hill (2002).

Kaufman, J. G., Schilling, P. E., and Nelson, F. G., *Met. Eng. Quart.* 9(3), 39 (1969).

Kay, G. W. C., and Laby, T. H., *Tables of Physical and Chemical Constants*, 14th ed., Longman Group Ltd., London, England (1973).

Kingery, W. D., Bowen, H. K., and Uhlmann, D. R., *Introduction to Ceramics*, 2nd ed., John Wiley & Sons, New York (1976).

Kinney, G. F. *Engineering Properties and Applications of Plastics*, John Wiley & Sons, New York (1957).

Kittel, C., *Introduction to Solid State Physics*, 3rd ed., John Wiley & Sons, New York (1967).

Kondepepudi, D., and Prigogine, I., *Modern Thermodynamics*, John Wiley & Sons, New York (1998).

Kraus, J. D., *Electromagnetics*, McGraw-Hill, New York (1953).

Kwok, H. L., *Electronic Materials*, PWS Publishing, Boston, MA (1997).

Larminie, J., and Dicks, A., *Fuel Cell Systems Explained*, John Wiley & Sons, New York (2000).

Lewis, G., *Selection of Engineering Materials*, Prentice Hall, Englewood Cliffs, NJ (1990).

Lindemann, F. A., *Z. Phys.* 11, 609 (1910).

Lowenheim, F. A., ed., *Modern Electroplating*, 2nd ed., John Wiley & Sons, New York (1963).

Lutjering, G., and Williams, J. C., *Titanium*, Springer, Berlin (2007).

Mark, J. E. ed., *Physical Properties of Polymers Handbook*, AIP Press, Woodbury, NY (1996).

Massa, W., *Crystal Structure Determination*, 2nd ed., Springer, Berlin (2004).

Massalski, T. B., editor, *Binary Alloy Phase Diagrams*, 2nd ed., ASM International, Metals Park, OH (1990).

Mayer, J. W., and Lau, S. S., *Electronic Materials Science: For Integrated Circuits in Si and GaAs*, Macmillan Publishing Co., New York (1990).

McCafferty. E., *Introduction to Corrosion Science*, Springer, New York (2010).

McCrum, N. G., Buckley, C. P., and Bucknall, C. B., *Principles of Polymer Engineering*, Oxford Science Publications, Oxford, England (1997).

Moffatt, W. G., Pearsall, G. W., and Wulff, J., *The Structure and Properties of Materials*, Vol. 1, John Wiley & Sons, New York (1964).

Myers, M. A., and Chawla, K. K., *Mechanical Behavior of Materials*, Cambridge University Press, Cambridge, England (2009).

Neto, A. H. C., The carbon new age, *Materials Today*, March, 13(3), 12 (2010).

Osswald, T. A., and Menges G., *Materials Science of Polymers for Engineers*, Hanser Publishers, Munich (2003).

Porter, D. A., and Easterling, K. E., *Phase Transfromations in Metals and Alloys*, Chapman and Hall, New York (1992).

Ragone, D. V., *Thermodynamics of Materials*, Vols. 1 and 2, John Wiley & Sons, New York (1995).

Ralls, K. M., Courtney, T. H., and Wulff, J., *Introduction to Materials Science and Engineering*, John Wiley & Sons, New York (1976).

Raymond, R., *Out of the Fiery Furnace: The Impact of Metals on the History of Mankind*, Penn State University Press, State College, PA (1990).

Reed-Hill, R. E., *Physical Metallurgy Principles*, 2nd ed., D. van Nostrand Co., New York (1973).

Reed-Hill, R. E., and Abbaschian, R., *Physical Metallurgy Principles*, 3rd ed., PWS, Boston, MA (1994).

Reese, R. L., *University Physics*, Brooks Cole Publishing Co., Pacific Grove, CA (2000).

Reimer, L., *Transmission Electron Microscopy: Physics of Image Formation and Microanalysis*, 4th ed., Springer, Berlin (1997).

Rochow, T. G., and Tucker, P. A., *Introduction to Microscopy by Means of Light, Electrons, X-Rays, or Acoustics*, 2nd ed., Plenum Press, New York (1994).

Rose, R. M., Shepard, L. A., and Wulff, J., *The Structure and Properties of Materials*, Vol. 4, *Electronic Properties*, John Wiley & Sons, New York (1966).

Rubenstein, M., and Colby, R. H., *Polymer Physics*, Oxford University Press, Oxford, England (2003).

Rudin, A., *The Elements of Polymer Science and Engineering*, 2nd ed., Academic Press, San Diego (1990).

Savage, G. *Carbon-Carbon Composites*, Chapman & Hall, London (1993).

Schaffer, J. P., Saxena, A., Antolovich, S. D., Sanders Jr., T. H., and Warner, S. B., *The Science and Design of Engineering Materials*, 2nd ed., WCB McGraw-Hill, Boston (1999).

Schultz, J. M., *Polymer Materials Science*, Prentice-Hall, Englewood Cliffs, NJ (1974).

Schwartz, M., *Handbook of Structural Ceramics*, McGraw-Hill Inc., New York (1992).

Schwartz, M. M., *Composite Materials, Properties, Nondestructive Testing, and Repair*, Prentice Hall, Upper Saddle River, NJ (1997).

Shewmon, P. G., *Diffusion in Solids*, McGraw-Hill, New York (1963).

Shewmon, P. G. "Diffusion", *Physical Metallurgy*, Cahn, R. W., ed. North-Holland, Amsterdam, 383 (1970).

Shull, P. J., ed. *Nondestructive Evaluation*, Marcel Decker, New York (2002).

Simmons, R. O., and Balluffi, R. W., *Phys. Rev.*, 119, 600 (1960).

Sims, C., Stoloff, N., and Hagel, W., *Superalloys II: High Temperature Materials for Aerospace and Industrial Power*, John Wiley & Sons, New York (1987).

Smith, W. F., and Hashemi, J., *Foundations of Materials Science and Engineering*, 5th ed., McGraw-Hill, New York (2010).

Smith, W. F., *Foundations of Materials Science and Engineering*, 3rd ed., McGraw-Hill, New York (2004).

Smithell's Metals Reference Book, eds. Gale, W. F., and Totemeier, T. C., Elsevier Butterworth-Heinemann, Oxford, England (2004).

Strong, A. B., *Plastics–Materials and Processing*, 3rd ed., Pearson-Prentice Hall, Upper Saddle River, NJ (2006).

Swalin, R. A., *Thermodynamics of Solids*, John Wiley & Sons, New York (1962).

Sze, S. M., ed., *Modern Semiconductor Device Physics*, John Wiley & Sons, New York (1998).

Sze, S. M., and Kwok, K. N., *Physics of Semiconductor Devices*, John Wiley & Sons, New York (2006).

Tee, B. C-K., Wang, C., Allen, R., and Bao, Z., "An electrically and mechanically self-healing composite with pressure- and flexion-sensitive properties for electronic skin applications", *Nature Nanotechnology*, 7 (2012) p. 825.

Tetelman, A. S., and McEvily, Jr., A. J., *Fracture of Structural Materials*, John Wiley & Sons, New York (1967).

Timings, R. L., *Engineering Materials*, 2nd ed., Addison Wesley Longman Ltd., Essex, England (1998).

Turton, R., *The Quantum Dot*, Oxford University Press, Oxford, England (1996).

Uhlig, H. H., and Revie, R. W., *Corrosion and Corrosion Control*, John Wiley & Sons, New York (1985).

van Krevelen, D. W., *Properties of Polymers: Correlations with Chemical Structure*, Elsevier, Amsterdam (1972).

Wang, F. E., *Bonding Theory for Metals and Alloys*, Elsevier, Amsterdam (2005).

Waterman, N. A., *The Fulmer Materials Optimizer*, Stokes Poges, Slough Berks, England (1974).

Waterman, N. A., and Ashby, M. F., *Materials Selector*, Chapman & Hall, London, England (1996).

Weertman, J., and Weertman, J. R., *Elementary Dislocation Theory*, Oxford University Press, New York (1992).

Wen, J., "Some Mechanical Properties of Typical, Polymer-Based Composites", *Physical Properties of Polymers Handbook*, Mark, J. E., ed., AIP Press, Woodbury, NY (1996).

Wilkes, P. *Solid State Theory in Metallurgy*, Cambridge University Press, Cambridge, England (1973).

Winding, C. C., and Hiatt, G. D., *Polymeric Materials*, McGraw-Hill, New York (1961).

Wood, W. A., *The Study of Metal Structures and Their Properties*, Pergamon Press, Elmsford, NY (1971).

INDEX

Unit Abbreviations

A = ampere
C = coulomb
°C = degrees celsius
cal = calorie
eV = electron volt
°F = degrees fahrenheit
g = gram
Hz = hertz

in = inch
J = joule
K = kelvin
lb_f = pound force
m = meter
N = newton
P = poise
Pa = pascal

psi = pounds per square inch
s = second
T = tesla
V = volts
W = watt
Wb = weber
Ω = ohms

Commonly Used Prefixes for Powers of 10

Power	Prefix	Abbreviation	Power	Prefix	Abbreviation
10^{-12}	pico	p	10^{-2}	centi	c
10^{-9}	nano	n	10^{3}	kilo	k
10^{-6}	micro	μ	10^{6}	mega	M
10^{-3}	milli	m	10^{9}	giga	G

Values of Selected Physical Constants

Physical constant	Symbol	Value
Avogadro's number	N_A	6.022×10^{23} particles/mole
Boltzmann's constant	k	1.38×10^{-23} J/atom·K; 8.62×10^{-5} eV/atom·K
Bohr Magneton	μ_B	9.27×10^{-24}A·m^2
Electron charge	e	1.602×10^{-19} C
Electron mass	m_e	9.11×10^{-23} kg
Permeability of a vacuum	μ_0	$4\pi \times 10^{-7}$ Wb/A·m
Permittivity of a vacuum	ε_0	8.85×10^{-12} F/m
Planck's constant	h	6.63×10^{-34} J·s
Speed of light	c	2.998×10^{8} m/s
Universal gas constant	R	8.31 J/mole·K; 1.987 cal/mole·K

Unit Conversion Factors

Length: 1 m = 10^{10} Angstrom (Å)   1 cm = 0.394 in   1 in = 2.54 cm

Mass: 1 kg = 2.205 lb_m 1 lb_m = 0.4536 kg

Density: 1 kg/m^3 = 10^{-3} g/cm^3 1 g/cm^3 = 0.0361 lb_m/in^3 1 lb_m/in^3 = 27.7 g/cm^3

Force: 1 N = 0.2248 lb_f 1 lb_f = 4.448 N

Stress: 1 MPa = 145 psi   1 psi = 6.90×10^3 Pa

Stress Intensity: 1 MPa(m)$^{1/2}$ = 910 psi(in)$^{1/2}$ 1 psi(in)$^{1/2}$ = 109.9 Pa(m)$^{1/2}$

Energy: 1 J = 6.24×10^{18} eV   1 eV = 1.602×10^{-19} J
1 J = 0.239 cal   1 cal = 4.184 J
1 J = 0.738 ft·lb_f 1 lb_f = 1.356 J

Viscosity: 1 Pa·s = 10 P   1 P = 0.1 Pa·s

Periodic table and properties of the elements

Legend:

Element
Atomic number
Atomic mass — Values in ( ) are for the most stable isotope.
Electronegativity
Cohesive energy in eV are for zero Kelvin, values in [ ] are for 298.15 K or for the melting point, whichever temperature is lower.
Melting temperature in K
Density at 300 K - Solids and liquids are in 10^3 kg/m³, gasses are at STP in kg/m³.

Color key: Metal, Semimetal, Nonmetal

Group	Element	At. no.	At. mass	Electroneg.	Cohesive energy	Melting temp (K)	Density
IA	H	1	1.008	2.1	4.48	14.03	0.09
0	He	2	4.003	-	0.009	0.95	0.179
IA	Li	3	6.939	0.9	1.65	453.7	0.53
IIA	Be	4	9.012	1.5	3.33	1560	1.85
IIIA	B	5	10.81	2.0	5.81	2300	2.34
IVA	C	6	12.01	2.5	7.36	4000	2.62
VA	N	7	14.01	3.0	[4.94]	63.14	1.251
VIA	O	8	16.00	3.5	[2.60]	50.35	1.429
VIIA	F	9	19.00	4.0	[0.87]	53.48	1.696
0	Ne	10	20.18	-	0.036	24.55	0.901
IA	Na	11	22.99	0.9	1.13	371	0.97
IIA	Mg	12	24.31	1.2	1.53	922	1.74
IIIA	Al	13	26.98	1.5	3.34	933	2.70
IVA	Si	14	28.09	1.8	4.64	1685	2.33
VA	P	15	30.97	2.1	[3.42]	317	1.82
VIA	S	16	32.06	2.5	2.86	388	2.07
VIIA	Cl	17	35.45	3.0	[1.39]	172	3.17
0	Ar	18	39.95	-	0.062	83.8	1.784
IA	K	19	39.10	0.8	0.94	336	0.86
IIA	Ca	20	40.08	1.0	1.825	1112	1.55
IIIB	Sc	21	44.96	1.3	3.93	1812	3.00
IVB	Ti	22	47.88	1.5	4.855	1943	4.51
VB	V	23	50.94	1.6	5.30	2175	5.80
VIB	Cr	24	52.00	1.6	4.10	2130	7.19
VIIB	Mn	25	54.94	1.5	2.98	1517	7.43
VIIIB	Fe	26	55.85	1.8	4.29	1809	7.86
VIIIB	Co	27	58.93	1.8	4.39	1768	8.90
VIIIB	Ni	28	58.71	1.8	4.435	1726	8.90
IB	Cu	29	63.54	1.9	3.50	1358	8.96
IIB	Zn	30	65.37	1.6	1.35	693	7.14
IIIA	Ga	31	69.72	1.6	2.78	303	5.91
IVA	Ge	32	72.59	1.8	3.87	1210	5.32
VA	As	33	74.99	2.0	3.0	1081	5.72
VIA	Se	34	78.96	2.4	2.13	494	4.80
VIIA	Br	35	79.91	2.8	1.22	266	3.12
0	Kr	36	83.80	-	0.130	115.8	3.74

Element	Z	Atomic weight	χ	value	m.p. (K)	density
Rb	37	85.47	0.8	0.86	313	1.53
Sr	38	87.62	1.0	[1.60]	1041	2.6
Cs	55	132.9	0.7	0.83	302	1.87
Ba	56	137.3	0.9	1.86	1002	3.5
Fr	87	(223)	0.7	–	300	–
Ra	88	(226)	0.9	1.86	973	5

Element	Z	Atomic weight	χ	value	m.p. (K)	density
Y	39	88.91	1.2	4.39	1799	4.5
Zr	40	91.22	1.4	6.32	2125	6.49
Nb	41	92.91	1.6	7.47	2740	8.55
Mo	42	95.94	1.8	6.81	2890	10.2
Tc	43	(99)	1.9	7.53	2473	11.5
Ru	44	101.1	2.2	6.615	2523	12.2
Rh	45	102.9	2.2	5.75	2236	12.4
Pd	46	106.4	2.2	3.94	1825	12.0
Ag	47	107.9	1.8	2.96	1234	10.5
Cd	48	112.4	1.7	1.16	594	8.65
In	49	114.8	1.7	2.6	430	7.31
Sn	50	118.7	1.8	3.12	505	7.30
Sb	51	121.8	1.9	2.7	904	6.86
Te	52	127.6	2.1	2.0	723	6.24
I	53	126.9	2.5	[1.11]	387	4.92
Xe	54	131.3	–	0.176	161.4	5.89

La–Lu 57–71 (χ = 1.1–1.2)

Element	Z	Atomic weight	χ	value	m.p. (K)	density
Hf	72	178.5	1.3	6.35	2500	13.1
Ta	73	181.0	1.5	8.09	3287	16.6
W	74	183.8	1.7	8.66	3680	19.3
Re	75	186.2	1.9	8.10	3453	21.0
Os	76	190.2	2.2	[8.10]	3300	22.4
Ir	77	192.2	2.2	6.93	2716	22.5
Pt	78	195.1	2.2	5.85	2045	21.4
Au	79	197.0	2.4	3.78	1338	19.3
Hg	80	200.6	1.9	[0.69]	234.3	13.53
Tl	81	204.4	1.8	1.87	723	11.85
Pb	82	207.2	1.8	2.04	600.6	11.4
Bi	83	209.0	1.9	2.15	544.5	9.8
Po	84	(210)	2.0	[1.49]	527	9.4
At	85	(210)	2.2	–	575	–
Rn	86	(222)	–	0.208	202	9.91

Lanthanides

Element	Z	Atomic weight	χ	value	m.p. (K)	density
La	57	138.9		4.49	1193	6.7
Ce	58	140.1		4.77	1071	6.78
Pr	59	140.9		3.9	1204	6.77
Nd	60	144.2		3.35	1289	7.00
Pm	61	(145)		–	1204	6.48
Sm	62	150.4		2.11	1345	7.54
Eu	63	152.0		1.80	1090	5.26
Gd	64	157.2		4.14	1585	7.89
Tb	65	158.9		4.1	1630	8.27
Dy	66	162.5		3.1	1682	8.54
Ho	67	164.9		3.0	1743	8.80
Er	68	167.3		3.3	1795	9.05
Tm	69	168.9		2.6	1818	9.33
Yb	70	173.0		1.6	1097	6.98
Lu	71	175.0		4.4	1936	9.84

Actinides

Element	Z	Atomic weight	χ	value	m.p. (K)	density
Ac	89	(227)		–	1323	10.07
Th	90	232.04		5.93	2028	11.7
Pa	91	(231)		5.46	–	15.4
U	92	238.03		5.41	1405	18.9
Np	93	(237)		4.55	910	20.4
Pu	94	(242)		4.0	913	19.8
Am	95	(243)		2.6	1268	13.6
Cm	96	(247)		–	1340	13.51
Bk	97	(247)		–	–	–
Cf	98	(249)		–	900	–
Es	99	(254)		–	–	–
Fm	100	(253)		–	–	–
Md	101	(256)		–	–	–
No	102	(254)		–	–	–
Lr	103	(257)		–	–	–

(1) The elements under this line are transition elements that do not correspond to the group numbers at the top of the page.

PRINCIPAL UNITS USED IN MECHANICS

Quantity	International System (SI)			U.S. Customary System (USCS)		
	Unit	Symbol	Formula	Unit	Symbol	Formula
Acceleration (angular)	radian per second squared		rad/s^2	radian per second squared		rad/s^2
Acceleration (linear)	meter per second squared		m/s^2	foot per second squared		ft/s^2
Area	square meter		m^2	square foot		ft^2
Density (mass) (Specific mass)	kilogram per cubic meter		kg/m^3	slug per cubic foot		$slug/ft^3$
Density (weight) (Specific weight)	newton per cubic meter		N/m^3	pound per cubic foot	pcf	lb/ft^3
Energy; work	joule	J	$N \cdot m$	foot-pound		ft-lb
Force	newton	N	$kg \cdot m/s^2$	pound	lb	(base unit)
Force per unit length (Intensity of force)	newton per meter		N/m	pound per foot		lb/ft
Frequency	hertz	Hz	s^{-1}	hertz	Hz	s^{-1}
Length	meter	m	(base unit)	foot	ft	(base unit)
Mass	kilogram	kg	(base unit)	slug		$lb\text{-}s^2/ft$
Moment of a force; torque	newton meter		$N \cdot m$	pound-foot		lb-ft
Moment of inertia (area)	meter to fourth power		m^4	inch to fourth power		$in.^4$
Moment of inertia (mass)	kilogram meter squared		$kg \cdot m^2$	slug foot squared		$slug\text{-}ft^2$
Power	watt	W	J/s $(N \cdot m/s)$	foot-pound per second		ft-lb/s
Pressure	pascal	Pa	N/m^2	pound per square foot	psf	lb/ft^2
Section modulus	meter to third power		m^3	inch to third power		$in.^3$
Stress	pascal	Pa	N/m^2	pound per square inch	psi	$lb/in.^2$
Time	second	s	(base unit)	second	s	(base unit)
Velocity (angular)	radian per second		rad/s	radian per second		rad/s
Velocity (linear)	meter per second		m/s	foot per second	fps	ft/s
Volume (liquids)	liter	L	$10^{-3} \, m^3$	gallon	gal.	$231 \, in.^3$
Volume (solids)	cubic meter		m^3	cubic foot	cf	ft^3